V[th] International Congress on X-Ray Optics and Microanalysis

V. Internationaler Kongreß für Röntgenoptik und Mikroanalyse

Ve Congrès International sur l'Optique des Rayons X et la Microanalyse

Tübingen, September 9th-14th, 1968

Edited by

G. Möllenstedt and K. H. Gaukler

With 558 Figures

Springer-Verlag Berlin Heidelberg GmbH 1969

Editores: Professor Dr. G. Möllenstedt and Dr. K. H. Gaukler
Institut für angewandte Physik der Universität, 74 Tübingen

ISBN 978-3-662-12110-8   ISBN 978-3-662-12108-5 (eBook)
DOI 10.1007/978-3-662-12108-5

The picture on the dust cover shows Fig. 1, p. 583, Contribution "The Application of Microprobe Analysis to Biology" by T. A. Hall and H. J. Höhling

Originally published by Springer-Verlag Berlin Heidelberg New York in 1969
Softcover reprint of the hardcover 1st edition 1969

Library of Congress Catalog Card Number 74-94153.

Preface

The Fifth International Congress on X-Ray Optics and Microanalysis was organized by the Institute of Applied Physics at Tübingen University in Western Germany from September 9th through 14th, 1968. Since 1956, when the First Conference was arranged in Cambridge, England by one of the pioneers in this field, V. E. COSSLETT, the experts in the fields of X-Ray Optics and Microanalysis have met every third year to exchange their scientific experiences. Later meetings were held at Uppsala, Sweden in 1959, at Stanford, California in 1962, and at Orsay, France in 1965. The participants in the 1968 Conference came from the following countries: Germany 140, France 60, Great Britain 55, USA 20, Netherlands 16, Switzerland 12, Austria 9, Sweden 7, Belgium 6, Japan 5, Italy 4, two each from Israel, Yugoslavia, Canada, Norway, Hungary and one each from Argentine, Poland, South Africa.

As at the latest congress in Paris the following central topics were treated: General problems of X-ray optics, physical bases of electron beam microanalysis, quantitative problems of X-ray microanalysis, instrumentation, microdiffraction, applications to metallurgy, mineralogy, and biology.

An exhibition showing some of the most modern instruments formed an important part of the conference. The Springer-Verlag, Heidelberg, deserves thanks for the careful and speedy work they have performed in printing these conference proceedings. We are further indebted to all contributors of this volume for their kind cooperation.

Tübingen, August 1969

G. MÖLLENSTEDT and K. H. GAUKLER

Table of Contents

X-Ray Optics

Electron Probe Microanalysis. Physical Bases

Electron Probe Microanalysis. Quantitative Analysis

Instrumentation

Microdiffraction

Metallurgical and Mineralogical Applications

Biological Applications

List of Authors

X-Ray Optics

Present State of X-Ray Interferometry

U. BONSE

Physikalisches Institut der Universität Münster, BRD

In optical interferometry, the generation and subsequent recombination of two (or more) coherent beams can be accomplished in many different ways, most of which cannot be applied to X-rays because the refractive index of all materials is too close to unity. It is not difficult, however, to deviate or to divide a beam of X-rays by means of diffraction by a crystal lattice. Therefore in existing X-ray interferometers extensive use is made of Laue and Bragg case diffraction.

Evidently the degree of lattice perfection of crystals used in interferometers will play a major role with regard to the phases of the various beams that are generated by diffraction. With faulted crystals the calculation of phase relationships can be a formidable problem. The situation is much simplified if perfect or practically perfect crystals can be used. Fortunately artificially grown silicon and germanium crystals and also some quartz and copper crystals with a high degree of perfection have become available in recent years. Diffraction by these crystals follows very closely the predictions of the well-known dynamical theory of X-ray diffraction.

1. Perfect Crystals as Diffraction Mirrors and Beam Splitters

a) Laue Case

In the so called Laue case of diffraction an incoming wave with wave vector K_0^i generates behind a crystal plate thin enough not to absorb all X-rays two diffracted waves with wave vectors K_0^d and K_h^d, which, by way of their formation, are phase coherent. The standing $|D|^2$ pattern (D is electric induction) set up by these waves has a spacing of the $|D|^2$ maximum and minimum planes that equals the Bragg planes spacing d of the particular reflection used. The position of this pattern with respect to the lattice and the ratio of amplitudes D_0^d and D_h^d of the waves in it depend in a characteristic way on the exact angle of incidence ψ_0 of K_0^i and on the thickness t of the crystal.

Behind a "thick" crystal plate, i.e. if $\mu t \approx 5$ or larger, where μ is the normal absorption coefficient, the planes of maximum $|D|^2$ of the outside pattern point *between* the atomic planes of the crystal (Fig. 1a), because of the two wavefields inside the crystal excited per state of polarization only the weakly absorbed Borrmann wavefield with antinodes between the atomic planes can penetrate the crystal. For anomalously small absorption to occur the angle of incidence ψ_0 must not deviate from the exact Bragg angle Θ_B by more than about one sec of arc in the case of Cu K_α and silicon (220) reflection. The thick crystal plate may preferably be used as beam divider because over the whole range $2\Delta\psi_0$ of anomalous transmission the waves K_0^d and K_h^d have practically equal intensity.

The relative phases of the transmitted waves can be learned from the position of the standing $|D|^2$ pattern behind the crystal. Fig. 1b illustrates the behaviour of this pattern if $\psi_0 \neq \Theta_B$. Although slightly tilting the antinodal planes continue to point between the atomic planes. It is worth noting that for this to be possible the rotation $\Delta\psi_h$ of the transmitted wave K_h^d has to have the *same* sign as the rotation $\Delta\psi_0$ of K_0^d but slightly *different* magnitude. With the light optical analogue, the semitransparent glas plate, always $\Delta\psi_h = -\Delta\psi_0$, what implies a distance

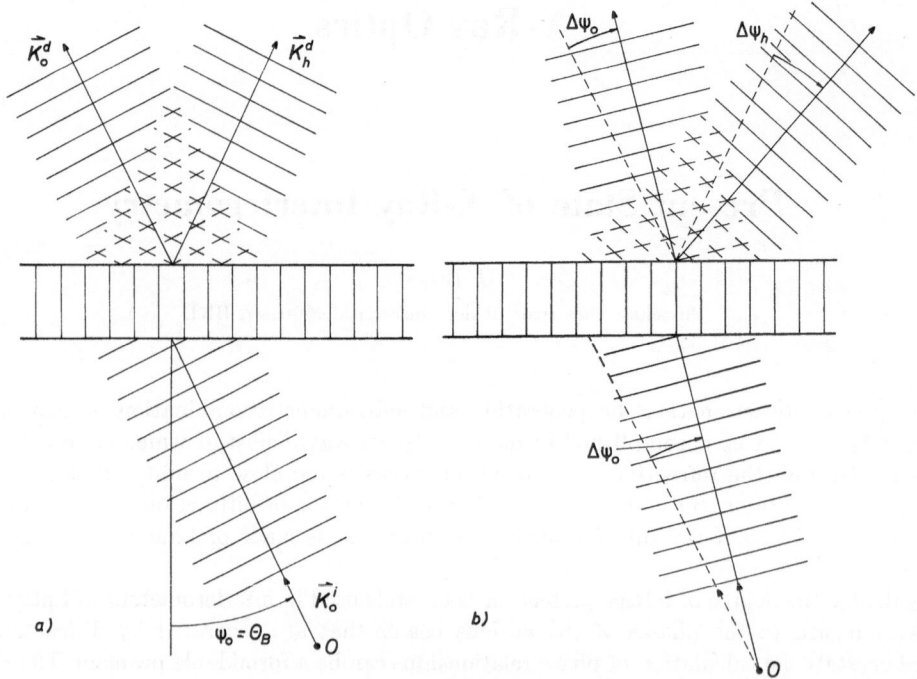

Fig. 1a and b. Outside waves in the case of Laue diffraction by a thick crystal. Note position of standing wave pattern. (a) Exact Bragg position; (b) slightly off Bragg position

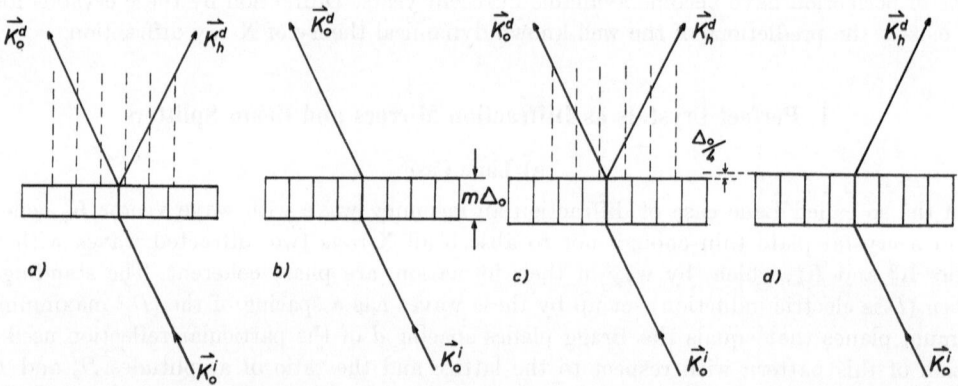

Fig. 2a—d. Outside waves in the case of Laue diffraction by a thin crystal. Crystal thickness varies in steps of the extinction distance Δ_0. Incidence at exact Bragg angle Θ_B. Note the position of standing wave pattern which is different for a and c and also different from the thick crystal case Fig. 1. m is an integer 0, 1, 2, 3, ...

variation of the antinodal planes with ψ_0. Differences of this kind have to be kept in mind whenever a comparison is made of phases and of directions of propagation vectors in light and X-ray interferometers of "corresponding" design.

With a "thin" crystal plate, i.e. if $\mu t \lesssim 1$, also the high absorption wavefield with antinodes of $|D|^2$ on the atomic planes penetrates the crystal and sets up a second standing pattern. The two outside patterns combine to a single pattern with just one pair of waves $\boldsymbol{K}_0^d$ and $\boldsymbol{K}_h^d$ with amplitudes D_0^d and D_h^d. The ratio D_h^d/D_0^d varies with incident angle ψ_0 and plate thickness t. For incidence at the exact Bragg angle this variation is illustrated in Fig. 2a—d, where a sequence of four crystal plates with increasing thickness t is considered. The thickness varies in steps of $\Delta_0/4$. Δ_0 is the well-known extinction distance. Let m be an integer 0, 1, 2, 3, ... For $t = m\Delta_0$ only the transmitted wave $\boldsymbol{K}_0^d$ in the direction of the incident wave is excited (Fig. 2b). If $t = (m - \frac{1}{4})\Delta_0$ or $t = (m + \frac{1}{4})\Delta_0$, both transmitted waves $\boldsymbol{K}_0^d$ and $\boldsymbol{K}_h^d$ occur and have equal intensity (Fig. 2a and 2c). The $|D|^2$ patterns set up by $\boldsymbol{K}_0^d$ and $\boldsymbol{K}_h^d$ point with their antinodal planes unto locations that are $\frac{1}{4}d$ right and left of the sites of the atomic planes respectively, as is indicated by the dashed lines in Fig. 2a and 2c. If $t=(m+\frac{1}{2})\varDelta_0$, then only the reflected wave $\boldsymbol{K}_h^d$ and no forward transmitted wave $\boldsymbol{K}_0^d$ is present. The changeover of the intensities between O- and H-waves is known as Pendellösung solution of the dynamical wave equations.

With a thin crystal either an ideal mirror or an ideal beam divider can be realized simply by choosing a suitable thickness. In practice it is necessary to manufacture crystal plates some few hundred microns thick with thickness tolerances of a few microns, which, with silicon and germanium crystals, is just about possible. An important advantage of thin crystals is the higher transmitted intensity. If however a considerable range $\varDelta\psi_0$ of incident angles or spherical waves instead of plane ones have to be used then a unique thickness to obtain either a mirror or a beam divider does not exist because the extinction distance varies with incident angle.

b) Bragg Case

In the Bragg case the orientation of diffracting planes is such that a wave $\boldsymbol{K}_h^b$ is generated in front of the crystal by surface reflection (Fig. 3). Since the surface reflection has 80 to 95% reflectivity a Bragg case crystal makes a very good mirror. However there is also a wavefield *entering* the crystal. If the surface is cut with an asymmetry angle $\varphi \neq 0$ against the Bragg planes as shown in Fig. 3 the energy flow can follow the low absorption direction parallel to the net planes and reach the rear surface of a thin crystal plate, thereby emitting a transmitted wave $\boldsymbol{K}_0^d$ which again by way of formation is phase coherent with the surface reflected wave $\boldsymbol{K}_h^b$. Usually the wave $\boldsymbol{K}_0^d$ is considerably less intense than $\boldsymbol{K}_h^b$ because of absorption along the path inside the crystal. The phase between $\boldsymbol{K}_h^b$ and $\boldsymbol{K}_0^d$ varies more with angle of incidence than in the case of a thick Laue crystal. Nevertheless Bragg case diffraction mirrors and beam dividers have successfully been used.

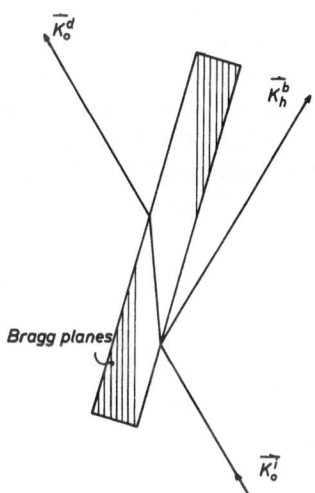

Fig. 3. Beam splitting by Bragg case diffraction

c) Phase Shift Due to Shift of Diffracting Crystal

Displacements of the diffracting crystal with respect to the source that have components normal to the Bragg planes affect the phases. A simple analysis shows that the phases of all *reflected* waves are altered by $2\pi\boldsymbol{c}\cdot\boldsymbol{h}$, where $\boldsymbol{c}$ is the displacement vector and $\boldsymbol{h}$ the diffraction vector. Phases of *transmitted* waves stay unchanged. *Qualitatively* the situation is the same as in light optics with a semitransparent mirror. Regarding *quantities*, however, it is very much different. In order to keep phases stationary all crystals of an X-ray interferometer should never move against each other by more than a small fraction of the Bragg plane spacing $d=1/|\boldsymbol{h}|$, i.e. some 0.1 angstroms. Thus the geometric instability of an X-ray interferometer has to be some thousand times smaller. An easy way to overcome this problem is to manufacture all instrument components as part of one monolithic crystal block.

The extreme sensitivity of the phases to minute shifts of the components of an X-ray interferometer has been successfully utilized to measure angstrom scale displacements with high accuracy.

2. Types of Working Interferometers

A variety of X-ray interferometers has been designed so far and put successfully to operation. Denoting by L or B whether a Laue crystal or a Bragg crystal is used as mirror and/or beam splitter the various interferometer types can be named as follows.

a) *L-L-L* Interferometer

In the first interferometer [1] Laue diffraction by thick crystals is used for beam splitting (splitter crystal S), beam reflection (mirror crystal M), and combined reflection-transmission (analyzer crystal A), see Fig. 4. Interferometric fringes are produced in the plane of A even without A being present. However the spacing of these fringes is too narrow to be observable because it is of the order of the Bragg plane spacing d. The function of A is to transform the atomic scale fringe pattern into a macroscopic one that is observable in either of the exit beams O and H. The macroscopic pattern can best be interpreted as being the moire pattern formed by superposition of the atomic scale standing wave pattern in front of A with the atomic planes of A. Inhomogenous phase shifts introduced in paths I or II will alter either the spacing or the orientation of the standing wave pattern and hence result in a dilational or a rotational moire pattern in the exit beams.

In Fig. 5 a useful modification is shown where the mirrors are no longer positioned in the same plane but placed some distance apart from each other [2]. As a result the two interfering paths contain very long and separated sections. This makes it possible to put even thicker objects in the beams without entering their overlap regions.

Usually, S, M, and A are parts of one single crystal block that have been left standing on a thicker crystalline base. With presently available crystals the length of phase objects is limited to about 8 cm. Larger objects require the mounting and alignment of separate crystals. Stability

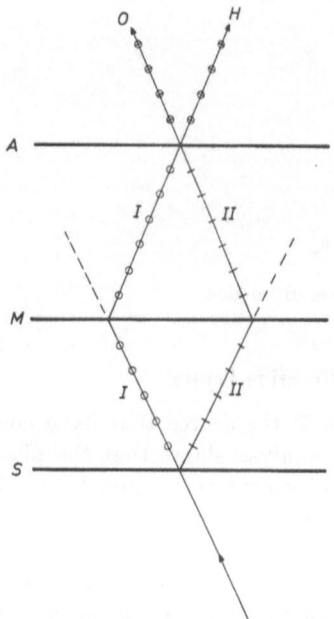

Fig. 4. Symmetric *L-L-L* interferometer

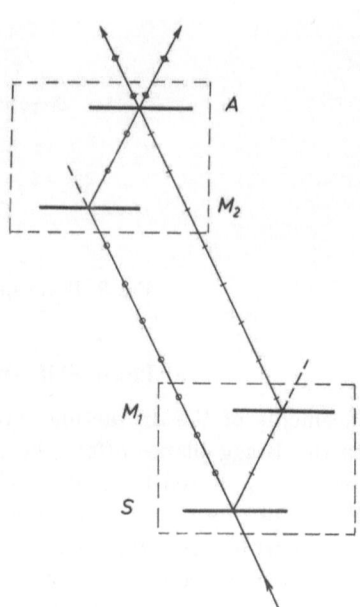

Fig. 5. Asymmetric *L-L-L* interferometer

problems are not so severe if the interferometer of Fig. 5 is used and if S and M_1 are made as one crystal and M_2 and A as a second crystal as illustrated in Fig. 5. Phase shifts caused by a relative displacement of the two crystals cancel in first order over path I and path II. This type has not yet been tried experimentally.

Shifts of the source are of no influence with either single or multiple crystal $L\text{-}L\text{-}L$ interferometers. The total transmitted intensity can be increased by using thin Laue crystals of appropriate thickness for S, M, and A.

b) *B-B-B* Interferometers

A $B\text{-}B\text{-}B$ interferometer [3] is shown in Fig. 6. A thin lamella separating two grooves acts as beam splitter and analyser. The outer walls of the grooves serve as mirrors. The grooves are cut at an angle φ with the Bragg planes in order to reduce absorption losses by shortening the path of the wavefield inside the crystal. High contrast fringes have been obtained with this type of interferometer [3].

Another possible design is the split lamella interferometer shown in Fig. 7. Like the split mirror interferometer of Fig. 5 it could be set up out of two crystals if extra long interfering paths are required. Again relative shifts between the crystals cancel if S and M_1 belong to one and M_2 and A to the other crystal. $B\text{-}B\text{-}B$ interferometers appear to be slightly more difficult to manufacture than $L\text{-}L\text{-}L$ interferometers.

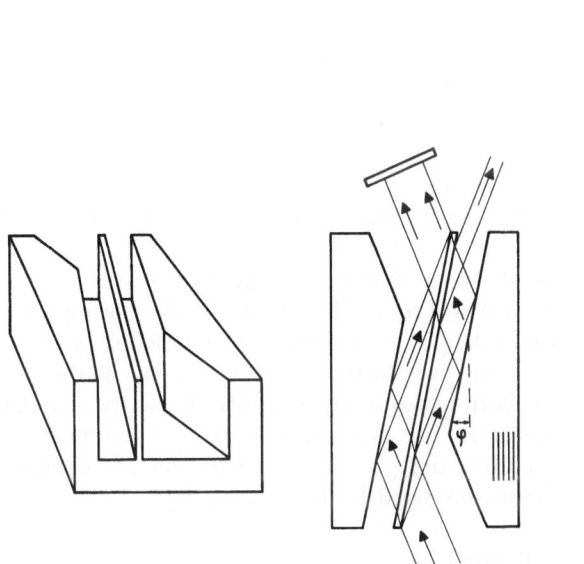

Fig. 6. Symmetric *B-B-B* interferometer

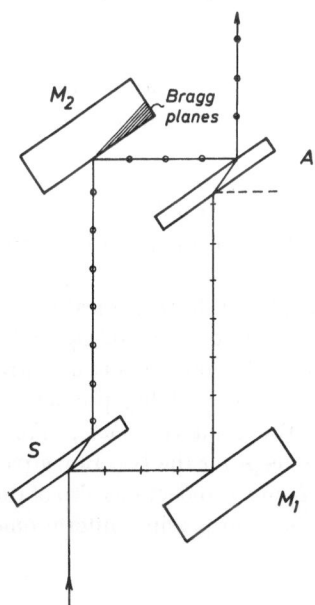

Fig. 7. Asymmetric *B-B-B* interferometer

c) *L-B-B-L* Interferometers

Thick Laue crystals are the most simple beam dividers and Bragg crystals are excellent mirrors. Thus a combination of both appears to be worth trying, particularly because a higher transmitted intensity should be attainable. There are however two restrictions. One is that the peaks of maximum diffraction do not occur at precisely the same angle in the Bragg and Laue cases but are separated by roughly the angular width of the Bragg peak, i.e. a few seconds of arc. Maximum transmitted intensity is obtainable only by "tuning" Bragg parts with Laue parts by appropriate tilts or temperature differences. The other is that in recombining the interfering beams exact overlap of coherent regions in them seems to be possible for one wavelength only, what is in contrast to pure L and pure B interferometers which are automatically coherent for a wide wavelength range.

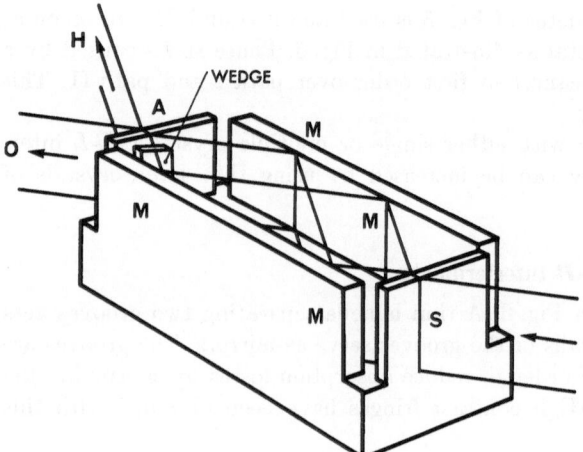

Fig. 8. Single crystal L-B-B-L interferometer

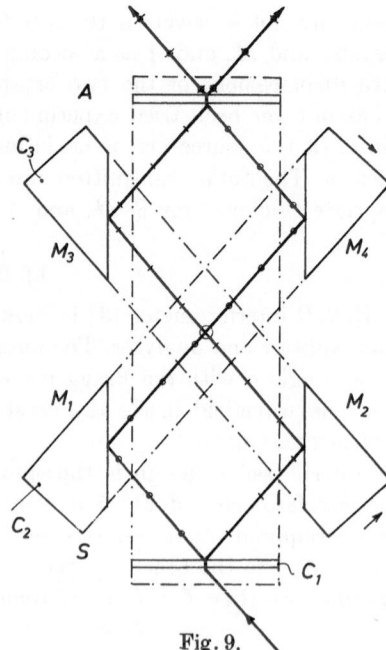

Fig. 9. Three-crystal L-B-B-L interferometer. Tuning is possible by rotating crystals C_2 and C_3 about central axis as indicated by small arrows

Fig. 9.

A study of the various thinkable combinations of L and B components reveals [4] that a promising kind is the L-B-B-L type shown in Fig. 8. Two Bragg reflections are interposed between the Laue diffractions because then defocussing effects due to nonparallellity of Bragg planes and mirror surfaces cancel to first order. Tuning can be achieved by heating crystals S and A. With silicon and Cu K_α radiation a temperature difference of about 17° C is necessary for the 220 reflection. With an interferometer of this type fringes have been observed without tuning [4].

A triple-crystal modification is show in Fig. 9, where A and S are parts of crystal C_1, M_1 and M_4 are parts of crystal C_2, and M_2 and M_3 are parts of crystal C_3. Tuning can be accomplished by rotating C_2 and C_3 about a central axis normal to the plane of the figure in a direction indicated by the arrows. Stability problems should be the same as with the split mirror Laue or the split lamella Bragg interferometer. This type has not been tested experimentally. Recently a similar device except for the two Bragg parts was suggested for use as X-ray resonator by COTTERIL [5]. Mixed L and B reflections also of different sets of diffracting planes for use with monochromators, resonators and possibly interferometers have been considered by DESLATTES [6].

3. Application

The possibility of making spacial phase measurements with the X-ray interferometer opens the way to X-ray phase contrast topography. It should be useful particularly with objects consisting of light elements like carbon, nitrogen, and oxygen that do not give considerable absorption contrast but would give phase contrast. Nothing is known about work actually done in this direction.

Of homogenous objects with throughout constant index of refraction thickness contours can be measured. KATO and TANEMURA [7] have applied the interferometer to measure the thickness of crystal wedges.

In the following some examples of more detailed studies with X-ray interferometers will be described.

a) Refractive Index Measurements

A method of performing precise measurements of the refractive index is illustrated in Fig. 10. The L-L-L interferometer of Fig. 4 is used with two wedges of the material to be measured located in the interfering paths as shown. The wedges are linked together and moved downwards.

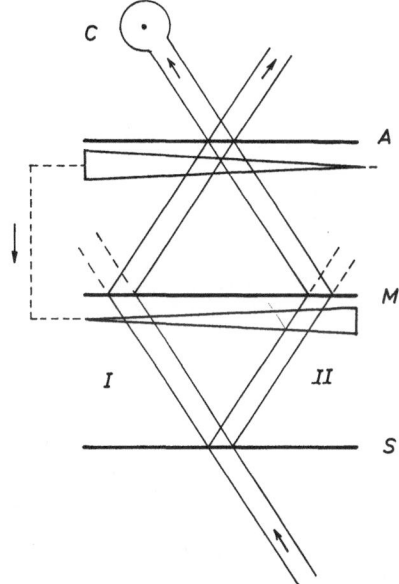

Fig. 10. Method for precision measurement of refractive index with L-L-L interferometer. C is a counter for counting moire fringe passages while wedges travel downwards as indicated

It is easy to see that they introduce a phase shift between paths I and II that is always constant over the cross section of the beams and varies linearly with wedge position. Over a full wedge travel a large number of high contrast fringes can be measured by means of the counter C. Knowing the wedge angle the deviation $\delta = 1 - n$ of the refractive index n from one can be evaluated very accurately.

So far lucite, beryllium, lithiumfluoride, and sodiumfluoride have been measured [8]. For Cu K_α radiation the following results were obtained:

	$\delta \times 10^6$ experiment	Theory	Deviation exp.-theory
Lucite	$4.135 \pm 0.13\%$	4.1353	-0.09%
Be	$5.296 \pm 0.16\%$	5.2947	-0.05%
LiF	$7.980 \pm 0.18\%$	7.9219	$+0.68\%$
NaF	$8.702 \pm 0.12\%$	8.675	$+0.27\%$

As can be seen the relative experimental errors are all below 2×10^{-3}, which demonstrates the precision of the method. The theoretical values have been corrected for dispersion. Very good agreement is obtained for lucite and Be. The cause of the larger discrepancy in the case of LiF is unknown.

b) Lattice Defects

As was pointed out, in the case of a L-L-L interferometer (Fig. 4), the intensity pattern in the exit beams O and H is essentially a moire pattern formed by the superposition of the Bragg planes of the analyser crystal A and the standing wave in front of it. Therefore, if crystal A contains lattice defects such as dislocations or stacking faults or simply strains due to local variations of the impurity concentration, they become visible in the moire pattern. Moreover, also defects within any of the other crystals, S or M, are imaged because they deform the standing wave pattern before A. By this method defects have been studied in silicon [1, 9], germanium [9], and quartz [9] crystals.

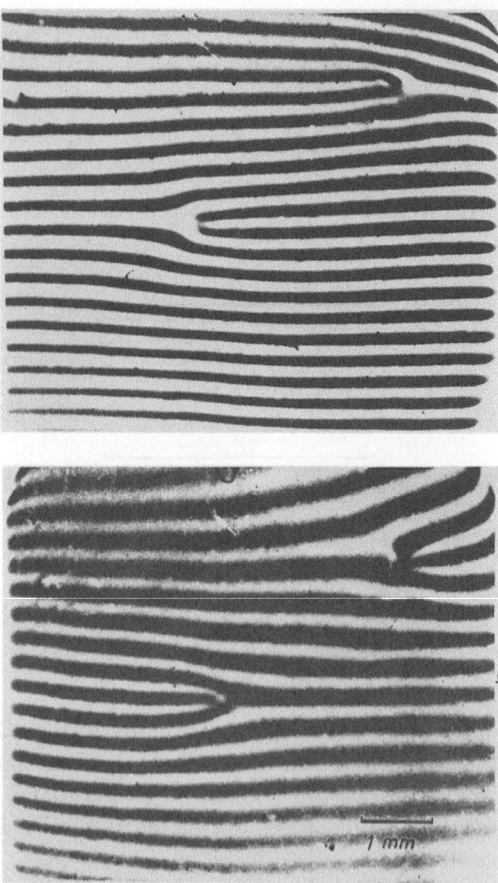

Fig. 11. Moire pattern of pair of dislocations with opposite Burger's vectors. The Bragg planes being vertical the pattern is a *rotational* moire pattern. Sign of relative rotation between perfect crystal and crystal containing dislocations is reversed between top and bottom picture. Note change of sides of moire half planes at dislocations. Silicon 220 reflection Cu K_α radiation

In Fig. 11 the rotational moire pattern of a pair of dislocations with opposite Burgers vectors is shown. It is seen that the moire half planes change sides when the sign of the relative rotation between the dislocated and the perfect lattice is reversed (Fig. 11 top and bottom).

It is worth noting that large crystal areas can be imaged at a time just as with ordinary X-ray topographic methods like the scanning method, the parallel beam, or the double crystal method, which can all be combined with the interferometric moire technique.

The sensitivity to lattice rotations and lattice dilations may be estimated as follows. With an imaged area that is 2 cm wide the maximum observable fringe distance Λ is also 2 cm. If Bragg planes of a spacing $d \approx 2$ angstroms (silicon 220 or germanium 220) are used then the smallest detectable strain $\Delta d/d = d/\Lambda = 10^{-8}$ and the smallest detectable rotation $\varrho = d/\Lambda = 10^{-8}$, what corresponds to about 2×10^{-3} sec of arc.

Generally speaking moire interferometry is a powerful tool for the investigation of defects in nearly perfect crystals.

c) Angstrom-Scale Length Measurements

The fringes of a moire pattern move by one fringe distance if the two gratings generating the pattern shift by one lattice parameter with respect to each other. Thus with a L-L-L interferometer the analyser A of which can be moved while S and M remain stationary, the Bragg plane spacing d can be measured by counting the number N of moire fringe passages for a macroscopically measureable analyser shift Δs: $d = \Delta s/N$. d can serve as a secondary length standard

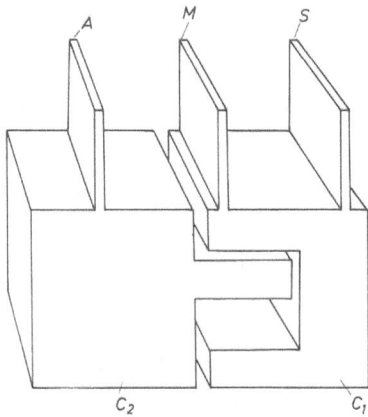

Fig. 12. Shape of crystals C_1 and C_2 of two-crystal interferometer. The groove below M is for fine adjustment about a horizontal axis by means of a multiply reflected X-ray beam passing the groove

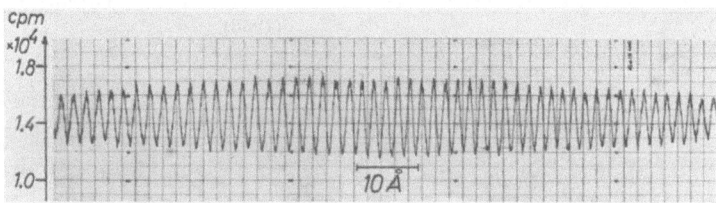

Fig. 13. Recorded moire fringe passages while crystal C_2 of Fig. 12 moves with respect to C_1 normal to Bragg planes

of the order of 10^{-10} m if Δs is directly measured by a coupled light interferometer in units of the internationally adopted standard light wave length. It may be pointed out that such a measurement of d does *not* depend on the precise knowledge of the wave length of the X-rays used. This is in contrast to the traditional d spacing measurement involving the measurement of a Bragg angle.

The main difficulty arising with an interferometric d measurement is to shift the analyser (or a corresponding crystal) *without* introducing misalignment and disturbance. Two slightly different experimental approaches to this end have been made recently. In either case the L-L-L interferometer is used.

BONSE and TE KAAT [10] mount the analyser A as a second independent crystal thus setting up a true two-crystal X-ray interferometer. The shape of the two crystals is illustrated in Fig. 12. Alignment of crystal C_1 with respect to crystal C_2 is achieved with the help of three X-ray beams that are diffracted by different sets of Bragg planes. The necessary stability is maintained by means of a temperature enclosure and an antivibration mount. Several hundred fringes could be counted in either exit beam when the analyser was shifted. An example of recorded fringes is shown in Fig. 13. The successful observation and control of moire fringes in a two crystal interferometer also demonstrates the feasibility of setting up larger multiple crystal interferometers that might be needed.

In the other approach by HART [11] the analyser crystal is part of the same crystal block as are crystals S and M but can be displaced by elastically deforming its link with the unit of S and M. The link consists of a pair of skillfully shaped leaf springs of crystalline material that represent a folded version of the spring strip traverse of JONES [12]. Also several hundred fringes could be observed. The device is considerably less sensitive to thermal and mechanical disturbances because it is much smaller and also more rigid than the above two-crystal interferometer. On the other hand it is size limited and alignment parameters cannot always be varied independently. Alignment is not too difficult because the initial single crystal orientation is maintained after cutting to some extent.

For a precision measurement of d some 10^6 fringes have to be counted. To accomplish this within a reasonable period of time a higher X-ray intensity is needed. As Hart [11] has shown some increase of intensity can be achieved by using thin Laue crystals.

The author likes to thank E. Te Kaat for valuable discussions. The Deutsche Forschungsgemeinschaft gave financial support which is gratefully acknowledged.

References

1. Bonse, U., and M. Hart: Appl. Phys. Letters 6, 155 (1965); — Z. Physik 188, 154 (1965); 190, 455 (1966).
2. — — Appl. Phys. Letters 7, 99 (1965).
3. — — Z. Physik 194, 1 (1966).
4. — — Acta Cryst. A 24, 240 (1968).
5. Cotterill, R. M. J.: Appl. Phys. Letters 12, 403 (1968).
6. Deslattes, R. D.: Appl. Phys. Letters 12, 133 (1968).
7. Kato, N., and S. Tanemura: Phys. Rev. Letters 19, 22 (1967).
8. Bonse, U., u. H. Hellkötter: Z. Physik 223, 345 (1969).
9. Hart, M.: Science Progress (Oxf.) 56, 429 (1968).
10. Bonse, U., and E. Te Kaat: Z. Physik 214, 16 (1968).
11. Hart, M.: Brit. J. Appl. phys. (J. phys. D), Ser. 2, 1, 1405 (1968).
12. Jones, R. V.: J. Sci. Instr. 28, 38 (1951). — Jones, R. V., and I. R. Young: J. Sci. Instr. 33, 11 (1956).

X-Ray Reflection Optics* (Recent Developments)

J. F. McGee, D. R. Hesser[1] and J. W. Milton[2]
Saint Louis University, Saint Louis, Missouri, U.S.A.

Abstract

Early X-ray microscopes of the reflection type were beset with geometrical aberrations which seriously limited their resolution. In addition, the problem was clouded by the role of surface defects to which an intolerable amount of non-symmetrical broadening of the image was attributed. Solution of the Kirchhoff diffraction integral shows that diffraction maxima are to be expected on the high angle side of the line image when spherical aberration is present and the magnification is greater than unity. A similar type of diffraction integral analysis leads to the discovery of the figure of an aspherical surface which is represented by a third order polynomial for which the spherical aberration is reduced to a negligible amount. The new figure increases the attainable resolution by an order of magnitude over that of spherical or cylindrical surfaces. A visible-light test of the aspherical figure at normal incidence is suggested by a solution of the diffraction integral for the case of reflection at normal incidence instead of grazing incidence as required for X-rays.

Introduction

As early as 1929 Jentzsch [1] showed that a concave surface would focus X-rays by total external reflection. It was by nature, a very astigmatic system, a point object giving rise to a line image. Much later Kirkpatrick [2] at Stanford University and Montel [3] at the University of Paris removed the astigmatism by placing two concave reflectors at right angles thereby accomplishing point to point imaging. Kirkpatrick placed one mirror behind the other; Montel juxtaposed them after making appropriate angular cuts. Since the astigmatism is readily removed by the right angle arrangement, the following discussion will be concerned with the optical properties of a single mirror in the meridional plane only. Ehrenberg [4] of the University of London earlier had noted that the line focus broadened considerably with an increase of exposure time. Although he attributed the broadening to scattering by surface defects, it is now clear that the excess broadening is a pure diffraction effect in the presence of a large amount of spherical aberration. The scattering observed in the neighborhood of a focused line image formed by X-rays totally reflected from a concave spherical or cylindrical polished reflector at grazing incidence has been theoretically and experimentally investigated [5]. It is assumed that all of the observed radiation is reflected. Because of the large wave-number involved, the Kirchhoff diffraction-integral may be used to evaluate the diffraction pattern produced by such a non-symmetrical optical system. Advantage is taken of the method of stationary phase in part of the analysis. The resulting asymmetric diffraction pattern is characterized at magnification $M > 1$ by a central maximum which drops off exponentially on the low-angle side and has secondary maxima on the high-angle side which are modulated by Fresnel terms. For magnification $M < 1$ the image inverts with the exponential decay side on the high-angle side of the central maximum.

* This work was in part supported by grants from: The National Institutes of Health, General Medical Science Division and the American Cancer Society.
1. Present address: McDonnell Douglas Corporation, St. Louis, Missouri.
2. Present address: Bishop McNamara High School, Kankakee, Illinois.

The diffraction results have proved useful in the understanding of previous results of Kirkpatrick and Baez as well as those of Ehrenberg and others.

A number of optical experiments are performed to illustrate and verify the results of the diffraction analysis.

The subject of resolving power is treated at length. Finally, the diffraction analysis has led to the discovery of the exact figure of an aspherical surface which effectively removes spherical aberration thus permitting an order of magnitude increase in resolving power. Applications of the results are possible in X-ray microscopy, X-ray telescopes and in medical radiography.

Theory

The diffraction of a beam of X-rays by a segment of a concave cylindrical mirror is to be formulated as a boundary value problem, considering only the meridian plane. We consider the diffraction of an electromagnetic wave polarized with its electric vector in the z direction (out of the paper) as shown in Fig. 1. The electric vector Ψ must satisfy the scalar Helmholtz equation

$$\nabla^2 \Psi + k^2 \Psi = 0 \tag{1}$$

and the Sommerfeld radiation condition; both conditions are contained in the integral equation

$$\Psi(\vec{r}) = \frac{1}{4\pi} \int_S [G(\vec{r_0^s}, \vec{r}) \, V_0 \Psi(\vec{r_0^s}) - \Psi(\vec{r_0^s}) \, V_0 G(\vec{r_0^s}, \vec{r})] \cdot dS \tag{2}$$

where $G(\vec{r_0}, \vec{r})$ is the Green's function, which is a function of the coordinates $\vec{r_0}$ of the source point (a, φ) and of the coordinates $\vec{r}$ of the point of observation (r, φ). The source point $\vec{r_0}$ is the cylindrical surface on which the boundary value of Ψ is prescribed. If the mirror, a metallic-coated glass-surface, is assumed to be a perfect conductor, then the boundary condition on the surface of the mirror is

$$\Psi(a, \varphi') = 0; \quad 2\pi - \alpha \leqq \varphi' \leqq \alpha \tag{3}$$

where 2α is the aperture angle of the mirror, measured from the center of curvature, and a is the radius of the cylindrical mirror. Outside of this range of φ', the boundary condition is unknown. Hence the integral equation is

$$\Psi(\vec{r}) = \frac{1}{4\pi} \int_0^{2\pi} G(\vec{r_0}, \vec{r}) \frac{\partial \Psi}{\partial a} a \, d\varphi' - \frac{1}{4\pi} \int_\alpha^{2\pi - \alpha} \Psi(\vec{r_0}) \frac{\partial G(\vec{r_0}, \vec{r})}{\partial a} a \, d\varphi'. \tag{4}$$

Since α is a small parameter in X-ray microscopes, a perturbation solution is sought in the form $\Psi = \Psi_0 + \Psi_1$, where Ψ_0 satisfies the unperturbed equation, i.e., Eq. (4) with $\alpha = 0$. Substituting

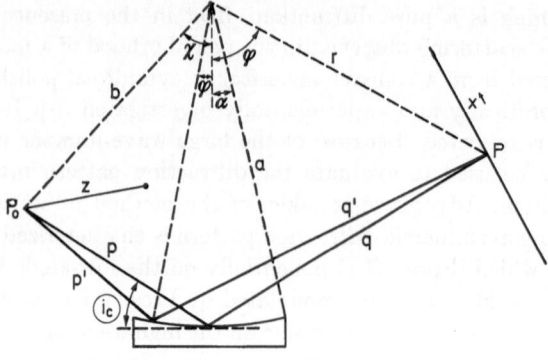

Fig. 1. Polar coordinate system used to locate the source $P_0(b, \chi)$ of a cylindrical wave diffracted by a mirror of aperture 2α, whose center is located a distance p from the source. The observation plane is located a distance q from the center of the mirror and is perpendicular to the chief ray (p, q). The angle i_c made by the chief ray and the tangent plane at the center of the mirror may be taken equal to the critical angle

this into Eq. (4) yields the exact integral equation for Ψ_1

$$\Psi_1 = \frac{1}{4\pi} \int\limits_{-\alpha}^{+\alpha} \Psi_0(\vec{r}_0) \frac{\partial G}{\partial a} a \, d\varphi' + \frac{1}{4\pi} \int\limits_{0}^{2\pi} G(\vec{r}_0, \vec{r}) \frac{\partial \Psi_1}{\partial a} a \, d\varphi' - \frac{1}{4\pi} \int\limits_{\alpha}^{2\pi-\alpha} \Psi_1(\vec{r}_0) \frac{\partial G}{\partial a} a \, d\varphi'. \tag{5}$$

This equation can be solved approximately by using the Born approximation, which neglects the terms in Ψ_1 on the right hand side of Eq. (5), and uses the unperturbed (free space) Green's function

$$G = \pi i H_0(k q'), \quad q'^2 = r^2 + a^2 - 2 a r \cos(\varphi' - \varphi) \tag{6}$$

in the remaining integral. We take Ψ_0 to be a cylindrical wave originating at P_0 (see Fig. 1),

$$\Psi_0 = H_0(k p'); \quad p'^2 = a^2 + b^2 - 2 a b \cos(\varphi + \chi). \tag{7}$$

Performing the indicated differentiation in Eq. (5) yields

$$\Psi_1 = -\frac{i a k}{4} \int\limits_{-\alpha}^{+\alpha} H_0(k p') H_1(k q') \frac{a - r \cos(\varphi' - \varphi)}{q'} d\varphi' \tag{8}$$

which reduces to

$$\Psi_1 = \frac{i a}{2\pi} \frac{a - r \cos\varphi}{p^{\frac{1}{2}} q^{\frac{3}{2}}} \int\limits_{-\alpha}^{+\alpha} e^{ik(p'+q')} d\varphi' \tag{9}$$

by using the asymptotic form of the Hankel functions, and by taking the slowly varying functions outside of the integral. The integral

$$J = \int\limits_{-\alpha}^{+\alpha} e^{ik(p'+q')} d\varphi' \tag{10}$$

appearing in Eq. (9) is just the Kirchhoff diffraction integral. It is evaluated by expanding the quantity $(p'+q')$ in a power series about $\varphi'=0$,

$$(p'+q') = \beta_0 + \beta_1 \varphi' + \beta_2 \varphi'^2 + \beta_3 \varphi'^3 + \cdots. \tag{11}$$

It is desirable to express the β's in terms of i_c, x, p, q, and a rather than r and φ (Fig. 2). The coefficients are, with $\beta_i = c_i + g_i x$,

$$
\begin{aligned}
c_0 &= p + q; & g_0 &= 0 \\[4pt]
c_1 &= 0; & g_1 &= \frac{2f}{q} \\[4pt]
c_2 &= 2f^2 \left(\frac{1}{p} + \frac{1}{q} \right) - 2f; & g_2 &= \frac{a \cos i_c}{2q} \left(\frac{4f}{q} - 1 \right) \\[4pt]
c_3 &= -2f^2 a \cos i_c \left(\frac{1}{p^2} - \frac{1}{q^2} \right) + f a \cos i_c \left(\frac{1}{p} - \frac{1}{q} \right) \\[4pt]
g_3 &= -\frac{4f^3}{q^3} + \frac{2f}{q^3} a^2 \cos^2 i_c + \frac{2f^2}{q^2} - \frac{a^2 \cos^2 i_c}{2q^2} - \frac{f}{3q}
\end{aligned}
\tag{12}
$$

where

$$f = \frac{a}{2} \sin i_c$$

is the meridian focal length. When

$$\frac{1}{p} + \frac{1}{q} = \frac{1}{f}$$

then

$$
\begin{aligned}
g_1 &= \frac{2}{M+1}; & c_2 &= 0; & g_2 &= \frac{a \cos i_c}{2q} \frac{3-M}{M+1} \\[4pt]
c_3 &= -a \cos i_c \frac{M-1}{M+1}; & & & g_3 &= \frac{a^2 \cos^2 i_c}{2q^2} \frac{3-M}{M+1} - \frac{1}{3} \frac{M^2 - 4M + 7}{(M+1)^3}.
\end{aligned}
\tag{13}
$$

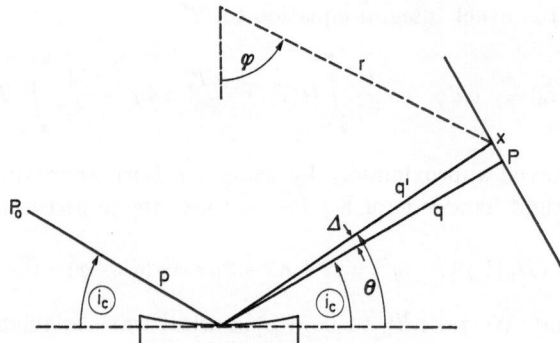

Fig. 2. New coordinates (q, θ) are used instead of (r, φ) to locate a point in the observation plane

The Kirchhoff integral is now, to third order in φ'

$$J = \int_{-\alpha}^{+\alpha} \exp[i\, k(\beta_1 \varphi' + \beta_2 \varphi'^2 + \beta_3 \varphi'^3)]\, d\varphi'. \tag{14}$$

The square term may be eliminated by the substitution

$$\varphi' = \varphi'' - \frac{\beta_2}{3\beta_3} \tag{15}$$

so that

$$J = e^{ikC} \int_{-\alpha+B}^{\alpha+B} \exp[i\, k(A\,\varphi'' + \beta_3 \varphi''^3)]\, d\varphi'', \tag{16}$$

where

$$A = \beta_1 - \frac{\beta_2^2}{3\beta_3}; \quad B = \frac{\beta_2}{3\beta_3}; \quad C = \frac{2}{27}\frac{\beta_2^3}{\beta_3^2} - \frac{\beta_1 \beta_2}{3\beta_3}.$$

The integral of Eq. (16) may now be evaluated by the method of stationary phase [6, 7]. The stationary points are given by

$$\varphi_0 = \pm\, \delta = \pm\, \sqrt{\frac{-A}{3\beta_3}}. \tag{17}$$

The substitution of $\varphi'' = \delta t$ in Eq. (16) yields

$$J = e^{ikC}\, \delta \int_{t_1}^{t_2} \exp[i\, k\beta_3 \delta^3 (t^3 - 3t)]\, dt. \tag{18}$$

The J integral may now be written as

$$J = e^{ikC}\, \delta \left[2\int_0^\infty \cos k\beta_3 \delta^3 (t^3 - 3t)\, dt - \int_{t_2}^\infty \exp i\, k\beta_3 \delta^3 (t^3 - 3t)\, dt - \int_{-t_1}^\infty \exp i\, k\beta_3 \delta^3 (t^3 - 3t)\, dt \right] \tag{19}$$

where the first integral is the Airy integral, and the remaining two integrals are shown to be Fresnel integrals by the method of stationary phase. In the last two integrals, the substitution

$$\frac{\pi}{2} u^2 = k\beta_3 \delta^3 (-t^3 + 3t + 2) \tag{20}$$

is made, and, by using the Lagragian inversion formula, it follows that

$$\frac{dt}{du} = \sqrt{\frac{\pi}{3z}} + \cdots, \quad z = 2k\beta_3 \delta^3. \tag{21}$$

The integral for J becomes

$$J = e^{ikC}\, \delta \left[\int_0^\infty \cos\frac{z}{2}(t^3 - 3t)\, dt - \sqrt{\frac{\pi}{3z}}\int_{\omega_2}^\infty e^{iz - i\frac{\pi}{2}u^2}\, du - \sqrt{\frac{\pi}{3z}}\int_{\omega_1}^\infty e^{-iz + i\frac{\pi}{2}u^2}\, du \right] \tag{22}$$

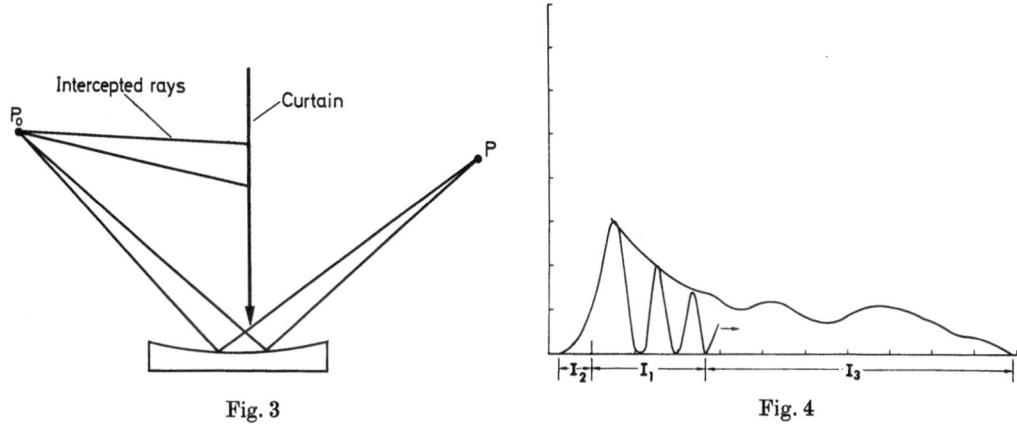

Fig. 3   Fig. 4

Fig. 3. A curtain or stop is used to insure that only the reflected rays reach the observation or image plane

Fig. 4. A typical intensity distribution given by Eq. (27). For x slightly greater than zero the Bessel functions $J_{\pm\frac{1}{3}}$ control the intensity as shown in region I_1. For x less than zero an approximate form of Eq. (27) shows that the intensity falls off essentially in an exponential manner as shown in region I_2. Finally for large positive values of x all terms of Eq. (27) are important. The modulation of the maxima of the intensity function by the Fresnel terms can be seen in region I_3 where only the envelope of the intensity distribution is drawn

where

$$\omega_1 = \sqrt{\frac{z}{\pi}(t_1^3 - 3t_1 + 2)}, \quad \omega_2 = \sqrt{\frac{z}{\pi}(-t_2^3 + 3t_2 + 2)}.$$

Finally,

$$J = e^{ikC}\,\delta\left[\frac{2\pi}{3}\left(J_{\frac{1}{3}}(z) + J_{-\frac{1}{3}}(z)\right) - \sqrt{\frac{\pi}{3z}}\sin\left(z + \frac{\pi}{4}\right) - \sqrt{\frac{\pi}{3z}}\left(e^{iz}\,\overline{\Phi}(\omega_2) + e^{-iz}\,\Phi(\omega_1)\right)\right] \qquad (23)$$

where

$$\Phi(\omega) = C(\omega) + i\,S(\omega) \qquad (24)$$

is the complex Fresnel integral, and the bar signifies the complex conjugate.

The intensity is usually found from

$$I \sim (\Psi_0 + \Psi_1)(\overline{\Psi}_0 + \overline{\Psi}_1) \qquad (25)$$

but, because of the "curtain" shown in Fig. 3, the incident wave, Ψ_0, is taken as zero in image space. This is approximated here by determining the intensity solely from

$$I \sim \Psi_1 \overline{\Psi}_1 \qquad (26)$$

or

$$\begin{aligned}
\frac{I}{I_0} = {}& \frac{\delta^2}{4\alpha^2}\left[\frac{2\pi}{3}\left(J_{\frac{1}{3}}(z) + J_{-\frac{1}{3}}(z)\right) - \sqrt{\frac{2\pi}{3z}}\sin\left(z + \frac{\pi}{4}\right)\right. \\
&\left. + \sqrt{\frac{\pi}{3z}}\{(C_1 + C_2)\cos z + (S_1 + S_2)\sin z\}\right]^2 \\
&+ \frac{\delta^2}{4\alpha^2}\left[\sqrt{\frac{\pi}{3z}}\{(S_1 - S_2)\cos z - (C_1 - C_2)\sin z\}\right]^2
\end{aligned} \qquad (27)$$

in which the constant I_0 is chosen so that when the Fraunhofer pattern is obtained, its maximum intensity is unity. The behavior of a typical intensity distribution is illustrated in Fig. 4. For a magnification $M = 3$, the intensity distribution can be evaluated in terms of the two dimensionless parameters

$$y = \frac{\alpha x}{6\lambda}$$

$$\varepsilon = \frac{a\,\alpha^3 \cos i_c}{2\lambda} \qquad (28)$$

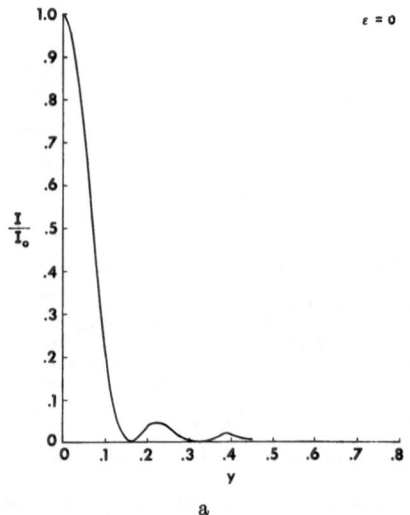

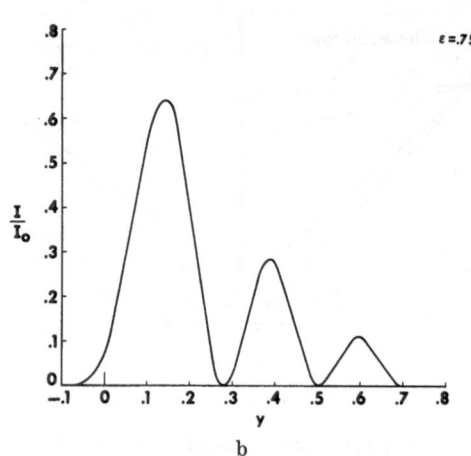

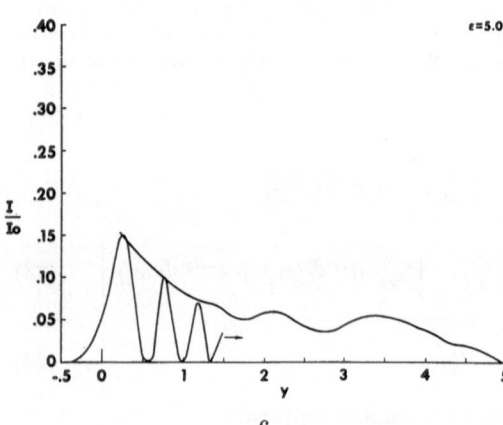

Fig. 5a. A theoretical intensity distribution with $M = 3$ as a function of the dimensionless parameters y and ε of Eq. (28) with $\varepsilon = 0$. The negative half of the symmetrical Fraunhofer pattern is not shown

Fig. 5b. Changing from $\varepsilon = 0$ to $\varepsilon = 0.75$ causes the intensity distribution to become asymmetrical as described under Fig. 4. The intensity of the principal maximum is lowered and shifted to a higher angle while the relative intensity of the secondary maxima increases

Fig. 5c. Changing from $\varepsilon = 0.75$ to $\varepsilon = 5.0$ causes a further decrease in the intensity and a shift to a still higher angle by the principal maximum. The relative intensity of the secondary maxima again increases while the overall breadth of the diffraction pattern increases and the modulation by the Fresnel terms becomes obvious

where ε is the number of wavelengths of spherical aberration. At $\varepsilon = 0$, the typical Fraunhofer pattern is obtained as in Fig. 5a. The remaining Figs. 5b, 5c for $\varepsilon = 0.75$ and $\varepsilon = 5$ respectively show that as ε increases, the principal maximum decreases and shifts to the high angle side. The secondary maxima on the low-angle side of the Fraunhofer ($\varepsilon = 0$) pattern (not shown) flatten out so that the intensity decreases exponentially. The secondary maxima on the high angle or $+ y$ side increase, and finally, as shown for $\varepsilon = 5$, the amplitude of the secondary maxima are modulated by Fresnel like fringes. Any further increase in ε will obviously widen the image even further.

The type of diffraction image to be obtained in an X-ray microscope can be determined by substituting typical values into (28). For instance, with $a = 325$ cm, $i_c = 55$ milliradian, $s = 2$ cm, $\lambda = 8.34$ Å, then one unit length, y, in the image plane corresponds to 16,263 Å, while ε is 57. It is apparent that, for the values used here, the image will not be good.

For a fixed radius of curvature, an increase in ε corresponds to an increase in the length of the mirror, since the angle of grazing incidence cannot exceed the critical angle. This indicates that one method of reducing the spherical aberration is to use aperture stops or slits to reduce the length of mirror illuminated by the X-ray beam. To obtain a quarter wavelength of spherical aberration, s should be 0.33 cm. This method was used by McGee and Milton [8] in connection with the simultaneous removal of the obliquity aberration.

A second method of improving the resolution concerns the figure of the mirror. This has been studied by several authors, Baez and Weissbluth [9], Dyson [10], and Baez [11]. At

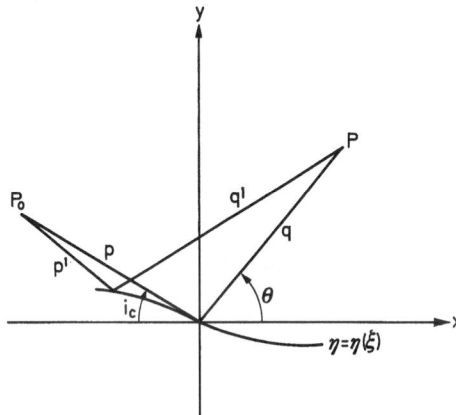

Fig. 6. The surface whose equation is to be determined under the condition $c_3 = 0$ is represented by $\eta = \eta(\xi)$. The object point is $P_0(x_0, y_0)$ and the image point is $P(x_i, y_i)$. The Cartesian coordinate system is so chosen that the surface passes through $(0, 0)$ and is tangent to the x axis at $(0, 0)$

normal incidence, the standard procedure to remove spherical aberration is to replace the spherical surface by a parabolic surface, as in a telescope. BAEZ and WEISSBLUTH [9] have made ray-tracing calculations for several conic sections. For point objects, they found that an ellipse gives the least spherical aberration at grazing incidence. However, as the object field increases, the blurring of the image soon becomes greater than that of a sphere.

BAEZ [11] had also made ray tracing calculations based on a cubic reflecting surface. Although his results were incomplete, it appeared possible that a cubic surface would produce a smaller blur than a conic section. The present diffraction analysis can also be applied to the problem of determining the best surface, resulting in an analytic equation for the surface in terms of the magnification, angle of incidence, etc.

The diffraction pattern spreads out principally because of the term c_3. This is usually very large at grazing incidence. To eliminate the spreading, a surface should be found for which $c_3 = 0$. The procedure for doing this is given below.

The starting point is again the Kirchhoff diffraction integral

$$J = \int e^{ik(p'+q')} dS. \tag{29}$$

In Fig. 6 is shown an arbitrary surface $\eta = \eta(\xi)$ whose equation is to be determined, and the cartessian coordinate system used. The expansion of $(p'+q')$ has the form

$$p' + q' = \sqrt{(\xi - x_0)^2 + (\eta - y_0)^2} + \sqrt{(\xi - x_i)^2 + (\eta - y_i)^2} = \beta_0 + \beta_1 \xi + \beta_2 \xi^2. \tag{30}$$

The equation of the surface is assumed to be written in the form

$$\eta = a_0 + a_1 \xi + \frac{a_2}{2!} \xi^2 + \frac{a_3}{3!} \xi^3 + \cdots. \tag{31}$$

It is obvious that the coordinate system can be chosen so that the surface passes through $(0, 0)$. This gives $a_0 = 0$. Also the coordinate system may be oriented so that the surface is tangent to the x-axis at the origin. This requires that $a_1 = 0$. The surface equation is now

$$\eta = \frac{a_2}{2!} \xi^2 + \frac{a_3}{3!} \xi^3 + \cdots. \tag{32}$$

The radius of curvature at the origin is

$$R = \frac{[1 + \eta'^2(0)]^{\frac{3}{2}}}{\eta''(0)} \tag{33}$$

and from Eq. (32)

$$R = \frac{1}{a_2} \quad \text{or} \quad a_2 = \frac{1}{R}. \tag{34}$$

With this equation, a focal length for the mirror can be defined in terms of the radius of curvature at the origin,

$$f_c = \frac{R}{2} \sin i_c. \tag{35}$$

The equation of the surface is now written as

$$\eta = \frac{\xi^2}{2R} + \frac{a_3}{6}\xi^3 + \cdots. \tag{36}$$

The following are the results of expanding Eq. (30) using Eq. (35), and the lens equation

$$\frac{1}{p} + \frac{1}{q} = \frac{1}{f_c}. \tag{37}$$

Then, the expansion coefficients are

$$\beta_0 = p + q; \quad \beta_1 = g_1 x; \quad \beta_2 = c_2 + g_2 x; \quad \beta_3 = c_3 + g_3 x$$

$$g_1 = \frac{2f_c}{qR}$$

$$c_2 = 0; \quad g_2 = \frac{\cos i_c}{qR}\left(\frac{2f_c}{q} - \frac{1}{2}\right) \tag{38}$$

$$c_3 = -\frac{f_c}{R^2}\cos i_c\left(\frac{1}{p} - \frac{1}{q}\right) - \frac{2}{3}\frac{f_c}{R}\eta'''(0)$$

$$g_3 = \frac{4f_c}{qR}\frac{\cos^2 i_c}{q^2} - \frac{4f_c^3}{q^3 R^3} - \frac{1}{2}\frac{\cos^2 i_c - \sin^2 i_c}{q^2 R} - \frac{\cos i_c}{6q}\eta'''(0)$$

and, from Eq. (36)

$$\eta'''(0) = a_3 \equiv 6\varepsilon_0. \tag{39}$$

The term required to remove c_3 is then found from Eq. (38) by setting $c_3 = 0$ and solving for a_3,

$$a_3 = -\frac{3}{2}\frac{\cos i_c}{R}\frac{q-p}{pq}$$

$$a_3 = -\frac{3}{2}\frac{\cos i_c}{R}\frac{M-1}{Mp} \tag{40}$$

$$\varepsilon_0 = -\frac{\cos i_c (M-1)}{4RMp}.$$

Hence, a cubic surface can be found to remove c_3 and thus eliminate a large amount of the spherical aberration. Some aberration is still contained in g_3. Two facts should be emphasized here. First, the conic sections do not contain a cubic term in their expansion, and second, that it is unnecessary to use higher order surfaces to remove the spherical aberration.

It is interesting to note that Baez [11] with the aid of a ray tracing program determined the geometrical "blur" for various values of ε_0 of a third-order surface. Using a $R = 440$ cm, $p = 2.0$ cm, $q = 147$ cm and $i_c = 8.97 \times 10^{-3}$ radians he plotted "blur" versus ε_0 over a range of positive and negative values of ε_0. Unfortunately he terminated his calculations when the "blur" was heading downward for an $\varepsilon_0 = -8 \times 10^{-6}$. If he had continued his calculations further into the negative region of ε_0 he would no doubt have found that the "blur" is a minimum for $\varepsilon_0 = -2.8 \times 10^{-4}$, the value calculated from Eq. (40) on the basis of the present diffraction analysis.

The diffraction integral now becomes

$$J = C\int_{\xi_1}^{\xi_2} e^{ik(\beta_1\xi + \beta_2\xi^2 + \beta_3\xi^3)} d\xi \sqrt{1 + \eta'^2}. \tag{41}$$

The square root may be taken outside, since it is a slowly varying function. The resulting integral may be integrated by parts immediately

$$J = \frac{C}{ikx} \frac{e^{ikx(g_1\xi + g_2\xi^2 + g_3\xi^3)}}{g_1 + 2g_2\xi + 3g_3\xi^2}\bigg|_{\xi_1}^{\xi_2}. \tag{42}$$

If $\xi_1 = -\xi_2$ and if $2g_2\xi$ can be neglected, then

$$\frac{I}{I_0} = \left(\frac{g_1\xi_2 + g_3\xi_2^3}{g_1 + 3g_2\xi_2^2}\right)^2 \frac{\sin^2 kx\xi_2(g_1 + g_3\xi_2^2)}{[kx\xi_2(g_1 + g_3\xi_2^2)]^2} \tag{43}$$

which is just the Fraunhofer intensity pattern, multiplied by a constant factor in front. The Rayleigh criterion may be applied to this directly to yield the resolvable distance

$$\text{R.D.} = \frac{1}{M}\frac{\lambda}{2\xi_2(g_1 + g_3\xi_2^2)}. \tag{44}$$

This is better than that given by a cylinder, and since spherical aberration will be absent a larger aperture can be used. In the limit of large magnification, the R.D. is given by

$$\text{R.D.} = \frac{\lambda}{4\xi_2}R. \tag{45}$$

If $\xi_2 = \dfrac{R\theta}{2}$, then this reduces to the standard result, except for the coefficient in front. An exact calculation for the case of normal incidence given this coefficient as 0.62. Eq. (45) also shows that better resolution can be obtained by using shorter radii of curvature.

Optical Experiments

a) Images Formed by Spherical Reflectors

Several optical experiments were performed to confirm the analysis presented above. Light from a gas laser ($\lambda = 6{,}328$ Å) was focused onto a pinhole in a metal sheet, and reflected by a spherical aluminized mirror at grazing-incidence angles of approximately ten degrees. The reflected beam is observed either on a screen or photographic film. An experimental diffraction pattern is shown in Fig. 7, in which both the sharp fringes of the diffraction pattern and the modulating Fresnel-like fringes are visible as predicted by Eq. (27). A densitometer trace made from an original negative is shown in Fig. 8. Note the lack of reproduction of the Fresnel-like

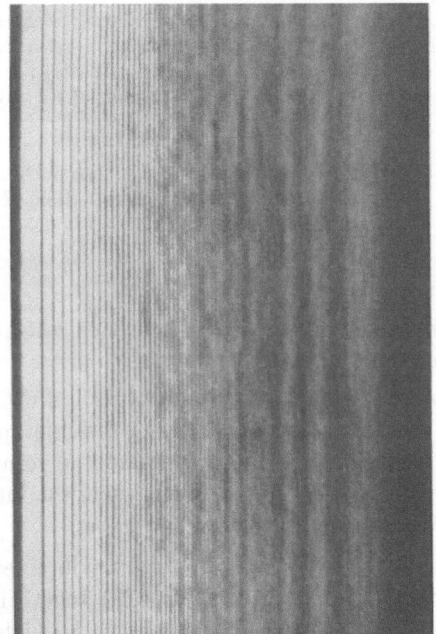
Fig. 7

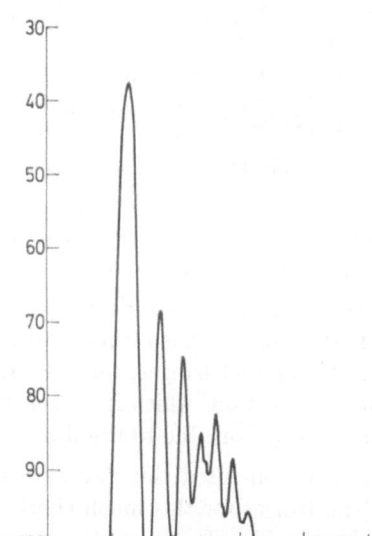
Fig. 8

Fig. 7. Diffraction pattern formed by a single spherical reflector illuminated at grazing incidence with light ($\lambda = 6{,}328$ Å) from a point source. The light distribution should be compared with the theoretical distribution shown in Fig. 4, calculated using Eq. (27). A large amount of spherical aberration is evident

Fig. 8. A densitometer tracing made from an original negative of a diffraction pattern similar to that shown in Fig. 7. Note the failure of the densitometer to reproduce the Fresnel modulating fringes

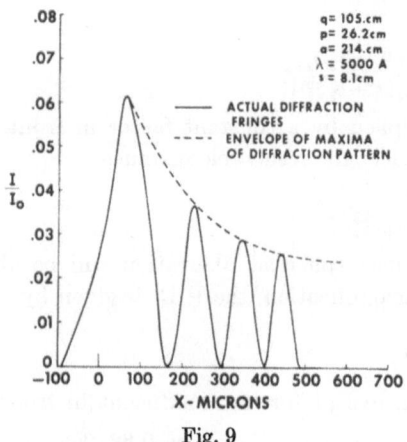

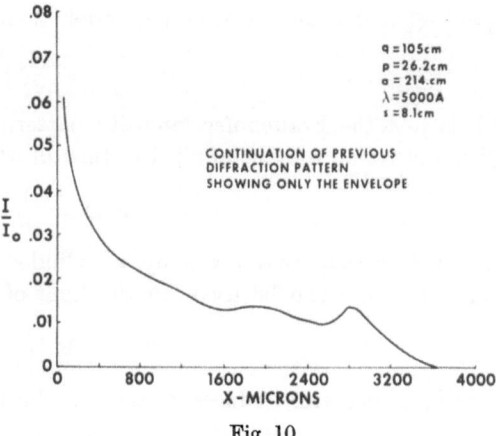

Fig. 9   Fig. 10

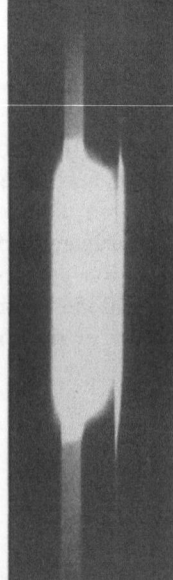

Fig. 9. Incomplete intensity distribution computed from Eq. (27) using the following parameters. $q = 105$ cm; $p = 26.2$ cm; $a = 214$ cm; $\lambda = 5,000$ A and $s = 8.1$ cm

Fig. 10. Since the width of the mirror $s = 8.1$ cm is large, a considerable amount of spherical aberration is present, resulting in a very broad diffraction pattern. Only the envelope of the diffraction pattern is drawn. Compare the x scale in microns of Fig. 9 with that of Fig. 10

Fig. 11. An X-ray reflection image of a slit formed by a six inch diameter concave glass reflector of 3 m radius with $\lambda = 1.54$ A radiation. Exposure time was four hours

Fig. 11

fringes by the densitometer. Another general conclusion of the analysis is that the pattern changes direction from $+ x$ of Fig. 2 to $- x$ for magnifications less than one.

Corresponding theoretical diffraction patterns are shown in Figs. 9 and 10. The first, Fig. 9 shows a portion of the sharp fringes, while Fig. 10 shows the envelope of the diffraction pattern including the Fresnel fringes, for $M = 4$. Densitometry data used to compare the results are shown in Table 1, from which it is seen that the fringe spacings agree well, while the intensity agreement is only fair, due to the difficulty in obtaining good densitometer traces.

The illustrations used thus far were for visible light. Fig. 11 shows a result obtained with X-rays. The fringes for this much shorter wavelength ($\lambda = 1.54$ Å) are too close together to be resolved by the film. However, the expected spreading out of the envelope on the high angle side can be seen quite well in this four-hour exposure. The weaker nonmeridional rays outline the slit image on the right and the left and indicate a lower angular limit to the image. A series of increasingly longer exposures, starting from one second, show an ever widening image as more of the diffraction pattern in the presence of severe spherical aberration is recorded with increasing time.

b) Images Formed by Aspheric Reflectors

A mirror whose contour in the meridian plane is a cubic curve will have the effect of reflecting rays from the source in a way that directs them more toward the Gaussian image-point than a spherical or cylindrical reflector will. Fig. 12 compares a cubic and a spherical reflector having the same value of R.

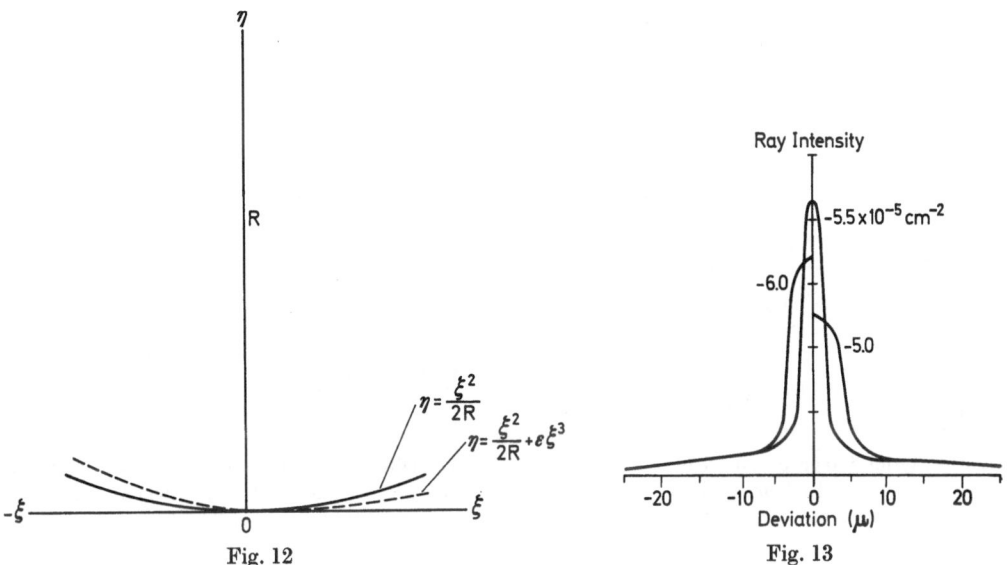

Fig. 12   Fig. 13

Fig. 12. Graphical difference between a cylinder of large radius R and a cubic surface with a negative ε_0. At one cm from the origin the difference in η amounts to 5.5×10^{-5} cm for a surface whose $\varepsilon_0 = -5.5 \times 10^{-5}$ cm^{-2}

Fig. 13. Results of a ray tracing study showing the effect of using a surface whose value of ε_0 is incorrect for given values of p, q, R and i according to Eq. (46)

A computer ray-tracing for a particular system was repeated for a number of values of the cubic coefficient ε_0. For a mirror with $R = 325$ cm, $i = 0.055$ rad, $p = 10.88$ cm, $q = 50.00$ cm, the value of ε_0 which gave the smallest deviation of rays was about $\varepsilon_0 = -5.5 \times 10^{-5}$ cm^{-2}.

Fig. 13 shows the geometrical ray-density or intensity diagram for three values of ε_0. It can be seen that for ε_0 too small in magnitude, the intensity is predominantly to the high angle side, and for ε_0 too large in magnitude, the ray densities are greatest in the low angle direction. For the value of ε_0 which gives the smallest deviation, the ray densities are distributed evenly on either side of the center; this suggests the envelope of a Fraunhofer type pattern.

The previous diffraction analysis shows that the spherical aberration can be almost totally removed by the use of a reflector whose contour is a cubic curve with the value of ε_0 given by the relation

$$\varepsilon_0 = -\frac{(M-1)\cos i}{4RMp}. \tag{46}$$

The value of ε_0 calculated from this expression using the same values of R, i, p and q as in the ray-tracing described above is $\varepsilon_0 = -5.54 \times 10^{-5}$ cm^{-2}. The same agreement was found for widely different system parameters.

An aspheric reflector [12] was produced with the following dimensions: $R = 243$ cm, $\varepsilon_0 = -2.54 \times 10^{-5}$ cm^{-2}, half-width $\xi_2 = 2$ cm. Images of a point source ($\lambda = 6,328$ Å) were obtained for a grazing incidence angle of 8° and various values of p and q. The parameters of the optical system were varied in such a way as to require different values of ε_0 to eliminate the spherical aberration. Fig. 14a—e shows some of the images obtained for the object and image distances of Table 1. The values of ε_0 required to remove theoretical the spherical aberration for these distances, best matches that of the actual aspheric reflector for Fig. 14c.

Table 1

Figure	p (cm)	q (cm)	ε_0 (10^{-5} cm^{-2})
14a	32.5	32.5	0.00
14b	25.0	45.0	−1.92
14c	23.5	51.5	−2.48
14d	22.0	61.5	−3.14
14e	21.0	69.5	−3.58

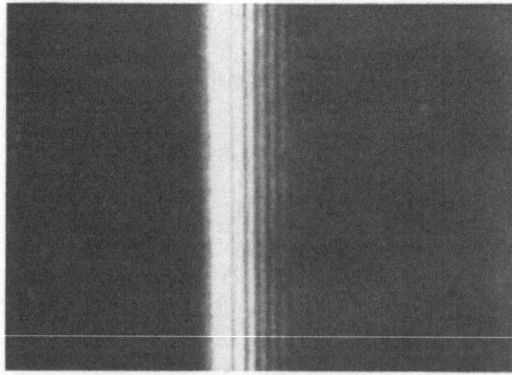

Fig. 14a. Diffraction image produced by a cubic surface whose $\varepsilon_0 = -2.5 \times 10^{-5}$ cm^{-2} when Eq. (46) requires that $\varepsilon_0 = 0.00$

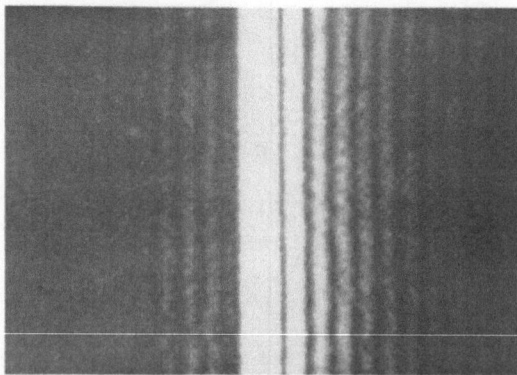

Fig. 14b. Diffraction image produced by a cubic surface whose $\varepsilon_0 = -2.5 \times 10^{-5}$ cm^{-2} when Eq. (46) requires that $\varepsilon_0 = -1.92 \times 10^{-5}$ cm^{-2}

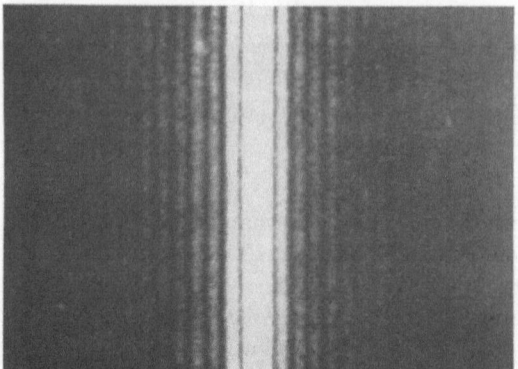

Fig. 14c. Diffraction image produced by a cubic surface whose $\varepsilon_0 = -2.5 \times 10^{-5}$ cm^{-2} when Eq. (46) requires that $\varepsilon_0 = -2.48 \times 10^{-5}$ cm^{-2}. Note the almost symmetrical diffraction pattern indicating a great reduction in spherical aberration

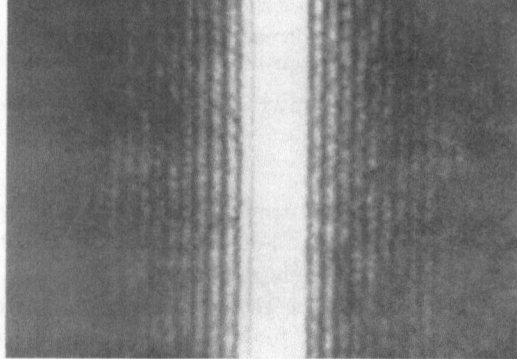

Fig. 14d. Diffraction image produced by a cubic surface whose $\varepsilon_0 = -2.5 \times 10^{-5}$ cm^{-2} when Eq. (46) requires $\varepsilon_0 = -3.14 \times 10^{-5}$ cm^{-2}. Note the change in the direction of the asymmetry in agreement with the ray tracing study shown in Fig. 13

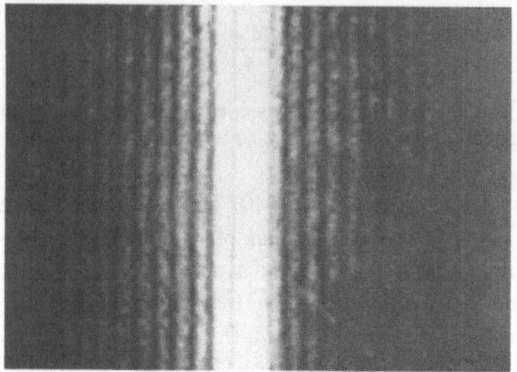

Fig. 14e

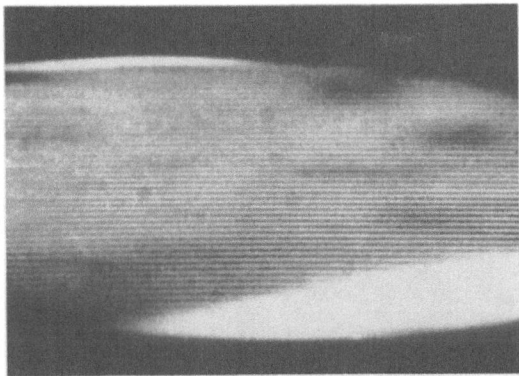

Fig. 15. Single mirror grazing incidence image of a wire mesh. Bars are 14 microns wide. The spacing between the edges of adjacent bars is 36 microns; $\lambda = 6{,}328$ A

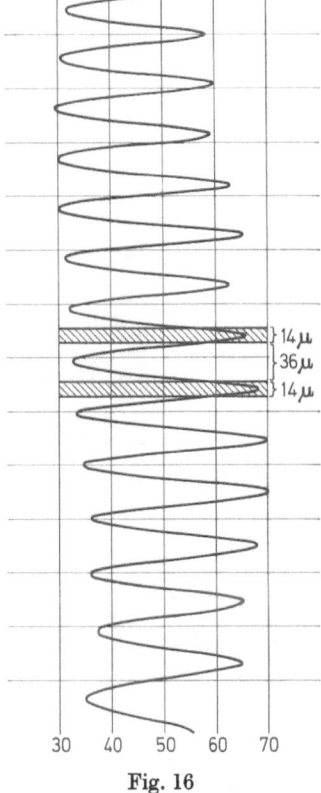

}14 μ
}36 μ
}14 μ

Fig. 16. A densitometer trace of the original negative of Fig. 15 showing a few superimposed bars

30   40   50   60   70

Fig. 16

The asymmetry of the diffraction images shifts in the direction predicted by the ray-tracing results of Fig. 13.

For the dimensions used to produce Fig. 14c the resolving distance computed from Eq. (44) is 26.6 μ. The resolving distance measured from Fig. 14c is 28.6 μ.

Values of p and q not too far from those used to obtain Fig. 14c still result in a nearly symmetrical Fraunhofer type pattern. This implies that points on an extended object field will be imaged rather well by a mirror designed to remove spherical aberration for only one point.

A 500 bar/in wire mesh with bar width $\approx 14\,\mu$ was imaged first with a single aspherical reflector, then with two aspheric mirrors. In the latter case an image was obtained of both the horizontal and vertical bars of the grid; the second mirror was employed in the Kirkpatrick manner to remove the extreme astigmatism resulting from the use of only one reflector. Fig. 15 shows the astigmatic image of the mesh for the case of $i = 8°$, $p = 23.5$ cm, $q = 54.5$ cm. The mirrors were 2 cm in length.

Fig. 16 is a densitometer trace of the original negative of Fig. 15. The bar spacing is superimposed. The magnification varies slightly as expected over the 500 μ field of the mesh.

Fig. 17 shows the image of the grid formed by two aspheric mirrors mounted at right angles to one another. The dimensions for this case are given in Table 2.

Table 2

i	p (cm)	q (cm)	$M(q/p)$	
8°	23.5	54.5	2.32	1st mirror
10°	27.5	50.5	1.84	2nd mirror

The lack of sharpness in the horizontal bars is due to the unequal magnifications of the two mirrors. This effect would be reduced considerably in systems with large magnification.

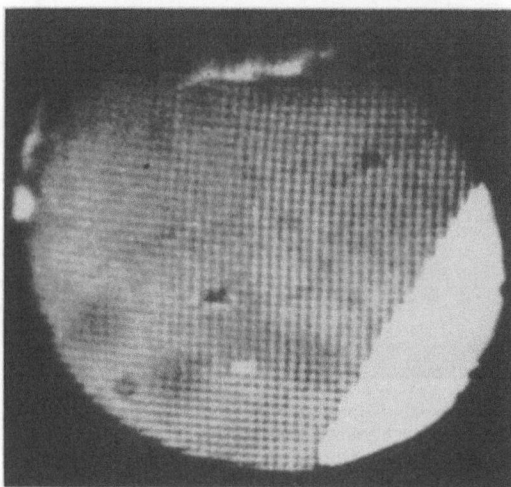

Fig. 17. A two mirror image of the wire mesh formed by cubic mirrors mounted at right angles in the Kirckpatrick manner; $\lambda = 6{,}328$ Å

c) A Normal Incidence Test of Aspherical X-ray Mirror Surfaces

It is possible that aspherical reflectors designed for use at grazing incidence may be tested at *normal* incidence using visible light. A simple test to determine the variation in curvature of the figured surface involves masking the mirror so that only a small aperture is illuminated normally by light from a point source. The reflected light will form a line image at the source if the source is separated from the mirror by a distance equal to the average radius of curvature of the illuminated strip. A measurement of curvature vs position can be used to determine the cubic coefficient ε_0 and the principal radius of curvature R. Further details of this method are given by McGee and Arrazola [12].

If R is known, then another kind of normal incidence investigation can be made. In this case the entire surface is illuminated normally by a point source located a distance R from the mirror, as shown in Fig. 18. Fig. 19 shows the highly assymmetrical diffraction pattern obtained; it closely resembles the image of a point source obtained from a spherical mirror at *grazing* incidence.

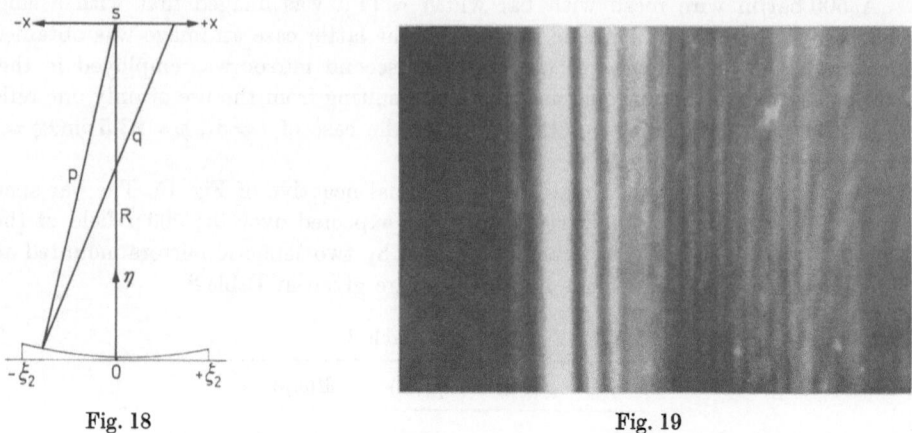

Fig. 18   Fig. 19

Fig. 18. Coordinate system used in establishing the intensity distribution of Eq. (47)

Fig. 19. A typical asymmetrical diffraction image observed when a cubic surface ($\varepsilon_0 = -2.5 \times 10^{-5}$ cm^{-2}) was illuminated at *normal* incidence by a point source at a distance R from the mirror. Note the similarity with that of the uncorrected cylinder at *grazing* incidence shown in Fig. 7

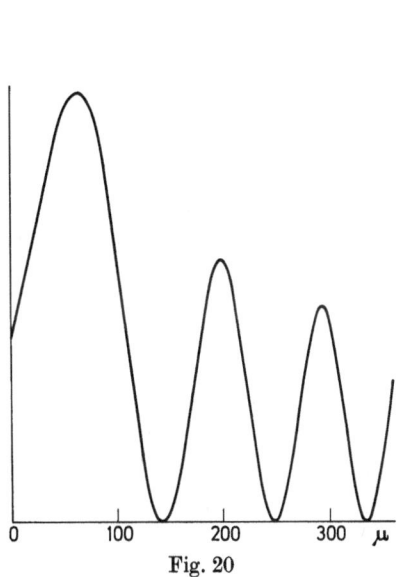
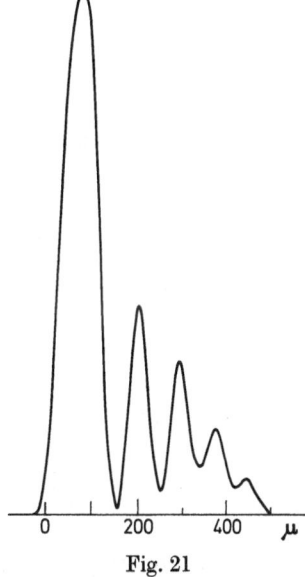

Fig. 20   Fig. 21

Fig. 20. A graph of the first few diffraction fringes as computed from Eq. (47) for the parameters used in making Fig. 19

Fig. 21. A densitometer trace made from the original negative of Fig. 19

The Kirchhoff diffraction integral can be evaluated for this case, using the method of stationary phase. The intensity to the right of the origin in the image plane is

$$I \sim p_0^2 \left\{ \frac{2\pi}{k^{\frac{2}{3}}} A_i(-\beta) - \sqrt{\frac{\pi}{2 k \zeta^{\frac{1}{2}}}} \left[(\cos z + \sin z) - 2(C_2 \cos z + S_2 \sin z)\right] \right\}^2.$$

$$p_0 = (-6\varepsilon_0)^{-\frac{1}{4}}, \quad \beta = \frac{k^{\frac{2}{3}} x}{R} (-6\varepsilon_0)^{-\frac{1}{3}}$$

(47)

$$\zeta = \frac{x}{R} p_0^2, \qquad z = \frac{2}{3} k \zeta^{\frac{1}{2}}$$

$$u_2 = (-6\varepsilon_0)^{\frac{1}{2}} \xi_2, \quad \omega_2 = \left[\frac{2k}{\pi} \left(\frac{1}{3} u_2^2 - \zeta u_2 + \frac{2}{3} \zeta^{\frac{3}{2}}\right)\right]^{\frac{1}{2}}.$$

The predominant term is the Airy function term involving $A_i(-\beta)$; the argument β is defined above. The first few fringes of Eq. (47) are shown in Fig. 20; Fig. 21 shows the corresponding experimental densitometer trace. The theoretical and experimental results are compared in Table 3.

Table 3

Fringes	Separation		Relative intensities	
	theor.	exp.	theor.	exp.
1—2	135	131	1.64	2.56
1—3	230	228	1.97	3.46
2—3	95	94.5	1.21	1.35
1—4	311	304	2.24	6.41
1—5	384	377	2.45	—

It can be seen that the positions of fringe maxima and minima are functions of the Airy function maxima and minima. If β_n and β_{n+j} are the arguments of the n-th and $(n+j)$-th maxima, and x_n and x_{n+j} the corresponding positions, then from the defining equation for β,

$$\varepsilon_0 = \frac{k^2}{6\,R^3(\beta_n - \beta_{n+j})^3}\,(x_{n+j} - x_n)^3\,. \tag{48}$$

Measurements of the maxima positions of Fig. 21 give an average value of $\varepsilon_0 = -2.56 \times 10^{-5}$ cm^{-2}. It should be noted that the dependence of ε_0 is on the third power of the fringe separation, so that the result is quite sensitive to small errors in the measurement of the separation. Nevertheless, there is confirmation from other measurements [12] which indicate that the reflector in question is indeed aspherical and can be described by the third order Eq. (36) with $\varepsilon_0 = -2.5 \times 10^{-5}$ cm^{-2}, where $\varepsilon_0 \equiv \frac{a_3}{6}$.

References

1. Jentzsch, F.: Physik. Z. **30**, 268 (1929).

2. Kirkpatrick, P., and A. V. Baez: J. Opt. Soc. Am. **38**, 766 (1948).

3. Montel, M.: Rev. Optique **32**, 585 (1953); — Opt. Acta **1**, 117 (1954); — Compt. Rend. **239**, 39 (1954).

4. Ehrenberg, W.: Nature **160**, 330 (1947); — J. Opt. Soc. Am. **39**, 741, 746 (1949).

5. Hesser, D. R., and J. F. McGee: J. Opt. Soc. Am. **53**, 525 (1963).

6. Copson, E. T.: Asymptotic expansions, p. 27, 107. London: Cambridge University Press 1965.

7. Chester, C., B. Friedman, and F. Ursell: Proc. Cambridge Phil. Soc. **53**, 599—611 (1957).

8. McGee, J. F., and J. W. Milton, in: X-ray microscopy and X-ray microanalysis, edit. by Engstrom et al. New York: Elsevier Publ. Co. 1960.

9. Baez, A. V., and M. Weissbluth: Phys. Rev. **93**, 942 (1954).

10. Dyson, J.: Proc. Phys. Soc. (London) B **65**, 580 (1952).

11. Baez, A. V., in: X-ray microscopy and microradiography, edit. by Cosslett, et al., p. 186. New York: Academic Press Inc. 1957.

12. McGee, J. F., and Ignacio Arrazola: Proceedings of the Fifth Internat. Congr. on X-Ray Optics and Microanalysis, Tübingen, Germany. Berlin-Heidelberg-New York: Springer 1969.

The Figuring of an Aspherical X-Ray Lens*

J. F. McGee and I. M. Arrazola[1]

Saint Louis University, Saint Louis, Missouri, U.S.A.

Abstract

A detailed account of the manufacture of an X-ray mirror whose cylindrical surface is approximately described by a third order polynomial is presented. Theoretical analysis shows that such a surface removes most of the spherical aberration at grazing angles of incidence. Because the coefficient of the required third order term is small, the cubic surface is treated as a perturbation of a second order cylinder. A grinding method has been found which produces this type of figure over a large portion of the glass work. Essentially, the method consists in the asymmetrical perturbation of the process used in producing a high quality right circular cylinder. Testing procedures and computational methods for determining the coefficients of the polynomial are discussed.

Introduction

Modern day demands upon optical components make it necessary not only to develop new manufacturing techniques, but have shown as well, the need to develop, from a rigorously analytical point of view, the fabrication aspect of optical components. Although great advances have been made in the design and testing of elaborate optical systems and glass surfaces, very little has been done through the centuries to put the fabrication of optical components on a scientific basis. There is no theory on the micro and macrogeometrical development of glass surfaces during the grinding and polishing processes, that would enable us to design an experiment so as to produce an optical surface with a given geometry and with sufficient optical quality. The fabrication of optical components, especially with regard to aspherical surfaces, is artisan based. The manufacturing aspect of optical engineering is still much of an art. For very judicious comments in this regard, the reader is referred to an excellent article by LARMER and GOLDSTEIN [1].

Workers in reflection X-ray optics have not been unaware of the advantages inherent in the use of reflecting surfaces other than spherical. C. M. LACHT and D. HARKER [2], for example, have experimented with bending optical flats so as to obtain mirrors of adjustable curvature. DuMOND, LIND, and COHEN [3] developed a method of generating circular cylindrical surfaces of large radii of curvature, on blocks of steel and cast iron. They report, however, that under optical and X-ray tests, the errors of focus of the surfaces were too large to be accepted for curved-crystal spectrometer work, and the surfaces had to be modified by subsequent lapping. In particular, D. R. HESSER and J. F. McGEE [4] have studied in detail the reflection of electromagnetic waves by a cylindrical mirror and developed a diffraction theory of image formation for this type of mirror. They have also shown that a surface which obeys the equation

$$y = x^2/2R + \varepsilon_0 x^3 + O(\text{higher}) \tag{1}$$

---

* This work was in part supported by grants from: The National Institutes of Health, General Medical Science Division and the American Cancer Society.

1. *Present address:* University of Bogota, Bogota, Colombia, S.A.

eliminates spherical aberration to a large degree and that no higher order terms are needed. J. F. McGee and J. W. Milton, private communication, have developed the diffraction theory of image formation by a cubic mirror, that is, a mirror whose surface obeys Eq. (1). They studied the image forming properties of the cubic surface at grazing incidences and also at normal incidence. It is with the fabrication and testing of a cubic mirror as described by Eq. (1) that this paper is concerned.

Apparatus

The same natural processes working in the generation of spherical surfaces make possible the fabrication of cylinders. Scott [5] has described the most common method applied in the manufacture of optical circular-cylinders of short radii of curvature. Larmer and Goldstein, loc. cit., have described the fabrication of an off-axis cylindrical-parabola, also of short radius of curvature. In both methods, however, a metal tool, previously figured to the desired geometry, is employed. In what follows a method is described for producing right circular cylinders in a "natural" way; as the grinding proceeds, the surface, as in the spherical case, tends more and more towards a right circular cylindrical.

It is known that in the grinding process, when the motion of the system possesses three degrees of freedom the shapes of the surfaces have a tendency to become segments of a sphere. It is natural to expect that by limiting the system motion to only two degrees of freedom we may be able to generate a right circular cylinder. By designing the tool holder in such a way that (a) the motion of the stroke is linear, and (b) the tool holder is free to rotate about an axis perpendicular to the stroke, a right circular cylinder may be generated, if the glass work is held fixed.

The Atlas universal-shaper of Fig. 1 was adapted to provide the driving mechanism for the stroke. The tool holder consists of an aluminum block 9.5 cm long × 7.0 cm wide × 1.9 cm thick. These dimensions are not critical. Two holes drilled at equal distances from the leading edges receive bolts for holding lead weights. The tool holder is coupled to the ram of the shaper by means of two identical brass arms 11.0 cm long. Each arm contains two small bearings with their axis of rotation 9.0 cm apart. The vise with which this type machine is ordinarily supplied to hold the metal work was removed and replaced by a steel platform to hold the glass work. The shaper is an ideal machine for the type of operation being described here, as it allows practical adjustment of stroke length and vertical height of the work and of the ram.

Fig. 1. An Atlas universal-shaper adapted for the grinding of cylindrical and aspherical surfaces. The tool holder is coupled to the ram of the shaper by means of two identical arms 11.0 cm long

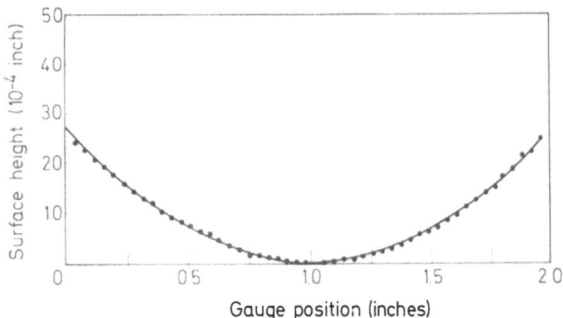

Fig. 2. Experimental points as measured with a precision dial-gauge coupled to a comparater carriage. The mirror under test was ground to be cylindrical. The smooth curve was determined by a root-mean-square fit of the points to a second-order polynomial with the aid of an IBM computer which yielded

$$y = (2.7309 - 2.1021\,x + 0.4085\,x^2) \cdot 10^{-3}$$

The optical material used for the mirrors was Pyrex, which has good dimensional stability, takes a good polish and is very stable with temperature and age. The mirrors were made from rectangular blanks 16.1 cm $\times 3.0$ cm $\times 0.5$ cm thick on the average. The material chosen for the tool was plate glass, which is much softer than Pyrex. A hard tool leaves long linear gouges in the direction of the stroke. By using a plate glass tool the grinding and the fining operations produce uniform results with a remarkable absence of bad scratches. The tools used had dimensions of 5.08 cm $\times 5.08$ cm $\times 1.27$ cm thick. A stroke length of 5.67 cm with a stroke frequency of 76 strokes per minute was used. The lead weights were distributed so that a uniform pressure would result on the initially flat surface. The total mass of the tool holder, plus arms and glass tool was approximately 1.75 kilograms.

To obtain some information as to how the glass surface developed during the grinding process, two identical glass blanks were ground individually under the same conditions. The abrasive used in both cases was Alundum No. 160. Equal quantities of abrasive were introduced manually every two hundred strokes. After every thousand or two thousand strokes, a measurement of what is essentially "y" vs. "x" was made on the surface. The measurements were carried out by linearly scanning the glass surface with a precision dial-gauge, used as a depth gauge. This instrument, manufactured by Brown & Sharpe (Model 2), permitted exact readings to $^1/_{10,000}$ of an inch. The gauge was mounted on a microscope comparator-carriage, manufactured by the Gaertner Scientific Corporation, and could be displaced a total distance of 5 centimeters. The advancement is effected by means of a micrometer screw, permitting readings to 0.5×10^{-2} millimeters. Fig. 2 shows a typical plot of a set of such measurements. For large R, $x^2/2R$ is a very good approximation to a circle. The data was fitted by the root mean square method to a second order polynomial, an IBM computer being used for the latter calculations.

Experimental Results — Right Circular Cylinder

Fig. 3 shows a plot of the curvature $1/R$ vs. N, the number of strokes. The linear behavior of the curvature with respect to the grinding time is evident. It should be noted that a theoretical curve describing the process should pass through the origin, for at $N = 0$, both the tool and the work are flat surfaces, and the curvature of a plane surface is zero. Thus we may write

$$1/R = kN \tag{2}$$

where k is a constant that depends upon the composition of the glass, the type of abrasive, and the pressure on the glass work. On the average, the root mean square error in the determination of the polynomial coefficients for the sets of measurements fluctuated between 11×10^{-5} and 6×10^{-5}.

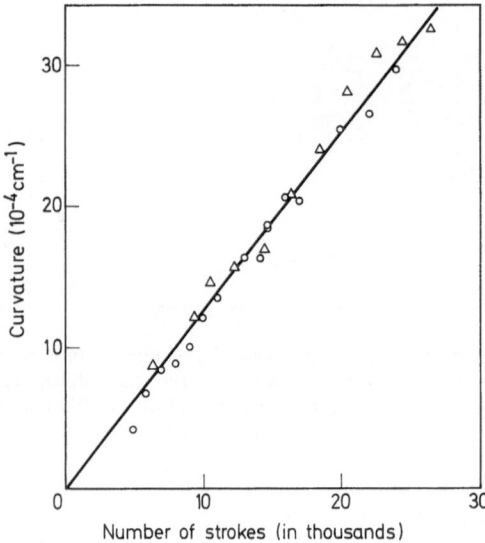

Fig. 3. A plot of the curvature $1/R$ vs. N, the number of strokes

The polishing of the cylinders was done by simply replacing the plate-glass tool with a soft pitch lap. The polishing agent was Red Rouge No. 25 (Universal Supply & Shellac Co.), although cerium oxide was found to also give good results. The only variation in the procedure consisted in advancing with each stroke the glass work perpendicularly to the direction of the stroke motion, at intervals of $1/100$ of an inch. It was found that some transverse curvature had been introduced especially at the edges of the blank, but on the whole this effect was negligible. Good mirrors were obtained after some 10 hours of polishing. For a particular case, a difference of 2.2% for the radius of curvature was found between the mechanical measurement previously described and a measurement made on the optical bench.

Experimental Procedure — The Aspheric Reflector

As mentioned earlier, a surface that satisfies Eq. (1) removes to a great extent spherical aberration. For practical purposes, the constant ε_0 is a small negative number compared with $1/2 R$ and thus we may look upon Eq. (1) as a perturbation upon a right circular cylinder. Obviously, the effect of the term $+ \varepsilon_0 x^3$ is to substract from the term $x^2/2 R$ for positive x, and to add to the term for negative values of x. In a previous section, a process for the grinding of right circular cylinders was described. To produce a cubic mirror we simply perturb the grinding of the circular cylinder in an asymmetric manner, by an uneven distribution of weights on the tool holder. The perturbation then, consists in placing more weight on one side than on the other of the tool holder. Using the latter weight arrangement, two initially flat glass surfaces were ground together, tool and work having the same dimensions as specified before. It was found that the glass surface of the tool was worn considerably more on the side favored by the weights. With each stroke the leading side wore away more and more. Another effect of the weight distribution was to rotate the glass tool very slowly, with each stroke, about an axis defined by the shaft coupled to the arms of the tool holder. The net effect of the slow rotation is that, as the grinding proceeds, less and less area toward the back of the glass tool is in contact with the glass work, with a subsequent decrease in the effective length of the stroke. This type of grinding is obviously not symmetric, and does in principle what we need to produce a cubic surface. To produce an aspherical surface, it is better to start by grinding a circular cylinder; the perturbation is then introduced when the radius of curvature of the cylinder is somewhere near the desired mean radius for the aspherical mirror. The polishing should be carried out with the perturbing weights in place. The grinding and polishing times depend principally on dimensions, total weight of perturbation, and polishing time needed to produce acceptable optical quality. It should be clear that not all attempts can be successful, and that the length of usable surface will vary from mirror to mirror.

Experimental Testing Methods

The main purpose for carrying out a series of tests on a lens is to abstract useful data which will allow judgments to be made on the acceptability or usefulness of the entire lens in an X-ray microscope. In our particular case we will be interested in determining to what extent, if any, is the method described in this paper successful in generating a cubic surface.

The lens equation for any concave mirror, as given by JENTZSCH [6], is

$$1/p + 1/q = 1/f_c \tag{3}$$

where $f_c = \frac{1}{2}(R \sin i_c)$, R is the radius of curvature at the origin, and i_c is the grazing angle; p and q are respectively the object and image distances. At normal incidence, Eq. (3) reduces to the well known formula

$$1/p + 1/q = 2/R. \tag{4}$$

At normal incidence, and for a segment of the cubic surface, one may define a focal length in terms of the average curvature of the mirror segment:

$$1/f = 2 \langle 1/r \rangle = 1/p + 1/q. \tag{5}$$

The curvature ϱ is given by

$$\varrho = 1/r = y''/[1 + (y')^2]^{\frac{3}{2}}. \tag{6}$$

For large R and small ε_0, as is the case in Eq. (1),

$$\varrho = 1/r \approx y'' = 1/R + 6\varepsilon_0 x. \tag{7}$$

So that within the approximations made, the curvature of the cubic surface is a linear function of x.

The experimental measurement used to determine variations in focal length required the covering of the silvered surface of an aspherical mirror with a window one centimeter wide. The mirror aperture was then illuminated normally by a point source to which was attached a calibrated eyepiece. The composite system of source and eyepiece could be positioned on an optical bench. Since $p = q$ at all times, the system is in focus when $p = q = r$, according to Eq. (5). If we let "a" and "b" be the x coordinates of the edges of the aperture, we may write

$$\langle \varrho \rangle = \langle 1/r \rangle \approx \frac{1}{b-a} \int_a^b (1/R - 6\varepsilon_0 x)\, dx$$

$$\langle \varrho \rangle \approx 1/R - 3\varepsilon_0 (b + a). \tag{8}$$

Defining the position of the center of aperture by

$$\langle x \rangle = \frac{1}{2}(b + a) \tag{9}$$

we obtain

$$\langle \varrho \rangle = \langle 1/r \rangle \approx 1/R - 6\varepsilon_0 x. \tag{10}$$

Fig. 4 is a plot of the experimental curvature vs. position of the center of aperture, for a particular mirror. The linear behavior of the curvature shows that this mirror is very approximately a cubic mirror over a length of five centimeters. The solid line represents the root mean square fit of the data to a straight line. The cubic coefficient ε_0 can be readily calculated from the slope of the line. For this mirror $\varepsilon_0 = -2.4 \times 10^{-5}$ cm^{-2}.

According to HESSER and McGEE, loc. cit., a cubic surface will remove spherical aberration to a large extent when the following equation is satisfied:

$$\varepsilon_0 = \frac{\cos i_c}{4R} \left[\frac{p - q}{pq} \right]. \tag{11}$$

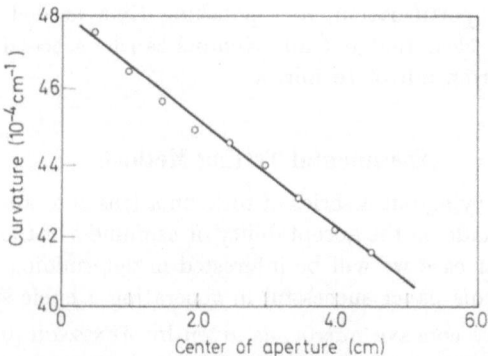

Fig. 4. Curvature $1/R$ vs. x the position of the center of a defining window covering a small section of the mirror should theoretically yield a straight line according to Eq. (10) from the slope of which ε_0 can be measured

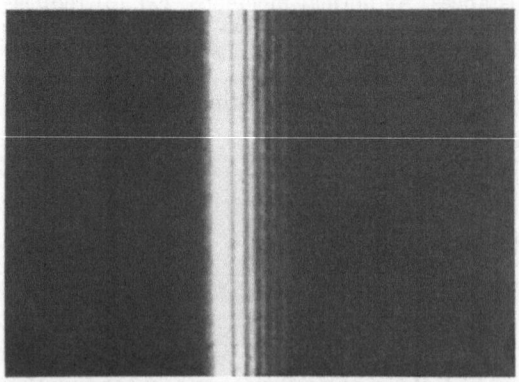

Fig. 5. The line image of a point source formed by a circular cylinder of radius $R = 260$ cm

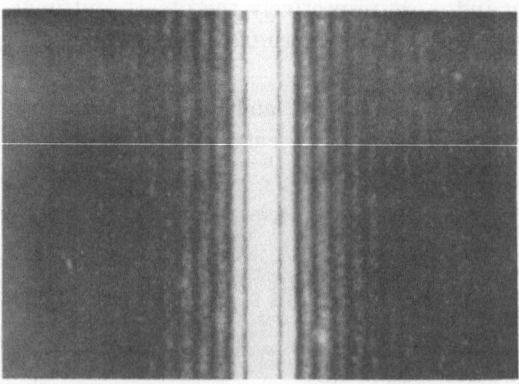

Fig. 6. The line image of a point source formed by a cubic mirror with an ε_0 of -2.5×10^{-5} cm^{-2}

Fig. 5 shows the line image of a point source formed by a circular cylinder of radius $R = 260$ cm. The spherical aberration is clearly evidenced by the secondary fringes lying on the right side of the principal maxima. For this photograph the grazing angle was $i_c = 10°$, $p = 30.5$ cm, $q = 92.5$ cm and $\lambda = 6,328$ Å. The source was a He-Ne laser. Exposure time was $^1/_{1,000}$ of a second. Fig. 6 shows the images of a point source formed by the particular cubic mirror we have been considering. For the image of Fig. 6 the parameters were $i_c = 8°$, $p = 23.5$ cm, $q = 51.5$ cm, $\lambda = 6,328$ Å and $R = 231$ cm. The Fraunhofer pattern that is observed indicates the removal, to a very large extent, of spherical aberration. The calculated cubic coefficient varies from 2.4 to 2.5×10^{-5} cm^{-2} because of the uncertainties in R and i_c.

References

1. Larmer, J. W., and E. Goldstein: Appl. Opt. 5, 676 (1966).
2. Lacht, C. M., and D. Harker: Rev. Sci. Instr. 22, 392 (1951).
3. DuMond, J. W., D. A. Lind, and E. R. Cohen: Rev. Sci. Instr. 18, 617 (1947).
4. Hesser, D. R., and J. F. McGee: J. Opt. Soc. Am. 53, 525 (1963).
5. Scott, R. M.: Optical manufacturing. In: Applied optics and optical engineering (Rudolph Kingslake, ed.), vol. III. New York: Academic Press 1965.
6. Jentzsch, F.: Physik. Z. 30, 268 (1929).

Untersuchung zur ASR (Anomalous Surface Reflection) von Röntgenstrahlen mit einer Mikrofokus-Röntgenröhre

K. Beck und G. Kühnen

Physikalisches Institut der Universität Würzburg, BRD

Einleitung

Fällt Röntgenstrahlung unter einem kleinen Glanzwinkel i auf eine uns eben erscheinende Grenzfläche zweier Medien, so stellen wir fest, daß die reflektierte Röntgenstrahlung das aus dem sichtbaren Spektralgebiet bekannte Reflexionsgesetz $i = i'$ erfüllt. Ist das erste Medium Luft, dann ist im Wellenlängengebiet der Röntgenstrahlen der Brechungsindex des zweiten Mediums stets kleiner als 1, und es kann bei hinreichend schwacher Absorption des zweiten Mediums bei kleinen Glanzwinkeln Totalreflexion auftreten. 1963 beobachtete Yoneda [1], daß für Glanzwinkel i, die nur wenig größer waren als der Grenzwinkel der Totalreflexion i_g, reflektierte Röntgenstrahlung nicht nur unter dem Winkel $i' = i$ zu beobachten war, sondern auch unter einem Winkel $i_a < i$ (Abb. 1). Yoneda nannte dieses Phänomen, daß Röntgenstrahlung anscheinend dem Reflexionsgesetz nicht gehorcht, ,,anomalous surface reflection" oder kurz ASR. Die bisherigen Versuche zur Deutung der ASR gehen davon aus, daß die einfallende Röntgenstrahlung gestreut (s. u.) und ein Teil der Streustrahlung auch dann totalreflektiert wird, wenn $i > i_g$ ist. Nach den entwickelten Vorstellungen handelt es sich entweder um Kleinwinkelstreuung an Verunreinigungen und Unregelmäßigkeiten der Spiegel (Warren und Clarke [2], Guentert [3]) oder zusätzlich um Kantenstreuung an Blenden (Nigam [4]). Gegenüber den bisher benutzten Anordnungen zur Untersuchung der ASR, bei denen mit einem kollimierten Strahl gearbeitet wurde, bietet die Mikrofokus-Röntgenröhre zwei Vorteile: einmal wird bei allen Glanzwinkeln stets der Spiegel in seiner ganzen Länge mit Strahlung beaufschlagt, und zum anderen können Blenden, die Kantenstreuung mit möglicher nachfolgender Totalreflexion ergeben, vermieden werden. Im folgenden werden einige experimentelle Ergebnisse von Reflexionsmessungen mitgeteilt, und es wird diskutiert, wieweit sie mit den bisher versuchten Deutungen der ASR verträglich sind.

Abb. 1. Schematische Skizze zur Reflexion von Röntgenstrahlen

Meßanordnung

Die Untersuchungen erstrecken sich auf drei Glasarten (BK 7, SF 3, Ultrasil) und verschiedene auf Glas BK 7 aufgedampfte Metalle mit Ordnungszahlen zwischen $Z = 12$ und $Z = 47$. Die Meßanordnung (Abb. 2) wurde bereits früher ausführlich beschrieben [5, 6]. Strahlungsquelle ist eine Röntgenröhre mit Cu-Membranantikathode nach Hink [7]. Der Fokus hat einen Durchmesser von 7 μm (Halbwertsbreite der gaußförmigen Verteilung der Strahlungsdichte). Die Strahlung ist über einen Winkelbereich von 0,17 rad isotrop. Die untere Kante der Bleiblende (9),

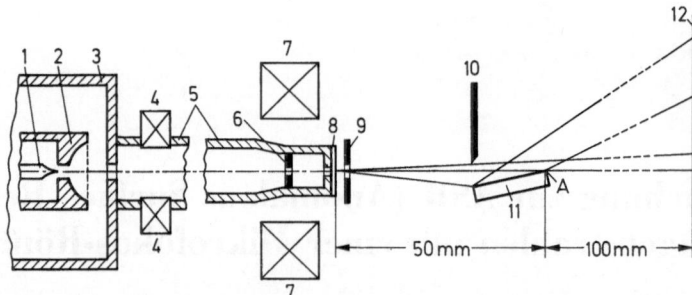

Abb. 2. Schematischer Aufbau der Mikrofokus-Röntgenröhre. *1* Haarnadelkathode; *2* Wehneltzylinder; *3* Anode; *4* magnetisches Strahlablenkungssystem; *5* Anodenrohr; *6* Aperturblende der magnetischen Polschuhlinse; *7* Polschuhlinse; *8* Membranantikathode; *9* Spaltblende aus Blei zur Ausblendung eines Röntgenstrahlbündels; *10* Bleiblende zur Abschirmung des direkten Strahls; *11* Spiegel, um Achse *A* drehbar; *12* Registrierebene

die die Röntgenstrahlung grob ausblenden soll, ist von der reflektierenden Fläche des Spiegels (*11*) tatsächlich nicht zu sehen, wie es nach dem Bild den Anschein hat. Als Spiegel dienen Glasquader mit den Abmessungen 20 mm × 20 mm × 5 mm. Ihre Oberfläche ist nach Angabe des Herstellers auf $\lambda/10$ der grünen Hg-Linie eben. Die Metallschichten werden durch Aufdampfen im Hochvakuum hergestellt. Die Registrierung der Strahlung erfolgt hauptsächlich fotografisch (Belichtungszeiten ca. 20 min), in einigen Fällen mit einem Proportionalzählrohr und nachfolgendem Einkanal-Impulshöhenanalysator mit schreibender Anzeige.

Meßergebnisse

Zunächst konnte sichergestellt werden, daß die noch im Strahlengang befindlichen Blenden — (*9*) und (*10*) in Abb. 2 — keinen Einfluß auf die Entstehung der ASR haben. Mit und ohne Blende (*9*) und bei unterschiedlichen Positionen der Kantenblende (*10*) tritt ASR unter gleichem Winkel und mit gleichbleibender maximaler Intensität auf.

Zwei mit dem Zählrohr registrierte Intensitätsverteilungen der an einer 1040 Å dicken Cu-Schicht reflektierten Strahlung zeigt die Abb. 3. i_{min}, der minimale Glanzwinkel, unter dem direkte Strahlung von der Röntgenröhre den Spiegel trifft, ist bei beiden Registrierungen größer als der Grenzwinkel der Totalreflexion i_g. Es tritt ASR auf, ein Intensitätsmaximum der ASR ist jedoch erst in der unteren Hälfte der Abb. 3 (bei größerem i_{min}) zu erkennen. Die aus den Untersuchungen von GUENTERT bekannte typische Form der ASR — der langsame Anstieg auf der Seite kleiner Winkel und der relativ schnelle Abfall auf der Seite großer Winkel mit dem Maximum in der Nähe von $i'/i_g = 1$ — ist deutlich zu erkennen. Bei der Berechnung von $i' = i_a$ wurde die Spiegelmitte als Entstehungsort der ASR angenommen, was einen maximalen Fehler für i_a von ca. 10% bedeutet. Wie i_a von der Spiegelstellung abhängt, zeigen die nächsten beiden Abbildungen. Die Spiegelstellung wird dabei durch den kleinsten Glanzwinkel i_{min} charakterisiert. Alle Winkel werden auf den Grenzwinkel der Totalreflexion i_g bezogen. In Abb. 4 sind für Glas BK 7 die aus fotografischen Aufnahmen ermittelten Beobachtungswinkel i' des Maximums und der beiden Begrenzungen der ASR für verschiedene minimale Glanzwinkel eingetragen. Es ist zu erkennen, daß sich die Lage des Maximums und die Begrenzung zu größeren Winkeln hin nur wenig ändern, die Begrenzung zu kleineren Winkeln hin dagegen ändert sich stark. Für die auf Glas BK 7 aufgedampften Elemente ist die Abhängigkeit der Lage des Maximums i_a/i_g über i_{min}/i_g in Abb. 5 dargestellt. Hier lassen sich deutlich zwei Gruppen mit verschiedener Abhängigkeit unterscheiden: Gruppe 1, zu der auch die drei untersuchten Gläser, die in der Darstellung fehlen, gehören, mit Cu, Mg, Ni, Zn und Ti zeigt nur eine schwache Abhängigkeit und Werte in der Nähe von $i_a/i_g = 1$. Gruppe 2, zu der auch eine Ti-Schicht gehört, zeigt hingegen eine starke Abhängigkeit mit einem Anstieg $i_a/i_{min} \approx 1$. Die Ti-Schichten, die ein unterschiedliches Verhalten zeigen, sind, soweit das unserer Kontrolle zugänglich war, unter gleichen Bedingungen hergestellt worden. Bei manchen Elementen der Gruppe 2 ist eine ASR schon zu beobachten,

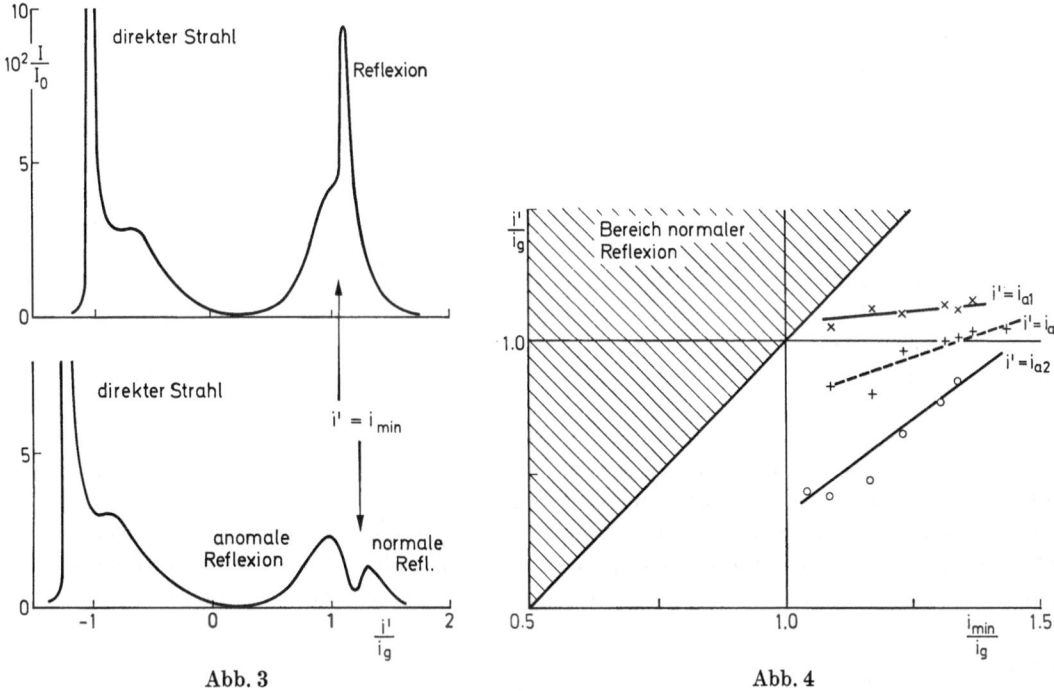

Abb. 3

Abb. 4

Abb. 3. Zwei für verschiedene Spiegelstellungen mit dem Zählrohr registrierte Intensitätsverteilungen der an einer 1040 Å dicken Cu-Schicht reflektierten Cu-K_α-Strahlung

Abb. 4. Abhängigkeit der Lage der anomalen Reflexion vom minimalen Glanzwinkel der Strahlung bei der Reflexion von Cu-K_α-Strahlung an Glas BK 7. $i' = i_{\min}$: Begrenzung der „normalen" Reflexion; $i' = i_{a1}$: Begrenzung der anomalen Reflexion auf der Seite großer Winkel; $i' = i_{a2}$: Begrenzung der anomalen Reflexion auf der Seite kleiner Winkel; $i' = i_a$: Winkel, unter dem die maximale Intensität der anomalen Reflexion auftritt; $i' = i_g$: Grenzwinkel der Totalreflexion

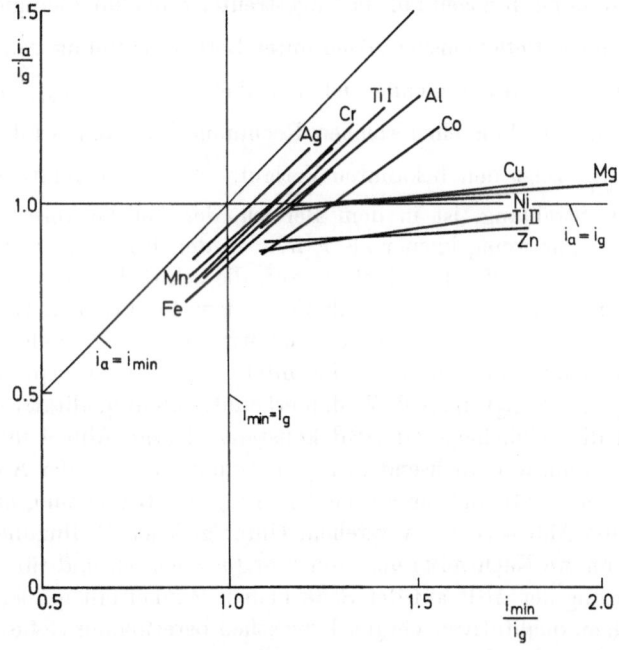

Abb. 5. Abhängigkeit der Winkellage i_a des Maximums der ASR von der Stellung des Spiegels für verschiedene Metallaufdampfschichten. Schichtdicken: Al 1175 Å, Ti I 800 Å, Ti II 1100 Å, Cr 600 Å, Mn 1480 Å, Ag 1060 Å, Fe 1420 Å, Co 340 Å, Ni 380 Å, Cu 980 Å, Zn 990 Å, Mg 1160 Å

wenn i_{min} noch kleiner ist als i_g. In der Nähe von $i_{min}/i_g = 1,2$ durchdringen sich die beiden Geradenscharen, und bei dieser Spiegelstellung ist für alle untersuchten Elemente $i_a/i_g \approx 1$. Die Intensität der ASR nimmt mit wachsendem i_{min} ab.

Diskussion

Eine Deutung für das Auftreten der ASR als Kantenstreuung an Blenden mit nachfolgender Totalreflexion nach Nigam scheidet aus. Die Streustrahlung der Kanten der Blenden (9) und (10) trifft nur unter größeren Glanzwinkeln auf den Spiegel als die direkte Strahlung vom Mikrofokus und kann daher die ASR nicht verursachen. Auch die endliche Ausdehnung des Fokus kann, wie eine Abschätzung zeigte, die ASR nicht erklären. Nach Warren und Clarke sollte an Verunreinigungen und Unregelmäßigkeiten der Spiegel gestreute Strahlung, die anschließend totalreflektiert wird, die ASR verursachen. Um in diesen Vorgang quantitativ Einblick zu gewinnen, wurde für ein einfaches Modell das winkelabhängige Reflexionsvermögen $\varrho^2(i')$ für Glas BK 7 berechnet. Hierbei wurde angenommen, daß alle ankommende Strahlung vor Auftreffen auf den Spiegel einmal gestreut wird und daß die Streuung durch die von Elliott [8] gemessene Winkelverteilung der Intensität in der Nähe der 1:1-Abbildung des Fokus einer Cosslett-Nixon-Röntgenröhre gegeben ist. Diese Streuung läßt sich durch eine Überlagerung einer Gaußkurve (Streuung σ) und einer Exponentialkurve (Konstanten K, γ) annähernd beschreiben. Man erhält:

$$\varrho^2(i') = \varrho_F^2(i') \cdot \frac{I(i')}{I_0}$$

$$= \varrho_F^2(i') \cdot \frac{1}{\sqrt{2\pi}\,\sigma^2 + \dfrac{2K}{\gamma}} \int\limits_{i_0 = i_{min}}^{i_0 = i_{max}} \left\{ \exp\left[-\frac{(i_0 - i')^2}{2\sigma^2} \right] + K \cdot \exp\left[-\gamma |i_0 - i'| \right] \right\} di_0. \qquad (1)$$

$\varrho_F^2(i')$: Aus den Fresnelschen Gleichungen berechnetes winkelabhängiges Reflexionsvermögen.

$\dfrac{I(i')}{I_0}$: Die durch Streuung modifizierte Winkelverteilung der Intensität der auf den Spiegel fallenden Strahlung, bezogen auf die ungestreute, winkelunabhängige Intensität I_0.

$\varrho^2(i')$: Winkelabhängiges Reflexionsvermögen unter Berücksichtigung von Streuung.

Integriert wird über alle vorkommenden Glanzwinkel i_0 zwischen i_{min} und i_{max}.

Die Abb. 6 zeigt das Ergebnis einer solchen Rechnung. Die Glanzwinkel i' wurden wieder auf i_g normiert. $\varrho_F^2(i'/i_g)$ zeigt den bekannten Verlauf. $\dfrac{I(i'/i_g)}{I_0}$, die relative Intensität der auf den Spiegel fallenden Strahlung, ist in dem Bereich, der üblicherweise vorkommt, nämlich zwischen i_{min} und i_{max}, nur wenig kleiner als 1, fällt an den Enden sehr schnell auf etwa 10^{-2} und weist im Anschluß daran Kantenschwänze auf, die schwächer und exponentiell abfallen. $\varrho^2(i'/i_g)$ ergibt sich durch punktweise Multiplikation beider Funktionen. Eine ASR ist deutlich zu erkennen, auch deren typische Form, die bedingt wird durch den steilen Abfall von $\varrho_F^2(i'/i_g)$ in der Nähe des Grenzwinkels der Totalreflexion infolge geringer Absorption des Spiegels. Den Einfluß von i_{min} auf $\varrho^2(i'/i_g)$ zeigt die Abb. 7. Man erkennt, daß in qualitativer Übereinstimmung mit dem Experiment die Winkellage der ASR konstant ist (vgl. Abb. 4 und z.T. Abb. 5) und die Intensität im Maximum mit wachsendem i_{min} abnimmt. Anhand der Abb. 7 lassen sich das Abnehmen der Breite der ASR und die starke Änderung der Begrenzung auf der Seite kleiner Glanzwinkel, wie es die Abb. 4 zeigte, verstehen. Unter gleichen Bedingungen (Intensität, Belichtungszeit) ist die untere Nachweisgrenze konstant ($\varrho^2 = const$), und für eine solche wandert in Abb. 7 die Begrenzung der ASR auf der Seite kleiner Winkel mit größer werdenden i_{min} zu größeren Winkeln. Einen qualitativen Vergleich zwischen berechneten Reflexionskurven $\varrho^2(i'/i_g)$ und den Winkelbereichen der Schwärzung fotografischer Registrierungen der an Glas BK 7 reflektierten Strahlung zeigt die Abb. 8. Wir finden gute Übereinstimmung zwischen Rechnung und Messung.

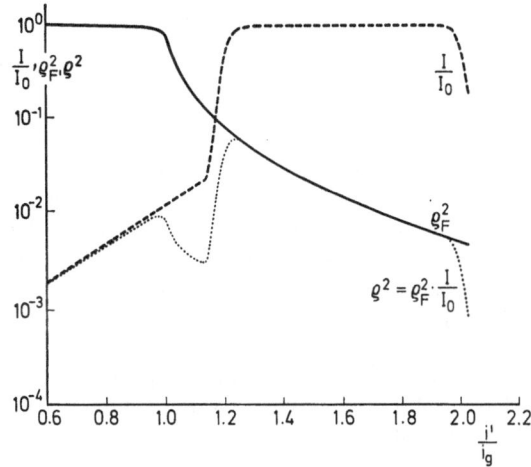

Abb. 6. $\varrho^2 = \varrho^2(i'/i_g)$, nach Gl. (1) berechnetes winkelabhängiges Reflexionsvermögen eines Spiegels aus Glas BK 7. $\varrho_F^2 = \varrho_F^2(i'/i_g)$, winkelabhängiges Reflexionsvermögen des gleichen Spiegels, berechnet nach den Fresnelschen Formeln mit dem komplexen Brechungsindex $n_{BK\,7} = 1 - \delta - jk = 1 - 8{,}05 \cdot 10^{-6} - j \cdot 1{,}41 \cdot 10^{-7}$.

$\dfrac{I}{I_0} = \dfrac{I(i'/i_g)}{I_0}$, die durch Streuung modifizierte Winkelverteilung der Intensität der auf den Spiegel fallenden Strahlung, bezogen auf die ungestreute, winkelunabhängige Intensität I_0. Berechnet mit $\sigma = 10^{-4}$ rad, $K = 10^{-2}$, $\gamma = 1{,}51 \cdot 10^{-3}$ rad, $i_{min} = 1{,}2\,i_g$

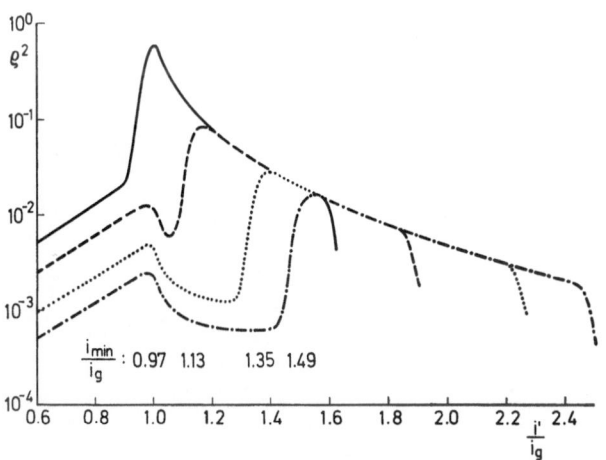

Abb. 7. Nach Gl. (1) berechnetes winkelabhängiges Reflexionsvermögen eines Spiegels aus Glas BK 7 für 4 verschiedene Spiegelstellungen

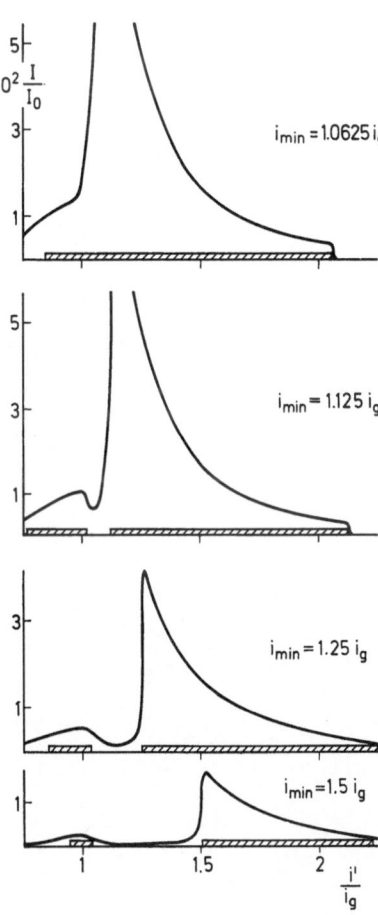

Abb. 8. Intensitätsverteilung der von Glas BK 7 reflektierten Cu-K_α-Strahlung bei vier verschiedenen Spiegelstellungen, berechnet nach dem Modell von Warren-Clarke, dazu zum Vergleich: Schwärzung der Fotoplatte ▨

Abb. 8

Zusammenfassung

In der vorgelegten Arbeit wurde der Versuch unternommen, die ASR unter Zugrundelegung eines einfachen Modells nach dem Vorschlag von Warren und Clarke zu berechnen. Für eine erste Gruppe von Metallen und Gläsern, für die $i_a \approx i_g$ ist, ergibt sich gute qualitative Übereinstimmung zwischen Rechnung und Messung bezüglich der Intensitätsänderung, Winkellage und Linienbreite der ASR. Bei einer zweiten Gruppe von Metallen wurde experimentell $i_a \approx i_{min}$ gefunden. Für die Metalle der zweiten Gruppe gibt es daher keine Übereinstimmung zwischen Rechnung und Messung. Bemerkenswert ist, daß zwei Spiegel desselben Metalls verschiedenen Gruppen angehören können. Eine Erklärung hierfür sowie für die Existenz zweier Gruppen von Spiegeln überhaupt konnte bisher nicht gegeben werden.

Die Autoren danken Herrn Professor Dr. W. Hink für die Ermöglichung der Arbeit an seinem Lehrstuhl und für deren Förderung. Die Rechnungen wurden auf der elektronischen Rechenanlage EL X 8 des Instituts für angewandte Mathematik durchgeführt. Dessen Vorstand, Herrn Professor Dr. W. Velte, gilt unser Dank für die zur Verfügung gestellte Rechenzeit.

Literatur

1. Yoneda, Y.: Phys. Rev. **131**, 2010 (1963).
2. Warren, B. E., and J. S. Clarke: J. Appl. Phys. **36**, 324 (1965).
3. Guentert, O. J.: J. Appl. Phys. **36**, 1361 (1965).
4. Nigam, A. N.: Phys. Rev. **138** A, 1189 (1965).
5. Hink, W., u. W. Petzold: Z. Angew. Phys. **10**, 135 (1958).
6. Kühnen, G.: Diss. FU Berlin, 1965.
7. Hink, W.: X-ray microscopy and X-ray micro-analysis. Proc. 2. Int. Symp., Stockholm, 1960, S. 83.
8. Elliott, S. B.: X-ray optics and X-ray micro-analysis. Proc. 3. Int. Symp., Stanford, 1962, S. 215.

Grazing Incidence X-Ray Telescopes

W. P. Reidy, R. Giacconi, G. Vaiana, L. van Speybroeck and T. Zehnpfennig

American Science and Engineering, Inc., Cambridge, Ma., U.S.A.

This paper reviews some of the work related to X-ray imaging which has been done at our laboratory over the past several years. Our interest has been to develop high resolution telescopes for studying X-ray emission from the sun and from other celestial X-ray sources. The systems which we have developed are based on the fact that X-rays are reflected with high efficiencies only at grazing angles of incidence.

Fig. 1 shows the calculated reflection efficiency for various materials (Be, Al, Ni, Au) at various grazing angles as a function of wavelength [1]. The telescopes which we are using now have a Kanigen [2] coated beryllium reflecting surface. The reflecting surface is tilted approximately 1 degree with respect to the optical axis; therefore, these systems have a short-wave cutoff on the order of 3 Å. The long-wave cutoff is determined by a filter placed in the optical path.

The requirement for shallow angles of incidence precludes the use of normal mirror geometries. Our mirror system is a double reflection system shown schematically in Fig. 2. The two reflecting surfaces are a paraboloid and, confocal to it, a hyperboloid. This design was first suggested by H. Wolter [3]. A single paraboloid used at grazing incidence is subject to severe coma; the second reflecting surface, the hyperboloid, corrects for this aberration. The mirror design requires that the optical axes of the paraboloid and hyperboloid coincide and that the focus of the paraboloid coincide with the focus of the other sheet of the hyperboloid. A mirror system of this type will provide high angular resolution over a narrow field of view. To maximize the reflection efficiency for a given mirror diameter and focal length, our mirrors are designed to have paraxial rays strike both reflecting surfaces at the same angle of incidence; therefore, the average

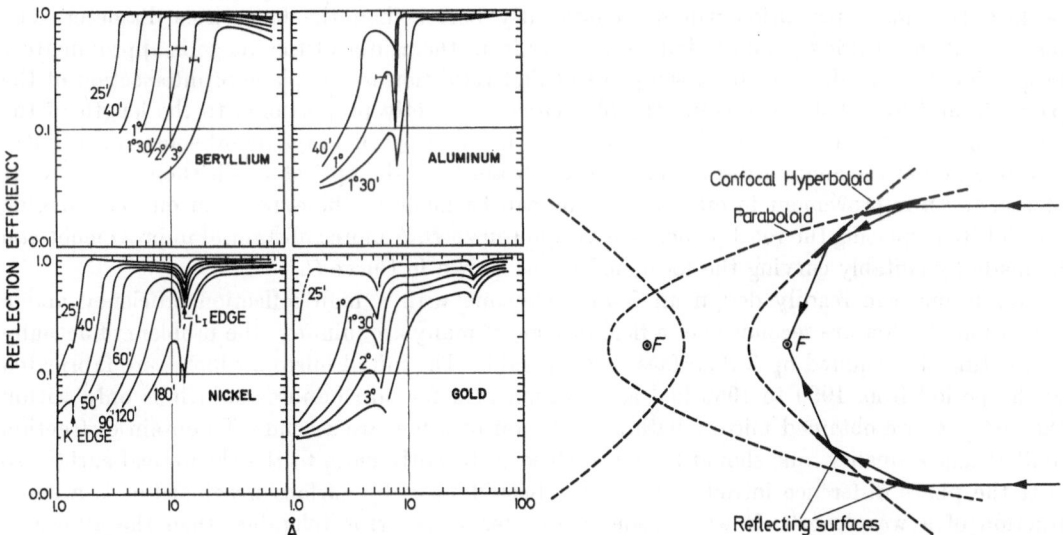

Fig. 1. Theoretical reflection efficiency as a function of wavelength

Fig. 2. Conceptual drawing of double reflection imaging system

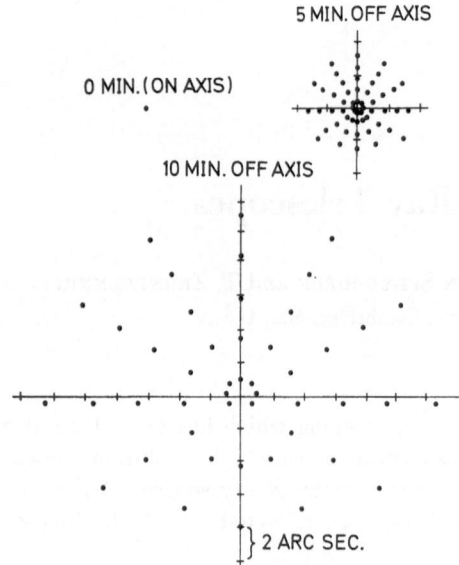

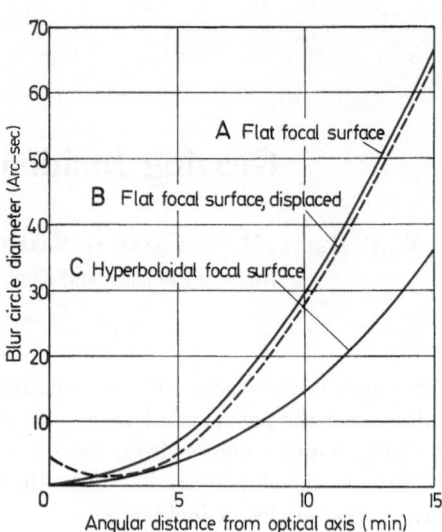

Fig. 3. Blur circle plots for rays incident 0, 5, and 10 minutes from axis

Fig. 4. Blur circle diameter as a function of angular distance from optical axis

hyperboloid slope is three times that of the paraboloid. Diameters of mirror systems we have built range from 7.5 cm to 23 cm. The geometrical image quality is not affected by a uniform scaling of the mirror but by changes in the average slope (i.e. ratio of diameter to focal length) and by changes in the ratio of diameter to length of the mirror element. For practical reasons, to minimize polishing, the mirror slope will be as large as possible consistent with the acceptable short wavelength cutoff.

Calculated blur circle plots for a mirror with a focal length to diameter ratio of 10:1 and a reflecting element length to diameter ratio of 1:1 are shown in Fig. 3. A bundle of 60 parallel rays evenly spaced over the telescope aperture was used in the ray tracing program. On-axis rays image to a point; off-axis rays image in a sharply peaked intensity distribution. The blur circle diameter is approximately 6 arc seconds for rays incident five minutes off axis and 18 arc seconds for rays incident 10 minutes off axis. Since the image intensity distribution is sharply peaked, the image resolution will be significantly better than the blur circle diameter. The distance of the striking points of individual rays from the center of the image is approximately proportional to the distance from the point of first reflection to the plane of intersection of the parabola and hyperbola. Therefore, the blur circle is directly proportional to the length of the reflecting elements. The blur circle diameter for a beam of parallel rays (point source at infinity) as a function of the angular displacement of the beam from the optical axis is shown in Curve A of Fig. 4. An improvement in off-axis imaging can be made at the expense of on-axis imaging by slightly displacing the focal plane, as shown in Curve B. A more substantial improvement can be made by suitably curving the focal surface, as shown in Curve C.

While one can readily design an X-ray telescope with a high reflection efficiency and a resolution of a few arc seconds over a field of view of many arc minutes, the problem of actually fabricating the required optical surfaces is formidable. The initial mirrors which were fabricated in the period from 1960 to 1962 had a resolution of a few arc minutes, and it is only within the last year we obtained mirrors with a resolution of a few arc seconds. To obtain diffraction limited image quality, one should have a surface sufficiently close to the theoretical surface so that the phase difference introduced in the reflected wave by surface irregularities is a small fraction of a wavelength. If we consider a quarter wave error tolerable, then the allowable height of a surface deviation is 85 Å for a mirror with an average slope of 1° and an X-ray wavelength of 12 angstroms. This is approximately 1/60 wave in visible light.

Fig. 5a and b. Photographs of X-ray telescope. a 44 cm² nested mirror for Apollo Telescope Mount. b 34 cm² mirror for rocket studies of solar corona

Since it is presently beyond the state of the art to fabricate large aspheric surfaces to this tolerance, we have examined the effects of various surface imperfections and thereby obtained a set of achievable tolerances which would assure adequate mirror performance. These tolerances are on mirror roundness, surface contour, and local surface finish. This basic requirement is that deviations in roundness and contour should not introduce local slope deviations in excess of a few arc seconds and that the height of irregularities in the local surface finish should be on the order of 100 Å or less. The requirements for roundness can be met. However, it has not yet been possible to eliminate small scale deviations in the surface contour which contribute to a spreading in the image plane of many arc minutes. Since the image distribution is still sharply peaked at the center, present systems have high angular resolution but not high image contrast. Surface finish on a local scale has been studied by electron microscopy techniques and the observed surface irregularities are on the order of 100 Å high and 1,000 Å wide. We believe the primary effect of these irregularities in finish is to decrease the observed efficiency

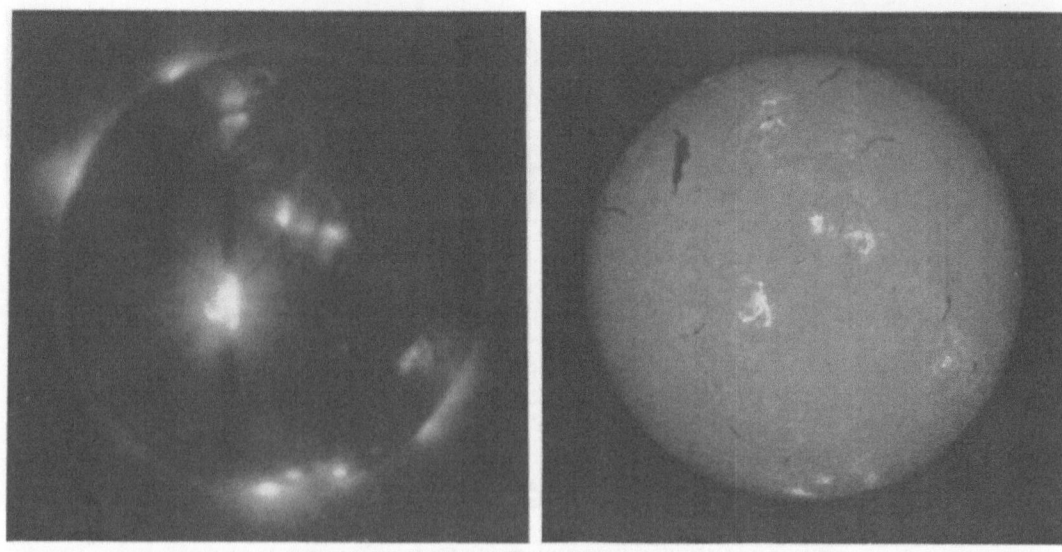

4 Micron aluminized mylar filter   Simulatanious H Photograph (courtesy of Essa)

E ↙↗ N

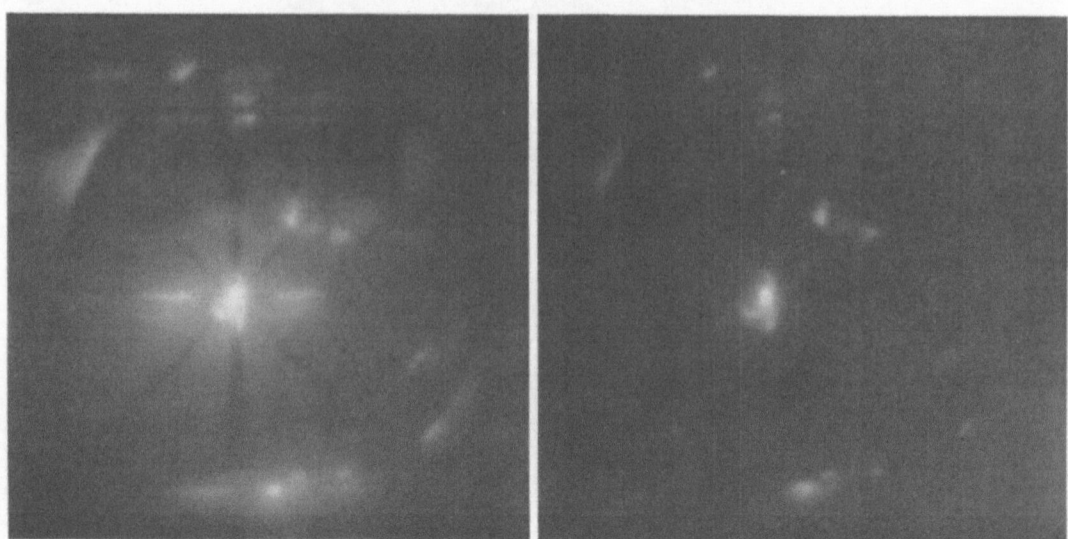

Slitless spectrogram, 12 micron beryllium filter   12 micron beryllium filter

Fig. 6. X-ray and $H\alpha$ photographs of a solar flare, 8 June 1968

rather than the resolution, since such imperfections should introduce large angle scattering which is not efficient at X-ray wavelengths.

Fig. 5 shows two X-ray telescope mirrors which have recently been fabricated [4]. One of the telescopes (Fig. 5a) has been designed for flight on the Apollo Telescope Mount (ATM) which will be the first manned orbiting solar observatory. This satellite will carry a variety of instruments to obtain high resolution X-ray and vacuum ultraviolet spectroheliograms and white light coronal observations during a two-month observational period. The ATM X-ray telescope consists of two nested confocal and coaxial imaging mirrors, each consisting of a paraboloid and an hyperboloid. The internal diameters of the two nested mirrors are 22.5 cm and 30 cm. The total geometric collecting area is 44 cm² and the focal length is 213 cm.

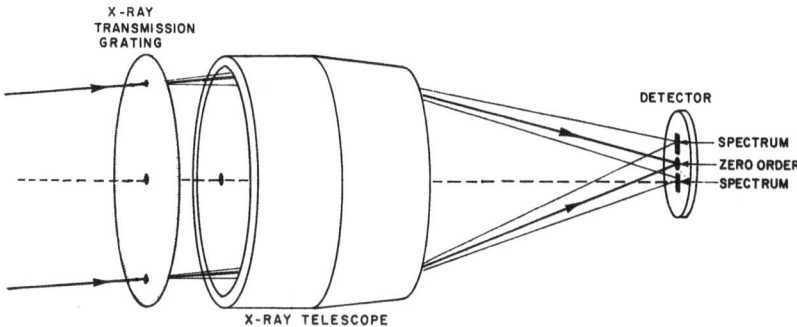

X-RAY
TRANSMISSION
GRATING

DETECTOR

SPECTRUM
ZERO ORDER
SPECTRUM

X-RAY TELESCOPE

Fig. 7. Schematic of slitless spectrometer

The telescope in Fig. 5b has been flown in two rocket flights to study solar X-ray emission. The telescope is 22.5 cm in diameter, has a focal length of 130 cm, and a collecting area of 33 cm². The reflecting efficiency of this telescope for a point source has been measured in the laboratory at 8.3 Å. Twenty percent of the energy in the image plane fell within 15 arc seconds of the image center. The total energy within 10 arc minutes of the image center is between 10 and 20 percent of the incident energy. This is to be compared to the theoretical double reflection efficiency of 50%. Laboratory tests of the X-ray resolution are difficult because of the problems involved in obtaining mechanical stability over the required exposure times. In tests made in a 21 meter long vacuum chamber, this telescope could clearly resolve two X-ray sources separated 7.4 seconds center-to-center. The best resolution measured to date was obtained during a rocket flight to study solar X-ray emission on 8 June of this year [5]. Three X-ray photographs obtained during that flight are shown in Fig. 6. An $H\alpha$ photograph ($\lambda = 6,562$ Å) taken at the same time from the ground is shown for comparison. The intense region in the center of the solar disk is a flare. Although $H\alpha$ emission must originate in regions significantly cooler than those responsible for the X-ray emission, there is a detailed correspondence between the $H\alpha$ and the X-ray features on the disk which indicates that the two emissions are physically linked. In addition, the X-ray photographs clearly show a three dimensional looping structure not visible in $H\alpha$. On the flight negatives of the X-ray photographs, filaments in the flare region 2 arc seconds wide and separated by 5 arc seconds can be resolved.

The photograph in Fig. 6 labeled Slitless Spectrogram contains in addition to the X-ray image the dispersed spectra of each of the emitting regions. This was obtained using the technique of X-ray slitless spectroscopy [6] developed at our laboratory. The technique consists of placing an X-ray transmission grating in front of the telescope as shown in Fig. 7. For a single point source, the image would consist of a zero order image, and bracketing it, spectra of various orders. When there are several X-ray emitting regions, separate spectra of each of the features are recorded simultaneously.

The transmission gratings, which are thin plastic replicas of conventional ruled gratings, have been shadowed at a shallow angle with gold. Since the plastic replica retains the impression of the grating surface, the gold will be deposited in thin lines. We have tested gratings ruled from 600 to 2,880 lines per mm. For such gratings, the dispersion varies from 0.25 to 1 arc minute per angstrom. The wavelength response of the gratings is determined on the long wavelength side by the absorption of the plastic substrate and on the short wavelength side by the transparency of the evaporated gold. The substrate is typically 10,000 Å of parylene (a thermoplastic polymer) which will transmit 93% at 8.3 Å. We have attempted to fabricate gratings with a gold line width equal to one-half the grating spacing. Theoretically for such a grating, 50% of the transmitted power will be in the zero order image and 20% in each of the first order spectra. The second order spectra will be supressed. We have fabricated 1,440 lines per mm gratings with as much as 18% of the transmitted power in the first order spectrum. Fig. 8 shows the resolution of two tungsten M lines at 6.74 and 6.97 Å. Strong first order and weak second

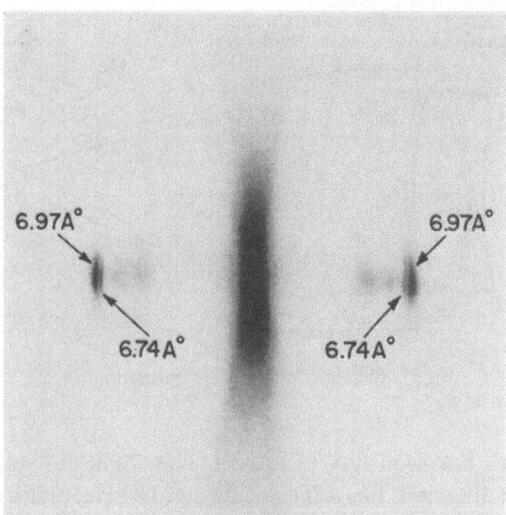

Fig. 8. Spectrum of laboratory source showing resolution of the tungsten $M\alpha$ and $M\beta$ lines at 6.97 Å and 6.74 Å

order spectra can be seen in the picture. The measurement indicates a spectral resolving power $\lambda/\Delta\lambda$ of approximately 50.

The development of the X-ray telescope from a laboratory curiosity to its present position as a major tool in X-ray astronomy has taken place in the last eight years. This rapid evolution will continue in the next decade. We are now designing telescopes an order of magnitude larger than presently available mirrors for satellite studies of celestial X-ray sources. By extending the nested mirror concept developed for ATM (Fig. 5a), it is feasible to design a telescope with a collecting area of 1,000 cm² and a focal length of 6 meters.

This work has been supported by the National Aeronautics and Space Administration under Contracts NAS 5-9041, NASW-1555 and NASW-1700.

References

1. Zehnpfennig, T., R. Giacconi, R. Haggerty, W. Reidy, and G. Vaiana: A laboratory program to develop improved grazing incidence X-ray optics. NASA CR-717, February, 1967.
2. Kanigen is a chemically deposited alloy consisting of 91% nickel and 9% phosphorous. Because it is amorphous, it is suitable for optical polishing.
3. Wolter, H.: Ann. Physik 10, 94, 286 (1952).
4. These mirrors were polished by Diffraction Limited, Inc., Bedford, Massachusetts, U.S.A.
5. Vaiana, G. S., W. P. Reidy, T. Zehnpfennig, L. van Speybroeck, and R. Giacconi: Science 161, 564 (1968).
6. Gursky, H., and T. Zehnpfennig: Appl. Opt. 5, 875 (1966). — Zehnpfennig, T.: Appl. Opt. 5, 1855 (1966).
7. Giacconi, R., W. P. Reidy, G. S. Vaiana, L. P. van Speybroeck, and T. F. Zehnpfennig: Space Sci. Rev., to be published (1969).

Application of X-Ray Optical Methods in Solar Astronomy

U. MAYER

Astronomisches Institut der Universität Tübingen, BRD

The possibility of using soft X-rays to image the sun has been studied in the Astronomical Institute at Tübingen. This activity was stimulated by previous theoretical work on solar X-rays by G. ELWERT of Tübingen University and by experimental work of the microminiaturisation group of the Institute of Applied Physics who could provide Fresnel-Microzoneplates [1] for soft X-ray optics.

In the wavelength range between 10 and 100 Å, in which we are mainly interested, the solar corona emits a great number of spectral lines of highly ionized atoms together with a weak continuum which may partly or totally consist of unresolved emission lines. Most of the energy flux is emitted from narrow zones on the solar disk, the coronal condensations. These active regions are spatially connected with solar plages, an attribute of solar activity observable in the visible range. Therefore an estimate of the dimension of X-ray plages and of the existence and location of active zones during a rocket experiment lying ahead can be obtained from earth bound observations. Depending on ionisation equilibrium the highest ionisation states occur in hot active regions only. Consequently the most energetic emission lines are emitted from plages only, whereas the soft radiation is also emitted from the quiet corona.

Fig. 1 shows X-ray spectra of the quiet sun between 11.8 and 25 Å, obtained by the U.S. Naval Research Laboratory [2] with a rocket borne KAP spectrometer. Two plages appear separately at short wavelengths, whereas the O VII lines at 21.6 and 22.8 Å are smeared over the solar disk. Fig. 2 shows the adjacent spectral region between 30 and 130 Å as obtained by J. E. MANSON [3] of Airforce Cambridge Research Laboratory with a grazing incidence spectrograph. At wavelengths shorter than 60 Å only a few bright lines appear, whereas the region of maximum emission of the quiet sun is crowded with intense lines.

In order to obtain information on the physical parameters of solar active regions high resolution X-ray images of the sun should be taken in narrow spectral bands. As solar activity is a highly variable phenomenon only long time or repeated observations can provide reliable knowledge.

Let us now estimate whether Fresnel zoneplates focused for a selected emission line and combined with a filter of suitable transmission or with a microminiaturized transmission grating can be used in a multicolor rocket spectroheliograph. The table shows a survey of recently published data on the solar emission between 13 and 100 Å (J. E. MANSON; K. EVANS and K. A. POUNDS [3, 4]). To make things clearer only the most intense lines in the long wavelength part of the table are included. The values given for the energy flux are instantaneous or short time averages. They can vary by an order of magnitude or even more with solar activity, especially at short wavelengths. The flux data below 20 Å are related to a single plage with the exception of two O VII lines (EVANS and POUNDS).

For a rocket experiment which we are preparing we plan to use zoneplates of 30 cm focal length, focused at 50, 34, 21.7, 19, 17, 15 and 14 Å, and combined with different filters. For zoneplates consisting of a hundred zones the diameters range from 0.4 to 0.8 mm. A zoneplate of 100 zones and 30 cm focal length at 15 Å has an outermost transparent zone of about 1 μm in width. Such plates can be made with a thickness of the opaque zones of about 1 μm. The transmission of a zone plate related to the primary focus is about 10% and the filter transmission

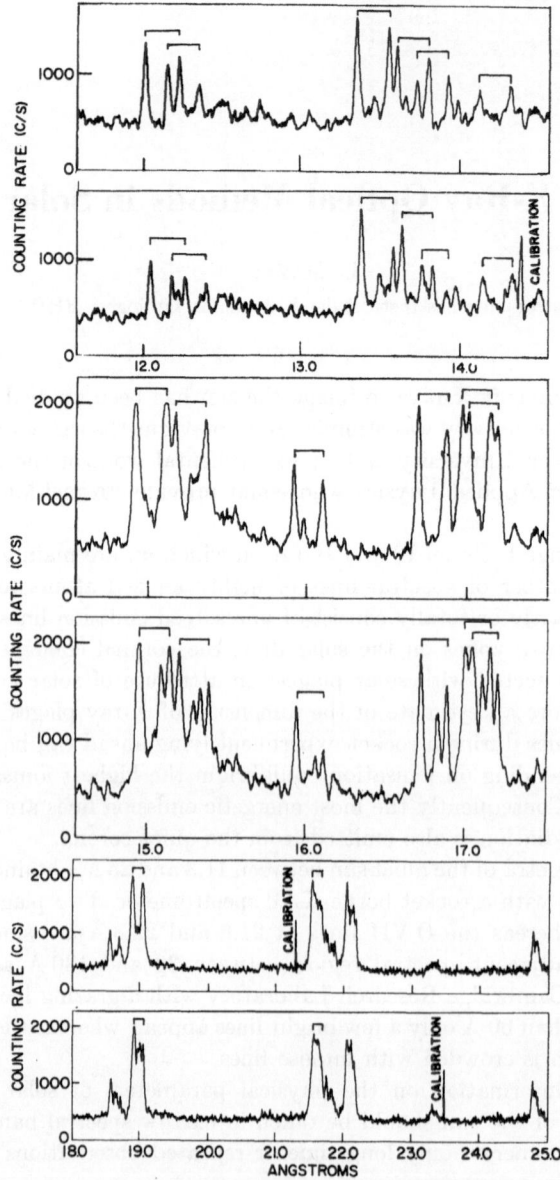

Fig. 1. Solar X-ray spectra from 11.8 to 25.0 Å. The upper plot of each pair is the ascent scan; the lower plot of each pair is the descent scan. The brackets indicate one solar diameter and identify the same spectrum lines from different active regions on the disk. (After G. FRITZ et al. [2])

can be adjusted to a value of approximately 50%. Taking the photon flux around 15 Å wavelength, given in the table and assuming a transmission of 5% and an exposure time of 200 sec about 10^4 photons would contribute to the image of a bright plage. Assuming further that the area of the plage is of the order of 1×1 minute of arc, the exposure of the film would be between 10^7 and 10^8 photons/cm². So that it can well be recorded on high speed soft X-ray films like Kodak Pathé SC 5 or Ilford special G (P. A. ATKINSON, K. A. POUNDS [7]; W. P. REIDY, G. S. VAIANA, T. ZEHNPFENNIG, R. GIACCONI [8]). For longer wavelengths the flux is higher and the area of the zoneplates is larger so that shorter exposure times or less transparent filter combinations with more favourable transmission curves can be chosen.

The filters to be used in this experiment are self-supporting foils of different materials with a free diameter of 2.5 mm to be mounted in front of the zoneplates. They have to serve two

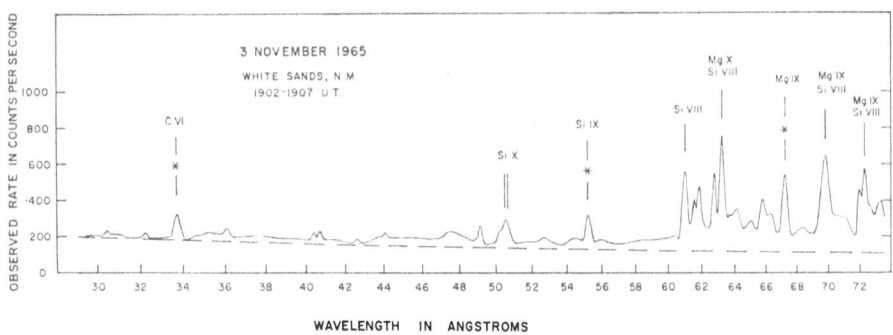

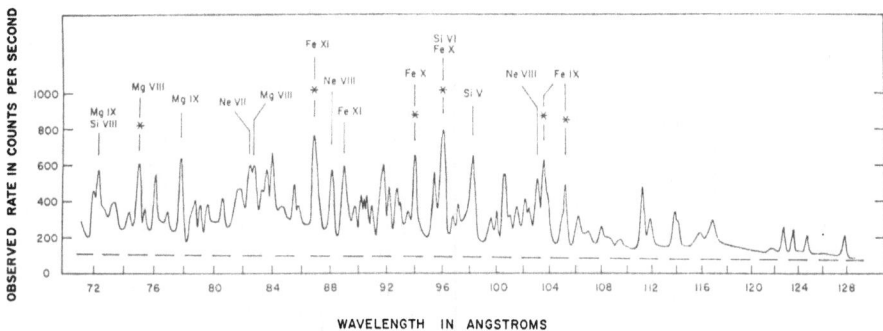

Fig. 2. Averaged observed solar spectrum after J. E. MANSON [3]

different purposes, to prevent visible and near ultraviolet radiation from reaching the film and to limit the transmitted spectral band. At 50 Å a thin carbon foil would meet these requirements but it was found that carbon foils, although they can be prepared opaque and self-supporting in the desired size tend to break under the mechanical stress occuring during a rocket start. A more reliable filter for wavelengths at the long wavelength side of the carbon K edge is a combination of a plastic foil, e.g. polystyrene, and 1,000 Å aluminum. In the wavelength range between 10 and 30 Å the L edges of the metals of atomic number 29 to 22 can be used for wavelength discrimination. It proved to be possible to prepare self-supporting foils of 2.5 mm in diameter and 2,000 to 2,500 Å in thickness of copper, nickel, iron and titanium which displayed considerable resistance to mechanical stress up to several g acceleration and several kc/sec vibration frequency. Only very few data are available on absorption coefficients of these metals on the long wavelength side of the L-edges. From experimentally supported values given by COMPTON and ALLISON [9] for copper, we conclude that the absorption of a copper foil 5,000 Å in thickness at 15 Å is of the order of 50%. The L absorption jump for the above mentioned metals is of the order of 5 so that a considerable discrimination against shorter wavelengths can be achieved. At the long wavelength side of the transmitted band the increasing absorption of the filter and the defocusing due to the chromatic aberration of the zoneplate will contribute to the suppression of unwanted lines.

The L_{III}-edges are for copper 13.3 Å   manganese 19.4 Å
                              nickel 14.5 Å   chromium 21.6 Å
                              cobalt 15.9 Å   titanium 27.3 Å
                              iron 17.5 Å

The investigation of the transmission curves of these metal foils is continued in our laboratory.

The theoretical limit of resolution of our zone plates is of the order of 1 sec of arc, practically we expect about 3″. This corresponds to 3 µm on the film in the focal plane which is better than the graininess of the film, so that the resolution will be limited by the film or by the pointing jitter of the payload. The next step in the development of small and inexpensive X-ray

Table 1. *Solar XUV- and ultrasoft X-ray spectrum. 30—100 Å major line intensities after* J. E. Manson [3]. *13—22 Å emission of a solar active region after* K. Evans *and* K. A. Pounds [4]

Wavelength λ [Å]	Ion	10^{-3} erg/cm² sec	10^6 photons/ cm² sec	Wavelength λ [Å]	Ion	10^{-3} erg/cm² sec	10^6 photons/ cm² sec
100.6 100.5		} 3.30	16.7	61.06	Si VIII (T)	2.90	8.9
96.07 95.87		} 4.50	21.8	55.28	Si IX (T)	2.92	8.1
95.38 93.98	Fe X (Z)	1.68 2.08	8.1 9.8	50.7 50.5	Si X (T) Si X (T) }	} 3.42	8.7
92.95 92.72		} 2.55	11.9	33.70	C VI (T)	2.54	4.3
91.8 91.6		} 3.22	14.9	21.80 21.60	} O VII	1.93 3.63 } 5.56	6.2
87.2 86.90	Fe XI (Z) }	3.76	16.5	18.97 18.63	O VIII O VII	1.30 0.56	1.25 0.53
82.78 82.50		} 3.78	15.7	17.77	O VII	0.18	0.16
81.8 81.6	Mg VIII (Z) Si VII (Z) }	2.66	10.9	17.05 16.77	Fe XVII Fe XVII	0.84 0.32 } 1.16	1.0
73.4 73.2		} 1.95	7.2	15.45 15.26 15.01	} Fe XVII	0.22 0.48 0.74	
69.9 { 69.7	Fe XV (Z) Mg IX (T) } Si VIII (Z)	3.67	12.9	16.01 15.17 14.82	} O VIII	0.20 0.12 0.02 } 1.78	1.35
66.4 66.3 66.1	Fe XVI (T) Fe XVI (T) }	1.95	6.5	14.40 14.25	} Fe XVIII	0.045 0.027	
65.9 65.6	Mg IX (Z) }	1.77	5.8	14.03 13.77	} Ni XIX	0.029 0.058	
62.80 63.25	Si VIII (T) Si VIII (T)	1.68 } 4.19 2.51	5.3 } 13.3 8.0	13.82 13.65	Fe XVII Ne VIII	0.088 0.03 } 0.53	0.37
				13.55 13.45	} Ne IX	0.074 0.18	

T: R. Tousey et al. (1965) [5]. Z: H. Zirin (1964) [6].

spectroheliographs is the combination of a microminiaturized transmission grating of 1,000 Å spacing, being prepared at the Institute of Applied Physics, with a Fresnel zoneplate. The dispersion of such a grating is 3.4′/Å in the wavelength range from 10 to 100 Å. As the angular diameter of the sun is about 32 minutes of arc the images of the sun in neighboring emission lines would overlap. It can be expected however that bright active zones would be separated even for closely spaced lines. E.g., an image focused to the strong carbon VI line at 33.7 Å would only overlap at one limb with the 50.5 Å image. The plus and minus first order of the grating-zoneplate combination are equivalent. Therefore at least two monochromatic images at the same wavelength but with different additional filtering or in different focal planes can be obtained with a single instrument. Another possibility, at present investigated in this laboratory, is to use the light in one of the two equivalent images for photoelectric intensity measurements.

Photoelectric methods, employing open structure photomultipliers, are frequently used in the extreme ultraviolet range above 100 Å wavelength. At shorter wavelengths the ordinary metallic photocathodes are relatively ineffective. Therefore gas flow counter tubes have been employed for intensity measurements although the gas supply system, the fragile window and the lower pulse amplification in the filling gas are disadvantages in space applications. At present the so called channeltron multipliers supplied by Bendix in the U.S.A. and by Mullard Ltd. in England display the most favourable pulse characteristics and pulse amplification factors for photon counting. However the quantum efficiency of this device has to be improved by the

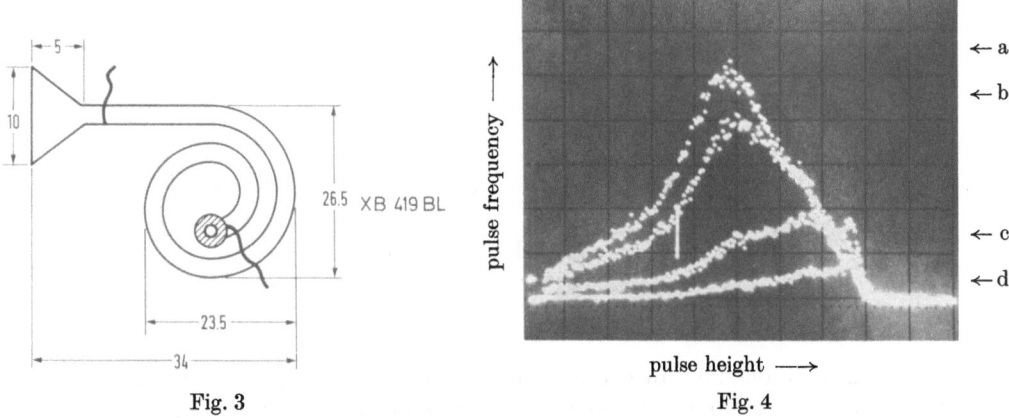

Fig. 3   Fig. 4

Fig. 3. Schematic drawing of a Mullard channeltron with funnel shaped orifice

Fig. 4. Pulse height distribution for a Mullard flare channeltron multiplier XB 419 BL. CuK radiation incident on the central capillary, a and b; and on the funnel wall, c and d. 250 counts/vertical division, 256 channels full scale, accumulation time 2 minutes

introduction of nonmetallic photocathodes to make them comparable with gas flow counters. Channeltron multipliers are available in different shapes, beginning with capillaries with an input diameter of less than 1 mm. For simultaneous intensity measurements in a transmission grating spectroheliograph flare channeltrons would be preferred (Fig. 3). Here the input of the capillary is widened into a funnel of up to 10 mm in diameter. This type of photomultiplier with a comparatively large cathode surface allows to monitor the total monochromatic flux of the solar image with an unusually small and ruggedized detector. It shall be mentioned, however, that the photoelectric effect at short wavelength depends strongly on the input angle of the radiation. Therefore a variation in quantum efficiency was expected in flare channeltrons between the central tube and the inclined funnel wall. Fig. 4 shows four pulse frequency distributions obtained with CuK radiation and a Mullard flare channeltron at different positions of the incident beam 1 mm in diameter. The scales of all four curves are the same. It is apparent that not only the pulse frequency, i.e. the quantum efficiency depends on the input angle, but also the shape of the pulse frequency curve, i.e. the pulse amplification. The investigations are continued.

The image integration method of H. J. EINIGHAMMER [10], which will be presented in a separate paper, can be used to improve the speed of a transmission grating spectrograph at short wavelength where the photon flux at low solar activity is critically small for this device or when the focal length has to be extended to obtain better resolution on the photographic film.

References

1. GROTE, K. H. v., G. MÖLLENSTEDT u. R. SPEIDEL: Optik **22**, 252 (1965).
2. FRITZ, G., R. W. KREPLIN, J. F. MEEKINS, A. E. UNZICKER, and H. FRIEDMAN: Astrophys. J. **148**, L 133 (1967).
3. MANSON, J. E.: Astrophys. J. **149**, 703 (1967).
4. EVANS, K., and K. A. POUNDS: Astrophys. J. **152**, 319 (1968).
5. TOUSEY, R., W. E. AUSTIN, J. D. PURCELL, and K. G. WIDING: Ann. Astrophys. **28**, 755 (1965).
6. ZIRIN, H.: Astrophys. J. **140**, 1332 (1964).
7. ATKINSON, P. A., and K. A. POUNDS: J. Phot. Sci. **12**, 302 (1964).
8. REIDY, W. P., G. S. VAIANA, T. ZEHNPFENNIG, and R. GIACCONI: Astrophys. J. **151**, 333 (1968).
9. COMPTON, A. H., and S. K. ALLISON: X-rays in theory and experiment, sec. ed., p. 526 ff. Toronto-New York-London: D. van Nostrand Comp., Inc. 1954.
10. EINIGHAMMER, H. J.: This publication, p. 50.

Detection of Weak X-Ray Sources by Image Integration

H. J. EINIGHAMMER

Astronomisches Institut der Universität Tübingen, BRD

In this paper we denote a radiation emitting source as weak, when the speed of the image forming system does not suffice to produce an image of the desired quality.

For the purpose of forming X-ray images of the sun the following systems have been applied or tested up till now: the pinhole [1], the zoneplate [2, 3], and the mirror system of WOLTER and GIACCONI [4]. If one wishes to investigate radiation with wavelengths smaller than 2—3 Å the grazing incidence mirror and the zoneplate cannot be employed to advantage. In this region one can only use the pinhole. This ceases to be useful with very hard radiation.

The pinhole camera, however, is not very satisfactory, because it is not possible to improve its very low speed. The speed of the zoneplate is also limited because an increase in the number of rings leads very soon to spacings which are no longer realisable.

In order to get more image information with pinholes and zoneplates in spite of these restrictions it is possible to apply the principle of signal integration, a method which is often used in electronics to detect weak recurrent signals. In electronics, however, weak temporal signals are analyzed over a noise level. But in our case we are concerned with underexposed photographic images, that is to say with weak spatial signals over a spatial noise level.

In integrating photographs we proceed as follows: Instead of one image with one pinhole we produce simultaneously N images with N pinholes. The same procedure applies to zoneplates. The images are superimposed to form a composite image, which contains the combined information of the single ones. Since in this way the image signal is multiplied by the factor N and the rms-noise only by the factor $\sqrt{N}$, there is a $\sqrt{N}$-increase of the signal to noise ratio. That means, that the detection threshold for weak signals is decreased, and the dynamic range of the X-ray film, which is not very large, can be better utilized.

Two optical techniques are presented here by means of which numerous images on a photoplate can be superimposed simultaneously[1]:

1. X-Ray Integral Photography [5]

The X-ray camera has N imaging systems side by side, in this case either zoneplates or pinholes (Fig. 1). We call this two-dimensional array of systems an objective matrix. After exposure with the matrix camera and after the usual development of the film, N images are obtained, that is to say, the corresponding image matrix is obtained.

In a second imaging step the single images are superimposed by projecting them upon each other by means of a lens matrix (LM) and a telescope-lens of large aperture. In order to do this it is necessary to arrange the elements of the lens matrix in the same way as those in the objective matrix. This can be achieved by adjusting the elements of one of the matrices in a way, that the element pattern of the other is reproduced.

To demonstrate the efficiency of this technique, a pilot experiment with visible light was carried out. The first step of the imaging process, the exposure with X-rays, was simulated — the lens matrix and visible light were used instead of the objective matrix and X-rays, to produce the image matrix.

---

1. For special considerations in respect of starlike objects see: DICKE [9].

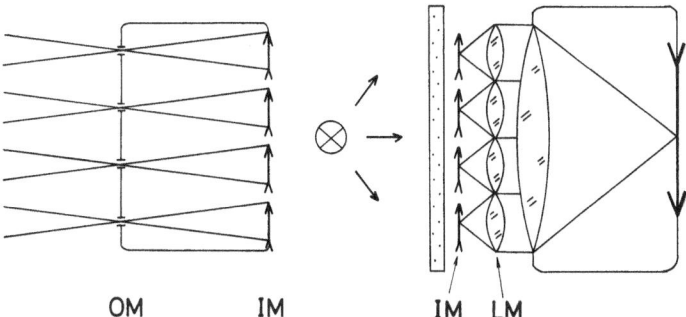

Fig. 1. On the left: X-ray camera with objective matrix OM and image matrix IM. On the right: Superposition of the images by means of a lens matrix LM and a telescope lens

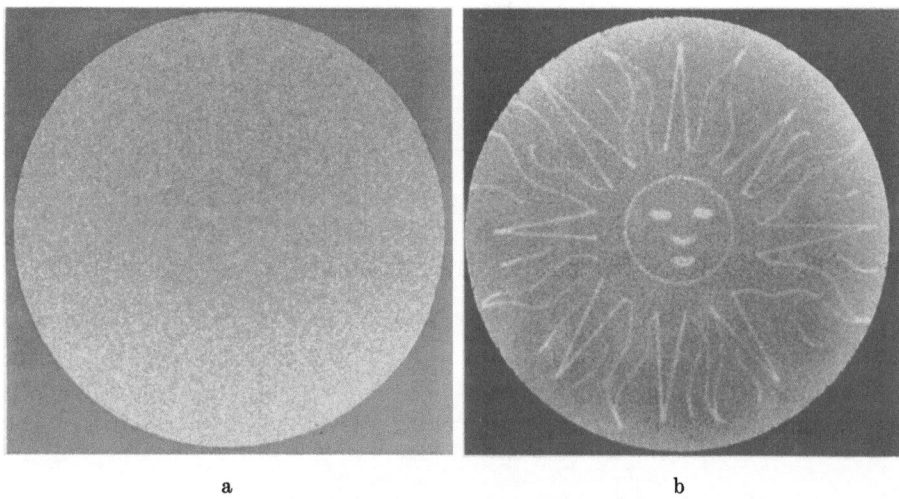

a   b

Fig. 2. a Single image of an underexposed image matrix. b Superposition of 85 single images, showing the increased signal to noise ratio

Fig. 2 shows the result. The left side shows one of the single images and the right the composite image consisting of 85 single ones. The effective speed of a matrix camera compared with that of an ordinary camera would in this case increase by almost 10 times. The image matrix here has been so much underexposed that almost no information can be seen in the grain noise. Only after the superimposition the signal is raised above the detection threshold.

Photographing an object and reproducing the image, both with the same lens matrix, corresponds to the first stages of Lippmann's Integral Photography [6] a method to produce stereoscopic images in the visible region. Using the proposed X-ray technique even three dimensional X-ray images can be made, but this clearly can not be exploited in X-ray astronomy.

If one wishes to proceed to a very large N, it becomes troublesome to match the two matrices mentioned above. This disadvantage is avoided by the following holographic technique, which is however at the present time more applicable to pinholes as imaging systems.

2. Holographic Image Integration [7]

Here, in the same way as before, an image matrix is produced by means of the matrix camera. For the integration procedure of course the distribution of the images or that of the pinholes must be known. If this information is stored by recording a hologram of the pinhole-matrix, this hologram can be used directly for an optical integration process.

4*

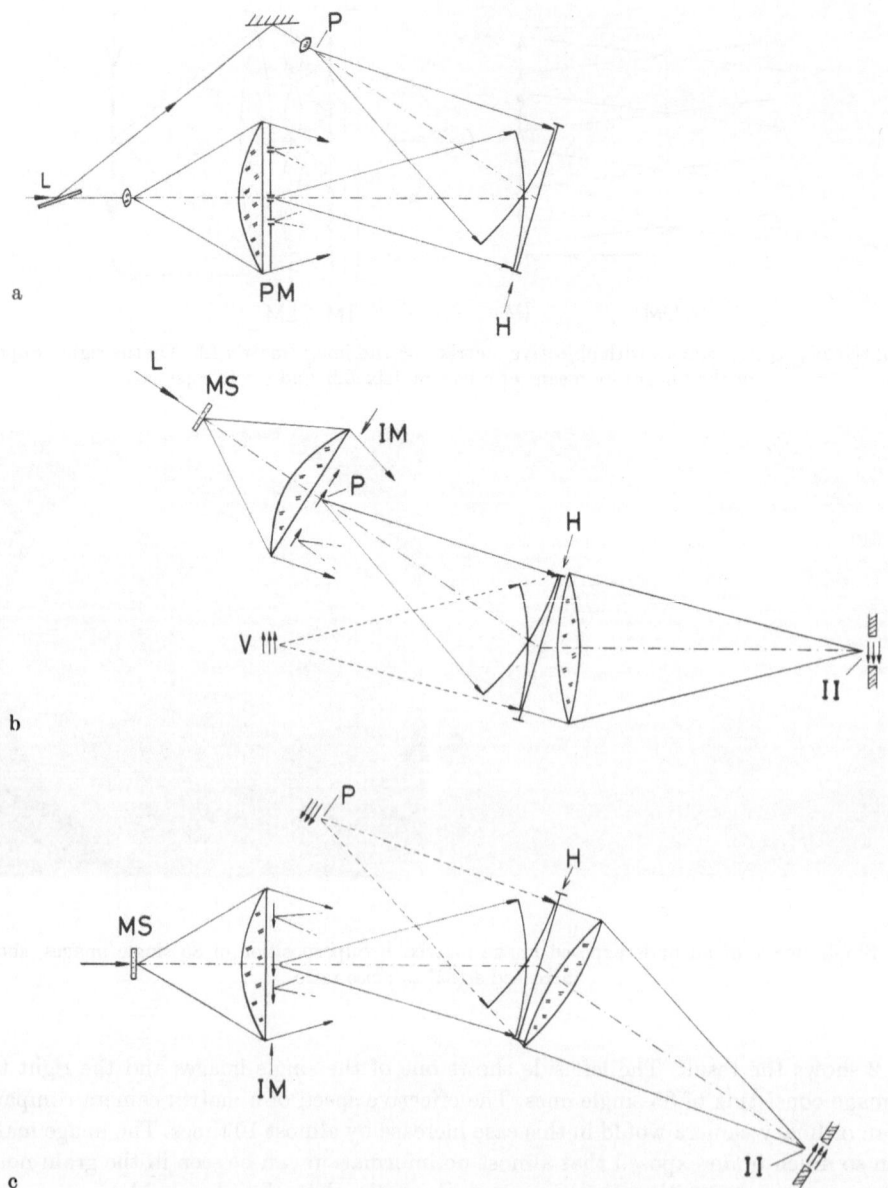

Fig. 3a—c. Holographic procedure for $N = 3$: Hologram recording (a). Image integration by hologram illumination with a modified reference beam (b) and by hologram illumination with a modified object beam (c). (*L* Laser beam, *PM* Pinhole matrix, *P* Reference point, *H* Hologram, *MS* Moving scatterer, *IM* Image matrix, *V* Virtual image plane, *II* Integrated image)

On Fig. 3a the optical arrangement for recording the hologram is shown. The hologram type selected is similar to the Fourier type [8] with interfering spherical waves of equal radius of curvature.

The pinhole matrix *PM* is illuminated via a condensor lens so that the diffraction patterns of the holes are superimposed on the photoplate *H*. The reference beam originates from the reference point *P*, *P* being the same distance from the plate *H* as the pinhole matrix.

To reconstruct the object in the usual manner, we could illuminate the hologram with the reference beam and find a virtual image of the pinhole matrix in the position where the real pinhole matrix was located during the hologram exposure.

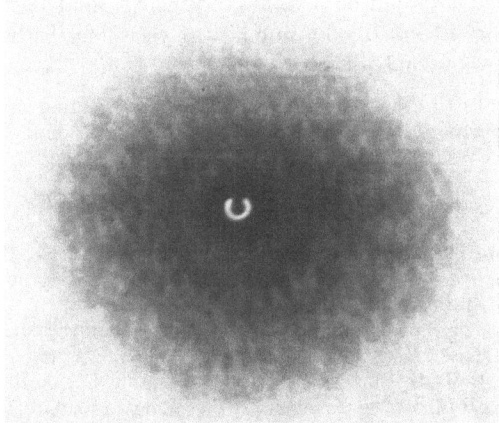

Fig. 4. Image of a bulb filament, taken with visible light by means of a matrix camera (100 pinholes) and holographic integration. The composit image in the center is surrounded by others which do not disturb the central one

In order to obtain the desired composite image the usual reconstruction process is modified by using the image matrix in the holographic arrangement. There are two ways to get the composit image:

1. The reference source P in Fig. 3a is replaced by the image matrix IM, which is illuminated using the condensor and a moving scatterer MS as shown in Fig. 3b.

Doing this, we have to arrange the image matrix so that its pattern is rotated by 180° relative to the virtual pinhole matrix, which would become visible in the normal case. The virtual image plane V of the hologram now contains N^2 images instead of the pinhole matrix. N of them are superimposed exactly. This is the composit image which can be photographed through the hologram. The formation of the integral image can be described by convolution of the image matrix with the pinhole matrix.

2. The second method gives identical results[2]. In this case the pinhole matrix PM in the recording arrangement (Fig. 3a) is replaced by the image matrix having the same orientation.

If the image matrix IM is illuminated by means of the condensor (see Fig. 3c) we find the wanted composit image, looking through the hologram, in the near vicinity of the reference point P.

The integration effect here can be understood in the following way:

Considering, that in holography the reference beam and the object beam are equivalent, we could also reconstruct the point P from the hologram by using the pinhole matrix as a reference source. Furthermore it is possible to obtain P when incoherent illumination of the pinhole matrix is applied (moving scatterer) or when the holes vary in brightness or even when only a single pinhole is illuminated, for the diffraction gratings of the hologram, which correspond to the single pinholes, can be assumed to work independently.

Essentially the pinhole matrix already represents an image matrix of an X-ray point source. A second object point corresponds to a second point matrix, displaced parallel to the first. If matrix and reference point P have the same distance from the hologram, and if the diffraction angles at the hologram remain small, then the corresponding second image point P' will be displaced in the same way from the first point P. This consideration holds for every image point.

Fig. 4 shows a picture of the glowing filament of a bulb, which was taken using the holographic procedure. The integrated image in the center is surrounded by others, which are not of interest and which do not disturb the central one. The matrix camera used had 100 pinholes, distributed over an area of 9×12 cm.

---

2. However the procedure is simplified because less adjustment of the elements is required.

An estimate shows, that it should be possible to image a solar flare of type II by means of a rocket-borne camera having about 10,000 pinholes at wavelengths between 1 and 3 Å with an angular resolution of several seconds of arc.

I wish to express my thanks to Professor G. MÖLLENSTEDT for his kind support. This research is wholly supported by the Bundesministerium für wissenschaftliche Forschung, Bonn.

References

1. BLAKE, R. L., T. A. CHUBB, H. FRIEDMANN, and A. E. UNZICKER: Astrophys. J. **137**, 3 (1963).
2. MÖLLENSTEDT, G., H. J. EINIGHAMMER, K. H. v. GROTE and U. MAYER: Optique des rayons X et microanalyse, p. 14. Paris: Hermann 1966. See also: EINIGHAMMER, H. J., G. ELWERT, and U. MAYER: Space research VII, p. 1336. Amsterdam: North-Holland Publ. Comp. 1967.
3. BOL RAAP, B. E., J. B. LE POOLE, J. H. DIJKSTRA, W. DE GRAAF, and L. J. LANTWAARD: Small rocket instrumentation techniques, p. 203. Amsterdam: North Holland Publ. Comp. 1969.
4. REIDY, W. P., R. GIACCONI, G. VAIANA, L. VAN SPEYBROECK, and T. ZEHNPFENNIG: This publication, p. 39.
5. EINIGHAMMER, H. J.: Naturwissenschaften **54**, 641 (1967).
6. LIPPMANN, G.: J. Phys. **7**, 821 (1908); — Compt. Rend. **146**, 446 (1908).
7. EINIGHAMMER, H. J.: Naturwissenschaften **55**, 295 (1968).
8. STROKE, G. W.: An introduction to coherent optics and holography. New York and London: Acad. Press 1966.
9. DICKE, R. H.: Astrophys. J. **153**, L 101 (1968).

Resolution Error Correction of Small Angle Scattering Data Using Hermite Functions

F. Hossfeld

Institut für Festkörper- und Neutronenphysik, Kernforschungsanlage Jülich, BRD

1. Introduction

In small angle scattering the determination of characteristic parameters has become a useful tool for the evaluation of scattered intensity of dilute systems containing identical particles [1]. But in most cases, especially in problems of solid state physics, the scattering systems are of more complicated structure. The most general information we can draw from elastic isotropic small angle scattering is contained in the correlation function $g(r)$ as the spatial Fourier transform of the scattering function $S(\varkappa)$, where $\varkappa = 4\pi \cdot \sin(\vartheta/2)/\lambda$. ϑ is the scattering angle and λ the wavelength.

In addition to principal limitations of the experimental information due to the fact of statistical errors and the merely finite $\varkappa$-range, the scattering distributions unfortunately are often substantially distorted by systematic errors caused by collimation effects and nonmonochromatic radiation. Thus in many cases no direct evidence of $S(\varkappa)$ can be obtained.

2. The Resolution Errors

We consider the intensity distribution i_0 of the primary beam in the detector plane which we describe by a coordinate system (h, u). h means the reduced angular variable $h = 2p \cdot \sin(\vartheta/2)$ in scanning direction, where p may be a scaling parameter determined by the experimental device. u describes the axis perpendicular to h. Restricting the discussion to slit arrangements we postulate that the primary intensity may be separated with respect to h, u, and λ. Thus, introducing instead of the wavelength distribution the distribution in "momentum" space $z(k)$, the relation may hold:

$$i_0(h, u, k) = z(k) \cdot i_B(h) \cdot i_H(u), \qquad k = 2\pi/(p\,\lambda). \tag{1}$$

Then we can formulate the resolution errors in small angle scattering by three successive integral equations [2]:

a) the wavelength distortion

$$J(h) = \int\limits_0^\infty z(k) \cdot S(k \cdot h)\,dk; \tag{2}$$

b) the slit width effect

$$V(h) = \int i_B(h') \cdot G(h - h')\,dh'; \tag{3}$$

c) the slit height effect

$$G(h) = \int i_H(u) \cdot J(\sqrt{h^2 + u^2})\,du. \tag{4}$$

Since small angle scattering implies approximation of the sine function by its argument: $\sin(\vartheta/2) \approx \vartheta/2$, we may formally extend the integration range in these equations to infinity.

The smearing effect by polychromatic radiation may be important in neutron scattering, whereas in X-ray small angle scattering usually wavelength and slit width distortion can be neglected compared to the slit height effect. Therefore in the following we shall concentrate on the latter. But we should emphasize that Hermite functions can be applied to correct all of these resolution errors in an adequate manner by series expansions leading to simple recurrence relations [3].

3. Correction Method Using Hermite Functions

The Hermite orthonormal functions are explicitely given [4] by

$$\psi_{2n+s}(x) = \left[\sum_{\nu=0}^{n} b_{2\nu+s}^{(2n+s)} x^{2\nu+s}\right] \cdot \exp(-x^2/2), \quad s = 0 \text{ or } 1; \quad n = 0, 1, \ldots. \tag{5}$$

The expansion of experimental scattering distributions for any functions must be truncated at a finite index because of the limited information available due to the finite number of sampling points and the statistical errors. Because of the completeness of the Hermite functions an expansion for this orthonormal system is consistent with the method of least squares, reducing all experimental information to a few expansion coefficients.

Whereas closed analytic solutions of the slit height equation could be found for the special cases of infinite slit height by GUINIER and FOURNET [5] and DuMOND [6] and for Gaussian shaped slit height weighting functions $i_H(u)$ by KRATKY, POROD and KAHOVEC [7] the Hermite function method yields a practical general solution of this problem independent of the special type of the weighting function [2]. After inserting the series expression for the solution

$$J(h) = \sum_{n=0}^{N} j_{2n} \psi_{2n}(h) \tag{6}$$

into the integral equation and expanding the experimental curve $G(h)$ for Hermite functions

$$G(h) = \sum_{n=0}^{N} g_{2n} \psi_{2n}(h), \tag{7}$$

comparison of coefficients of equal power of h yields a system of linear equations with a triangular matrix coupling the coefficients j_{2n} with the known coefficients g_{2n}. This system is easily solved for the j_{2n} yielding the undistorted function $J(h)$. The special shape of the slit height weighting function is involved in the triangular matrix.

It should be mentioned that error propagation may be discussed within the frame of this formalism [2].

4. Optimal Scaling Factors and Numerical Results

In the Hermite expansion of a Gaussian $f(x) = \exp(-a^2 x^2/2)$ all coefficients excepting the zero order one vanish because of orthonormality if the x-scale is transformed to $a = 1$. Thus we can improve series convergence by taking Gaussian shape as a first approach to the distributions of small angle scattering according to Guinier's Approximation. For adaption of the scale we may use properly the variance

$$\sigma_f^2 = \int f(x) \cdot x^2 dx / \int f(x) \, dx. \tag{8}$$

Transforming $x^+ = x/\sigma_f$ yields $\sigma_{f+}^2 = 1$ in agreement with the variance of $\psi_0(x)$. Since the distributions more or less deviate from Gaussian shape, one should modify this transformation by introducing an optimizing factor φ_{opt} according to

$$\sigma_{\mathrm{opt}} = \varphi_{\mathrm{opt}} \cdot \sigma_f. \tag{9}$$

The factor φ_{opt} can be determined by a least squares criterium for the deviations of the numerical results from the theoretically expected solution in testing the method. We got for the scattering function of spherical particles $\varphi_{\mathrm{opt}} = 1.45$ whereas for Lorentzian curves $\varphi_{\mathrm{opt}} = 0.50$. For Gaussians, of course, $\varphi_{\mathrm{opt}} = 1$.

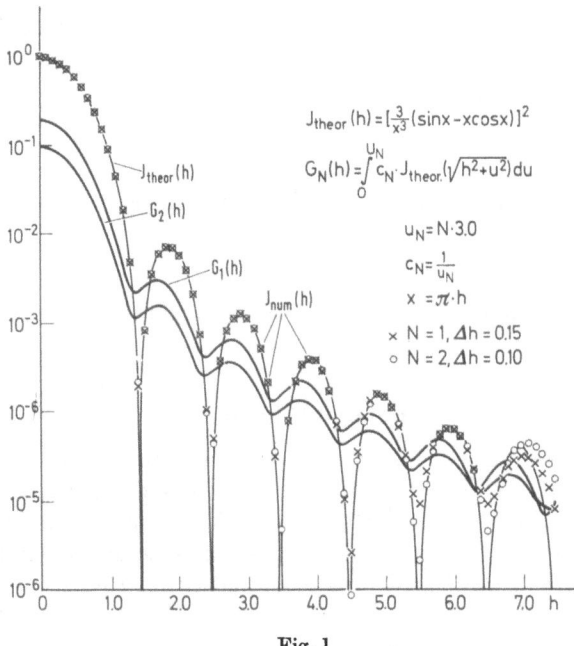

Fig. 1

Using FORTRAN IV programs the efficiency of the method was tested for various types of functions. On IBM 360/75 the execution time per slit height correction including the calculation of the correlation function takes 19 seconds using 76 sampling points for $G(h)$, 121 sampling points for $i_H(u)$ as input and 25 Hermite functions to approximate $G(h)$, for instance.

The Fig. 1 gives some results concerning the scattering by spheres. The full curves named $G_1(h)$ and $G_2(h)$ correspond to the distributions resulting from the distortion by rectangular slit height weighting functions of various heights applied to the theoretical scattering function of spheres, $J_{\text{theor}}(h)$. The crosses and circles represent the numerical results of the slit height correction method using 20 Hermite functions and different spacing of the sampling points for $G(h)$ as input data. Although in order to get the distribution $G_2(h)$, a slit twice as high as for $G_1(h)$ was applied, the minima are much better resolved in the solution indicated by circles, thus demonstrating the advantage of the higher density of the input sampling points due to $\Delta h = 0.10$ compared to $\Delta h = 0.15$ for $G_1(h)$.

References

1. POROD, G., in: H. BRUMBERGER, Small-angle X-ray scattering. Proceedings of the conference held at Syracuse University, June 1965, p. 1. New York: Gordon and Breach.
2. HOSSFELD, F.: Thesis, Technische Hochschule Aachen, 1967.
3. — Acta Cryst. A **24**, 643 (1968).
4. SZEGÖ, G.: Orthogonal polynomials. Amer. Math. Soc., 1959.
5. GUINIER, A., et G. FOURNET: J. Phys. Radium 8, 345 (1947).
6. DuMOND, J. W. M.: Phys. Rev. **72**, 83 (1947).
7. KRATKY, O., G. POROD u. L. KAHOVEC: Z. Elektrochem. **55**, 53 (1951).

Electron Probe Microanalysis. Physical Bases

Principles and Limitations of Electron Probe Microanalysis

V. E. Cosslett

Cavendish Laboratory, University of Cambridge, England

General Principles

Electron probe microanalysis essentially involves a miniaturisation of the classical procedure of X-ray spectrometry, based upon the discovery by Moseley [1] that each element in the periodic table is characterised by a relatively simple X-ray spectrum. The basic experimental arrangement is outlined in Fig. 1. An electron beam (E) of moderate energy (5—40 kV) is directed on to the sample (S) to be investigated, in vacuo, and the X-rays (X) emitted from it (in all directions) are dispersively analysis by means of a Bragg spectrometer (C). The latter employs a crystal of known atomic lattice spacing to diffract X-rays of differing wavelengths into different angles, in direct analogy to the formation of a spectrum from visible light by a glass prism. The wavelength λ diffracted at a given angle θ (the glancing angle, measured from the crystal surface) is related to the lattice spacing d by the simple equation:

$$n \lambda = 2d \sin \theta$$

where n is the order of the spectrum, usually unity in most experimental conditions. Since the wavelength of an X-ray emission line from an element is little influenced by its state of chemical combination, the elementary composition of a sample is readily found qualitatively from a measurement of its X-ray spectrum. A quantitative determination is obtainable if a comparative experiment is made on a pure sample of each element present. If the recorded X-ray intensity from element A in the unknown sample is $(I_A)_s$ and that from a pure reference sample of A is $(I_A)_r$, in identical experimental conditions, then the weight fraction of A in the sample is given in first approximation by

$$C_A = (I_A)_s/(I_A)_r.$$

For more refined analysis, various corrections must be made to this simple relation.

X-ray spectrometry has long been established as a technique of analysis on a macroscopic scale, the primary electron beam covering an area of 1 mm² or more on the sample. Its application on the microscopical scale was the idea of Castaing and Guinier [2] and independently of Borovskii [3]. Castaing adapted an electrostatic electron microscope to focus a minute electron probe, a few μm in diameter, on to the sample and analysed the X-rays with a spectrometer situated outside the microscope (Fig. 2). The construction and operation of the apparatus is described in his thesis [4], together with a discussion of the necessary corrections to the raw results. By moving the sample stepwise under the probe, a point-by-point analysis could be made across its surface.

The remaining stage in the development of the modern electron probe microanalyser was the introduction of scanning techniques by Duncumb and Cosslett [5]. By scanning the probe in a regular raster across the sample, and feeding the amplified X-ray signal to a cathode ray tube operating synchronously, it is possible to display an image of the sample surface in terms of X-ray brightness, and thus in terms of the variation in concentration of the particular element for which the crystal spectrometer has been set. The system, diagrammatically shown in Fig. 3, was described in Duncumb's thesis [6]. It incorporates also a detector $(S.C.)$ for electrons scat-

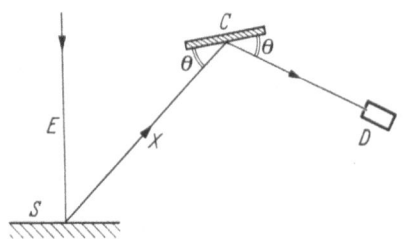

Fig. 1. Principle of microprobe analysis. *E* fine electron beam; *S* sample; *X* emitted X-rays; *C* crystal spectrometer; *D* detector

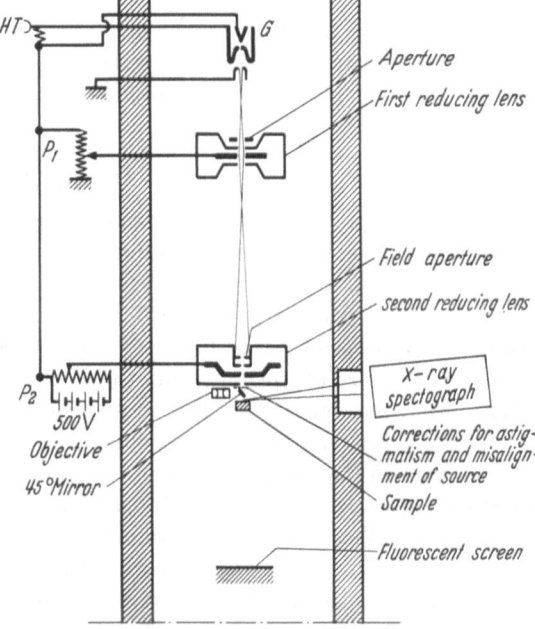

Fig. 2. Original microprobe analyser of Castaing and Guinier [2, 4] employing electrostatic lenses; the fluorescent screen was used for aligning them. *G* electron gun

Fig. 2

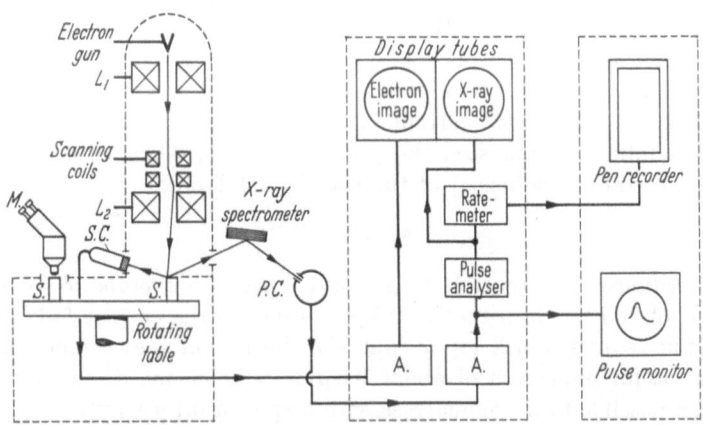

Fig. 3. Scanning electron microanalyser system [5, 6], with magnetic electron lenses, $L_1 L_2$. *A* amplifier; *M* optical microscope; *PC* proportional counter; *SC* scintillation counter + photomultiplier; *S* sample

tered back off the surface, which provides a signal several orders of magnitude greater than the X-ray signal and hence a much more readily visible image on a display tube. This technique, borrowed from scanning electron microscopy, greatly facilitates searching a specimen for areas of interest to be examined in detail. An optical viewing system is usually provided as well, as in Figs. 2 and 3, to permit direct comparison with normal metallurgical or mineralogical observations.

The microprobe method of analysis has found a great variety of applications in most branches of science and technology, primarily to metals and minerals, but recently to biological tissues also. To meet the demand, several different commercial models of microanalyser are now produced, descriptions of which are readily available. The various problems involved in the design and operation of the instrument, and in the evaluation of the results, have also been reviewed in the literature from time to time. For general surveys see, in particular, Castaing [7], Cosslett [8], Duncumb [9] and Heinrich [10]. The state of knowledge concerning correction procedures is reviewed in the present volume by Philibert [11]. Examples of applications can be found in the proceedings of the earlier symposia in this series, and of some other meetings, listed at the end of this paper. We shall do no more here than indicate the chief problems that have to be faced in the design of a microanalyser and the limitations that exist in our knowledge of the physical processes involved in the production of characteristic X-rays.

Technical Problems

An electron probe microanalyser consists of a probe-forming system, an X-ray analysing system, and between these some arrangement for inserting and positioning a sample. There may also be an optical system and an electron detection system for observing it directly or as a television image. The basic technical requirement is to focus the maximum possible electron current into a probe of given size, so as to excite the maximum X-ray output, and then to collect as much as possible of this output in the X-ray analysing system.

The probe-forming system has an electron gun to produce and accelerate a directed beam of electrons, a first electron lens to collimate it and a second lens (loosely called the "objective") to focus it on to the sample (see Figs. 2 and 3). The electron gun should have a small cross-over or virtual source, and a high brightness combined with reasonably long working life. The triode type of gun, with a biassed beam-forming electrode (Wehnelt cylinder) in addition to cathode and anode, is almost universally used because its properties are well-known, if still not fully understood. A tungsten hairpin serves as cathode, although attempts continue to adapt for routine use the pointed tip of tungsten developed for field-emission microscopy.

The current density in the beam emerging from such a gun is determined by the applied voltage as well as by the temperature of the filament. An ideal electron optical system would conserve the beam brightness, and on this basis one may calculate from the Langmuir relation the maximum current that can be delivered into a spot of given diameter. In fact even the magnetic electron lenses, now used in preference to the electrostatic type, have so great a spherical aberration that the actual current in a probe falls far short of this theoretical limit. If no other factors influence spot size, the probe current i_p depends on the spot diameter d and the spherical aberration coefficient of the lens C_s as follows:

$$i_p \propto (d^8 \cdot C_s^{-2})^{\frac{1}{3}}.$$

In the best circumstances it is possible to get only 1 μA into a probe of diameter 1 μm, even at a beam voltage of 30 kV. For a spot of 0.1 μm (1,000 A) the current falls to about 2.10^{-9} A, with a corresponding reduction in X-ray output. For this reason most commercial microanalysers have a minimum useful probe size of about 0.5 μm; only in one or two recent experimental models has it been possible to make analyses with a spot of 0.1 μA [12].

Clearly there is need for improvement in the design of the final electron lens, but this is largely dependent on success in finding some practicable way of correcting the spherical aberration of electron lenses generally. Meanwhile there is scope for the designer in trying to reconcile in the most efficient manner the contradictory requirements put upon this final lens: a long working distance to allow adequate space for large samples, and a short focal length to minimise spherical aberration. At the same time it has to allow the X-ray beam a free path to the spectrometer, as well as an optical path for a viewing microscope. It is not surprising that the commercial models differ widely in the design of their final electron lens. Use of a Le Poole minilens is a particularly interesting attempt at a solution [13, 14].

The X-ray analysing system also has to reconcile conflicting demands, and in particular high wavelength resolution with adequate collecting efficiency. It is in the nature of a crystal spectrometer that it can collect only a small fraction (10^{-3} to 10^{-4}) of the X-rays coming from the point of impact of the probe. The size of crystal is itself limited, as it is very difficult to grow big perfect crystals and even more difficult to bend them, as has to be done in order to bring the reflected X-rays to a focus. In addition, because of the need to rotate (and usually also to translate) it, its distance from the axis of the instrument cannot be kept small. The complicating factor is the need to keep constant the take-off angle, — the angle at which X-rays leave the sample in the direction of the crystal, — in order to simplify the correction which has to be made for absorption of X-rays in the sample. If this direction is to stay fixed, whilst the crystal is rotated into the right position to reflect each selected X-ray wavelength, the only solution is for it to move to and fro along this direction at the same time.

Here again there is great scope for ingenuity in designing compact yet precise mechanisms, and the various microanalysers differ considerably in the way the problem has been solved. An additional requirement is that the detector must move in conjunction with the to-fro and rotational motions of the crystal, so that it is always in the path of the reflected X-ray beam. Add to this the requirement for the whole system to be in vacuum, so as to eliminate the absorption in air of the longer wavelength X-rays, and it will be appreciated that we have here an engineering problem of considerable complexity.

Partly in consequence of these disadvantages of the crystal spectrometer and partly because of the difficulty it has in reflecting long wavelength X-rays, such as those from carbon and oxygen, increasing attention is being paid to the prospects of using two quite different detection systems: the ruled grating or a non-dispersive detector. The present state of their development is discussed in later papers in this volume. It is enough to say here that the higher reflectivity of gratings in the soft X-ray region is likely to be offset by even greater engineering difficulties in mounting and moving them to deal with a wide range of wavelengths. On the other hand the solid-state detector, which has a natural ability to discriminate one wavelength from another, has not yet been shown capable of doing so for a wavelength longer than that from calcium and so it cannot be used for elements of low atomic number. The gas-flow proportional counter can be so used, but has inadequate inherent wavelength resolution and therefore requires some electronic means of unscrambling a group of wavelengths. Research in this direction continues [15], mainly because a counter (like the solid state detector) can be brought up close to the sample to provide a very high collection efficiency, which is again especially desirable when analysing light elements, from which characteristic X-ray emission is very weak. The use of balanced X-ray filters, as originally devised by Ross for macroscopic spectrometry, in conjunction with a non-dispersive detector is an interesting new idea [16].

Further technical problems, of an entirely different nature, arise in connection with the recording, processing and presentation of the data derived from the X-ray signal. In early microanalysers it came out as a count rate or a graph from a pen recorder, or semi-quantitatively as a photograph of the image displayed on the cathode ray tube. There is an increasing tendency to present the data in a form suitable for feeding into a computer, where all the necessary corrections can be automatically made. So the output may appear on a punched tape or as a set of numbers printed by an electric typewriter. For visual assessment it may also be imaged in terms of contour lines showing different levels of concentration directly [10]. There is a strong tendency towards automation in the operation of the microanalyser itself, and this will become stronger as it is used more and more in routine industrial control processes.

Limitations in Our Knowledge of X-Ray Physics

Given an adequate X-ray signal, the ultimate accuracy of an analysis depends on the precision with which we can apply various corrections. These involve a knowledge of the distribution in depth of X-ray production in a target, of the absorption coefficients for X-rays of given wavelength in given materials and of the efficiency of secondary production of characteristic X-rays, since an element can be excited to emit its own wavelength by harder X-rays generated elsewhere in the target. Basic to these phenomena is a detailed understanding of how the incident electrons spread out as they penetrate into the sample: their mean free path, scattering angle and rate of energy loss. It is also important to know what fraction of them are scattered back through the surface and with what energy, since this amount of energy is no longer available for producing X-rays. For an element of high density 25—30% of the incident energy can be lost in this way. In all these respects there are wide gaps in our knowledge.

These corrections could be avoided altogether by preparing a sample which is as close as possible in composition and structure to the unknown. For a two-element system this may not be too difficult, although even here a process of successive approximation has to be followed. For a multicomponent system it becomes quite impracticable, since the distribution as well as the relative amounts of each element would have to be known and reproduced.

So we have to make the best use possible of the available knowledge in applying corrections in any particular case. A number of semi-empirical procedures have been devised, the relative merits of which are still under active discussion [11]. For the most part the models of electron penetration on which they are based are extremely crude. An analysis of X-ray production in a target at the atomic level of detail should be our aim, but the computational difficulties are great. For instance, the range of 25 kV electrons in aluminium is about 5 μm, whereas the mean free path for elastic scattering is initially 300 A and decreases as energy is lost. So several hundred scattering acts occur (on average) along the path of each electron, and the result of each collision is random in respect of angular deflection and energy loss. Hence a great number of separate paths must be computed if we wish to have a statistically accurate overall picture of the spatial and energy distribution of electrons in the target, from which to determine the corresponding X-ray distribution. Nevertheless progress is being made with this so-called Monte Carlo approach, good agreement having been obtained in respect of some experimentally measurable quantities: in the energy and distribution of backscattered electrons, and in the depth distribution of X-ray production [17].

At the same time, more experimental measurements are required of the parameters which enter into the correction procedures, especially of absorption coefficients at long wavelengths, of electron backscattering and of the variation in X-ray emission with take-off angle and voltage. As the demand for higher accuracy of analysis grows, and as increasingly complex systems have to be examined, so our methods of correction will have to become more sophisticated.

The Present State of the Art and Prospects for Further Improvement

With the commercial type of microprobe analyser it is now possible, in the surface of a solid sample, to analyse any element from the top of the periodic table down to and including boron. The analytical accuracy ($\Delta C/C$) depends greatly on the composition of the sample, because of the correction difficulties, but in favourable cases it can be better than 1%. The detection limit depends mainly on the counting time one is willing to allow, since it depends on the signal to noise ratio in the output. In a 10 sec count it can be below 100 parts per million, for instance. Many other analytical methods have a much higher sensitivity, but the great advantage of the microprobe is its power of localisation. The amount of sample analysed is roughly that within a cylinder of diameter equal to that of the probe (0.5—1 μm) and depth equal to rather less than the range of electrons of the incident energy, a few micrometers at most. So the analysed volume may be less than 10^{-12} cc, and in this a mass of 10^{-15} g of an element is readily detectable. In transmission, on an extraction replica containing precipitates or a biological section of mineralised tissue, a spot of diameter 0.1 μm can be analysed in a sample only 0.1—0.25 μm thick, i.e. a volume as small as 10^{-15} cc.

The prospects for any radical improvement in this performance are not great, at least in the immediate future. The spatial resolution is limited primarily by the depth of penetration of the electron probe into the specimen, which can be made less only by lowering the beam voltage, the result of which is a more than proportionate reduction in X-ray output. The X-ray signal might be restored if a gun of higher brightness or a greatly improved objective lens were to be devised, but these developments are likely to be slow in coming. So far as the analysis of solid samples is concerned, the best hope lies in detection systems of higher collecting efficiency, especially the solid state detector.

For transmission specimens the prospect is rather better since the effect of electron scattering on spot size is negligible in a film thickness small compared with the range. Here improvements in gun brightness and detector efficiency might well allow areas down to a few hundred Angström units in diameter to be analysed, an advance which would be particularly important in many biological applications [18].

Beyond this point it seems unlikely that microprobe X-ray analysis can proceed much further. Even if, in due course, a spherically corrected lens becomes available, any appreciable increase in current density at the specimen is ruled out by its heating effect, except on materials of high thermal conductivity. It may still be possible to advance beyond the present limits of analytical accuracy and spatial localisation, however, by detecting the electrons which have given rise to a characteristic X-ray line instead of detecting the X-ray line itself. These electrons pass on with very little deviation from their incident direction and may be recorded by means of an energy loss analyser positioned on the axis behind the specimen. Thus they are detectable with much greater efficiency than are X-ray lines, which are emitted over the whole sphere around an excited atom. Preliminary experiments [19] indicate that this microprobe electron-analyser should be capable of a spatial resolution an order of magnitude better than that at present attainable with the microprobe X-ray analyser. It is limited to thin films, of course, since the loss-electrons have to be detected in the forward direction.

In this and many other directions the subject of microprobe analysis is still in a state of very active development, as may be seen in greater detail from the latter contributions to the present volume.

References

1. Moseley, H. G. J.: Phil. Mag. 26, 1024 (1913); 27, 703 (1914).
2. Castaing, R., and A. Guinier: Proc. Conf. Electron Microscopy, Delft, p. 60. Delft: Nijhoff 1949.
3. Borovskii, I. B.: Collection of problems in metallurgy, p. 135. Moscow: A.N. 1953.
4. Castaing, R.: Thesis, University of Paris (1951).
5. Duncumb, P., and V. E. Cosslett: X-ray microscopy and microradiography, p. 374. New York: Academic Press 1957.
6. — Thesis, University of Cambridge (1957).
7. Castaing, R.: Advan. Electron. Electron Phys. 13, 317 (1960); — X-ray optics and X-ray microanalysis, p. 263. New York: Academic Press 1963.
8. Cosslett, V. E.: Rept. Progr. Phys. 28, 381 (1965); — J. Electronmicroscopy (Tokyo) 16, 51 (1967).
9. Duncumb, P.: Sci. Progr. (Oxford) 55, 511 (1967).
10. Heinrich, K. F. J.: National Bureau of Standards, Technical Note No. 278 (1967).
11. Philibert, J.: This volume, p. 114.
12. Cooke, C. J., and P. Duncumb: This volume, p. 245.
13. Fontijn, L. A.: This volume, p. 261.
14. Chapman, P. F.: This volume, p. 241.
15. Wardell, I. R. M., and V. E. Cosslett: The electron microprobe, p. 23. New York: Wiley 1966.
16. Pichoir, F.: This volume, p. 373.
17. Bishop, H. E.: Brit. J. Appl. Phys., Ser. II, 1, 673 (1968).
18. Hall, T. A.: Proceedings of Microprobe Symposium, National Bureau of Standards, Washington 1967, p. 269.
19. Wittry, D. B., R. P. Ferrier, and V. E. Cosslett: This volume, p. 293.

Conference Proceedings

Cambridge Symposium, 1956: X-ray microscopy and microradiography. New York: Academic Press 1957.
Stockholm Symposium, 1959: X-ray microscopy and X-ray microanalysis. Amsterdam: Elsevier 1960.
Stanford Symposium, 1962: X-ray optics and X-ray microanalysis. New York: Academic Press 1963.
Paris Conference, 1965: Optique des rayons X et microanalyse. Paris: Hermann 1966.
The electron microprobe. New York: Wiley 1966.
Advances in X-ray analysis. New York: Plenum Press; especially vol. 7 (1964), 8 (1965), 9 (1966) and 10 (1967).

Recent Progress of Electron Microprobe in Japan

G. Shinoda

Faculty of Engineering, Osaka University, Japan

Introduction

During three years after the Paris conference progress in the electron microprobe application becomes remarkable in Japan as well as other countries. On the otherhand in the instrumentation field, although many improvements have been done, outstanding changes have not been experienced. This is because the electron microprobe itself is nearly completed. However, from an urgent request of practical field some new improvements have been done. Therefore the next step or progress would be found in the related field.

Such tendencies are found in scanning microscope and electron velocity analyser.

Among the application fields, that of iron and steel research is most active. However, the way to application for biological field has been opened recently and such field becomes active by and by.

Instrumentation

Among progresses in instrumentation the most remarkable ones are those related to scanning images. However, the problem related to the analysis of light element should be mentioned at first. In the dispersive method, the spectrometers and crystals should be improved. Fukuo (Akashi Co.) has designed a goniometer of quite new type, however the details of it is not published yet.

In iron and steel industry analysis of oxygen is very important as well as carbon. Oxygen would be analysed using KAP crystal, however, in a certain commercial goniometer the sensibility becomes poor for wavelength range around oxygen. If we use lead stearate instead of KAP, for certain geometrical condition sensibility would be much improved. Shiraiwa and Fujino [1] succeeded using a Rowland circle of 11″ in diamater, as shown in Fig. 1.

Recently good crystals of lead stearate become commercially available. Okano [2] improved the method of preparation and obtained a good crystal. Stearate was built-up on a stepwise surface and Fig. 2a shows line shapes obtained by an ordinary built-up film, improved one called pseudo-Johansson type and diffraction grating respectively. The line shape of C-K_α obtained by it is as good as that of diffraction grating. As the intensity of X-ray diffracted by crystal is much stronger than that of diffraction grating, such improvement will provide a powerful tool to investigate important spectroscopic problems such as chemical shift of soft X-rays, as shown in Fig. 2b.

On the non-diopersive method, Okano [3] investigated various related phenomena both theoretically and experimentally, and clarified the factors to determine the limit of dectectability. Theoretical calculation on signal and noise ratio has agreed with the experimental result of Birks. Also, some improvements on the instrumentation such as elimination of level fluctuation in an amplifier have been done. For the practical application side, matrix method of Dolby was adopted and succeeded on the separation of two neighboring elements such as carbon and boron. Also the problem on unsymmetrical pulse height distribution has been investigated and found a linear relation between pulse height and incident energy.

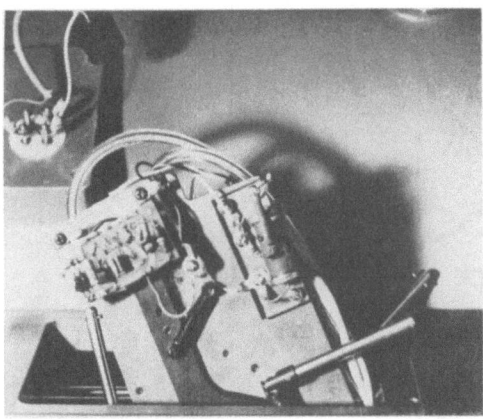

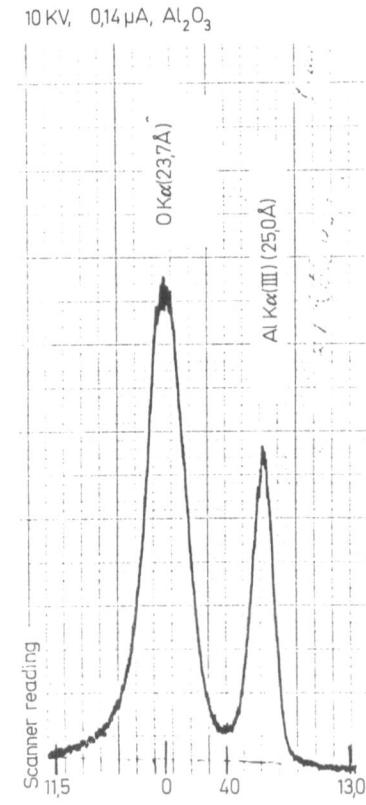

Fig. 1. Improved goniometer for oxygen analysis, by SHIRAIWA and FUJINO

When carbon is analysed by an ordinary instrument, contamination of carbon or its compound from the diffusion pump oil is experienced. OKANO obtained depth distribution curve of such contamination layer and compared it with SHINODA et al.'s result for carbon. Fig. 3 is the results and the latter was obtained by angular distribution method.

SHIMIZU [4] discussed the pulse height distribution from the standpoint of single electron spectrum and reached a conclusion that the newly obtained pulse height distribution for lithium is explained by WEIBULL function expression as well as more heavier elements already published.

In electron microprobe much information would be obtained from X-ray method and optimum conditions to use it should be utilization of generated X-ray. However, if we pay attention to the resolving power another method would be conceivable. From this standpoint of information theory SHINODA and SHIMIZU [5] discussed the problem and pointed out that with the specimen current or backscattered electron method much improvements of resolving power would be obtained. Fig. 4 is the result showing that resolving power for X-ray method has better information quantity and poorer resolving power. The condition will be pronounced in the secondary electron method and even in an ordinary electron microprobe much better resolution would be obtained. In HASHIMOTO and KOSUGE's [6] accessory for JEOLCO instrument resolution of 1,000 Å has been obtained.

Fundamental problems of scanning microscope using secondary electrons were already solved by Cambridge group and EVERHART. On the basis of these results KIMOTO [7] succeeded to develop new scanning microscope commercially available having resolving power of $100 \sim 200$ Å, and magnification of several hundreds to several hundreds hundred. SAKAKI and NAGATANI [8] also developed another type of scanning microscope with co-operation of Hitachi Co.

As the scanning microscope gives a flat image, sometimes determination of the contours of the image becomes necessary. KIMOTO et al. [9] developed a new method of determination of contours of image using stereographic photographs.

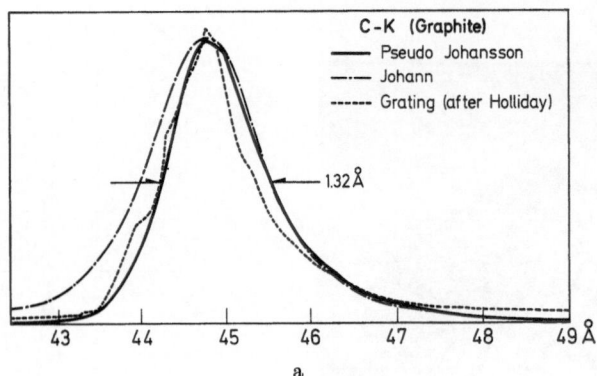

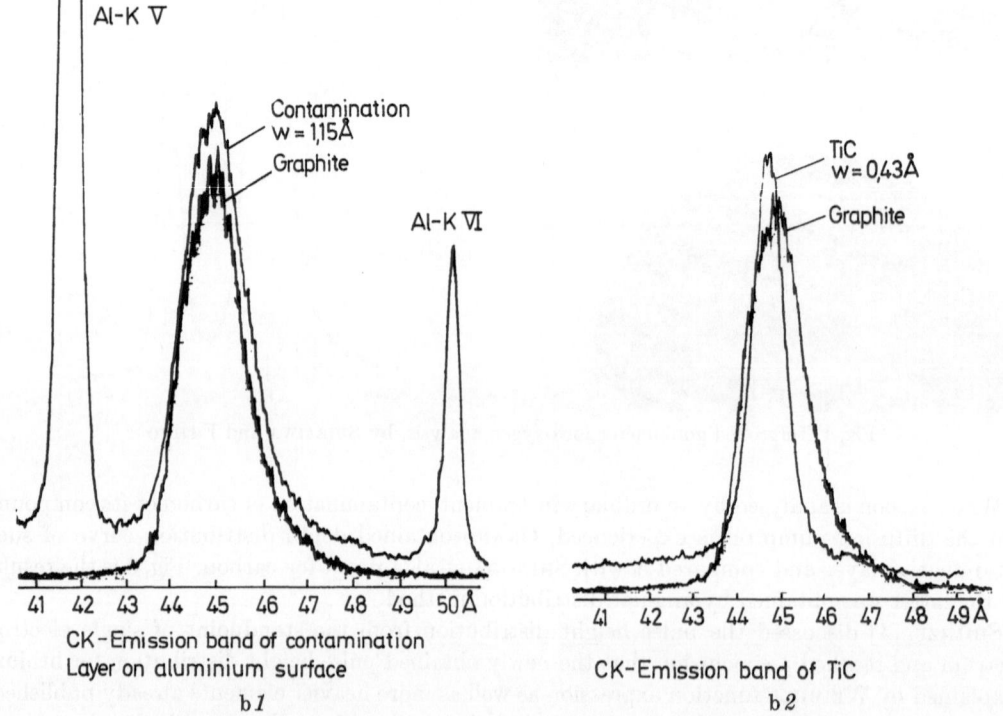

Fig. 2. a Spectra obtained by Johan (ordinary stearate film, OKANO), pseudo-Johansson (improved stearate film, OKANO) and diffraction grating (after HOLLIDAY). b Spectra of C-*K* (OKANO). *1.* Graphite and contamination layer. *2.* Comparison of graphite and TiC (chemical shift)

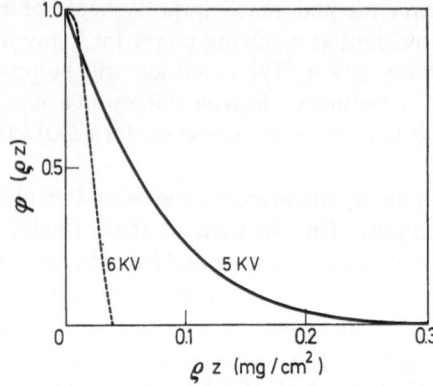

Fig. 3. Comparison of depth distribution curves obtained by contamination layer (OKANO) and calculated from angular distributions of X-ray (MURATA)

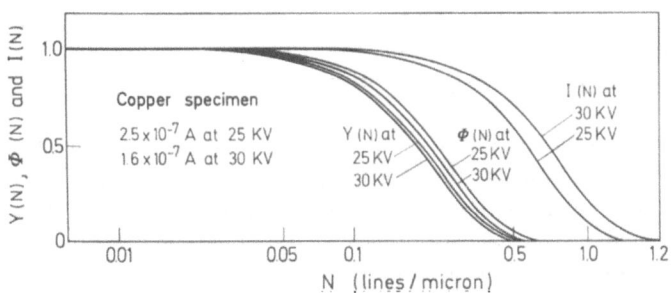

Fig. 4. Response functions for X-ray method $Y(N)$ and backscattered electron method $I(N)$. $\Phi(N)$ corresponds to the diffused X-ray source and $Y(N) = I(N) \cdot \Phi(N)$. (SHIMIZU)

a

Chemical Composition (%)				
Ta	O	N	C	H
Bal	0.08	0.005	0.004	0.0007

b

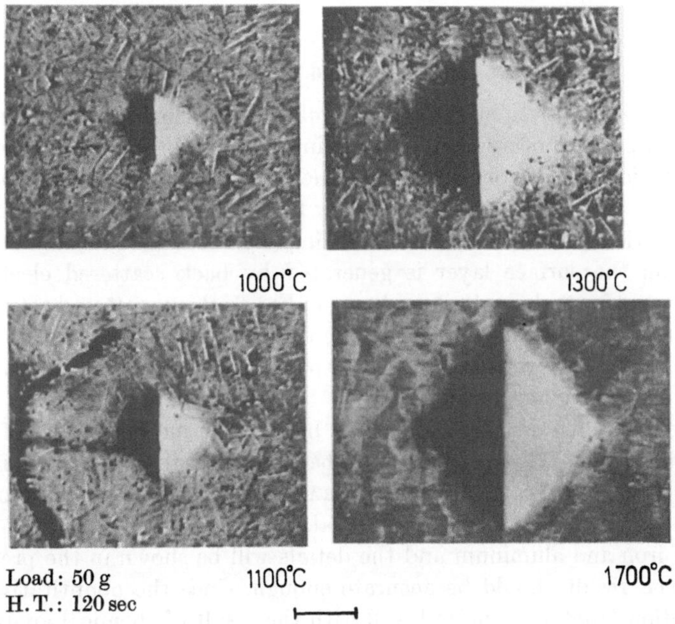

Fig. 5a and b. Scanning images of indentation for tantalum at various temperatures (a), and high temperature hardness of tantalum obtained by Akashi instrument (b)

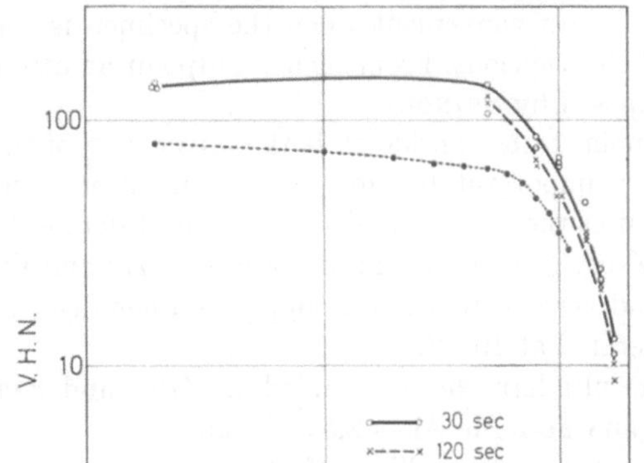

Another scanning instrument is a high temperature scanning device of Akashi Co. [10]. It has a tantalum heating device and can be raised up to 1,700°C. Also it has a diamond pyramid and high temperature hardness is measured by its indentation shown in the scanning image (Fig. 5).

By these scanning images quantitative informations are very difficult to obtain. Some attempts such as to use tri-chromatic filters are reported. Tomura et al. [11] using luminosity discrimination ability of human eye developed a special signal storing circuit to obtain an eight-step contour mapping image. Thus the scanning image becomes quantitative one.

Another method to improve resolution is an electron velocity analyser. Watanabe's analyser is well known and already published. Ichinokawa [12] (details will be shown by his contributed paper) also developed his original one having special magnetic poles attached to an ordinary electron microscope. By such instrument high resolution convenient to investigate microscopic phenomena such as the initial stage of phase change would be obtained.

Electronic Behaviors and Generation of X-ray

Resolving power in electron microprobe mainly depends on the X-ray source size and its diameter is greater than that of primary beam owing to the diffusion of electrons in the specimen. On the lateral diffusion of electrons in the specimen Shimizu and Shinoda [13] published their results several years ago. Further study especially for depth direction has been carried out using Monte Carlo calculation. This has been done mainly by Murata and the result is as shown in Fig. 6. X-rays from the surface layer is generated by back scattered electrons, because the population of electrons for each depth is as shown in Fig. 7. Back scattered electrons are composed of those scattered at a shallower depth than the depth of complete diffusion and therefore the informations from back scattered electrons correspond to the surface state. The condition will be nearly similar for the secondary electrons.

The depth distribution function is obtained by various methods. One of them, use of the angular distribution of characteristic ray, has been applied to aluminum and compared with that obtained by Monte Carlo calculation by Murata et al. Agreement was good, futher development to calcium has been done by Shimizu et al. Next angular distribution analysis has been done for alloys of iron and aluminum and the details will be shown in the present author's contributed paper. The result should be accurate enough, since the quantitative result corrected by above distribution function coincided well with the result of chemical analysis.

Among the methods to obtain distribution function, the angular distribution of characteristic ray method is most convenient when the specimen is composed of unknown elements and the structure is heterogeneous. From such standpoint an attempt to analyse heterogeneous specimen has been proposed by Shinoda.

Another point to be considered is the unisotropy of the material. For ions the effect of channelling is very important but for X-ray generation it is not so clear. However, at the grain boundary region the behavior of elections is different from the interior of the specimen and anomaly in X-ray generation will be observed. Izui and Furuno [14] compared X-ray intensities from large and very small grained copper and comfirmed that in the latter about 10% stronger X-ray is generated at 10 kV.

Similar results have been obtained by Oda and Satta in Si-K_α in Al—12% Si, C-K_α in Fe—0.5% C and Fe-K_α in Al — 2.5% Fe alloys.

Even though the channelling effect in an unisotropic specimen is not so clear, it has been postulated by anomalies of specimen current in a copper-zinc alloy. Such anomalies were already reported by Shirai et al. [15] in γ-phase of that alloy system. It should be expected that the specimen current of γ-phase is larger than that of ε-phase, however the result with contradiction has been always obtained. Afterwards, this has been attributed to the second order effect of the orientation by Shinoda et al. [16] and this is an evidence of channelling effect in electron penetration. Such evidence should exist in X-ray generation, however as it is a second order effect to obtain reliable result would not be so easy.

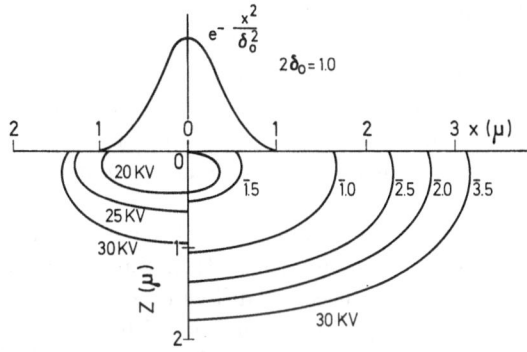

Fig. 6. Shape of X-ray source for copper. Upper: intensity distribution of electron beam, lower: intensity contours, lower left: contours for $1/e$ of maximum intensity and lower right: contours of X-ray intensity for 30 kV in logarithmic scale. (MURATA et al.)

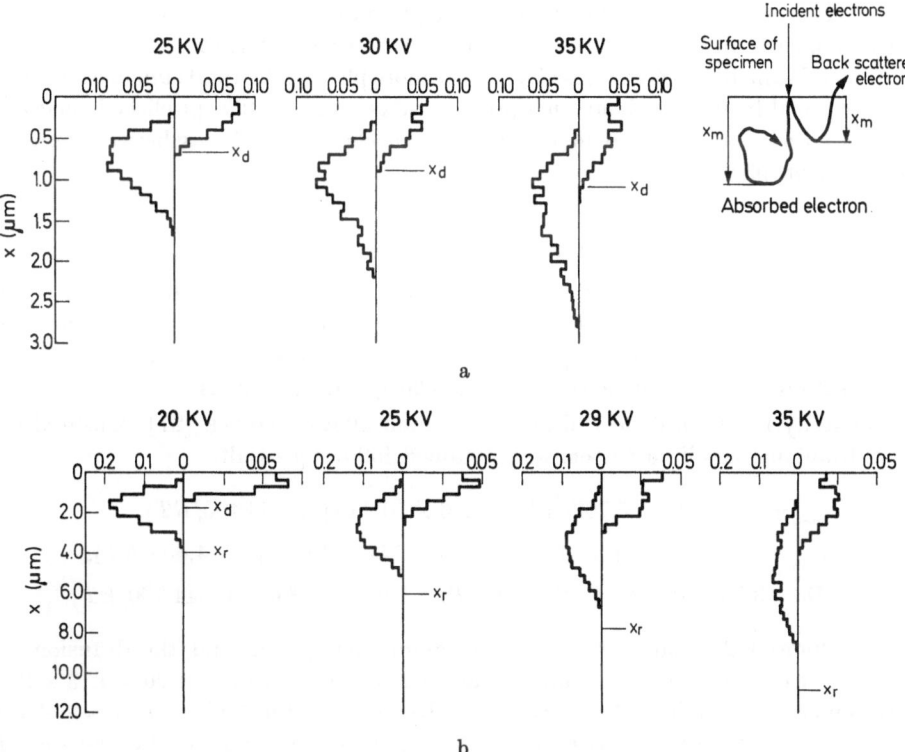

Fig. 7a and b. Distribution of maximum penetration depth for backscattered (right hand side) and absorbed left hand side) electrons. a copper, b aluminum. (MURATA et al.)

Quantitative Procedures

Quantitative procedure is very important from the practical standpoint. Many investigators proposed their own procedures and recently the result becomes quite satisfactory for homogeneous specimen. However for inhomogeneous specimen, such as non-metallic inclusions with complex structures in steel, agreement between the results of chemical analysis and electron-probe analysis is quite poor. The reason should be partly attributable to unsatisfactory theoretical considerations, however main reasons should be found elsewhere.

Among kinds of quantitative correction, theoretical background of atomic number effect is not complete yet. SAWATANI and MUROTA [17] proposed a new method, modifying Archard-Mulvey method. The energy loss in the specimen has been calculated by full-Bethe formula.

As the relation between the effective ionization cross section and electron energy he used Green-Cosslett relation and as the depth of complete diffusion Cosslett-Thomson formula has been adopted. At first he applied his correction method to gold-copper binary alloy system and obtained good result and then applied to iron-aluminum system. The agreement was satisfactory.

In the quantitative procedure, fluorescence correction is also important. Shiraiwa and Fujino [18] investigated this problem and after tedious calculation using logarithmic intergral they have reached an exact solution.

Thus for homogeneous specimen the problem has been nearly completely solved and the procedure became automatic using electronic computers. But when a specimen is composed of heterogeneous components, the problem becomes difficult. Oda and Satta [19] investigated the X-ray intensities from heterogeneous specimen and reached following conclusions. If a specimen is a heterogeneous mixture of elements A and B, and $\mu_A \varrho_A > \mu_B \varrho_B$ the intensity of X-ray from B element decreases with the increase of particle diameter and vice versa. Thus the X-ray generation of heterogeneous specimen is very complicated and therefore the present author proposed use of angular distribution curve for the very specimen. However if the dispersion curve of grain diameter is determined or known both the grain diameter and its composition would be determined from such measurement, as pointed out by Oda and Satta.

If a non-metallic inclusion is composed of two or more phases, the total sum of each component element is not equal to 100%. About such problems Sasaki et al. [20] published their elaborate work, and attributed it on the difference of generation mechanisms of X-ray between homogeneous and inhomogeneous phases.

Metallurgical Problems

In the early stage of electron microprobe application problems of diffusion and equilibrium diagrams had been active. Recently, such field has been extended to the study of nonequilibrium reaction. Study on segregation or crystallization mechanism also belongs to this category and the electron microprobe can provide the means to clarify such problems.

Diffusion study has been done mainly on titanium alloys. Inouye [21] measured diffusion constant in titanium-vanadium system and obtained following result.

$$D_{N_{v=0}} = 1.24 \exp(-5{,}700/RT) + 4.46 \times 10^{-4} \exp(-33{,}400/RT).$$

$$D_{Ti} = 3.22 \times 10^4 \exp(-97{,}300/RT) + 3.56 \times 10^2 \exp(-4{,}600/RT).$$

$$D_V = 3.73 \times 10^3 \exp(-90{,}200/RT) + 1.67 \times 10^{-2} \exp(-44{,}700/RT).$$

Hirano and Ouchi [22] studied titanium and aluminum system and the diffusion constant increases with the concentration of aluminum, as for $2 \sim 12$ at % Al $D = 1.20 \sim 5.13 \times 10^{-13}$ cm²/sec. Activation energies are in α-phase $21.7 \sim 25.8$ Kcal/mol and in β-phase $18.7 \sim 31.1$ Kcal/mol for the same alloy compositions. Hirano also studied diffusion on Fe—Co, Fe—Al, Ti—Ni, Fe—Ti and Fe—Co—Ni systems. He determined precisely γ-loops of Fe—Al and Fe—Ti systems.

In aluminum-copper alloy, by a eutectic reaction α solid solution and $CuAl_2$ crystallize simultaneously. But it is under an equilibrium condition. Kimoto et al. [23] using their heating device and confirmed the existence of intermediate solid phase extremely rich, in copper, as shown in Fig. 8. That phase appears around α phase and afterwards disintegrates into $\alpha + CuAl_2$.

Even in the liquid state diffusion velocity is not so high and microscopic and macroscopic segregation will take place. Shinoda et al. [24] has investigated copper-tin alloys and found that the dendrite cores crystallizes step-wise i.e. the concentration gradient along radial direction of the core is not gradual but step wise. Also they found significant outward stream of liquid during the crystallization stage corresponding to the inverse segregation.

When an alloy composed of many components their solidification mechanism is very much complicated. Shiraiwa et al. [25] investigated nitrided silicon-killed steel and found a new mineral of which crystal structure is similar to AlN but contains manganese, silicon, aluminum and nitrogen and belongs to $P6_3$ mc.

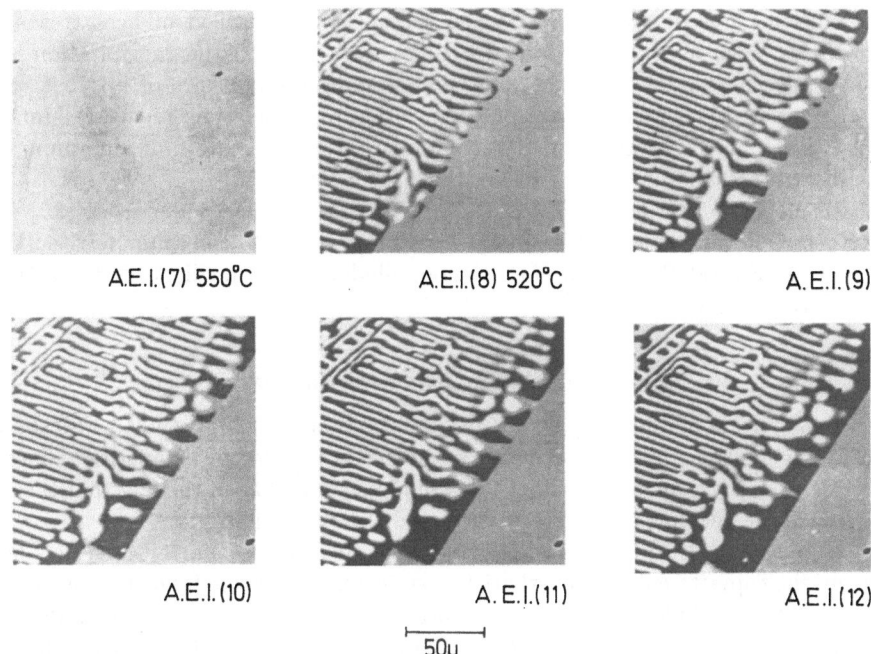

Fig. 8. Specimen current image for aluminum-10 wt.% copper alloy near the eutectic point, after KIMOTO et al., white; α-solid solution, grey; solid with eutectic composition and black; CuAl₂

Surface phenomena of metals are combined effects of diffusion and chemical reaction. Dezincation in brass is a kind of diffusion. NAGASAKI [26] determined the depths where the concentration of zinc decreases from 30 wt.% to 20 wt.% at 800°C and for 4 hrs. as 100 μ and for 8 hrs. as 200 μ. The electron microprobe is used to identify oxidation products. WATANABE et al. found four oxidatien layers on an 18—8 stainless steel and identified as the first and second layer: $(Fe, Cr)_2O_3$ and the third and fourth layers: spinel type $(Fe, Cr)_3O_4$. Also the enriching process in the surface layer of atmosphere exposed steel has been investigated. But it will take much times to reach definite conclusions.

In the solid state reaction only SHINODA et al.'s work [27] is found. They postulated that α phase precipitation from β-phase of copper-zinc system occurs martensitic transformation manner and at the growth stage diffusion takes place, however it takes many days to reach equilibrium state. Further details will be mentioned in the last chapter.

Non-Metallic Inclusion

As there are too many works reported in this field, their results would be stated only briefly.

First of all the method of experiment should be mentioned. Several years ago Yawata Tokyo Research Laboratory developed a method using sliced specimen for mineralogical microscopy and afterwards this method became popular. Also extraction replica method with joint use of electron microscope and electron diffraction became popular and as pointed out by KATO et al. [28] it is very convenient when the inclusion size is less than 1 μ. Also KATO et al. [29] developed a cathode ray luminescence method and found the color of luminescence of Al_2O_3 or SiO_2 changes sensibly by small addition of manganese and iron.

On the aluminum deoxidation SANO et al. [30] made an elaborate work, not only on the inclusion itself but also on its diffusion process and investigation from the standpoint of equilibrium diagram. When the steel contains silicon, manganese and aluminum main constituents in the bottom side are FeO and MnO and in the upper side SiO_2 and Al_2O_3 become rich.

On the effect of compound deoxidizer containing silicon and calcium, WATANABE and KUSAKAWA [31] studied on influence of holding time.

Such inclusion plays deterious effect when the specimen is rolled or heat treated. Several reports are found such as Shiraiwa and Fujino's work [32] on the precipitation velocity of rimmed steel inclusion (MnO—SiO_2—Al_2O_3) and softening temperature of vitreous component, Saito et al.'s work [33] on the change of composition of spinel type inclusion MnO, Cr_2O_3 in 18-8 stainless steel and Sawatani and Sakata's one [34] on formation of aluminum mono-oxy carbide, Al_2OC in the bandlike region of cold rolled sheet steel.

On the mechanism of formation of ferrite, we can find Kojima et al.'s work [35]. They confirmed existence of ternary calcium ferrite $4CaO \cdot FeO \cdot 4Fe_2O_3$ and also barium ferrite $2BaO \cdot Fe_2O_3$ and $BaO \cdot 6Fe_2O_3$. Sasaki [36] found new mineral, alkali aluminum silicate $K_2O \cdot Al_2O_3SiO_2Na_2O$ in the blast furnace wall.

Mineralogical and Biological Applications

Several papers are found in the study of ore. Fujiki and Imai [37] studied $FeO \cdot Fe_2O_3$—TiO_2 system oxide minerals. They found titanium in hematite but not in magnetite. This is because these minerals are formed by rapid cooling and would be verified from the fact that these minerals are derived from the decomposition of andesites and their pyroclastics.

Takeuchi and Fujiki [38] studied from the synthetical stand point on Cu—Fe—S minerals, especially on chalcopyrite-cubanitepyrrbatite system and obtained following result. Complete solid solution between chalcopyrite and pyrrhotite is only found above 800°C. Pyrrhotite can contain copper 2.5 wt.% at 600°C, 4% at 700°C and 6% at 800°C. Also the solubility of iron in chalcopyrite was studied.

In the biological field study of tooth becomes active recently especially by Japan Dental University Group. Another group's result, depth distribution of calcium for analysis of tooth enamel will be given in the present authors contribution paper.

Kossel Technique

Several years after discovery of Kossel lines of Prof. Kossel of University of Tübingen. Fujiwara developed his capillary X-ray tube. The method belongs to a pseudo-Kossel technique and during the war this method had been applied to the study of light alloys, especially study of fatigue in superduraluminum. After the electron microprobe became popular this technique has been much improved both in true- and pseudo-Kossel technique. This latter technique is used by many X-ray stress measurement workers and for instance problems on fatigue have been studied by them. In this paper only those concerning to the EPMA Kossel technique would be given.

On the nature of the difference of true- and pseudo-Kossel lines Umeno [39] investigated using germanium single crystal of various thickness. As shown in Fig. 9 intensity ratios of anomalous transmission and Kossel lines vary with change of thickness and it would be interpreted by the discussion on the kinematical and dynamical theories of diffraction. Also he found evidences of lattice defects in Kossel lines.

One application of Kossel technique is the precise determination of lattice constant and many methods such as use of lens form diffraction lines, or utilization of two osculating lines and etc. However such figures do not appear always. In the case of iron such conditions never fulfilled. Therefore Morimoto et al. [40] proposed a method to use intersection of the straight line connecting the intersections of $(0\bar{1}1)$, $(\bar{1}12)$ and (022), $(\bar{2}11)$, and the diffraction line (022).

As the EPMA-Kossel technique provides a mean to investigate very small area, it would be useful to study behaviors of microcrystals or grain boundary rather than single crystal. From this standpoint Umeno and Shinoda [41] studied plastic deformation of macro- and microcrystals of aluminum. Especially attentions are paid on the grain boundary region.

When an aluminum plate consists of several macrocrystals is stretched, deformation near the grain boundary is influenced by the direction of stress induced by the slip of neighboring crystals. If grain boundary is parallel to the direction of tension, remarkable splitting or tearing of Kossel

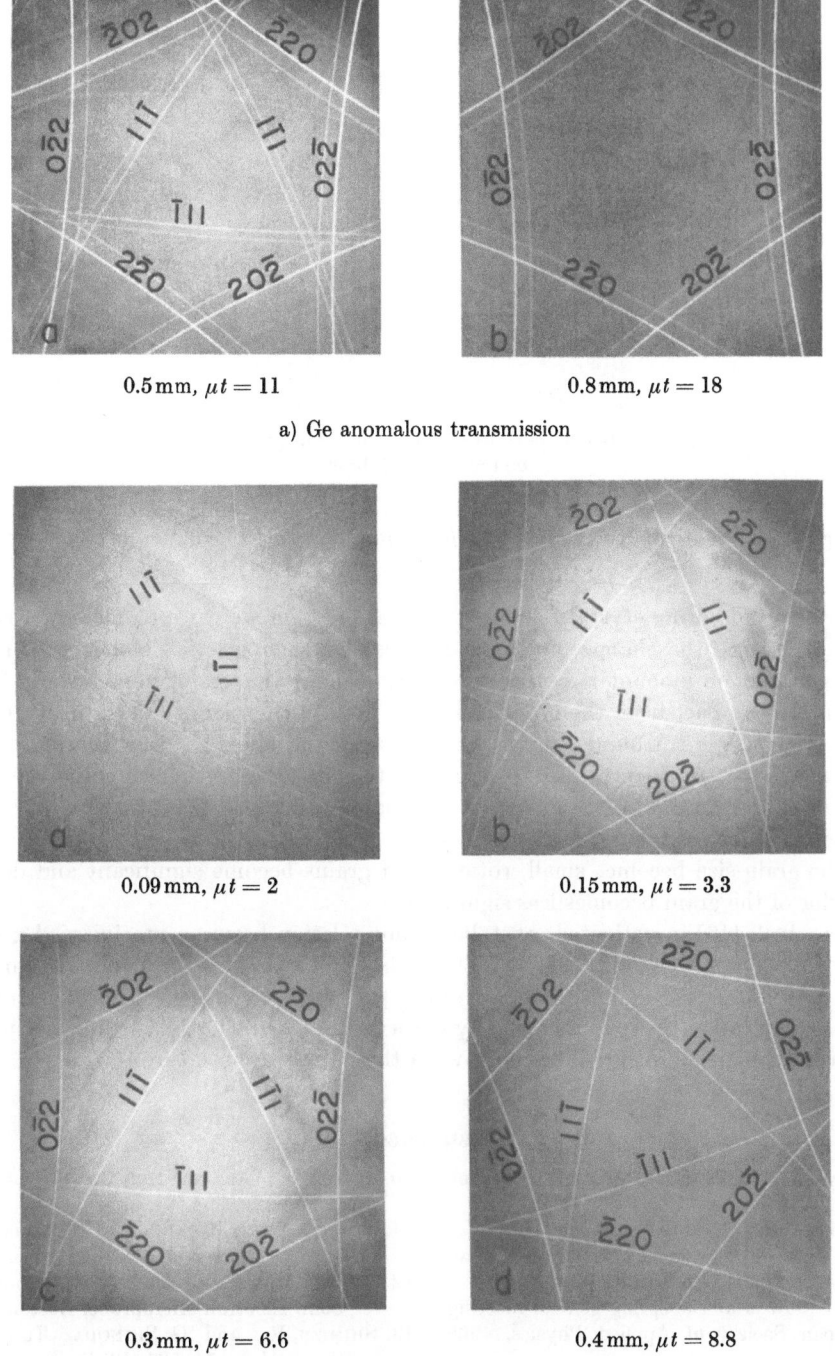

0.5 mm, $\mu t = 11$ 0.8 mm, $\mu t = 18$

a) Ge anomalous transmission

0.09 mm, $\mu t = 2$ 0.15 mm, $\mu t = 3.3$

0.3 mm, $\mu t = 6.6$ 0.4 mm, $\mu t = 8.8$

b) Ge Kossel pattern

Fig. 9a and b. Kossel lines of germanium; after UMENO

lines due to a slip plane parallel to the grain boundary is observed while for those perpendicular to the grain boundary no significant change in diffraction lines is observed. When the direction of tension is perpendicular to the grain boundary remarkable changes of Kossel lines are observed for a slip plane perpendicular to the grain boundary. A slip induced by the slip of the neighboring crystal would be predicted, if we calculate Schmidt factors for the direction of stress induced by

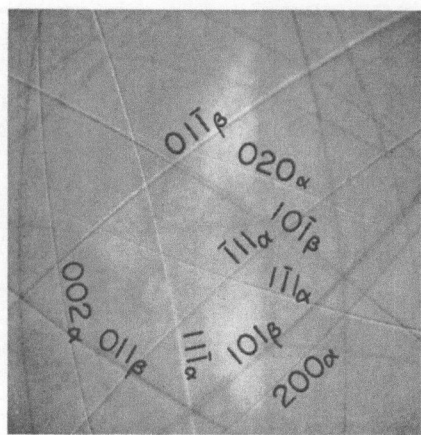

Kossel pattern of α-precipitate at grain
boundaries of β-brass
$\{110\}_\beta \| \{111\}_\alpha \cdot \langle 111 \rangle_\beta \| \langle 110 \rangle_\alpha$

Fig. 10. Martensitic transformation from β to α in copper-zinc alloy, after Shinoda et al.

the slip of the neighboring crystal. Thus by such calculation we can predict how the next slip occurs and how large the change of Kossel lines is. By such plastic deformation, rotation of crystals occurs and its amount is not constant throughout the grain, namely it is larger in a region, about 100 μ apart from the grain boundary than in the vicinity of grain boundary.

If a macrocrystal is surrounded by several crystals, the effects of neighboring crystals are different for each part of the crystal. If one part of the crystal can slip easily and the other part is difficult to slip, formation of deformation twin is plausible. Such twins have been observed several times in aluminum (Umeno et al. [42]).

When the grain size becomes small, rotations of grains become significant and deformation in the interior of the grain becomes less significant.

Shinoda et al. [16] investigated crystallographic relation between precipitated α phase and matrix β in copper—45 wt.% zinc alloy. As shown in Fig. 10, the relation was quite similar to the martensitic transformation in steel namely $[110]_\alpha \| [111]_\beta$, $(111)_\alpha \| (110)_\beta$. However in the above relation scattering of less than 1° was usually observed. Also when α phase appears in the grain boundary it always grows in the direction toward the center of curvature.

References

1. Shiraiwa, T., and N. Fujino: Jap. J. Appl. Phys. (in press).
2. Okano, K.: Thesis Osaka Univ. 1967.
3. — Read at the spring general meeting of Japan Society of Applied Physics, 1968 April.
4. Shimizu, R.: Read at the spring general meeting of Japan Society of Applied Physics, 1968 April.
5. — Thesis Osaka Univ. 1964.
   Shinoda, G.: To be published shortly.
6. Hashimoto, H., and T. Kosuge: Private communication.
7. Kimoto, S.: Read at the Second National Conf. Electron Microprobe, Boston 1967, June.
8. Sakaki, Y., and R. Nagatani: Nagoya Univ. Private communication.
9. Kimoto, S.: Read at the Spring General Meeting of Japan Society of Applied Physics, 1968 April.
10. Fukuo, S.: Akashi Seisakusho Ltd., Private Communication.
11. Tomura, T., H. Okano, K. Hara, and T. Watanabe: Advance in X-ray analysis. In press.
12. Ichinokawa, T.: Read at the Second National Conf. Electron Microprobe, Boston June 1967.
13. Shimizu, R., and G. Shinoda: Technol. Rept. Osaka Univ. 14, 897 (1964).
14. Izui, K., and S. Furuno: Jap. J. Appl. Phys. 7, 184 (1967).
15. Shirai, S., A. Onoguchi, and T. Ichinokawa: Jap. J. Appl. Phys. 7, 277 (1967).
16. Shinoda, G., K. Isokawa, and M. Umeno: Paper presented to the Denver Conference 1968.
17. Sawatani, T., and S. Murota: Read at the Spring General Meeting of Japan Inst. Metals 1968.
18. Shiraiwa, T., and N. Fujino: Read at the Pittsburgh Conf. Anal. Chem. and Appl. Spect. 1967, March.

19. Oda, Y., and K. Satta: Read at the Spring General Meeting of Japan Society of Applied Physics, 1968 April.
20. Sasaki, M., T. Takahari, and H. Hammada: Trans. ISIJ 1968 (in press).
21. Inouye, K.: Read at the Spring General Meeting of Jap. Inst. Metals 1968.
22. Hirano, K., and K. Ouchi: Read at the Spring General Meeting of Jap. Inst. Metals 1968.
23. Kimoto, S., H. Hashimoto, and K. Tada: Solid State Phys. **22**, No. 12 (1967) [in Japanese].
24. Shinoda, G., M. Umeno, and S. Fuchikawa: To be published in Trans. Japan Inst. Metals.
25. Shiraiwa, T., Y. Sakamoto, and N. Fujino: Jap. J. Appl. Phys. **6**, 1025 (1967).
26. Nagasaki, K., S. Haruyama, and I. Ito: Read at the Spring General Meeting of Jap. Inst. Metals 1968.
27. Shinoda, G., K. Isokawa, and M. Umeno: Read at the Denver Conference, Aug. 21, 1968.
28. Kato, K., T. Watanabe, and S. Yoshida: Denki Seiko **38**, 350 (1968) [in Japanese].
29. — S. Tokunaga, and Y. Takano: Denki Seiko **36**, 10 (1965) [in Japanese].
30. Members of the group are K. Sano, Y. Wanibe, K. Mukai and H. Sakao and the results will be published in J. Jap. Inst. Met. [in Japanese].
31. Watanabe, Y., and R. Kusakawa: Read at the Spring General Meeting of the Iron and Steel Inst. Japan 1968.
32. Shiraiwa, T., N. Fujino, and N. Matsuno: Read at the Spring General Meeting of the Iron and Steel Inst. Japan 1968.
33. Saito, T., I. Uchiyama, and T. Araki: Read at the Spring General Meeting of the Iron and Steel Inst. Japan 1968.
34. Sawatani, T., and S. Sakata: Tetsu To Hagane **54**, 455 (1968) [in Japanese].
35. Kojima, K., K. Nagano, and T. Inazumi: Read at the Spring General Meeting of the Iron and Steel Inst. Japan 1968.
36. Sasaki, M.: Private communication.
37. Fujiki, Y., and H. Imai: Nippon Kogyo-kai Shi **81**, 2 (1965) [in Japanese].
38. Takeuchi, S., and Y. Fujiki: Nippon Kogyo-kai Shi **84**, 1 (1968) [in Japanese].
39. Umeno, M.: Private communication.
40. Morimoto, H., O. Kamikaito, and K. Satta: Private communication.
41. Umeno, M., and G. Shinoda: J. Mat. Sci. **3**, 120 (1968).
42. — N. Gennai, and G. Shinoda: Proc. Int. Conf. Strength Metals, Suppl. to the Trans. Japan Inst. Metals **9**, 499 (1968).

Measurements of Backscattered Electron Energy Spectra for Primary Beam Energies of 10 to 30 keV

E. F. H. St. G. Darlington

Cavendish Laboratory, University of Cambridge, England

This work was carried out to examine the effects of beam energy, take-off angle and specimen atomic number on the energy spectra of backscattered electrons. The effect these spectral differences have on the backscattering corrections in microanalysis is also being examined.

The energy spectra were measured using the apparatus shown in Fig. 1. An electron beam of 0.1 mm diameter was incident normally on thick, polycrystalline metallic specimens. The backscattered electrons were energy analysed with a 127° cylindrical electrostatic analyser and Faraday cage. The analyser had an entry aperture subtending approximately 10^{-3} steradians at the specimen and an energy resolution of 2%. The ultimate vacuum of the system used was only 10^{-5} torr which made it necessary to traverse the specimen perpendicular to the direction of the incident beam. This reduced the effects of contamination by continually presenting a fresh area of the specimen to the electron beam.

Energy spectra from high purity specimens of Al, Cu, Ag and Au were measured for beam energies (E_0) of 10, 20 and 30 keV, and at backscattering angles (θ) of $112\frac{1}{2}°$, 135° and $157\frac{1}{2}°$ measured relative to the forward beam direction. The Cu, Ag and Au specimens were mechanically polished (1 μ diamond paste and fine commercial metal polish), while the Al specimen was spark planed and then polished in an electrolytic cell (18—20 volts, perchloric acid and ethanol, below 0°C).

The energy analyser used had a constant percentage energy resolution which meant that the current measured was proportional to the energy distribution ($\partial\eta/\partial E$) multiplied by the energy being analysed (E). Since the noise of the current measuring system was fairly constant the error in measuring the energy distribution became very large for small values of E and tended to infinity as E tended to 0. The results have to be integrated from $E = 0$ to $E = E_0$, the initial beam energy, to obtain the backscattering coefficient at one angle ($\eta_{(\theta)}$), so some method must be used to prevent the experimental curves from diverging at the origin. A third order polynomial was fitted to the first third of the experimental curves such that the energy spectra were constrained to pass through $\partial\eta/\partial E = 0$ at $E = 0$ with positive or zero slope.

Fig. 2. shows the curves of $\partial\eta/\partial E$ against E for Al at 135° and 30 keV from a number of different experimental runs on two different Al specimens, to show the repeatability of the measurements.

In all the experimental curves shown the area under the curve was normalised to unity, for convenience, because normalisation to η also requires integration over all backscattering angles.

The variation in energy spectra with beam energy (E_0) was measured for Cu and Al, at 135° backscattering angle and beam energies of 10, 20 and 30 keV. Fig. 3 shows the spectra for Cu which are almost identical regardless of energy. Fig. 4 shows the spectra for Al which are also very similar, particularly at 20 and 30 keV. All the curves overlap within the range of their experimental errors. The invariance with energy shown by these results is in agreement with the experimental measurements of Kulenkampff and Rüttiger [1] and with the theoretical treatment of Bishop [2].

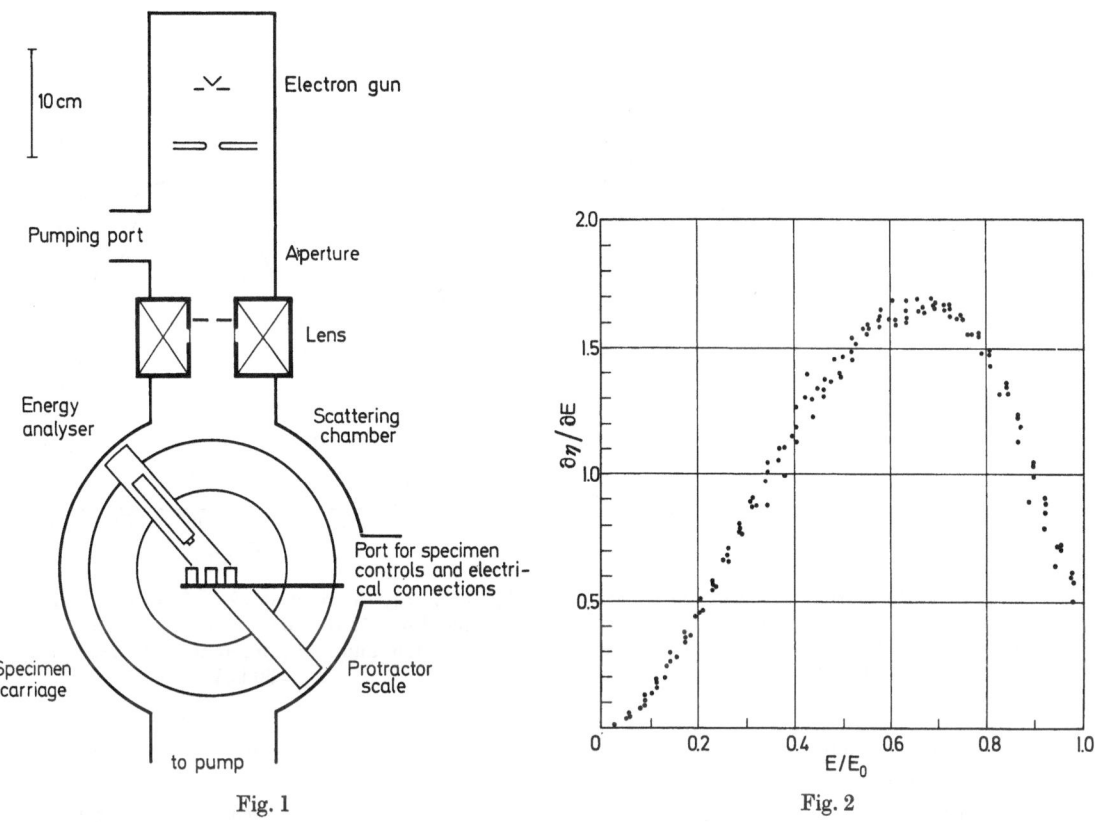

Fig. 1   Fig. 2

Fig. 1. Diagram of the apparatus

Fig. 2. Energy distribution $(\partial \eta / \partial E)$ of backscattered electrons from Al, recorded at an angle of 135° from the incident direction, for an incident beam energy $E_0 = 30$ keV, showing the experimental values from several measurements

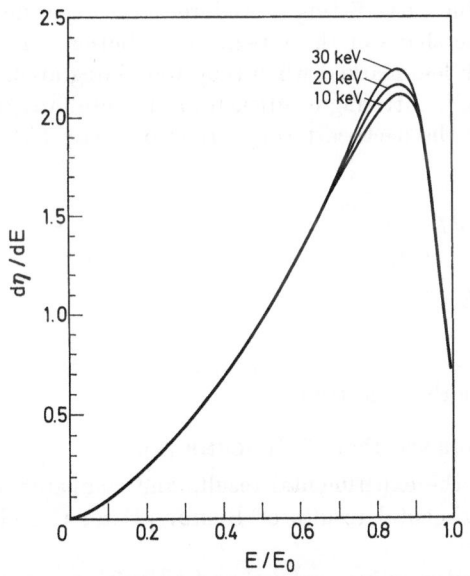

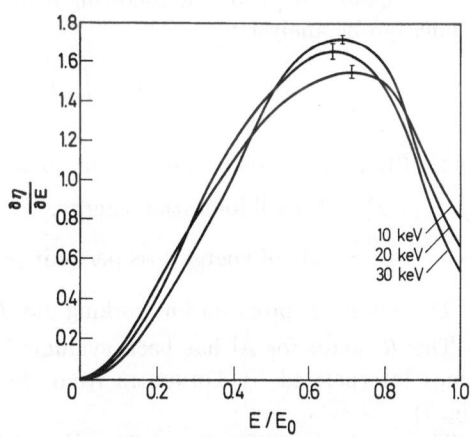

Fig. 3. Energy distribution $(\partial \eta / \partial E)$ of backscattered electrons from Cu recorded at a backscattering angle of 135° for incident beam energies (E_0) of 10, 20, and 30 keV

Fig. 4. Energy distribution $(\partial \eta / \partial E)$ of backscattered electrons from Al recorded at a backscattering angle of 135° for incident beam energies (E_0) of 10, 20 and 30 keV

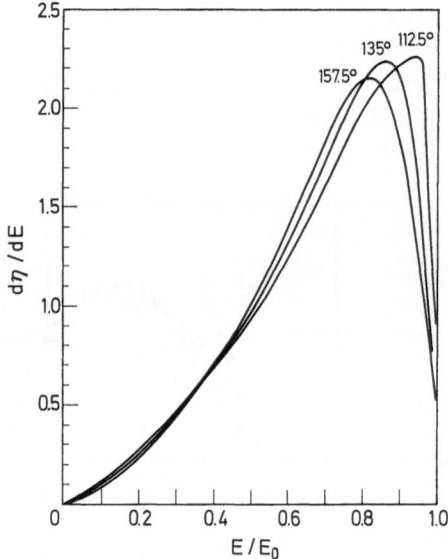
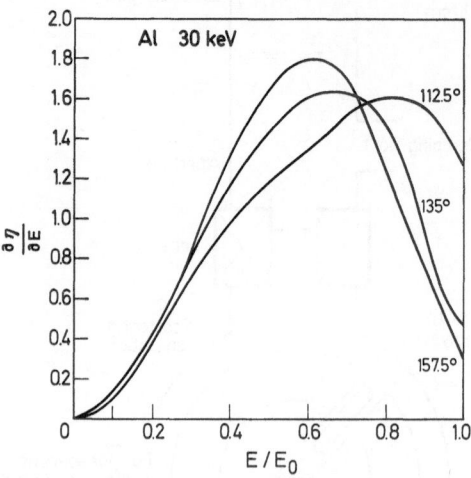

Fig. 5. Energy distribution $(\partial\eta/\partial E)$ of backscattered electrons from Cu recorded at backscattering angles of $112^{1}/_{2}°$, $135°$ and $137^{1}/_{2}°$ for an incident beam energy $E_0 = 30$ keV

Fig. 6. Energy distribution $(\partial\eta/\partial E)$ of backscattered electrons from Al recorded at backscattering angles of $112^{1}/_{2}°$, $135°$ and $157^{1}/_{2}°$ for an incident beam energy $E_0 = 30$ keV

The variation of energy spectra with backscattering angle is shown in Fig. 5 and 6 for Cu and Al at 30 keV, for angles of $112\frac{1}{2}°$, $135°$ and $157\frac{1}{2}°$.

The shift in peak energy with take-off angle is quite pronounced for both elements and is in a direction which would be expected from a model similar to that used by Everhart [3]. A combination of Rutherford scattering with a continuous energy loss approximation leads qualitatively to the conclusion that at smaller backscattering angles the peak and mean energies of backscattered electrons should be higher than at higher angles.

The integral spectra of $\eta(E, \theta)$ against fractional energy $W = E/E_0$ were obtained by numerical integration. For comparatively large values of W the curve fitting procedure to the experimental results affects only the normalisation and not the shape of the integral distribution. The differences between differential spectra became much less marked when they were integrated.

If these integral curves are normalised to the backscattering coefficient (η) the resultant graph of $\eta(E)$ against E yields the following formula for the backscattering correction factor (R) used in microprobe analysis:

$$R = 1 - \int_{E_0}^{E_x} \eta(E) \frac{\Phi(E)}{dE/dS} dE \left/ \int_{E_0}^{E_x} \frac{\Phi(E)}{dE/dS} dE \right.$$

where $\Phi(E) =$ ionisation cross section for energy E,

$\quad E_x =$ X-shell ionisation energy,

$\quad \dfrac{dE}{dS} =$ rate of energy loss per unit path length of electron.

The computer program for working out R factors was that of Duncumb [4].

The R factor for Al has been evaluated from the experimental results and compared with values interpolated by Duncumb from the experimental results of Bishop [2] for C and Cu (Fig. 7).

The maximum difference at 30 keV and $135°$ between interpolated and experimental results is 0.44%. The maximum difference between results at $112\frac{1}{2}°$ and $135°$ at 30 keV is 1.2%. The difference between the spectrum integrated over the angles measured by assuming a Lambertian cosine distribution and the $135°$ spectrum alone is 0.16%.

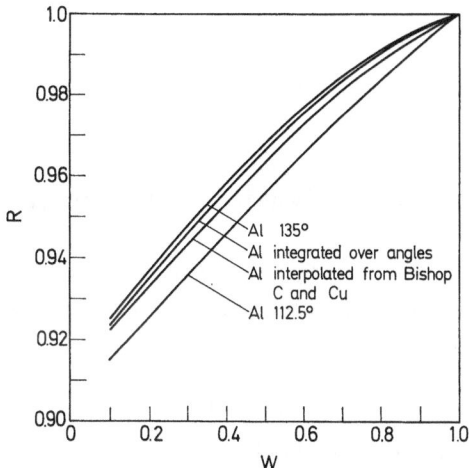

Fig. 7. "R factor" backscattering correction for aluminum plotted against fractional energy W $(=E_x/E_0)$ and evaluated from different energy distributions

For gold, the corresponding results of the present work show a maximum difference of 0.24% from those of Bishop at 135° and 30 keV. The maximum difference between present results at 10 and 30 keV at 135° is 3.2%. The maximum difference between angular integrated results and 135° only results is 0.95% at 30 keV and 0.67% at 10 keV.

From these results it is clear that at least for low atomic numbers the 135° curve alone gives a sufficiently close approximation to the total distribution. A fairly simple interpolation procedure is adequate to provide corrections for atomic numbers between those experimentally determined, at least for atomic numbers less than 39, provided that spectra from a reasonable number of well spaced elements have been measured.

A graph of R factor against E/E_0, at least for Al, will show almost no variation for beam energies between 10 and 30 keV arising from variations in energy spectrum $(\partial \eta/\partial E)$.

Acknowledgements. This work was done during the tenure of a research studentship, for which my thanks are due to the Science Research Council. I am grateful to Dr. DUNCUMB of Tube Investment Research Laboratories for providing computer facilities and a program for working out the R factor quoted, as well as providing data for comparison.

References

1. KULENHAMPFF, H., u. K. RÜTTIGER: Z. Physik **137**, 426 (1954).
2. BISHOP, H. E.: Thesis University of Cambridge 1966.
3. EVERHART, T. E.: J. Appl. Phys. **31**, 1483 (1960).
4. DUNCUMB, P., and S. REED: Tube Investments Technical Report No. 221.

Peak to Background Ratio in Microprobe Analysis

S. J. B. REED

Department of Mineralogy, British Museum (Natural History), London, England

Abstract

The peak to background ratio governs the limit of detection and the accuracy of the micro-probe analysis of very low concentrations. Theoretical formulae are derived for the fundamental peak to background ratio, assuming that the continuous X-ray spectrum is the only source of background. The effect of the spectrometer resolution and natural line width is considered. Some experimental data for the $\mathrm{Cr}\,K_{\alpha 1}$ and $\mathrm{Cd}\,L_{\alpha 1}$ lines are compared with theory.

1. Introduction

The limit of detection in electron-probe microanalysis is governed by the possibility of distinguishing the peak of a characteristic X-ray line above the level of the background. Usually the limit is considered to be set by the counting statistics of the measurement of peak and background intensities. For a recent discussion of this topic, see ZIEBOLD (1967).

The minimum detectable concentration is proportional to $(I_P R_{PB})^{-\frac{1}{2}}$, neglecting the effect of absorption and other corrections, which ZIEBOLD has calculated. Here I_P is the peak intensity from the pure element and R_{PB} is the peak to background ratio, defined as the ratio of I_P from the pure element to the background level on the specimen.

In practice several sources of counts contribute to the total background intensity. Of these, one is fundamental, namely the continuous X-ray spectrum. The others are dependent on instrumental factors, and may be minimised by suitable design and operation.

In this paper the theoretical framework for calculating the peak to background ratio for $K_{\alpha 1}$, $L_{\alpha 1}$ and $M_{\alpha 1}$ lines, in relation to the continuous spectrum only, is considered. Uncertainties in the physical data are such that it is necessary to check theory against experiment. Some experimental measurements are described and their implications are discussed.

2. Spectrometer Resolution and Natural Line Width

The resolution of an X-ray spectrometer can be defined in terms of the "window width", ΔV_1 (volts), which is the full width at half maximum of the "window" function, or response curve of the spectrometer as a function of X-ray quantum energy. Natural line width, ΔV_0, may be defined in similar terms. The observed line width, assuming both line and window functions are Gaussian, is given by $\Delta V^2 = \Delta V_0^2 + \Delta V_1^2$.

The observed peak to background ratio, R_{PB}, is related to the ideal peak to background ratio, $R_{PB}(0)$, which would be attained if $\Delta V_1 \ll \Delta V_0$, by the equation:

$$R_{PB} = \left(\frac{\Delta V_0}{\Delta V}\right) R_{PB}(0). \tag{1}$$

Plotting the profile of a line, by scanning the spectrometer and recording the X-ray intensity, enables ΔV to be determined. In measuring the peak width, the effect of the adjacent α_2 line must be allowed for, if it is not completely resolved. The width, $\Delta(2\theta)$, measured in degrees of 2θ ($\theta=$ Bragg angle) can be converted into $\varDelta V$ by the equation:

$$\varDelta V = 8.73 \times 10^{-3} V_1 \quad \cot\theta \, \varDelta(2\theta) \tag{2}$$

where V_1 is the quantum energy of the line.

COMPTON and ALLISON (1935) list experimental $K_{\alpha 1}$ line widths, which fit the relation $\varDelta V_0 = 4.4 \times 10^{-4} V_1$ quite well, for atomic numbers $20-40$. Calculated on the basis of the later data given by PARRATT (1935, 1936), the constant has the lower value 3.1×10^{-4} for atomic numbers $16-31$. There is also some evidence that a smaller value is required for lower atomic numbers. Applying the same formula to the experimental $L_{\alpha 1}$ and $M_{\alpha 1}$ line widths also given by COMPTON and ALLISON, the optimum values of the constant are 9.1×10^{-4} and 1.1×10^{-3} respectively.

3. Intensity of Continuous Spectrum

Observation of the continuous spectrum shows that the total intensity is proportional to the atomic number, Z, of the target, and the distribution of energy in the spectrum is a nearly linear function of the X-ray quantum energy V, falling to zero at the cut-off $V=V_0$. The intensity expressed as quantum rate is thus proportional to $Z\left(\frac{V_0}{V}-1\right)$. Experimental values of the constant of proportionality are rather variable, but using an average value, the intensity in the continuous spectrum in quanta per incident electron, per unit quantum energy interval, is given by:

$$I_c = 2.2 \times 10^{-9} Z\left(\frac{V_0}{V}-1\right). \tag{3}$$

It is well known that the distribution of intensity of the continuous spectrum emitted by a thin film target is highly anisotropic. However, with solid targets the anisotropy is blurred by the effect of electron scattering in tending to randomise the directions of the electrons. The effect of scattering is greatest for heavy elements, but DYSON (1959) showed experimentally that even light elements show only moderate anisotropy for $V < 0.5 V_0$, which is the region of interest here. It is therefore not necessary to include an extra factor in the intensity expression: any small anisotropy effect can be allowed for empirically by adjusting the constant.

GREEN and COSSLETT (1968) suggest that the mean depth of production of X-rays in the continuous spectrum is approximately the same as that of characteristic X-rays. If so, absorption can be neglected in calculating peak to background ratios. However, measurements by BIRKS et al. (1964, 1965) show R_{PB} increasing with decreasing X-ray take-off angle, indicating that more absorption is suffered by the continuous spectrum than by the characteristic line. This effect was only appreciable in pure elements when the take-off angle was lower than is usual in practice, but larger mass absorption coefficients can occur with compound specimens, so the absorption effect may not always be negligible.

4. Intensity of Characteristic Lines

GREEN and COSSLETT (1961) derived a formula for the number of K quanta per incident electron generated in a solid target. This formula contains a constant from the Thomson-Whiddington law used to give the rate of loss of energy of electrons with distance, which can be replaced by the appropriate expression from the more accurate Bethe law. With this modification, and adding a factor 0.6 for the ratio of $K_{\alpha 1}$ intensity to total K intensity, the formula becomes:

$$I_{K_{\alpha 1}} = 0.13 \frac{\omega_K R}{B}\left(\frac{V_0}{V_K}-1\right)^{1.67}. \tag{4}$$

Here B is the "stopping number": $B = Z \ln\left\{\frac{1.166\,\overline{V}}{J}\right\}$, where J is the mean ionisation potential, and $\overline{V}$ is the mean electron energy, which may be calculated with adequate accuracy from the relation: $\overline{V} = \frac{V_0 + V_K}{2}$. R is the correction factor for electron backscattering. Data for evaluating B and R are given in several papers on the atomic number correction, including that used here, by Duncumb and Reed (1968). The contribution of K ionisation by X-rays in the continuous spectrum is small and is neglected here. The K shell fluorescence yield, ω_K, can be calculated from the formula: $\omega_K = \dfrac{Z^4}{a_K + Z^4}$ due to Burhop (1952), putting $a_K = 1.0 \times 10^6$.

Green and Cosslett (1968) have shown that substitution of a theoretical expression for the L_{III} shell ionisation cross-section due to Burhop (1940) gives a formula for intensity of L lines identical to that for the K shell, except for the value of the constant. However measured L intensities were found to be in poor agreement with the theory, and required the constant to be increased by a factor of approximately 2. Experiments by Hink (1965) and Reed (1965) support this conclusion, so this factor has been used here. If the factor 0.65 for the ratio of $L_{\alpha 1}$ intensity to total L_{III} intensity is included, the constant in Eq. (3) for the $L_{\alpha 1}$ line becomes 0.40. The L_{III} shell fluorescence yield can be calculated from the Burhop formula, setting the constant equal to 1.0×10^8. The L_{III} shell excitation potential $V_{L_{III}}$ should be substituted for the K shell excitation potential V_K.

Usable data on the absolute intensity of M lines are few, however, Green and Cosslett (1968) give results for Au M_α. These have been used to calculate the value 0.64 for the constant in the equation for the intensity of $M_{\alpha 1}$ lines, assuming that the M_V shell fluorescence yield is given by the Burhop equation with a constant of 1.4×10^9 (based on data collected by Fink et al. (1968).

5. Theoretical Peak to Background Ratio

Eqs. (3) and (4) can be combined to give a theoretical formula for the fundamental peak to background ratio of the $K_{\alpha 1}$ line:

$$R_{PB}(0) = 5.9 \times 10^7 \frac{\omega_K R \left(\dfrac{V_0}{V_K} - 1 \right)^{1.67}}{BZ \left(\dfrac{V_0}{V_{K_{\alpha 1}}} - 1 \right) \Delta V_0}. \tag{5}$$

For the $L_{\alpha 1}$ and $M_{\alpha 1}$ lines, the appropriate values of the constant are respectively 1.8×10^8 and 2.9×10^8, and ω_K, V_K and V_{K_α} are replaced by $\omega_{L_{III}}$, $V_{L_{III}}$, $V_{L_{\alpha 1}}$ and ω_{M_V}, V_{M_V}, $V_{M_{\alpha 1}}$.

Eq. (5) can be simplified by substituting for ΔV_0, assuming $\Delta V_0 = \text{const } V_1$, where the constant has values 3.1×10^{-4}, 9.1×10^{-4} and 1.1×10^{-3} for $K_{\alpha 1}$, $L_{\alpha 1}$ and $M_{\alpha 1}$ lines:

$$R_{PB}(0) = 1.9 \times 10^8 \frac{\omega_K R \left(\dfrac{V_0}{V_K} - 1 \right)^{1.67}}{BZ(V_0 - V_{K_{\alpha 1})}}. \tag{6}$$

The constant in Eq. (6) takes values 2.0×10^8 and 2.6×10^8 for $L_{\alpha 1}$ and $M_{\alpha 1}$ lines. V_0 and $V_{K_{\alpha 1}}$ are expressed in keV.

Eqs. (1), (2) and (6) give the predicted practical peak to background ratio:

$$R_{PB} = 6.8 \times 10^6 \frac{\omega_K R \left(\dfrac{V_0}{V_K} - 1 \right)^{1.67} \cdot \tan\theta}{BZ(V_0 - V_{K_{\alpha 1}}) \Delta(2\theta)}. \tag{7}$$

A given fundamental peak to background ratio can be converted to the predicted practical value, using the relation:

$$R_{PB} = \text{const } \frac{\tan\theta}{\Delta(2\theta)} R_{PB}(0). \tag{8}$$

So far, only pure elements have been considered. At the limit of detection, the analysed element is present in a very small concentration in a usually compound material. Eqs. (5), (6) and (7) can be used for compounds, if mean values of B, R and Z are substituted.

6. Experimental Results

Measurements of the peak to background ratio for the Cr $K_{\alpha 1}$ and Cd $L_{\alpha 1}$ lines were carried out, with accelerating voltages of 15 to 35 kV. The instrument used was a Cambridge Instrument Co. "Geoscan", with normal electron incidence and an X-ray take-off angle of 75°.

For the Cr measurements, a lithium fluoride crystal was used in the spectrometer, with Johannson geometry (Rowland circle radius 25 cm) and a slit 30 μm wide in front of the counter. Pulse height analysis was employed, with a threshold of 4 V and a channel width of 3.5 V.

At each kilovoltage the probe current was adjusted to give about 6,000 cps on the peak, and the intensity was measured on pure Cr and corrected for dead-time. The beam was then positioned on a specimen of pure Mn and the background count-rate measured for a total of 200 s. The background count was then repeated with the crystal removed, and this count was subtracted from the first background count to give the corrected background intensity. This was then multiplied by the ratio of the atomic numbers of Cr and Mn. The difference in mass absorption coefficients of Cr and Mn for the $Cr K_{\alpha 1}$ line is very small, so that differences in absorption are unlikely to have been significant.

The peak profile was plotted, and a correction for the effect of the $K_{\alpha 2}$ line was estimated and applied to the peak intensity measurement. The full width at half intensity of the $K_{\alpha 1}$ line was measured.

A similar procedure was adopted for the Cd measurements, except that a quartz crystal was used, and the peak count-rate was about 4,000 cps. The background was measured on a specimen of pure Ag. A correction was made for the effect of the extreme tail of the $Ag K_{\beta 1}$ line.

The table gives the results, together with values of R_{PB} and $R_{PB}(0)$ calculated from Eqs. (6) and (7).

Table. *Theoretical and experimental peak to background ratios*

	V_0 (kV)	Calculated		Measured		V_0 (kV)	Calculated		Measured
		$R_{PB}(0)$	R_{PB}	R_{PB}			$R_{PB}(0)$	R_{PB}	R_{PB}
a) $Cr K_{\alpha 1}$	15	3,320	630	770	b) $Cd L_{\alpha 1}$	15	710	340	180
	20	4,170	790	1,110		20	810	390	230
	25	4,950	940	1,460		25	900	430	290
	30	5,610	1,070	1,740		30	990	480	330
	35	6,200	1,180	1,890		35	1,080	520	360

7. Discussion

Theory and experiment agree to within a factor of 2, as the table shows. The theory predicts peak to background ratios that are too high for $Cd L_{\alpha 1}$ and too low for $Cr K_{\alpha 1}$. The most likely sources of error are the constant in Eq. (3) for the intensity of the continuous spectrum, which is probably not strictly a constant, and the natural line width data.

References

BIRKS, L. S., R. E. SEEBOLD, A. P. BATT, and J. S. GROSSO: J. Appl. Phys. **35**, 2578 (1964).
— — B. K. GRANT, and J. S. GROSSO: J. Appl. Phys. **36**, 699 (1965).
BURHOP, E. H. S.: Proc. Cambridge Phil. Soc. **36**, 43 (1940).
— The auger effect. Camb. Univ. Press 1952.
COMPTON, A. H., and S. K. ALLISON: X-rays in theory and experiment. Macmillan 1935.
DUNCUMB, P., and S. J. B. REED: Quantitative electron probe microanalysis (ed. K. F. J. HEINRICH). N.B.S. Spec. Publ. **298** (1968).

DYSON, N. A.: Proc. Phys. Soc. (London) **73**, 924 (1959).
FINK, R. W., R. C. JOPSON, H. MARK, and C. D. SWIFT: Rev. Mod. Phys. **38**, 513 (1966).
GREEN, M., and V. E. COSSLETT: Proc. Phys. Soc. (London) **78**, 1206 (1961).
— — Brit. J. Appl. Phys. (J. Phys. D) Ser. 2, **1**, 425 (1968).
HINK, W.: Z. Physik **182**, 227 (1965).
PARRATT, L. G.: Rev. Sci. Instr. **6**, 387 (1935).
— Phys. Rev. **50**, 1 (1936).
REED, S. J. B.: Brit. J. Appl. Phys. **16**, 913 (1965).
ZIEBOLD, T. O.: Anal. Chem. **39**, 858 (1967).

A Study of the Deadtime Correction in Electron Probe Microanalysis

D. R. Beaman*, R. Lewis** and J. A. Isasi*

*The Dow Chemical Company, Midland, Michigan, U.S.A.
** Consolidated Electrodynamics Corporation, Monrovia, California, U.S.A.

Introduction

The magnitude of the deadtime correction must be known in quantitative electron microprobe analyses since failure to apply the correction can lead to errors in the measured relative X-ray intensity ratio, k, which often are of the same magnitude as the precision in the determination of k; even greater errors will result when analyzing dilute alloys while using pure element standards. Heinrich, Vieth and Yakowitz [1] developed a simple method of determining the deadtime, τ, in the electron microprobe based on the model proposed by Ruark and Brammer [2] where the true counting rate, N_t^0, is given by $N^0/(1-\tau N^0)$ and N^0 is the measured X-ray intensity from a pure element. Several investigators [3—7] have reported that τ is a function of the counting rate and becomes essentially zero below intensities varying from 4,000 to 14,000 cps [3—6]; in the Ruark [2] model, τ is independent of intensity. In an attempt to resolve this problem, a careful comparison was made between the current method described by Heinrich et al. [1], and the ratio method, which gives a range of zero deadtime, or negligible coincidence loss, described by Short [3], and Borile, Short and Tabock [4]. The deadtimes of the electronic components were measured, both individually and as complete counting systems, in an attempt to identify the deadtime source and to correlate X-ray (current and ratio methods) and electronic deadtimes.

Experimental Method

There are several different means of performing the ratio technique measurement based simply on the requirement of having available two X-ray intensities, one presumably low enough to be unaffected by deadtime when losses in the higher intensity are detectable. The method can be used as either a single or double spectrometer technique, with the two intensities being provided by any of several possibilities: K_α and K_β; two different orders of K_α; K_α and filtered K_α; K_α and spectrometer slit reduced K_α; K_α from a pure element and an alloy. An alloy method was used herein where a pure Fe and a homogeneous Fe—37% Ni alloy or an Fe—10% Ni alloy were mounted in a single mount in juxtaposition with adjacent edges about 40 microns apart. Both alloys were stable in the electron beam and gave equivalent results. Intensities from the pure and alloy materials were recorded consecutively on a single spectrometer by moving the sample 75—100 μ under the beam between measurements. Optical refocusing between measurements was unnecessary. This single spectrometer method was selected in preference to several others, which were also studied, primarily because it involved a single detector and counting channel. This simplified the interpretation of the results and facilitated the correlation of X-ray and electronic measurements to such an extent that the two spectrometer methods were rejected notwithstanding the fact that they have the distinct advantage of measuring the two intensities simultaneously, therein minimizing drift problems. The sample stage could be repositioned with better precision than the peaks on the spectrometer. Counting times of 50 to 100 seconds were used at each intensity level and anticontamination was provided by a cold plate attached to the lower portion of the objective lens. Control of contamination was essential to provide the precision being sought for the 2—3 hours required to complete an experimental run. The pure $Fe\,K_\alpha$ intensity was varied from 300 cps to 60,000 cps, and an equal number of measurements were made while increasing and decreasing the beam current. A 25 μ incident beam size was utilized. The acceleration potential was 25 kV and specimen currents, i, ranged between 1 and 250 nanoamperes. Recording of the specimen current, i, while not required in the ratio procedure, made it possible to extract data for the current technique (N^0 and i) from the information collected in any given run of the ratio method.

When independent current measurements were made, experimental runs were accomplished in 35—55 minutes and higher currents were often used in conjunction with a spectrometer slit to provide low intensities at moderate currents in an attempt to improve the precision in the measurement of current at low counting rates. This procedure avoided the measurement of currents below 3 nanoamperes and resulted in good precision in the range of intensities where the electron microprobe is normally operated.

All measurements were made using a Cameca/CEC electron probe microanalyzer operated at 25 kV. The specimen current was monitored continuously with a digital voltmeter. Simultaneous measurements of the $Fe\,K_\alpha$ intensity were always made using two separate counting channels and spectrometers; one spectrometer contained a LiF and the other an $10\bar{1}1$ quartz crystal. The flow proportional X-ray detectors were normally operated at 3 atmospheres pressure with P-10 gas at an anode potential of about 2,100 volts. Preamplifiers and amplifiers of different manufactures were used to study the effect of pulse shape. A double pulse generator, two single pulse generators and two calibrated dual trace oscilloscopes were used in studying the electronic components.

Results and Discussion

X-Ray Measurements. Six different experimental determinations of the deadtime, using the current method, produced values of 3.9 ± 0.3 and 4.3 ± 0.3 μsecs for the spectrometers with the LiF and $10\bar{1}1$ quartz crystals respectively. The N^0/i versus N^0 curves (Fig. 1 is typical) were linear to about 30,000 cps, beyond which baseline intensity reduction [1] was often noted due to the pulse amplitude decrease at higher intensities presumably caused by detector anode wire contamination [8]. In all cases the good linear fit to the data confirmed the Ruark model.

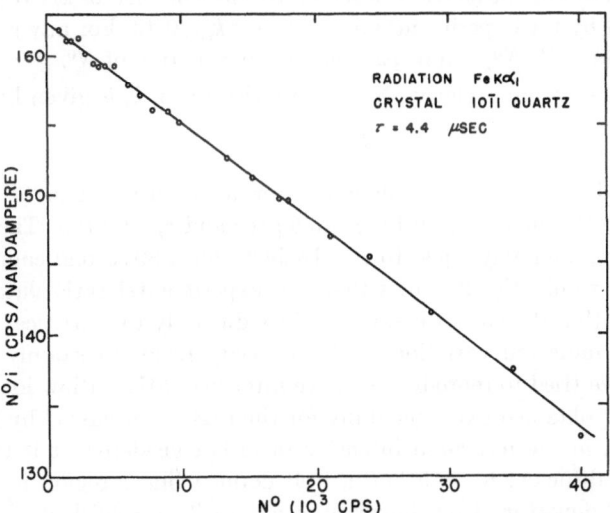

Fig. 1. Example of the data collected using the current method [1] and a single spectrometer equipped with a spectrometer slit

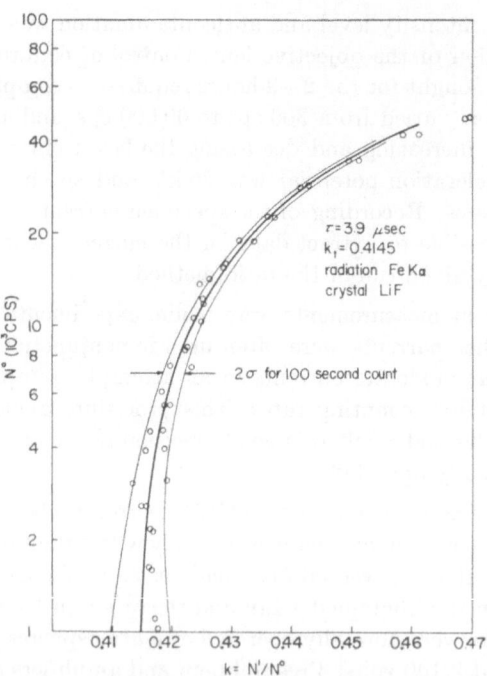

Fig. 2. Example of the data collected using the ratio method [3] and a single spectrometer. k is the ratio of the FeK_α X-ray intensity from an Fe—37% Ni alloy to that from pure Fe

The experimental data for two different runs, using the ratio technique for the spectrometer containing the LiF crystal, are shown in Fig. 2. The heavy solid curve was calculated for the deadtime of $3.9\,\mu sec$ determined from current measurements using the following expression, derived assuming that the Ruark model applies:

$$k_t = N_t'/N_t^0 = N'(1-\tau N^0)/N^0(1-\tau N'). \qquad (1)$$

k_t is the true X-ray intensity ratio in the absence of any deadtime effects and is assumed to be the value obtained at low intensity values. N is the measured X-ray intensity from an alloy; N_t' and N_t^0 are the true counting rates for the alloy and pure material respectively. Only the position of the curve, and not the shape, is affected by the selection of k_t over the possible range of k_t values established by the experimental data: $0.41 < k_t < 0.42$. For any given value of τ and k_t, N' and therefore k ($k = N'/N^0$) can be calculated as a function of N^0.

The variance in the measured relative X-ray intensity ratio, k, is given by

$$\sigma_k^2 = k(k+1)/N^0. \qquad (2)$$

The $\pm 2\,\sigma_k$ error band in Fig. 2 was calculated for a counting time of 100 seconds using Eq. (2) and imposed on the curve drawn for $\tau = 3.9\,\mu sec$ and $k_t = 0.4145$. The $\pm 2\sigma$ band for an 100 second count is approximately equal to a $\pm 1\sigma$ band for a 30 second count. The variance in N^0 is small compared to σ_k^2. Fig. 2 shows that the experimental ratio data is well described by the Ruark model within the variance expected in k due only to statistical fluctuations in the X-ray intensities. The indicated variation in the intensity ratio is certainly familiar to microprobe analysts who have tried to reproduce relative intensity ratios at low intensity levels and is generally less than that obtained experimentally for the reasons discussed by Ziebold [9].

When the precision in the measured intensity ratio is considered, it is clear that an exact determination of k_t is difficult; in addition, a well defined linear region is not apparent and, therefore, the point of departure from linearity is not easily established. Since σ_k is not very sensitive to τ, the less the system deadtime is, the greater will be the difficulty encountered in trying to determine whether or not a region of linearity exists; e.g., if τ is constant at 1 μsec,

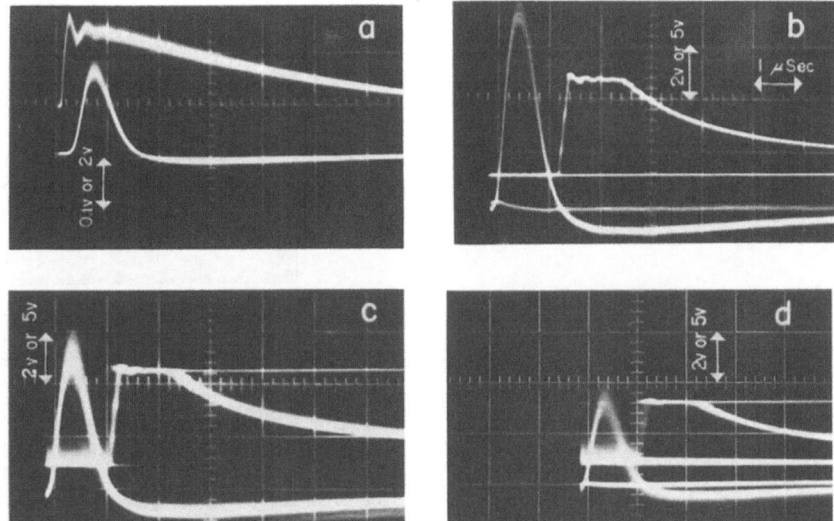

Fig. 3 a—d. Oscilloscope photographs of a system consisting of a Cameca high pressure detector, an Acton preamplifier, an Ortec shaping linear amplifier, and a Mesco PHA. For the linear amplifier and PHA outputs, the vertical scales are 2 volts/large division and 5 volts/large division respectively. The sweep rate is 1 μsec/large division. The horizontal line represents a baseline voltage of 1 volt. Fe K_α X-radiation and a LiF crystal are used. a Upper trace is the preamplifier output with a vertical scale of 0.1 volt/large division and lower trace is the linear amplifier output; b trace on left is the amplifier output and trace on right is the PHA output, intensity = 1,500 cps; c intensity = 20,000 cps; d intensity = 55,000 cps

k will change by 0.0025 as N^0 increases from 1,000 cps to 10,000 cps, while $2\sigma_k$ over the same range will vary from ± 0.0088 to ± 0.0028 for a 30 second count and from 0.0048 to 0.0016 for an 100 second count.

In order to evaluate the experimental data and methods, it is necessary to know the expected precision in τ for the two techniques. For the current method HEINRICH et al. [1] have derived

$$\tau = \{1 - [(N^0/i)/K]\}/N^0 \tag{3}$$

where K is the value of N^0/i at $N^0 = 0$. The most probable total error in τ, $d\tau/\tau$, is given by the square root of the sum of the squares of the component errors due, in the current method, to variation in N^0, K, and N^0/i. Differentiating Eq. (3) gives component errors of $-dN^0/N^0$, $(N^0/i)dK/K(K - N^0/i)$, and $d(N^0/i)/(K - N^0/i)$ for variations in N^0, K and N^0/i respectively.

If the Ruark model holds, τ for the ratio method calculated from Eq. (1) with $k = N'/N^0$ is given by

$$\tau = (k_t - k)/k N^0 (k_t - 1). \tag{4}$$

Differentiating this expression gives component errors of $(k - 1)dk_t/(k_t - 1)(k_t - k)$, $-dN^0/N^0$ and $k_t dk/(k - k_t) k$ for variations in k_t, N^0 and k respectively.

Maximum and minimum possible values of $d\tau/\tau$ were estimated using variations $(1\,\sigma)$ extracted from the experimental curves, and for a 30 second count interval the calculated precision of the current method was from 2 to 8%. This agrees favorably with the value of 7% found in six experimental runs. For the same count interval, the calculated precision for the ratio method was 20—38%; for 100 seconds counting time, this improved to 14—23%. The precision of the ratio method is lower for equivalent count intervals, because τ depends upon two relative intensity ratios. The precision of this method can apparently be improved by using longer counting intervals [4, 10]. BORILE et al. [10], performed experimental runs that lasted over 8 hours.

Electronic Measurements. The output of the various electronic components was observed on a dual trace oscilloscope with the sweep triggered by selected input signals. Fig. 3 illustrates the behavior of a system consisting of a Cameca high pressure proportional counter detecting Fe K_α

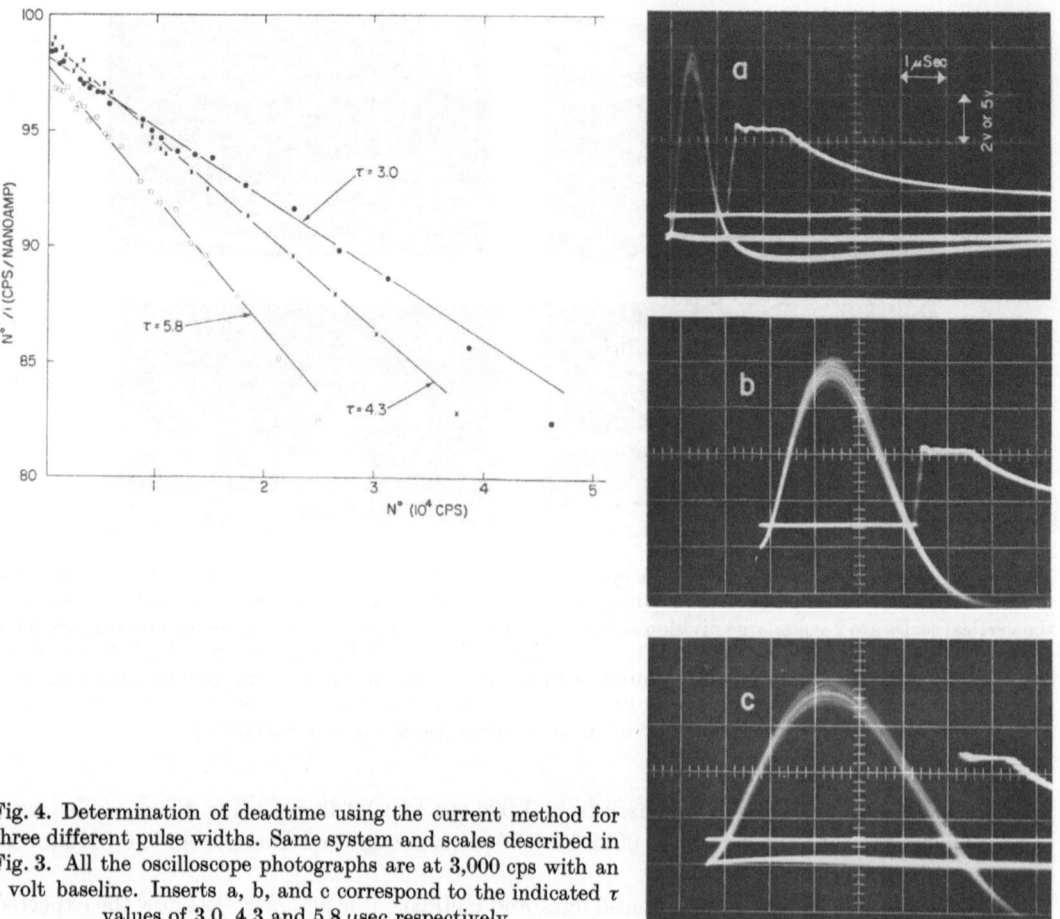

Fig. 4. Determination of deadtime using the current method for three different pulse widths. Same system and scales described in Fig. 3. All the oscilloscope photographs are at 3,000 cps with an 1 volt baseline. Inserts a, b, and c correspond to the indicated τ values of 3.0, 4.3 and 5.8 μsec respectively

X-radiation, an Acton preamplifier, an Ortec shaping linear amplifier and a Mesco pulse height analyzer (PHA). All of the signals except one have negative polarities which have been inverted at the oscilloscope. The positive output of the Acton preamplifier was inverted by the shaping amplifier. In Fig. 3a, where the preamplifier output has been used to trigger the oscilloscope sweep, it is apparent that the amplifier is triggered on the leading edge of the preamplifier pulse. The rapid rise time in the preamplifier leads to a negligibly small contribution to circuit deadtime (< 0.2 μsec). In Fig. 3b, where the oscilloscope sweep is triggered by the amplifier output, the leading edge of the PHA output pulse is seen to coincide with the point where the trailing edge of amplifier pulse intersects the baseline. In this system the PHA is triggered on the leading edge of the amplifier pulse, but the PHA output pulse is not released until the amplifier pulse voltage decays to the baseline level. The Saphymo scaler used triggers on the leading edge of the PHA output, but will not accept another pulse until the PHA output decays to 6.8 volts. Figs. 3c and d show a serious pulse amplitude shift at higher intensities, but a negligible effect on system deadtime. These measurements suggested that the electronic deadtime might be related to the time delay between the triggering of the amplifier and the point where the PHA output decays to 6.8 volts plus the 0.1 μsec time lag between preamplifier and amplifier triggering. It should be emphasized that the magnitude of the delay times measured on the oscilloscope is independent of the mode of operation of the electronic components.

The effect of the amplifier pulse width on system deadtime was studied by varying the pulse width using three different shaping times. In Fig. 4 the oscilloscope photographs for the different pulse widths are shown, in addition to the corresponding current measurement data. For the narrow pulse the electronic and X-ray (current method) deadtimes agree, but as the pulse width increases the total electronic values become greater than the X-ray values (Table and Fig. 4).

Table. *A comparison of X-ray and electronic deadtimes showing the effect of amplifier pulse width*

Detector Preamplifier Amplifier	Amplifier and generator pulse width in μsec	X-ray deadtime in μsec		Electronic deadtime in μsec	
		ratio	current	pulse pair resolution	component total
Cameca Cameca Mesco	4.2 ± 0.3[a]	3.7 ± 0.4	3.9 ± 0.3	4.2	6.7 ± 0.2
Cameca Acton Ortec	1.2 1.2 3.6 5.8	3.4 ± 0.5	3.5 ± 0.4 3.0 4.1 5.8	2.2 2.2 3.6 5.8	3.2 ± 0.2 3.2 5.2 7.1

[a] 1σ variation in cases where several experimental runs were made.

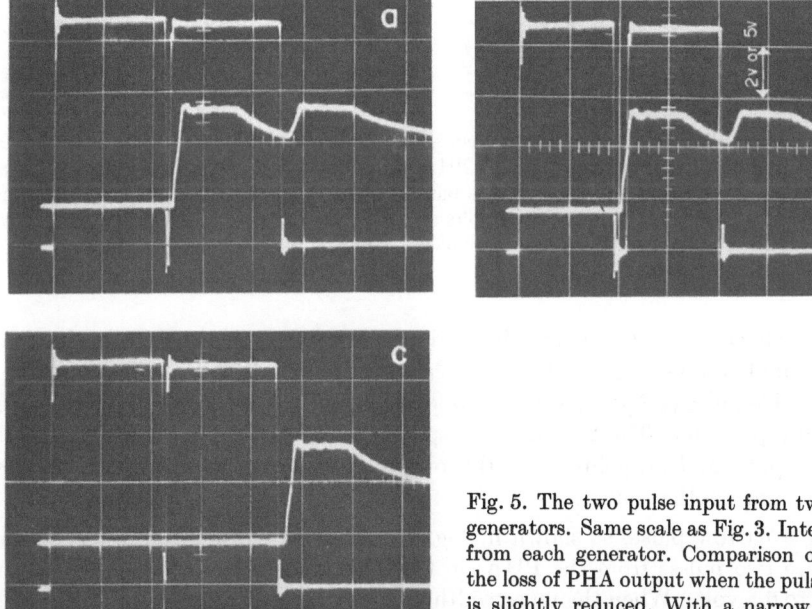

Fig. 5. The two pulse input from two separate pulse generators. Same scale as Fig. 3. Intensity = 5,000 cps from each generator. Comparison of a and c shows the loss of PHA output when the pulse pair separation is slightly reduced. With a narrow pulse, the scaler just fails to resolve two pulses in b

The system deadtime was next measured using a method in which square wave pulses from two single pulse generators were utilized. The trigger output of the first generator was used to trigger the second generator and the combined output of the two was used as the input to the system at various locations; namely, at the preamplifier after attenuation, at the linear amplifier, at the PHA and at the scaler. The pulse separation was controlled by the pulse delay of the first generator and the oscilloscope sweep was triggered by the trigger output of the first generator. This approach was preferred over a double pulse generator where pulse shrinkage of the second pulse occurred at narrow pulse separations which corresponded to high generator frequencies. Whenever ambiguities due to pulse shrinkage could be avoided the results obtained with the double pulse generator and a triple pulse were the same as with the two pulse input. The electronic deadtime was taken to be the pulse pair resolution, measured from leading edge to leading edge, when the measured counting rate halved. The table shows agreement, within experimental error, between this electronic deadtime and the current measurements for the wider pulse widths.

When using a wide pulse as shown in Figs. 5a and c, pulse pair resolution is reached when the PHA fails to detect two pulses from the amplifier; i.e., the PHA output from the first pulse disappears. The implication here is that if the pulse width is greater than 2.2 μsec, which is the

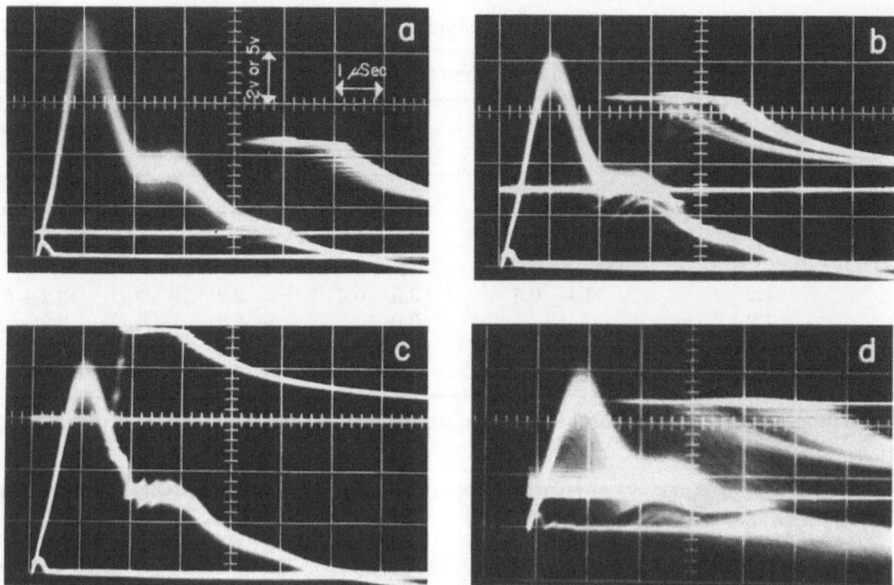

Fig. 6a—d. Oscilloscope photographs of a system consisting of a Cameca high pressure detector, a Cameca preamplifier, a Mesco linear amplifier and a Mesco PHA. Same scale as in Fig. 3. Fe K_α radiation and a LiF crystal are used. This series shows the effect of the baseline setting on the PHA output. a baseline $=1$ volt and intensity $=6,000$ cps; b baseline $=3$ volts and intensity $=6,000$ cps; c baseline $=7$ volts and intensity $=6,000$ cps; d baseline $=1$ volt and intensity $=40,000$ cps

experimentally determined value where the leading PHA pulse first vanishes, the deadtime is given by the amplifier pulse width. The fact that system deadtime measured by the current method using a wide pulse (4.2 μsec) was the same with the PHA as when the PHA was bypassed supports this interpretation. The 2.2 μsec corresponds to the time delay between the release of the PHA output pulse and the point where this output has decayed to a level (6.8 volts) at which the scaler can accept another pulse.

When using a narrow pulse, as shown in Fig. 5b, pulse pair resolution is reached when the scaler fails to see two pulses from the PHA; in this case the PHA output of the leading pulse does not decay to 6.8 volts. When the pulse width was less than 2.2 μsec, the pulse pair resolution was constant at 2.2 μsec. The implication here is that the deadtime is 2.2 μsec regardless of pulse width. This was not observed in the experimental current measurements where τ was 3.5 ± 0.4 μsec for a pulse width of 1.2 μsec.

Fig. 6 shows the effect of varying the baseline voltage when using the Cameca preamplifier and the Mesco amplifier. The most significant observation is that an increasing baseline reduces the effective pulse width by permitting earlier release of the PHA output pulse, and thereby leads to slightly reduced deadtimes. The poorly defined trailing edge in this preamplifier is troublesome and can lead to additional PHA output as shown in Fig. 6b. Comparison of Figs. 6a and d shows a large pulse amplitude shift. The broader and lowered distribution at higher intensities has led to additional output and an ambiguous deadtime. Thus, the only intensity dependence detected was a result of the pulse amplitude sensitivity to intensity which led to reduced pulse widths for a fixed baseline. However, this effect accounted for less than 0.2 μsec of deadtime at intensities below 30,000 cps when using a shaped amplifier pulse.

In the present investigation only Fe K_α radiation was used and the dependence of τ on wavelength was not studied. The results obtained herein indicate that τ depends primarily on the amplifier pulse width. At constant gain this width varies slowly with atomic number; however, if the pulse amplitude is maintained constant the change in pulse width is negligible between 1.4 and 7.1 Å.

Deadtime Model. The controversy concerning the intensity sensitivity of τ led to an examination of the Ruark formulation which is indeed a surprisingly simple one in view of the difficulties now apparent. RUARK states that if the deadtime is τ, the true number of counts during a time interval τ is $N_t^0 \tau$; there are N^0 count intervals every second; there then are $N^0 N_t^0 \tau$ counts lost every second through coincidence losses; the true number of counts is the sum of the actual measured count rate and the lost counts: $N_t^0 = N^0 + N^0 N_t^0 \tau$. This is easily recognized as $N_t^0 = N^0/(1 - \tau N^0)$.

Our measurements of electronic deadtime have led us to believe that the amplifier does accept pulses in such a manner as to extend the system deadtime when a second pulse arrives at some time dt $(dt < \tau)$ after the original pulse. The Ruark formulation does not account for such a possibility because the system is assumed to be inoperative for a fixed time τ. We have considered this effect in deriving a new expression for τ and preliminary calculations indicate that the Ruark formulation applies at the intensities normally encountered in microanalysis; however, at high intensities and deadtimes, deviation from the Ruark model occur and increased deadtimes, which increase with intensity, are predicted. This effect is usually small in the range where we have detected deviation from linearity in the current measurements ($N^0 > 30,000$ cps) prior to the baseline intensity reduction which occurs above 60,000 cps. This phenomenon will be discussed in a later report.

Summary

While we have not been able at this time to explain all of the observed phenomena, several preliminary conclusions, applicable to the X-ray detection systems studied in this investigation, may be made.

1. N^0/i versus N^0 (current method) data were collected in 30 different experimental runs and in all cases a good linear fit to the data was obtained in the interval from 1,000 to 30,000 cps.

2. When data for the current and ratio methods (18 experimental runs) were collected concurrently and the deadtime extracted from the current measurements using the RUARK [2] model was used to calculate k versus N^0 data, a fit to the experimental ratio data was obtained which fell within the limits of statistical counting error ($\pm 2\sigma_k$). All of the experimental ratio data could be treated with the Ruark model and the τ values extracted in this manner were in good agreement with those determined using the current method (see the table). The ratio data collected herein did not in any case suggest a region of negligible coincidence loss.

3. Careful observation of the preamplifier, amplifier and PHA outputs failed to reveal any significant intensity sensitivity for intensities ranging up to 30,000 cps. Even in the systems which exhibited large pulse amplitude shifts the effect on pulse width was small when using a 1.0 volt baseline setting. Higher baseline values could be troublesome.

4. For the above conditions of measurement (pulse width > 3.0 µsec) the electronic deadtime measured using a two pulse input at the preamplifier was, within experimental errors, the same as the X-ray deadtime, implying a negligible detector deadtime.

5. In view of 1—4 we conclude that, for the electronic components and detector studied, the deadtime is constant and non zero over the entire intensity interval investigated and, therefore, must be applied to all intensities less than 40,000 cps.

6. In view of the above, the current method rather than the ratio method is recommended for measuring deadtime because: a) in the range of intensities normally encountered in microanalysis the model describing the method appears to be valid; b) it is possible to collect good data in a relatively short period (45 minutes versus 3—8 hours for the ratio method) of time; c) the expected precision in the determination of τ is sufficiently good for quantitative microanalysis and, for equal count times, significantly better than in the ratio technique; d) the deadtime and, therefore, the correction are easily and directly obtained from the experimental data without the need for the construction of correction curves. This is important because the deadtime is associated with the electronic components, whose characteristics alter with age, making it necessary to periodically measure τ.

When making current measurements it is necessary to accurately measure low currents, which is relatively easy with good equipment, but it is not necessary to use currents below 3 nanoamperes if the instrument has a spectrometer slit system. The difficulties described by Heinrich et al. [1] must be avoided: a) the specimen current must be proportional to the beam current — in our instrument the fractional standard deviation in the beam current/specimen current ratio for 18 readings in the range of 1—500 nanoamperes was 0.2%; b) sample biasing at high currents should be accounted for; c) the intensity at which baseline intensity reduction arises must be determined prior to collecting current data.

7. In our system where the PHA output is released on the trailing edge of the linear amplifier output, the system deadtime appears to be given by the amplifier pulse width if this width is greater than the time delay (2.2 μsec) between the release of the PHA output and the point where the scaler can accept another pulse. For such wide pulses the deadtime is related to the failure of the PHA to distinguish the pulses. An increase in the X-ray deadtime with increasing pulse width was shown. For narrow pulse widths (1.2 μsec) it was not possible to correlate the X-ray deadtimes with pulse pair resolution data. In this case the deadtime appears to be related to the failure of the scaler to distinguish pulses.

8. There is a deviation from the Ruark [2] model at intensities in the range of 30,000 to 60,000 cps which can not be attributed to baseline intensity reduction. This effect may be due to a measurable intensity sensitivity in this range caused by an extended deadtime resulting from pulse interaction.

9. For the Cameca instrument and Mesco-Saphymo electronics we found the following: a) the deadtime depends on the counting channel used, but is close to 4 μsec (3.9 ± 0.3 μsec for spectrometer and count channel number 1 — 4.3 ± 0.3 seconds for spectrometer and count channel number 2); b) the deadtime was not affected by varying the detector pressure between 2 and 3 atmospheres.

Acknowledgement. The authors wish to thank H. K. Birnbaum of the University of Illinois and K. F. J. Heinrich of the National Bureau of Standards for their helpful comments. Thanks are also due L. Solosky for his assistance in performing the experimental measurements.

References

1. Heinrich, K. F. J., D. Vieth, and H. Yakowitz: In: Advances in X-ray analysis, vol. 9 (G. Mallett, M. Fay and W. Mueller, eds.), p. 208. New York: Plenum Press 1966.
2. Ruark, A., and F. E. Brammer: Phys. Rev. **52**, 322 (1937).
3. Short, M. A.: Rev. Sci. Instr. **31**, 618 (1960).
4. Borile, F., M. A. Short, and J. Tabock: In: Transactions of The Third National Conference on Electron Microprobe Analysis, Chicago, Illinois, July 31—August 2, 1968, paper No. 32.
5. Clayton, D. B.: Private communication concerning the activities of the Midlands Microanalysis Group 1967.

6. Beaman, D. R., and T. P. Schreiber. In: Transactions of The Third National Conference on Electron Microprobe Analysis, Chicago, Illinois, July 31—August 2, 1968, paper No. 49.
7. Wolf, R. C., and V. G. Macres: Stanford University, Stanford, California, Department of Materials Science, Report No. 63—18 (1963).
8. Spielberg, N.: In: Transactions of The Third National Conference on Electron Microprobe Analysis, Chicago, Illinois, July 31—August 2, 1968, paper No. 34.
9. Ziebold, T. O.: Anal. Chem. **39**, 859 (1967).
10. Borile, F.: Private communication.

The Measurement of Total Mass per Unit Area and Elemental Weight-Fractions along Line Scans in Thin Specimens

T. A. HALL and P. WERBA

Cavendish Laboratory, Free School Lane, Cambridge, England

As reported earlier [1—3], one can measure elemental weight-fractions and total mass per unit area in thin specimens in the microprobe by combining observations of characteristic line intensities with simultaneous observations of the bremsstrahlung. The bremsstrahlung counting rate also provides a means to subtract background instantaneously from a characteristic peak, with no need for background measurements off the Bragg angle, thus permitting quantitative line-scans even when the background is large and varies from point to point.

In this paper we shall describe tests of the validity and range of application of the method. We shall discuss only relative measurements, as absolute measurements have been discussed elsewhere [3].

Fig. 1 shows the arrangement of components in the column. Besides the conventional X-ray spectrometer, an additional counter receives X-rays directly from the specimen, and its count is restricted to a suitable range of quantum energies by pulse-height analysis.

In order to avoid excessive background, the specimen must be mounted on a very thin support. In our laboratory the support is usually a Nylon film weighing approximately 40 μg/cm². The film in turn is stretched over the end of an aluminium tube.

The continuum counting rate serves two distinct purposes: It is a measure of total mass per unit area, and it also furnishes a means of correction for background in the X-ray spectrometer. The principle of the correction is shown schematically in Fig. 2. The background W consists

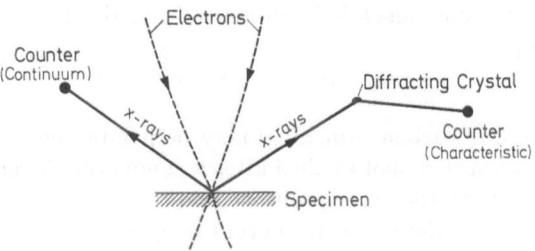

Fig. 1. Disposition of components

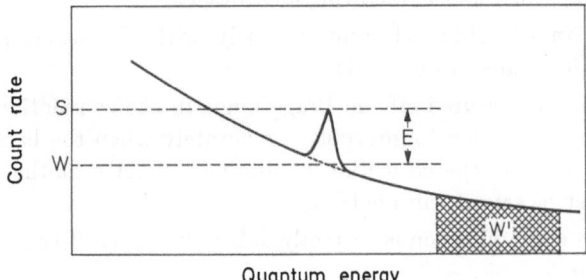

Fig. 2. Schematic, principle of background correction by monitored band of white radiation

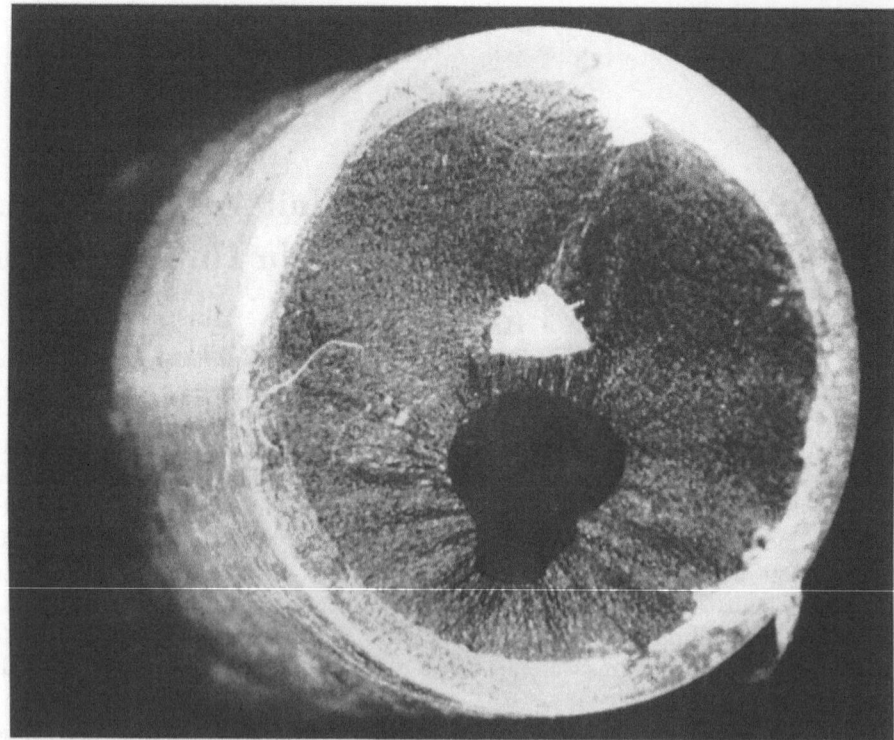

Fig. 3. Thin specimen (white) on thin nylon support, aluminium-coated, plus painted Aquadag spot (black)

almost entirely of continuum radiation very near the wave length of the characteristic line. From basic theory [1], it is expected that this background and the signal in the direct counter, W', will be in a constant ratio r. The only necessary conditions for this are that the energy-band of W' must not include a significant amount of characteristic radiation, and that this band (and the band around the characteristic line) must be below the energy of the radiating probe electrons. Then the desired characteristic signal E is obtained from the total spectrometer signal S by means of the relationship

$$E = S - W = S - r\,W'. \tag{1}$$

In practice, a thin spot of carbon (Aquadag) may be painted on the supporting film next to the specimen. Fig. 3 shows such a spot applied after a conducting layer of aluminium was evaporated onto the specimen. With the probe positioned on the Aquadag, a fraction of the ratemeter output W' is substracted from the ratemeter output S by means of an analogue computer, and the fraction is adjusted to make the computer output *nil*. The output from any spot in the specimen should then be the characteristic signal E.

This method of correction has the following advantages:

1. The corrected signal is obtained simultaneously with the spectrometer signal, so that background-corrected line scans can be made.

2. When background is measured off the Bragg angle in the conventional way, two or more off-angle measurements are needed to interpolate accurately when the background curve is not "flat". In the present method the correction is obtained directly at the Bragg angle and the spectrometer need never be set off this angle.

3. Background from contamination is correctly subtracted even if the amount of contamination varies during a run.

4. Instrumental drifts between on-angle and off-angle runs are avoided.

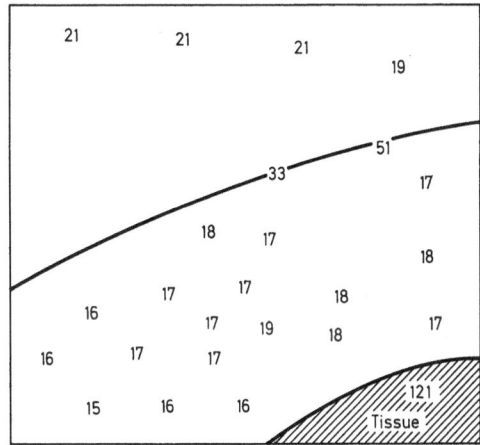

Fig. 4. Variation of white count from aluminized Nylon support. Field 250×250 μm

The use of the signal W' as a measure of local mass per unit area is based on the equation

$$W' = k M \sum_r C_r Z_r + B. \tag{2}$$

Here k is a constant, M is local mass per unit area, C_r and Z_r are the weight-fraction and atomic number of constituent r, the sum is taken over all constituent elements, and B is the presumably constant background from the supporting film. The equation in fact merely states that for sufficiently thin specimens, W' depends linearly on M, and that the efficiency of bremsstrahlung production is proportional to Z. If necessary, one may determine the weight-fractions C_r by the measurement of several characteristic radiations [3] and thus evaluate the factor $(\sum_r C_r Z_r)$ in Eq. (2). However, we shall restrict our attention here to the case, common in the study of biological soft tissues, where only a few elements contribute significantly to the sum and the mean atomic number $(\sum_r C_r Z_r)$ is virtually constant[1]. Then the quantity $(W' - B)$ is a direct measure of local mass per unit area. In practice, a constant is subtracted in the analogue computer from the ratemeter signal W', and the difference-output is adjusted to *nil* when the probe is positioned on the bare support; this output then gives a continuous record of $(W' - B)$, relative mass per unit area, as the probe moves over the specimen.

We shall now consider some assumptions and limitations of the method.

1. The band of white radiation must exclude characteristic X-rays. Since all of the requirements of the method are best satisfied at high column voltages, it is generally not difficult to select a band comfortably between the shortest characteristic wave length and the X-ray cutoff determined by the column voltage.

2. For the measurement of mass per unit area, the supporting film should be uniform. (This is not required for accurate background correction!) Fig. 4 shows the signal W' from a 5-μm section of soft tissue and from many points in the adjacent supporting film. The film is badly non-uniform at occasional folds or creases which are clearly seen in an electron scanning image, but otherwise local non-uniformity usually imposes an uncertainty of a few per cent (a few μg/cm²).

---

1. It is interesting to make a comparison with the microradiographic method of measuring mass per unit area. Because X-ray absorption coefficients vary approximately as Z^4, the quantity measured by the microradiographic method is essentially $(M \sum_r C_r Z_r^4)$. Even the co-factor of M in this expression scarcely varies in soft tissues, as shown exhaustively by LINDSTRÖM [4].

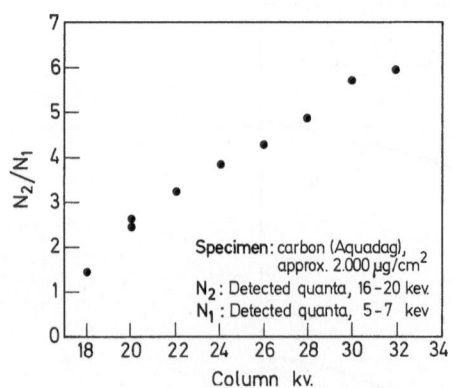

Fig. 5. Non-proportionality of white radiation in different energy bands in thick specimens. Dependence on column kV

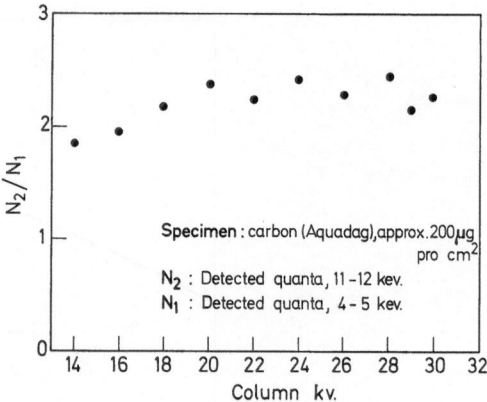

Fig. 6. Proportionality of white radiation in different energy bands in thin specimens. Independence of column kV

3. The observed continuum radiation should come only from the specimen. There are several extraneous sources:

a) Electrons are scattered sideways by the specimen and generate continuum as they strike the walls of the tube on which the film is mounted. In our instrument this background seems negligible until the probe is moved more than $\frac{2}{3}$ of the way from the centre of the tube towards a wall "seen" by the continuum-detector. (The tube diameter is 1 cm.)

b) Backscattered electrons strike the objective lens and other objects near the specimen. We evaluated the amount of the resulting radiation seen by the continuum-detector by comparing the intensity from a specimen presenting a flat surface (predominantly legitimate) with the intensity recorded when the probe struck the bottom of a well (all illegitimate). The illegitimate continuum arising from nearby surfaces was of the order of 1% of the signal.

c) Cooke and Duncumb [5] have discussed the severe difficulties arising from extraneous radiation when one analyses very thin specimens non-diffractively in an electron microscope-microanalyser. Although at high quantum energies the extraneous background is less than at the energies they recorded, it is not yet clear how adequately extraneous continuum can be suppressed in such instruments.

4. For background correction, the ratio of continuum intensities in the monitor channel and in the narrow band accepted by the crystal should be independent of atomic number. From theory no dependence is expected, and we have not detected any dependence in our limited observations to date.

5. For the accurate measurement of local mass per unit area, the specimen must be thin enough so that the bremsstrahlung intensity is linearly proportional to local mass per unit area. Published data [3] indicate that for 30-kV electrons, thicknesses up to at least 600 µg/cm² are safe. In this range of thickness, X-ray absorption is usually negligible for the selected high-energy band of the continuum.

6. For accurate background correction, the restrictions on thickness are that the incident electrons must not be degraded below the highest observed X-ray quantum energy, and that absorption of the observed X-ray wave lengths in the specimen must be low. The range of adequate operating conditions is indicated in the following data.

Fig. 5 shows the dependence on column voltage of the ratio of intensities generated in a relatively thick specimen (2,000 µg/cm²) in two energy bands of the continuum. (Both bands were observed in the same counter, the "continuum-detector".) The ratio is not satisfactorily constant because electrons are being degraded in the specimen to energies below the top of the upper window, an effect which becomes worse as the column voltage is lowered.

Fig. 6 shows a similar test under better conditions. With a thickness of 200 µg/cm² and an upper window of 11—12 kV, the effect of electron degradation is not seen until the column voltage is lowered to approximately 18 kV.

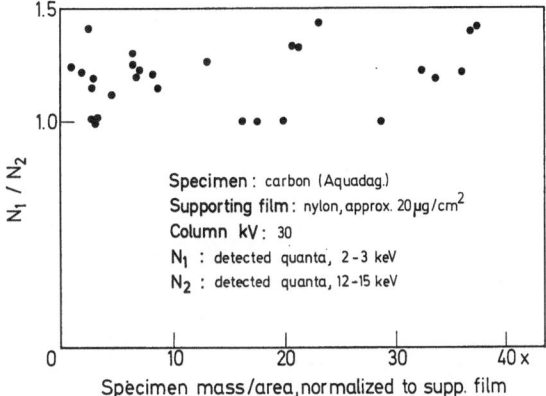

Fig. 7. Proportionality of white radiation in different energy bands in thin specimens. Independence of specimen thickness

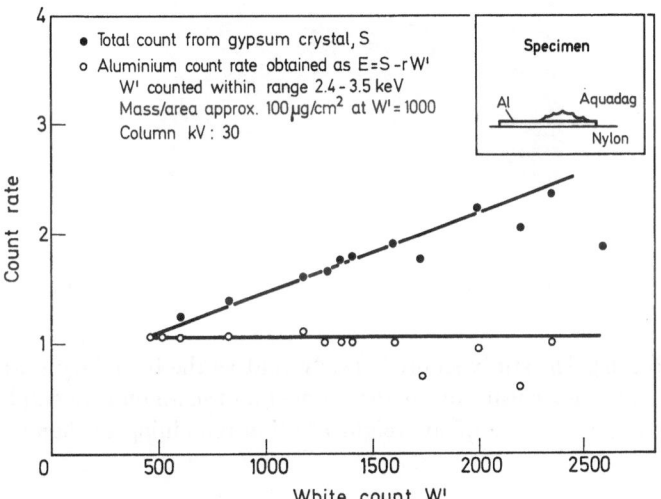

Fig. 8. Subtraction of background from varying Aquadag to get signal from constant aluminium layer

Fig. 7 shows the ratio as a function of specimen thickness (i.e., as a function of the counting rate in the high-energy band) at constant column voltage. However, in this case both counters were used; the spectrometer counter received X-rays directly, with the diffractor moved out of the way, and it was set to a low energy band of 2—3 kV. While no overall dependence of the ratio on thickness is seen, the points are not well consistent. We attribute this to lumpiness in the Aquadag specimen, resulting in unequal absorption paths to the two detectors and occasional high absorption of the 2—3 kV radiation. Thus the painted Aquadag film can lead to error.

Fig. 8 shows the method of background correction applied to a uniform layer of aluminium over which Aquadag was painted non-uniformly. The specimen is much thinner than in the preceding Figure and the results are more consistent. It is seen that the high continuum background from the Aquadag can be successfully substracted to give the correct aluminium signal even when the background is larger than the signal. However, absorption enters again as a limiting factor at the higher thicknesses, as expected for the soft radiations which are involved.

Just as the signal W' is proportional to mass per unit area in thin specimens, the corrected signal E is proportional to amount of element per unit area [3]. Hence the ratio of these two signals E/W' is proportional to local weight-fraction. The display of weight-fraction (E/W') and amount of element per unit area (E) along a line of scan is illustrated in Fig. 9, taken from a study of the distribution of sulphur and total dry mass near the boundary of viable epidermis

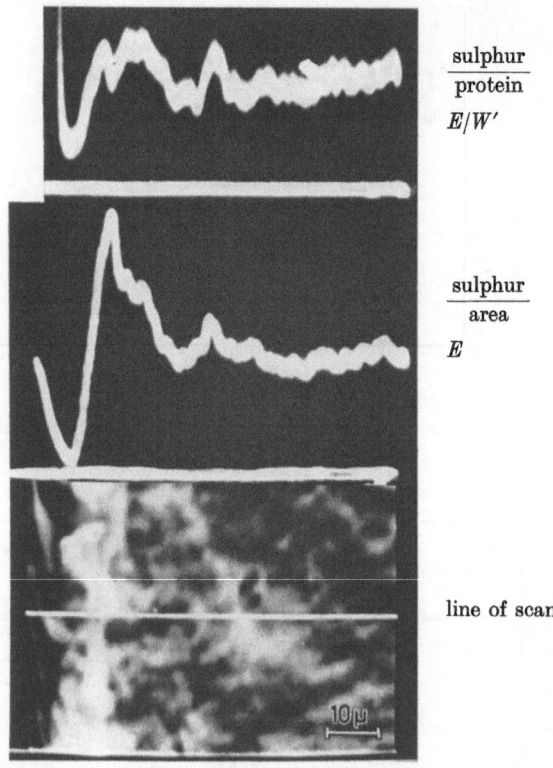

sulphur
―――――
protein

E/W'

sulphur
―――――
area

E

line of scan

10 μ

Fig. 9. Quantitative displays of weight-fraction and amount per unit area along a line of scan

and corneum in skin [6]. The study showed clearly that as the boundary is approached from the side of the living tissue, the density of the dry tissue and the amount of sulphur per unit volume increase steeply in parallel, the sulphur weight-fraction remaining unchanged.

References

1. HALL, T. A., A. J. HALE, and V. R. SWITSUR: In: The electron microprobe (eds. HEINRICH, MCKINLEY and WITTRY), p. 805. New York: John Wiley & Sons 1966.
2. MARSHALL, D. J., and T. A. HALL: In: X-ray optics and microanalysis (eds. CASTAING, DESCAMPS and PHILIBERT), p. 374. Paris: Hermann, 1966.
3. — — Brit. J. Appl. Phys. 1, 1651 (1968).
4. LINDSTRÖM, B.: Acta Radiol., Suppl. 125 (1955).
5. COOKE, C. J., and P. DUNCUMB: In: X-ray optics and microanalysis (eds. CASTAING, DESCAMPS and PHILIBERT), p. 467. Paris: Hermann 1966.
6. SIMS, R. T., and T. A. HALL: J. Cell Science 3, 563 (1968).

A Method for Composition Determination
of Alloy Thin Films

B. Djurić and D. Cerović

Institute of Nuclear Sciences "Boris Kidrič" Vinča, Beograd — Yugoslavia

Abstract

To determine the composition of alloy thin films it is necessary to know their thickness. In the present work the possibility of composition determination without knowing the exact thickness is theoretically discussed. A method based on determination of the ratio of intensities of the elements present in the film is proposed. The results were experimentally tested on nickel-chromium thin films.

Introduction

In the electron probe microanalysis of thin films the intensities of the characteristic X-rays will be a function not only of the composition but also of the thickness of the specimen; consequently, for the exact determination of concentration it is necessary to known the thickness of the specimen.

Methods have been evolved for the composition determination of thin films with known thickness, and conversely, for the thickness determination of films of known composition by means of the electron probe microanalysis. So far, no satisfactory procedure has been developed to determine the composition of thin films whose exact thickness is not known. In the present work a method for the determination of the composition of thin films whose thickness is not known is suggested.

Previous Work

Numerous authors have studied the question of thickness determination of thin films with known composition, mainly on pure element films.

Sweeney et al. [1] employed the simplest method: they measured the intensity of the characteristic X-rays in pure metal thin films of known thickness. On the basis of this, calibration curves were constructed facilitating the determination of unknown specimen thickness by measuring the intensity of characteristic X-rays under identical experimental conditions. Schumacher et al. [2, 3] evolved a method for the experimental determination of the electron accelerating voltage needed for the electrons to just penetrate the films. Here too, film thickness can be determined by means of calibration curves relating the penetrating voltage and film thickness. Cockett and Davis [4] compared both of these methods and found that the former is more suitable for thinner and the latter for thicker films. They further showed that the calibration curves may be also mathematically computed if the intensity distribution curves are experimentally determined.

Marshall and Hall [5] described a method of composition determination of films whose thickness is unknown by measuring the characteristic X-rays and the continual X-ray spectrum of the film. In its present form, this method produces only semi-quantitative results.

Theory

According to the well-known CASTAING's expression [6], the intensity of characteristic X-rays generated by electron incidence in a thin film of pure element A is

$$(I_A^A)_F = \int_0^{\varrho x} \phi(\varrho_A x) e^{-X_A^A \varrho_A x} d(\varrho_A x) \tag{1}$$

where x is the thickness of the film, ϱ_A the density of pure element A, and X_A^A the absorption coefficient of the characteristic radiation A in pure element A. $\phi(\varrho_A x)$ is a complex function representing the distribution in depth of the characteristic emission of element A.

The characteristic X-ray intensity of element A from an alloy film $A B$ in which the weight fraction of element A is equal to c_A, can be written

$$(I_A^{AB})_F = c_A \int_0^{\varrho x} \phi(\varrho_A x) e^{-X_A^{AB} \varrho_A x} d(\varrho_A x) \tag{2}$$

where X_A^{AB} represents the absorption coefficient of radiation A in the alloy.

In order to find the unknown concentration c_A by measuring the characteristic X-ray intensities from the specimen and standard it is also necessary to compute the integrals contained in Eqs. (1) and (2). This cannot be done by the usual methods applicable to the massive specimens. A possibility to calculate the integrals was given by HUTCHINS [7], who showed that the expression on the right side of Eq. (1) can be expanded to a series

$$\int_0^{\varrho x} \phi(\varrho_A x) e^{-X_A^A \varrho_A x} d(\varrho_A x) \propto \varrho_A x + (\varrho_A x)^2 \left(\frac{B}{2} - \frac{X}{2} \right) - (\varrho_A x)^3 \left(\frac{BX}{3} - \frac{X^2}{6} \right) + \cdots$$
$$+ (\varrho_A x)^n (-1)^n \frac{B X^{n-2}}{(n-1)!} - \frac{B X^{n-2}}{n!} - \frac{X^{n-1}}{n!}. \tag{3}$$

Parameter B is here the measure of electron diffusion in the film. Its values for different experimental conditions were calculated by HUTCHINS.

According to HUTCHINS, the X-ray intensity from a film, normalized in respect to the intensity of the massive standard can be expressed in the following way

$$\frac{I_F}{I_M} = \frac{\varrho x + (\varrho x)^2 \left(\frac{B}{2} - \frac{X}{2} \right) - \cdots}{\int_0^\infty \phi(\varrho x) e^{-X \varrho x} d(\varrho x)}. \tag{4}$$

For defined experimental conditions the integral in Eq. (4) has a constant value, denoted hereafter by $1/K$.

For films of pure elements A and B of the same thickness x Eq. (4) yields the following intensity ratios

$$\frac{(I_A^A)_F}{(I_A^A)_M} = K_A \left[\varrho_A x + (\varrho_A x)^2 \left(\frac{B}{2} - \frac{X_A^A}{2} \right) - \cdots \right], \tag{5}$$

$$\frac{(I_B^B)_F}{(I_B^B)_M} = K_B \left[\varrho_B x + (\varrho_B x)^2 \left(\frac{B}{2} - \frac{X_B^B}{2} \right) - \cdots \right]. \tag{6}$$

When measuring the intensity of thin film characteristic radiation, account must be taken of the influence of the substrate to which the films have been deposited. The substrate continual spectrum and the high energy electrons which are back-scattered from the substrate may augment the intensity of the characteristic X-radiation generated by the primary electron beam. Since the intensity of the continual spectrum and the amount of back-scattered electrons increase with the atomic number of substrate Z, substrate correction is given by the expression [7]

$$\frac{I_{FS}}{I_F} = 1 + f(Z) \tag{7}$$

where I_{FS} and I_F are the intensities of the X-ray radiations of thin films with and without substrate, respectively. Experimentally determined values of function $f(Z)$ [7, 8] indicate that substrate correction depends also on working conditions, first of all on the electron accelerating voltage.

By dividing Eq. (5) by Eq. (6) and introducing the correction for the substrate, $I_F = I_{FS}/(1+f(Z))$, which is the same for both films, the following ratio denoted by a is obtained

$$\frac{\dfrac{(I_A^A)_{FS}}{(I_A^A)_M}}{\dfrac{(I_B^B)_{FS}}{(I_B^B)_M}} = \frac{K_A\left[\varrho_A\,x + (\varrho_A\,x)^2\left(\dfrac{B}{2}-\dfrac{X_A^A}{2}\right)-\cdots\right]}{K_B\left[\varrho_B\,x + (\varrho_B\,x)^2\left(\dfrac{B}{2}-\dfrac{X_B^B}{2}\right)-\cdots\right]} = a. \tag{8}$$

For a thin film composed of alloy $A\,B$, with weight fractions of components c_A and c_B, an analogous ratio denoted by b may be written

$$\frac{\dfrac{(I_A^{AB})_{FS}}{(I_A^A)_M}}{\dfrac{(I_B^{AB})_{FS}}{(I_B^B)_M}} = \frac{K_A\left[\varrho_A\,x + (\varrho_A\,x)^2\left(\dfrac{B}{2}-\dfrac{X_A^{AB}}{2}\right)\cdots\right]c_A}{K_B\left[\varrho_B\,x + (\varrho_B\,x)^2\left(\dfrac{B}{2}-\dfrac{X_B^{AB}}{2}\right)\cdots\right]c_B} = b. \tag{9}$$

The last two equations can be used for calculation of film composition.

Depending on film thickness two cases can be considered.

a) For small film thicknesses experimental data show that the intensity of characteristic X-rays is a linear function of the thickness. In that case Eqs. (8) and (9) reduce to

$$\frac{\dfrac{(I_A^A)_{FS}}{(I_A^A)_M}}{\dfrac{(I_B^B)_{FS}}{(I_B^B)_M}} = \frac{K_A\,\varrho_A\,x}{K_B\,\varrho_B\,x} = a_1 \tag{10}$$

and

$$\frac{\dfrac{(I_A^{AB})_{FS}}{(I_A^A)_M}}{\dfrac{(I_B^{AB})_{FS}}{(I_B^B)_M}} = \frac{K_A\,\varrho_A\,x\,c_A}{K_B\,\varrho_B\,x\,c_B} = b_1. \tag{11}$$

By combining them we obtain

$$a_1\frac{c_A}{c_B} = b_1. \tag{12}$$

Since the relation $c_A + c_B = 1$ is also valid here, it is possible to determine the unknown concentrations.

In order to determine the constant a_1 it is necessary to have films of pure elements A and B of known thickness. The films must be either without substrate or on the same substrate. After constructing the I_F/I_M versus $\varrho\,x$ diagrams, it is possible to determine a_1 by finding the ratio of slopes of the straight lines in the diagrams. Once determined, the value of constant a_1 can be used for all subsequent calculations carried out under identical conditions. The value of b_1 is obtained by measuring the characteristic radiation intensities of elements A and B on the alloy film with the parallel intensity measurements on pure element massive standards.

This procedure, applicable within the range where the dependence between X-ray intensities and the film thickness is linear, eliminates the need of knowing the thickness of investigated films.

b) In thicker films, where the intensity is not a linear function of the thickness, approximative concentration measurements are possible only in a limited number of cases, provided that approximate film thickness can be assessed and that the absorption of characteristic radiations does not differ much in both elements. In that case x in Eq. (9) is replaced by the approximately known thickness of the investigated film and the absorption coefficients of the alloy by average values of pure element absorption coefficients. The constants K_A and K_B are found from the slopes of the calibration curves and the intensity ratios are determined experimentally. As the ratio of the two expanded series does not vary too much with the film thickness x, the results, although less accurate, may still be adequate for practical purposes.

In principle, this method may be also used to determine the composition of multicomponent thin films.

Experimental Results

A few measurements were carried out with nickel-chromium thin films obtained by vacuum deposition under different conditions, starting from a solid alloy with 80 wt.% nickel and 20 wt.% chromium. The films were evaporated on to a ceramic substrate. A SEM 2 electron probe microanalyser (AEI-Manchester) was used in the experiments. The electron accelerating voltage was 20 kV and the electron current was 50 μA. Nickel and chromium K_α lines were used. Measurements were carried out at several points of each specimen, and measurement times were elected so as to warrant a sufficient statistical reliability of results. The radiation intensity of both chromium and nickel was determined at each point. Before and after thin film measurements, the characteristic radiation intensity was measured on standard nickel and chromium massive specimens.

To obtain the constants needed for calculation of the film compositions calibration curves were determined by measuring the characteristic X-line intensities of chromium and nickel obtained from pure metal films of known thickness. It was found that a linear dependency between the characteristic X-radiation intensity and film thickness exists up to about 1,000 Å. The calibration curves are shown in Fig. 1. The intensities were normalized in respect to the intensities of pure massive nickel and chromium standards. Constant a_1 [Eq. (10)] for the linear part of the diagram was determined from the slop ratio of nickel and chromium lines, amounting to 1.41. The calculated curves according to Eq. (4) are also presented in the Fig. 1.

The compositions of thin films with different thickness and composition were determined. The values of the measured nickel and chromium concentration in some of the investigated films are given in the table.

For the purpose of testing the method, beside composition determination of thin films, thickness measurements of the investigated films were carried out by means of a quartz oscillator with resonant frequency, with an accuracy of ±5%. The measured thicknesses are given in the table. On the other hand, the thicknesses of the films were calculated on the basis of the determined nickel and chromium concentration in the film and with the aid of the diagram

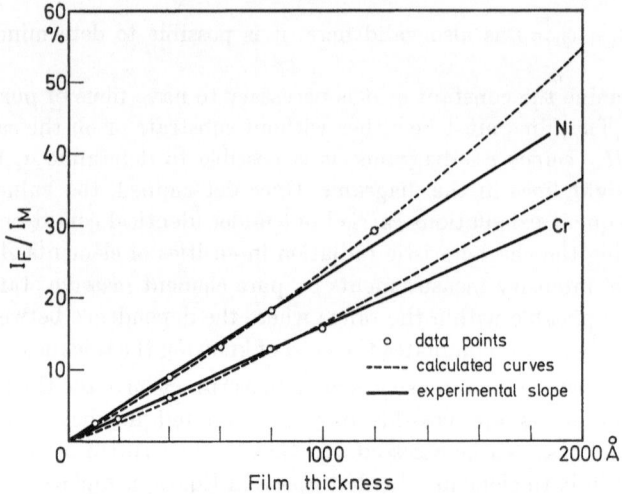

Fig. 1. Intensity data from pure nickel and chromium films

Table. *Experimental results*

Film	c_{Ni}(%)	c_{Cr}(%)	Measured thickness (Å)	Calculated thickness (Å)
1	43.2	56.8	750	765
2	54.0	46.0	900	920
3	12.5	87.5	1150	1110

in Fig. 1. Film thicknesses determined in this manner are also quoted in the table. The comparison of the measured film thicknesses with those calculated on the basis of the film composition shows that the deviations are within the range of the thickness measurement error.

For thicker films it is necessary to know the approximate thickness in order to calculate the ratio of expanded series in Eq. (9). This ratio varies from 1.29 to 1.44 when the film thickness changes from 100 to 5,000 Å. To show that only rough estimate of the film thickness is needed the following example can be taken: for a film of NiCr alloy whose thickness lies somewhere between 2,000 and 3,000 Å the difference between the calculated compositions will not exceed 1% (abs. 0.5 wt.%) when assuming the thickness of 2,000 or 3,000 Å.

The main factor affecting the accuracy of concentration determination according to the method under review are the error obtained during intensity measurements, and the error in the thickness determination of pure element thin filmes which served for the construction of calibration curves.

Conclusions

A method of the composition determination of alloy thin films of unknown thickness from the ratio of the normalized intensities of the characteristic X-rays of the individual alloying components is suggested. The method can be employed with a satisfactory accuracy for film thicknesses within the range of the linear dependence of thin film intensity on film thickness. In a limited number of cases the method can be used with a lesser accuracy also for films whose thickness exceedes that range. The method does not require any correction for the substrate to which the film was deposited.

Acknowledgements. We are grateful to T. NENADOVIĆ and T. DIMITRIJEVIĆ for providing the thin film specimens.

References

1. SWEENEY, W. E., R. E. SEEBOLD, and L. S. BIRKS: J. Appl. Phys. **31**, No. 6, 1061 (1960).
2. KRIEGLER, R., and B. W. SCHUMACHER: Plating 393 (1960).
3. SCHUMACHER, B. W., and S. S. MITRA: Proc. Electronic Components Conf. Washington 1962, p. 152.
4. COCKETT, G. H., and C. D. DAVIS: Brit. J. Appl. Phys. **14**, 813 (1963).
5. MARSHALL, D. J., et TH. A. HALL: Optique des rayons X et microanalyse (CASTAING, DESCHAMPS, PHILIBERT, eds.), p. 374. Paris: Hermann 1966.
6. CASTAING, R.: Advan. Electron. Electron Phys. **13**, 317 (1960).
7. HUTCHINS, G. A.: The electron probe (McKINLEY, HEINRICH, WITTREY ed.), p. 390. New York: John Wiley & Sons 1966.
8. HEINRICH, K. F. J.: Advan. in X-Ray Anal. **7**, 325 (1964).

An Experimental Method for Determining the Depth Distribution of Characteristic X-Rays in Electron Microprobe Specimens

U. Schmitz, P. L. Ryder and W. Pitsch

Max-Planck-Institut für Eisenforschung, Düsseldorf, Germany

Abstract

An experimental method is described to determine the depth distribution of characteristic X-rays produced in electron microprobe specimens. The results obtained by measurements on Cu are described and compared with results of other methods given in the literature.

1. Introduction

Exact knowledge on the depth distribution of characteristic X-rays in a solid specimen provides the basis on which concentrations are determined quantitatively by electron probe microanalysis. Several methods are known to achieve this knowledge either experimentally (by the "tracer method" [1—3] and by the "analysis of angular distribution of characteristic X-rays" [3]) or theoretically (by the "Monte-Carlo calculations" [3—6]). Due to certain limitations and difficulties inherent in the experimental technique of the first of the above methods it is felt that another approach could provide useful additional information on the production of X-rays in solid targets, although it should be emphasized that the new method described here does not in principle produce more reliable results than the former but rather complementary information.

2. Experimental Principle of the Method

In order to determine the depth distribution of e.g. the K_α-radiation of element A a specimen is prepared as shown in Fig. 1. (The preparation technique of such a specimen is described in a subsequent section.) The specimen contains three parts, being made of elements A, B and again

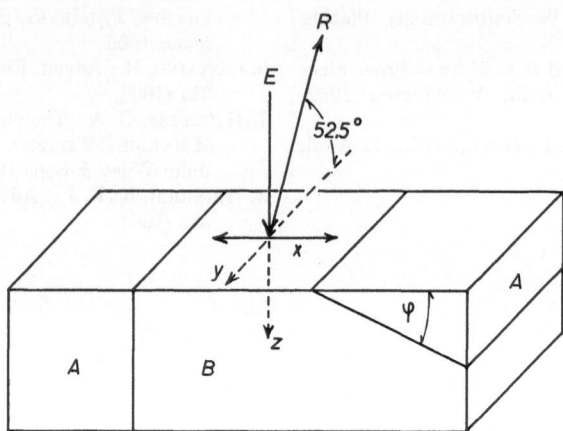

Fig. 1. Schematic drawing of a specimen with two interfaces A/B and B/A, the first being perpendicular and the second inclined to the free specimen surface

A in such a way that the interface A/B on the left side is perpendicular to the upper free specimen surface whereas the other interface B/A makes an angle φ ($\varphi \approx 3-5°$) with that surface. The specimen is positioned in the microanalyser in such a way that the direction z of the electron beam E and the direction y of the surface traces of the interfaces A/B and B/A are coplanar with the spectrometer circle and the X-ray beam R which is measured by the spectrometer; ϑ is the take-off angle.

The electron beam is deflected perpendicular to the interfaces A/B and B/A, that is along the coordinate x and the intensity $I(x)$ of the A-K_α-radiation is measured as a function of x across each interface. These $I(x)$-curves provide the main experimental information on which the analysis of the A-K_α-depth distribution is based.

3. Theoretical Principle of the Method

Theoretically the A-K_α depth distribution in a solid A-specimen is obtained from a function $\psi(x, y, z)\, \varrho\, dx\, dy\, dz$ (ϱ is the density of A) which describes the intensity of the A-K_α radiation produced in the specimen by primary electron excitation and to a less extent by fluorescence from the white X-ray radiation also produced in the specimen. The total intensity I_∞ of the X-ray beam R measured by the spectrometer is then:

$$I_\infty = C\, p_0\, \varrho \iiint dx\, dy\, dz\, \psi(x, y, z) \exp\{-\chi\, \varrho\, z\} \tag{1}$$

with

$$\chi = (\mu/\varrho)\, \mathrm{cosec}\, \vartheta$$

where C is some constant (depending e.g. on the experimental conditions), p_0 is the density of the specimen current, μ is the absorption coefficient of A-K_α-radiation and the integral is to be taken over the volume of the "excitation region", i.e. the region in which appreciable X-ray emission is excited. This region will be treated for simplicity as having a square cross-section b^2 with a size parameter b which is treated as being independent of z. Furthermore the distribution of X-rays along the coordinates x and y in each z-layer is replaced by an x, y-independant average value ϕ (which will be described in the usual manner as a function of ϱz, rather than of z); hence

$$\int\limits_0^b dx \int\limits_0^b dy\, \psi(x, y, z) = b^2 \phi(\varrho z). \tag{2}$$

For I_∞ then the usual expression follows from (1) and (2):

$$I_\infty = C\, p_0\, b^2 \int\limits_0^\infty \phi(\varrho z) \exp\{-\chi\, \varrho z\}\, \varrho\, dz \tag{3}$$

where $P_0 = p_0\, b^2$ is the specimen current; the integral will be briefly labelled as F_∞ in the following.

The choice of the element B in the specimen is governed by similar considerations to those governing the choice of the tracer element in the tracer method [1—3] (i.e. the atomic number of B should be as little as possible less than the atomic number of A, in order to avoid fluorescence from characteristic B-radiation and further to have almost the same behaviour of electrons in B as in A). Then the $I(x)$-curve from the wedge-shaped A-crystal can be analysed by the following, simplified geometrical considerations:

The position of the excitation region relative to the wedge-shaped A-crystal is described by the coordinate x, where $x = 0$, when this region just touches the tip of the wedge. Two situations than have to be treated: first the wedge penetrates partly into the region (see Fig. 2a) and second the wedge penetrates the region totally (see Fig. 2b). In each case the A-volume in the region will be treated as a rectangular parallelepiped, with the edges x, b and $x/2 \cdot \tan \varphi$ in the first case and b, b and $\left(x - \dfrac{b}{2}\right) \cdot \tan \varphi$ in the second case. Here, the slope of the bottom of the A-wedge is neglected, which is justified by the smallness of φ.

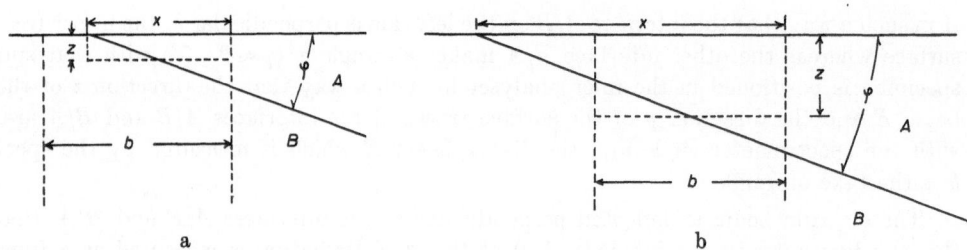

Fig. 2. Diagram of the geometrical relationship between the A-wedge and the volume in which X-ray are excited when the wedge penetrates this volume partly (a) or totally (b)

Then the A-K_α-intensity curve may be written

$$I(x) = C\, p_0\, x\, b \int_0^{\frac{x}{2} \tan \varphi} \phi(\varrho z) \exp\{-\chi \varrho z\}\, \varrho\, dz \tag{4a}$$

for $0 \leqq x \leqq b$ (see Fig. 2a), and

$$I(x) = C\, p_0\, b^2 \int_0^{\left(x - \frac{b}{2}\right) \tan \varphi} \phi(\varrho z) \exp\{-\chi \varrho z\}\, \varrho\, dz \tag{4b}$$

for $b \leqq x < \infty$ (see Fig. 2b).

In order to eliminate the unknown constant C Eqs. (4a) and (4b) are divided by (3); then by differentiating with respect to x it follows that

$$\frac{\phi(\varrho z) \exp\{-\chi \varrho z\}}{F_\infty} = \frac{2b}{x^2 \varrho \tan \varphi} \left[x\, \frac{d}{dx} \left(\frac{I(x)}{I_\infty} \right) - \frac{I(x)}{I_\infty} \right] \tag{5a}$$

for $0 \leqq x \leqq b$ with $z = \frac{x}{2} \tan \varphi$, and

$$\frac{\phi(\varrho z) \exp\{-\chi \varrho z\}}{F_\infty} = \frac{1}{\varrho \tan \varphi}\, \frac{d}{dx} \left(\frac{I(x)}{I_\infty} \right) \tag{5b}$$

for $b \leqq x < \infty$ with $z = \left(x - \frac{b}{2} \right) \tan \varphi$.

These two Eqs. (5a) and (5b) provide the basis for determining the depth distribution $\phi(\varrho z)$ from the experimental curves $I(x)$. However, it was found that Eq. (5a) could not be used with sufficient accuracy to determine $\phi(\varrho z)/F_\infty$. Therefore in practice $\phi(\varrho z)/F_\infty$ was determined mainly from Eq. (5b) in the range $b \leqq x < \infty$, that is in the range $\frac{b}{2} \cdot \tan \varphi \leqq z < \infty$. Eq. (5a) was used only to determine $\phi(0)/F_\infty$ at $x = z = 0$ in the following way: Since $I(0) = 0$ and $(dI/dx)_{x=0} = 0$, the first approximation of the $I(x)$-curves at low x-values is $I(x) \approx I_2 \cdot x^2$. Hence

$$\lim_{x \to 0} \frac{2b}{x^2 \varrho \tan \varphi} \left[x\, \frac{d}{dx} \left(\frac{I(x)}{I_\infty} \right) - \frac{I(x)}{I_\infty} \right] = \frac{2b\, I_2}{\varrho \tan \varphi\, I_\infty}$$

which gives

$$\phi(0)/F_\infty = 2b\, I_2/\varrho \tan \varphi \cdot I_\infty. \tag{6}$$

Since the value I_2/I_∞ can be obtained from the experimental $I(x)$-curves as described in the next section, $\phi(0)/F_\infty$ is determined by this procedure.

4. Application of the Method

In order to test the method described in this paper the depth distribution of Cu-K_α-radiation in pure Cu was measured and the results were compared with theoretical predictions [3, 4] and equivalent measurements by the tracer method [1]. The specimen used in this case was composed of $A = $ Cu and $B = $ Ni, and was prepared as follows.

A pure Ni-specimen was polished on two mutually perpendicular surfaces PQ and QR (see Fig. 3). This specimen was electroplated with Cu to a Cu layer thickness of about 0.5 mm (using a bath containing per liter 20 g Na_2CO_3, 8 g $NaHSO_4$, 20 g $(CH_3COO)_2Cu$, 12 g concentrated

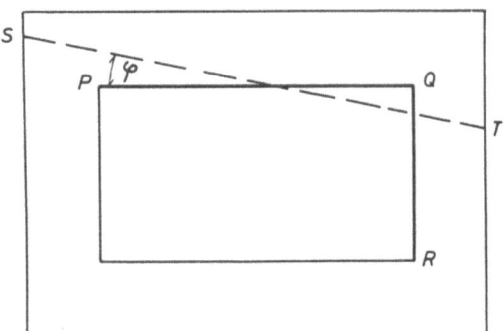

Fig. 3. Schematic representation of the successive stages in the preparation of the wedge specimen

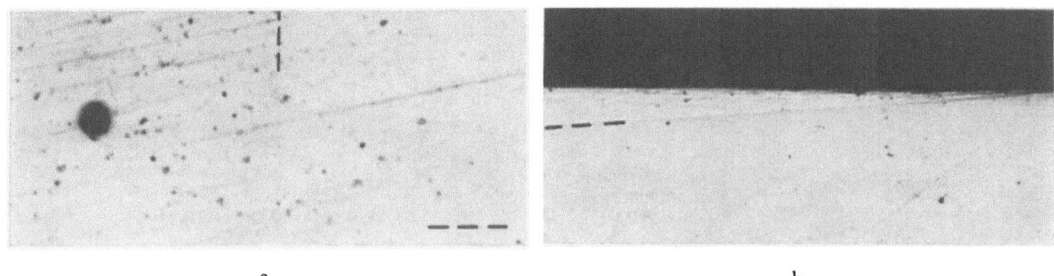

a   b

Fig. 4a and b. Optical micrographs of a Cu-wedge on Ni: a) free specimen surface with a contamination spot on the Cu-wedge (1,000:1); b) cross-section perpendicular to this surface at the tip of the wedge (200:1)

NH_4OH, 28 g KCN, 2—10 mg gelatine at a current density of 10 mA/cm². The specimen was then polished parallel to the plane ST which made a small angle φ to the plane PQ. In Fig. 3 ST indicated the final position of the specimen surface on which the measurements were made. Since the angle φ is small (3—5°) the interface QR between Ni and Cu is almost perpendicular to the specimen surface ST.

The specimen surface at the Ni/Cu interface of the wedge shaped Cu-crystal is shown in Fig. 4a, where the interface trace is marked by the perpendicular dotted line. Since the cross-section of the electron probe is treated as if it were square, the electron beam should not be astigmatic; this can be checked by means of a contamination spot. Such a spot is shown in Fig. 4a (left) on the Cu-wedge. After the $I(x)$-curves had been measured a cross-section of the specimen was observed almost along the line of measurement (see horizontal dotted line in Fig. 4a). This cross-section is shown in Fig. 4b, where the Cu-wedge is indicated by a dotted line. The value of $\tan\varphi$ determined from the original photographic plates was 0.073. In order to test the accuracy of the determination of $\tan\varphi$ the measurement was repeated on another cross-section about 1 mm from the first one; in this case the value obtained was $\tan\varphi = 0.075$. In the following analysis the average value of 0.074 has been used.

The value of b was determined from the $I(x)$-curve taken at the perpendicular Cu/Ni-interface by the approximate expression

$$0.6 \cdot b = (x_2 - x_1)$$

where $(x_2 - x_1)$ is deduced from the intensity values $I(x_1) = 0.2\, I_\infty$ and $I(x_2) = 0.8\, I_\infty$, where I_∞ is the total difference of Cu-K_α-intensities produced on pure Cu and on pure Ni.

An experimental $I(x)/I_\infty$-curve taken at the Cu-wedge at 30 kV, with $P_0 = 10$ nA and $b = 1.3\ \mu$m, is shown in Fig. 5a. For the purpose of analysis this curve is replaced by a smooth averaged curve as shown in Fig. 5b. Since the position of $x = 0$ on the x-coordinate could not be determined with sufficient accuracy by observing the specimen surface during the $I(x)$-measurements, this position was obtained by the following consideration: Since the first approximation of the $I(x)$-curve at small x-values is $I(x) \approx I_2 \cdot x^2$, several parabolic curves were

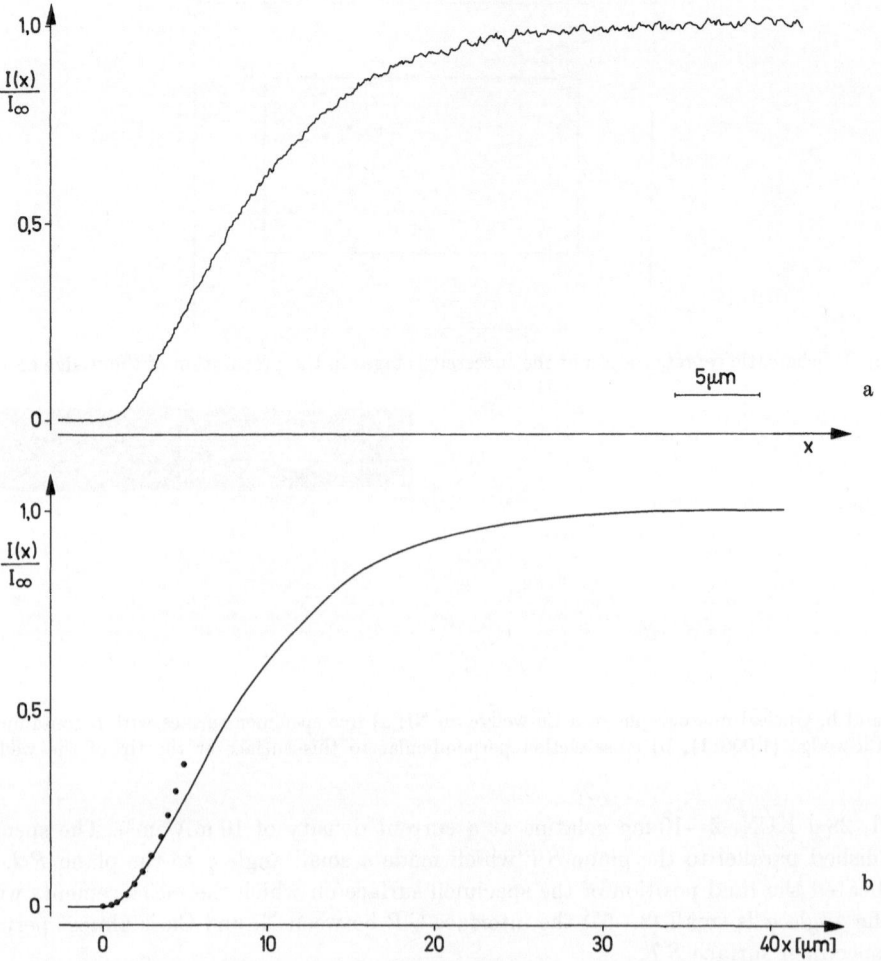

Fig. 5a and b. Cu-K_α-intensity $I(x)$ at 30 kV as a function of distance x perpendicular to the Cu-wedge/Ni junction expressed as a fraction of the intensity I_∞ of a thick Cu specimen: a) experimental curve; b) smooth averaged curve

drawn with different I_2-constants and compared with the experimental $I(x)$-curve. The curve with the best fit was selected (see dotted line in Fig. 5b) and from this curve the position $x = 0$ in the experimental curve was determined; it was found that the position $x = 0$ was not very sensitive to the value of the I_2-constant.

Table 1

Accelerating voltage of the electron beam [kV]	Specimen current P_0 [nA]	Average diameter of the excitation region b [μm]	Depth of 95% X-ray excitation z_{95} [μm]
30	5	1.35	1.55
30	10	1.3	1.55
30	30	1.6	1.55
25.5	5	1.1	1.30
25.5	10	1.45	1.30
25.5	30	1.5	1.30
20	5	1.7	0.90
20	10	1.6	1.00
20	50	2.35	0.90

Nine different $I(x)$-curves have been measured at the Cu-wedge on Ni under the experimental conditions summarized in Table 1. The maximum depth at which the electrons still have sufficient energy to excite X-ray emission may be determined as a function of the accelerating voltage directly from these curves. In Table 1, the measured values of z_{95}, the depth at which 95% of the total X-ray intensity has been excited, are listed in the last column for comparison.

From these nine $I(x)$-curves the functions $\phi(\varrho z) \cdot \exp\{-\chi\,\varrho z\}/F_\infty$ were obtained by graphical differentiation and the use of Eq. (5b) for $b \leq x < \infty$. The absorption factor $\exp\{-\chi\,\varrho z\}$ was calculated with the value $\vartheta = 52.5°$ and with the absorption coefficient of Cu-K_α-radiation in Cu which is $\mu/\varrho = 54$ cm²/g and was found to be almost equal to unity in this instance. Thus finally one obtains the functions $\phi(\varrho z)/F_\infty$, which are plotted with respect to the depth coordinate ϱz in Figs. 6a, 6b and 6c for 30 kV, 25.5 kV and 20 kV respectively. Further the value $\phi(0)/F_\infty$ is indicated in the figures, as determined from Eq. (6).

Since the depth distribution $\phi(\varrho z)/F_\infty$ is independent of the values of P_0 and b in Table 1, the $\phi(\varrho z)/F_\infty$-curves determined at the same acceleration voltage should also be same. Therefore the scatter of the experimental points in Fig. 6 is a direct measure of the uncertainty of this method. Average curves have been fitted to these points, as is shown by solid lines in Fig. 6.

In order to be able to compare these three curves with one another, and with previous results in the literature, it is necessary to normalise all the curves in the same way. From Eq. (3) it is seen that the measured K_α intensity on a thick specimen of the pure element is proportional to the instrument parameter C, the specimen current P_0 and the integral

$$F_\infty = \int_0^\infty \phi(\varrho z) \exp\{-\chi\,\varrho z\}\, \varrho\, dz.$$

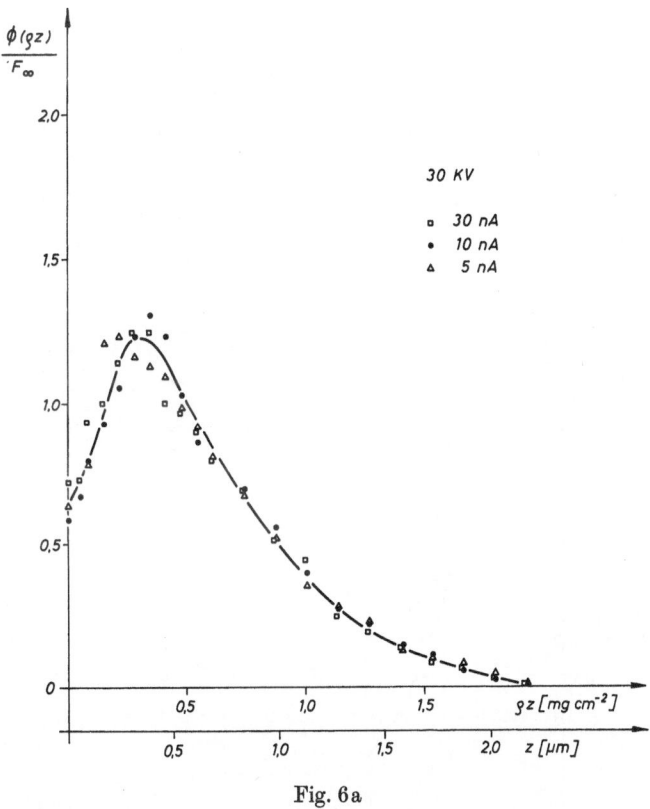

Fig. 6a

Fig. 6a—c. The depth distribution curves $\phi(\varrho z)/F_\infty$ in arbitrary units determined from the experimental $I(x)$-curves a at 30 kV, b at 25,5 kV, c at 20 kV. The measurements at different specimen currents (see Table 1) are indicated by different symbols as follows: triangle at 5 nA, filled circle at 10 nA, square at 30 nA, open circle at 50 nA

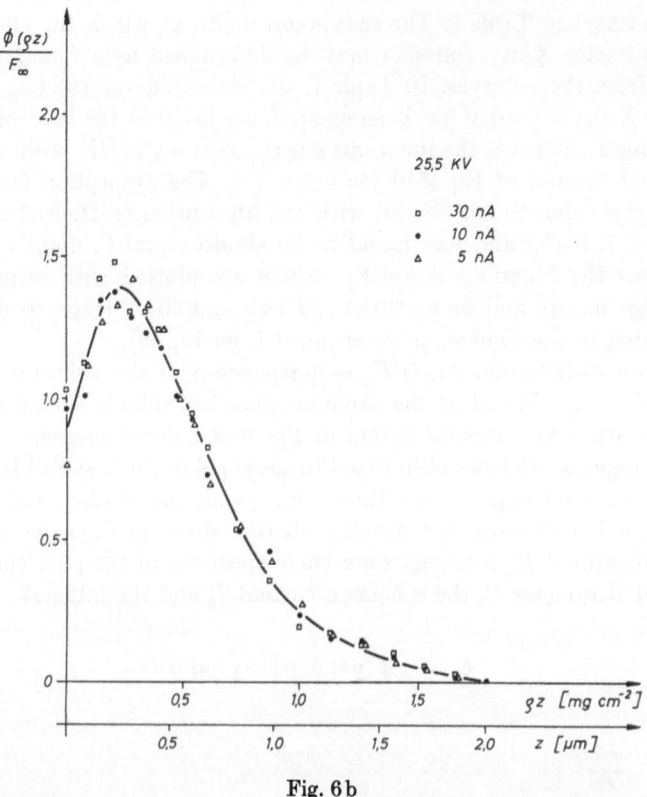

Fig. 6 b

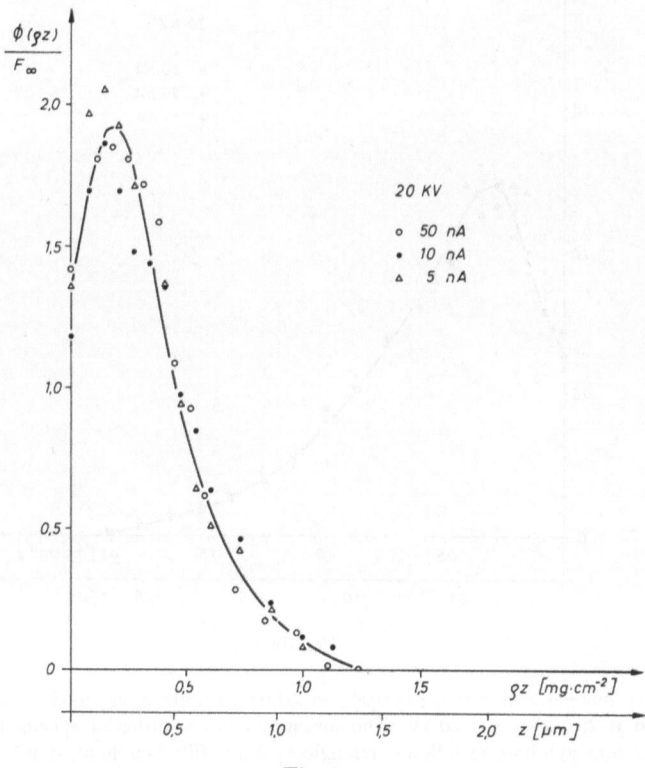

Fig. 6 c

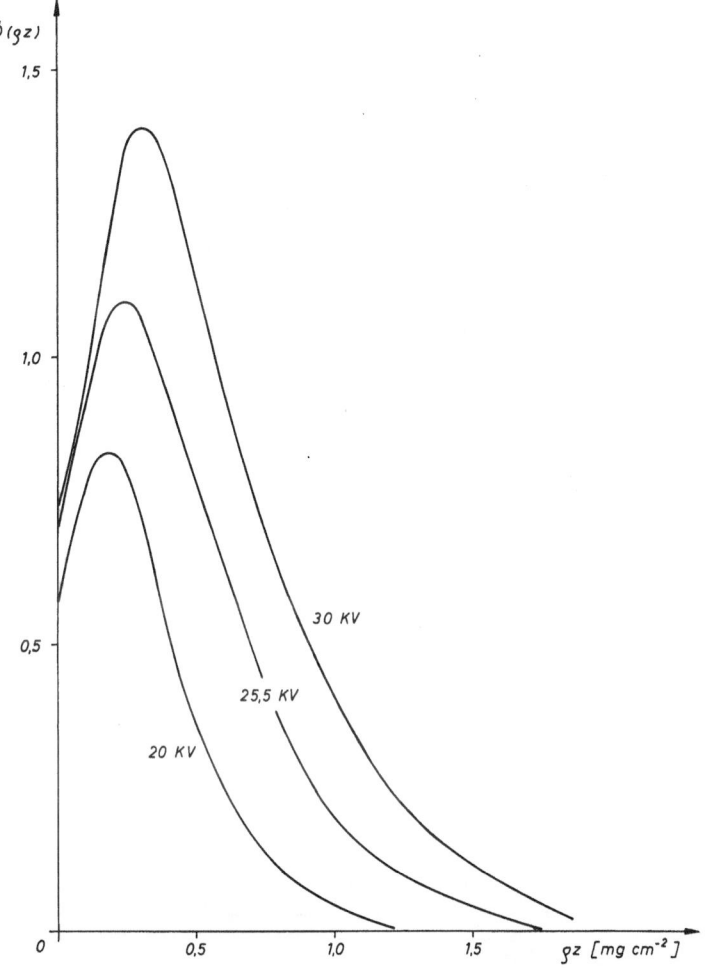

Fig. 7. The three experimental $\phi(\varrho z)$ curves of Fig. 6, after normalising in such a way that the integral F_∞ of Eq. (3) is proportional to the total K_α intensity per unit current measured on a thick copper specimen at the appropriate accelerating voltage (see Table 2)

If C is regarded as a purely geometrical constant, dependent only on the instrument and the state of alignment, the function $\phi(\varrho z)$ should be normalised in such a way that the value of F_∞ is proportional to the measured characteristic X-ray intensity per unit specimen current (I_∞/P_0), i.e. $C = I_\infty/P_0 F_\infty =$ constant. For the ARL instrument used in the present investigation, the values of I_∞/P_0 have been measured on a pure copper target (Cu-K_α radiation) for a series of accelerating voltages. The values obtained, after subtraction of the background intensity, which was measured on nickel, are listed in Table 2. With the help of these data, the three curves of Fig. 6 were normalised to a constant value of C. It was found that, in calculating the values of F_∞ from the $\phi(\varrho z)$-curves, the effect of the absorption factor $\exp(-\chi \varrho z)$ was so small in comparison to the experimental scatter that it could be neglected.

The three normalised $\phi(\varrho z)$ curves are plotted together for comparison in Fig. 7. In Fig. 8 the 30 kV curve is compared with the experimental results of CASTAING and DESCAMPS [1] and the theoretical curves due to BISHOP [4] and SHIMIZU, MURATA and SHINODA [3], all curves having been normalised as described above. Except for slight deviations for small values of z and in the tail of the curves, the agreement is good, particularly with the theoretical curve of BISHOP. The agreement of the other two curves with the literature was found to be equally good, allowing for a somewhat greater experimental scatter at lower accelerating voltages.

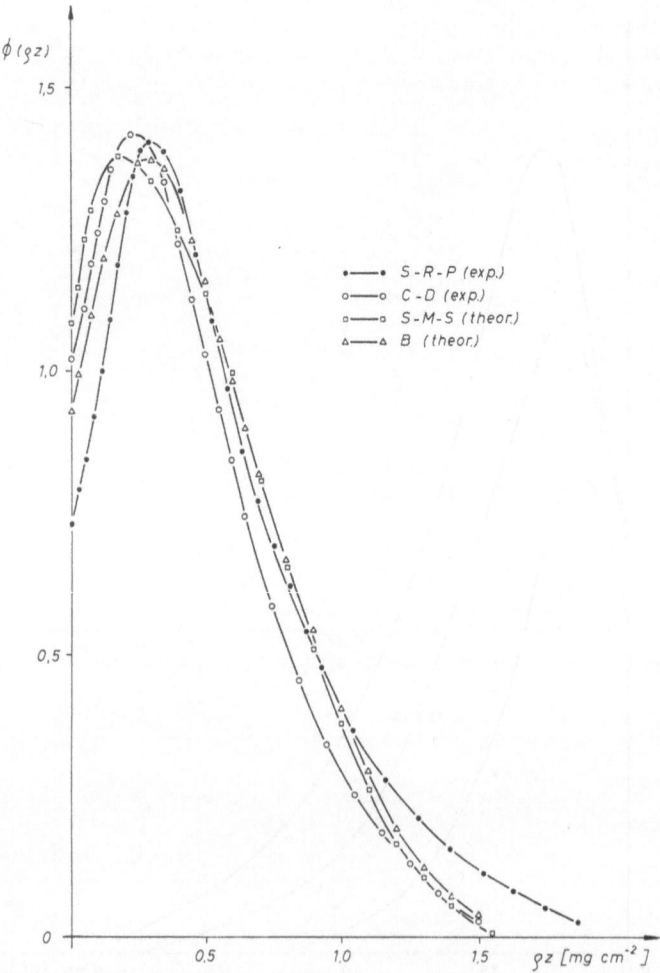

Fig. 8. Comparison of the results of the present investigation at 30 kV (S—R—P) with the measurements of Castaing and Descamps (C—D) and the theoretical curves derived by Bishop (B) and Shimizu et al. (S—M—S). The curve C—D was measured at 29 kV; all other curves refer to 30 kV

Table 2

Accelerating voltage [kV]	Cu-K_α intensities per unit specimen current [counts/sec nA]
30	1180
29	1080
25.5	810
20	450

5. Discussion

The validity of the simplifying assumptions made in the theoretical treatment of this method is basically confirmed by the comparatively small scatter of the experimental points in Fig. 6 and the good agreement with the other curves shown in Fig. 8. Further, the replacement of the actual excitation region (see Fig. 2) by a rectangular parallelepiped, which ignores the slope of the wedge over the distance b, was justified by calculating the $I(x)$-curve from the depth distribution curves of CASTAING and DESCAMPS [1] with and without this simplification. The two curves thus obtained could not be distinguished from one another.

Although the general features of the $\phi(\varrho z)$ curves obtained from the wedge curves are in good agreement with the literature, minor deviations are observed at small values of z and at greater depths in the tails of the curves. It is possible that these deviations may be due, at least in part, to systematic errors in the determination of the experimental conditions (accelerating voltage, wedge angle, density of the material etc.). These effects, and the whole question of the reproducibility of the curves, must therefore be investigated thoroughly before a decision can be made as to whether the observed deviations are significant or not. The results of this investigation will be reported in a future paper.

In conclusion it may be stated that the experimental technique described in the present paper provides a convenient, alternative method for determining the depth distribution function $\phi(\varrho z)$ of the production of characteristic X-rays in solid targets bombarded with electron beams. Some refinement of the experimental technique may, however, be necessary before the $\phi(\varrho z)$ curves so obtained may be regarded with confidence as quantitatively accurate in all respects.

Acknowledgement. Thanks are due to Dipl.-chem. H. HÄLBIG for experimental assistance and to the Landesamt für Forschung des Landes Nordrhein-Westfalen for financial support.

References

1. CASTAING, R., and J. DESCAMPS: J. Phys. Radium 16, 304—317 (1955).
2. —, and J. HENOC: X-ray optics and microanalysis, p. 120. Paris: Hermann 1966.
3. SHIMIZU, R., K. MURATA, and G. SHINODA: Reference [2], p. 127.
4. BISHOP, H. E.: Reference [2], p. 112.
5. GREEN, M.: Proc. Phys. Soc. (London) 82, 204 (1963).
6. BISHOP, H. E.: Proc. Phys. Soc. (London) 85, 855 (1965).
7. SHINODA, G.: Reference [2], p. 97.

Electron Probe Microanalysis. Quantitative Analysis

Etat actuel des méthodes quantitatives d'analyse par sonde électronique

J. PHILIBERT

IRSID — Saint-Germain-en-Laye, France

I. Introduction

Le document fondamental pour l'étude de l'analyse quantitative par émission X excitée par une sonde électronique demeure toujours la thèse de CASTAING [1]. En effet le traitement original que CASTAING présentait en 1952 constitue la base de tous les développements ultérieurs. De ceux-ci, qui se sont multipliés depuis quelques années, on a parfois tenté de dresser des arbres généalogiques, pour les classer en vue de mieux discuter leur validité [2, 3]. On peut s'interroger sur l'origine de tous ces traitements et sur les raisons qui ont provoqué la publication, parfois confidentielle! d'un tel nombre de méthodes de calculs.

L'opération essentielle depuis le travail de CASTAING reste la comparaison des intensités d'une même raie émise par l'échantillon et le témoin. Plaçons-nous dans le cas le plus simple où celui-ci est un élément pur A. Traditionnellement le rapport des intensités mesurées est écrit $k_A = I_A/I(A)$. Le problème consiste à relier ce rapport à la concentration locale de l'élément A, C_A (exprimée en masse %).

Pour résoudre ce problème, il faut pouvoir passer des intensités *mesurées*, aux intensités *engendrées* dans la cible, et plus précisément aux intensités engendrées *primaires*, c'est-à-dire dues aux ionisations par les électrons du faisceau et non par les rayonnements X produits (fig. 1).

La théorie doit alors établir la relation entre la concentration C_A et le rapport des intensités primaires engendrées. Dans la première approximation de CASTAING, on admet que cette relation est une égalité. Les calculs de correction sont alors limités aux effets d'absorption et de fluorescence; mais il apparut vite que ces corrections n'étaient pas suffisantes, autrement dit que la première approximation n'était pas en général valide. Ce fut le mérite particulier de POOLE et THOMAS en 1961 [4] de montrer sur un grand nombre d'analyses l'importance de la «correction de numéro atomique» ou «d'effet de (Z)». Ces dénominations quelque peu impropres, mais maintenant familières, ont pour origine la perte de validité de la première approximation lorsque les numéros atomiques des espèces constitutives sont trop différents.

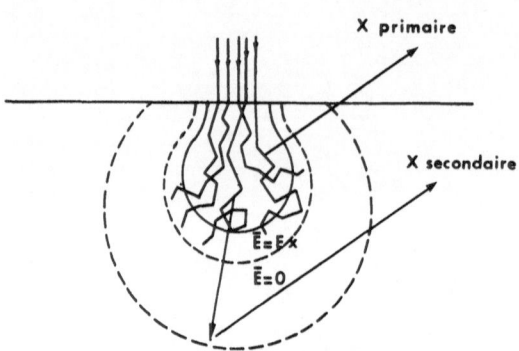

Fig. 1. Schéma des volumes excités pour les émissions primaire et secondaire

De ce point de vue, il y a donc lieu de distinguer soigneusement la correction d'effet de numéro atomique (Z) qui permet de relier concentration et intensités engendrées primaires, des corrections d'absorption (A) et de fluorescence (F) qui relient celles-ci aux intensités émergentes ou, à un facteur près, aux intensités mesurées. Cette distinction est schématisée par la séquence ZAF ou FAZ suivant le sens dans lequel on procède [5].

En fait toutes ces relations ne sont pas indépendantes, car elles reposent sur les mêmes bases physiques: la diffusion et la perte d'énergie des électrons incidents dans la cible. La connaissance de ces phénomènes doit permettre le calcul des intensités engendrées et de leur distribution en profondeur. Cette dernière détermine l'importance de l'absorption, et par conséquence de l'émission secondaire de fluorescence. C'est pourquoi nombre d'auteurs ont proposé des formules universelles dans lesquelles au moins les effets (Z) et (A) étaient bloqués. Mais cette approche ne paraît pas la meilleure tant sur le plan théorique que pratique:

1. Sur le plan théorique, toute théorie générale qui décrit la distribution énergétique et spatiale des électrons dans la cible, comme la méthode de Monte-Carlo par exemple, permet évidemment de calculer aussi bien les intensités engendrées que les intensités émergentes. Mais les facteurs de correction d'absorption $f(\chi)$ sont relativement insensibles à la forme exacte des courbes de distribution en profondeur des ionisations $\varphi(\varrho z)$, ce qui facilite le calcul de cette correction quel que soit le modèle utilisé. Mais celui-ci peut devenir insuffisant pour le calcul du nombre total d'ionisations, et par conséquent, de l'effet de numéro atomique. Or la mise au point de formules analytiques n'est possible qu'au prix de modèles très simplifiés, qui de ce fait ne sauraient prévoir l'effet de Z. D'*ailleurs* même avec les modèles les plus raffinés, la précision se semble pas encore tout à fait suffisante pour le calcul du nombre total d'ionisations (surtout à cause des effets de rétrodiffusion).

2. Sur le plan pratique, plusieurs raisons militent en faveur d'une séparation claire des effets (A) et (Z):

— la relation (Z) ne dépend que de l'énergie des électrons, du taux d'excitation et de l'angle d'incidence, alors que la relation (A) dépend en outre de l'angle d'émergence des rayonnements reçus dans les spectromètres.

— une formule «bloquée» ne révèle pas clairement l'importance relative des divers effets. Or celle-ci peut être fort variable; c'est en particulier le cas des rayonnements de très grande longueur d'onde: l'épaisseur de la couche d'où émerge le rayonnement mesuré est beaucoup plus faible que la pénétration des électrons, de sorte que les effets (Z) et (A) paraissent ici quasi-indépendants.

— une formule bloquée n'offre pas la meilleure voie d'optimisation des conditions expérimentales, c'est-à-dire de choix des conditions qui minimisent tel effet plutôt que tel autre (par exemple celui pour lequel l'incertitude prévue est la plus grande).

Il faut cependant noter qu'une formule «bloquée» ou universelle peut se révéler en pratique extrêmement utile, car sa simplicité permet une grande rapidité de calculs avec des moyens limités.

Les remarques précédentes ne s'appliquent pas aux modèles théoriques, tels que la méthode de Monte Carlo ou l'équation de transport de Boltzmann, qui donnent directement les intensités émergentes en fonction de la concentration. Mais ces modèles permettent a fortiori la décomposition de la relation globale en effet de Z et absorption.

On se trouve donc en présence de deux types d'approche, auxquels correspondent deux modes de calcul: l'un conduit à des formulations analytiques suivant les schémas ZAF, l'autre essentiellement numérique est fondé sur des modèles beaucoup plus fins.

Comme toutes ces théories reposent sur la connaissance des processus d'interaction des électrons et de la cible, il nous paraît utile de revoir brièvement les bases physiques de la microanalyse.

II. Pénétration et diffusion des électrons dans la cible

Lorsqu'un électron pénètre dans la cible, il peut subir plusieurs types d'interactions. Les plus importantes dans le domaine d'énergie qui nous intéresse (2—50 kV) sont la diffusion élastique par les noyaux atomiques et la diffusion inélastique par les électrons.

Par suite de ces diffusions la trajectoire affecte une forme de zigzag; à chaque interaction l'électron est dévié dans une certaine direction. Il s'agit donc d'un problème de mouvement aléatoire. En outre, au cours des interactions inélastiques les électrons perdent une certaine quantité d'énergie par ionisation ou excitation des électrons atomiques ou de conduction.

La diffusion élastique est caractérisée par sa section efficace. Des formules plus exactes que l'expression initiale de Rutherford ont été calculées. Le résultat essentiel est que la section efficace varie comme Z^2/E^2. Cette variation avec le carré du numéro atomique est très importante. Dans une cible de faible Z les trajectoires seront plus rectilignes que dans une cible de Z élevé. En outre, par suite de la variation en $1/E^2$, la trajectoire sera de plus en plus zigzagante au fur et à mesure de la pénétration de l'électron. D'où la forme du volume irradié, suivant le numéro atomique Z, forme que rappellent les classiques schémas de la fig. 2. Mais toutes les formules sont moins précises pour les valeurs élevées de Z, ce qui les rend en particulier impropres à prévoir correctement les processus de rétrodiffusion élastique.

Il est assez facile de tenir compte des diffusions *inélastiques* pour étudier la trajectoire électronique, sachant que la section efficace est à peu près Z fois plus petite que la précédente. L'importance de ce type de collision réside dans les pertes d'énergie de l'électron incident: comme il s'agit d'évènements aléatoires, après un parcours donné, les énergies des électrons sont distribuées statistiquement (dispersion ou «straggling»), mais par suite des fortes pertes d'énergie par ionisation des niveaux profonds K, L, la perte d'énergie moyenne est plus élevée que la perte la plus probable. Pour des raisons de commodité, comme les fortes pertes sont relativement rares, on néglige la dispersion en remplaçant ces évènements discrets et aléatoires par une perte continue caractérisée par la loi de décroissance de l'énergie moyenne E en fonction du parcours s. L'expression de dE/ds la plus usitée est celle de Bethe. Outre son caractère fondamentalement très approché, elle contient un terme dit de potentiel d'ionisation moyen (J) dont les valeurs ne sont pas très bien connues dans le domaine d'énergie qui nous intéresse.

L'intégration de la loi de Bethe permet de calculer le parcours moyen des électrons dans la cible, dit «parcours de Bethe», ou plus improprement «pénétration de Bethe» (Bethe's range)[1]. Celui-ci est proportionnel à E^2; Exprimé en termes de masse par unité de surface, il augmente avec le numéro atomique Z.

La pénétration des électrons dans la cible peut être évaluée de diverses manières, suivant, en gros, que l'on considère l'énergie moyenne ou l'énergie la plus probable des électrons transmis ou le nombre de ces électrons. La pénétration P varie peu sensiblement avec Z, mais elle croît avec l'énergie suivant une loi en E^n ($n \sim 1,5$ à 2). Aussi lorsque l'énergie des électrons incidents augmente, le parcours moyen et la pénétration augmentent en E_0^n alors que la section efficace de diffusion diminue comme E_0^{-2}; en d'autres termes plus l'électron peut pénétrer dans la cible

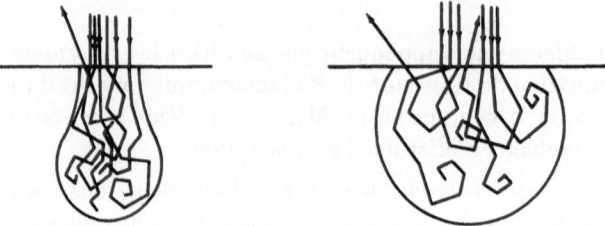

Fig. 2. Variation du volume excité avec le numéro atomique

---

1. Le parcours est mesuré le long de la trajectoire, alors que la pénétration est mesurée suivant la coordonnée z normale à la surface (profondeur). L'importance de la diffusion fait que, pour les électrons, ces deux quantités peuvent être différentes (cf. fig. 2).

moins il peut être diffusé. Il s'ensuit que la forme du volume irradié ne change pas de manière sensible avec l'énergie; mais ses dimensions croissent avec E_0; par contre sa forme est très sensible à Z (cf. fig. 2).

Ces brefs rappels nous permettent d'analyser rapidement les modèles théoriques de pénétration des électrons dans les cibles.

III. Modèles théoriques de pénétration diffuse des électrons

Une théorie complète permettrait le calcul de la distribution spatiale et énergétique des électrons, de leurs pertes d'énergie, ainsi que des ionisations. Le problème physique de base, à savoir la détermination de la distribution des électrons dans les trois directions de l'espace, n'a pas été résolu de manière rigoureuse, essentiellement par manque d'une théorie de la diffusion multiple; c'est pourquoi des modèles simplifiés ont été proposés. Il faut remarquer à ce sujet que les difficultés sont différentes selon que l'on s'intéresse à la totalité du rayonnement engendré, ou à sa distribution spatiale suivant les trois dimensions ou suivant la profondeur.

1. Modèle simple: trajectoires moyennes

Ce modèle permet de calculer très simplement le nombre d'ionisations par trajectoire dans un milieu infini. Soit ds l'élément de trajectoire, $\psi_A^X(E)$ la section efficace d'ionisation du niveau atomique $X(A)$ considéré; le nombre d'ionisations sera:

$$dn_X = n_A \psi_A^X(E)\, ds$$

où n_A est le nombre d'atomes par unité de volume, soit $N \varrho C_A / A$, si N est le nombre d'Avogadro, ϱ la densité, A la masse atomique; d'où:

$$dn_X = C_A \frac{N}{A} \psi_A^X(E)\, d\varrho\, s$$

en introduisant les distances en masse par cm². Pour toute la trajectoire, on trouve le nombre d'ionisations n_X en intégrant de E_0 à E_X, E_0 désignant l'énergie des électrons incidents:

$$n_X = C_A \frac{N}{A} \int_{E_0}^{E_X} \psi_A^X(E)\, d\varrho\, s.$$

Pour faire le calcul, il faut avoir une loi de correspondance énergie-parcours. Le modèle suppose cette loi universelle, en considérant une perte d'énergie moyenne continue, situation assez éloignée de la réalité comme nous l'avons signalé. On écrit alors:

$$n_X = C_A \frac{N}{A} \int_{E_0}^{E_X} \frac{\psi_A^X(E)}{dE/d(\varrho\, s)}\, dE.$$

En utilisant pour $dE/d\varrho s$ la loi de Bethe, pour ψ_A^X la section efficace d'ionisation de Bethe, l'intégrale se calcule bien pour un corps pur ou un alliage. Le résultat peut s'écrire, pour un électron:

$$n_X = C_A \frac{N}{A} \cdot \frac{1}{S_A}$$

où S est fonction de E_0 et des caractéristiques des niveaux atomiques $X(A)$.

En comparant témoin A pur et échantillon, il vient immédiatement:

$$k_A = \frac{n_X \text{ éch.}}{n_X \text{ témoin}} = C_A \frac{S(A)}{S_A}.$$

La première approximation n'est pas vérifiée. Il y a une correction due à la différence des pouvoirs de ralentissement des deux cibles, différence qui croît avec l'écart des numéros atomiques de celles-ci.

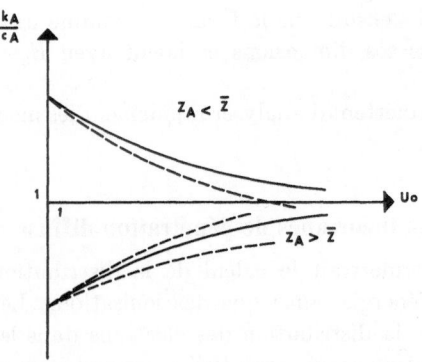

Fig. 3. Effet de numéro atomique seul (courbes en trait plein) et combiné avec l'absorption (courbes en tirets)

Indépendamment des hypothèses simplificatrices relatives à l'application de la loi de Bethe, qui peuvent être gênantes pour les éléments légers ou pour des échantillons minces, ce modèle en considérant des trajectoires dans un milieu infini, calcule un nombre d'ionisations trop élevé. En effet, dans une cible semi-infinie, par suite de processus de retrodiffusion élastique ou de diffusion multiple, une fraction des électrons ressort de la cible: ce sont les électrons rétrodiffusés, caractérisés par leur spectre énergétique $d\eta/dE$ et leur nombre total η.

Dans le modèle précédent on introduit un facteur correctif R pour tenir compte des pertes d'ionisations du fait de la rétrodiffusion d'électrons d'énergie supérieure à E_X en écrivant:

$$n_X = C_A \frac{N}{A} \cdot \frac{1}{S_A} \cdot R_A$$

d'où finalement:

$$k_A = C_A \frac{R_A/S_A}{R(A)/S(A)}.$$

Le facteur R doit alors être évalué par une autre voie. On a utilisé des moyens simplifiés à partir des distributions énergétiques expérimentales des électrons rétrodiffusés (voir par ex. P. M. Thomas [6]). On peut de manière plus élaborée appliquer le modèle des trajectoires moyennes à ces électrons et, connaissant leur distribution énergétique, calculer le nombre d'ionisations perdues $1-R$. Ces calculs ont été faits par P. Duncumb [7] d'une part, H. E. Bishop [8] d'autre part, soit à partir de distributions expérimentales, soit à partir de distributions calculées par la méthode de Monte-Carlo [9]. Il est très remarquable que les courbes ainsi calculées ($1-R$ en fonction de Z et de l'inverse du taux d'excitation $U_0 = E_0/E_X$) soient en accord raisonnable avec les déterminations expérimentales directes de Dérian et Castaing sur Al, Cu, Au [10, 11].

Une méthode différente a été utilisée par Brown [17]: il compare au nombre total d'ionisations prédit le nombre calculé théoriquement à l'aide de l'équation de transport (cf. plus bas), nombre égal à l'intégrale de la fonction $\varphi(\varrho z)$ calculée.

Il faut noter que l'effet de numéro atomique est d'autant plus faible que le taux d'excitation est plus élevé, par suite d'une compensation des effets de la loi de ralentissement et de la rétrodiffusion. Soit, par exemple, l'analyse d'un élément A dans un métal de base B, tel que $Z_A < Z_B$ (analyse de l'élément léger). Le témoin est A pur; S éch. $< S$ tém., mais R éch. $< R$ tém. Il y a donc une certaine compensation des facteurs du terme (R éch./R tém.). (S tém./S éch.), compensation d'autant meilleure que les pertes par rétrodiffusion sont plus importantes, c'est-à-dire l'excitation plus forte. Néanmoins le rapport des S l'emporte toujours sur le rapport des R, de sorte que l'élément léger est dosé par excès, l'élément lourd par défaut (fig. 3).

Mais le modèle des trajectoires moyennes ne permet pas le calcul de la distribution spatiale des pertes d'énergie ni des ionisations. Par suite de la diffusion multiple, il y a une différence énorme entre le parcours s et la pénétration z dans la cible et il n'existe pas de loi de diffusion en z satisfaisante pour un nombre de collisions élevé. Cependant des lois empiriques de perte d'énergie dE/dz ont été proposées pour l'énergie moyenne ou l'énergie la plus probable du faisceau d'électrons. Leur validité a été discutée de manière détaillée par Cosslett et Thomas [12, 13].

2. Modèles de trajectoires individuelles

Comme il serait trop compliqué, et impossible dans l'état actuel de nos connaissances théoriques, de reconstituer les trajectoires réelles d'un grand nombre d'électrons, on doit calculer des trajectoires individuelles virtuelles. Il s'agit là d'une méthode de simulation, dite de Monte-Carlo. La première tentative de ce genre fut celle de Archard et Mulvey (1962) [14] fondée sur un modèle très simple: les trajectoires sont réduites à deux segments rectilignes, une seule diffusion se produisant à la profondeur dite de «complète diffusion». Malgré ses simplifications drastiques, ce modèle donna des résultats très intéressants pour l'effet Z, la distribution en profondeur du rayonnement, le coefficient de rétrodiffusion et permit d'établir des courbes de correction pour les effets Z et A.

Des modèles plus réalistes ont été utilisés par la suite (Green [15], Bishop [9], Shinoda et col. [16]). Bishop par exemple considère 5000 trajectoires; chacune de celles-ci est constituée de 25 éléments rectilignes, égaux à 1/25e du parcours de Bethe. La liaison entre éléments successifs est calculée à l'aide d'une loi de diffusion multiple. Ce modèle est encore assez simplifié, puisque le nombre d'éléments de trajectoire est petit et constant quel que soit le parcours effectué, et que l'on utilise une loi de perte d'énergie continue, sans dispersion.

La méthode permet de calculer la distribution énergétique et spatiale des électrons $n(s, z)$ ou $n(E, z)$, en particulier les divers moments de ces distributions. Elle permet également de calculer la distribution des pertes d'énergie, le nombre total d'ionisations X et leur distribution en profondeur $\varphi(\varrho z)$. Malgré les simplifications du modèle, les précisions obtenues sont encourageantes, en particulier en ce qui concerne la rétrodiffusion ou les courbes $\varphi(\varrho z)$. La méthode permet en principe d'établir des courbes d'étalonnage donnant les valeurs des intensités engendrées en fonction de C_A et du taux d'excitation $U_0 = E_0/E_X$ (c'est-à-dire de déterminer l'effet de Z), ainsi que les valeurs de celles-ci en fonction des intensités émergentes (calcul de l'absorption par l'intermédiaire de $\varphi(\varrho z)$). Elle s'est révélée utile pour l'analyse des éléments très légers où les méthodes de correction habituelles ne s'appliquent pas: Bishop [8] a établi des courbes d'étalonnage k_A en fonction de E_0 pour quelques cas intéressants (Raie K du carbone dans SiC, ou Fe_3C). Mais bien entendu la méthode de Monte-Carlo ne fournit pas d'expression analytique des diverses corrections.

3. Modèle collectif

Par suite du grand nombre de diffusions qui se produisent le long de la trajectoire, la pénétration peut être considérée comme un mouvement brownien et traitée par les mêmes équations que la théorie cinétique des gaz, en particulier par l'équation de transport de Boltzmann. En général il n'est pas possible de résoudre cette équation dans le cas qui nous intéresse. Cependant Lewis a pu calculer les moments des distributions angulaires et spatiales des électrons dans une cible infinie, mais non dans la cible semi-infinie de l'expérience. Les expressions des moments permettent d'évaluer les fonctions de distribution spatiale et énergétique des électrons, ainsi que le taux de dissipation d'énergie avec la pénétration, toujours pour une cible infinie.

Cependant pour une cible semi-infinie, Bethe, Rose et Smith ont réussi à résoudre l'équation de Boltzmann à condition de négliger la diffusion aux grands angles et de supposer une loi continue de perte d'énergie. Leur équation simplifiée a été résolue numériquement par Brown et Ogilvie [17] à l'aide d'une méthode de différences finies. Ces auteurs ont ainsi pu calculer la distribution des électrons en fonction de leur parcours et de leur énergie. Ils en ont déduit les distributions d'électrons rétrodiffusés, le facteur R, ainsi que la loi de distribution spatiale des ionisations $\varphi(\varrho z)$. Le calcul a été raffiné par la suite.

Là encore la méthode ne fournit pas d'expression analytique pour les corrections. Elle donne des graphiques du facteur correctif R et des corrections d'absorption $f(\chi)$.

Un modèle collectif très approché a été utilisé par divers auteurs [18, 27—29] pour calculer la distribution $\varphi(\varrho z)$. Il se fonde sur une description collective de la pénétration diffuse des électrons. On sait qu'il y a lieu de distinguer, pour des profondeurs croissantes et suivant le nombre moyen de collisions subies, des régions dites de diffusions simple, plurale, multiple, complète. Les caractéristiques des deux premiers processus se laissent analyser assez facilement à l'aide de la section efficace de Rutherford. Mais à partir d'une vingtaine de diffusions élémentaires, il n'en va plus de même: la distribution angulaire des vitesses des électrons suit une loi d'allure gaussienne dont la théorie ne prévoit pas très bien l'écart-type (loi de Bothe). Pour des pénétrations plus grandes la diffusion est complète, les vitesses étant distribuées suivant une loi de Lambert. La loi de transmission des électrons, nombre d'électrons transmis par une épaisseur z, est dans ces conditions fonction exponentielle de z (loi de Lenard), l'angle le plus probable de déviation des vitesses est maximal et constant. Nous renvoyons le lecteur aux publications de Cosslett et Thomas [13] pour la discussion de ces théories et leur comparaison aux résultats expérimentaux.

Dans le modèle collectif simplifié (cf. Philibert [18]) dit «modèle exponentiel», on admet que la loi de Lenard est valable à toutes pénétrations, et que le taux de diffusion augmente exponentiellement avec la profondeur z jusqu'à la limite de complète diffusion. On calcule ainsi le nombre d'électrons en fonction de la profondeur. Comme on ne connaît pas la loi de perte d'énergie moyenne en fonction de cette profondeur, on suppose que la section efficace d'ionisation est indépendante de l'énergie, ce qui constitue une approximation passable pour des énergies $E > 1,5 E_X$. Le modèle permet de calculer des fonctions $\varphi(\varrho z)$, qui renferment trois paramètres, ajustables par comparaison avec les courbes expérimentales. Il est très facile d'en déduire la correction d'absorption. Certains auteurs ont cru pouvoir en tirer également l'effet de numéro atomique, mais ceci ne nous paraît pas très correct du fait de la structure même du modèle. En effet notre mauvaise connaissance des lois de la pénétration et de la diffusion des électrons ne permet pas un calcul rigoureux de $\varphi(\varrho z)$. L'accord des lois théoriques avec l'expérience paraît souvent fortuit, car ces lois ne se vérifient jamais sur un large domaine de numéro atomique ou d'énergie. Un calcul plus rigoureux combinerait la fonction d'absorption différentielle des électrons avec la profondeur z et une loi de perte d'énergie moyenne pour obtenir le taux de dissipation d'énergie en fonction de z et finalement le nombre d'ionisations. Dans l'état actuel de nos connaissances et du développement des méthodes de calcul sur machine, cette approche ne paraît guère prometteuse.

IV. Méthodes de calcul de correction d'effet de numéro atomique et d'absorption

En dépit de leur caractère distinct nous traiterons ensemble ces deux corrections, puisqu'elles ont souvent été associées, renvoyant la correction de fluorescence à un dernier paragraphe.

1. Méthodes ZAF

Dans celles-ci les corrections (Z) et (A) sont effectuées séparément. Le traitement général est grosso modo le même pour toutes le méthodes.

a) — Correction (Z)

α) — Méthode de Poole et Thomas [3, 6]

Poole et Thomas partent de la relation:

$$k_A = C_A \frac{R_A/S_A}{R(A)/S(A)}.$$

Ils écrivent $R(A)/S(A) = \alpha_A$ et ils supposent pour le calcul relatif à l'échantillon que:

$$R_A/S_A = C_A \alpha_A + C_B \alpha_B.$$

De la sorte le calcul est réduit aux termes R/S relatifs aux éléments purs. R a été évalué en fonction de Z et du taux d'excitation U_0 à partir de données expérimentales; ce qui a permis d'établir une famille de courbes $R(Z, U_0)$. S est pris dans les tables de Nelms pour Z et $E = (E_0 + E_X)/2$.

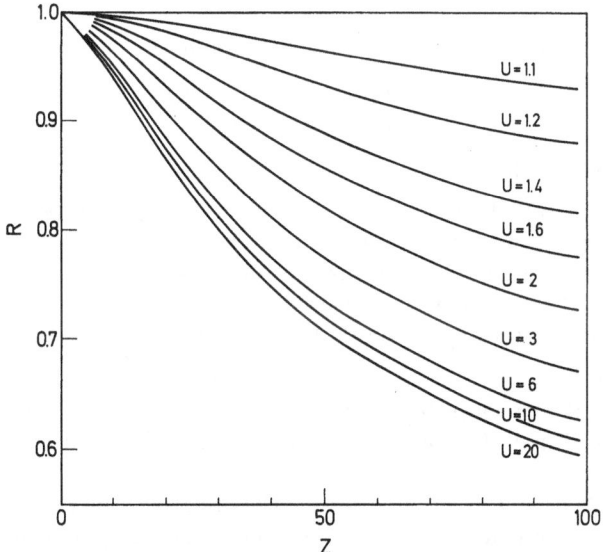

Fig. 4. Facteur de rétrodifusion R, d'après Duncumb

Rappelons que les tables de NELMS [19] sont établies à partir de la loi de ralentissement de BETHE. Les hypothèses ainsi faites ne sont justifiées que par un argument heuristique.

β) — *Méthode de* DUNCUMB *et al.*

Divers auteurs DUNCUMB, REED, DA CASA [7, 21], BISHOP [8], TIXIER et PHILIBERT [20] ont proposé une application plus rigoureuse des équations de base.

Le facteur R a été évalué à partir des distributions énergétiques normalisées des électrons rétrodiffusés, et les résultats sont en accord satisfaisant avec les déterminations expérimentales directes de DERIAN et CASTAING (cf. plus haut). Les phénomènes de rétrodiffusion étant relativement insensibles à l'énergie incidente E_0, R ne dépend que de deux paramètres Z et U_0 et peut, de ce fait, être facilement tabulé (fig. 4).

Pour un alliage, on suppose que

$$R_A = \sum C_i \, R(i)$$

les valeurs $R(i)$ étant prises pour les éléments purs, mais pour le taux d'excitation U_0 de l'élément A analysé. Cette approximation semble justifiée dans les cas les plus défavorables par un calcul direct à l'aide de la méthode de Monte-Carlo [BISHOP [8]].

Le terme S peut être évalué directement à partir de la loi de Bethe, en faisant le calcul pour une valeur moyenne de l'énergie $E = \frac{1}{2}(E_0 + E_X)$, approximation n'entraînant pas d'erreur supérieure à 1% selon DUNCUMB et REED et en prenant pour l'échantillon: $S_A = \sum_i C_i \, S(i)$.

Cependant TIXIER et PHILIBERT [20] ont montré que S pouvait être calculé par intégration analytique et ils ont fourni les expressions de ce terme.

Les principales incertitudes de cette méthode ont deux origines:

— rétrodiffusion: R_A devrait varier avec E_0 pour des cibles de numéro atomique élevé, alors que l'on suppose seulement une variation avec U_0 (sans distinguer les ionisations K ou L). L'interpolation linéaire pour un alliage est mal justifiée et n'a été que peu vérifiée;

— loi de ralentissement. Outre le problème de la validité de la loi de Bethe, subsiste l'incertitude sur le terme J, potentiel moyen d'ionisation. On prend souvent $J = 11{,}5\,Z$, mais DUNCUMB et DA CASA [21] ont préféré déterminer une loi $J(Z)$ empirique de façon à optimiser les précisions sur les concentrations calculées.

b) — *Correction* (A)

La correction d'absorption s'exprime sous la forme simple

$$k_A = \left[\frac{I_A}{I(A)}\right]_{\text{engendré}} \cdot \frac{f(\chi)\ \text{éch.}}{f(\chi)\ \text{tém.}}$$

où χ a sa signification habituelle, $\chi = (\mu/\varrho)\ \operatorname{cosec}\theta$.

Les fonctions $f(\chi)$ sont différentes pour l'échantillon et le témoin par suite des différences de distribution en profondeur $\varphi(\varrho z)$ du rayonnement engendré.

Il est possible d'utiliser des courbes $f(\chi)$ expérimentales mais leur nombre est limité et l'interpolation peu commode, d'autant plus que pour un niveau excité X de l'élément analysé A, la courbe $f(\chi)$ varie suivant la composition de l'alliage. D'où diverses méthodes:

— *méthodes approchées*

Brown et Ogilvie [17] ne considèrent pour un alliage AB que les courbes limites relatives à l'excitation de A dans A pur et B pur, de B dans A pur et B pur et effectuent pour l'alliage une interpolation linéaire.

Poole et Thomas [3, 6] n'utilisent que les courbes relatives au témoin (calculées à l'aide de la formule de Philibert pour $F(\chi)$, voir ci-dessous), mais ils prennent pour l'alliage un χ efficace:

$$\chi_{A\,\text{eff.}} = \frac{\chi_A}{\Sigma\, C_i\, \dfrac{\alpha_{iA}}{\alpha_{AA}}} \cdot$$

— *méthode de* Philibert-Duncumb-Shields

A l'aide de l'expression analytique de $\varphi(\varrho z)$ fondée sur le modèle exponentiel, il est facile de calculer sa transformée de Laplace $F(\chi)$ [15]. Cette expression peut être simplifiée pour des raisons purement pratiques afin de ne contenir que deux paramètres:

$$F(\chi) = \frac{1}{\left(1+\dfrac{\chi}{\sigma}\right)\left[1+h\left(1+\dfrac{\chi}{\sigma}\right)\right]}$$

on en déduit la fonction d'absorption:

$$f(\chi) = \frac{F(\chi)}{F(0)} = \frac{1}{\left(1+\dfrac{\chi}{\sigma}\right)\left(1+\dfrac{h}{1+h}\,\dfrac{\chi}{\sigma}\right)}$$

avec $h = 1,2\,A/Z^2$ et σ un «coefficient de Lenard» qui varie avec E_0. Il y a lieu de ne pas confondre les deux fonctions F et f. Seule $f(\chi)$, rapport des intensités émergente et engendrée, corrige l'absorption pure. L'importance de celle-ci dépend évidemment du numéro atomique Z (facteur h de la formule), puisqu'elle est fonction de la distribution en profondeur de l'émission, qui varie avec Z. Par contre $F(\chi)$ tient compte d'un effet de pénétration diffuse, dont le lien avec la correction d'effet de Z n'apparaît pas très clairement. Il importe à ce sujet de distinguer les différences de facteurs de ralentissement (c'est-à-dire de parcours) qui conditionnent les émissions totales (facteur S des formules), et les différences de pénétration et de profondeur d'émission dont dépend l'absorption.

La formule ci-dessus ne rendant pas compte de la valeur finie $f(\chi) = 1$, lorsque $E_0 \to E_X$, Duncumb et Shields [22] ont proposé une loi empirique pour σ:

$$\sigma = \frac{\sigma_0}{E_0^n - E_X^n}$$

avec $n = 1,5$ et $\sigma_0 = 2,39 \cdot 10^5$ (E en keV). Les valeurs de $f(\chi)$ ont été tabulées par Adler et Goldstein [23]. Ces expressions de σ et de $f(\chi)$ rendent bien compte de la variation de $f(\chi)$ pour une cible donnée suivant la nature de l'élément excité (cf. fig. 5).

D'autres auteurs ont proposé diverses modifications de la formule initiale. Par exemple, pour la variation de σ, Heinrich [24] choisit $n = 1,65$ et $\sigma_0 = 4,5 \cdot 10^5$.

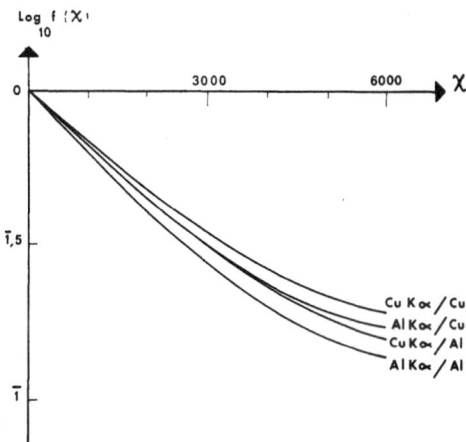

Fig. 5. Courbes $f(\chi)$ limites dans le système Al—Cu

Pour un alliage on utilise la même formule avec une valeur moyenne de $h : \bar{h} = \sum c_i h_i$, bien que cette relation n'ait pas reçu de justification satisfaisante. Enfin la formule ne s'applique plus aux très fortes absorptions $\left(f(\chi) < 0,2 \right)$ et de toute façon il est conseillé de choisir des conditions opératoires telles que $f(\chi) > 0,5$, ou même 0,8 [39].

2. Méthodes simplifiées

Différents auteurs se sont efforcés de simplifier les expressions précédentes pour faciliter le calcul. C'est ainsi que BELK [25] partant des formules de POOLE et THOMAS et de PHILIBERT écrit:

$$C_A = k_A \left[1 + \frac{Z_A - \bar{Z}}{100} \right] \left[1 + \frac{\overline{\mu/\varrho} - (\mu/\varrho)_A}{P + (\mu/\varrho)_A} \right]$$

où $P = \sigma \sin \theta (1 + h)/(1 + 2h)$; $\bar{Z}$ est le numéro atomique moyen.

Dès que $Z > 40$, un Z efficace doit être utilisé. La formule de BELK n'est applicable que si l'effet d'absorption est faible (conditions précisées dans la publication de BELK [25]).

Une autre expression simplifiée a été proposée par ZIELBOLD et OGILVIE [26]. Ces auteurs remarquent que, pour des alliages binaires $A\,B$, une représentation linéaire à l'aide d'un paramètre α peut être utilisée:

$$\frac{1 - k_A}{k_A} = \alpha \, \frac{1 - C_A}{C_A}$$

avec

$$\alpha = 0,95 \left(\frac{\sigma + \chi_B}{\sigma + \chi_A} \right) \left(\frac{Z_A}{Z_B} \right)^{0,28}.$$

La méthode a l'avantage de traiter aisément les ternaires connaissant les coefficients α des binaires.

3. Méthodes combinées

Par méthodes combinées nous entendons celles qui reposent sur un traitement où les effets Z et A sont analysés à l'aide du même modèle, en l'espèce le modèle collectif simplifié ou exponentiel.

Ces calculs ont été développés par TONG [27] et par THEISEN [28]; ils conduisent à une expression voisine de celle de PHILIBERT, mais sans la limiter à la seule correction d'absorption. Dans une première étape ces auteurs ont calculé une expression identique, aux notations près, à $F(\chi)$ donnée ci-dessus, avec $h = 3,5\,A/Z^2$ et $\sigma = \sigma_0/V^2$.

Plus récemment Theisen [29] a reconsidéré son expression et posé:

$$k_A = C_A \frac{R_A}{R(A)} \frac{f(\chi)\,\text{éch.}}{f(\chi)\,\text{tém.}} .$$

Le $f(\chi)$ est semblable à celui de Philibert, avec un coefficient dit le Lenard

$$\sigma = \frac{8{,}9 \cdot 10^5}{(E_0 - E_X)^2}$$

et

$$h = 1{,}72 \cdot 10^{-6} E_0^2 \cdot A/Z^2 .$$

Cette expression tient compte de l'effet de rétrodiffusion R (calculé d'après Thomas), mais semble négliger l'effet de ralentissement des électrons en ne considérant que le facteur d'absorption pure $f(\chi)$.

De son côté Tong [30] a développé, à la suite de calculs plus raffinés sur le modèle collectif, une expression de $F(\chi)$ universelle:

$$F(\chi) = \frac{1}{Z/A} \left[\frac{1}{1 + \dfrac{\chi}{\sigma}} - \frac{1 - \phi(0)/K}{1 + \dfrac{\chi}{\sigma} + \dfrac{Z}{m}} \right] .$$

Dans un alliage on calcule des moyennes pondérées par les C_i pour Z/A, Z et χ. Le coefficient de Lenard σ varie avec $(Z/A)\, E_0^{-1{,}7}$ $K = 4$, $m = 7$ à 5 kV et 10 à 20 kV.

V. Correction de fluorescence

Comme l'on sait il y a lieu de distinguer les émissions secondaires excitées par le spectre continu et par des raies caractéristiques. La première se produit dans tous les objets, alors que la seconde ne concerne que les cibles composées de plusieurs éléments. Le rayonnement secondaire est lui-même absorbé et le rapport entre les intensités émergentes et engendrées dépend de la distribution en profondeur de cette dernière.

Les formules de correction utilisées en pratique bloquent les deux effets: émission de fluorescence et absorption du rayonnement secondaire, de sorte qu'elles s'expriment en termes d'intensités secondaires émergentes I_f^A rapportées à l'intensité primaire émergente I_A — d'où le sens de la séquence FAZ quand on doit «remonter» des intensités émergentes globales, aux intensités primaires émergentes par la correction de fluorescence, puis aux intensités primaires engendrées par la correction d'absorption, et enfin à la concentration par la correction d'effet de numéro atomique.

1. Fluorescence par raie caractéristique

Cette émission avait été calculée par Castaing, moyennant certaines approximations sur $\varphi(\varrho z)$. Diverses évaluations (Wittry [31], Duncumb et Shields [32]) ont montré que des expressions analytiques plus exactes que la simple exponentielle choisie initialement n'affectaient pas sensiblement le résultat du calcul. En effet la profondeur moyenne d'émission secondaire est plus élevée que celle de l'émission directe, d'où une absorption plus forte qui rend inutile une meilleure approximation sur $\varphi(\varrho z)$. On trouve ainsi, avec les notations traditionnelles, pour la fluorescence de A excitée par une raie de l'élément B:

$$\frac{I_f^A}{I^A} = \frac{1}{2}\,\omega_A \cdot z^A \left(\frac{r-1}{r} \right)_A \cdot C_A \cdot \frac{\mu_B^A}{\bar{\mu}_B} \cdot \left[\frac{\log(1+u)}{u} + \frac{\log(1+v)}{v} \right] .$$

$$\left(\frac{f_B(\chi^A)}{f_A(\chi^A)} \right)_{\text{éch.}} \cdot \left(\frac{I_B}{I_A} \right)_{\text{engendré}}$$

où les fonctions f sont relatives aux raies A et B dans l'échantillon, et I^A et I^B sont les intensités engendrées dans l'échantillon. Ce rapport peut être calculé directement, suivant un mode très voisin de celui choisi pour l'effet de numéro atomique [44]. Moyennant certaines approximations sur les fonctions $f(\chi)$ et sur les sections efficaces d'ionisation, on déduit aisément la formule originale de Castaing, ou sa variante suggérée par REED [33].

Cependant la formule ci-dessus a une signification physique plus immédiate et se prête mieux aux cas plus complexes que la fluorescence $K(B) \to K_\alpha(A)$ (par exemple $K \to K_\beta$, ou $K \to L$, $L \to L$, ...). Les calculs leur paraissant quelque peu lourds, divers auteurs ont proposé des formules simplifiées [26, 34].

2. Fluorescence par le spectre continu

Cette émission s'effectue dans un volume encore plus grand que l'émission de fluorescence précédente et que, a fortiori, l'émission directe. Aussi peut elle être calculée en admettant une distribution de l'émission de spectre continu très simple (source réduite à un point à la surface). Pour les mêmes raisons, la profondeur moyenne de l'émission secondaire est grande, et l'absorption élevée. De ce fait la correction est généralement faible dans une cible homogène. Une formule de correction a été établie par HÉNOC [36]; elle devrait être plus largement utilisée car il n'est pas certain que la correction soit toujours négligeable.

Remarque

Les corrections de fluorescence, en particulier celle de spectre continu, peuvent être très importantes dans le cas de cibles non homogènes, par exemple de petites phases précipitées dans une matrice. L'effet est bien mis en évidence par l'étude de placages A/B dits «couples non diffusés» (voir par exemple MAURICE, HÉNOC [37, 44]).

VI. Pratique et test des calculs

Après ce bref rappel des principales méthodes de calcul, (un grand nombre de variantes n'ont pu être citées) examinons maintenant leur application pratique.

1. Pratique des calculs

Dans de nombreux cas plusieurs corrections sont nécessaires, sinon toutes. Il est maintenant courant d'utiliser à cet effet un ordinateur. Plusieurs raisons justifient cette pratique. Outre la rapidité des calculs, il faut signaler:

— la précision nécessaire. Notons par exemple que la formule de correction d'absorption nécessite une bonne précision des calculs, que souvent ne fournit pas la règle;

— l'utilisation de méthodes de correction les plus exactes. L'usage de formules simplifiées n'apparaît plus guère justifié;

— l'itération. En effet, les formules de correction s'écrivent d'une manière générale:

$$C_A = k_A \cdot F(C_i)$$

où les C_i sont les concentrations inconnues des divers éléments.

Une première approximation consiste à prendre pour C_i les valeurs mesurées k_i, mais en général la grandeur des corrections rend une ou plusieurs itérations nécessaires.

Afin d'accélérer la convergence des calculs, diverses méthodes d'itération ont été essayées. La plus adaptée semble être celle de WEGSTEIN [cf. 38].

De nombreux programmes de calculs de correction ont déjà été publiés; cependant nombre de ces programmes sont fort incomplets ou utilisent des méthodes de correction certainement peu exactes. Ils diffèrent en outre par la manière suivant laquelle sont introduits les paramètres physiques nécessaires tels que coefficients d'absorption, énergies d'ionisation, etc. Parfois des sous-programmes sont utilisés pour calculer ces paramètres, afin d'éviter de mettre toutes ces données en mémoire, ou d'avoir à les entrer à chaque calcul.

Tous les nouveaux développements dans le domaine des calculs de correction devront tenir compte de l'utilisation de l'ordinateur pour la pratique courante. Les différents groupes de «microsondeurs» ont un rôle utile à jouer pour la mise au point et la communication des programmes.

Il importe de souligner ici l'impact de l'ordinateur sur la pratique des mesures. Il peut sembler trivial de noter que le calcul sur machines n'a de sens qu'en fonction de la précision des données de base. D'où, en particulier, la nécessité d'un contrôle strict de la précision statistique des mesures des intensités, l'augmentation du nombre des mesures qui implique une automatisation de la technique expérimentale (déplacement automatique de l'échantillon, sorties sur ruban perforé ou sur cartes), le choix des conditions optimales pour minimiser les corrections, toutes conditions qui peuvent exiger une certaine reconversion de l'opérateur.

2. Précision des calculs

Nous n'insisterons pas sur la précision relative, c'est-à-dire celle des valeurs expérimentales des intensités, ni sur toutes les causes d'erreurs accidentelles, systématiques ou statistiques qui peuvent affecter les mesures. La précision absolue dépend par contre des calculs de correction. Il faut tout d'abord signaler l'importance des indéterminations sur les corrections dues aux erreurs affectant les paramètres utilisés pour effectuer les calculs, tout spécialement les coefficients d'absorption, les rendements de fluorescence. La propagation des erreurs a été particulièrement dénoncée par Heinrich [39]: selon cet auteur la correction d'absorption ne pourrait plus être calculée dès que $f(\chi) < 0,8$ si une précision de 1% est demandée sur la concentration. Un gros effort a été fait depuis quelque temps pour mesurer des coefficients d'absorption (la microsonde n'est pas l'appareil le mieux adapté à ce genre de mesures, notons-le en passant), pour systématiser les valeurs expérimentales et établir des tables de valeurs cohérentes. Signalons les tables de Heinrich [40], Vollath et Theisen [41], Frazer [42].

3. Test des calculs

De très nombreux chercheurs ont voulu vérifier et comparer les diverses méthodes de correction. Les difficultés sont nombreuses; il faudrait en effet trouver des cas où une seule correction devrait être appliquée. Si ceci est possible pour l'absorption ou la fluorescence, il n'en va pas de même pour l'effet de numéro atomique. On ne peut étudier qu'un ensemble de corrections ($Z + A$, par exemple) et des effets de compensation entre celles-ci sont toujours possibles; a l'intérieur d'une seule correction d'ailleurs, de tels effets de compensation sont probables soit entre les divers termes qui la composent, soit dans le rapport échantillon/témoin. Il en résulte que les résultats pratiques sont généralement meilleurs que ne le ferait croire la fragilité des bases théoriques des formules utilisées.

Comme souvent les valeurs vraies des compositions chimiques ne sont pas connues, certains auteurs ont utilisé le test $\Sigma C_i = 100\%$. Il s'agit là d'une condition nécessaire, mais nullement suffisante, toujours du fait d'éventuelles compensations. Ce test est particulièrement insuffisant dans le cas de mélanges d'oxydes, l'oxygène n'étant pas dosé (ou du moins pas dosé avec une précision équivalente à celle des autres éléments), de sorte que des hypothèses doivent être faites sur le degré d'oxydation des éléments métalliques.

Poole et Thomas ont développé les premiers la méthode des histogrammes en se fondant sur les analyses de 150 alliages [6], puis plus récemment de 229 [3]. Un grand nombre de méthodes de correction ont été appliquées à ces alliages, et les fréquences des écarts relatifs aux concentrations «vraies» ont été comparées à l'aide d'histogrammes (fig. 6). Cette méthode a l'avantage de considérer un grand nombre d'alliages et de noyer ainsi les cas particuliers et certaines erreurs systématiques dans la dispersion statistique. Elle permet de chiffrer la qualité de la méthode par l'écart de la moyenne à la valeur vraie et par la dispersion statistique des résultats. Elle peut mettre en évidence l'insuffisance de telle ou telle correction si l'on compare à l'ensemble des résultats les cas où l'absorption est faible, ceux relatifs à l'analyse de l'élément lourd, ou léger,

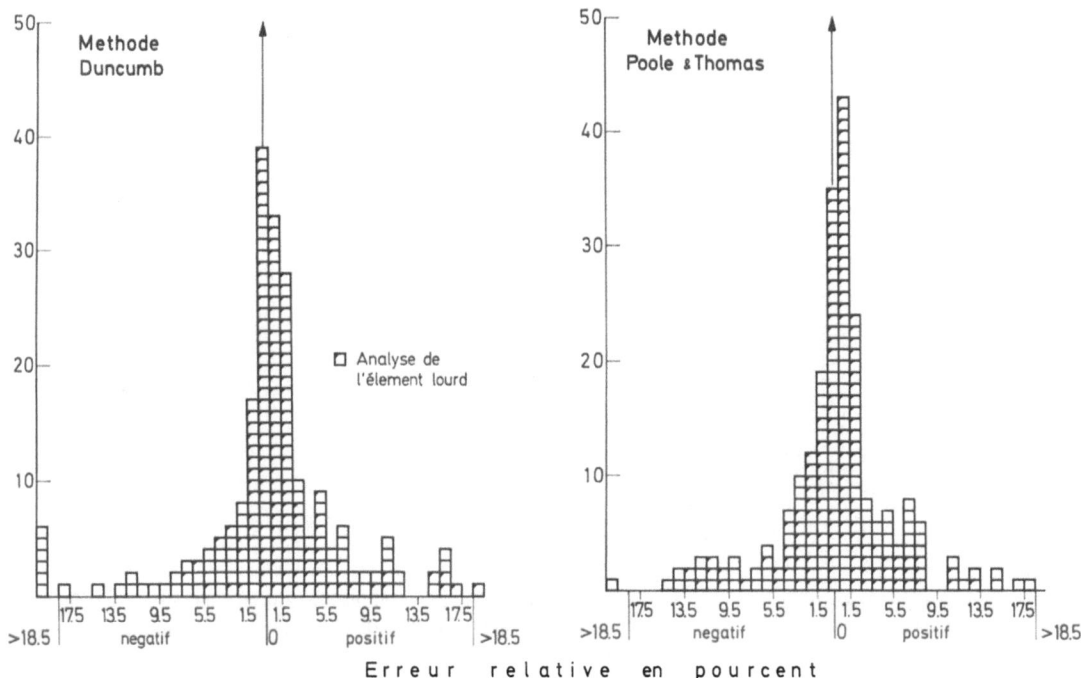

Fig. 6. Histogrammes établis par Poole après calculs de correction sur 229 cas

etc. ... les dissymétries observées décèlent alors l'existence de sur-ou sous-corrections et indiquent quelle correction semble déficiente. Le tableau suivant emprunté à Poole [3] montre le résultat de ce test:

Tableau 1

Méthode	Pourcentage de résultats tombant dans l'intervalle d'erreur $\pm 2,5\%$	
	150 alliages	229 alliages
Archard et Mulvey	22	—
Theisen 1	27	—
Theisen 2	< 27	—
Birks	13	—
Ziebold et Ogilvie	37	39
Poole et Thomas	55	61
Belk	42	47
Duncumb — Da Casa	53	55

Les calculs n'ont été effectués sur les 79 alliages supplémentaires que pour les 4 meilleures méthodes.

Dans une étude statistique portant sur 48 alliages, choisis pour la faiblesse des effets d'absorption et de fluorescence (correction $< 2\%$, analyse d'un élément lourd dans une matrice légère), Duncumb et col. [7] trouvent 81% des résultats avec une erreur à $\pm 1,5\%$. Pour les numéros atomiques moyens, le facteur correctif varie de 0,75 à 1,35, et l'erreur sur la concentration ne saurait dépasser 3% dans les pires cas. La précision semble moins bonne aux forts numéros atomiques. Dans une comparaison portant sur 240 alliages, dont 59 présentaient une

Methode Philibert–Tixier J=11.5 Z

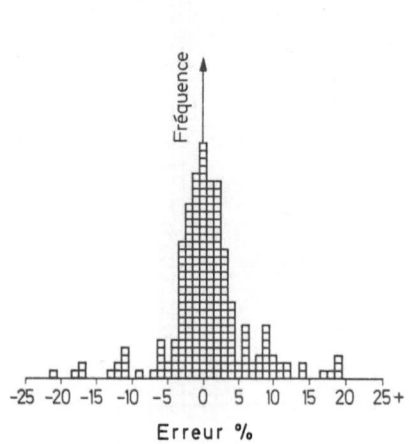

Methode Duncumb–Da Casa avec J ajusté

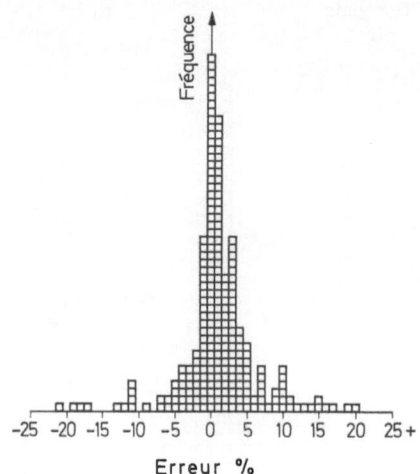

Methode Philibert–Tixier avec le J de Duncumb–Da Casa

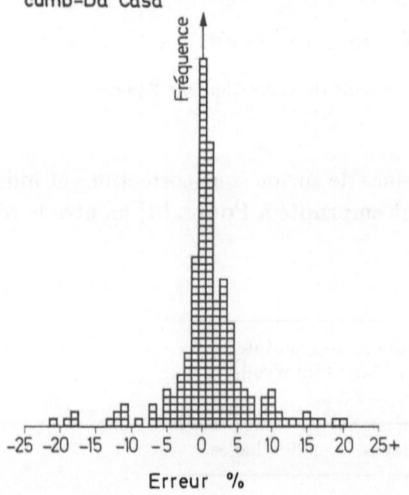

Fig. 7. Histogrammes portant sur 240 alliages, influence du calcul de S et de J dans l'effet de numéro atomique

Tableau 2

Méthode	Pourcentage de résultats tombant dans l'intervalle d'erreur	
	$\pm 2,5\%$	$\pm 5\%$
I	56,22	77,68
II	56,48	78,66
III	57,32	79,07

très faible correction d'absorption (en gris sur la fig. 7), les mêmes auteurs ont comparé la méthode de Duncumb et Da Casa (I), la méthode de Tixier, Philibert avec $J = 11,5Z$ (II) ou le J proposé par Duncumb et Da Casa (III) (fig. 7 et Tableau 2).

Les différences ne semblent pas significatives. Le succès égal de ces variantes et de la méthode de Poole et Thomas est un nouveau signe de la compensation de diverses erreurs intrinsèques, soit entre les divers termes des formules, soit dans le rapport échantillon/témoin (voir par exemple Duncumb et Reed [7]).

A titre d'exemple pratique, nous indiquons le résultat d'un test circulaire effectué en 1968 parmi 25 utilisateurs du groupe français. L'échantillon était un alliage de fer contenant un peu moins de 1% d'aluminium. Les mesures devaient être effectuées d'une part dans des conditions imposées (10 kV, 50 nA), d'autre part dans les conditions jugées optimales par l'opérateur. Si les valeurs expérimentales (dans les conditions fixées) révèlent un accord satisfaisant entre les divers laboratoires, il n'en va plus de même pour les résultats corrigés (fig. 8). Les causes de cette dispersion sont à rechercher dans le mauvais choix ou de la méthode de calcul, ou des divers paramètres utilisés (coefficients d'absorption, etc.).

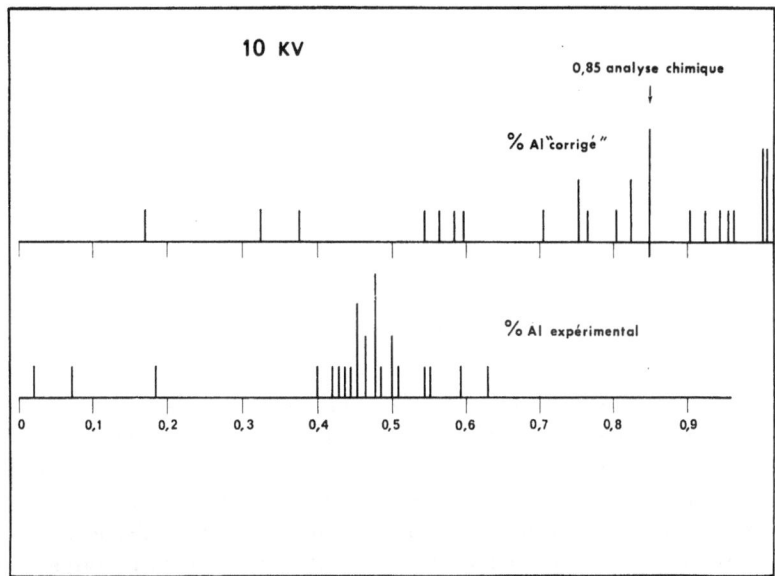

Fig. 8. Résultats d'un test circulaire parmi 25 utilisateurs français

4. Cas particuliers

a) Oxydes. Ce type d'échantillons est fréquent en métallurgie (inclusions non métalliques, laitiers, réfractaires ...) et en minéralogie. Les histogrammes dont nous avons parlé comportent un certain nombre d'oxydes simples, et ceux-ci ne se manifestent pas par de plus forts écarts que les métaux. Il semble cependant que certains oxydes les plus isolants donnent lieu à des résultats moins précis (variation de J avec l'état chimique ? charge électrostatique ?), mais que ce défaut disparaisse dans le cas d'échantillons complexes si les témoins utilisés sont des oxydes simples, tels que le quartz, le corindon, le périclase, ou un peu plus complexes comme la wollastonite [45].

b) Rayons X mous. Dans le domaine des très grandes longueurs d'onde, l'absorption devient considérable de sorte que seul le rayonnement engendré dans une couche superficielle émerge de la cible. La formule de correction d'absorption n'est plus valable car ses hypothèses de base (fonction d'ionisation constante, $\varphi(0) = 1$) sont trop grossières. D'autre part la correction de numéro atomique devient également imprécise aux faibles énergies utilisées, les valeurs de R n'étant pas calculables puisque la rétrodiffusion n'est pas connue aux faibles énergies. Deux méthodes ont été utilisées:

— méthode de Monte-Carlo: les calculs ont été effectués par BISHOP [8] pour la raie $K_\alpha C$ dans SiC et Fe_3C entre 5 et 25 kV et l'accord satisfaisant avec l'expérience démontre l'utilité du modèle dans ce domaine.

— méthode de MELFORD-DUNCUMB [43] dans l'approximation de la couche superficielle (très forte absorption), ces auteurs ont montré que les corrections Z et A peuvent être bloquées:

$$k_A = C_A \frac{\varphi(0)_{\text{éch.}}}{\varphi(0)_{\text{témoin}}} \cdot \frac{\chi_{\text{témoin}}}{\chi_{\text{éch.}}} .$$

Les valeurs de $\varphi(0)$ peuvent être déterminées expérimentalement ou calculées soit par la méthode de Monte-Carlo, soit, comme le facteur R, à partir des distributions énergétiques d'électrons rétrodiffusés.

5. Échantillons minces

C'est le cas des lames minces ou des répliques avec extraction préparées pour la microscopie électronique. Comme il n'est pas pratique ou possible de préparer des témoins minces de même masse superficielle que les échantillons, on est conduit à utiliser les témoins massifs habituels.

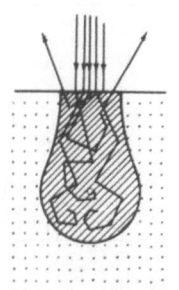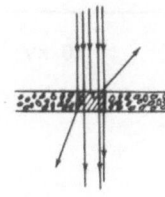

Fig. 9. Volume excité dans un échantillon massif et une lame mince

La masse analysée n'étant pas connue, il faut comparer les émissions de deux éléments au même point. Pour l'échantillon seul intervient le rapport des sections efficaces d'ionisation des deux éléments, mais les effets (Z) et (A) y sont en première approximation négligeables alors qu'ils sont forts pour les témoins (fig. 9). Les corrections sont donc de signe inverse du signe habituel caractéristique des échantillons massifs. Des méthodes très voisines ont été développées par Tixier et Philibert [20] et par Duncumb et col. [21]. On a, pour le couple d'éléments $A B$:

$$\frac{k_A}{k_B} = \frac{C_A}{C_B} \cdot \frac{P_{(A)}}{P_{(B)}} \cdot \frac{f(\chi)_{(A)}}{f(\chi)_{(B)}}.$$

Les $f(\chi)$ sont calculées pour A et B purs à la manière habituelle. Des formules très simples permettent le calcul de $P(A)$ et $P(B)$.

Conclusion

Les méthodes de calcul de corrections ont connu depuis quelques années un développement tel que leur précision et leur généralité ont été nettement améliorées. En toute rigueur les meilleures approches se fondent sur la méthode de Monte-Carlo, sur l'équation de transport, ou sur d'autres modèles statistiques. Malheureusement l'insuffisance de notre connaissance des lois de base ne permet pas de les utiliser aussi efficacement que le permettrait leur puissance. En outre la complexité des calculs nécessite la libre disposition d'un très gros calculateur et le prix de revient du calcul semble encore, pour le moment, prohibitif. Aussi, à l'heure actuelle, la méthode la plus simple et la plus économique paraît-elle être fondée sur l'évaluation séparée des effets de numéro atomique, d'absorption et de fluorescence (ZAF). Cependant cette situation pourrait évoluer rapidement dans les prochaines années. Quoi qu'il en soit les conditions opératoires (excitation spécialement) doivent être optimisées, de façon à réduire l'importance des corrections sans nuire à la précision statistique des mesures. Il serait souhaitable que de nouvelles études expérimentales et théoriques fussent entreprises afin de compléter notre connaissance des paramètres physiques avec une précision telle que l'erreur absolue sur le calcul de correction reste, de leur fait, inférieure à 1 % ; citons: la loi de ralentissement, la distribution énergétique des électrons rétrodiffusés, les sections efficaces d'ionisation, les coefficients d'absorption, les rendements de fluorescence.

Bibliographie

1. Castaing, R.: Thèse — Publication Onera No 55 (1952).
2. Duncumb, P., and P. K. Shields: Brit. J. Appl. Phys. **14**, 617 (1963).
3. Poole, D. M.: N B S Seminar, Washington 1967.
4. —, and P. M. Thomas: J. Inst. Metals **90**, 228 (1961—62).
5. Philibert, J.: Metaux Corrosion **40** (1964), Publication IRSID No B51. — J. Microscopie **6**, 889 (1967).
6. Thomas, P. M., and D. M. Poole: The electron microprobe, p. 269. New York: John Wiley & Sons 1966.
7. Duncumb, P., and S. J. B. Reed: NBS Seminar, Washington 1967.
8. Bishop, H. E.: Brit. J. Appl. Phys. Sér. 2, 1, 673 (1968).
9. — Brit. J. Appl. Phys. **18**, 703 (1967).
10. Derian, J. C., et R. Castaing: Optique des rayons X et microanalyse, p. 193. Paris: Hermann 1966.

11. DERIAN, J. C., et R. CASTAING: Thèse, Rapport CEA 1966, R 3052.
12. COSSLETT, V. E.: Optique des rayons X et microanalyse, p. 85. Paris: Hermann 1966.
13. —, and R. N. THOMAS: Brit. J. Appl. Phys. **15**, 235, 883, 1283 (1964); **16**, 779 (1965).
14. ARCHARD, G. D., and T. MULVEY: X ray optics and microanalysis, p. 393. Stanford: Academic Press 1963.
15. GREEN, M.: Proc. Phys. Soc. (London) **82**, 204 (1963).
16. SHINODA, G.: Optique des rayons X et microanalyse, p. 97. Paris: Hermann 1966. — SHIMIZU, R., K. MURATA, and G. SHINODA: Optique des rayons X et microanalyse, p. 127. Paris: Hermann 1966.
17. BROWN, D. B., and R. E. OGILVIE: J. Appl. Phys. **37**, 4429 (1966). — Optique des rayons X et microanalyse, p. 139. Paris: Hermann 1966.
18. PHILIBERT, J.: X ray optics and microanalysis, p. 379. Stanford: Academic Press 1963.
19. NELMS, A. T.: NBS Circular No 577 (1956) et supplément (1958).
20. PHILIBERT, J., and R. TIXIER: Brit. J. Appl. Phys., Sér. 2, **1**, 685 (1968).
21. DUNCUMB, P., et K. DA CASA: communication particulière 1967.
22. —, and P. K. SHIELDS: The electron microprobe, p. 284. New York: John Wiley & Sons 1966.
23. ADLER, I., et J. GOLDSTEIN: Optique des rayons X et microanalyse, p. 210. Paris: Hermann 1966.
24. HEINRICH, K. F. J.: Second conf. electron microprobe analysis, Boston 1967.
25. BELK, J. A.: Optique des rayons X et microanalyse, p. 214. Paris: Hermann 1966.
26. ZIEBOLD, T. O., and R. E. OGILVIE: Ann. Chem. **35**, 621 (1963); **36**, 322 (1964).
27. TONG, M.: Non publié.
28. THEISEN, R.: Rapport Euratom (1961), I—1.
29. — Quantitative Electron Probe Microanalysis. Berlin-Heidelberg-New York: Springer 1965.
30. ROUBEROL, J. M., M. TONG, and C. CONTY: Rev. GAMS No. 4, 334 (1966).
31. WITTRY, D.: Thèse Cal. Tech. Pasadena 1967.
32. DUNCUMB, P., and P. K. SHIELDS: X ray optics and microanalysis, p. 329. Stanford: Academic Press 1963.
33. REED, S. J. B.: Brit. J. Appl. Phys. **16**, 913 (1965).
34. WITTRY, D.: Advances X ray analysis. Plenum Press **7**, 395 (1964).
35. COLBY, J. W., et D. K. CONLEY: Optique des rayons X et microanalyse, p. 263. Paris: Hermann 1966.
36. HENOC, J.: Rapport CNET, Paris 1962, PCM 655.
37. MAURICE, F., R. SEGUIN et J. HENOC: Optique des rayons X et microanalyse, p. 357. Paris: Hermann 1966.
38. REED, S. J. B., and P. K. MASON: Second Conf. Electron Microprobe Analysis, Boston 1967.
39. HEINRICH, K. F. J.: N.B.S. Seminar 1967.
40. — The electron microprobe, p. 296. New York: John Wiley & Sons 1966.
41. THEISEN, R., and D. VOLLATH: Table of X-ray mass attenuation coefficients. Düsseldorf: Stahleisen 1967.
42. FRAZER, J. Z.: A computer fit to mass absorption coefficient data. University of California 1967, SIO Ref. No. 67—29.
43. DUNCUMB, P., et D. MELFORD: Optique des rayons X et microanalyse, p. 240. Paris: Hermann 1966.
44. HENOC, J., et F. MAURICE: Communication à ce congrès 1968.
45. SWEATMAN, T. R., et J. V. P. LONG: Communication à ce congrès 1968.

Présentation d'un programme de calcul sur ordinateur des diverses corrections à appliquer aux analyses à la microsonde électronique

C. Zeller et F. Zeller

C.N.R.S., Laboratoire de Minéralogie et de Cristallographie, Faculté des Sciences, 54 — Nancy — France
I.N.R.A., Station de Biométrie, 54 — Champenoux — France

Résumé

Dans le but d'écrire un programme de calcul sur ordinateur des diverses corrections à appliquer aux analyses à la microsonde électronique, les auteurs ont été amenés à mettre ces calculs sous une forme nouvelle dans laquelle les inconnus C_i (concentrations vraies) sont nettement mises en évidence. Des expressions analytiques classiques ont été employées sauf en ce qui concerne le facteur de rétrodiffusion R dont la formule est originale. Ce formalisme de calcul, vérifié expérimentalement, permet d'effectuer les corrections d'analyses d'échantillons contenant jusqu'à 12 éléments, avec une convergence rapide vers la solution.

1. Introduction

En microanalyse par émission X on écrit d'habitude que l'intensité caractéristique émise par un élément A d'une anticathode de composition donnée est de la forme :

$$I_A^{K_\alpha} = C_A \frac{R}{S} (1+F) f(\chi) = C_A X_A \tag{1}$$

où C_A est la concentration de l'élément considéré dans l'anticathode. Rappelons que R est le facteur de rétrodiffusion, $1/S$ le nombre de photons engendrés par électron incident perdant toute son énergie dans la cible, F le renforcement relatif de la raie analysée par effet de fluorescence et $f(\chi)$ l'effet d'autoabsorption. Ces quatre fonctions dépendent de la concentration respective des différents éléments constituant l'anticathode, de leur nature et des conditions expérimentales de l'analyse. Chaque variable dont dépendent ces fonctions est une combinaison linéaire des concentrations de la forme :

$$M = \sum C_j M_j. \tag{2}$$

Il ne faut pas oublier que le but de la microanalyse est de déterminer ces concentrations. Dans le cas d'échantillons complexes (3 éléments et plus) le calcul des relations entre les variables devient rapidement inextricable et lors de la résolution numérique il peut se produire des instabilités, voire des divergences, dans les solutions. C'est pour cette raison que nous avons cherché une approximation valable qui permette de séparer les variables C_i pour des échantillons comportant plus de deux éléments.

2. Formulation du problème

Dans un travail précédent (Zeller, 1968) nous avons montré que l'on peut mettre la grandeur X définie par (1) sous la forme d'une moyenne du type :

$$X_i^r = \sum_{j=1}^n C_j X_{ij}^r \tag{3}$$

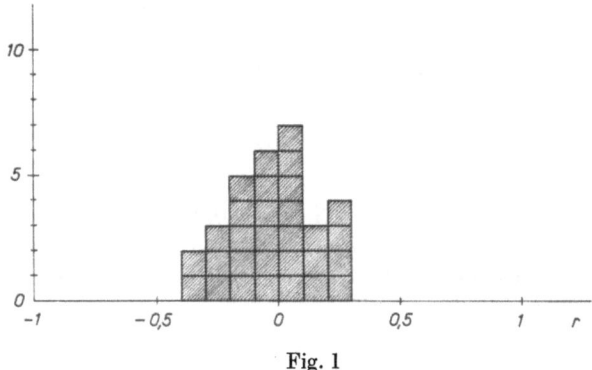

Fig. 1

où r est une constante comprise entre -1 et $+1$. i étant l'indice de l'élément émetteur, la somme est effectuée sur les n composants j de l'anticathode, y compris l'émetteur. Les X_{ij} sont définies tout naturellement et leurs valeurs sont celles de X pour des échantillons binaires dans lesquelles la concentration C_i de l'élément émetteur tend vers zéro, la concentration C_j de l'élément absorbant tendant évidemment vers l'unité.

Les formules analytiques employées pour calculer $1/S$, F et $f(\chi)$ sont classiques; par contre celle du facteur de rétrodiffusion R est originale (cf. Annexe): $1/S$ s'obtient par intégration du rapport de la section efficace d'ionisation et du ralentissement des électrons. F est la correction de fluorescence de Castaing employée sous la forme donnée dans un travail précédent (ZELLER, 1966). $f(\chi)$ est la correction d'autoabsorption de Philibert avec le coefficient de LENARD modifié par DUNCUMB et SHIELDS.

Nous avons alors comparé les valeurs de X calculées et mesurées pour 30 substances analysées à la microsonde et dont nous possédions l'analyse chimique. Ce sont des oxydes, des silicates et des sulfures contenant 3 éléments ou plus. Pour chaque échantillon nous avons calculé la valeur de r pour laquelle la formule (3) est vérifiée au mieux. L'histogramme de la fig. 1 montre que r est pratiquement constant et que sa valeur moyenne est peu différente de zéro. Ceci nous conduit à admettre pour X une moyenne logarithmique. En effet si l'on fait un développement limité de la formule (3) et que l'on fasse tendre r vers zéro on obtient:

$$\mathrm{Log}\, X_i = \sum_{j=1}^{n} C_j \, \mathrm{Log}\, X_{ij}. \tag{4}$$

3. Pratique des calculs de correction

On calcule au moyen de la formule (4) la valeur de X_i relative au témoin, soit X_i^*, en effectuant la somme sur tous les éléments constitutifs de ce témoin. En incluant X_{ij}, X_i^* et C_i^* la concentration en élément i du témoin dans un même facteur G_{ij} soit:

$$G_{ij} = \frac{X_{ij}}{C_i^* X_i^*} \tag{5}$$

nous obtenons le système suivant de n équations à n inconnues C_i, où k_i est le rapport des intensités X émergentes mesurées sur l'échantillon et sur le témoin:

$$k_i = C_i \exp\left(\sum_{j=1}^{n} C_j \, \mathrm{Log}\, G_{ij}\right), \quad i = 1, \dots, n. \tag{6}$$

La résolution de ce système est facile sur ordinateur et les calculs ne nécessitent que la connaissance de la matrice $[G_{ij}]$. Les coefficients de cette matrice sont obtenus à partir des données relatives d'une part à la nature des éléments constitutifs de l'échantillon étudié et des témoins, d'autre part aux conditions expérimentales de l'analyse.

Un problème se pose pour les oxydes parce qu'il n'est pas possible d'analyser aisément l'oxygène à la microsonde. De manière à conserver un système de n équations à n inconnues on remplace l'équation (6) relative à l'oxygène par l'équation:

$$\sum_{j=1}^{n} (\alpha_j - 1)\, C_j = 0$$

où α_j est l'inverse de la proportion de métal j dans l'oxyde correspondant. Pour les éléments à valence multiple il y a une incertitude et il faut effectuer les calculs pour les nombres de valences extrêmes. L'expérience montre que malgré cette incertitude les rapports cationiques restent bien définis.

4. Conclusion

Ces calculs sont effectués à l'aide d'un programme écrit en langage Algol pour un ordinateur CAE 510 8 K, référencé «*Corrections Sonde*» (Zeller, Thèse, 1968).

Ce programme a été testé et utilisé pour des échantillons de minéraux isolants (oxydes, sulfures, silicates) contenant jusqu'à 12 éléments. Les effets de numéro atomique, de fluorescence et d'autoabsorption varient chacun considérablement d'un échantillon à l'autre.

L'un des intérêts de cette méthode est que pour une série d'échantillons comportant les mêmes éléments et analysés dans les mêmes conditions expérimentales, il suffit de calculer une fois pour toutes la matrice $[G_{ij}]$ et pour chaque analyse il ne reste plus qu'à résoudre le système (6) (environ 20 minutes pour un système d'ordre 12).

5. Annexe — Facteur de rétrodiffusion

Généralement on pose pour la limite R_M du facteur de rétrodiffusion l'expression:

$$1 - R_M = \eta\, \xi_M$$

où η est le coefficient de rétrodiffusion pratiquement indépendant de la haute tension et ξ_M l'efficacité maximum des électrons rétrodiffusés. En utilisant un modèle de pénétration électronique moyenne nous sommes arrivés aux résultats suivants:

$$\eta = \frac{K}{K + 2\sigma}$$

$$\xi_M = 1 - 2\left(\frac{1}{2\sigma} + \frac{1}{2\sigma + K}\right)$$

avec

$$K = \frac{Z}{7,5} \quad \text{et} \quad \sigma = 4\left(1 + \frac{Z^2}{14400}\right)$$

Z étant le numéro atomique. Les grandeurs K et σ sont liées respectivement à la loi de diffusion et à la loi de ralentissement des électrons. On montre ensuite que la variation de R avec le taux d'excitation U_0 est donnée par l'expression approchée:

$$\frac{1-R}{1-R_M} = 1 - \frac{1}{5}\,\frac{U_0 - 1 - (1 - 1/U_0^4)/4}{U_0 \log U_0 - U_0 + 1}.$$

Ces résultats, pour des éléments de numéro atomique moyen ($12 < Z < 30$), coïncident à une bonne approximation avec les valeurs expérimentales de Derian (1966). Les calculs du facteur de rétrodiffusion R font l'objet d'une publication en préparation.

Bibliographie

Derian, J. C.: Thèse, Faculté des Sciences Paris 1966.
Zeller, C.: C. R. Acad. Sci., Paris **263**, 1050 (1966).
Zeller, C.: Thèse, Faculté des Sciences Nancy 1968.
— J. microscopie (sous presse) (1968).

Programme de calcul de l'intensité
des rayons X émis en microanalyse à sonde électronique et analyse quantitative

B. Pascal

Electricité de France, Centre de Renardieres, Ecuelles, France

1. Introduction

Le calcul est basé sur la connaissance de la diffusion en profondeur des électrons du faisceau incident, déterminée par simulation par la méthode de Monte-Carlo de Bishop [1]. On en déduit la fonction de distribution du rayonnement X engendré $\varphi(\varrho z)$. La transformée de Laplace de $\varphi(\varrho z)$ donne l'intensité primaire émergente. Une correction de fluorescence par les raies caractéristiques inspirée de celle de Castaing, conduit à l'intensité totale émergente. On déduit par itération la concentration massique d'un échantillon en comparant la grandeur ainsi calculée et l'intensité réellement mesurée.

2. Organigramme de calcul (fig. 1)

Le programme de calcul comprend les étapes suivantes:

1 — Lecture des rapports I/I_0 des intensités de l'échantillon et des témoins réellement mesurées et des autres caractéristiques de l'analyse.

2 — Calcul par la méthode de Monte-Carlo, ou lecture si ce calcul a été effectué auparavent, des intensités I_0' correspondant aux témoins.

3 — Chois d'une composition de première approximation.

4 — Calcul de la répartition des électrons incidents dans la cible et déduction des intensités I' correspondant à chaque élément.

5 — Comparaison des rapports I'/I_0' calculés et I/I_0; déduction d'une nouvelle approximation par interpolation parabolique entre le point calculé et les points ($c = 0$, $I/I_0 = 0$; $c = 1$, $I/I_0 = 1$).

6 — Itération.

3. Calcul de la distribution en profondeur des électrons du faisceau incident

3.1. Section efficace de diffusion. Bishop suppose que la diffusion angulaire est essentiellement due aux interactions élastiques et que la loi de ralentissement ne dépend que des interactions inélastiques. Il utilise la section efficace de diffusion de Rutherford écrantée valable dans la première approximation de Born, multipliée par un facteur $\frac{Z+1}{Z}$ cherchant à tenir compte des effets inélastiques (Spencer, 1959 [2]).

$$\sigma(\theta) = \frac{Z^2 e^4}{16 E^2} \frac{1}{\left(\sin^2 \frac{\theta}{2} + \frac{\theta_0^2}{4}\right)^2} \frac{Z+1}{Z}$$

$$\theta_0 = \frac{\lambda}{2\pi R}, \quad R = a_H Z^{-\frac{1}{3}}, \quad a_H = 0{,}529 \, \text{A}°. \tag{1}$$

Programme de calcul

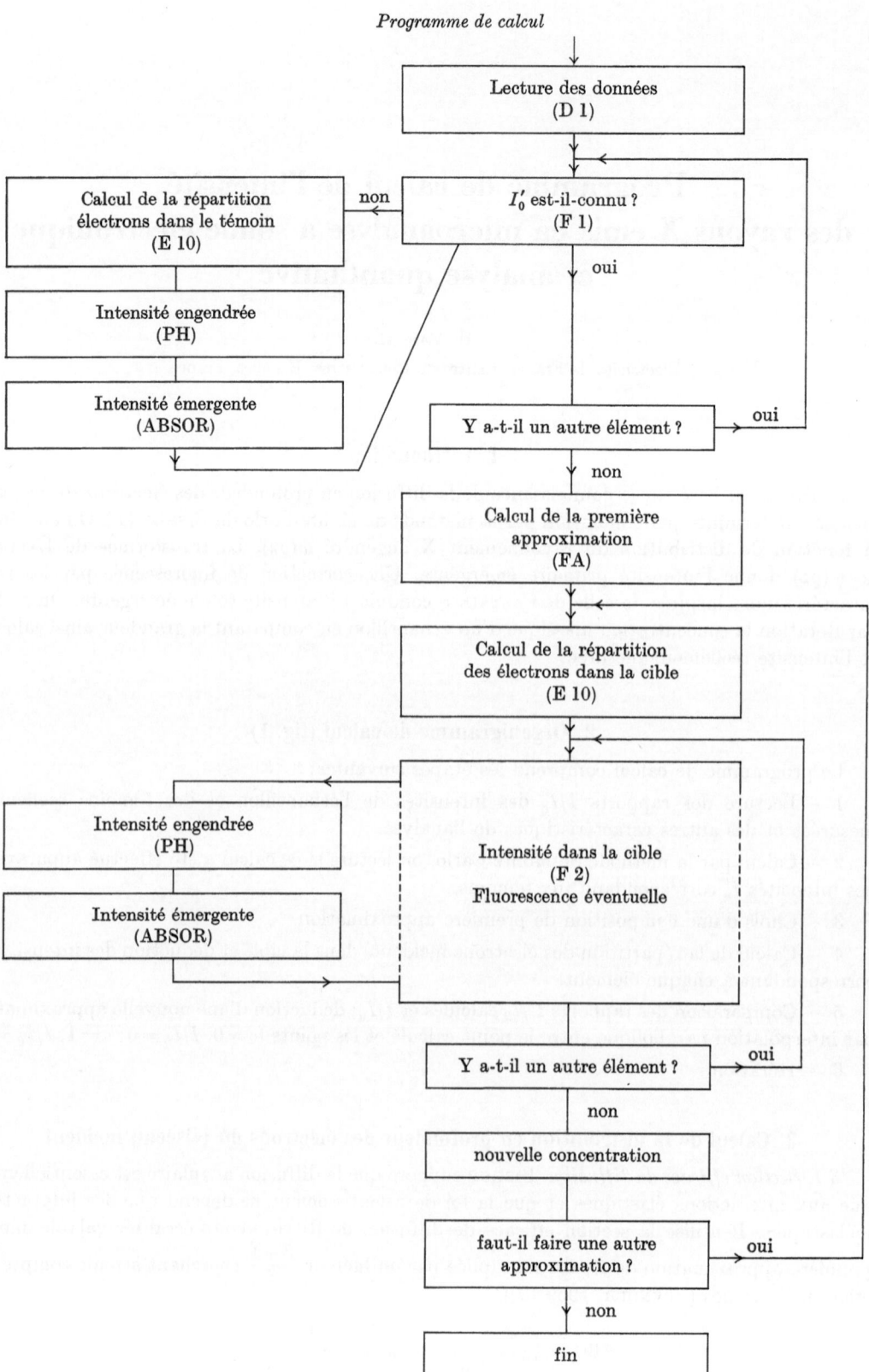

Fig. 1

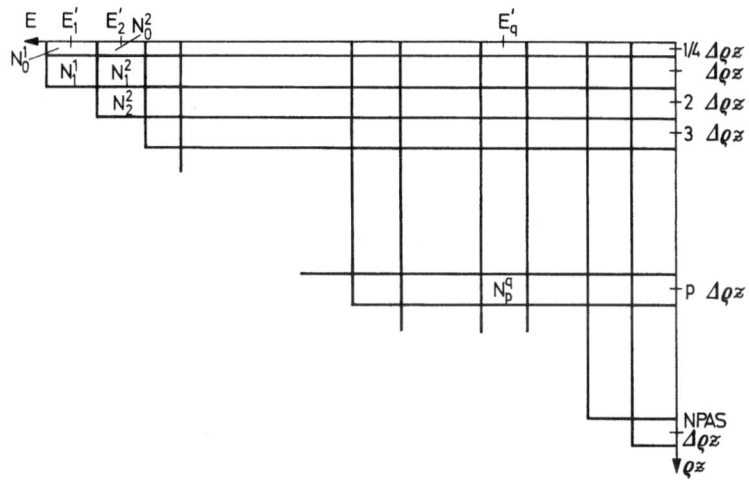

Fig. 2. Diagramme de la diffusion des électrons dans la cible $N_p^q = N f(E, pz)$

La perte d'énergie est supposée suivre la loi de BETHE:

$$\frac{dE}{d \varrho s} = - \frac{2\pi N e^4}{E} \frac{Z}{A} \log 1{,}166 \frac{E}{J}.$$

J est le potentiel moyen d'ionisation, de l'ordre de $13 \times Z$ eV; nous employons les valeurs numériques de CALDWELL [3].

3.2. Diffusion multiple. La trajectoire des électrons est divisée en NPAS étapes d'égale longueur $\varDelta \varrho z$, leur nombre peut aller jusqu'à 45.

A chaque étape est associée une énergie moyenne E, la distribution des angles de diffusion est déduite de (1) par la méthode de GOUDSMIT et SAUNDERSON [4]; on obtient un certain nombre (128) d'angles β équiprobables, qui sont tabulés dans la mémoire de l'ordinateur.

On suit chacun des N électrons incidents pas à pas, trois nombres de hazard déterminent le point où a lieu la diffusion, l'angle β de diffusion par rapport à la direction initiale et l'angle d'azimuth par rapport à la verticale. Les résultats peuvent être schématisés par le tableau de la fig. 2. Le rapport N_p^q/N est considéré comme la densité des électrons ayant une énergie égale à l'énergie E_q' de la fin de la $q^{\text{iéme}}$ étape et se trouvant à la profondeur $p\varDelta \varrho z$. Ces électrons perdent à cette profondeur une énergie $E_{q-1} - E_q$.

4. Calcul de l'intensité primaire émergente

Soit $n(E, \varrho z)$ la densité d'électrons à la profondeur ϱz, à un facteur de proportionnalité près, égal au poids de la raie analysée dans son niveau quantique, l'intensité primaire émergente s'écrit:

$$I_1 = \omega_A \frac{C_A}{A} \int \int n(E, \varrho z) \frac{\psi(E)}{dE/d\varrho z} e^{-\mu \varrho z/\sin \theta} \, dE \, d\varrho z \qquad (2)$$

(ω_A est le rendement quantique du niveau considéré).

On a choisi pour la fonction d'ionisation $\psi(E)$ l'expression donnée par la loi de Bethe modifiée par WORTHINGTON et TOMLIN [5]. L'équation (2) est approchée par l'expression:

$$I_1 = \omega_A \frac{C_A}{A} \frac{1}{N} \Sigma_p \{ \Sigma_q N_p^q \psi(E_q') \} e^{-\mu p \varDelta \varrho z/\sin \theta} \varDelta \varrho z.$$

Le terme entre accolades correspond à la fonction $\varphi(\varrho z)$, répartition des ionisations en profondeur; il est calculé pour les points $\frac{1}{4}\varDelta \varrho z$, $\varDelta \varrho z$, $2\varDelta \varrho z$, ..., NPAS $\varDelta \varrho z$; il est ensuite assimilé à un polynome du 4e degré pour le calcul de la transformée de LAPLACE. Le résultat obtenu tient compte des effets d'absorption et de numéro atomique.

5. Correction de fluorescence

On a négligé l'influence du fond continu. Supposons qu'une raie caractéristique du niveau K de B excite la raie du niveau K de A analysée. Si la fonction $\varphi(\varrho z)$ est supposée de la forme $e^{-\sigma \varrho z}$ la valeur de l'intensité de fluorescence se déduit des calculs de Castaing [6]:

$$I_A^f = K \omega_A C_A I_{1A} \left\{ \frac{\text{Log}(1+x)}{x} + \frac{\text{Log}(1+y)}{y} \right\} \frac{j_B}{j_A}.$$

j_B et j_A sont les intensités des raies de B et de A engendrées.

$$K = \frac{1}{2} \frac{r_A - 1}{r_A} \frac{\mu_B^A}{\mu_B}; \qquad x = \frac{\mu_A}{\mu_B \sin \theta}; \qquad y = \frac{\sigma}{\mu_B}.$$

j_B / j_A et σ peuvent être calculés à l'aide de la fonction $\varphi(\varrho z)$:

$$\frac{j_B}{j_A} = \frac{C_B}{C_A} \frac{A}{B} \frac{\int \varphi_B(\varrho z) \, d\varrho z}{\int \varphi_A(\varrho z) \, d\varrho z}$$

$$\sigma = \frac{\varphi_B(0)}{\int \varphi_B(\varrho z) \, d\varrho z}.$$

En fait la fonction $\varphi_B(\varrho z)$ présente un maximum au voisinage de l'origine. On écrit:

$$\varphi(\varrho z) = \varphi_0 \, e^{-\sigma \varrho z} + \varphi'$$

φ' est tel que $\varphi'(0) = 0$; $\int \varphi' = 0$. On a choisi l'expression

$$\varphi' = \alpha \, \varphi(0) \, [e^{-\sigma' \varrho z} - 4 e^{-2\sigma' \varrho z} + 3 e^{-3\sigma' \varrho z}]$$

α et σ' sont tels que le maximum de φ coïncide avec celui de $\varphi(0) \, e^{-\sigma \varrho z} + \varphi'$:

$$\sigma' = \frac{0{,}303}{\varrho z_m}$$

$$\alpha = \frac{\varphi(\varrho z_m) - \varphi_{(0)} \, e^{-\sigma' \varrho z_m}}{\varphi_{(0)} (e^{-\sigma' \varrho z_m} - 4 e^{-2\sigma' \varrho z_m} + 3 e^{-\sigma' \varrho z_m})}$$

ϱz_m est l'abscisse du maximum de $\varphi - \varphi_{(0)} \, e^{-\sigma \varrho z}$. L'intensité de fluorescence s'écrit alors:

$$I_A^f = K \omega_B C_B I_{1A} \frac{A}{B} \frac{\int \varphi_B(\varrho z)}{\int \varphi_A(\varrho z)} \left(\frac{\text{Log}(1+x)}{x} + \frac{\text{Log}(1+y)}{y} + \frac{\alpha}{y'} \text{Log} \left(\frac{1 + 4y' + 3y'^2}{1 + 4y' + 4y'^2} \right) \right) \qquad (3)$$

avec

$$y' = \frac{\sigma'}{\mu_B}.$$

Dans le cas de fluorescence des raies L nous utilisons la méthode de Reed [7]: on multiplie, suivant les cas, (3) par un coefficient P_{KL}, P_{LL} ou P_{LK} égal à 0,24, 1 ou 4,2.

6. Discussion

6.1. Précision du calcul. Les erreurs du programme sont de deux ordres; le unes sont dues au procédé de calcul:

1. Emploi d'une loi de diffusion multiple qui surestime la longueur des trajectoires réelles, substituant à des éléments curvilignes des segments de droite.

2. Erreur statistique due au nombre fini de trajectoires.

Les autres sont dues aux lois physiques employées:

3. Emploi d'une section efficace de diffusion valable uniquement dans la première approximation de Born.

4. Emploi d'une loi de perte d'énergie continue et abstraction de la dispersion des pertes d'énergies (Straggling).

5. Emploi d'une section efficace d'ionisation $\psi(E)$ approchée.

A ces erreurs il faut ajouter celles dues à l'incertitude sur les coefficients expérimentaux et particulièrement les coefficients d'absorption.

On diminuera les erreurs 1 et 2 en augmentant le nombre de pas et de trajectoires mais le temps de calcul sera d'autant plus long; la première sera minimisée en prenant le même nombre de pas pour la diffusion dans le témoin et dans l'échantillon. La Tableau 1 donne les coefficients de rétrodiffusion calculés pour des matrices de Carbone, d'Aluminium, de Cuivre et d'Or, pour un nombre de pas égal à 15, 25, 45 et 65, à 30 kV.

La précision statistique sur le résultat final d'une analyse est difficile à évaluer théoriquement; une série d'essais a montré que l'erreur quadratique moyenne $\sqrt{\dfrac{1}{n}\sum_1^n\left(\dfrac{\Delta I}{I}\right)^2}$ est de l'ordre de $\dfrac{1}{4\sqrt{N}}$ pour 25 étapes et un taux d'excitation de 5; l'erreur augmente lorsque le nombre d'étapes ou le taux d'excitation diminue.

6.2. Pratique d'un calcul de correction. On cherchera à réduire les écarts à la loi de proportionnalité: dans le cas de fortes corrections d'absorption on ne choisira pas un taux d'excitation trop grand, en effet, la forme de la fonction $\varphi(\varrho z)$, au voisinage de l'origine, est alors très importante or elle dépend beaucoup de la section efficace d'ionisation $\psi(E)$, dont l'expression théorique employée n'est qu'approchée.

Le nombre de pas peut être réduit (10 à 15) si les éléments ont des numéros atomiques voisins (Fe, Cr, Ni), car les erreurs se compensent.

La première approximation du calcul pourra se faire, soit en employant d'autres méthodes de corrections moins longues, soit, plus simplement, en prenant le rapport des intensités; un premier calcul avec un petit nombre de trajectoires (500) donne généralement une valeur suffisamment approchée pour que, dans le cas d'un mélange binaire, le résultat de l'approximation

Tableau 1

Coefficient expérimental de BISHOP		Coefficient calculé par le programme E-9				Calculé par BISHOP
		65 pas	45 pas	25 pas	15 pas	
Au	0,521		0,548	0,561	0,581	0,568
			0,562			
N			10000	10000	20000	5000
Cu	0,319		0,3226	0,3358	0,3421	0,339
			0,3264	0,3368	0,3412	
N			10000	10000	10000	5000
AL		0,154				
	0,155	0,1545	0,1581	0,1638	0,1668	0,166
		0,1547				
N		10000	50000	50000	100000	5000
C	0,06		0,0539	0,0575	0,0604	0,06
N			10000	10000	20000	5000

Tableau 2. *Analyse d'un sulfure de Cuivre, Nickel, Cobalt et Fer. $V_1 = 19\,kV$; $NPAS = 25$; pour les témoins $N = 5000$; coefficients d'absorption de* HEINRICH

	Première approximation	Itérations				Analyse chimique
		n° 1: $N = 500$	n° 2: $N = 500$	n° 3: $N = 1000$	n° 4: $N = 5000$	
Cu	0,3	0,357	0,353	0,353	0,352	0,4
Ni	27,7	32,147	31,875	31,852	31,760	32,3
Co	0,4	0,441	0,440	0,440	0,438	0,5
Fe	32,8	33,533	33,808	33,709	33,574	34
S	28	31,522	33,768	33,819	33,846	33,3
			100,244	100,173	99,970	

suivante soit dans les limites de l'erreur statistique autour de la concentration finale. Dans le cas de plusieurs éléments 3 ou 4 approximations sont nécessaires. La Tableau 2 représente les résultats successifs de l'analyse d'un sulfure mixte de Fer, Chrome, Cobalt, Nickel et Cuivre.

7. Conclusion

— Le programme de calcul permet d'effectuer l'analyse d'un échantillon ou d'une série d'échantillons contenant jusqu'à 10 éléments.

— On peut espérer une précision relative de 1% si les analyses sont faites dans des conditions convenables (taux d'excitation compris entre 1,5 et 5).

— Une bonne connaissance des coefficients expérimentaux et spécialement des coefficients d'absorption est indispensable pour être sûr de la signification de l'analyse.

— Si on a calculé antérieurement, une fois pour toute, l'intensité correspondant aux témoins, une précision statistique de 1% est atteinte en 10 secondes sur ordinateur CDC 6600.

Bibliographie

1. Bishop, H. E.: Calculations of electron penetration and X-ray production in a solid target. Optique des Rayons X et microanalyse. Paris: Hermann 1966.
2. Spencer, L. V.: Energy dissipation by fast electrons. N.B.S. Monograph 1 (1959).
3. Caldwell, D. O.: Phys. Rev. 100, 291 (1955).
4. Goudsmit, S., et J. L. Saunderson: Multiple scattering of electrons. Phys. Rev. 57, 24 (1940).
5. Worthington, C. R., et S. G. Tomlin: Proc. Roy. Soc. A 69, 401 (1956).
6. Castaing, R.: Thèse université de Paris, publication O.N.E.R.A. No. 54 (1951).
7. Reed, S. J. B.: Characteristic fluorescence corrections in electronprobe microanalysis. Brit. J. Appl. Phys. 16, 913 (1965).

Formulae for Absorption Correction with Regard to Indirect Excitation of *K*, *L* and *M* Lines

E. Preuss

KFA-Jülich, ITP, Jülich, BRD

When the electron beam of a microprobe strikes the surface of a specimen, considering in this case a single element, one part of the electrons is reflected or back scattered. The other part which enters the specimen can excite X-rays. The kinetic energy of the electrons decreases with greater penetration depths. The profile of the surfaces with constant energy of the electrons and photons is demonstrated in Fig. 1. There are shown in a schematic drawing the boundaries

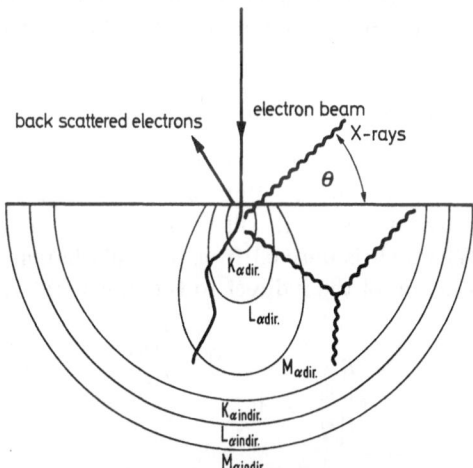

Fig. 1. The boundaries for K_α, L_α and M_α radiation excited by electrons and X-ray photons in a schematic drawing

for K_α, L_α and M_α radiation which are excited directly by electron impact and indirectly by X-ray photons. PHILIBERT [1] considers in his calculations for absorption correction only that part of the X-rays which is excited by direct electron impact. Eq. (1) is his well known expression for absorption correction

$$f(\chi)_{\text{dir.}} = \frac{1+h}{\left(1+\dfrac{\chi}{\sigma}\right)\left[1+h\left(1+\dfrac{\chi}{\sigma}\right)\right]} \qquad \begin{aligned} &\chi = \frac{\mu}{\varrho}\,\text{cosec }\theta, \\ &\theta = \text{take-off angle}, \\ &\frac{\mu}{\varrho} = \text{mass absorption} \\ &\qquad \text{coefficient}. \end{aligned} \qquad (1)$$

When he develops this relative expression the atomic constants, e.g. the excitation cross sections are lost. Therefore his formula is valid for all excitation levels for X-rays. The Lenard coefficient σ in Eq. (2) used by PHILIBERT is inversely proportional to the square of the acceleration voltage V of the electrons.

$$\sigma = \frac{\sigma_0}{V^2}; \qquad h = 1.2\,\frac{A}{Z^2} \qquad \begin{aligned} &A = \text{atomic weight}, \\ &Z = \text{atomic number}. \end{aligned} \qquad (2)$$

By this assumption results the simple expression for h in Eq. (2). Theisen [2] considers that the electrons must have at least a kinetic energy of the height of the excitation levels for X-rays V_C and he finds the expression for σ respectively for h in Eq. (3).

$$\sigma = \frac{8.9 \cdot 10^5}{(V - V_C)^2}; \qquad h = 1.5308 \, \frac{V^2}{(V - V_C)^2} \, \frac{A}{Z^2} \, . \tag{3}$$

Some other good approximation for the experimental values of $f(\chi)$ is achieved by using the dependence of the Lenard coefficient from the acceleration voltage given by Duncumb and Shields [3], Eq. (4).

$$\sigma = \frac{2.39 \cdot 10^5}{V^{1.5} - V_C^{1.5}}; \qquad h = 1.2 \, \frac{A}{Z^2} \, . \tag{4}$$

This is used here also in the following calculations.

Green [4] divides the function $f(\chi)$ in two parts, one part regards to the directly and the other part to the indirectly excited X-ray intensity. His formula is some little transformed and shown in Eq. (5).

$$f(\chi) \atop \text{total} = \left[1 - S \left(1 - \frac{f(\chi) \atop \text{indir.}}{f(\chi) \atop \text{dir.}} \right) \right] f(\chi) \atop \text{dir.} \, . \tag{5}$$

In this expression is S the fraction of the indirect and the total number of X-ray photons generated in the specimen. The analytical expressions for these S values of the K_α, L_α and M_α radiation are given in Eq. (6).

$$S_K = \frac{Z^4}{10^7 + Z^4}; \qquad S_L = \frac{Z^4}{10^8 + Z^4}; \qquad \left(S_M = \frac{Z^4}{10^9 + Z^4} \right) . \tag{6}$$

It must be mentioned that the equation for the M_α radiation in broken brackets is arbitrary since no experimental values were known. But this expression gives best approximation to experimental values of $f(\chi)$ by Green, as it will be demonstrated in the following part.

It can be seen in Eq. (7) which is derived from Eq. (5) that in choosing S_M, S must have a smallest value since the fraction of the indirect to the direct $f(\chi)$ value can never get negative.

$$\frac{f(\chi) \atop \text{indir.}}{f(\chi) \atop \text{dir.}} = 1 - \frac{1 - \dfrac{f(\chi) \atop \text{total}}{f(\chi) \atop \text{dir.}}}{S} \, . \tag{7}$$

Thus the simple expression for S_M in Eq. (6) was formed.

In Fig. 2 the corresponding curves for these S values are plotted. Comparing experimental results of others e.g. of Webster [5], Castaing [6], Descamps [6] etc. one can find that these values differ partially very much from these curves. Furthermore it can be seen from which atomic number fluorescent excitation corrections for the different lines must be regarded, e.g. for K_α radiation from about atomic number 35, for L_α radiation from atomic number 55 and for M_α radiation from atomic number 75.

We consider again Eq. (7). With this expression the indirect contribution of $f(\chi)$ can be calculated by using the experimental values of $f(\chi)$ given by Green [4], Eq. (6) for S and Eq. (1) resp. (4) by Duncumb and Shields for the direct part of $f(\chi)$. This is performed for L_α and M_α radiation by using experimental values for gold. For K_α radiation theoretical curves by Webster and Green [4] were available which are to be seen in Fig. 3. These curves were approximated by the analytical function Eq. (8) shown in this Fig. 3.

$$\frac{f(\chi) \atop K_{\text{indir.}}}{f(\chi) \atop K_{\text{dir.}}} = \exp \left[- \frac{(A \chi)^{\frac{0.483}{U_0^{0.2}}} - 0.552 + 0.056 \, U_0}{1.25 / U_0^{0.32}} \right]$$

$$U_0 = \frac{V}{V_C}; \qquad A = \frac{\varrho}{\mu_K} \, . \tag{8}$$

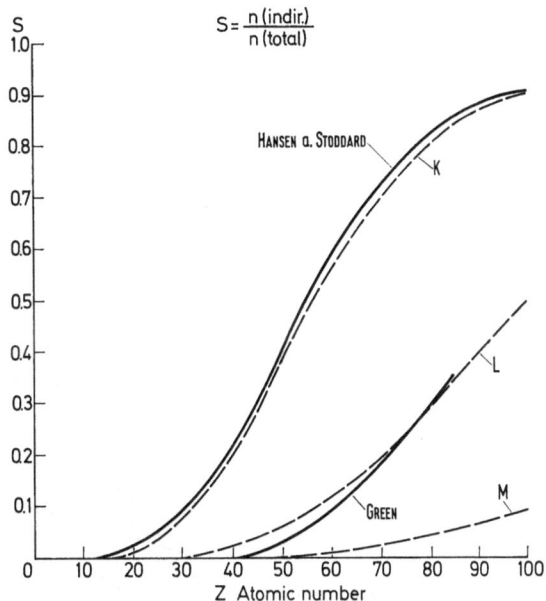

Fig. 2. The ratio of indirect to total production of K_α, L_α and M_α radiation

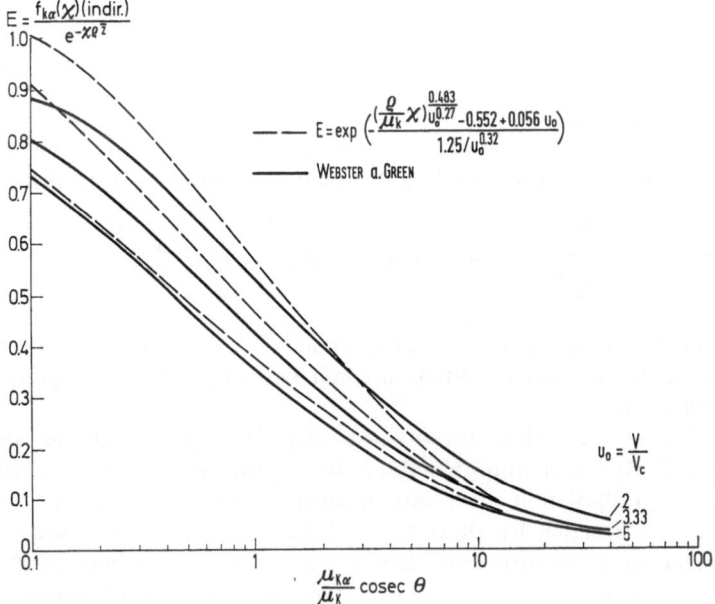

Fig. 3. Theoretical $f(\chi)$ curves for indirect production of K_α radiation

Here is ϱ the density of the specimen and μ_K the absorption coefficient at the high side of the K absorption edge. It must be mentioned that this expression and the following equations for L_α and M_α indirect $f(\chi)$ contribution are choosen in such a manner that they can be calculated with the computer type programma 101 from Olivetti.

In Fig. 4 the corresponding curves for L_α radiation are to be seen. Eq. (9) is the derived expression for this indirect $f(\chi)$ contribution.

$$\frac{f(\chi)_{L\text{indir.}}}{f(\chi)_{L\text{dir.}}} = \exp\left(-U_0 \cdot 10^{-3} \chi^{\sqrt[4]{U_0}}\right); \quad U_0 = \frac{V}{V_C}. \tag{9}$$

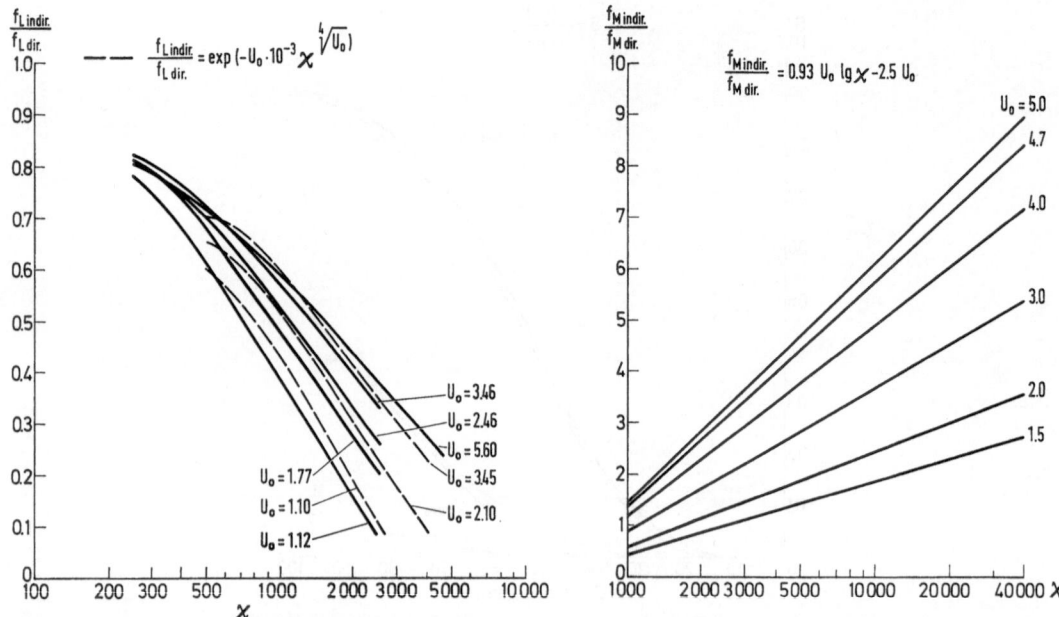

Fig. 4. Experimental $f(\chi)$ curves for indirect production of L_α radiation

Fig. 5. Experimental $f(\chi)$ curves for indirect production of M_α radiation

In Fig. 5 the corresponding curves for M_α radiation are plotted. It must be noticed that in contrast to the curves for K_α and L_α indirect $f(\chi)$ contribution here this part increases with decreasing take-off angle. A quantitative explanation for this is not achieved. The derived curves are identical with those calculated by the analytical function Eq. (10).

$$\frac{\underset{M\text{indir.}}{f(\chi)}}{\underset{M\text{dir.}}{f(\chi)}} = 0.93\,U_0 \lg \chi - 2.5\,U_0; \qquad U_0 = \frac{V}{V_C}. \tag{10}$$

This equation could be compared only with experimental $f(\chi)$ values for gold given by GREEN, while the expressions for K_α and L_α radiation could be compared with experimental values of several different elements.

Using the analytical expressions Eq. (8) resp. Eq. (9) resp. Eq. (10) in combination with Eqs. (1), (4), (5) and (6) better approximations to experimental values are achieved. This is shown in Fig. 6 for K_α radiation of silver with experimental curves given by GREEN, theoretical curves for absorption correction by DUNCUMB and SHIELDS and for the better approximation of those curves given by these corrections here for indirect contribution. In Fig. 7 resp. Fig. 8 the corresponding curves for L_α radiation of tantalum resp. for M_α radiation of gold are demonstrated.

It was the aim of this work to derive equations for absorption corrections with regard to indirect excitation which must be conceived as an effect by that element itself induced. The expressions were set up to a simple form for practical calculations. Better approximations to experimental results can be expected. This was examined for many elements with experimental curves given by GREEN. For M_α corrections comparisons with experimental results of other elements than gold are necessary and the measurement of indirectly generated X-ray contribution is desirable. In application to quantitative analysis of composed specimens better results may be expected in cases where this indirect excitation is considerable, e.g. when the elements in the specimen have almost the same atomic number. Of course other effects, e.g. fluorescent excitation by lines or by continous radiation of an element with a greater atomic number must be considered and cannot be replaced by these corrections.

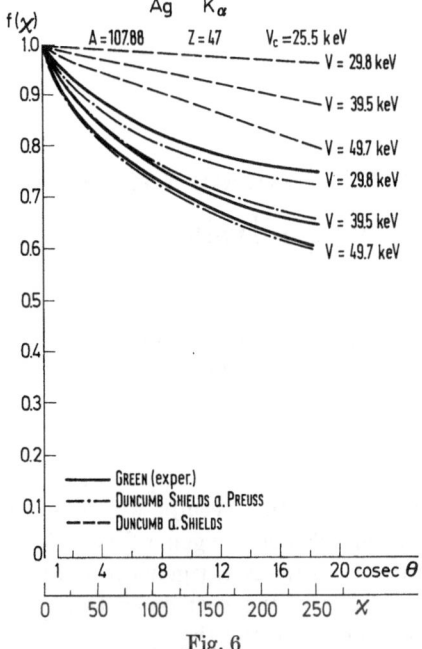

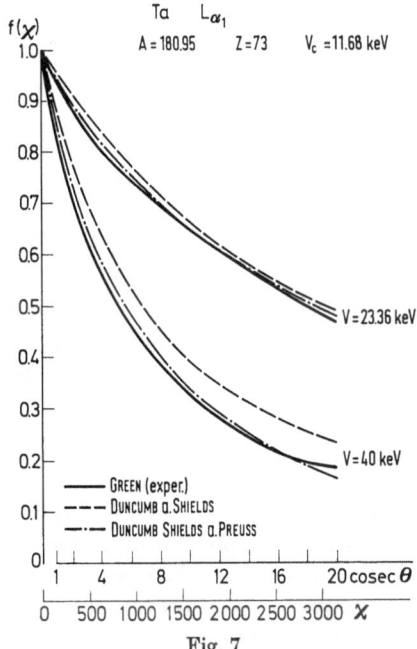

Fig. 6

Fig. 7

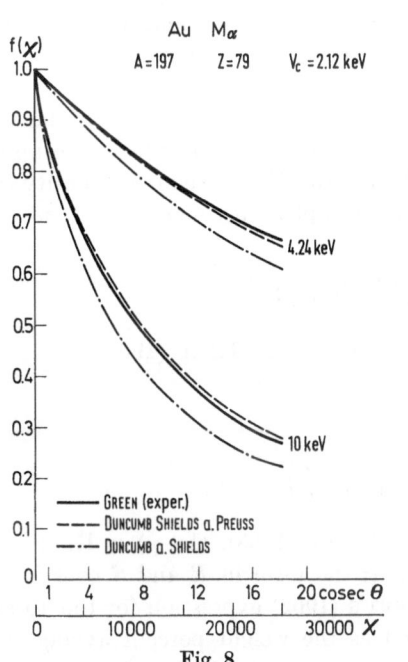

Fig. 8

Fig. 6. Experimental and theoretical curves of $f(\chi)$ for Ag K_α

Fig. 7. Experimental and theoretical curves of $f(\chi)$ for Ta $L_{\alpha 1}$

Fig. 8. Experimental and theoretical curves of $f(\chi)$ for Au M_α

References

1. PHILIBERT, D.: X-ray optics and X-ray-microanalysis, p. 379. New York and London: Academic Press 1963. — L'analyse quantitative en microanalyse par sonde électronique, IRSID, B-No. 51, Février 1965.

2. THEISEN, R.: Quantitative electron microprobe analysis. Berlin-Heidelberg-New York: Springer 1965.

3. DUNCUMB, P., and P. K. SHIELDS: The electron microprobe. New York-London-Sidney: John Wiley & Sons, Inc. 1966.

4. GREEN, M.: X-ray optics and X-ray microanalysis, p. 361. New York and London: Academic Press 1963.

5. WEBSTER, D. L.: Proc. Natl. Acad. Sci. U.S. **13**, 445 (1927); **14**, 331 (1928).

6. CASTAING, R., et J. DESCAMPS: J. Phys. Radium 304 (Mars 1955).

Accuracy of Atomic Number and Absorption Corrections in Electron Probe Microanalysis

P. Duncumb*, P. K. Shields-Mason** and C. da Casa*

* Tube Investments Research Laboratories, Hinxton Hall, Cambridge, England
** National Physical Laboratory, Teddington, Middlesex, England

Theory

The practising probe analyst requires a method for correcting results which is accurate, easy to apply, and applicable to a wide variety of specimens. To achieve generality, it is essential that the procedure should be based on a good physical theory of the X-ray generation-absorption-fluorescence processes. However, on account of the complexity of the electron scattering process, this necessarily contains approximations, and some parameters in the theory are best obtained from an empirical fit to the analysis of specimens of known composition. If a large number of such measurements are made on different specimens under varying conditions, the effect of random experimental errors can be distinguished from systematic trends arising from imperfections in the theory.

In this paper, we use the following correction procedure, which is based on the work of several authors, as described by Duncumb and Reed (1967), and Bishop (1968). It is applicable to the analysis of all elements from atomic number 11 upwards. The measured intensity ratio k_A between specimen and standard for an element A in a complex specimen $AB\ldots$, is given in terms of the true concentration C_A by:

$$k_A = C_A \cdot \frac{R_{AB}\,\bar{S}_A}{R_A\,\bar{S}_{AB}} \cdot \frac{f(\chi)_{AB}}{f(\chi)_A} \cdot (1+\gamma)$$

where R is the backscatter coefficient, tabulated by Duncumb and Reed (1967)

$$\bar{S} = \frac{Z}{A} \cdot \log_e\left(\frac{1.17\,\bar{E}}{J}\right)$$

$$f(\chi) = 1 \Big/ \left(1 + \frac{\chi}{\sigma}\right)\left(1 + \frac{h}{1+h} \cdot \frac{\chi}{\sigma}\right)$$

and Z, A, $\bar{E}$, χ and h have the usual meaning, as in Bishop (1968). $(1+\gamma)$ is Reed's (1965) correction for fluorescence from characteristic radiation: the term in R and $\bar{S}$ constitutes the atomic number correction, and that in $f(\chi)$ is Philibert's (1962) expression for the absorption correction. R_{AB}, $\bar{S}_{AB}$, χ_{AB}, and h_{AB} are all calculated as the weight percent average for the individual elements present.

This leaves unspecified σ, the modified Lenard coefficient, and J, the mean ionisation potential. To select optimum expressions for these, we have processed some 450 probe analyses on specimens of known composition with various forms of σ and J, and selected those which give the best overall fit. For each analysis C_A is calculated from k_A iteratively and the error ε obtained from

$$\varepsilon = \frac{C_A(\text{calculated}) - C_A(\text{true})}{C_A(\text{true})} \times 100\,\%.$$

For each group of analyses, a measure of the overall error can then be defined as the RMS average of ε, and used as a criterion for the selection of σ and J.

Experimental Data

The experimental data is made up of 333 measurements on binary compounds from 11 different laboratories, and 121 analyses of complex silicates made with high precision by SWEATMAN and LONG (this conference). Because the latter form a critical test of the absorption correction, it is convenient to consider them separately.

Binary analyses. These include the 229 measurements collected by POOLE (1967) to which have been added some further analyses of oxides, sulphides and intermetallic compounds giving a wide coverage of the periodic table. Elements from atomic number 12—92, in the presence of others of atomic number 6—92, have been analysed by their K_α, L_α, M_α, or M_β radiation at beam energies of 10—40 kV with take-off angles ranging from 15°—75°. Some early measurements are of doubtful accuracy but are included for completeness; more recent data is expected to have a relative precision of 1%, both in probe measurement and chemical or other determination of true composition.

It is convenient to recognise two sub-groups of the binary data: — (a) a set of 101 measurements on specimens exhibiting a high atomic number effect (referred to as Z alloys) and (b) 42 analyses of aluminium compounds showing a high absorption correction (Al K_α alloys). The Z alloys consist of uranium compounds of elements of atomic number 6—77 (analysed by U L_α radiation), metal-aluminium compounds of elements 22—78 (by M K_α or L_α) and oxides of metals 12—40 (by M K_α). The Al K_α alloys cover a more limited range of atomic number from 22—29. In all the binary analyses the standard consisted of the pure element.

The effect of fluorescence from the continuum was calculated by the method described by SPRINGER (1967a) for 254 of the binary analyses. The correction exceeded 1% in only 38 cases (mainly Cu—Au and Au—Cu), reaching a maximum of 4.4%, and was not thought to justify including in the procedure described above. For the oxides (and hence silicates) it was negligible.

Silicates. 9 different silicates were analysed at energies of 10—20 kV with take-off angles of 40° and 75° (SWEATMAN and LONG, this conference). The major constituents were Mg, Al, Si, Ca, Ti and Fe (not all present together), all considered as oxides. In addition there were trace amounts ($<1\%$) of Nb, Mn or Na. The silicates were therefore treated as 8-component mixtures, for which probe data was used to determine the 6 major constituents; the trace element composition was assumed from chemical analysis and the oxygen content obtained by difference from 100%. The corrected probe data was then compared with the true concentrations of the major constituents, obtained by chemical analysis and further checked by X-ray fluorescence analysis. After rejecting all probe analyses of elements present in less than 5% true concentration, 121 useful error value measurements remained, in which the contribution from random experimental errors was believed to be less than 1%. The standards used in the probe analysis were MgO, Al_2O_3, SiO_2, Wollastonite, Ti and Fe.

Atomic Number Effect — Selection of J

Values of the mean ionisation potential J reported in the literature (see, for example, SPRINGER, 1967b), fall in the range 9—15 eV per unit atomic number, a commonly used approximation being

$$J/Z = 11.5 \text{ eV}.$$

Although J appears under the logarithm in the formula for $\bar{S}$, it seems that for the purpose of probe analysis it is not sufficient to take J/Z as constant. More complex functions have been given by SPRINGER (1967b) to fit the data of BAKKER-SEGRÉ and of CALDWELL, together with a suggested compromise between the two, but none of these appear to bring about any significant difference in conditions of probe analysis. All agree, however, in indicating an increase in J/Z for low values of Z.

In an attempt to obtain values more directly related to probe analysis, DUNCUMB and DA CASA (1967) found that a unique J/Z vs. Z curve could be plotted empirically from the analysis

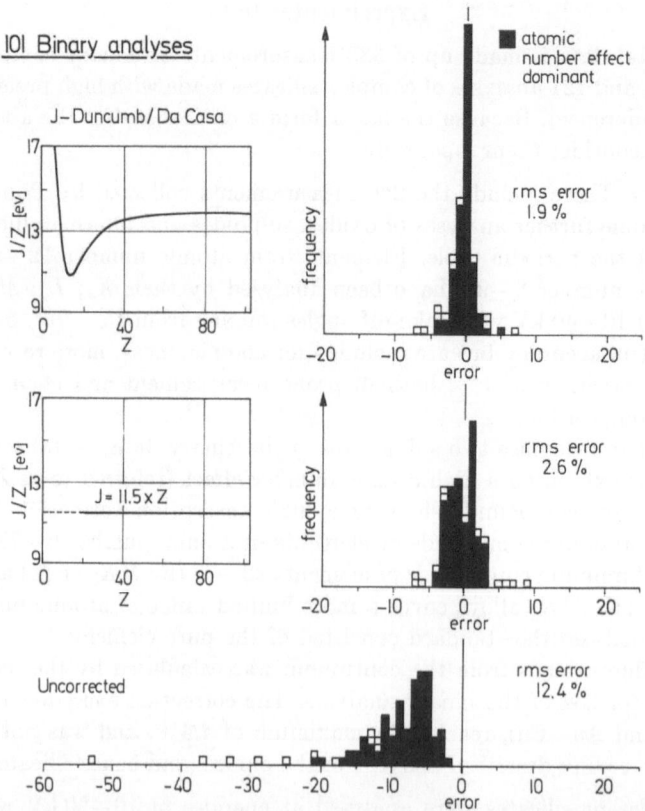

Fig. 1. Error histograms of 101 analyses of known binary alloys chosen for high atomic number effect, showing how RMS error is affected by choice of J/Z. Lower: Uncorrected analyses. Middle: Corrected using $J/Z=11.5$ eV. Upper: Corrected using Duncumb/da Casa J/Z

of some 40 alloys of known composition measured in the microprobe. This curve is described by the expression

$$J/Z = 14.0\,(1-e^{-0.1Z}) + 75.5/Z^{Z/7.5} - Z/(100+Z)$$

and shows a minimum in the region of $Z=12$, below which it rises sharply for elements of lower atomic number. Some examples of its use have been given by Duncumb and Reed (1967). The present results confirm the improvement to be obtained in using this expression as compared to the constant of 11.5 eV or to any of the alternatives given by Springer.

Fig. 1 shows error histograms of the "Z alloys" referred to in the previous section: with no correction applied at all, corrected using $J/Z=11.5$ eV, and corrected using Duncumb and da Casa's expression. The successive improvement is apparent from the narrowing histograms and is further quantified by the RMS error, which reduces from 12.4% (uncorrected) to 2.6% (constant J/Z) to 1.9% (Duncumb/da Casa J/Z). The shaded squares represent those cases where the atomic number effect is stronger than the absorption correction, from which it is seen that the dominant correction is indeed the former in this group of alloys. The absorption correction was carried out as indicated in the next section, though the particular form of σ does not influence the choice of J to any appreciable extent.

We therefore conclude that the fitted J values of Duncumb and da Casa offer a real improvement in accuracy, not only on the 40 analyses for which they were originally obtained, but also on the more varied collection of 101 analyses described here. Furthermore, since a unique value of J is still assigned to each element, the use of these values is little more complicated than with J/Z constant.

Absorption Correction — Optimisation of Sigma

PHILIBERT's (1962) expression describes well the variation of $f(\chi)$ with χ, providing the Lenard coefficient σ is chosen properly. A modified form of this parameter, taking into account the critical excitation potential E_c as well as beam voltage E_0, was proposed by DUNCUMB and SHIELDS (1964):

$$\sigma = \frac{C \times 10^5}{E_0^n - E_c^n}$$

where C and n were constants chosen to fit Green's experimental $f(\chi)$ curves for pure elements, having values of 2.39 and 1.5 respectively. HEINRICH (1967) subsequently proposed that these constants be altered to 4.5 and 1.65 respectively, though no indication was given as to how critical this choice was. It therefore seemed worthwhile to investigate the optimisation of σ, by studying the effect of varying C and n on the RMS error in the analysis of a number of known alloys exhibiting a strong absorption correction.

Three groups of alloys were chosen for this study: the 333 binaries referred to above, the 42 AlK_α alloys and the 121 silicate analyses. Results are shown in Fig. 2, in which RMS error is plotted against constant C for values of n of 1.5, 1.65 and 1.8.

Looking first at the curves for all 333 binaries, we see that shallow minima occur at each value of n, differing in level only slightly between the three. Because of the strong absorption correction, the AlK_α curves show sharper minima, which decrease slightly as n increases, but the RMS error is still strong. The most critical test is given by the silicate analyses, in which the higher precision of probe and chemical analysis results in RMS errors some 5 times smaller than those for the binaries. Furthermore, the sharpness of the minima is such that the use of HEINRICH's σ in place of DUNCUMB and SHIELDS' σ is sufficient to halve the RMS error; HEINRICH's value of C almost coincides with the minimum for $n = 1.65$. There is still no significant change of minimum RMS error with n, at least between $n = 1.5$ and 1.8. Whatever the power n one can always choose the constant C to give an RMS error no more than 1.3%. We therefore conclude that, on the present data, there is no improvement to be gained by further adjusting C and n, and that HEINRICH's values are to be recommended. This conclusion is unaltered by the use of THEISEN's (1967) mass absorption coefficients instead of HEINRICH's (1964), and is not affected by the choice of J in the atomic number correction.

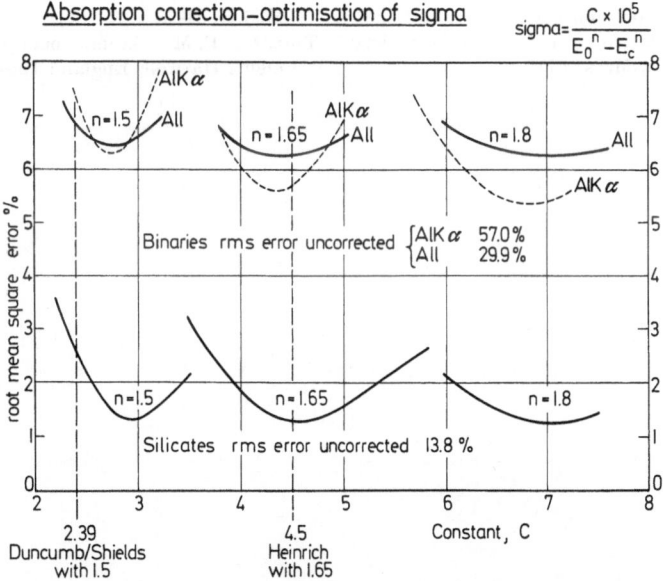

Fig. 2. RMS error for 3 groups of data having strong absorption corrections, plotted for various forms of modified Lenard coefficient σ. The 333 binaries analyses (marked "all") show shallow optimum values, the 42 aluminium alloys (marked "AlK_α") are sharper, and the 121 silicates best of all, indicating that HEINRICH's form of σ is to be preferred to that of DUNCUMB and SHIELDS

Conclusions

We have shown that the use of DUNCUMB and DA CASA'S J values and HEINRICH'S σ values give a useful improvement in accuracy over previously used forms of these parameters. A detailed comparison has not been attempted of the method as a whole with other correction procedures, but present indications are that it compares favourably with that of THOMAS (1964), which in turn has been compared with a number of other approaches by POOLE (1967). As a measure of the overall effectiveness of the method, it is noteworthy that the RMS errors of the uncorrected values of the selected groups of binaries and of the silicate analyses are all reduced 5—10 times by use of the recommended correction procedure. Some uncertainty still exists in the correction for fluorescence from the continuum, which has been ignored, and which may reach 5% in extreme cases. Nevertheless, this "order-of-magnitude" reduction in error seems to be a fair measure of what may now be expected in general.

Acknowledgments. The authors gratefully acknowledge the provision of the silicate data by Dr. T. R. SWEATMAN and Dr. J. V. P. LONG of the University of Cambridge, and the assistance in computer programming given by Miss E. M. IBBOTSON of Tube Investments. They are indebted to Dr. A. FRANKS of the National Physical Laboratory for his encouragement, and to the Chairman of Tube Investments for permission to publish this paper.

References

BISHOP, H. E.: Brit. J. Appl. Phys., Ser. 2, **1**, 673 (1967).

DUNCUMB, P., and C. DA CASA: Conference on Electron Probe Microanalysis, Institute of Physics and Physical Society, London (1967).

—, and S. J. B. REED: Seminar on Quantitative Electron Probe Microanalysis, National Bureau of Standards, Washington D.C. 1967, ed.; NBS Special Publication 298, 1968.

—, and P. K. SHIELDS: Proc. Symposium on The Electron Microprobe 1964. New York: John Wiley & Sons 1966, p. 284.

HEINRICH, K. F. J.: Proc. Symposium on The Electron Microprobe 1964. New York: John Wiley & Sons 1966, p. 296.

— Second National Conference on Electron Microprobe Analysis, Electron Probe Analysis Society of America, Boston, Mass. 1967.

PHILIBERT, J.: Proc. Third International Symposium on X-ray Optics and X-ray Microanalysis 1962. New York: Academic Press 1963, p. 379.

POOLE, D. M.: Seminar on Quantitative Electron Probe Microanalysis, National Bureau of Standards, Washington D. C. 1967, ed.; NBS Special Publication 298, 1968.

REED, S. J. B.: Brit. J. Appl. Phys. **16**, 913 (1965).

SPRINGER, G.: Neues Jahrb. Mineral., Abhandl. **106**, 3, 241 (1967a); — Neues Jahrb. Mineral., Monatsh. **9/10**, 304 (1967b).

THEISEN, R., and D. VOLLATH: Tables of X-ray mass attenuation coefficients. Düsseldorf: Stahleisen M.B.H. 1967.

THOMAS, P. M.: Atomic energy research establishment, Harwell, England 1964. Report R 4593.

Propagation of Errors in Correction Models for Quantitative Electron Probe Microanalysis

K. F. J. HEINRICH and H. YAKOWITZ

Institute for Materials Research, National Bureau of Standards, Washington, D.C., U.S.A.

Abstract

Lack of appropriate standards frequently forces the analyst to use elemental standards. The usefulness of correction models is limited by the accuracy to which the input parameters are known. Uncertainties in presumably known quantities (mass absorption coefficients, fluorescence yield, mean ionization potentials, etc.) are in many cases the limiting factors. The resulting analytical errors can be minimized by judicious choice of experimental conditions. This paper will give examples involving the corrections for absorption, fluorescence by characteristic lines, and atomic number effects.

Key words: Microprobe analysis, error propagation, absorption of X-rays, fluorescence, quantitative analysis, atomic number effects.

Introduction

Quantitative electron probe microanalysis is performed by comparing the X-ray emission of the specimen with that of a standard reference material. If the compositions of specimen and standard are similar, accuracy depends mainly on the precision of the measurements and upon the accuracy to which the composition of the standard is known.

If matching standards are not available, one must use the "absolute method", employing standards of simple composition, such as elements and oxides. However, several properties affecting the X-ray emission vary with target composition. The resulting "matrix effects" must be corrected in calculating the result of the analysis. This is usually done by means of algebraic formulae or "models" obtained by a combination of theory and empirical adjustment.

The error distribution of a large group of analyses is frequently used [1, 2] to evaluate the efficiency of a correction method. However, as shown in an analysis of Thomas' atomic number effect study [3] by one of us [4], such an evaluation is of limited value unless all sources of error are recognized.

Errors may arise in the *measurement* of X-ray intensities, in the models for *corrections*, and in the estimated composition of *standards*. In the correction, errors may be due to failure of the *model* itself, and to errors in the *input parameters* (X-ray attenuation coefficients, fluorescent yields, operating voltage, emergence angle, etc.). However, the literature concerns itself almost exclusively with the discussion of models, largely ignoring other causes of error. This is most serious and disturbing. No estimate of accuracy of a method is possible without considering all errors, and their effect on the final result.

Errors, Uncertainties, and Error Propagation

Random errors, which can be handled by the familiar statistical techniques, are usually of minor importance, since the accuracy of electron probe microanalysis is seldom close to the precision of the measurements. Systematic errors may arise from all sources mentioned above.

To estimate these errors, we must first estimate a range of uncertainty within which we believe each potential source of error (measurement, model, input parameter, standard) to be known. Then we must determine the uncertainty in the analytical result caused by the estimated uncertainty in the sources of error.

Let us assume that we wish to know the effect on the analysis of the uncertainty of mass attenuation coefficients. An inspection of reported values of mass attenuation coefficients, and their discrepancies, may lead us to believe that we know these parameters with an uncertainty not larger than 5%. (Since this statement implies a weighting of experimental data its validity cannot be supported or refuted by statistical methods.) We now perform the correction calculation twice, using for the value of the mass attenuation coefficient in question first the upper boundary value, and then the lower boundary value of the uncertainty range. The difference between the results indicates the resulting uncertainty.

If error propagation within a relatively simple model is to be tested over a wide range of conditions, a general (algebraic) solution can be obtained by the propagation of error formulae [5]. It should be noted that the choice of the model is in general not very important for error propagation of the parameters, since high accuracy is not needed in these calculations. Therefore, it is useful to simplify models for this specific purpose. To illustrate this method, let us consider the usual technique of correction calculation using the multiplicative factors:

$$C = k \cdot k_F \cdot k_A \cdot k_Z \tag{1}$$

where C is the calculated weight-fraction, k the intensity ratio corrected for coincidence losses and background, k_F a correction factor for fluorescent emission, k_A a correction factor for X-ray absorption, and k_Z a correction factor for the "atomic number correction". By applying the error propagation formulae we obtain:

$$\Delta C / C = \Delta k / k + \Delta k_F / k_F + \Delta k_A / k_A + \Delta k_Z / k_Z. \tag{2}$$

Hence, a relative error in any of the k-factors will produce a relative error of same magnitude and sign in the calculated weight-fraction.

Errors in the Measurement

Although random errors in the X-ray intensity measurement can easily be recognized, systematic errors due to coincidence losses of the detector system, defocusing of the X-ray optics, and faulty background correction are factors that should not be overlooked. More research is warranted, particularly in the problem of background correction. It is important that pulse height shrinkage at high counting rates [6] should not be allowed to alter the value of the dead time, τ [7]. If this is achieved by appropriate adjustment of detector and electronics, the dead-time losses in the measurement of the standard will be partly compensated by those of the specimen, particularly at high concentrations:

If $r = N_1 / N_2$ is the ratio of the count rates of specimen to standard, and τ is the dead-time,

$$\frac{\Delta r}{r} = (N_1 - N_2) \Delta \tau. \tag{3}$$

Absorption Correction [8]

The effects of an error in the absorption function $f(\chi)$ — henceforth called f in this paper — can be obtained by the error propagation formulae:

$$k_F = \frac{f_{\text{std}}}{f_{\text{spec}}} \tag{4}$$

$$\frac{\Delta C}{C} = \frac{\Delta k_F}{k_F} = \frac{\Delta f_{\text{std}}}{f_{\text{std}}} - \frac{\Delta f_{\text{spec}}}{f_{\text{spec}}}. \tag{5}$$

We see that a relative error in f_{spec} produces a relative error of the same magnitude and changed sign in the result[1]. To observe the effects of errors on input parameters in the absorption correction, the Philibert model [9] can be simplified as follows:

$$\frac{1}{f} = \left(1 + \frac{\chi}{\sigma}\right)\left(1 + \frac{h}{1+h} \cdot \frac{\chi}{\sigma}\right) \cong 1 + \frac{\chi}{\sigma}. \tag{6}$$

The symbols used in this and the following equations have the conventional meanings. Error propagation yields:

$$\frac{\Delta f}{f} = (1 - f)\left[\frac{-\Delta(\mu/\varrho)}{\mu/\varrho} + \cot\theta\,\Delta\theta + \frac{\Delta\sigma}{\sigma}\right]. \tag{7}$$

Using Duncumb's model for σ [10] we obtain $(\Delta\sigma/\sigma)$ as a function of error in the operating voltage:

$$\sigma = \frac{C}{E^n - E_c^n} \tag{8}$$

$$\frac{\Delta\sigma}{\sigma} = \frac{n\,E^{n-1}}{E^n - E_c^n}\,\Delta E \tag{9}$$

hence,

$$\frac{\Delta f}{f} = (1 - f)\left[\frac{-\Delta(\mu/\varrho)}{\mu/\varrho} + \cot\theta\,\Delta\theta + \frac{n\,E^{n-1}}{E^n - E_c^n}\,\Delta E\right]. \tag{10}$$

We must now set bounds of uncertainty on the input parameters μ/ϱ (the mass attenuation coefficient), θ (the X-ray emergence angle), and E (the operating voltage). For μ/ϱ, we believe that an uncertainty of 5% is a rather optimistic estimate, in view of disagreements in literature. The X-ray emergence angle depends not only on the instrument geometry, but also on the relative efficiency of different parts of the crystal, and of the flatness and positioning of the specimen. An uncertainty of one degree is a conservative estimate. In the estimation of the uncertainty of the operating voltage, we must include the possible effects of specimen coating, and also the fact that in the conventional gun design, the nominal high voltage is connected not between filament and specimen but between the gridcap (Wehnelt cylinder) and the specimen. Under typical operating conditions, the potential of the filament may differ from that of the gridcap by several hundred volts. We assume an error of 250 V in the value of operating potential.

With these assumptions, the effect on f of the uncertainty in the mass attenuation coefficient becomes:

$$\frac{\Delta f}{f} = -0.05\,(1 - f). \tag{11}$$

The effect of the uncertainty in the X-ray emergence angle is

$$\frac{\Delta f}{f} = \frac{\cot\theta}{57.3}\,(1 - f). \tag{12}$$

Assuming that in Eq. (8), $n = \frac{5}{3}$ (4), we obtain the effect of the uncertainty in the operating voltage as:

$$\frac{\Delta f}{f} = -\frac{0.4\,E^{\frac{2}{3}}}{E^{\frac{5}{3}} - E_c^{\frac{5}{3}}}\,(1 - f). \tag{13}$$

All these effects increase rapidly with decreasing value of f. If we are not willing to accept an error in C — and hence in f — larger than 1%, it follows from Eq. (11) that the absorption function f must be 0.8 or larger. It should be noted too that the models for f tend to diverge for values of f below 0.8. This imposes a severe restriction in the use of soft X-rays for quantitative analysis. Such analysis, if at all possible, must be performed using low operating voltages, and high X-ray emergence angles, in order to keep f tolerably large.

---

1. Whether the relative systematic errors in the specimen and standard combine by addition or difference in Eq. 5 (and in similar subsequent equations) depends on the sign of the error in the input parameters, if known [5]. Except for the atomic number correction, compensation of errors from specimen to standard is almost always negligible.

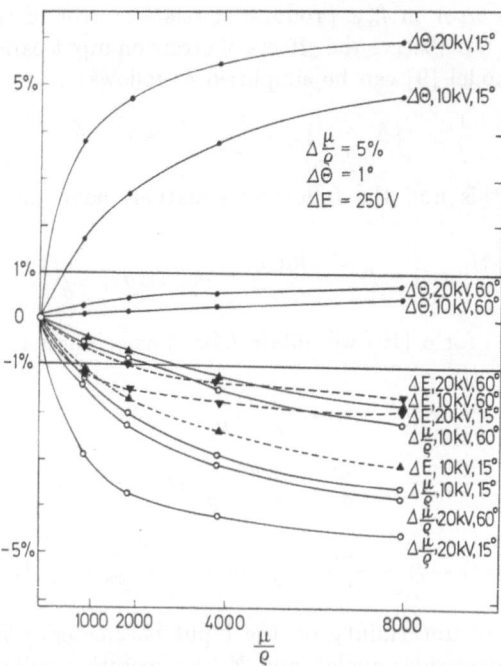

Fig. 1. Error of the absorption correction factor, k_E, caused by errors in the mass attenuation coefficient of the specimen, of the X-ray emergence angle, or the operating voltage E

By combining Eqs. (6) and (11)—(13), and assuming that $E_c = 0$, we can obtain the data shown in Fig. 1, which shows the effects of errors of μ/ϱ, θ, and E, as a function of the mass attenuation coefficient of the target. The curves are drawn for two emergence angles (15° and 60°), and for two operating voltages (10 kV and 20 kV). Within the needed accuracy, this graph can be considered valid for all target compositions. The condition ($E_c = 0$) does not impose serious restrictions to the use of this graph, since for radiation of high minimum excitation potential the absorption correction is usually small.

In view of the form of Eq. (4), we must consider how much compensation can be expected in case of errors affecting both f_{std} and f_{spec}. If pure elements are used as standards, the absorption coefficient of the standard is usually considerably smaller than that of the specimen. For instance, the absorption coefficient of aluminum for aluminum K_α radiation is approximately 400, while those of most other elements are 3 to 20 times higher. Hence, very little if any, compensation will occur in cases of severe absorption correction.

For the analysis with very soft radiation, Duncumb's model [11]

$$C = K \frac{\Phi(0)_{std}}{\Phi(0)_{spec}} \cdot \frac{\chi_{spec}}{\chi_{std}} \tag{14}$$

gives the following expression for error propagation:

$$\frac{\Delta C}{C} = \frac{\Delta(\mu/\varrho)_{spec}}{(\mu/\varrho)_{spec}} - \frac{\Delta(\mu/\varrho)_{std}}{(\mu/\varrho)_{std}} \tag{15}$$

equivalent to Eq. (7), with ($f \to 0$). This seems to preclude the possibility of quantitative analysis with very soft radiation by the "absolute method", in view of the uncertainties of the attenuation coefficients in the long wavelength region.

Since the accuracy of models for the absorption correction is mainly judged by their agreement with experimental f curves, some comment on these is warranted. We find for these curves the same uncertainties as previously discussed for the correction of electron probe data. Few authors have given numerical data — not to mention error bars or uncertainty bounds —, and at least some of the published curves do not show experimental points and are difficult to read. Consequently, the information presently available is of limited value for quantitative purposes. Further measurements of high accuracy are urgently needed.

Fluorescence Correction

A detailed analysis of error propagation in the fluorescence correction can be found in reference [12]. In view of the greater complexity of the model, the numerical method of error propagation was employed. The results are shown in Figs. 2 and 3. Besides uncertainties of the model (particularly for excitation of K-lines by L-lines and vice versa), only the uncertainty of the fluorescent yield, ω, is of importance. The uncertainty increases moderately with operating

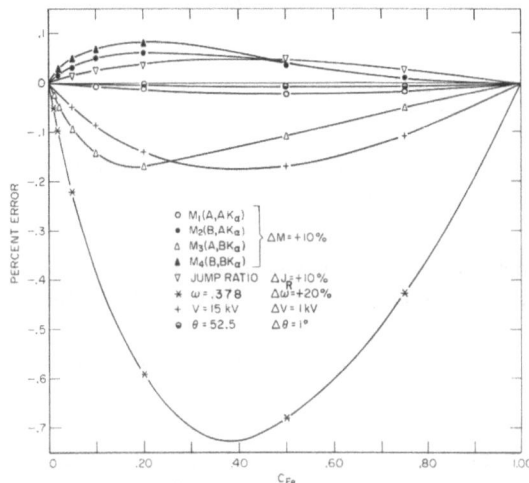

Fig. 2. Error in percent of the iron concentration, caused in Fe—Ni alloys by various input parameters for Reed's equation. "M" stands for μ/ϱ. Note that J_R here is the absorption jump ratio

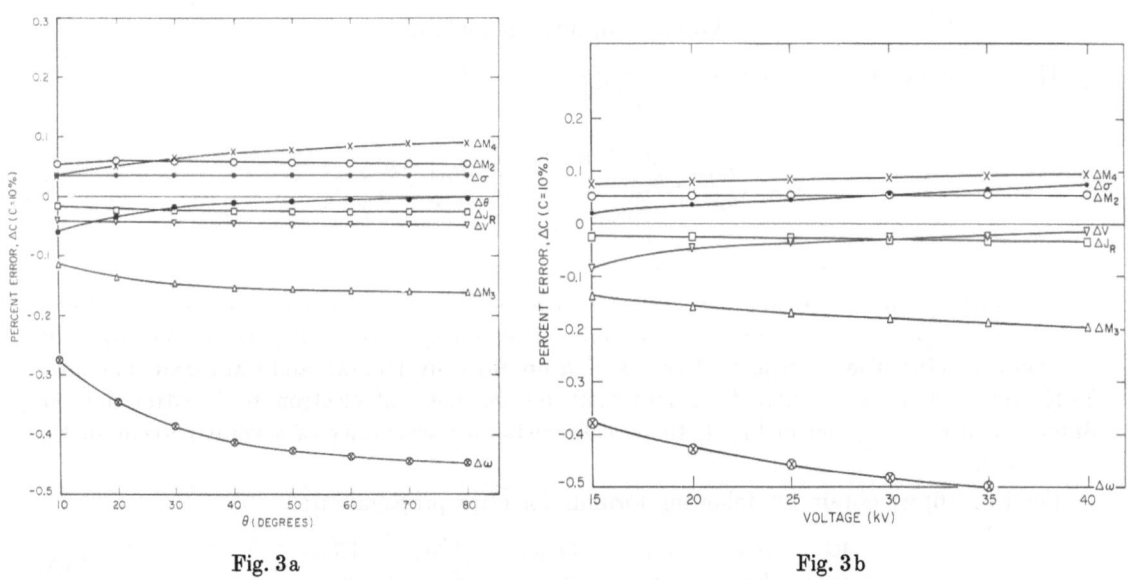

Fig. 3a   Fig. 3b

Fig. 3a. Error in percent of the iron concentration, in a 0.1 Fe—0.9 Ni alloy, as a function of X-ray emergence angle. Input errors are the same as in Fig. 2

Fig. 3b. Error in percent of the iron concentration in 0.1 Fe—0.9 Ni alloy as a function of accelerating voltage

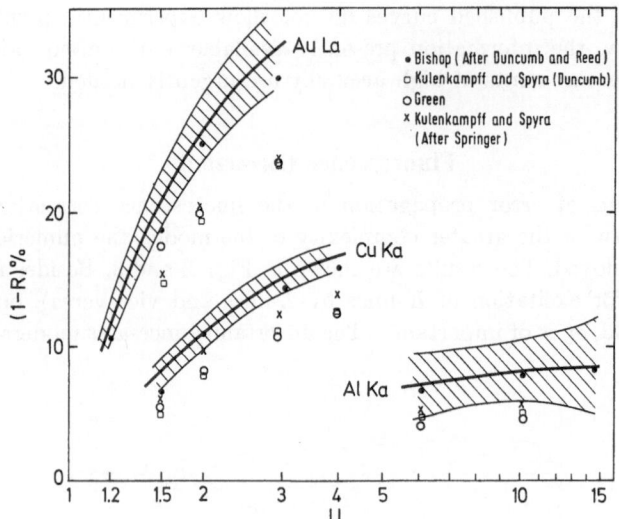

Fig. 4. Uncertainty of the backscatter correction coefficient, R. Solid line and error bounds after Dérian and Castaing [17]

voltage and with X-ray emergence angle. Let us assume a specimen of the composition 0.1 Fe — 0.9 Ni, analyzed at 20 keV. The estimated uncertainty in the value of ω is taken to be 20%. The resulting uncertainty in the calculated value or iron is 0.33% for an X-ray emergence angle of 15°, and 0.44% for 60°. But the estimate of an uncertainty of 20% in the value of ω is probably pessimistic.

As pointed out by Castaing [13], the fluorescence uncertainties become severe when local concentration changes are large. Unless the geometry of such specimens is well known, quantitative analysis is impossible under any operating conditions.

We have not as yet studied the error propagation in the correction of fluorescence due to continuous radiation.

Atomic Number Correction

The atomic number correction can be expressed as follows:

$$k_Z = \frac{R_{std}}{R_{spec}} \cdot \frac{\int_{E_c}^{E_0} \left(\frac{Q}{S}\right)_{std} dE}{\int_{E_c}^{E_0} \left(\frac{Q}{S}\right)_{spec} dE} \simeq \frac{R_{std}}{R_{spec}} \cdot \frac{\bar{S}_{spec}}{\bar{S}_{std}}. \tag{16}$$

In this equation, R denotes the backscatter correction factors, Q the ionization cross-sections, and S the stopping powers. Values for R were calculated by Green [14], Duncumb [15], and by Springer [16]. The experimental values of R obtained by Dérian and Castaing [17] agree fairly well with those calculated by Duncumb on the basis of electron backscatter data of Bishop. However, as seen in Fig. 4, there still exists an uncertainty of several percent in the values of R.

For Eq. (16) we obtain the following formula for error propagation:

$$\frac{\Delta C}{C} = \frac{\Delta K_Z}{K_Z} = \frac{\Delta R_{std}}{R_{std}} - \frac{\Delta R_{spec}}{R_{spec}} - \frac{\Delta \bar{S}_{std}}{\bar{S}_{std}} + \frac{\Delta \bar{S}_{spec}}{\bar{S}_{spec}}. \tag{16A}$$

Fig. 4 thus shows that we still do not know the R factors with accuracy sufficient for quantitative analysis of specimens which differ considerably in atomic number from the standards. The additivity of the R factors (in mass-proportion) for multi-element targets seems reasonably

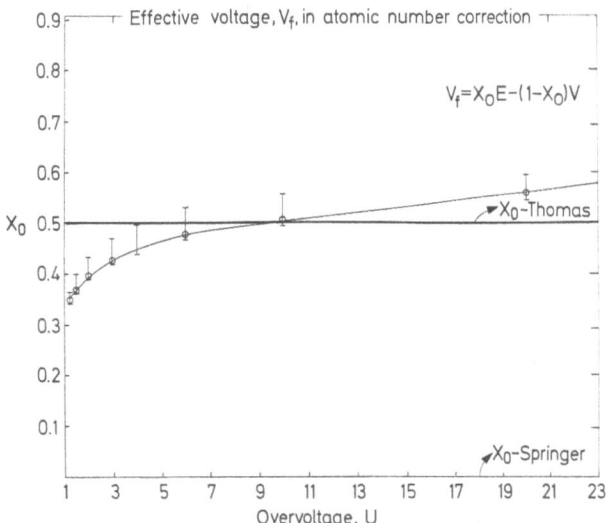

Fig. 5. Effective voltage, V_f, which gives the same result as the full integration in Eq. (16). Bars indicate ranges of binaries of elements from Be to Au. Values insensitive to J

well established. It is not generally recognized, however, that electron backscatter coefficients, and therefore probably also R factors, do not vary smoothly with atomic number [21].

Expressions for the ionization cross-section were formulated by GREEN and COSSLETT [18], and by POCKMANN et al. [19]. We have performed the integrations shown in Eq. (16) for elements from $Z = 4$ to $Z = 79$, using both expressions and we find no effects of the difference of these expressions on the atomic number correction. It was in fact proposed by THOMAS [3] and by SPRINGER [20] that an average value of the stopping power S can be used, thus eliminating Q [Eq. (16), right side]. The value of the stopping power is usually calculated by Bethe's equation as used by DUNCUMB and REED [15]:

$$S = K \cdot \frac{Z}{A} \cdot \frac{1}{E} \cdot \ln\left(\frac{1.166\,E}{J}\right). \tag{17}$$

Here, the constant K cancels in the calculation of k_Z; J is the mean ionization potential, other symbols have the same significance as in previous equations. Using this equation, with various sets of values of J shown in Fig. 7, we have determined for several combinations of elements and overvoltage an effective voltage, V_f, which gives the same k_Z values as the full integrations indicated in Eq. (16). As shown in Fig. 5, the proposed usage of a potential half way between the operating voltage and the minimum excitation potential leads in most cases to little error. However, as shown in Fig. 6, for large atomic number differences, errors exceeding 1% may arise.

There is also some doubt about the stopping power of multi-element targets. THOMAS [3] used the method of averaging (R/S), equivalent to the α-correction of CASTAING [13]. DUNCUMB and REED [15] prefer the separate averaging of the values of R and S, in mass-proportions. The resulting differences in the k_Z values amount to more than 3% for large differences in atomic number.

The accuracy of Bethe's law at low operating voltages is probably poor. Furthermore, there is considerable discrepancy concerning the values of the mean ionization potential J (Fig. 7), particularly for low atomic number elements. It is also known that J varies with the chemical state of the respective element. The effects of the uncertainty in J on the result are shown in Fig. 8.

The values of J proposed by DUNCUMB and REED [15] were obtained by empirical manipulation, using the error distribution of a series of analyses as a criterion for fit. There is nothing wrong with such an empirical adjustment, but it must be recognized that this adjustment

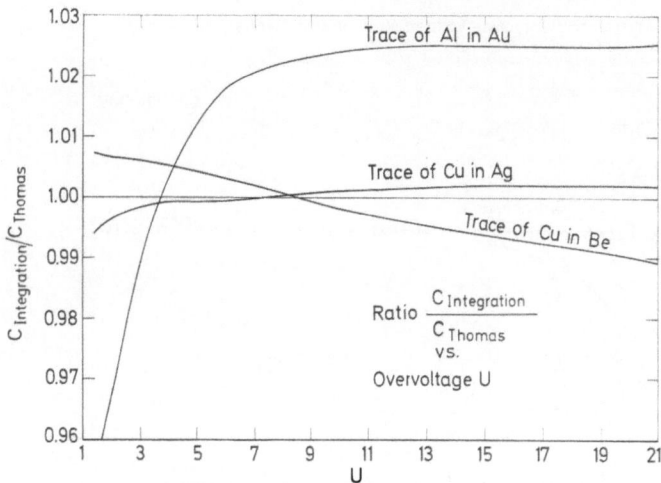

Fig. 6. Error due to the use of Thomas' effective voltage in calculation of stopping power, as a function of overvoltage

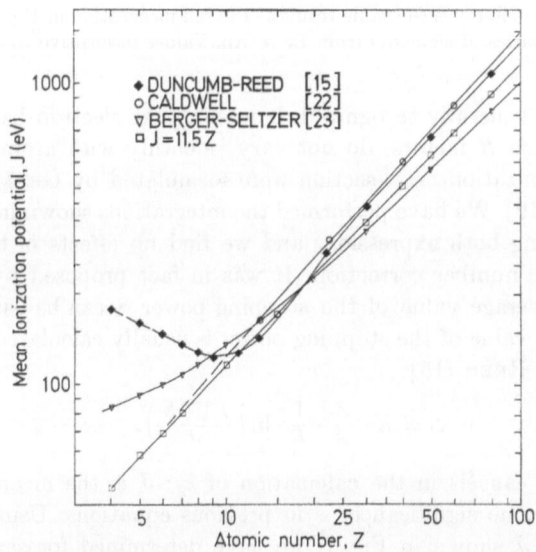

Fig. 7. Proposed values for the mean ionization potential J

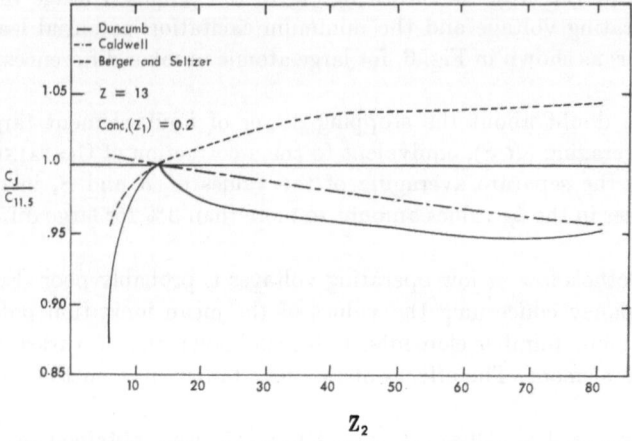

Fig. 8. Effect of varying expressions for the mean ionization potential J. C_J is the result of the analysis of a binary containing 0.2 Al and 0.8 of the element Z_2. $C_{11.5}$ is the result of using $J = 11.5\,Z$. Operating voltage was assumed to be 20 kV

probably compensates for other factors as well, and that it should be revised whenever any other factor is readjusted.

If the atomic number correction is over- or underestimated, errors of similar magnitude and opposite sign are produced in the determination of components of binary specimens. Therefore, it is incorrect to consider summation of components to close to 100% as a sufficient criterion of analytical accuracy, as proposed recently [25].

Some significant sources of errors may not have been recognized yet. For example, noticeable effects may arise from varying crystal orientation, and from different stopping power in metals and non-metallic materials.

It should also be pointed out that most of the uncertainties discussed in this paper cannot be overcome by the more detailed and complex Monte-Carlo calculations or by the use of transport equations. We believe that a proper recognition of all sources of error is necessary in order to achieve a significant advance in quantitative electron probe microanalysis. Meanwhile, the consideration of error propagation permits choosing the operating conditions in such a way as to minimize the errors of analysis.

References

1. POOLE, D. M., and P. M. THOMAS: The electron microprobe (McKINLEY, HEINRICH, WITTRY, eds.), p. 269. New York: John Wiley & Sons 1966.
2. Quantitative electron probe microanalysis. NBS Special Publ. **298**, 93, 133 (1968).
3. THOMAS, P. M.: AERE Report 4593 U. K. At. Energy Authority, 1964.
4. HEINRICH, K. F. J.: Advanc. X-ray Microanalysis, **11**, 40 (1968).
5. KU, H. H.: J. Res. Natl. Bur. Std. C **70**, 263 (1966).
6. SPIELBERG, N.: Rev. Sci. Instr. **37**, 1268 (1966).
7. HEINRICH, K. F. J., D. VIETH, and H. YAKOWITZ: Advanc. X-Ray Analysis **9**, 208 (1966).
8. YAKOWITZ, H., and K. F. J. HEINRICH: Mikrochim. Acta (1) 182 (1968).
9. PHILIBERT, J.: X-ray optics and X-ray analysis (PATTEE, COSSLETT, ENGSTRÖM, eds.), p. 379. New York: Academic Press 1963.
10. DUNCUMB, P., and P. K. SHIELDS: The electron microprobe (McKINLEY, HEINRICH, WITTRY, eds.), p. 284. New York: John Wiley & Sons 1966.
11. —, et D. A. MELFORD: Optique des rayons X et microanalyse (CASTAING, DESCHAMPS, PHILIBERT, eds.), p. 240. Paris: Hermann 1966.
12. HEINRICH, K. F. J., and H. YAKOWITZ: Mikrochim. Acta 905 (1968).
13. CASTAING, R.: Advanc. Electron. Electron Phys. **13**, 317 (1960).
14. GREEN, M.: Thesis. Cambridge University 1964.
15. DUNCUMB, P., and S. J. B. REED: Quantitative electron probe microanalysis. NBS, Special Publ. **298**, 133 (1968).
16. SPRINGER, G.: Neues Jahrb. Mineral., Monatsh. 9/10, 304 (1967).
17. DÉRIAN, J. C., et R. CASTAING: Optique des rayons X et microanalyse (CASTAING, DESCHAMPS, PHILIBERT, eds.), p. 193. Paris: Hermann 1966.
18. GREEN, M., and V. E. COSSLETT: Proc. Phys. Soc. (London) **78**, 1206 (1961).
19. POCKMANN, L. T., D. L. WEBSTER, P. KIRKPATRICK, and K. HARWORTH: Phys. Rev. **71**, 330 (1947).
20. SPRINGER, G.: Neues Jahrb., Mineral., Monatsh. (4), 113 (1966).
21. HEINRICH, K. F. J.: Quantitative electron probe microanalysis. NBS Special Publ. **298**, 8 (1968).
22. CALDWELL, D. O.: Phys. Rev. **100**, 291 (1955).
23. BERGER, M. J., and S. M. SELTZER: N. Sc. Sci., Nat. Res. Council Publ. **1133** (Washington, D. C., 1964), p. 205.
24. MULVEY, T.: Quantitative electron probe microanalysis, NBS, Special Publ. **298**, 81 (1968).
25. SALTER, W. J. M.: Brit. J. Appl. Phys. Ser. 2, **1**, 541 (1968).

Prüfung der Korrekturen für Ordnungszahl, Absorption und sekundäre Fluoreszenz an metallischen Zweistofflegierungen mit nur je einem Matrixeffekt

CH. THOMA und P. HÖLLER

Stranski-Institut für Metallurgie der Hüttenwerk Oberhausen AG, Oberhausen, BRD

Experimentelle Überprüfung der Korrekturformeln für Ordnungszahl, Absorption und sekundäre Fluoreszenz an metallischen Zweistofflegierungen der Systeme Mn—Fe, Ni—Fe, Al—Mg, Al—Fe, Si—Fe, V—Si, Nb—V bei Anregungsspannungen zwischen 10 und 40 kV. Mittlere Abweichungen der korrigierten Meßwerte von den Gehalten für Mn—Fe, Ni—Fe, V—Si, Nb—V $\leqq 5\%$. Für Al—Mg, Al—Fe, Si—Fe (starke Absorption) systematische Abweichungen bis zu 20%. Vergleich mit den Ergebnissen einer eigenen früheren Untersuchung, bei der der Entnahmewinkel der Röntgenstrahlung variiert wurde, sowie mit Messungen anderer Autoren. Diskussion der im Falle starker Absorption auftretenden systematischen Fehler anhand der Philibertschen Formel.

Bei der quantitativen Analyse an der Mikrosonde ist man auf die Benutzung von Korrekturverfahren angewiesen, da in der Regel Eichproben mit gleicher Absorption, sekundärer Fluoreszenz bzw. gleichem Ordnungszahleffekt nicht zur Verfügung stehen. Bei der Ableitung der bekannten Korrekturformeln wurden starke Vereinfachungen gemacht. Ihre Anwendbarkeit muß daher an typischen Proben mit bekannter Zusammensetzung überprüft werden. Besonders geeignet sind Proben, in denen einer der drei Matrixeffekte gegenüber den anderen stark hervor tritt. Zusätzlich müssen die Anregungsspannung und der Entnahmewinkel der charakteristischen Röntgenstrahlung, die in die Korrekturformeln als Parameter eingehen, variiert werden.

Der Einfluß des Entnahmewinkels wurde von den Verfassern in einer früheren Arbeit [1] an Proben der Zweistofflegierungen Fe—Mn, Fe—Si und Fe—Ni geprüft. Während bei den Proben der Systeme Fe—Mn und Fe—Ni sowohl bei $\theta = 20°$ als auch 75° die korrigierten Meßwerte um nicht mehr als 3% von den chemisch ermittelten Gehalten abweichen, ist dies bei Fe—Si nur bei den nach PHILIBERT [2] korrigierten Meßwerten für $\theta = 20°$ der Fall. Für $\theta = 75°$ liegen die korrigierten Meßwerte im Mittel um 6% über den chemisch ermittelten Gehalten. Die Tatsache, daß die Analyse ein und derselben Probe bei stark unterschiedlichem Entnahmewinkel der Röntgenstrahlung und Anwendung der gleichen Korrekturformel auf die Meßergebnisse unterschiedliche Gehalte ergibt, zeigt, daß das angewandte Verfahren nicht richtig sein kann. In den Untersuchungen, über die hier berichtet wird, wird der zweite der in die Korrekturen eingehenden Parameter, die Anregungsspannung, in einem großen Bereich variiert.

Versuchsdurchführung

Es wurden metallische Zweistofflegierungen ausgewählt, bei denen einer der Matrixeffekte — Ordnungszahl, Absorption oder sekundäre Fluoreszenz — gegenüber den anderen stark überwiegt. Dadurch war es möglich, die Korrekturverfahren einzeln zu prüfen. Zusätzlich wurden Proben untersucht, bei denen sowohl Ordnungszahl- als auch Absorptionseffekt zu berücksichtigen sind.

Die Proben wurden im Lichtbogenofen erschmolzen und nach Abschalten des Lichtbogens durch schnelles Niederdrücken eines Kupferstempels abgeschreckt [3]. In den nach dieser Methode zweimal umgeschmolzenen Proben waren die Legierungspartner homogen verteilt.

Die Untersuchung wurde an einem Gerät mit 75° Austrittswinkel* durchgeführt. Jede Probe wurde an 10 Stellen analysiert; die gemessenen Impulszahlen wurden gemittelt.

Korrekturverfahren

Auf die Meßwerte, das sind die Verhältnisse der nach Untergrund und Zählverlust korrigierten Impulszahlen von Probe und reinem Metall, wurden folgende Korrekturverfahren angewandt:

Für Ordnungszahl das von DUNCUMB [4] angegebene Verfahren.

Zur Absorptionskorrektur die Philibertsche Formel

$$f(\chi) = \frac{1+h}{(1+\chi/\sigma)+h(1+\chi/\sigma)^2},$$

wobei für σ sowohl die von PHILIBERT [2] als auch die von DUNCUMB-SHIELDS [5] angegebenen Werte eingesetzt wurden.

Für die Fluoreszenzkorrektur die von REED und LONG [6] angegebene Formel, die eine Modifizierung der Castaingschen Korrekturformel darstellt. Die Massenabsorptionskoeffizienten wurden einer Zusammenstellung von HEINRICH [7] entnommen.

Versuchsergebnisse

Mn in Fe. Da die Korrekturfaktoren für Ordnungszahl, Absorption und sekundäre Fluoreszenz das Meßergebnis um nicht mehr als ca. 1% beeinflussen, wurden sie in diesem Falle nicht berücksichtigt. Unabhängig von der Anregungsspannung ergeben die Meßwerte (Abb. 1) einen Gehalt von im Mittel 9,6%, wobei die Einzelwerte zwischen 9,7 und 9,4% liegen. Chemisch wurde ein Gehalt von 9,4% bestimmt. Der Variationskoeffizient für die Abweichung der Meßwerte vom Mittelwert errechnet sich zu 1,5%.

Ni in Fe. Wegen der starken Absorption von $Ni K\alpha$ in Eisen ergibt sich ein mit der Anregungsspannung zunehmender Abfall der unkorrigierten Meßwerte (Abb. 2). Auf diese Werte wurde

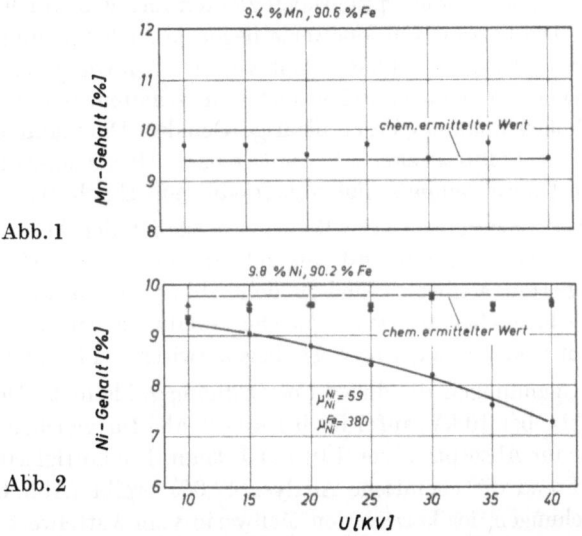

Abb. 1

Abb. 2

Abb. 1. Gemessene Mn-Gehalte in einer Legierung mit 9,4% Mn, 90,6% Fe

Abb. 2. Gemessene und korrigierte Ni-Gehalte in einer Legierung mit 9,8% Ni, 90,2% Fe

● Meßwerte
▲ Absorption korrigiert nach PHILIBERT   } Ordnungszahl korrigiert
■ Absorption korrigiert nach PHILIBERT-DUNCUMB-SHIELDS } nach DUNCUMB

* Geoscan der Fa. Cambridge Instrument Company Ltd.

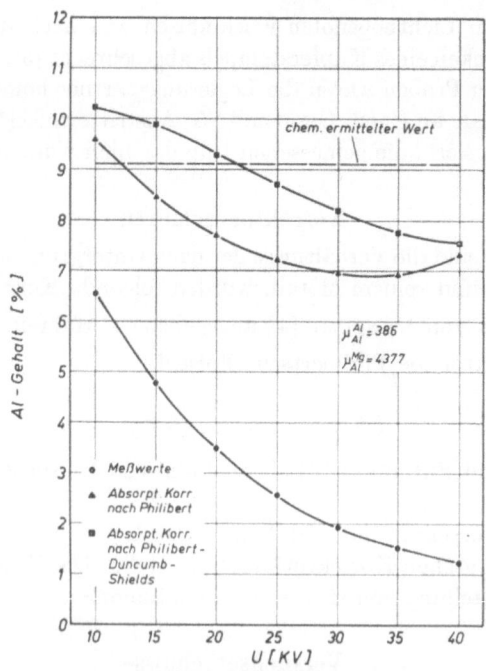

Abb. 3. Gemessene und korrigierte Al-Gehalte in einer Legierung mit 9,1% Al, 90,9% Mg

die Philibertsche Absorptionskorrektur angewandt mit σ nach PHILIBERT sowie nach DUNCUMB-SHIELDS. Lediglich bei 10 kV ergibt sich ein merklicher Unterschied zwischen beiden Verfahren, da bei dieser Spannung σ nach PHILIBERT 9600, nach DUNCUMB-SHIELDS dagegen 30250 wird. Der Mittelwert der korrigierten Meßwerte liegt um 0,2% absolut unter dem Gehalt von 9,8%. Der Variationskoeffizient ist 1,2%.

Al in Mg. Der Abfall der gemessenen Konzentration mit steigender Anregungsspannung macht sich wegen des extrem hohen Massenabsorptionskoeffizienten hier wesentlich stärker bemerkbar als bei Ni in Fe (Abb. 3). Die korrigierten Meßwerte liegen je nach Spannung bzw. Korrekturverfahren zwischen 6,9 und 10,2%. In Abhängigkeit von der Anregungsspannung ergeben sich systematische Abweichungen von den chemisch ermittelten Gehalten. Bei niedrigen Spannungen erhält man zu hohe, bei hohen Spannungen zu niedrige Gehalte. Die maximalen Abweichungen vom Sollwert sind mit σ nach PHILIBERT 24%, mit σ nach DUNCUMB-SHIELDS 18% relativ. Beide Korrekturverfahren liefern demnach nur unbefriedigende Ergebnisse.

Al in Fe, Si in Fe. Diese Legierungen verhalten sich bezüglich der Absorption wie Al in Mg. Hinzu kommt die Berücksichtigung des Ordnungszahleffektes wegen $\Delta Z = 13$ bzw. 12. Die unkorrigierten und korrigierten Meßwerte sind in den Abb. 4 und 5 aufgetragen. Qualitativ ergibt sich trotz der unterschiedlichen Matrix der gleiche Verlauf wie bei Al in Mg. Die maximalen Abweichungen vom Gehalt sind etwas geringer, sie liegen zwischen 14 und 17%.

V Si$_2$. Bei niedrigen Spannungen dominiert der Ordnungszahleffekt. Der zugehörige Korrekturfaktor sinkt von 1,14 bei 10 kV auf 1,09 bei 40 kV ab. Im gleichen Spannungsbereich steigt der Korrekturfaktor für Absorption von 1,01 auf 1,08 an. Die korrigierten Meßwerte liegen im Mittel bei 49,0%, während die chemische Analyse 47,0% ergibt (Abb. 6). Der Variationskoeffizient für die Abweichungen der korrigierten Meßwerte vom Mittelwert beträgt 1,6%.

Nb in V. Dieser Fall ist ähnlich dem vorhergehenden, auch hier handelt es sich um ein schweres Element in leichter Matrix. Bei niedrigen Spannungen dominiert der Ordnungszahleffekt. Die korrigierten Meßwerte fallen von 8,6% Nb bei 10 kV auf 8,1% Nb bei 40 kV ab, der Gehalt ist 8,0%.

Fe in Ni. Bei dieser Legierung ist nur die sekundäre Fluoreszenz zu berücksichtigen. Die nach der Formel von REED und LONG korrigierten Meßwerte (Abb. 8) liegen im Mittel bei 8,0%, chemisch wurden 7,9% Fe bestimmt. Der Variationskoeffizient ist ca. 1% (Abb. 8).

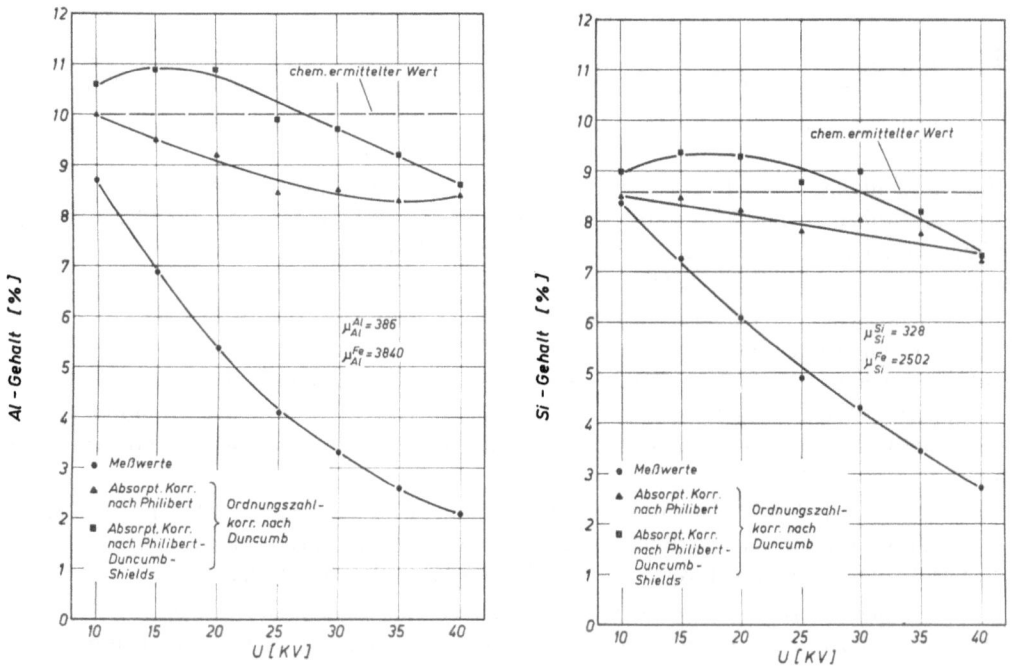

Abb. 4. Gemessene und korrigierte Al-Gehalte in einer Legierung mit 10,0% Al, 90,0% Fe

Abb. 5. Gemessene und korrigierte Si-Gehalte in einer Legierung mit 8,6% Si, 91,4% Fe

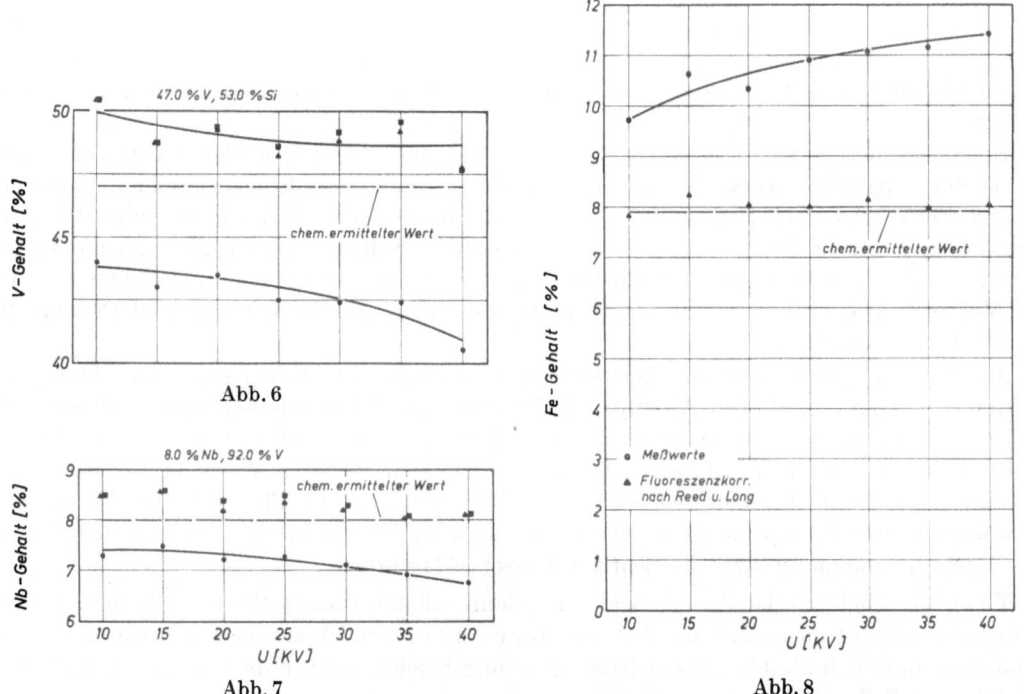

Abb. 6

Abb. 7

Abb. 8

Abb. 6. Gemessene und korrigierte V-Gehalte in einer Legierung mit 47,0% V, 53,0% Si

Abb. 7. Gemessene und korrigierte Nb-Gehalte in einer Legierung mit 8,0% Nb, 92,0% V

● Meßwerte
▲ Absorption korrigiert nach PHILIBERT
■ Absorption korrigiert nach PHILIBERT-DUNCUMB-SHIELDS
} Ordnungszahl korrigiert nach DUNCUMB

Abb. 8. Gemessene und korrigierte Fe-Gehalte in einer Legierung mit 7,9% Fe, 92,1% Ni

Diskussion und Ergebnisse

Es soll versucht werden, aus den Ergebnissen dieser und der vorangegangenen Untersuchung [1] Aussagen über den Gültigkeitsbereich der angewandten Korrekturformeln abzuleiten. Dabei werden auch Ergebnisse einiger anderer Autoren, soweit sie in diesen Rahmen passen, berücksichtigt.

Für den Fall, daß keine Korrekturen anzuwenden sind, liegen die Ergebnisse von Messungen an Mn—Fe-Proben im Bereich von 0—50% Mn für $\theta = 20$ und 75° [1] sowie die hier mitgeteilten Ergebnisse für eine Probe mit 9,4% Mn bei 10—40 kV Anregungsspannung vor. Die mittleren Abweichungen vom Gehalt sind stets $\leq 2\%$.

Die gleichen Untersuchungen wurden zur sekundären Fluoreszenz an Fe—Ni-Proben durchgeführt. Die mittleren Abweichungen der nach REED und LONG korrigierten Meßwerte vom Gehalt sind auch hier $\leq 2\%$. COLBY und CONLEY [8] haben die sekundäre Fluoreszenz am System Cr—Fe geprüft. Wendet man auf die für eine Probe mit 10% Cr zwischen 15 und 40 kV angegebenen Meßwerte die hier benutzte Korrektur an, dann ergibt sich eine mittlere Abweichung vom Gehalt von $< 2\%$.

Die Ordnungszahlkorrektur wurde durch Messungen an Proben der Systeme V—Si sowie Nb—V geprüft. Im unteren Spannungsbereich, wo die Absorptionskorrektur zu vernachlässigen ist, liegen die korrigierten Meßwerte um ca. 5% bzw. 7% über den Gehalten.

Zur Absorptionskorrektur liegen eigene Messungen an den Systemen Ni—Fe, Al—Mg, Al—Fe, Si—Fe vor. Für Ni in Fe erhält man mittlere Abweichungen von $\leq 2\%$. Der Untersuchungsumfang ist der gleiche wie bei Mn in Fe bzw. Fe in Ni. Bei den Proben der übrigen 3 Systeme ergeben sich dagegen Abweichungen bis zu 24%. Qualitativ verhalten sich die errechneten Gehalte von Al bzw. Si in den 3 Legierungen in Abhängigkeit von der Anregungsspannung gleich, die mit σ nach PHILIBERT bzw. nach DUNCUMB-SHIELDS korrigierten Werte weichen jedoch im mittleren Spannungsbereich stark voneinander ab. Für jede Legierung und jedes σ läßt sich jeweils eine Spannung angeben, bei der der korrigierte Meßwert mit dem chemisch ermittelten Gehalt übereinstimmt. Unterhalb dieser Spannung errechnet man zu hohe, oberhalb zu niedrige Gehalte.

Ergebnisse der Fe-Bestimmung in Fe—Cr-Proben sind ebenfalls in der Arbeit von COLBY und CONLEY angegeben. Dieser Fall ist mit Ni in Fe zu vergleichen. Korrigiert man die zwischen 15 und 40 kV erhaltenen Meßwerte für eine Probe mit ca. 10% Fe nach DUNCUMB-SHIELDS, dann liegen sämtliche Gehalte um 5% relativ unter dem Sollwert. Im Gegensatz dazu ergeben sich systematische Abweichungen von mehr als 10% nach oben, wenn man die gleiche Korrektur auf Meßwerte von Proben des Systems Ni—Fe anwendet, die von ZIEBOLD und OGILVIE [9] veröffentlicht wurden.

Aus den Darlegungen folgt, daß besonders große systematische Fehler bei starker Absorption auftreten. Die an den Proben Al—Mg, Al—Fe und Al—Si festgestellten systematischen Abweichungen der korrigierten Meßwerte von den chemisch ermittelten Gehalten lassen die Vermutung zu, daß die in die Absorptionskorrekturformel eingesetzten Werte für σ falsch sind. Man kann aus den Meßergebnissen der Al—Mg-Legierung und der Philibertschen Formel σ in Abhängigkeit von der Spannung errechnen. Mit diesen σ-Werten erhält man natürlich für die Al—Mg-Legierung und $\theta = 75°$ die richtigen Korrekturfaktoren.

Wenn die Philibertsche Formel richtig ist, dann müßten diese σ-Werte auch für Al in Fe gültig sein. Dies ist aber nicht der Fall. Nur bei ca. 10 und 40 kV erhält man geringere Abweichungen vom Gehalt als 3%. Im mittleren Spannungsbereich weichen die korrigierten Meßwerte um 8% vom Sollwert ab.

Es läßt sich auch leicht zeigen, daß in den 3 aufgeführten Fällen andere Massenabsorptionskoeffizienten zu keiner besseren Übereinstimmung führen können. Dadurch würden die korrigierten Meßwerte nach oben oder unten verschoben, und zwar bei hohen Spannungen stärker als bei niedrigen. Insgesamt ergibt sich eine Verschiebung und Drehung der korrigierten Meßkurven. Rechnet man z.B. im Falle Al—Mg mit einem Massenabsorptionskoeffizienten von 5500 statt 4377, dann ergeben sich für die nach DUNCUMB-SHIELDS korrigierten Meßwerte bei

10 kV 11,4% statt 10,2% und bei 40 kV 9,8% statt 7,5%. Erst bei $\mu = 7000$ erhält man einen von der Spannung unabhängigen Wert, der allerdings mit ca. 14% um mehr als 50% vom chemisch ermittelten Gehalt abweicht.

Diese Befunde zeigen, daß die systematischen Fehler, die bei starker Absorption auftreten, nicht durch in die Korrekturformel eingehende falsche Parameter wie σ oder μ hervorgerufen werden, sondern dadurch, daß die Formel nicht richtig ist. Zur weiteren Prüfung werden die Proben, an denen die hier diskutierten Messungen durchgeführt wurden, z.Z. von 3 anderen Instituten untersucht, die über Geräte mit 20 und 50° Austrittswinkel verfügen.

An den Mikrosonden in den Laboratorien der Stahlindustrie werden hauptsächlich nicht-metallische Einschlüsse, Schlacken, Rost- und Zunderschichten sowie Steigerungen untersucht. Die im Falle starker Absorption auftretenden großen systematischen Fehler machen sich besonders bei der quantitativen Analyse von Einschlüssen und Schlacken bemerkbar. Dazu kommen noch die Unsicherheiten in der Ordnungszahlkorrektur. Das bedeutet, daß in diesem Falle eine geforderte Genauigkeit von ca. 5% ohne die Benutzung von Eichproben nicht möglich ist. Dies ist natürlich im Hinblick auf die Vielzahl und die Schwierigkeiten bei der Herstellung solcher Proben unbefriedigend.

Literatur

1. HÖLLER, P., u. C. THOMA: Z. Physik. Chem. (Frankfurt) **53** (1—6), 286 (1967).
2. PHILIBERT, J.: X-ray optics and X-ray microanalysis, p. 379. New York and London: Academic Press 1963.
3. HÖLLER, P.: AEH **37**, 483 (1966).
4. DUNCUMB, P.: Seminar on quantitative microprobe analysis. NBS, Washington, 1967.
5. —, and P. K. SHIELDS: X-ray optics and X-ray microanalysis, p. 329. New York and London: Academic Press 1963.
6. REED, S. J. B., and J. V. P. LONG: 3. Intern. Symp. X-Ray Opt. and X-Ray Microanalysis, Stanford 1962.
7. HEINRICH, K. F. J.: The electron microprobe, p. 295. New York: John Wiley & Sons 1966.
8. COLBY, J. W., and D. K. CONLEY: X-ray opt. and X-ray microanalysis, p. 263. Paris: Hermann 1966.
9. ZIEBOLD, T. O., and R. E. OGILVIE: Anal. Chem. **35**, 621 (1963).

Eine empirische Methode
zur quantitativen chemischen Analyse von Mikroteilchen
mit der Mikrosonde

H.-J. Hoffmann, J. H. Weihrauch und H. Fechtig

Max-Planck-Institut für Kernphysik, Heidelberg, BRD

Solange das von einem Elektronenstrahl zur charakteristischen Röntgenstrahlung angeregte Volumen vollständig in der zu analysierenden Probe enthalten ist, gilt in erster Näherung die von Castaing [1] gefundene Beziehung:

$$C_A = \frac{I_A}{IStdA} . \tag{1a}$$

C_A: Konzentration des Elementes A in der Probe;
I_A: Intensität einer ergiebigen Röntgenlinie des Elementes A, die von der Probe emittiert wird;
$IStdA$:   Intensität derselben Röntgenlinie, die von der Standardprobe, die zu 100% aus dem Element A besteht, emittiert wird.

Dieses Meßprinzip läßt sich in abgewandelter Form auch auf Teilchen anwenden, deren Dicke kleiner als die maximale Anregungstiefe $x_{m,A}$ der zur Analyse verwendeten Röntgenlinie ist. Sie ist gegeben durch [1]:

$$x_{m,A} = 0{,}033 \left(U^{1,7} - U_K^{1,7}(A) \right) \frac{\bar{A}}{\bar{\varrho}\,\bar{z}} . \tag{2}$$

U: Beschleunigungsspannung der Elektronen;
$U_K(A)$:   Mindestspannung, die erforderlich ist, um die Röntgenlinie des Elementes A anzuregen;
$\bar{A}$:   mittleres Atomgewicht der Elemente in der Probe;
$\bar{\varrho}$:   mittlere Dichte der Probe;
$\bar{z}$:   mittlere Ordnungszahl der Elemente in der Probe;
C_i: Konzentration der Elemente in der Probe

mit $\qquad \bar{A} = \sum_i C_i A_i; \quad \bar{\varrho} = \sum_i C_i \varrho_i; \quad \bar{z} = \sum_i C_i z_i.$

Die maximale Anregungstiefe in einer Probe, die mehrere Elemente E (nur das Element A) enthält, sei mit $x_{m,A}(E)$ $(x_{m,A}(A))$ bezeichnet.

Die Konzentration eines Elementes ergibt sich analog zu Gl. (1a) aus einem Intensitätsvergleich der charakteristischen Röntgenstrahlung, die von dem zu untersuchenden Teilchen und einem dazu „äquivalenten" Teilchen emittiert wird.

Anstatt auf die Standardprobe wird nun die Röntgenintensität auf die des äquivalenten Teilchens bezogen, welches den gleichen prozentualen Verlust an Röntgenstrahlung gegenüber einer Standardprobe aufweisen soll, wie das zu analysierende Teilchen gegenüber einer von anregungsfähigen Elektronen undurchdringbaren Probe desselben Materials.

Solche äquivalente Teilchen können zwar nicht hergestellt werden, aber ihre Röntgenintensität läßt sich bei Kenntnis des prozentualen Verlustes der Röntgenstrahlung näherungsweise berechnen.

Wegen der Ortsinstabilität und den Durchmesserschwankungen des Elektronenstrahls über eine längere Meßzeit wird die Punktanalyse durch ein Rasterverfahren ersetzt, wobei der Elektronenstrahl zeilenförmig eine vorgegebene Fläche abtastet.

Die Rasterfläche wird so dimensioniert, daß der Abstand zwischen Teilchenrand und Rasterflächenbegrenzung immer größer als der Elektronenstrahldurchmesser ist. Dadurch wird jede Stelle des Teilchens pro Rasterzyklus von gleich vielen Elektronen getroffen und die Strahlinstabilitäten mitteln sich bei mehrmaligem Abfahren der Rasterfläche heraus. Das heißt, die Analyse hängt nur noch vom Teilchen ab. Die Gleichung lautet dann:

$$C_A = \frac{R_A^n\, T(E)/R^{n'} StdA}{R_A^1\, T(A)/R^{n'} StdA} \equiv \frac{R_{A,\mathrm{rel}}\, T(E)}{R_{A,\mathrm{rel}}\, T(A)}. \tag{1b}$$

$R_A^n\, T(E)$: Anzahl der Röntgenimpulse des Teilchens nach n Rasterzyklen, diviert durch die Meßzeit;

$R_A^1\, T(A)$: Anzahl der Röntgenimpulse des äquivalenten Teilchens nach einem Rasterzyklus, dividiert durch die Meßzeit;

$R^{n'} StdA$: Anzahl der Röntgenimpulse von der Standardprobe nach n' Rasterzyklen, dividiert durch die Meßzeit.

n bzw. n' werden so groß gewählt, daß der statistische Fehler der Anzahl der Röntgenimpulse kleiner als 1% ist.

Damit die Messung unabhängig vom Elektronenstrom ist, wurde die Anzahl der Röntgenimpulse des Teilchens und des äquivalenten Teilchens auf die der Standardprobe bezogen.

Bisher wurden noch keine näheren Angaben über die Form und Größe eines äquivalenten Teilchens gemacht. Sie ergeben sich aus 2 Forderungen:

1. Damit der Röntgenintensitätsverlust experimentell untersucht werden kann, muß seine Dicke konstant sein.

2. Damit Teilchen und äquivalentes Teilchen denselben Prozentsatz an Röntgenstrahlung verlieren, müssen sie durch folgende 3 Bedingungen miteinander korreliert sein:

$$\frac{\bar{x}_T}{x_{m,A}(E)} = \frac{x}{x_{m,A}(A)}, \tag{3a}$$

$\bar{x}_T$: mittlere Teilchendicke,
x: konstante Dicke des äquivalenten Teilchens;

$$\frac{\bar{F}_T}{x_{m,A}^2(E)} = \frac{F}{x_{m,A}^2(A)} \tag{3b}$$

mit

$$\bar{x}_T \cdot \bar{F}_T = V_T(E),$$

$\bar{F}_T$: mittlere Teilchenfläche,
F: Fläche des äquivalenten Teilchens,
$V_T(E)$: Volumen des zu analysierenden Teilchens;

$$\frac{U_T}{x_{m,A}(E)} = \frac{U}{x_{m,A}(A)}, \tag{3c}$$

U_T: Teilchenumfang,
U: Umfang des äquivalenten Teilchens.

Das heißt, das äquivalente Teilchen ist eine Platte, deren Form durch die des Teilchens bestimmt ist.

Der Nenner von Gl. (1b) ist gegeben durch:

$$R_{A,\mathrm{rel}}\, T(A) = \frac{F}{F_0}\, f(x)\, f(\mathrm{Rand}). \tag{4}$$

Ist $x > x_{m,A}(A)$ und würden die Elektronen nicht aus ihrer Einfallsrichtung gestreut, so wäre $R_{A,\mathrm{rel}}\, T(A)$ durch F/F_0 gegeben. Da aber $x < x_{m,A}(A)$ ist, wird die Platte einerseits von anregungsfähigen Elektronen an ihrer Unterseite durchdrungen, woraus ein Röntgenintensitätsverlust

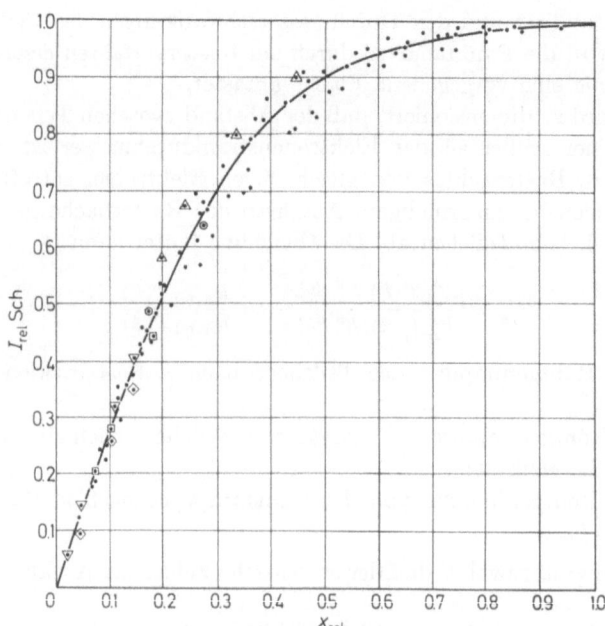

Abb. 1. Das Röntgenintensitätsverhältnis $I_{rel}Sch$ von Schichten und Standardprobe als Funktion der relativen
Schichtdicke x_{rel}

	Schichten	Unterlage
●	Aluminium	Kupfer
●	Gold	Kupfer
☐	Germanium	Silber
◇	Germanium	Kupfer
▽	Kupfer	Silber
△	Silber	Kupfer
○	Gold	Silber

resultiert, der durch die Funktion $f(x)$ korrigiert werden soll. Andererseits treten infolge der Streuprozesse anregungsfähige Elektronen auch am Rande der Platte aus. Dieser Röntgenintensitätsverlust soll durch die Funktion f(Rand) korrigiert werden.

Die Funktion $f(x)$ wurde empirisch bestimmt. Dazu wurden Metallschichten verschiedener Elemente hergestellt und deren Röntgenintensität bei verschiedenen Beschleunigungsspannungen gemessen.

In Abb. 1 ist das Röntgenintensitätsverhältnis $I_{rel}Sch$ von Schichten und Standardprobe als Funktion der relativen Schichtdicke $x_{rel} = x/x_{m,A}$ aufgetragen. Das Intensitätsverhältnis ist unabhängig vom Element und allen Beschleunigungsspannungen $U > 1{,}7\,U_K$. $f(x)$ ist durch $I_{rel}Sch$ für den der Plattendicke x entsprechenden Wert von x_{rel} gegeben.

f(Rand) wird aus einer experimentell bestimmten Kurve abgeleitet.

$$f(\text{Rand}) = (0{,}31 \pm 0{,}01) - (0{,}008 \pm 0{,}001)\,\frac{x_{m,A} \cdot U}{F}\,. \tag{5}$$

Diese Gleichung gilt für $0{,}2 \leqq x_{rel} \leqq 2$. Das heißt, die Plattendicke kann bis zu $^1/_5$ kleiner als die maximale Anregungstiefe sein. Die Konstanten in Gl. (5) wurden an Hand von Kügelchen ermittelt, deren Maße durch die eines Zylinders ersetzt wurden, dessen Volumen gleich dem Kugelvolumen und dessen Höhe gleich der mittleren Kugeldicke ist. Der dabei entstehende Fehler ist vernachlässigbar, verglichen mit dem Gesamtfehler.

Die Konzentration C_A eines Elementes A in einem kleinen Teilchen T erhält man also in erster Näherung aus:

$$C_A = \frac{R_{A,\,\mathrm{rel}}\, T(E)}{\dfrac{F}{F_0} \cdot f(x) \cdot \left(0{,}31 - 0{,}008 \dfrac{x_{m,A} \cdot U}{F}\right)} \,. \tag{1c}$$

Die Teilchenmaße ($\bar{x}_T$, U_T, $\bar{F}_T$) werden in einem Rasterelektronenmikroskop gemessen und in die entsprechenden Plattenmaße (x, U, F) umgerechnet.

Da die Konzentrationen C_i noch nicht bekannt sind, wird für x, U und F zunächst $\bar{x}_T$, U_T und $\bar{F}_T$ eingesetzt. Aus Gl. (1c) lassen sich daher die Konzentrationen zunächst nur in grober Näherung berechnen. Mit diesen Näherungswerten werden die Mittelwerte der Materialkonstanten ($\bar{A}$, $\bar{\varrho}$, $\bar{z}$) gebildet. Damit werden nun die Konzentrationen nochmals genauer berechnet. Anschließend werden die Korrekturen für Massenabsorption der Röntgenstrahlung und Röntgenfluoreszenzanregung vorgenommen.

Der prozentuale systematische Fehler der Konzentrationen läßt sich für Kügelchen mit $\pm 20\%$ abschätzen. Die bei Stahl- und Nickel-Chromkügelchen gefundenen Abweichungen vom Sollwert bestätigen diese Abschätzung.

Oft ist das Verhältnis der Konzentrationen von größerem Interesse als der absolute Gehalt eines Elementes in einer Probe. Dieser Wert ist wesentlich genauer. Experimentell wurde bei den obengenannten Kügelchen keine Abweichung der gemessenen Verhältnisse von den wahren Verhältnissen größer als $\pm 4\%$ gefunden.

Literatur

1. Castaing, R.: Electron probe microanalysis. Electron. Electron Physics 13, 317 (1960).

The Magnitude of the "Continuous" Fluorescence Correction in Electronprobe Analysis

G. SPRINGER und B. ROSNER

Institut für Mineralogie und Lagerstättenlehre, Technische Hochschule Aachen, BRD

I. Introduction

In mathematical procedures for relating the X-ray intensity measured with an electron-probe to concentration values the fluorescence due to the continuous spectrum ("continuous" fluorescence) is generally neglected. Only the fluorescence by characteristic lines is taken into account in addition to absorption, and atomic number effects. This seems to be justified since several authors in particular KIRIANENCO et al. (1962) have demonstrated that the influence of the continuous radiation is fairly small. This is so even in cases where it might be expected to be considerable, namely where there is a big difference in the average atomic number between specimen and standard. But there are definitely instances where this effect is appreciable and should not be ignored. The present study sets out to evaluate the continuous fluorescence correction for a large number of specimens in order to establish to what extent the omission of this correction affects the accuracy of the microprobe determinations.

II. The Correction Formula

The formula which has been applied to determine the seize of the continuous fluorescence has been derived earlier (SPRINGER, 1967a) and for K-radiation it is as follows:

$$\frac{I'_{fcA}}{I'_A} = 4.34 \cdot 10^{-6} \cdot \frac{r_K - 1}{r_K} E_K \cdot \bar{Z} \cdot A \cdot \frac{\mu_K^A}{\mu_K} \cdot h(g, U_0);$$

$$g = (\mu_A/\mu_K) \cdot \mathrm{cosec}\,\theta; \qquad U_0 = E_0/E_K;$$

$$h(g, U_0) = (U_0 \ln U_0 - U_0 + 1)^{-1} \int_1^{U_0} \frac{U_0 - U}{U} \cdot \frac{\ln(1 + g\,U^3)}{g\,U^3}\,dU \approx \frac{\ln(1 + g\,U_0)}{g\,U_0}.$$

This expression gives the ratio of the X-ray intensity I'_{fcA} excited by the continuum to the primary intensity I'_A excited by the electrons. Both intensities are those from a particular element A and are measured at the target surface. To eliminate the effect of the continuous fluorescence from the measured intensity $(= I'_A + I'_{fcA})$ it is necessary to multiply the latter with the factor $1/(1 + I'_{fcA}/I'_A)$. The other symbols employed in the formula have the usual meaning namely:

r_K K-edge absorption jump ratio for element A;

E_K critical excitation energy of K radiation for element A in keV;

E_0 electron excitation energy in keV;

$\bar{Z}$ mean atomic number of target;

A atomic weight of element A;

μ_K^A, μ_K mass absorption coefficients on high energy side of K absorption edge, for element A and compound specimen respectively;

μ_A mass absorption coefficient of measured K-radiation from element A in compound;

θ X-ray take-off angle.

$h(g, U_0)$ takes account of the absorption which the secondary characteristic radiation suffers in the target. The integration to evaluate this factor can be carried out fairly easily by computer but the calculation is made much simpler by the use of the expression $\ln(1+g\,U_0)/g\,U_0$ which is an approximation that is hardly less accurate in the usual range of g and U_0. The integration has, of course, to be divided up into several ranges if there are absorption edges in the energy interval between E_K and E_0. Under these circumstances too, the simple expression for the absorption factor makes numerical evaluation easier.

The formula given applies to the case that K-radiation is measured and a similar expression can be derived for L-radiation (SPRINGER, 1967a). The absorption coefficients used for the present work are derived from the tables of HEINRICH (1964).

III. Results and Discussion

The calculations have been performed for specimens for which quantitative electronprobe determinations have been carried out previously. Most of these examples have been taken from the literature, in particular from the tabulation by POOLE and THOMAS (THOMAS, 1964) and an article by PEISSKER (1967). Some measurements, mostly on sulphidic and oxidic minerals, were supplied by the present authors. About 50 different specimens on which more than 200 microprobe measurements at various kilovoltages were made have been considered.

Fig. 1 shows a histogram dealing with these cases. The ratio of the corrected microprobe concentration to the true concentration is plotted on the abszissa and the frequency of occurrence on the ordinate. For the corrections the usual formulae have been employed, namely a DUNCUMB-SHIELDS modified absorption correction after PHILIBERT, a modified POOLE-THOMAS (DUNCUMB) atomic number correction with a constant J/Z-value and a REED modified CASTAING correction for the characteristic fluorescence as well as the above described correction for the continuous fluorescence. A detailed description of the correction procedure is given by SPRINGER (1967b). There should, of course, be complete agreement between the concentrations detected with the electronprobe and the true content. Obviously there are experimental errors and also deficiencies in the correction formulae and a certain amount of spreading on the histogram arises from this. It is interesting to notice, however, that the standard deviation from the theoretical value 100% is about $\pm 2\%$, this means that approximately $^2/_3$ of all the measurements fall within the range 98 to 102%. This agrees quite well with the experimental accuracy that can usually be obtained for the type of measurement under consideration.

It would be appropriate in the present context to draw a similar diagram but leaving out the correction for continuous fluorescence. From the variation of the shape of the frequency distribution curve it would then be possible to estimate the effect which the fluorescence by the

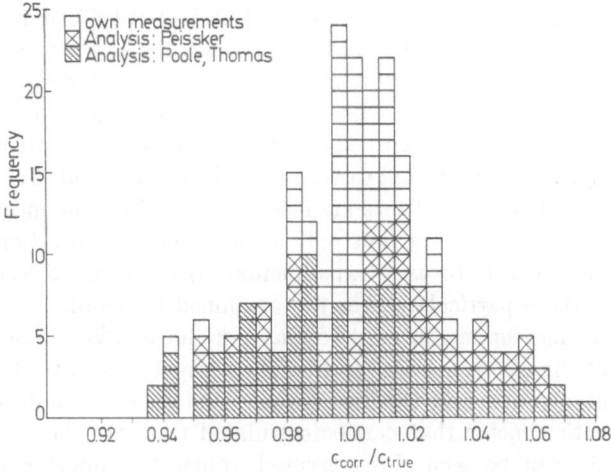

Fig. 1. Histogram showing agreement between corrected microprobe results and true concentrations

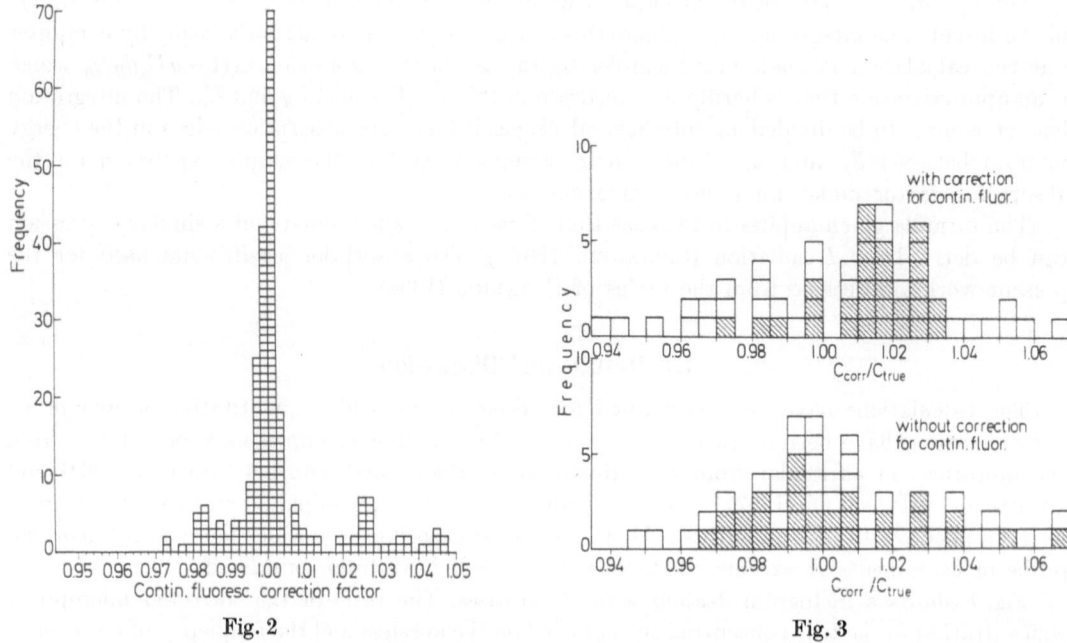

Fig. 2

Fig. 3

Fig. 2. Frequency with which "continuous fluorescence" correction factors of a certain magnitude occur

Fig. 3. Histograms showing agreement between corrected microprobe results and true concentrations for cases where "continuous fluorescence" correction is greater than 1%

continuum has. That such a straightforward approach is not quite suitable is demonstrated in Fig. 2. In this figure is plotted the frequency with which a continuous fluorescence correction of a certain magnitude occurs. It can be seen that some corrections can be as large as 5% but it is also obvious that in a vast majority of cases the correction is very small. In about 80% of all the instances considered it is less than $\pm 1\%$ which is below the limits of accuracy presently encountered in many microprobe measurements.

A better way of demonstrating the effect of continuous fluorescence is therefore to plot only those cases on the histogram for which the continuous fluorescence correction is larger than, say, 1% and this has been done in Fig. 3. The difference in the two histograms lies in the inclusion or omission of the continuous fluorescence correction. Altogether 17 different specimens have been considered. It can be seen that the maximum of the distribution curve lies at about 99.5% if the correction is not applied and at 101.5% if it is. Although in each of the two cases the deviation from the theoretical value 100% is within the presumed limits of experimental error it appears from this statistical treatment that it is preferable to leave the correction out altogether as is frequently done in many laboratories. An overcorrection may be produced otherwise. All that can possibly be said in favour of introducing the correction is that it leads to a slightly less scattered distribution. This is particularly true for the examples represented by the shaded squares. These are instances for which the measurements have been carried out at several kilovoltages and not only at one. The concentrations determined in this way appear to be more accurate because experimental errors are more readily apparent. In the following four figures these particular cases are examined in detail.

Fig. 4 shows $\mathrm{Au}\,L_\alpha$ measurements on two Cu—Au-alloys. Plotted on the abscissa is the excitation potential of the electron beam and on the ordinate the ratio between the measured and uncorrected concentration and the true value. Under ideal conditions this ratio should be equal to the correction factor that can be calculated from the theory of quantitative electronprobe analysis. As can be seen the measured values lie almost exclusively in between the curves that represent the correction factors with and without inclusion of the continuous

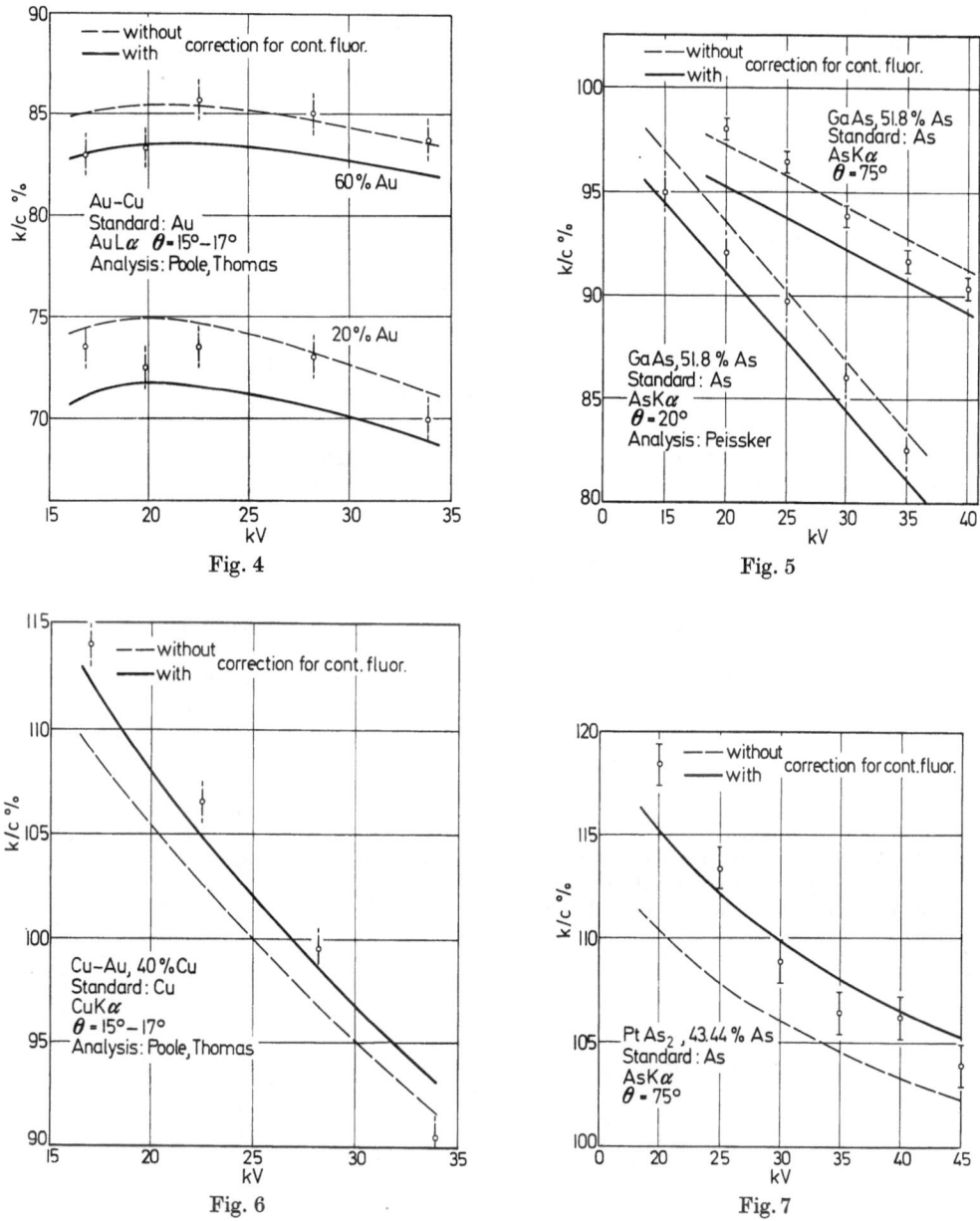

Fig. 4—7. Comparison of measured correction factors k/c with calculated ones at various excitation voltages. k = measured intensity ratio between specimen and standard, c = true concentration ratio

fluorescence. The inaccuracies introduced by the procedure of measurement and by the correction for other effects conceal the rather small influence of the continuum. The same is true for GaAs (Fig. 5). This compound has been analysed with two different instruments having X-ray take-off angles of 20° and 75°. The correction for continuous fluorescence is almost equal in the two cases and it is certainly not true to generalise that a high take-off angle leads to a larger continuous fluorescence correction. The magnitude of the correction has to be worked out in every case. As this example shows it is not only the difference in the mean atomic number between specimen and standard that determines it, but also the difference in the absorption properties. Fig. 6 shows the case of another Au—Cu alloy, but this time Cu has been determined. A definite improvement of the results is noticed when the continuous fluorescence is taken into account. This is even more apparent when the compound $PtAs_2$ is examined (Fig. 7). There is a fairly large continuous fluorescence correction of a about 4% and it would certainly lead to erroneous results if this correction were not considered. It should be mentioned that $Pt L_{\gamma 1}$ radiation causes a characteristic fluorescence contribution to the $As K_\alpha$ intensity. As the correction for this is difficult to evaluate and probably not greater than 0.5% it has been neglected.

IV. Conclusion

In conclusion it can be said that although in the great majority of cases the continuous fluorescence correction is neglegibly small there are instances where an improvement in the results can be attained if it is taken into consideration. This is particularly so where this correction amounts to more than about 2%. In cases where it is less, inaccuracies in the measurements and in the correction for other influences seem to conceal the effect of the continuous fluorescence. It appears that the presently used formulae for absorption and atomic number differences also correct for continuous fluorescence to some extent so that an overcompensation is produced if allowance for the latter influence is made specially. In the light of this, a revision of the data on which the present methods of absorption and atomic number correction are based seems desirable.

References

HEINRICH, K. F. J.: X-ray absorption uncertainty. The electron microprobe (T. D. McKINLEY et al., eds.) 1964. New York: John Wiley & Sons 1966.

KIRIANENKO, A., F. MAURICE, D. CALAIS, and Y. ADDA: Analysis of heavy elements ($Z > 80$) with the Castaing Microprobe: Application to the analysis of binary systems containing uranium. X-ray Optics and X-ray Microanalysis (H. H. PATTEE et al., eds.) 1962. New York: Academic Press 1963.

PEISSKER, E.: Vergleich verschiedener Korrekturverfahren für die quantitative Elektronenstrahl-Mikroanalyse. Mikrochim. Acta, Suppl. 2, 156—172 (1967).

SPRINGER, G.: The correction for "continuous fluorescence" in electronprobe microanalysis. Neues Jahrb. Mineral., Abhandl. 106, 241—256 (1967a).

— Die Berechnung von Korrekturen für die quantitative Elektronenstrahl-Mikroanalyse. Fortschr. Miner. 45, 103—124 (1967b).

THOMAS, P. M.: A method for correcting for atomic number effects in electronprobe microanalyses. U. K. Atomic Energy Research Establishment 1964, Report No. 4593.

An Attempt for Quantitative Procedure

Gunji Shinoda*

Osaka University, Faculty of Engineering, Yamada, Suita, Osaka, Japan

Introduction

Corrections for quantitative analysis in electron probe microanalyser (EPMA) has been developed since Castaing's pioneer work in 1951 [1] and many investigators attacked the problem. However, even nowadays existence of considerable amounts of errors between the values of chemical analysis and those of EPMA is repoted, sometimes. Usually, the error has been estimated as less than $\pm 1\%$, however some one claims $\pm 0.5\%$ and others say it is difficult to keep it less than $\pm 5\%$ always.

One reason for such divergency of opinions should be attributed to imcompleteness in the theoretical background of the quantitative procedures, especially on the method of corrections for the atomic number effects. However, it would not be so large. On the other hand if a specimen consists of heterogeneous mixture the compositions determined by chemical analysis means an average for a macrovolume. As the result of EPMA quantitative analysis is that of a microvolume it will be quite natural that there exists much discrepancies between these two values if the specimen is not homogeneous.

On the other hand we know that if the specimen is homogeneous the result of EPMA quantitative analysis agrees quite well with that of chemical analysis. This means that if there is no unknown factors in the behaviors of electrons in the specimen, present method of quantitative correction has enough accuracy even though there exists some incompleteness in the theoretical background. Therefore a method appropriate for heterogeneous specimen should be investigated. Present paper is an attempt along such lines.

Chemical and EPMA Analyses

Unless the specimen is uniform in electron microprobe scale the comparizon of the data of chemical analysis with those of EPMA analysis will be failed because in EPMA method the volume used for analysis is about 10^{-13} cm³ while that for chemical analysis is very much larger.

In a β-phase of copper-45 wt.% zinc alloy we can find about 1% fluctuation of concentration of zinc among each grain. This was obtained by a specimen quenched from 870° C and several times cold worked and tempered at 450° C, therefore it will be a well solution treated state. In a cast condition such fluctuation reaches about 2%. The composition of this alloy would be uniform at a higher temperature before quenching, however during cooling precipitation occurs and the compositions of the matrices become no more uniform. For a standard specimen used for quantitative analysis such homogeneity would be preserved. But in an industrial material such homogeneity is never expected. Therefore an accurate method of microscopical chemical analysis which is able to replace ordinary chemical analysis should be investigated.

When a specimen is composed of minute and heterogeneous grains, motion of electrons near the grain boundaries and manner of X-ray generation become very much complicated and consequently the X-ray intensity distribution would be different from more homogeneous or coarse grained specimen. For such case if depth distribution function for the very specimen is

---

* Present affiliation: Japan Women's University, Tokyo.

once obtained experimentally, a way to reach an exact absorption correction would be found. For such purpose our method using angular distribution of characteristic X-ray has been developed.

Angular Distribution of Characteristic X-ray and its Application to Quantitative Procedure

a) Homogeneous Specimen

Among the experimental methods to obtain depth distribution function, the measurement of angular distribution of characteristic X-ray is the only one which is able to obtain informations from the analysing point of the specimen. This method has been developed from such stand point and the depth distribution function obtained by this method has enough accuracy. As details of it is already published [2], here it will be explained only briefly.

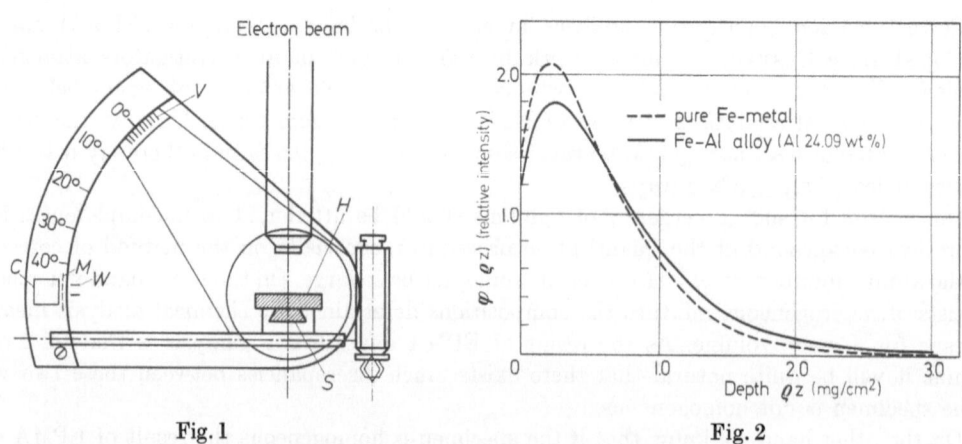

Fig. 1   Fig. 2

Fig. 1. General view of the angular distribution measuring apparatus. *C* counter; *H* specimen; *M* mylar film (60 μm); *S* specimen stage; *V* vernier; *W* slit (100 μm)

Fig. 2. Depth distribution functions of iron-aluminum alloys

The instrument, shown in Fig. 1, is constructed as an accessory of the JEX-31 EPMA and by a proportional counter angular distributions of X-ray is measured. If monochromatisation is necessary, the proportional counter should be replaced by a semiconducter transducer and a pulse height analyser. As the method of analysis of angular distribution of characteristic X-ray is applicable to both copper and aluminum, as reported already, the next step is to find a distribution function of an alloy composed of two elements having nearly equal atomic numbers with copper and aluminum. Thus iron and aluminum alloys were investigated. The alloys are homogeneous solid solution and the depth distribution function calculated from the angular distribution using minimum seeking method is shown in Fig. 2. The results of quantitative analysis corrected by above distribution function agreed quite well with those of chemical analysis. Therefore, as to the homogeneous specimen difficult problems concerning to the quantitative procedure are almost solved.

b) Heterogeneous Specimen

Next, consideration should be done for a specimen consists of heterogeneous mixtures. When each grain of the mixture is coarse, microanalysis for each individual grain is possible. On the other hand, if the specimen consists of very small heterogeneous grains in addition to the effect of beam size another unknown factors play an important role. The motion of electrons in the specimen is very much complicated in the grain boundary region and the manner of X-ray generation should be different from that of single crystal specimen [3]. It has been reported that the intensity of X-ray is greater in the small grained specimen. Therefore if the

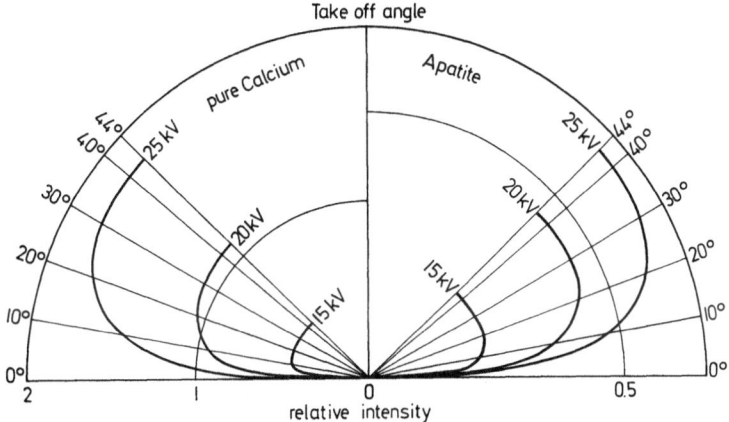

Fig. 3. Angular distribution of $\mathrm{Ca}\,K_\alpha$ for metallic calcium and apatite

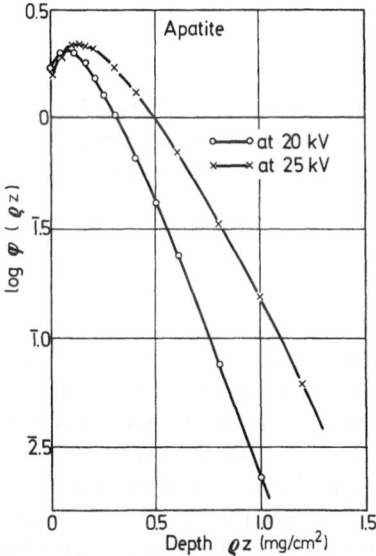

Fig. 4. Depth distribution function for apatite

specimen is composed of several elements, the depth distribution function would change with the condition of grain dispersion, such as grain size and its distributions, shapes and others. For such case the depth distribution function should be determined for the very specimen and the method to use angular distribution would be most convenient.

Form these standpoint, the analysis of calcium in human tooth enamels has been done. The followings are the result of SHIMIZU et al. As the calcium content of human tooth enamels is nearly equal to that of a natural mineral apatite the depth distribution curves have been obtained at first by metallic calcium and next by calcium in apatite $(\mathrm{Ca_5[FCl(PO_4)_3]})$. Fig. 3 shows the results of measurement of the intensities of angular distribution of characteristic X-ray for each case and Fig. 4 is the depth distribution of calcium for apatite.

Quantitative Analysis of Microvolume by Backward Diffraction

When a solid solution specimen has a microsegregation, we cannot expect agreement between the result of chemical analysis and that of electron microprobe quantitative analysis because the former gives mean value while the latter correrponds to a local value. For such case determination of accurate compositions of a microvolume at the analysis point is necessary. In other

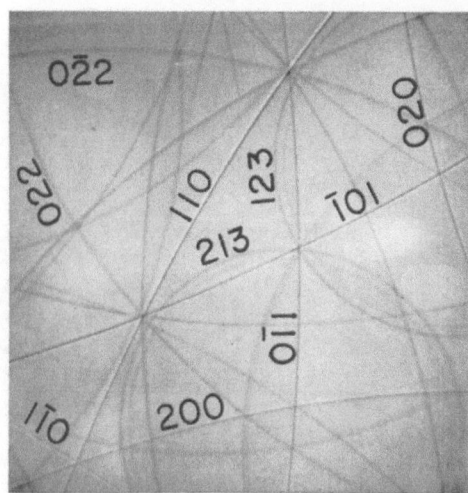

Fig. 5. Kossel patterns of matrix β of Cu—45% Zn alloy. β-phase: $a = 2.492(9)$ Å, bcc, 45.7 wt.% Zn

word an appropriate method of microvolume analysis which is able to replace ordinary chemical analysis is required. Fortunately if a specimen is a binary solid solution the relation between the compositions and lattice parameters are obtained with enough accuracy. Therefore, if the lattice parameters are determined for the analysis point, it will give an exact composition of that microvolume. Among many diffraction techniques the EPMA-Kossel technique would be most suitable. In this techniques precise lattice constant determination is done by back reflection method using special intersections of lines. There are two kinds of Kossel techniques, namely transmission and reflection. For back reflection, the latter is usually taken, however, even by transmission method dark lines due to back reflection are obtained and the lattice parameters would be determined precisely. Fig. 5 is an example obtained by copper-45 wt.% zinc alloy heat treated at 450° C after quenching from 870° C and cold worked. Well resolved $K_{\alpha 1}$ and $K_{\alpha 2}$ doublet of {321} back reflection pattern will be seen. The photograph has been taken using true Kossel technique and Zn K_α lines have been filtered off by a nickel filter. Though several sets of lens pattern are found in this photograph it was more convenient to use {321} ring diameters to determine lattice constant. The relation between lattice constant and concentration of zinc in β brass is given by Owen and Pickup [5] and from their data 45.7 wt.% zinc was obtained.

On the contrary from the intensity of copper and zinc K_α lines in the linear scan diagram the concentration of zinc at the same point has been determined as 45.9 wt.% zinc from Castaing's first approximation. Agreement between them is quite satisfactory. Because copper and zinc are neighboring elements in the periodic table. Therefore the manner of X-ray generation and its absorption would be nearly the same and also the atomic number correction would not be necessary.

However, such coincidence means that precise lattice constant measurement can provide a composition for a microvolume with enough accuracy. And for such case data of chemical analysis should be replaced by such compositions deduced from lattice constant measurement.

Use of Specimen Current

As pointed out already by present author et al. [6] the specimen current image has higher resolving power compared to the X-ray image. It was believed that the specimen current decreases monotonously with the increase of the mean atomic number and this property is utilized as a mean of quantitative analysis. However as pointed out by Ichinokawa et al. [7] there exists anomalies in specimen current in ε phase of copper-zinc alloy, i.e. in the ε phase with higher content of zinc increase of specimen current has been observed. Later, present author et al. [8] pointed out that as in this system the difference of atomic numbers of both elements is only one, the second order effect, such as the dependency of specimen current on the orientation, becomes predominant.

Therefore, for application of specimen current method to quantitative analysis care must be taken for such features, however in a crystal with higher isotropic nature, such as a usual body centered cubic crystal, these channelling effect should be insignificant.

To increase resolving power of an electron microprobe use of secondary electron emission would be advantageous. However it mainly depends on the surface properties and we cannot expect much informations related to quantitative analysis. The condition would be nearly the same as in the specimen current method.

References

1. CASTAING, R.: Thesis. University of Paris 1951; Publ. O.N.E.R.A. No. 55.
2. SHIMIZU, R., K. MURATA, and G. SHINODA: Technol. Rept. Osaka Univ. 17, 13 (1967).
3. IZUI, K., and S. FURUNO: Japan. J. Appl. Phys. 7, 184 (1968). — ODA, Y., and K. SATTA: Read at the Spring General Meeting of Japan Society of Applied Physics, March 31, 1968.
4. SHIMIZU, R., K. MURATA, Y. WATANABE, S. KATO, and S. YOSHIDA: Read at the Spring General Meeting of Japan Society of Applied Physics, March 31, 1968.
5. OWEN, E. A., and L. PICKUP: Proc. Roy. Soc. (London), Ser. A 137, 397 (1932).
6. SHIMIZU, R.: Thesis. Osaka University 1964. — SHINODA, G., H. KAWABE, K. MURATA and T. SHIRAI: Technol. Rept. Osaka Univ. 16, 423 (1966).
7. ICHINOKAWA, T., S. SHIRAI, and A. ONOGUCHI: Japan. J. Appl. Phys. 6, 277 (1967).
8. SHINODA, G., K. ISOKAWA, and M. UMENO: Kossel line micro-diffraction study on precipitation of α from β in copper-zinc alloys: Read at the Denver Conference on Application of X-ray Analysis, Aug. 21, 1968.

Analyse quantitative d'echantillons minces

R. Tixier et J. Philibert

Institut de Recherches de la Sidérurgie Française, St.-Germain-en-Laye, France

Resume

L'analyse des échantillons minces est un des domaines d'applications de la microanalyse à sonde électronique qui, déjà préconisé en 1951 par Castaing, n'a encore connu que des développements limités malgré son grand intérêt. Cette situation est en partie liée aux difficultés très particulières que pose l'analyse quantitative de tels échantillons. Leur solution dépend directement de la connaissance des phénomènes physiques mis en jeu, leur étude peut en outre apporter une contribution intéressante à certains aspects de l'analyse classique d'échantillons massifs.

On expose un modèle physique des corrections d'effets de numéro atomique et d'absorption qui permet l'analyse d'échantillons minces avec une précision raisonnable. Quelques exemples d'études faites à l'IRSID sont cités à l'appui de ce modèle.

I. Introduction

L'analyse quantitative d'échantillons minces est un des domaines d'application de la microanalyse par sonde électronique qui n'a encore connu que peu de développements malgré son grand intérêt. Ce genre d'étude complète des informations obtenues en microscopie électronique en fournissant l'analyse des précipités observés et permet souvent de mieux interpréter leurs diagrammes de diffraction électronique. En outre le pouvoir séparateur géométrique de la microanalyse est amélioré par la suppression de la diffusion des électrons dans l'échantillon et, dans le cas des répliques avec extraction, par la disparition de la matrice. Les difficultés pratiques de ces analyses sont discutées dans une autre communication à ce congrès, nous voudrions envisager ici essentiellement les problèmes théoriques qu'elles soulèvent.

II. Choix des temoins

La seule possibilité de supprimer complètement toutes les corrections serait, théoriquement, d'utiliser des témoins minces et d'épaisseur massique exactement égale à celle de l'échantillon. Cette solution est illusoire, puisque d'une part l'épaisseur massique de l'échantillon peut varier considérablement d'un point à l'autre et que d'autre part, si l'on connaît l'épaisseur massique on peut souvent calculer la composition, du moins pour un binaire (et dans ce cas la solution envisagée n'est plus nécessaire).

Il serait pourtant intéressant de disposer de témoins minces (< 500 Å) d'épaisseur connue; des corrections seraient encore nécessaires, mais leur magnitude serait réduite. Si l'on pouvait opérer en microanalyse à des tensions d'accélération comparables à celles de la microscopie électronique, cette solution serait sans doute indispensable. Il est possible qu'elle doive être envisagée pour l'analyse des éléments légers. Cependant il est pratiquement encore très difficile de fabriquer de tels témoins et de connaître leur épaisseur. C'est pourquoi nous envisagerons ici surtout le cas des analyses faites en utilisant des témoins massifs.

III. Mesure des concentrations apparentes

Rappelons que l'on ne peut pas faire d'analyse absolue de la teneur en un élément en un point, puisque la masse de matière analysée n'est pas connue. En effet le faisceau traverse l'échantillon, l'épaisseur de celui-ci est inconnue et le précipité analysé peut être d'une taille inférieure au diamètre de la sonde. On mesure des concentrations relatives d'un élément par rapport à un autre. En fait, pour chaque élément et au même point de l'échantillon, on mesure une concentration apparente $k_A = I_A$ (échantillon)$/I(A)$ (témoin) et le résultat de l'analyse est le rapport k_A/k_B. Ce rapport n'est généralement pas égal au rapport réel des concentrations C_A/C_B; il faut calculer des coefficients de correction pour obtenir la valeur réelle.

IV. Correction de l'effet de numero atomique

Cet effet est relatif aux intensités engendrées

1. Rétrodiffusion

Sur les témoins les facteurs de rétrodiffusion $R(A)$ et $R(B)$ sont ceux définis et tabulés [1—3]. Sur l'échantillon par contre, on ignore complètement les valeurs des facteurs de rétrodiffusion. Cependant, on sait que, pour un échantillon très mince, ces facteurs sont très voisins de 1; on supposera donc que le rapport R_A/R_B des facteurs de rétrodiffusion sur l'échantillon est pratiquement égal à un.

2. Pénétration

Nous avons schématisé ce qui se passe quand les numéros atomiques des deux éléments sont assez différents (fig. 1). Rappelons que les deux éléments sont analysés à une même tension et non à un même taux d'excitation pour des raisons pratiques. Les volumes excités dans les témoins sont de forme et de taille différentes, alors que dans la cible mince ces volumes sont les mêmes pour les deux éléments si ce n'est que ceux-ci sont excités inégalement en raison des différences de sections efficaces d'ionisation. L'effet de numéro atomique va donc être différent de celui qui caractérise les échantillons massifs. On va supposer que le ralentissement dans la cible mince peut être négligé pour le calcul du rapport des concentrations. On écrira donc que le nombre moyen d'ionisations sur le niveau j des atomes A rencontrés par un électron le long de sa trajectoire est:

$$d\,n_j = \left[C_A \,\varrho_A \cdot \frac{N}{A} \right] [\psi_A^j(E)\,d\,s]. \tag{1}$$

Le premier terme représente le nombre d'atomes de A, le second la section efficace.

E est l'énergie de l'électron,

s l'abscisse curviligne sur la trajectoire,

N est le nombre d'Avogadro,

A est la masse atomique de l'élément, ϱ_A sa masse spécifique.

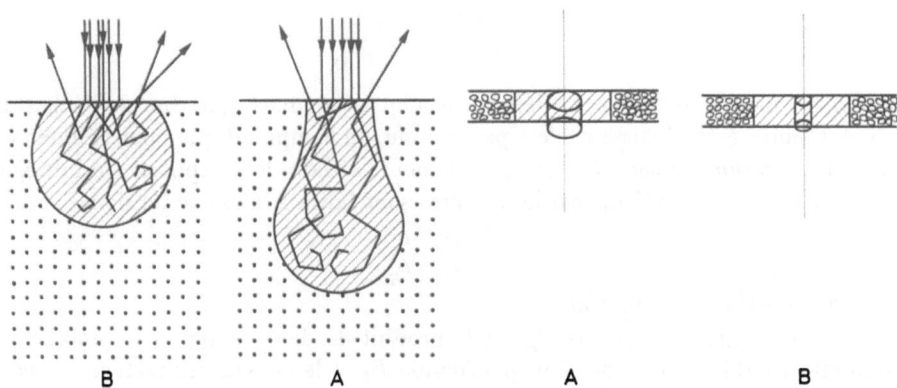

Fig. 1. Schéma des volumes ionisés dans un échantillon mince et dans les témoins massifs. A élement léger, B élement lourd

Dans le témoin massif on doit intégrer l'équation (1) sur toute la trajectoire. En introduisant la loi de ralentissement continu, $dE/d\varrho s$, on écrit:

$$n_j = C_A \cdot \frac{N}{A} \cdot \int_{E_0}^{E_j^A} \frac{\psi_A^j(E)\,dE}{dE/d\varrho s}. \tag{2}$$

E_j^A est l'énergie d'ionisation du niveau j, de l'atome A,
E_0 est l'énergie initiale de l'électron.

Supposons le témoin formé par l'élément pur, $C_A = 1$. S'il était composé, la forme du calcul resterait la même. Utilisons les lois de Bethe pour représenter la section efficace et la loi de ralentissement [4]:

$$d n_j = C_A \cdot \frac{N}{A} \cdot \frac{\pi e^4}{E_0 \cdot E_j^A} \cdot Z_j \cdot b_j \cdot \log U_0^A \, d\varrho s \tag{3}$$

pour l'échantillon mince et:

$$n_j = R_{(A)} \cdot \frac{N}{A} \cdot \frac{Z_j\,b_j}{2N} \cdot \int_1^{U_0^A} \frac{\log U}{M_A \cdot \log U \cdot W_A^j} \, dU \tag{4}$$

pour le témoin massif, en introduisant le facteur de rétrodiffusion et en notant

$$U_0^A = \frac{E_0}{E_j^A}, \qquad M_A = \frac{Z}{A}, \qquad W_A^j = \frac{1{,}166\, E_j^A}{J_A}.$$

Z_A est le numéro atomique de A, E_j^A est exprimée en électron-volt, J_A est le potentiel moyen d'ionisation des atomes A. Selon WILSON $J_A = 11{,}5\, Z_A$ [5], d'autres valeurs ont été proposées par DUNCUMB [6] et par TURNER [7]. On sait [8] que le terme intégral peut être écrit:

$$\frac{1}{S'_{(A)}} = \frac{1}{M_A} \left[U_0^A - 1 - \frac{\text{Log}\, W_A}{W_A} (li\, U_0 W_A - li\, W_A) \right] \tag{5}$$

$li\, x = \int_0^x dt/\log t$, $x \geqq 0$, est la fonction logarithmique intégrale (fonction Eulérienne). Cette fonction peut être calculée par son développement:

$$li\, x = C + \log|\log x| + \sum_{n=1}^{n=\infty} \frac{(\log x)^n}{n \cdot n!} \tag{6}$$

où $C =$ constante d'Euler.

La concentration apparente mesurée de l'élément A peut être écrite en considérant les intensités engendrées:

$$k_A = \frac{d n_j}{n_j} = C_A \cdot \frac{\pi e^4 2N}{E_0} \cdot \left[\frac{\text{Log}\, U_0^A}{E_j^A} \cdot \frac{1}{R_{(A)}/S'_{(A)}} \right] d\varrho s. \tag{7}$$

Nous appelerons P_A le terme entre crochets:

$$P_A = \frac{\text{Log}\, U_0^A}{E_j^A} \cdot \frac{1}{R_{(A)}/S'_{(A)}}. \tag{8}$$

L'équation (7) montre bien que k_A dépend de l'épaisseur massique $d\varrho s$ de l'échantillon au point analysé. Comme cette épaisseur n'est pas connue, on comprend que pour l'éliminer il faille mesurer les concentrations apparentes de deux éléments et faire leur rapport. On fait également disparaître ainsi le terme $\pi e^4 2N/E_0$, car les mesures sont faites à la même tension d'accélération. Il reste:

$$\frac{k_A}{k_B} = \frac{C_A}{C_B} \frac{P_A}{P_B} \tag{9}$$

Si $P_A < P_B$, on aura $(k_A/k_B) < (C_A/C_B)$.

On voit d'après l'équation (8) que P_A est le produit de deux facteurs; le premier provient de la section efficace d'ionisation de A pour l'énergie E_0 et le second des facteurs de correction de l'effet de numéro atomique dans le témoin A pur. Contrairement au cas des échantillons massifs, le facteur de correction ne dépend pas de la composition.

V. Correction d'absorption

Cette correction fait passer des intensités émergentes aux intensités engendrées. Dans l'échantillon mince le problème est différent de celui posé par un échantillon massif. En effet la fonction $\Phi(\varrho z)$ qui représente la distribution de l'ionisation en fonction de la profondeur n'est pas la même. Dans un échantillon massif, le maximum de $\Phi(\varrho z)$ se situe à une certaine profondeur dans l'échantillon [9], en raison de l'importance des effets de diffusion multiple et de rétrodiffusion en volume. Si l'échantillon est mince cet effet disparaît. Si l'échantillon était tellement mince qu'il n'y ait pas diffusion appréciable des électrons, la fonction d'ionisation, $\varphi(\varrho z)$, ne dépendrait pas de l'épaisseur traversée [10]:

$$\varphi(\varrho z) = k$$

d'où
$$F(\chi) = k\,\varDelta\varrho z - k \cdot \chi\,\frac{(\varDelta\varrho z)^2}{2!} + k\,\chi^2\,\frac{(\varDelta\varrho z)^3}{3!}\cdots$$

$$F(0) = k\,\varDelta\varrho z$$

d'où
$$f(\chi) = 1 - \chi\,\frac{\varDelta\varrho z}{2!} + \cdots$$

si
$$\chi\,\frac{\varDelta\varrho z}{2!} \ll 1 \quad \text{soit} \quad \chi\,\varDelta\varrho z < 0,1.$$

On pourra écrire qu'au premier ordre près, en $\varDelta\varrho z$, $f(\chi) \cong 1$.

Cependant les échantillons ne sont pas assez minces en général pour que cette hypothèse soit réaliste. Aux tensions d'accélération utilisées les précipités sont le plus souvent presque opaques. Montrons que jusqu'à des épaisseurs égales à la profondeur de complète diffusion des électrons la correction reste négligeable, au même ordre près. Dans ce cas:

$$\varphi(\varrho z) \propto e^{-\sigma\varrho z},$$

d'où
$$F(\chi) \propto \int_0^{\varDelta\varrho z} e^{-\sigma\varrho z} \cdot e^{-\chi\varrho z} \cdot d\varrho z$$

$$F(\chi) \propto \varDelta\varrho z - (\sigma + \chi)\,\frac{(\varDelta\varrho z)^2}{2!} + \cdots$$

$$F(0) \propto \varDelta\varrho z - \frac{\sigma}{2!}\,(\varDelta\varrho z)^2 + \cdots$$

Au premier ordre près en $\varDelta\varrho z$, si $\chi\varDelta\varrho z < 0,1$ (10)

$$f(\chi) \cong 1.$$

Si la condition (10) est respectée, il n'y a pas à calculer de correction d'absorption pour les mesures d'intensités dans l'échantillon mince. On voit bien l'intérêt de placer la réplique sur une grille et non sur un support massif.

Dans les témoins les intensités émergentes peuvent différer considérablement des intensités engendrées. En effet on est amené dans de telles analyses à étudier les deux éléments simultanément pour des raisons pratiques. Si de plus on observe l'échantillon par microscopie électronique, on est conduit à travailler à des tensions assez élevées pour exciter suffisamment l'élément le plus lourd et permettre l'observation sur l'écran. Il en résulte des taux d'excitation très élevés, surtout dans le cas des éléments légers, et par conséquent des facteurs d'absorption $f(\chi)$ non négligeables dans les témoins. Là encore on remarque que les facteurs de correction ne dépendent pas de la composition de l'échantillon mince.

VI. Corrections de fluorescence

Dans l'échantillon les corrections de fluorescence doivent être faibles: en effet elles tiennent essentiellement à l'émission en profondeur des rayons X. Si on utilise un témoin composé, ou on le choisit de telle sorte que ces corrections n'aient pas lieu d'être calculées, ou on calcule les corrections suivant les méthodes classiques.

VII. Calcul pratique de la correction

On a pour l'élément A:

$$k_A = C_A \cdot \frac{\pi\,e^4\,2N}{E_0} \cdot P_A \cdot \frac{\Delta\varrho\,z}{f(\chi)_{(A)}}$$

et pour le rapport de A sur B:

$$\frac{k_A}{k_B} = \frac{C_A}{C_B} \cdot \frac{P_A}{P_B} \cdot \frac{f(\chi)_{(B)}}{f(\chi)_{(A)}}. \tag{11}$$

Rappelons que

$$P_A = \frac{\mathrm{Log}\,U_0^A}{E_j^A} \cdot \frac{1}{R_{(A)}/S'_{(A)}}.$$

VIII. Discussion

Le modèle qui vient d'être décrit suppose l'échantillon « mince ». Il convient de préciser à partir de quelle épaisseur il peut s'appliquer. Pour les longueurs d'ondes moyennes et courtes, la pénétration des électrons dans le domaine d'énergie qui intéresse la microanalyse est faible comparativement aux épaisseurs que le rayonement X émis par l'élément peut traverser avant d'être complètement absorbé. Si l'échantillon est suffisamment mince pour être observé par microscopie électronique, on sera assuré qu'il est mince au sens des approximations faites dans la correction de l'effet de numéro atomique et a fortiori pour la correction d'absorption.

Par contre, pour les grandes longueurs d'onde, ce critère n'est pas suffisant, car la situation peut être inversée, l'absorption des rayons X dans l'échantillon peut devenir notable, alors que celui-ci est transparent aux électrons et reste donc facilement observable pour une épaisseur supérieure à 1000 Å (éléments de faible densité). Ceci est d'autant plus gênant que l'échantillon s'observera plus facilement avec une tension d'accélération élevée, ce qui entraîne une correction d'absorption dans le témoin pour laquelle les modèles actuels de calcul de $f(\chi)$ sont très peu précis. Il semble que pour résoudre en partie une telle difficulté, il faudrait utiliser des témoins minces d'épaisseur bien connue. Le modèle de calcul des corrections que l'on a décrit s'adapte facilement à ce cas d'analyse. Il deviendrait alors concevable de faire de la microanalyse à des tensions d'accélération de l'ordre de 100 kV ou plus. En outre, la valeur du modèle de correction proposé dépend également du choix des expressions analytiques de la section efficace d'ionisation et de la loi de ralentissement [4, 11]. Dans l'échantillon mince nous utilisons la section efficace en supposant qu'il n'y a pas diffusion inélastique[1]. Cette approximation se justifie car l'énergie la plus probable reste voisine de l'énergie incidente pour des épaisseurs déjà notables [10]. Dans les témoins massifs, la loi de ralentissement s'applique à une énergie moyenne de l'électron et non à l'énergie la plus probable, ce qui conduit à une approximation dans la correction de l'effet de numéro atomique.

Dans la microanalyse classique sur échantillons massifs, les imprécisions des formules de correction se compensent entre l'échantillon et le témoin. Dans la microanalyse des échantillons minces, il n'en va plus de même. En particulier on atteint souvent les limites de validité des formules de correction d'absorption. Les facteurs de correction sont élevés; aussi cette méthode pourrait être utilisée pour déceler certaines erreurs systématiques dues aux formules de correction.

P. Duncumb et K. Da Casa ont établi indépendamment de nous un modèle de correction pour l'analyse des échantillons minces [12]. Ce modèle est fondé sur des hypothèses physiques analogues aux nôtres, mais la méthode de calcul est différente. Nous avons pu comparer nos résultats et constater que les valeurs que nous trouvions étaient très voisines.

IX. Exemples d'applications

1. *Analyse à différentes tensions de fins précipités de Al_6Mn extraits sur film de carbone* (fig. 2). Pour les faibles tensions d'accélération le rapport mesuré Al/Mn est inférieur au rapport réel (2, 95). Ceci provient de l'importance de l'effet de numéro atomique dans les témoins qui, dans ce cas,

---

1. Nous nous proposons de déterminer quantitativement l'épaisseur massique limite pour laquelle cette approximation reste valide.

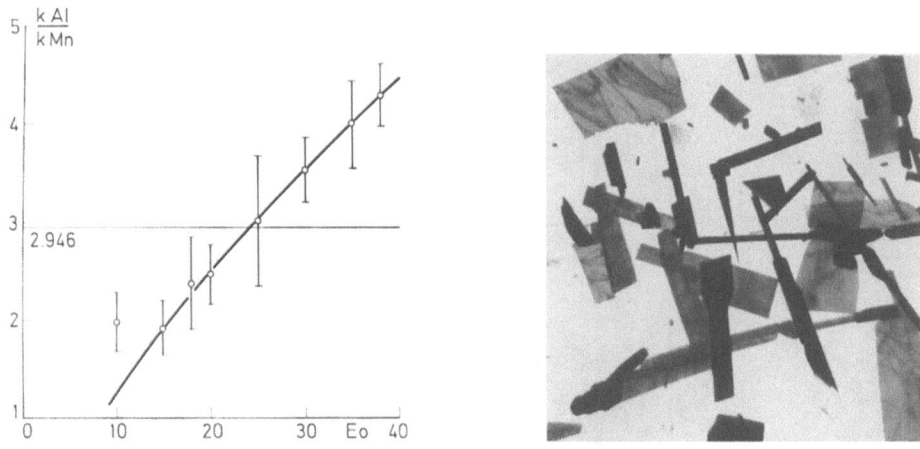

Fig. 2. Analyse à des tensions différentes de précipités Al₆Mn

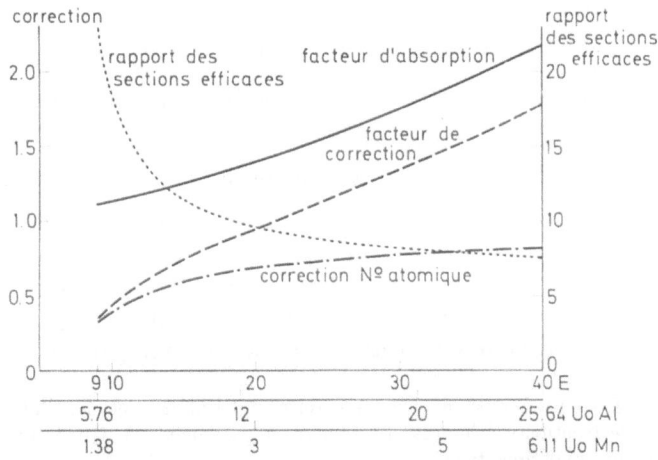

Fig. 3. Correction de l'analyse de Al₆Mn, influence des différents facteurs

produit une variation contraire de celle que l'on observe dans les analyses d'échantillons massifs. Au-dessus de 25 kV l'effet de l'absorption l'emporte et le rapport mesuré est plus grand que le rapport réel. On voit que les deux effets sont contrevariants. L'absorption a aussi un effet contraire de celui qui s'observe dans les analyses d'échantillon massif; de plus la correction est souvent bien plus importante.

La fig. 3 montre la variation de la correction pour cette analyse et l'influence des différents facteurs.

2. *Analyse de lames minces de microscopie électronique, afin d'examiner l'influence de l'épaisseur et de la tension* (fig. 4). Les mesures ont été faites sur une lame de Al—4% Cu et une lame de Ni₃Fe. Les mesures ont montré que, dans les parties minces des lames, les corrections décrites ici permettaient de retrouver avec une bonne précision les concentrations réelles. Dans les parties épaisses les corrections classiques étaient applicables. Mais pour les tensions les plus faibles ou pour les épaisseurs intermédiaires, il n'était possible d'appliquer aucune méthode, a priori, ni même de prévoir si le facteur de correction à appliquer serait inférieur ou supérieur à 1.

3. *Analyse à plusieurs tensions de précipités intermétalliques extraits sur réplique.* Dans une étude d'un acier inoxydable à durcissement contrôlé, l'analyse a conduit à donner aux précipités la formule Ni—Al. Nous avons pu confirmer cette mesure par diffraction électronique et diffraction de rayons X, en identifiant le composé NiAl β de structure cubique centrée du type ClCs, appelé aussi ferrite ordonnée.

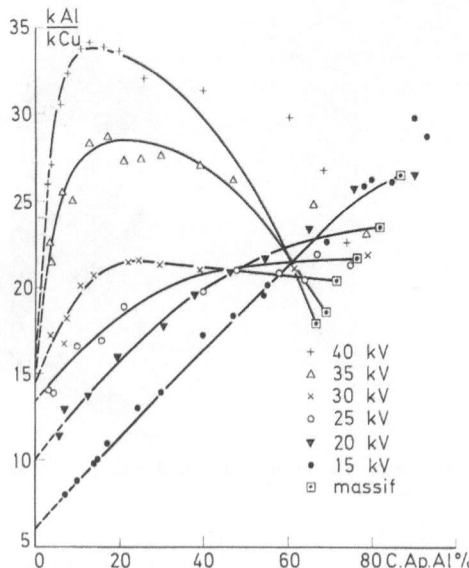

Fig. 4. Analyse de Al—4% Cu en fonction de l'épaisseur et de la tension

X. Conclusion

L'analyse quantitative des échantillons de microscopie électronique peut être faite à la microsonde de Castaing. Notre modèle de calcul de corrections met en évidence les causes des écarts entre les valeurs mesurées et les valeurs réelles. Les différences entre ce calcul et celui des corrections bien connues dans un échantillon massif permettent de mieux comprendre la nature physique de l'origine des corrections en microanalyse par sonde électronique.

Remerciements. Les auteurs tiennent à remercier P. Duncumb et K. Da Casa (maintenant Madame Nileshwar) pour de fructueux échanges. Ils remercient également A. Quennevat et D. Bryckaert pour de nombreuses mesures expérimentales.

Bibliographie

1. Duncumb, P.: Private communication unpublished.
2. Bishop, H. E.: Some electron backscattering measurements for solid targets. Optique des rayons X et microanalyse, Orsay 1965, p. 153—158. Paris: Hermann 1966.
3. Derian, J. C.: Détermination expérimentale du facteur de rétrodiffusion en microanalyse par émission X. Rapport C.E.A. — R 3052, 1966.
4. Bethe, H.: Zur Theorie des Durchgangs schneller Korpuskularstrahlen durch Materie. Ann. Physik 5, 325—400 (1930).
5. Wilson, R. R.: Range and ionization measurements on high speed protons. Phys. Rev. 60, 11, 749—753 (1941).
6. Duncumb, P., and S. J. B. Reed: Progress in the calculation of stopping power and backscatter effects. Quantitative electron probe microanalysis N.B.S. Special Publication 298 (1968).
7. Turner, J. E.: Values of I and I adj. suggested by the subcommitte. Studies in penetration of charged particles in matter — N.A.S. — N.R.C. Publication 133, P 99—101, Washington 1964.
8. Philibert, J., and R. Tixier: Electron penetration and the atomic number correction in electron probe microanalysis. Brit. J. Appl. Phys. 1, 685—694 (1968).
9. Castaing, R., et J. Henoc: Répartition en profondeur du rayonnement caractéristique. Optique des rayons X et microanalyse, Orsay 1965. p. 120—126. Paris: Hermann 1966.
10. Cosslett, V. E.: The absorption of electron energy in solid target. Optique des Rayons X et Microanalyse, Orsay 1965, p. 85—96. Paris: Hermann 1966.
11. Worthington, C. R., and S. G. Tomlin: The intensity of emission of characteristics X radiation. Proc. Phys. Sty, 69, 401—412 (1956).
12. Duncumb, P., and K. Da Casa: Private communication unpublished.

Phenomenes de fluorescence aux limites de phases

M. J. HENOC[1], Mlle F. MAURICE[2] et Mme A. ZEMSKOFF[2]

Gif-sur-Yvette, France

Résumé

Il faut toujours prendre en considération les rayonnements parasites de fluorescence, particulièrement aux limites de phases. Des expériences menées sur quelques couples non diffusés, complétant une étude antérieure, permettent de mettre en évidence les phénomènes secondaires que l'on peut prévoir par le calcul pour lequel une détermination théorique du nombre de photons émis est nécessaire. La précision des résultats expérimentaux dépend largement de la préparation des échantillons. De cette manière, il est possible de vérifier la méthode à l'aluminium et de donner une estimation des corrections de fluorescence.

Introduction

Quand on étudie des limites de phases à l'aide d'une microsonde, le rayonnement parasite de fluorescence rend particulièrement délicate l'interprétation des résultats expérimentaux, en effet le rayonnement X analysé par le spectrographe peut provenir soit directement de la zone de l'échantillon baignée par le faisceau d'électrons, soit encore des zones limitrophes qui sont le siège d'un rayonnement de fluorescence.

Dans un mémoire précédent [1] nous avons considéré l'importance du rayonnement de fluorescence excité par l'ensemble du spectre K, nous voudrions ici étudier séparément l'effet du spectre continu et de diverses raies caractéristiques.

Nous avons repris le même type d'études que précédemment sur des couples de deux métaux A et B n'ayant pas subi de traitement de diffusion [1]. Dans ces conditions il s'agit alors de calculer et de mesurer l'intensité du rayonnement X émis dans une raie caractéristique de A quand l'impact de la sonde se trouve dans B à une distance d de l'interface des deux métaux.

Résultats théoriques

1. Calcul du rayonnement de fluorescence dû au spectre continu

Si l'on admet que le rayonnement primaire est émis en surface, la situation est semblable à celle qui a été évoquée précédemment [1]. Il suffit de substituer un rayonnement polychromatique à un rayonnement monochromatique. Avec les notations de la fig. 1, l'intensité I_f du rayonnement secondaire émis dans la raie i de A excité par un rayonnement d'intensité $I_\lambda^c d\lambda$ de largeur spectrale $d\lambda$, en admettant qu'aucune discontinuité de A ou B ne se trouve dans l'intervalle (λ_0, λ_i), peut se mettre sous la forme:

$$I_{f(A)} = \int_{\lambda_0}^{\lambda_i} \int_{\phi=0}^{\pi} \int_{\psi=0}^{\pi/2} F(\phi, \psi, \lambda)\, d\phi\, d\psi\, d\lambda$$

1. C.N.E.T — Issy les Moulineaux.
2. C.E.A — CEN SACLAY.

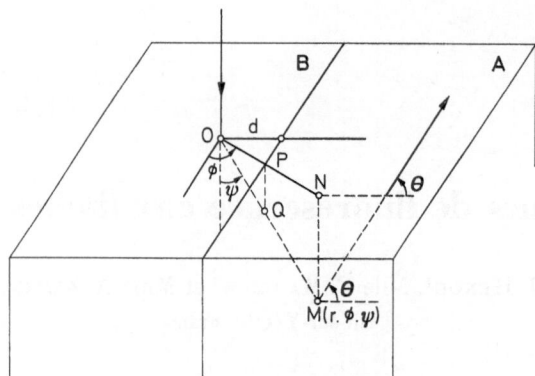

Fig. 1. Schéma de principe

où

$$F(\phi, \psi, \lambda) = \frac{1}{4\pi}\, \omega_i(A) \left(\frac{r_i - 1}{r_j}\right)_A z_{iA}\, I_\lambda^c \mu_\lambda^A \frac{\sin \phi}{\mu_A^A \operatorname{cosec} \theta \cos \psi + \mu_\lambda^A}$$

$$\cdot \exp - (\mu_\lambda^B \varrho_B + \mu_A^A \varrho_A \cos \psi \operatorname{cosec} \theta)\, \frac{d}{\sin \phi \sin \psi}$$

$$I_\lambda^c = 2{,}26 \cdot 10^{-13} Z_B \left(\frac{1}{\lambda_0} - \frac{1}{\lambda}\right)\frac{1}{\lambda} \quad \text{expression de Kramers} \quad (\lambda \text{ en cm})$$

$$\mu_\lambda^X = \mu_{\lambda_0}^X \left(\frac{\lambda}{\lambda_0}\right)^3.$$

Quand l'intervalle (λ_0, λ_i) contient plusieurs discontinuités de A ou B, il est divisé en sous intervalles limités par les différentes discontinuités soit (λ_0, λ_j), $(\lambda_j, \lambda_{j+1}) \dots (\lambda_{i-1}, \lambda_i)$. On obtient alors:

$$I_{f(A)} = \sum_{j=0}^{i-1} \int_{\lambda_j}^{\lambda_{j+1}} \int_{\phi=0}^{\pi} \int_{\psi=0}^{\pi/2} F(\phi, \psi, \lambda)\, d\phi\, d\psi\, d\lambda.$$

Les μ_λ^X, r_i, r_j, $w_i(A)$ sont calculés pour chaque bande suivant un procédé bien connu [2, 3].

Les intensités sont définies, par convention, en nombre de photons émis dans l'angle solide 4π stéradian.

2. Calcul du rayonnement de fluorescence dû aux raies caractéristiques

En admettant comme précédemment que le rayonnement primaire est émis en surface, c'est-à-dire que la source du rayonnement excitateur peut être assimilée à un point, l'intensité du rayonnement de fluorescence s'exprime de la façon suivante [1]:

$$I_{f(A)} = \int_{\phi=0}^{\pi} \int_{\psi=0}^{\pi/2} F(\phi, \psi)\, d\phi\, d\psi$$

avec

$$F(\phi, \psi) = \frac{\mu_B^A}{4\pi}\, \omega_i(A) \left(\frac{r_i - 1}{r_j}\right)_A I(B) \frac{\sin \psi}{\mu_B^A + \mu_A^A \operatorname{cosec} \theta \cos \psi}$$

$$\cdot \exp - [(\mu_B^B \varrho_B + \mu_A^A \varrho_A \operatorname{cosec} \theta \cos \psi)\, d/(\sin \psi \sin \phi)].$$

Cette expression présente deux différences par rapport à la formule plus élaborée qui fait intervenir la distribution du rayonnement primaire en fonction de la profondeur: tout d'abord elle majore l'effet du rayonnement dirigé vers le bas puisqu'elle ne tient pas compte du terme d'absorption du rayonnement secondaire le long du parcours $\varrho z \operatorname{cosec} \theta$ et d'autre part, elle néglige l'effet du rayonnement primaire dirigé vers le haut. D'une façon générale, ces deux quantités ont sensiblement la même importance si bien que l'erreur résultante est faible. La situation est toutefois beaucoup moins favorable quand le rayonnement secondaire a une faible longueur d'onde. Il est alors préférable d'évaluer la contribution du rayonnement dirigé vers le haut en admettant, par exemple, que l'on peut approcher la fonction $\varphi(\varrho z)$ par une expression de la forme $I_0 \exp(-\sigma \varrho z) - [4]$.

Enfin, la valeur du rayonnement secondaire est obtenue en fonction du rayonnement global $I(B)$ de l'élément B, ce fait n'est pas caractéristique de l'approximation employée. Le calcul complet nécessite par conséquent, la comparaison entre les intensités émises normalement par les éléments A et B.

Dans tous les cas, il est préférable de calculer ces intensités de la manière exposée au paragraphe suivant, plutôt que d'utiliser l'approximation de CASTAING [4] basée sur la fonction d'ionisation de ROSSELAND, qui est d'un emploi facile uniquement dans le cas d'excitation d'un rayonnement de la série K par d'autres raies de la même série et pour un taux d'excitation supérieur à 3.

3. Calcul de l'intensité du rayonnement primaire

Il s'agit d'évaluer l'intensité du rayonnement primaire I_A émergent d'un échantillon, contenant A en concentration massique C_A, lorsqu'il est soumis au bombardement du faisceau d'électrons. La méthode suivie ne comporte que des modifications minimes par rapport à celle que nous avons déjà utilisée [5]. Le nombre de photons émis dans la raie i appartenant à la série l issue de transition m s'écrit:

$$I_A^0 = w_l(A) \, z_{iA} \, C_A \, \frac{N}{A} \int_{E_0}^{E_m} \frac{Q}{dE/d\varrho \, s} \, dE$$

où

$$Q = \frac{\pi e^4}{E \, E_m} Z_{nl} \, b \, L \, \frac{4E}{B_m} \qquad \text{avec } B_m = 4 E_m$$

expression de WORTHINGTON et TOMLIN [6]

$$dE/d\varrho \, s = -\frac{2\pi e^4 N}{E} \sum_{A, B, \ldots} C_A \, \frac{Z_A}{A} \, L \, \frac{1{,}166 \, E}{J_A} \qquad \text{(avec } J_A = 11{,}5 Z_A)$$

expression de Bethe modifiée en admettant que l'effet des atomes de nature différente est additif.

Suivant PHILIBERT [7], en posant

$$M = \sum_{A, B, \ldots} C_A \, \frac{Z_A}{A}$$

et

$$M L W = \sum_{A, B, \ldots} C_A \, \frac{Z_A}{A} \, L \, \frac{1{,}166 \, E_m}{J_A}$$

on obtient:

$$I_A^0 = \omega_l(A) \, z_{iA} \, \frac{C_A}{A} \, \frac{Z_{nl}}{2} \, b \, \frac{1}{M} \left\{ (U_0 - 1) - [E i(L U_0 W) - E i(L W)] \, \frac{L W}{W} \right\}.$$

Pour tenir compte de la rétrodiffusion des électrons et de l'absorption du rayonnement émis avant l'émergence il faut introduire les facteurs R_A et $f(\chi)_A$ calculés respectivement par les méthodes de BISHOP [8] et PHILIBERT [9]; finalement on obtient:

$$I_A = \omega_l(A) \, z_{iA} \, \frac{C_A}{A} \, \frac{Z_{nl}}{2} \, b \, \frac{1}{M} \left\{ (U_0 - 1) - [E i(L U_0 W) - E i(L W)] \, \frac{L W}{W} \right\} R_A \, f(\chi)_A.$$

Dans le cas d'un élément pur, cette expression se réduit à:

$$I_{(A)} = \omega_l(A) \, z_{iA} \, \frac{Z_{nl}}{2} \, \frac{b}{Z_A} \left\{ (U_0 - 1) - \left[E i\left(L \, \frac{1{,}166 \, E_0}{J_A}\right) - E i\left(L \, \frac{1{,}166 \, E_m}{J_A}\right) \right] \frac{L \, \frac{1{,}166 \, E_m}{J_A}}{\frac{1{,}166 \, E_m}{J_A}} \right\} R_{(A)} f(\chi)_{(A)}.$$

La comparaison de quelques résultats expérimentaux [5—10] et de cette formule théorique permet de définir une valeur de la constante b.

On est amené à poser $b = 0,76$ pour les raies appartenant aux séries K et L.

Il faut noter que cette méthode est davantage justifiée par les résultats qu'elle a donné dans les calculs de correction que par une argumentation physique rigoureuse, notamment en ce qui concerne l'introduction du facteur R.

Tous les calculs théoriques ont été faits à l'aide d'un ordinateur IBM. Ces différents éléments permettent de redéterminer théoriquement les concentrations $I_{f(A)}/I_{(A)}$.

Tableau des couples étudiés

Couple	Raie analysée	Effet observé
V—Ti	V $K_{\alpha 1}$	F.C.
Cu—Ni	Cu $K_{\alpha 1}$	F.C.
Zr—Nb	Nb $K_{\alpha 1}$	F.C.
Au—Pt	Au $L_{\beta 1}$	F.C.
Ti—V	Ti $K_{\alpha 1}$	F.C. + V K_{β}
Cu—Ni	Ni $K_{\alpha 1}$	F.C. + Cu K_{β}
Zr—Nb	Zr $K_{\alpha 1}$	F.C. + Nb K_{β}
Y—Mo	Y $K_{\alpha 1}$	F.C. + Mo K_{α} + Mo K_{β}
Ni—Zn	Ni $K_{\alpha 1}$	F.C. + Zn K_{α} + Zn K_{β}
Ti—Cr	Ti $K_{\alpha 1}$	F.C. + Cr K_{α} + Cr K_{β}
Au—Pt	Pt $L_{\beta 1}$	F.C. + Au $L_{\gamma 1}$
Au—Pt	Au $L_{\alpha 1}$	F.C. + Pt $L_{\gamma 1}$
Au—Pt	Pt $L_{\alpha 1}$	F.C. + Au $L_{\gamma 1}$ + Au $L_{\beta 2}$ + Au $L_{\beta 3}$

Les fig. 2, 3, 4, 6, 7, montrent l'évolution de la concentration apparente de l'élément excité A dans l'élément excitateur B en fonction de la distance d du point d'impact des électrons à l'interface.

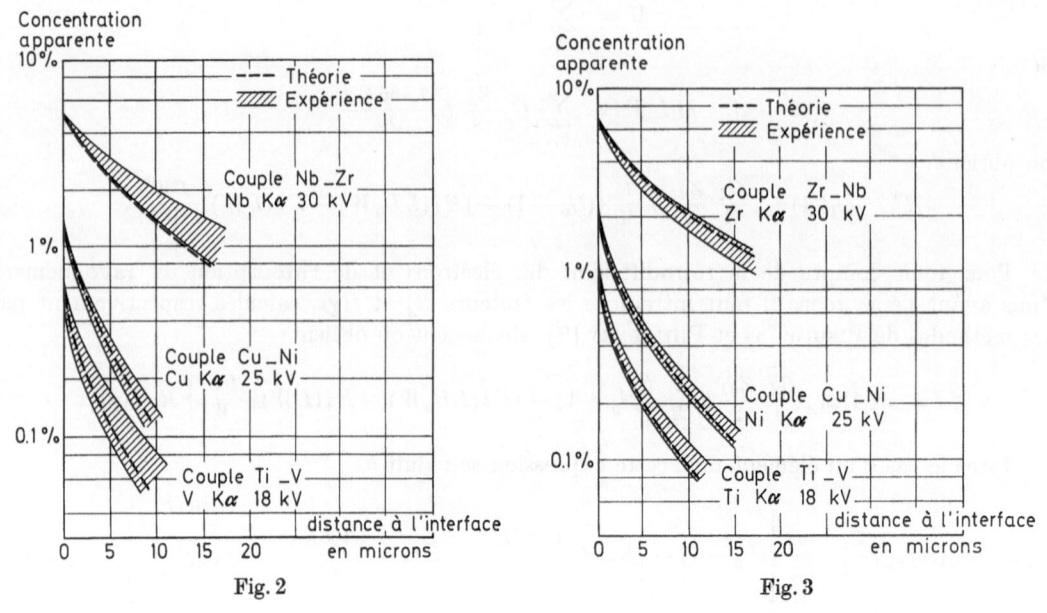

Fig. 2. Effet de fluorescence dû au spectre continu

Fig. 3. Effet de fluorescence dû au spectre continu, aux raies K_{β}

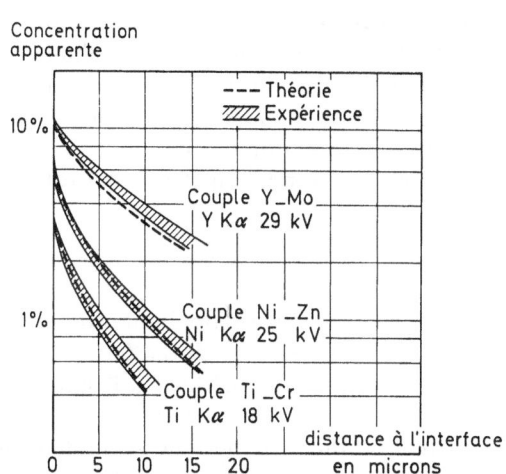

Fig. 4. Effet de fluorescence dû au spectre continu, aux raies K_β, aux raies K_α

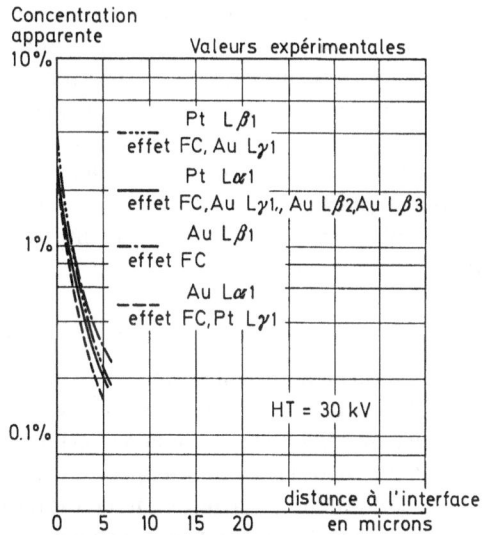

Fig. 5. Effet de fluorescence couple Au—Pt, valeurs expérimentales

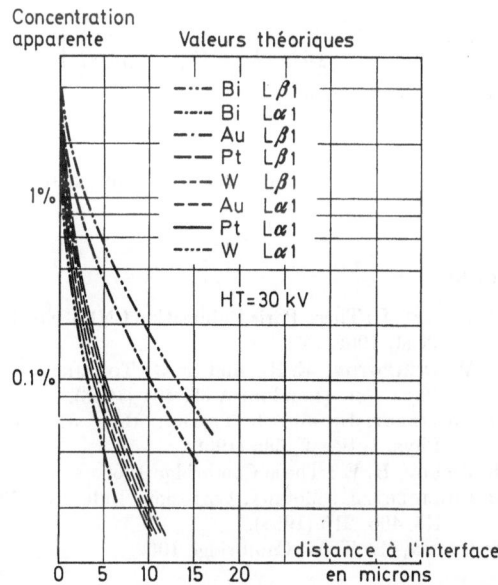

Fig. 6. Effet de fluorescence couple Au—Pt, valeurs théoriques

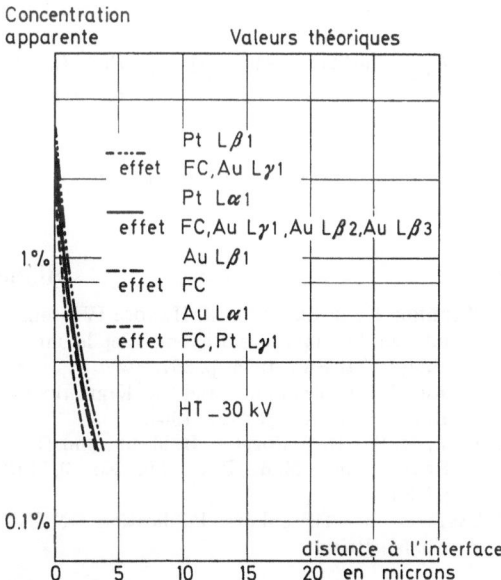

Fig. 7. Effet de fluorescence dû au spectre continu, raies L

Résultats expérimentaux

Les couples figurant au tableau ci-dessus ont été soigneusement préparés suivant la technique indiquée dans le mémoire précédent [1] les mesures effectuées sur ces échantillons sont reportées sur les fig. 2, 3, 4, 5. La microsonde utilisée est une microsonde CAMECA MS 46.

Discussion

Les couples doivent être choisis de manière à observer un phénomène de fluorescence nettement sensible. La sélection est assez facile pour les raies K elle est beaucoup plus délicate pour les raies L; non seulement à cause de la faible intensité du rayonnement secondaire mais aussi par suite des difficultés de préparation; pratiquement nous nous sommes limités à l'étude du couple or-platine.

Les résultats théoriques et expérimentaux sont en bon accord pour la plupart des couples étudiés dans le domaine des raies K. Les différences observées sur le seul couple Y—Mo peuvent être attribuées aux difficultés expérimentales (rapport pic sur fond et polissage). Pour les raies L la décroissance de l'intensité du rayonnement secondaire en fonction de la distance à l'interface semble être plus rapide que pour les raies K; étant donné les conditions expérimentales on peut considérer l'accord entre les valeurs théoriques et mesurées comme satisfaisant. Il faut noter que l'intensité du rayonnement secondaire dû au spectre continu pour $d = 0$ doit être deux fois plus faible que le résultat obtenu par la méthode à l'aluminium [2] ce qui est bien vérifié.

Conclusion

Cette méthode de calcul permet d'estimer la correction de fluorescence aux limites de phases qu'il s'agisse de métaux purs ou d'alliages, d'excitation de raie du spectre K ou L.

Parallèlement, il a été possible de vérifier les résultats obtenus par la méthode à l'aluminium sur l'effet de fluorescence dû au spectre continu.

Enfin l'approximation faite sur la forme de la répartition du rayonnement en profondeur ainsi que le calcul des intensités primaires émergentes permet de donner une estimation de l'importance du rayonnement secondaire de fluorescence dans le cas d'alliages homogènes à condition de modifier très légèrement l'expression de Castaing [4] soit:

$$\frac{I_{fA}}{I_{(A)}} = \frac{\omega(A)}{2}\left(\frac{r-1}{r}\right)_A \frac{\mu_B^A}{\mu_B^{AB}} C_A \frac{R_B}{R_A} \frac{I_B}{I_{(A)}} \frac{f^{AB}(\chi_A)}{f^A(\chi_{(A)})} \left[\frac{L\left(1 + \frac{\mu_A^{AB}\cosec\theta}{\mu_B^{AB}}\right)}{\frac{\mu_A^{AB}\cosec\theta}{\mu_B^{AB}}} + \frac{L\left(1 + \frac{\sigma}{\mu_B^{AB}}\right)}{\frac{\sigma}{\mu_B^{AB}}} \right].$$

Cette expression utilisée pour l'arséniure de gallium recoupe à moins de 2% les résultats expérimentaux.

Bibliographie

1. Maurice, F., R. Seguin et J. Henoc: IV° Congr. int. sur l'optique des Rayons X et la micro-analyse, ORSAY 1965, p. 357.
2. Henoc, J., F. Maurice, and A. Kirianenko: Rapport C.E.A.—R 2421, 1964.
3. Fink, R. W., R. C. Jopson, H. Mark, and C. D. Swift: Rev. Mod. Phys. 38, No. 3, 513 (1966).
4. Castaing, R.: Thèse Paris Publication ONERA No. 55, 1952.
5. Henoc, J.: Thèse Paris Publication CNET No. 655 PCM, 1962.
6. Worthington, E. R., and S. G. Tomlin: Proc. Phys. Soc. (London) A 69, 401 (1956).
7. Philibert, J., and R. Tixier: Brit. J. Appl. Phys. 1, Sér. 2, 685 (1968).
8. Bishop, H. E.: Thèse Cambridge 1966.
9. Philibert, J.: Metaux, Corrosion, Industries 40, No. 466, 216 (1964).
10. Green, M.: Thèse Cambridge 1962.

Die Massenschwächungskoeffizienten der Kohlenstoff-K_α-Linie in Abhängigkeit von der Ordnungszahl

E. Kohlhaas und F. Scheiding

Versuchsanstalt der Röchlingschen Eisen- und Stahlwerke GmbH, Völklingen (Saar), BRD

Für verschiedene Gruppen technisch interessanter Werkstoffe wie die Stähle, die hochwarmfesten Legierungen, die Hartmetalle u. a. m. ist der Kohlenstoff ein wichtiges Legierungselement. Seiner quantitativen Analyse kommt in den genannten Werkstoffen auch in Mikrobereichen eine besondere Bedeutung zu.

Die praktische Durchführung der Kohlenstoffanalyse mit der Elektronenmikrosonde stößt aber auf meßtechnische Schwierigkeiten, da wegen der extremen Langwelligkeit der Kohlenstoff-K_α-Linie — etwa 44 Å — Analysatorkristalle mit entsprechend großen Gitterkonstanten erforderlich sind. Die leichte Absorbierbarkeit der Linie verlangt zudem die Verwendung von Vakuumspektrometern und besonders durchlässigen Zählrohrfenstern. Weiterhin muß die Bildung von Kontaminationsschichten auf der Probenoberfläche verhindert werden. Bei den vorliegenden Untersuchungen wurde zu diesem Zweck sowohl ein Kühlfinger benutzt als auch Sauerstoff auf die Probe aufgeblasen. Die Abb. 1 zeigt schematisch die Meßanordnung.

Bezüglich der für die quantitative Kohlenstoffanalyse notwendigen Korrekturrechnungen bestehen ebenfalls noch einige Probleme. Abgesehen von der Frage, durch welche der zahlreichen in der Literatur angegebenen Korrekturformeln die Absorptionsvorgänge am besten beschrieben werden, ist es für die Durchführung der Rechnungen vor allem erforderlich, die Massenschwächungskoeffizienten der Kohlenstoff-K_α-Linie für die verschiedensten Elemente zu kennen. Diese Werte fehlen jedoch weitgehend für die Legierungselemente der technisch wichtigsten Werkstoffe, bzw. die bisher in der Literatur gemachten wenigen Angaben weichen beachtlich voneinander ab.

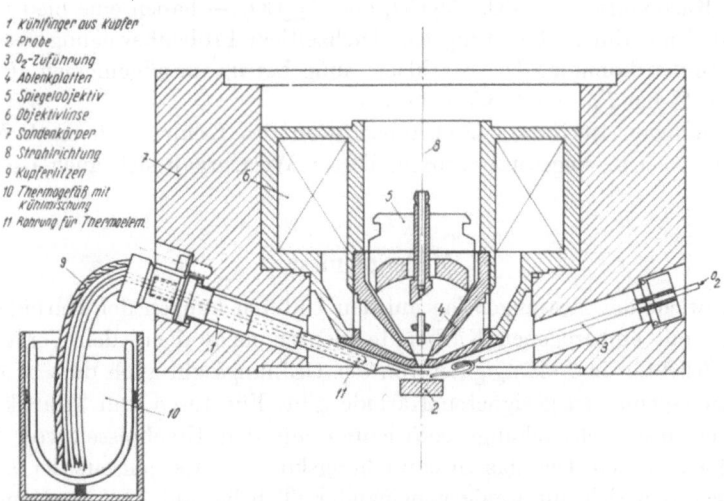

1 Kühlfinger aus Kupfer
2 Probe
3 O_2-Zuführung
4 Ablenkplatten
5 Spiegelobjektiv
6 Objektivlinse
7 Sondenkörper
8 Strahlrichtung
9 Kupferlitzen
10 Thermogefäß mit Kühlmischung
11 Rohrung für Thermoelem.

Abb. 1. Schematische Darstellung der Ausführung und Anbringung des Kühlfingers und des Röhrchens für die Sauerstoffzuführung in der Mikrosonde CAMECA MS46

B. L. Henke, R. White und B. Lundberg [1] berechneten unter Zugrundelegung der Tabellen von S. J. M. Allen [2] und J. A. Victoreen [3] die Massenschwächungskoeffizienten der leichten Elemente bis zur L-Kante, d.h. für die Kohlenstoff-K_α-Linie bis zum Element der Ordnungszahl 17 (Chlor). P. Duncumb und D. A. Melford [4] benutzten eigene Mikrosonden-messungen an SiC, TiC und Fe_3C, um mit einer Monte-Carlo-Methode nach H. E. Bishop die Massenschwächungskoeffizienten für Si, Ti und Fe zu ermitteln. Umfangreichere Mikrosonden-Untersuchungen führten D. E. Fornwalt und A. V. Manzione [5] an den meisten stabilen binären Karbiden aus und versuchten, mittels der Korrekturformel von T. E. Ziebold und R. E. Ogilvie [6] aus den Meßwerten die Massenschwächungskoeffizienten zu bestimmen. Ein Vergleich ihrer Ergebnisse mit den von P. Duncumb und D. A. Melford [4] erhaltenen Daten weist sehr große Unterschiede auf. Bei der Anwendung der Korrekturformel nach Ziebold und Ogilvie wird für die Meßwerte der Elemente Bor und Silizium — gemessen an B_4C und SiC — absolut keine Übereinstimmung mit den Rechnungen von Henke, White und Lundberg erzielt. Die Ursache für die genannten Abweichungen liegt wohl darin, daß die Korrekturformel nach Ziebold und Ogilvie eine auf die Praxis abgestimmte Vereinfachung darstellt und keineswegs auf binäre Systeme mit stark unterschiedlichen Ordnungszahlen angewendet werden darf und auch für benachbarte leichte Elemente die genannte Korrekturformel nicht mehr gültig ist.

Die Idee, eine der bekannten Korrekturformeln zur Berechnung bzw. Abschätzung der Massen-schwächungskoeffizienten heranzuziehen, wurde in der vorliegenden Arbeit aufgegriffen. Es wurde jedoch die Korrekturformel nach R. Theisen [7] benutzt, die eine weitere Verfeinerung des Ansatzes von J. Philibert [8] darstellt. R. Theisen setzt den Lenard-Koeffizienten in folgender Form an:

$$\sigma = \frac{8,9 \cdot 10^5}{(V - V_C)^2},$$

wobei V = Strahlspannung und V_C = Anregungsspannung des zu analysierenden Elementes bedeuten.

Für den effektiven Rückstreukoeffizienten R wurden die Werte von H. Kuhlenkampff und W. Spyra [9] eingesetzt.

Die Messungen erfolgten an Karbiden und einigen Karbonaten (Tabelle). Eisen wurde auch an einer Eichreihe mit bekannten gelösten Kohlenstoffgehalten [10] untersucht. Da es äußerst schwierig ist, reine Kobalt- und Nickelkarbide herzustellen, wurden die Messungen an $(Fe_2Co)C$ und $Ni_{5,6}W_{6,4}C_4$ ausgeführt. Mit Hilfe der vorher durch Untersuchungen an Fe_3C und WC errechneten Massenschwächungskoeffizienten für Fe und W konnten die entsprechenden Koeffizienten für Co und Ni ermittelt werden. Die Karbonatkristalle ließen sich in Sn-Pastillen ein-drücken. Einige Karbonate — Li_2CO_3, $MgCO_3$ und Ag_2CO_3 — haben eine niedrige Zersetzungs-temperatur, so daß nur durch Rasterung bei gleichzeitiger Probenbewegung verwendbare Meß-werte erhalten werden konnten. Die Strahlspannung betrug im allgemeinen 10 kV; zusätzlich wurde für Fe und Si bei 7,5 und 15 kV gemessen.

Bei den Rechnungen ist für die Selbstabsorption des Kohlenstoffs nach Messungen von M. Green [11] $f(\chi) = 0,45$ eingesetzt worden. Dieser Wert ergab sich auch nach der Theisen-schen Formel.

Meßergebnisse

Das Schwergewicht der Messungen lag auf den Untersuchungen der Karbide. Die ebenfalls in die Messungen mit einbezogenen Karbonate dienten mehr dazu, den Verlauf des Massen-schwächungskoeffizienten in Abhängigkeit von der Ordnungszahl auch dort zu erfassen, wo es keine für die Untersuchungen geeigneten Karbide gibt. Für Eisen und Titan konnten nahezu übereinstimmende Massenschwächungskoeffizienten mit den Ergebnissen von Duncumb und Melford gewonnen werden. Die Massenschwächungskoeffizienten jenseits der L-Kante von Ca bis zu Cr unterscheiden sich nur wenig voneinander (Tabelle und Abb. 2). Vom Chrom an ist ein starker Anstieg zu verzeichnen. Wie schon Fornwalt und Manzione durch Extrapolation ihrer Meßwerte vermuteten, zeigt die Einbeziehung weiterer Elemente in der vorliegenden Arbeit

Tabelle. *Zusammenstellung der mit der Mikrosonde CAMECA untersuchten Karbide und Karbonate*

Element	Z	chem. Formel	Impuls-ausbeute [Jmp./sec]	Metall $\Delta\mu$ Kohlenstoff eigene Meßwerte	Metall $\Delta\mu$ Kohlenstoff Henke† u. Duncumb/Melford²	Element	Z	chem. Formel	Impuls-ausbeute [Jmp./sec]	Metall $\Delta\mu$ Kohlenstoff eigene Meßwerte	Metall $\Delta\mu$ Kohlenstoff Henke† u. Duncumb/Melford²
Lithium	3	Li_2CO_3	95	14300	9400 [1]	Kobalt	27	Fe_2CoC	35	16700	—
Bor	5	B_4C	14	33000	32540 [1]	Nickel	28	$Ni_{3,6}W_{6+}C_4$	97	49800	—
Natrium	11	Na_2CO_3	44	17300	16650 [1]	Zink	30	$ZnCO_3$	6	68400	—
Magnesium	12	$MgCO_3$	45	31400	22850 [1]	Strontium	38	$SrCO_3$	2	49000	—
Aluminium	13	Al_4C_3	48	24400	24910 [1]	Zirkon	40	ZrC	28	20600	—
Silizium 7,5 KV	14	SiC	95	30500	33840 [1,2]	Niob	41	NbC	34	15800	—
Silizium 10KV	14	SiC	57	26200	33840 [1,2]	Molybdän	42	Mo_2C	23	12500	—
Silizium 15KV	14	SiC	43	36900	33840 [1,2]	Silber	47	$AgCO_3$	30	13300	—
Kalzium	20	Ca_2CO_3	98	6800	—	Barium	56	$BaCO_3$	43	6300	—
Titan	22	TiC	148	6900	6550 [2]	Hafnium	72	HfC	35	7600	—
Vanadium	23	VC	132	7500	—	Tantal	73	TaC	32	8500	—
Chrom	24	Cr_2C_3	86	7700	—	Wolfram	74	WC	30	10000	—
Eisen 7,5KV	26	Fe_3C	43	9100	11000 [2]						
Eisen 10KV	26	Fe_3C	32	10800	11000 [2]			Strahlspannung 10 KV			
Eisen 15 KV	26	Fe_3C	25	10000	11000 [2]			Impulsausbeute Diamant 1660 Jmp./sec $\mu\,^C_C = 2170$			

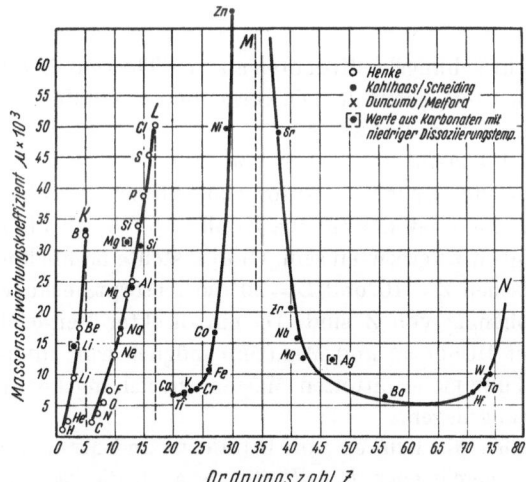

Abb. 2. Der Massenabsorptionskoeffizient der Kohlenstoff-K_α-Linie in Abhängigkeit von der Ordnungszahl Z

deutlich, daß etwa bei der Ordnungszahl 34 eine M-Absorptionskante auftritt. Des weiteren bestätigt sich eine Abnahme der Massenschwächungskoeffizienten mit steigender Ordnungszahl jenseits der M-Kante. Der Wiederanstieg der Massenschwächungskoeffizienten für Hf, Ta und W könnte auf das Vorhandensein einer N-Kante hindeuten. Von den Elementen, die im Bereich bis zur L-Kante liegen, werden nur drei stabile Karbide gebildet, nämlich B_4C, Al_4C_3 und SiC. Es konnten aber auch Messungen an Karbonaten dort vorgenommen werden: Li_2CO_3, Na_2CO_3 und $MgCO_3$. Wie aus der Abb. 2 zu ersehen ist, wird für die genannten Elemente eine befriedigende Übereinstimmung mit den berechneten Werten von HENKE, WHITE und LUNDBERG erzielt. Nur für Si zeigt sich eine größere Abweichung, obwohl die gemessenen Impulsausbeuten bei Strahlspannungen von 7,5, 10 und 15 kV recht gut mit den Werten von DUNCUMB und MELFORD vergleichbar sind.

Eine gewisse Berechtigung für die Anwendbarkeit der Korrekturformel nach R. THEISEN und damit für die Richtigkeit der berechneten Massenschwächungskoeffizienten ergaben Untersuchungen verschiedenster Komplexkarbide, wie z.B. von Karbiden in Schnellarbeitsstählen und Superlegierungen. Über eine Analyse der im Karbid enthaltenen Elemente wurde mittels

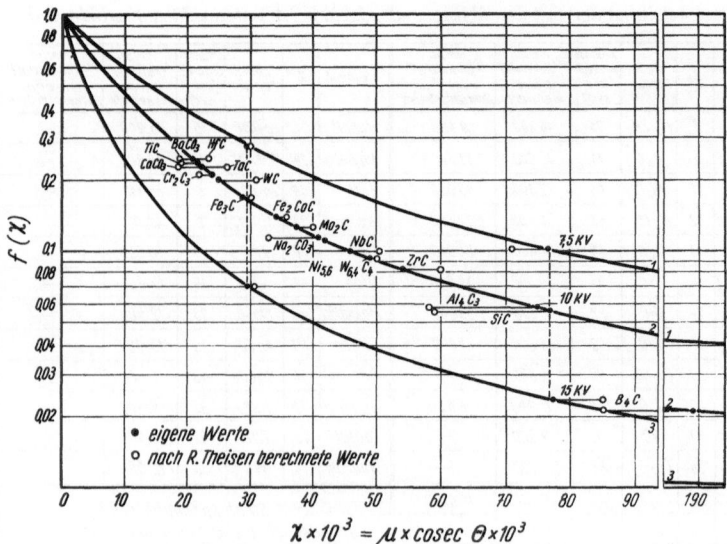

Abb. 3. $f(\chi)$-Kurven für die Kohlenstoff-K_α-Strahlung in Siliziumkarbid bei verschiedenen Strahlspannungen nach Duncumb und Melford

der berechneten Massenschwächungskoeffizienten aus der Theisenschen Formel die zu erwartende Impulsausbeute für Kohlenstoff ermittelt. Die vorgenommenen Messungen erbrachten stets eine Übereinstimmung mit den Erwartungswerten.

In der bereits mehrfach angeführten Arbeit von Duncumb und Melford ist eine nach einer Monte-Carlo-Methode erstellte Schar von Absorptionskurven in Abhängigkeit von $V - V_C$ veröffentlicht worden. Diese gilt zwar nur für die Kohlenstoff-K_α-Strahlung in SiC, es sollte aber eine weitgehendere Gültigkeit zu erwarten sein. Green stellte nämlich fest, daß für die Elemente mit Ordnungszahlen zwischen $Z = 10$ und $Z = 30$ die Absorptionskurven in Abhängigkeit von $V - V_C$ weitgehend unabhängig von Z sind. Da ein direkter Vergleich der eigenen Meßwerte nur für Si, Ti und Fe mit Duncumb und Melford möglich war, interessierte, wie alle übrigen nach Korrekturen von Theisen ermittelten Massenschwächungskoeffizienten im Vergleich zu einer Monte-Carlo-Rechnung liegen.

In der Abb. 3 sind daher die Kurven von Duncumb und Melford für SiC wiedergegeben, in die sowohl die eigenen gemessenen $f(\chi)$-Werte als auch die nach der Theisenschen Formel errechneten eingetragen wurden. Die Abweichungen zwischen Messung und Rechnung bewegen sich in solch engen Grenzen, daß damit die Beobachtungen von Green auch für Kohlenstoff ($Z = 6$) bestätigt werden. Die Differenzen dürften auf eine nicht ausreichende Erfassung des Atomzahleinflusses bei R. Theisen zurückzuführen sein. Die beste Übereinstimmung wird für die Elemente Ca bis Ni erreicht. Aber selbst im Bereich der schweren Elemente, für die Green keine Aussage macht ($Z > 30$), kann das Ergebnis als zufriedenstellend angesehen werden. Starke Abweichungen treten bei den leichten Elementen Bor, Aluminium und Silizium auf. Wie schon erwähnt, stimmen jedoch die Werte der vorliegenden Arbeit für Bor und Aluminium gut mit Henke, White und Lundberg überein.

Zusammenfassung

Aus Messungen an Karbiden und Karbonaten wurden nach der Korrekturformel von R. Theisen die Massenschwächungskoeffizienten der Kohlenstoff-K_α-Linie für zahlreiche Elemente berechnet und in Abhängigkeit von der Ordnungszahl dargestellt. Eine Bestätigung für die Richtigkeit dieses Vorgehens erbrachten Messungen an Karbiden komplexer Zusammensetzung. Des weiteren wurde eine befriedigende Übereinstimmung mit bisher bekanntgewordenen Werten einiger anderer Autoren erzielt.

Literatur

1. HENKE, B. L., R. WHITE, and B. LUNDBERG: J. Appl. Phys. **28**, 98—105 (1956).
2. ALLEN, S. J. M.: Handbook of chemistry and physics, 45. ed., E-69. Cleveland: Chemical Rubber Co.
3. VICTOREEN, J. A.: J. Appl. Phys. **20**, 1141 (1949).
4. DUNCUMB, P., and D. A. MELFORD: Proceedings of the 4th Int. Congr. on X-ray Optics and X-ray Microanalysis Paris, September 1965.
5. FORNWALT, D. E., and A. V. MANZIONE: Norelco-Reptr, **13**, No. 2, 39—63 (1966).
6. ZIEBOLD, T. O., and R. E. OGILVIE: Anal. Chem. **36**, No. 2, 322—327 (1964).
7. THEISEN, R.: Quantitative electron microprobe analysis. Berlin-Heidelberg-New York: Springer 1965.
8. PHILIBERT, J.: Métaux, Corrosion, Industries No. 466, 216—239 (Juni 1964).
9. KUHLENKAMPFF, H., u. W. SPYRA: Z. Physik **137**, 416 (1954).
10. KOHLHAAS, E., u. F. SCHEIDING: Arch. Eisenhuettenw. **40**, Nr. 1, 47—51 (1969).
11. GREEN, M.: Proc. Phys. Soc. (London) **83**, 435—451 (1964).

Messung des Massenabsorptionskoeffizienten in Kohlenstoff im Wellenbereich 10 bis 70 Å

W. Weisweiler

Institut für Chemische Technik der Universität Karlsruhe, BRD

Zusammenfassung. Dünne Folien aus reinem, ungeordnetem Kohlenstoff werden durch mehrmaliges Bedampfen hergestellt. Nach Ablösen von der Glasunterlage ergeben sich 2,7 cm² große, faltenfreie Folien, deren gleichmäßige Dicken mit einem Photometer, durch Wägung und mit dem Interferenzverfahren nach Tolansky zu 3500 bis 6500 Å bestimmt werden. Die freitragenden Folien gestatten nach Einbringen in den Röntgenstrahlengang des dispersiven Analysatorsystems einer Mikrosonde (JEOL JXA-3A) die Strahlungsabsorption in elementarem Kohlenstoff zu ermitteln. Zahlreiche Messungen für 7 Wellenlängen zwischen 10 und 70 Å bestätigen unterhalb der Kohlenstoff-K-Absorptionskante Massenabsorptionskoeffizienten, die bisher über die Gasphase gewonnen wurden, liegen jedoch nach der Absorptionskante höher als die bisher nur über Kohlenstoffverbindungen erhaltenen Werte. Für die Absorption von Kohlenstoff-K-Strahlung in Kohlenstoff, die in der Literatur zwischen 1720 und 4100 angegeben wird, ergibt sich $\mu/\varrho = 2535 \pm 150$ cm²/g. Weiterhin werden die Homogenität, die Emissionsspektren und das Kontaminationsverhalten von Kohlenstoffproben verschiedenen kristallinen Ordnungszustandes untersucht und festgestellt, daß sich Harzkohlenstoff als Standard eignet.

Summary. Thin foils were made of pure disordered carbon, employing vapor deposition. After separation from the glass substrate, the foils were found to be free of wrinkles. Their size amounted to 2.7 cm². The thickness as determined by photometer, weighing and interference procedures according to Tolansky proved to be uniform, the individual values of different foils ranging from 3500 to 6500 Å. After insertion into the X-ray path of the dispersive analyser system of an electron microprobe (JEOL JXA-3A), the self-supporting foils permitted determination of the mass attenuation coefficients of elementary carbon. Below the carbon K-absorption edge, numerous measurements carried out for 7 wavelengths between 10 and 70 Å yielded values in agreement with literature data derived from the gas phase. However, above the absorption edge, the results in this investigation obtained are higher than those known from determinations based on carbon compounds. For the absorption of the K_α-radiation in carbon, the mass attenuation coefficient μ/ϱ was found to be 2535 ± 150 cm²/g, the respective literature values ranging from 1720 to 4100 cm²/g. In addition, investigations were carried out regarding homogeneity, emission spectra, and contamination effect of carbon samples of various degrees of crystaline order. Glasslike carbon was found to be highly suitable as a standard.

Einführung

Die Bestimmung leichter Elemente ($_4$Be$-_9$F) mit Hilfe der Elektronenstrahl-Mikroanalyse konzentrierte sich bisher auf Verfahren zur Erlangung höherer Impulszählraten und besserer Signal/Untergrund-Verhältnisse. Eine qualitative Analyse der emittierten ultraweichen Röntgenstrahlung nach dem dispersiven Verfahren einerseits oder der Methode der Energiediskriminierung andererseits bietet heute keine grundsätzlichen Schwierigkeiten mehr. In günstigen Fällen können auch einige 100 ppm leichter Elemente einwandfrei aufgelöst werden; bei höheren Gehalten müssen jedoch Korrekturen für Absorption, Sekundärfluoreszenz und für Unterschiede in den Atomnummern berücksichtigt werden. Diese Korrekturen verlangen die genaue Kenntnis der Massenabsorptionskoeffizienten aller erzeugten Strahlungen in allen vorhandenen Elementen.

Die gebräuchlichen Tabellen der Absorptionskoeffizienten [1] erstrecken sich nur bis zur $Na K_\alpha$-Strahlung (11,9 Å), für längerwellige Strahlungen sind lediglich einzelne Werte in der Literatur verfügbar. Für quantitative Analysen sind neben Daten für Sauerstoff und Stickstoff besonders Absorptionswerte in Kohlenstoff erforderlich. Diese sind aber nur für Kohlenstoffverbindungen gemessen worden, wobei der Einfluß der chemischen Bindung nicht vernachlässigt werden darf.

In der vorliegenden Arbeit wird daher der Versuch unternommen, die Absorption langwelliger Röntgenstrahlen unter Mikrosondebedingungen in elementarem Kohlenstoff zu bestimmen. Die Messung der relativ hohen Absorption für Strahlungen ab 8 Å Wellenlänge verlangt eine geringe Massendicke des Absorbers, so daß zunächst dünne freitragende Kohlenstoff-Folien mit bekannter Dicke benötigt werden.

Herstellung der Kohlenstoff-Folien

In einer Vakuum-Bedampfungsanlage (JEOL JEE-4B) wird ähnlich der Methode von BRADLEY [2] zur Erzeugung von Replicas Kohle im Lichtbogen verdampft. Dazu sind Ströme bis 50 A zwischen einer angespitzten und stumpfen Elektrode (Spektralkohle RW 2 der Fa. Ringsdorff) nötig. In einem Abstand von 10 cm über dem Lichtbogen werden jeweils 2 Mikroskop-Objektträger 18×18 mm gleichzeitig bedampft. Zur Erzielung gleichmäßiger Schichtdicken müssen nach jedem der 4 Bedampfungsvorgänge das Objektträgerpaar horizontal um 180° gedreht werden. Eines der Plättchen wird nach Ankratzen seiner Ränder in destilliertes Wasser getaucht und öfters leicht angetippt. Nach ca. 1 min hat sich die Kohlefolie vom Glas abgetrennt, schwimmt frei auf dem Wasser und kann mit einem Polyäthylenrähmchen vorsichtig aufgefischt werden. Die faltenfreien Kohlenstoff-Folien haben eine freitragende Querschnittsfläche von 2,3 cm². Mit der Röntgenstrukturanalyse wird gefunden, daß der Kohlenstoff in besonders ungeordneter Form vorliegt [keine $(h k l)$-Reflexe mit $h \neq 0$, $k \neq 0$].

Dickenbestimmung der Folien

Mit den im folgenden beschriebenen Methoden wird bestätigt, daß die Forderung nach gleichmäßiger Schichtdicke für die beiden gemeinsam bedampften Plättchen sowohl untereinander als auch über ihren Querschnitt erfüllt ist. Eine der beiden Folien auf geeichtem Objektträger wird untersucht:

a) Prüfung der Homogenität auf gleichmäßigen Lichtdurchlaß mit einem empfindlichen Photometer. Für dünne Folien mit Dicken $d = 300$ bis 1500 Å ergeben sich Werte der optischen Dichte $D = \log(J_0/J) = 0,2$ bis 1,3 in Übereinstimmung mit den Meßergebnissen von TOYODA, AGAR-COSSLETT und GRAF [3], die alle einen nahezu linearen Zusammenhang zwischen d und D fanden. Im Dickenbereich $d = 1500$ bis 6000 Å kann aufgrund der vorliegenden Messungen die Linearität extrapoliert werden ($D = 1,3$ bis 5,5), allerdings mit abnehmender Genauigkeit.

b) Wägung der Kohlenstoff-Folie. Mittels einer 10^{-6} g noch auflösenden Mikroanalysenwaage ergeben sich Folienflächendichten von 63 bis 116 µg/cm². Über die ermittelte Xylol-Dichte der Folien bei 20° C von $\varrho = 1,81$ g/ml lassen sich die Schichtstärken errechnen.

c) Dickenbestimmung nach dem Interferenzverfahren in Reflexion nach TOLANSKY [4] über die grüne Thallium-Linie ($\lambda = 5350$ Å). Sofern die Versetzungsordnung der Interferenzlinien durch eine besondere Abdeckmethode bei der Bedampfung ermittelt worden ist oder die ungefähre Dicke durch das optische Verfahren oder durch Wägung bekannt ist, kann bei photographischer Aufnahme die Foliendicke auf ca. 20 Å genau vermessen werden. Für die eingesetzten Folien 1 bis 5 (Abb. 4) ergeben sich Dicken von $x = 3586$, 6398, 5817, 4543 und 4523 Å.

Anordnung und Messung

Die Untersuchungen werden an einem Elektronenstrahl-Mikroanalysator (JEOL JXA-3A) mit einem Zusatz zur Bestimmung leichter Elemente durchgeführt. Ein 10 keV-Elektronenstrahl wird auf eine Festkörperprobe gebündelt und die emittierte charakteristische Röntgenstrahlung über eine fokussierende Spektrometeranordnung im Vakuum analysiert. Zur Bestimmung der $B K_\alpha$-Strahlung mit $\lambda = 67{,}2$ Å Wellenlänge muß ein Lignoceratpseudokristall ($2d \simeq 128$ Å) eingesetzt werden, für die Strahlungen $\lambda = 44{,}8 - 13{,}3$ Å ein Bleistearat und für die kürzeren Wellenlängen die Glimmerunterlage des Stearats ($2d = 19{,}92$ Å). Die Auflösung der Spektrometer für langwellige Strahlungen liegt bei $\Delta\lambda/\lambda = 1$ bis $5 \cdot 10^2$. Als Detektoren eignen sich Proportionalzählrohre mit Gasdurchfluß (P 10) und dünnen Kollodiumfenstern auf Ni-Stützgittern. Das von außen bedienbare Mylarfenster zwischen Probenraum und Spektrometersystem wird durch eine Kohlenstoff-Folie ersetzt.

Der Meßvorgang ist folgender: Über den Zeitraum von 1 min wird die ausgeblendete Röntgenstrahlung J_0 [Imp./min] registriert. In den folgenden 15 sec wird die Folie in den Strahlengang eingeschwenkt und danach 1 min lang die verminderte Impulsrate J gezählt. Unmittelbar vor und nach der bis zu 30mal je Folie wiederholten Meßfolge werden 5mal die Untergrundintensitäten bestimmt. Für jede Messung wird aus dem Verhältnis der geschwächten zur ungeschwächten Strahlungsintensität nach $J/J_0 = \exp[-(\mu/\varrho)\varrho x]$ der Massenabsorptionskoeffizient (μ/ϱ) berechnet und die Gesamtergebnisse einer statistischen Auswertung unterzogen.

Standardproben und Emissionsspektren

Als Strahler werden 10 Materialien untersucht: die K_α-Strahlung von B, C, Al; die L_α-Strahlung von Ti, Fe, Cu; $N K_\alpha$ aus BN, $O K_\alpha$ aus SiO_2, $F K_\alpha$ aus LiF und $Na K_\alpha$ aus NaCl. Für jeden Strahler gibt das über den Bragg-Winkel 2Θ aufgezeichnete Intensitätsprofil Auskunft einerseits über die geeignete Winkellage der Untergrundintensitätsmessung, andererseits über die Wellenlänge der maximalen Zählrate. Die Justierung der Analysatorkristalle erfolgt mit $Cu K_\alpha$-Strahlung über die Glimmerunterlage der Seifenkristalle und wird mit der L-Strahlung der reinen Elemente Cu, Fe und Ti über die Seifenkristalle geprüft. TiL- und $F K$-Strahlung sind weitere Kontrollen, obwohl wegen geringen Impulsausbeuten hiermit keine Absorptionskoeffizienten bestimmt werden. Die über die aufgezeichneten Profile erhaltenen Wellenlängen maximaler Zählraten stimmen mit den tabellierten [z.B. 1] im Falle reiner Elemente als Strahler gut überein. Im Bereiche der ultraweichen Röntgenstrahlung macht sich der Einfluß der chemischen Bindung auf die Gestalt und das Maximum des Emissionsspektrums bemerkbar. FISCHER und BAUN [5] wiesen diesen Effekt an einigen Nitriden und Karbiden nach. Die Verschiebung des $O K_\alpha$- und $N K_\alpha$-Maximums beträgt etwa 0,4 Å zum längerwelligen (Tabelle 1). Bei den vorliegenden Untersuchungen werden zur Erzielung optimaler Signal/Untergrund-Verhältnisse diejenigen Wellenlängen ($\lambda_{gemessen}$) bzw. Bragg-Winkel eingestellt, die den maximalen Impulsraten bei differentieller Messung zuzuordnen sind.

In diesem Zusammenhang interessieren besonders reine Kohlenstoffproben. Mit einem optisch homogenen Standard muß an jeder Stelle der Oberfläche unter sonst gleichen Bedingungen dieselbe Impulsrate reproduzierbar sein. Drei Kohlenstoffproben (Abb. 1) von unterschiedlichem kristallinen Ordnungszustand werden unter einem 1 μm ⌀-Elektronenstrahl über eine Strecke von 30 μm zeitlich unmittelbar nacheinander abgetastet. Bei allen Messungen sind Probenstrom und Untergrundintensitäten konstant. Polierte Spektralkohle ergibt hier die meisten Impulse, abgesehen von der Zählstatistik aber auch erhebliche Schwankungen, verursacht durch Poren. Mit Spaltflächen von Maniwauki-Naturgraphit erhält man ein gleichmäßigeres Impulsbild, falls der Elektronenstrahl auf keine Stufe der Probenoberfläche trifft. Gute Ergebnisse gewinnt man mit Harzkohlenstoff, der einen nur diffusen 002-Reflex zeigt, eine Art Flüssigkeitsstruktur aufweist und im 10^{-3} bis 10^{-6} cm-Bereich keine Poren hat. Bei seiner Herstellung aus Furfurylalkoholharz bleibt die glatte Gußoberfläche bei der Verkokung und Wärmebehandlung erhalten, so daß ein Polieren des Standards sich erübrigt.

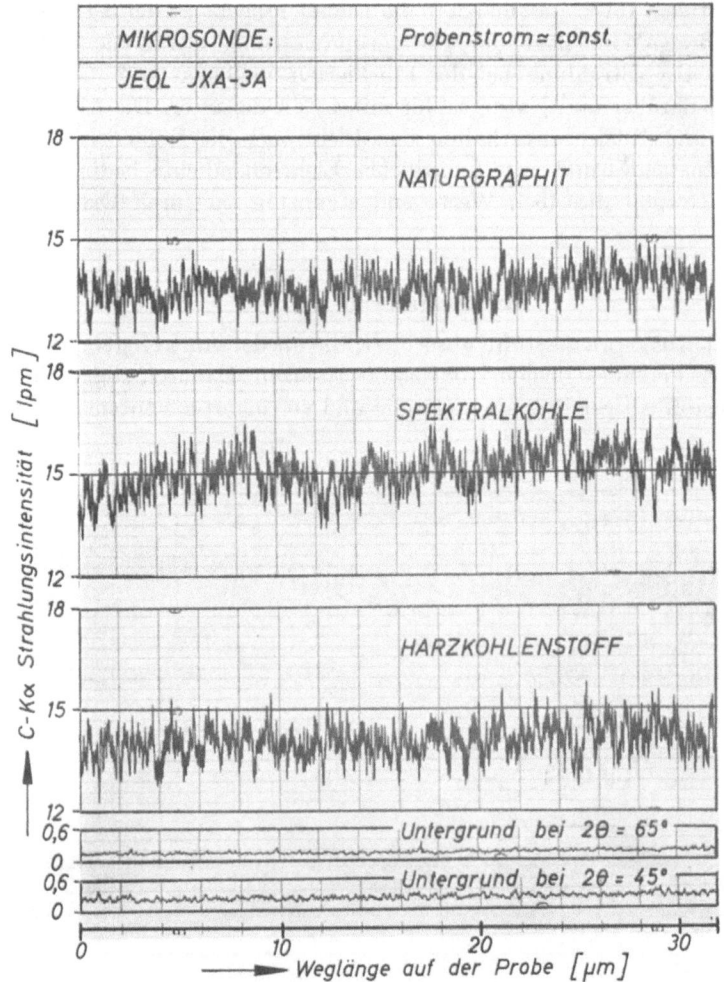

Abb. 1. Untersuchung von Kohlenstoffen als Standards

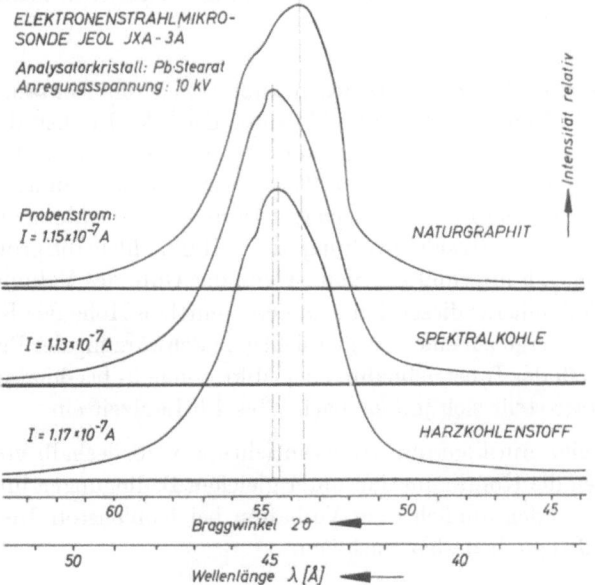

Abb. 2. Emissionsspektren von Kohlenstoffen

Abb. 2 zeigt in relativem Intensitätsmaßstab die Emissionsspektren der 3 Kohlenstoffproben aufgetragen bis zur Bragg-Winkellage der Untergrundmessung. Harzkohlenstoff bringt eine Maximalintensität der CK_α-Strahlung bei der tabellierten Wellenlänge $\lambda = 44{,}8\ \text{Å}$, Spektralkohle liegt mit $45{,}0\ \text{Å}$[1] darüber und Naturgraphit mit $44{,}3\ \text{Å}$ darunter. Die Abweichungen sind durch das Eindring- und Rückstreuverhalten der Elektronen in Kohlenstoff verschiedenen kristallinen Ordnungszustandes und unterschiedlicher Elektronendichte bedingt. FISCHER und BAUN [5] fanden für Graphit dieselben Werte und ergänzten sie durch Diamant- und Rußanalysen.

Kontamination

Auf der Probe sind Kohlenwasserstoffe absorbiert, die an der Auftreffstelle des Elektronenstrahls gecrackt werden und zusätzlichen Kohlenstoff absetzen. Dadurch steigt die CK_α-Strahlungsintensität kohlenstoffhaltiger Präparate ständig und verringert die Intensität anderer lang-

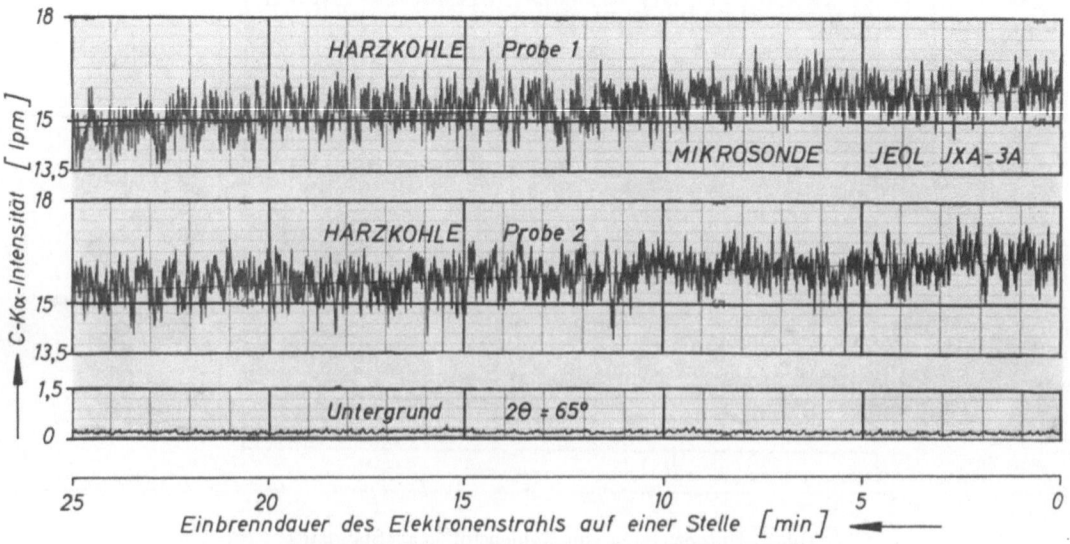

Abb. 3. Kohlenstoffkontamination auf einem Kohlenstoffstandard

welliger Strahlungen durch Absorption. Dagegen äußert sich die Kontamination bei der Bestrahlung eines Kohlenstoffstandards in einer Abnahme der CK_α-Intensität. Diese Erscheinung kann folgendermaßen gedeutet werden: Um den Auftreffort des Elektronenstrahls auf der Probenoberfläche bildet sich ein Kohlenstoffwall von Torusform, der das Eindringen der Elektronen nicht behindert, aber die aus dem Probeninneren emittierte CK_α-Strahlung zusätzlich absorbiert. In Abb. 3 ist das Ergebnis der Strahleinwirkung auf 2 Harzkohlenstoffproben dargestellt. Die Anfangsintensität fällt in $^1/_2$ h um rund 8% und ist von der Güte des Vakuums abhängig. Überschläglich errechnet sich damit für diesen Zeitraum eine mittlere Höhe des Kohlenstoffwalls von $700\ \text{Å}$ (bereits 100 bis $300\ \text{Å}$ ergeben eine deutlich sichtbare Schwärzung der Probe). Der Kontaminationseffekt ist kurz nach der Inbetriebnahme der Mikrosonde 3- bis 6mal größer als in Abb. 3. Eine konstante Impulsrate stellt sich jeweils nach 3 bis 4 h Laufzeit ein.

Unter Verzicht auf eine Antikontaminationseinrichtung wird deshalb vor jeder Meßfolge in mehreren Blindversuchen die Kontamination unter gleichen Bedingungen untersucht. Für jeden der anderen 9 Strahler werden ähnlich dem Verhalten bei Kohlenstoff Intensitätskorrekturen ermittelt, die annähernd reproduzierbar ausfallen.

---

1. Nach neueren Messungen liegt die Maximalintensität des Emissionsprofils rd. $0{,}4\ \text{Å}$ tiefer.

Absorptionskoeffizienten

Abb. 4 zeigt die Häufigkeitsverteilung der Massenabsorptionskoeffizienten in Kohlenstoff (μ/ϱ) [cm²/g] bei verschiedenen Strahlungen und den Folien 1 bis 5. Die (μ/ϱ)-Werte sind in äquidistante Klassen unterteilt. Die mittleren quadratischen Abweichungen $\pm\sigma$ sind als Parallelstriche zur Abszisse um den arithmetischen (μ/ϱ)-Mittelwert aufgetragen. Mit Ausnahme der Ergebnisse mit $\mathrm{Cu}\,L_\alpha$- und $\mathrm{C}\,K_\alpha$-Strahlung, für die relativ viele Messungen vorliegen (Tabelle 1), sind die Verteilungen breit und weichen von einer Normalverteilung stark ab. Die über 3 Zehnerpotenzen sich erstreckenden Absorptionswerte, aufgetragen über den Wellenlängen (Abb. 5, dicke Punkte), zeigen nur geringe Abweichungen vom interpolierten Kurvenzug.

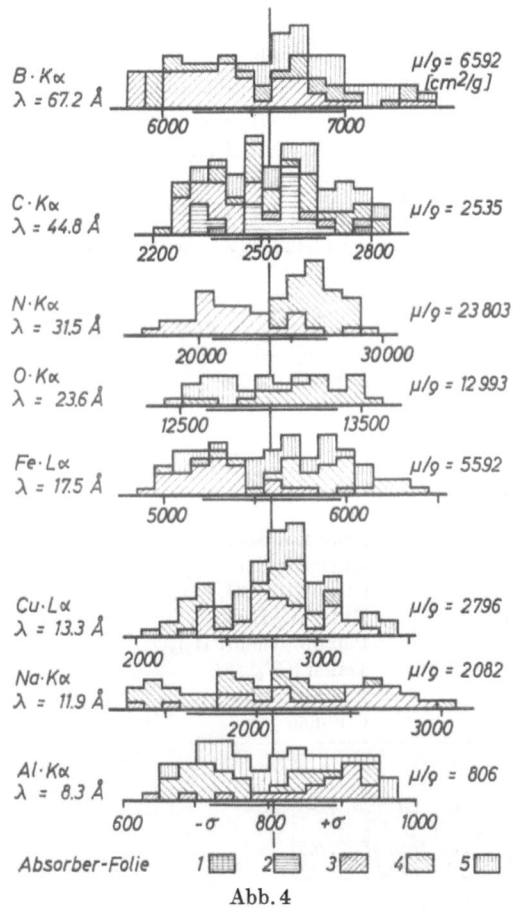

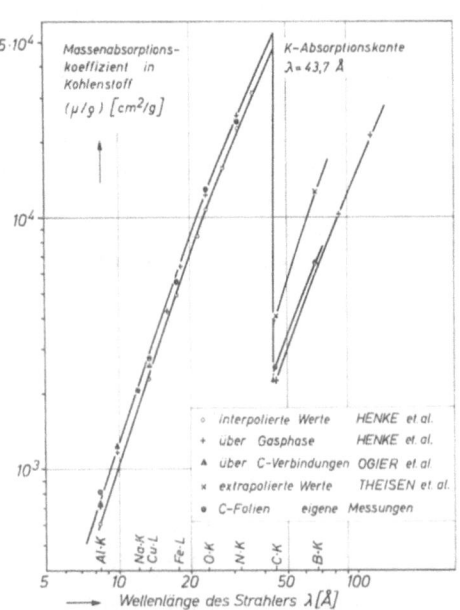

Abb. 5. Absorption in Kohlenstoff als Funktion der Wellenlänge

Abb. 4

Abb. 4. Häufigkeitsverteilungen der gemessenen Massenabsorptionskoeffizienten

Tabelle 1. *Meßergebnisse zur Absorption langwelliger Röntgenstrahlung in Kohlenstoff-Folien*

Wellenlängen		Strahlung	Strahler	Absorptionskoeffizient	Streuung σ	Auswahlmessungen
$\lambda_{\text{gemessen}}$ Å	$\lambda_{\text{Tab.}}$ Å	—	—	μ/ϱ cm²/g	cm²/g	n —
8,3	8,32	Al K_α	Al	806	90	77
11,9	11,89	Na K_α	NaCl	2082	470	55
13,4	13,33	Cu L_α	Cu	2756	300	86
17,6	17,57	Fe L_α	Fe	5591	380	76
18,4	18,32	F K_α	LiF	—	—	—
24,0	23,61	O K_α	SiO$_2$	12993	360	36
27,2	27,39	Ti L_α	Ti	—	—	—
32,0	31,57	N K_α	BN	23803	3140	57
44,8	44,78	C K_α	C	2535	155	103
68,0	67,20	B K_α	B	6592	421	97

Vergleich mit früheren Ergebnissen

Für ultraweiche Röntgenstrahlen $> 8\,\text{Å}$ Wellenlänge sind im Gegensatz zu harten Strahlen in der Literatur [6] nur wenige Angaben über die Absorption in Kohlenstoff vorhanden. Eine experimentell zugängliche Möglichkeit wird darin gesehen, die Absorption von Kohlenstoffverbindungen zu bestimmen und auf Kohlenstoff umzurechnen. Dazu müssen bei gesättigten oder ungesättigten Kohlenwasserstoffen die Absorption in Wasserstoff und bei Verwendung von Kunststoff-Folien zusätzlich die Absorption in Sauerstoff oder Fluor bekannt sein.

Messner [7] untersucht die Absorption in Kohlenstoff für BK_α- und CK_α-Strahlung an 5 Kohlenwasserstoffen und an Wasserstoff zu bestimmen (Abb. 6). Er bestätigt den noch nicht erfaßbaren Einfluß der chemischen Bindung und kann keinen Absorptionswert für elementaren Kohlenstoff angeben. Einige ältere Werte von Allen [8] werden von Henke, White und Lundberg [9] aufgegriffen und mittels einer semiempirischen Methode die Absorption der K- und L-Schalen-Elektronen berechnet. Für Kohlenstoff als Absorber ergibt sich in Abb. 5 die extrapolierte untere Kurve vor der K-Absorptionskante. In weiteren Messungen bestimmen Henke et al. [10] in zahlreichen Gasen (Tabelle 2) die Absorptionskoeffizienten, die bis zur Absorptionskante mit unseren Messungen übereinstimmen (obere Kurve in Abb. 5). Nach der Absorptionskante ermitteln sowohl Henke et al. [10] wie Ogier et al. [11] niederere Absorptionskoeffizienten, während die Extrapolation von Theisen und Vollath [12] zu höheren Werten

Tabelle 2. *Absorptionskoeffizienten (μ/ϱ) der Kohlenstoff-K_α-Strahlung in Kohlenstoff*

μ/ϱ [cm²/g]	Literatur	Ermittlung über
2000—2180	Messner [7]	C_2H_2, C_2H_4, C_2H_6
1720—2660	Messner [7]	CH_4, C_2H_6, C_3H_8
2170	Allen	—
	Henke, White und Lundberg [9]	semiempirische Interpolation
2280	Henke, Elgin und Lent [10]	CH_4, C_2H_6, CO, CO_2
2280	Ogier, Lucas und Park [11]	Mylar $(C_{10}H_8O_4)$ Polypropylen $(CH_2)_x$ Teflon $(CF_2)_x$
4092	Theisen, Vollath [12]	Extrapolation
2535 ± 150	diese Arbeit	C-Folien

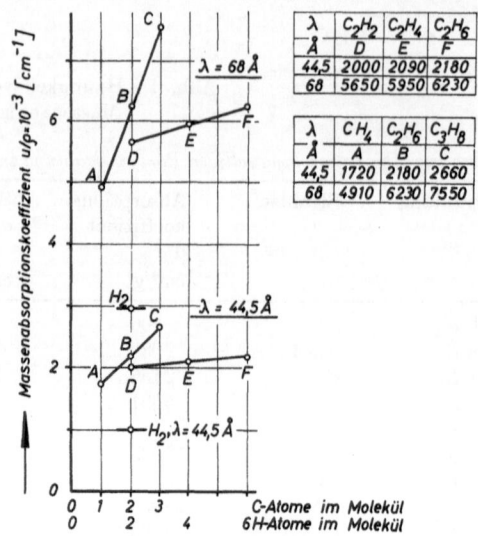

Abb. 6. Massenabsorptionskoeffizienten von Kohlenstoff aus Kohlenwasserstoffen nach Messner [7]

führt. In Tabelle 2 sind die eigenen Messungen an elementarem, ungeordnetem Kohlenstoff und CK_α-Strahlung den bisher ermittelten, meist kleineren Absorptionskoeffizienten (μ/ϱ) aus Kohlenstoffverbindungen gegenübergestellt.

Mit diesen gemessenen Absorptionsdaten für elementaren Kohlenstoff als Absorber kann die Anwendbarkeit der üblichen Ansätze [z.B. 13] zur Absorptionskorrektur bei Kohlenstoff überprüft werden.

Anmerkungen. Der Autor dankt Frl. B. Fischer und Herrn R. Kraft für die experimentelle Mitarbeit. Besonders sei Herrn Dr. B. Kegel (Kernreaktor Karlsruhe) für wertvolle Diskussionen und der Deutschen Forschungsgemeinschaft für die Zurverfügungstellung der Mikrosonde herzlich gedankt.

Literatur

1. Sagel, K.: Tabellen zur Röntgen-Emissions- und Absorptionsanalyse, S. 40, 47. Berlin-Göttingen-Heidelberg: Springer 1959.

2. Bradley, D. E.: Evaporated carbon films for use in the electron microscopie. Brit. J. Appl. Phys. **5**, 2 (1954).

3. Graf, K.: Optische Dichte und Dicke von aufgedampften Kohlenstoffschichten. Optik **18**, 3 (1961).

4. Tolansky, S.: Multiple-beam interferometry of surfaces and films. Oxford 1958.

5. Fischer, D. W., and W. L. Baun: Effect of chemical combination on the soft X-ray K emission bands of nitrogen and carbon. J. Chem. Phys. **43**, 6 (1965).

6. Gmelins Handbuch der anorganischen Chemie. Syst. Nr. 35, 8. Aufl., Teil B, Liefg 1, S. 150. Weinheim/Bergstr.: Verlag Chemie 1967.

7. Messner, R. H.: Der Einfluß der chemischen Bindung auf den Absorptionskoeffizienten leichter Elemente im Gebiete ultraweicher Röntgenstrahlen. Z. Physik **85**, 11/12 (1933).

8. Allen, S. J. M.: Tabellen im "Handbook of chemistry and physics", 41. ed., p. 2662/2667. Cleveland/Ohio 1959.

9. Henke, B. L., R. White, and B. Lundberg: Semiempirical determination of mass absorption coefficients for the 5 to 50 Angstrom X-ray region. J. Appl. Phys. **28**, 1 (1957).

10. — R. L. Elgin, R. E. Lent, and R. B. Ledingham: X-ray absorption in the 2-to-200 Å region. Norelco Reptr **9** (1967).

11. Ogier, W. T., G. J. Lucas, and R. J. Park: Ultrasoft X-ray absorption coefficients of Al, Be, C, O and F. Appl. Phys. Letters **5**, 7 (1964).

12. Theisen, R., u. D. Vollath: Tabellen der Massenschwächungskoeffizienten von Röntgenstrahlen. Düsseldorf: Verlag Stahleisen 1967.

13. — Quantitative electron microprobe analysis. Berlin-Heidelberg-New York: Springer 1965.

Instrumentation

Recent Advances in Instrumentation for Microprobe Analysis

D. B. WITTRY

Departments of Materials Science and Electrical Engineering
University of Southern California, Los Angeles, California, U.S.A.

Introduction

While the most important examples of microprobe analysis have come from the use of an electron probe to produce excited states in the specimen and an X-ray analyzer to detect the radiative decay of these excited states, other forms of local excitation have also been used, for example, a focussed ion beam or a focussed or collimated photon beam. The signals that result from the local excitation can also assume many additional forms such as backscattered electrons, secondary electrons or ions, and optical radiation. Some of these signals obtained with electron beam excitation have been discussed in W. C. NIXON's review paper at this conference [1].

A large number of instruments for microanalysis already exist in which different types of excitation or signal information are used. Before discussing some of these instruments in detail, it is useful to have some overall perspective to see how some of these instruments relate to each other. Such a perspective can be obtained by considering the various analytical techniques that have been used for microanalysis and the relationship of these techniques to the basic instruments from which they have evolved. This is indicated schematically in Fig. 1.

New instruments have been developed either by modifying the basic instruments so that more than one technique could be performed, or by combining the function of two or more instruments into a single instrument. For example, instruments for X-ray microanalysis have resulted from the combination of an electron probe with an X-ray analyzer, but similar instruments have also been derived from scanning microscopes with suitable modifications. Still newer instruments have come from the combination of an ion probe with an ion spectrometer or from the use of an ion spectrometer with an ion emission microscope. Higher orders of combination of basic instruments have also achieved some practical importance, for example, the combination of a transmission electron microscope with an electron probe microanalyzer (TEMMA) or scanning electron microscope with a microprobe analyzer (SEMMA).

While all of the techniques shown on Fig. 1 have had some impact on the development of new instruments, not all of these techniques have been equally suitable for microprobe methods of analysis. Microprobe methods are usually employed when the exciting radiation (that is photons or charged particles) can be focussed and the emitted or transmitted radiation (also photons or charged particles) cannot be focussed. For example, in X-ray microanalysis, the microprobe approach has been successful chiefly because the electrons can be focussed easily while the emitted X-rays can be focussed only with difficulty.

An alternative to the microprobe method is the selected-area method in which a large area of the specimen is irradiated and the signal information is detected from a small region of this area. This method can be used advantageously if the signal information can be focussed since an aperture can be introduced in the image plane to define the selected-area.

If both the incident radiation and the signal information can be focussed, either microprobe methods or selected area methods can be used. In principle, the results obtained by these two methods should be similar, but, in practice, they are usually different for the following reasons: a) aberrations of the illuminating system, b) aberrations of the signal focussing system, c) anomalies caused by surface effects such as contamination, oxidation, adsorption or desorption,

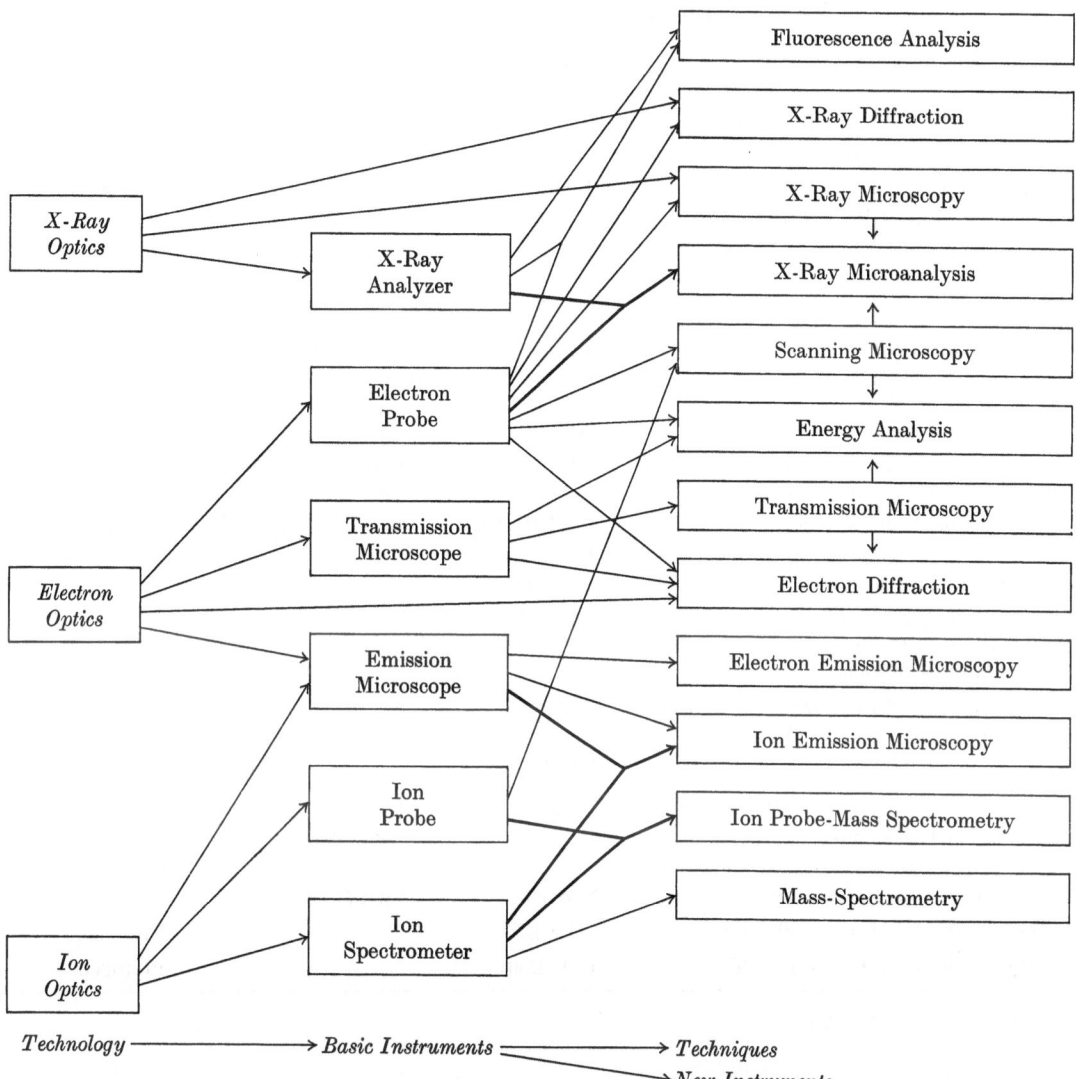

Fig. 1. Some techniques of microanalysis and their relationship to a few basic instruments

d) differences due to the finite lifetime of states responsible for the observed signal, e) differences in the efficiency of information recording, and f) differences in the ease of information processing. Some of these factors favor microprobe methods, others favor selected-area methods but unfortunately there is not sufficient information at the present time to know which method will prove to be best for all of the cases that are included in the scope of this conference. In the present paper, microprobe methods will be emphasized; nevertheless, it is important to remember that selected-area methods can provide alternative approaches or can be used as ancillary techniques.

Since electron probe X-ray microanalysis has become a widely accepted research tool [2], this paper will begin with a discussion of electron probe microanalyzers and recent developments in components for microprobe analysis. Next, we shall discuss accessories for microanalyzers, various approaches to instruments combining other techniques with electron probe microanalysis, and the use of computers with microprobe instruments for control or for information processing. Finally, two new analytical techniques will be mentioned, namely, electron spectroscopy using Auger electron emission excited by electron bombardment and mass spectrometry using ions sputtered from the specimen by an ion probe.

Electron Probe Microanalyzers

At the present time there are more than 11 manufacturers of instruments specifically intended for electron probe microanalysis and 16 different models available (Instruments made in the USSR were not included because of insufficient information). Eight of the 16 instruments now available have been announced within the past year; these new instruments have either been demonstrated at the instrument exhibition of this conference or have been the subject of papers presented in technical sessions. A summary of the instruments now available is given in the table.

Table. *Electron probe microanalyzers*[a]

Name of manufacturer	Brief designation	Current models (former)	New models
GEC-AEI (Electronics) Ltd.	AEI	SEM 2A (SEM 2)	
Applied Research Laboratories, Inc. (and Shimadzu Seisakusho Ltd., licensee)	ARL	EMX-SM, (EMX) AMX	
Cambridge Scientific Instruments, Ltd.	CAMBRIDGE INSTRUMENTS	(Microscan II) (Geoscan)	Microscan III Microscan V
Canal Industrial Corporation	CANALCO	Microsource	
Compagnie d'Applications Mechaniques a l'Electronique Au Cinema et a l'Atomistique	CAMECA	MS 46 DLC (MS 85)	
Société Française d'Instruments de Contrôle et d'Analyses (subsidiary of ARL)	FICA		AMX model 80,000
Hitachi, Ltd.	HITACHI	XMA S	
Japan Electron Optics Laboratory Co., Ltd.	JEOL	JXA 3A	JXA 5
Materials Analysis Company	MAC	Model 400	Model 400 S Model 450
Philips Electronic Instruments	NORELCO	AMR 3	
Siemens & Halske Aktiengesellschaft	SIEMENS		Elmisonde
Technisch Physische Dienst TNO-TH	TPD-Delft		Prototype

[a] Listed alphabetically.

Almost all commercial instruments employ a triode electron gun with a tungsten hairpin filament and two magnetic electron lenses. Exceptions are the JXA 3A which is normally provided with only one lens and the new MAC 400 S which employs a double condenser lens. For instruments with two lenses, the geometrical demagnification is typically between 200 and 500 and the smallest spot size varies between 0.1 and 0.7 micron. A useful criterion of the performance of the electron optics is the current that can be achieved on a focal spot of a given size. Values quoted for a 0.5 micron spot at 30 kV vary between 0.01 and 1 μa. The highest values are obtained with instruments having a small working distance; more typical values are 0.1 or 0.05 μa, but this can be as low as 0.01 μa in instruments with very large working distance.

Electron probe microanalyzers can be classified according to the angle of incidence of the electron beam on the specimen, the type of viewing system used and the X-ray take off angle. Normal incidence of the electron beam on the specimen and a high quality simultaneous optical viewing system usually imply the use of a reflecting optical objective coaxial with the electron beam. This approach has been used in the EMX of ARL, the Microsource of CANALCO, the MS 46 DLC of CAMECA, the XMA 5 of HITACHI, the JXA 3A and JXA 5 of JEOL, and the AMR 3 of NORELCO. Normal incidence of the electron beam can also be obtained with viewing systems consisting of refracting objectives of low numerical aperture and mirrors as in the AMX of ARL, the Microscan III and Microscan V of CAMBRIDGE INSTRUMENTS, and the Elmisonde of SIEMENS. In some cases, a second optical system is used for observation when the electron beam is not striking the specimen, as for example in the Microscan V and the Elmisonde. By sacrificing the normal incidence of the electron probe on the specimen, it is possible to use mirrors and refracting objectives as in the SEM 2 A of GEC-AEI or to view the specimen directly as in the MAC 400 S and the first model of TPD-Delft.

In the design of new instruments, higher X-ray take off angles seem to be preferred because of smaller absorption corrections and reduced effects due to surface roughness. If the X-rays are received below the electron optical objective lens, the maximum X-ray take off angle has been limited by the close proximity of this lens, particularly if the electron beam is normal to the surface of the specimen. For example, the AMR 3 has 15 degrees, the MS 46 DLC has 18 degrees, the JXA 3 A has 20 degrees and the Microscan III has 25 degrees. Higher "effective" X-ray take off angles are possible if the specimen is tilted with respect to normal incidence of the electron beam. Examples of this latter case and the actual X-ray take off angles employed are as follows: the SEM 2 A (26.5 degrees), the new instrument of TPD-Delft (30 degrees) and the MAC 400 S (34 degrees). High X-ray take off angles are also possible by using large working distance for the electron probe as in the JXA 5 (40 degrees) and the AMX (52.5 degrees). Finally, very high X-ray take off angles are possible by taking the X-rays back through the final electron lens or a part of this lens. This is done in the XMA 5 (38.5 degrees), the EMX (52.5 degrees), and the Microscan V (75 degrees).

X-ray detection systems vary mainly in the number of spectrometers and the spectrometer mechanism (linear in angle, linear in wavelength, variable focal circle radius, semifocussing, etc.). Some instruments have employed only one spectrometer with facilities for crystal changing inside the vacuum (SEM 2 A). Most instruments provide at least two spectrometers; sometimes these are available with crystal changers. The more elaborate instruments provide for the use of 3 fully scanning spectrometers (EMX and XMA 5) or 4 spectrometers (MS 46 DLC and Microscan III). In some cases, additional spectrometers are available as accessories, for example, newly-developed semiscanning spectrometers or semifocussing spectrometers.

The quality and efficiency of X-ray detection varies widely among the available instruments. These factors depend on the types of crystals used, the size of crystals, the radius of the focal circle, size of the slits, and the techniques used in the manufacture of crystals. Both the peak intensities and the line-to-background ratio are important, but the peak intensities will suffice for illustration. For a beam current of 10^{-8} amp and 30 kv the counting rates for CuK_α from pure Cu range from 800 to 14,000 counts/second. For softer X-rays, the efficiency is poorer, i.e. for 10^{-8} amp and 10 kv, the counting rates vary from 5 to 1,400 counts/second for CK radiation from graphite and 5—400 counts/second for OK radiation from quartz.

Often, poor performance of the X-ray detection system is offset by superior performance of the electron beam system (or vice-versa) so that both factors should be considered in evaluating any given instrument. However, a detailed comparison of the performance of specific instruments is difficult at the present time because of differences in methods used for measuring focal spot sizes and because of the variability of X-ray spectrometer crystals.

Components for Electron Probe Microanalysis

Basic components used in most electron probe microanalyzers today differ little from the components that were used a decade ago. The usual electron source is a tungsten hairpin filament in a triode electron gun, most electron lenses are based on the type studied by LIEBMANN [3], and the X-ray spectrometers are scanning curved crystal spectrometers as used in early versions. The most significant change in instrumentation over the past decade has been the incorporation of new crystals and detectors for analysis of light elements [4]. More significant changes are now beginning to take place because of the availability of new components for the electron optical and X-ray optical systems of the microanalyzer.

Several types electron sources have been suggested as an alternative to the conventional source. The pointed filament cathode [5] has been well known for more than a decade but for

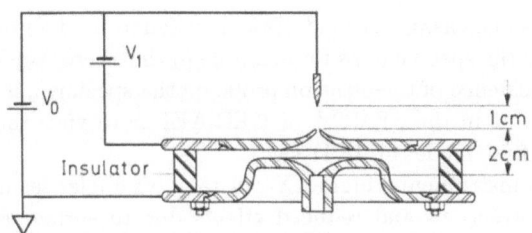

Fig. 2. Field emission gun developed by CREWE et al. [10]

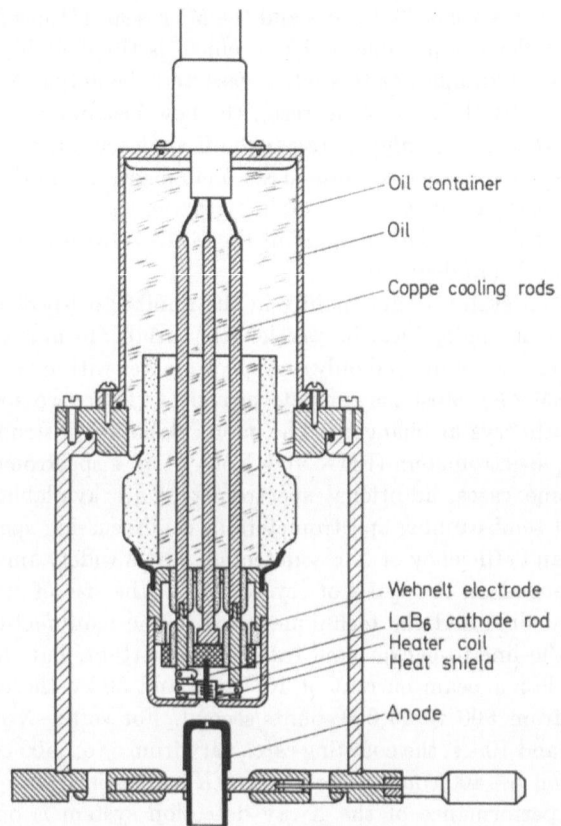

Fig. 3. Electron gun using an indirectly heated cathode of lanthanum hexaboride [11]

practical reasons has never been adopted by instrument manufacturers. It provides greater brightness and smaller source size but so far has been limited to special applications in a few laboratories because of shorter life than conventional cathodes. Field emission sources for electron probes were investigated many years ago [6—9] but were not successfully used until recently. By using oriented tungsten wire, high speed ion pumps, and ultrahigh vacuum techniques CREWE et al. [10] have been able to develop a practical field emission electron gun. This gun is shown in Fig. 2. When used in a probe forming system it provides no significant advantage for focal spot sizes of the order of 2,000 Å but for focal spot sizes of the order of 100 Å it gives $2^1/_2$ orders of magnitude greater current than conventional sources. Moreover, this gun has provided focal spot sizes of the order of 250 Å without the use of additional lenses.

An electron gun using a lanthanum hexaboride cathode was recently developed by BROERS [11]. This gun, shown in Fig. 3 may prove to be of more immediate practical importance than the field emission gun since it can operate with the typical vacuum conditions of microanalyzers. It is superior to the conventional gun in brightness, cathode life, and mechanical stability.

The use of miniature electron lenses to obtain large working distances with an electron probe system was suggested by LE POOLE [12]. Lenses of this type now being used consist of a carefully wound coil, usually of conical shape with no iron casing. Because of the high power dissipated in the small coil, the minimum focal length at high voltage is often determined by heat dissipation and water cooling is essential. Interest in lenses of this type is indicated by several papers at this conference [13]. Reduction in the size of conventional iron-shrouded lenses has also been possible by using a smaller volume for the coil windings and careful attention to heat dissipation. This approach [14] has been successfully used in the EMX-SM of ARL (see Fig. 5). No significant advantage is gained in working distance, but more space is made available for receiving signals from the specimen.

Recent advances in X-ray detection systems for electron probe microanalyzers fall into three categories, namely: a) modified versions of conventional curved crystal spectrometers, b) grating spectrometers for soft X-ray detection and c) non-dispersive detection using newly developed solid state detectors. Modified versions of conventional X-ray spectrometers have been recently announced by two manufacturers. Semi-scanning spectrometers have been developed for use in addition to the normal scanning spectrometers in the EMX. These are available with various crystals of two different radii and have a small scanning range so that peak and background intensities can be determined. A new version of the semifocussing X-ray spectrometer has also been developed for use on the Microscan III and the Stereoscan II A. The semi-focussing principle was used in order to fit two such spectrometers in a relatively small housing and to have the plane of the focal circle oriented so that the intensities would be least sensitive to changes in the vertical position of the specimen.

Grating spectrometers for analysis of soft X-rays have been under investigation in several laboratories. NICHOLSON [15] has been studying the use of concave blazed gratings and finds that both peak intensity and resolution are higher for the grating spectrometer than for crystal spectrometers for the ultrasoft X-ray region. FRANKS and his coworkers [16] have investigated the use of Siegbahn type gratings for X-ray analysis. They have recently developed a concave grating spectrometer [17] and have used it on an earlier version of the Microscan. The advantages

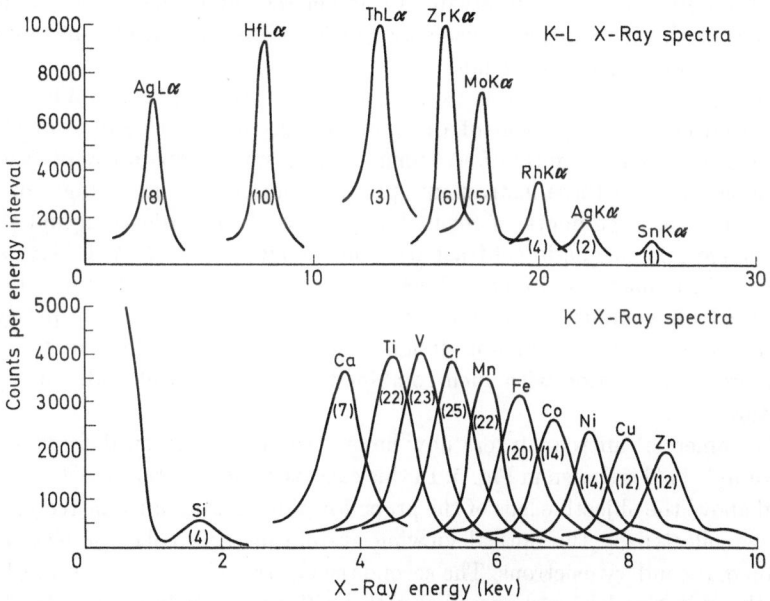

Fig. 4. Composite of pure-element spectra obtained with a solid state detector and a multichannel pulse height analyzer. Peak to background ratios are given in parenthesis. Integration time was 100 seconds; upper curves were obtained at 35 kV and 1 na for the L lines, 5 na for the K lines; lower curves were obtained at 20 kV and 1 na. From FITZGERALD et al. [18]. (Copyright 1968 by the American Association for the Advancement of Science)

of using grating spectrometers as indicated by the work of FRANKS and his coworkers and NICHOLSON can be summarized as follows: a) higher intensities than available with crystals in the 20—100 Å range, giving better sensitivity in microanalysis of light elements, b) higher resolution than available spectrometers so that more information can be obtained on the effects of chemical bonding on soft X-ray spectra, and c) larger wavelength range with the possible extension to 250 Å or more if suitable detectors are available.

Non-dispersive analysis using a cooled lithium-drifted Si detector has recently been described by FITZGERALD et al. [18]. Typical results for various characteristic X-ray lines from pure elements are shown in Fig. 4. Photon energies as low as 1.7 keV (Si K_α) can be separated from the noise, but the practical limit was 3.7 keV (Ca K_α) because of attenuation in the 0.125 mm Be window isolating the detector chamber from the main vacuum system. The advantages of non-dispersive detection are as follows: a) reduction in the time required for qualitative analysis, b) the possibility of using smaller electron probe currents, and c) the possibility of simultaneous analysis on many elements.

Accessories for Electron Probe Microanalyzers

Accessories for electron probe microanalyzers now available include units for concentration mapping or for linear and areal analysis [19], Kossel line cameras for both transmission [20] and back-reflection [21], programmable X-ray spectrometers [22] and special stages for heating [23] or tilting the specimen [24] or for using transmitted polarized light [25]. Anticontamination devices such as cold plates [26], cold fingers [27], or an air jet [28] are available for most instruments; in one instrument, the reduction of contamination was recently obtained by cooling the objective lens of the electron optical system [29].

Electron beam scanning systems [30] are now considered standard equipment for microanalyzers and detectors for backscattered electrons are available either as standard equipment or as accessories. In some cases, the backscattered electron detector is a simple collector plate or a surface-barrier semiconductor diode, but more often, a scintillation counter is used. In recent versions of scintillation counter detectors the scintillator is held at a positive potential of the order of 10 keV so that improved efficiency can be obtained when low beam voltages are used. With such a detector, it is also possible to detect secondary electrons, but unless special configurations are used, the resulting signal may consist of an undesirable mixture of secondary electron and backscattered electron signals.

For many applications, it is desirable to use a separate detector for secondary electrons and backscattered electrons. Incorporation of detectors for secondary electrons in electron probe microanalyzers has resulted in a form of scanning electron microscope-microanalyzer (SEMMA). Presently available forms of these instruments provide most of the advantages of the scanning electron microscope, i.e. large depth of focus, large range of magnification, and the possibility of obtaining voltage and magnetic contrast (see the review paper of W. C. NIXON). The best resolution of these instruments is 1,000 to 2,000 Angstrom or about 4—8 times poorer than the resolution specified for commercial scanning microscopes. However, this is partly due to the fact that most microanalyzers were not designed with sufficient attention to vibration and stray fields. Improved resolution with such combined instruments will undoubtedly be possible in the near future.

Almost all commercial microanalyzers now have secondary electron detection and SEMMA features. An example [14] is shown in Fig. 5. In this case, the backscatter detector is a scintillation counter located above the objective lens of the probe forming system; alternatively, backscattered electrons can be collected by the optical viewing system or by a collector biased to eliminate the contribution of secondary electrons. The secondary electron detector is a scintillation counter located below the objective lens and provided with baffles to exclude most of the backscattered electrons. With suitable potentials applied to the baffles or to the specimen, the efficiency of detecting secondary electrons can be optimized. Other types of secondary electron detectors that provide partial separation of secondary and backscattered electron signals have been used in conventional scanning electron microscopes, for example, a biased scintillation counter has

ELECTRON DETECTION SYSTEM

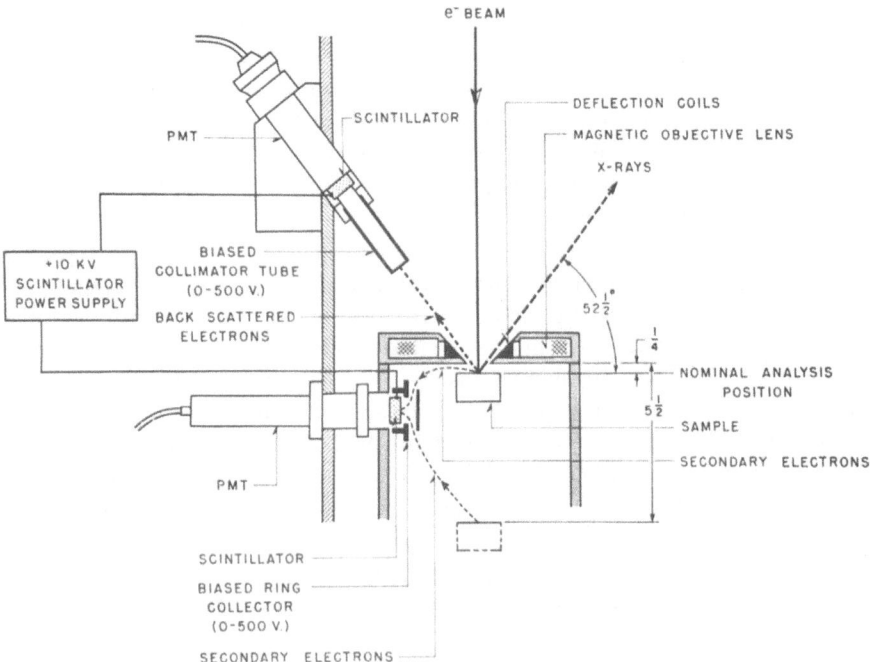

Fig. 5. Example of an electron detection system used in a scanning electron microscope-microanalyzer. From DAVIDSON and NEUHAUS [14]

been used with a grid (Stereoscan) or a quadrupole lens (JSM) for collimating the secondary electrons. The use of channel electron multipliers for secondary electron detection has also been investigated on an experimental basis [31, 32].

The addition of a transmission electron microscope as an accessory to the microanalyzer has resulted in a second form of combined instrument (TEMMA). At present, such an accessory is available on only one commercial microanalyzer, the MS 46 DLC of Cameca [33]. It uses two lenses, provides a resolution of 50 Å and has a maximum magnification of 10,000.

Other Instruments Combining Electron Microscopy and Microanalysis

In addition to accessories for conventional electron probe microanalyzers to provide TEMMA and SEMMA instruments, three other possibilities exist, the addition of accessories to transmission electron microscopes (TEM) the addition of accessories to scanning electron microscopes (SEM), and the design of completely new instruments.

Accessories to provide microprobe analysis in TEM are now available for use with the following electron microscopes: the EM 200 of Philips; the EM 6 G of GEC-AEI; the JEM 7, 7 A and 120 of JEOL; the HU-11 series of Hitachi; and the Elmiskop I and I A of Siemens. Possibilities of further improvement in the performance of microscope conversions to TEMMA by using the minilens have been discussed by COOKE and DUNCUMB [34] and CHAPMAN [35].

An X-ray spectrometer for use with the SEM has been announced recently. This spectrometer, provided as an accessory to the Stereoscan, is identical to the one used on the Microscan III. Other accessories for microanalysis in the SEM are also under development [36].

At the present time, there are only two combined instruments in use that were specifically designed to combine transmission electron microscopy and microprobe analysis. The first instrument of this type, designed by P. DUNCUMB [37], is still in operation at Tube Investments Research Laboratories. A second instrument, called a probescope has recently been described by SHIPPERT et al. [38].

Use of Computers with Electron Microprobes

The use of computers has had an impact on electron probe techniques in four principle areas, namely: 1. calculation of corrections to data in point-by-point analysis using conventional correction procedures, 2. calculation of the theoretical intensity-concentration relations using Monte-Carlo or transport models, 3. control of the electron probe or the specimen stage for microanalysis, 4. processing of image information obtained with scanning beam instruments. The first two of these examples will not be considered in this present paper because of limited space. The latter two cases are of great importance in the design of new instruments and in the modification of existing instruments.

Electron probe instruments already exist with memory systems of various types to control the spectrometers or the specimen stage during X-ray analysis. A sequential analysis for pre-selected elements can be performed by using optical readout of punched cards (as in the programmable spectrometers of ARL), a non-sequential analysis pre-recorded on punched tape can be performed by using servo control of the spectrometers (as in the Permma unit of AEI [39]) and movement of the stage to the position of standards can be accomplished automatically (as in the servo controlled stage of Cambridge Instruments). Most instruments are now provided with step scanning stages and automatic readout on electric typewriters or punched tape.

The control of spectrometers in commercial instruments has provided the possibility of pre-selecting the spectrometer position, the scaler range and the PHA threshold [39]. A further advance in automatic control of microprobe instruments would be the automatic determination of the appropriate variables for each element of interest. This features has been called an "element store" by LONG [40] who is developing a system for automated analysis for use on geological specimens. LONG is also attempting to realize further advantages from automatic control of the specimen stage. He has proposed a system that would make it possible to preselect various points on the specimen to be analyzed and the elements to be analyzed at these points. The appropriate conditions for the analysis would be automatically provided from the element store. Computer control of a microprobe instrument for automated analysis has been also described by LIFSHIN and HANNEMAN [41].

Computer processing of information obtained from scanning beam instruments will also be of increasing importance for inclusion counting or classification and for the determination of relative concentrations of different phases. A comprehensive survey of these techniques has been made by DORFLER [42]. The first applications of microprobe instruments to inclusion counting are relatively recent [43, 44]. MELFORD and WIDDINGTON [44] described an inclusion counter that employed signals from 2 X-ray spectrometers and a backscatter detector for classification and also used a memory system to avoid counting the same inclusion twice. A simple form of inclusion counter based only on the signal from backscattered electrons was recently developed by Vickers Limited and demonstrated at the 1968 Physics Exhibition in London. Accessories to electron probe microanalyzers to perform lineal analysis or phase integration have been described by TONG et al. [45] and by DORFLER [46]. Application of scanning electron probes to the determination of the shape of particles is also of current interest [47]. This is a more difficult problem than the use of signal levels to classify particles or phases since special computer techniques for pattern recognition are required.

Auger Electron Spectrometry

Within the past few years, it has been demonstrated that a chemical analysis can be obtained by measuring the energy of Auger electrons ejected during the non-radiative decay of excited atoms. The energy of the Auger electron is determined by the difference between the initial state having an inner shell vacancy and the final state which usually has two vacancies in a shell of lower ionization energy. Thus the energy of the Auger electrons varies in a systematic manner with increasing atomic number, similar to the variation in energy of characteristic X-rays. Owing to energy loss during escape, the Auger emission lines are not sharp; moreover,

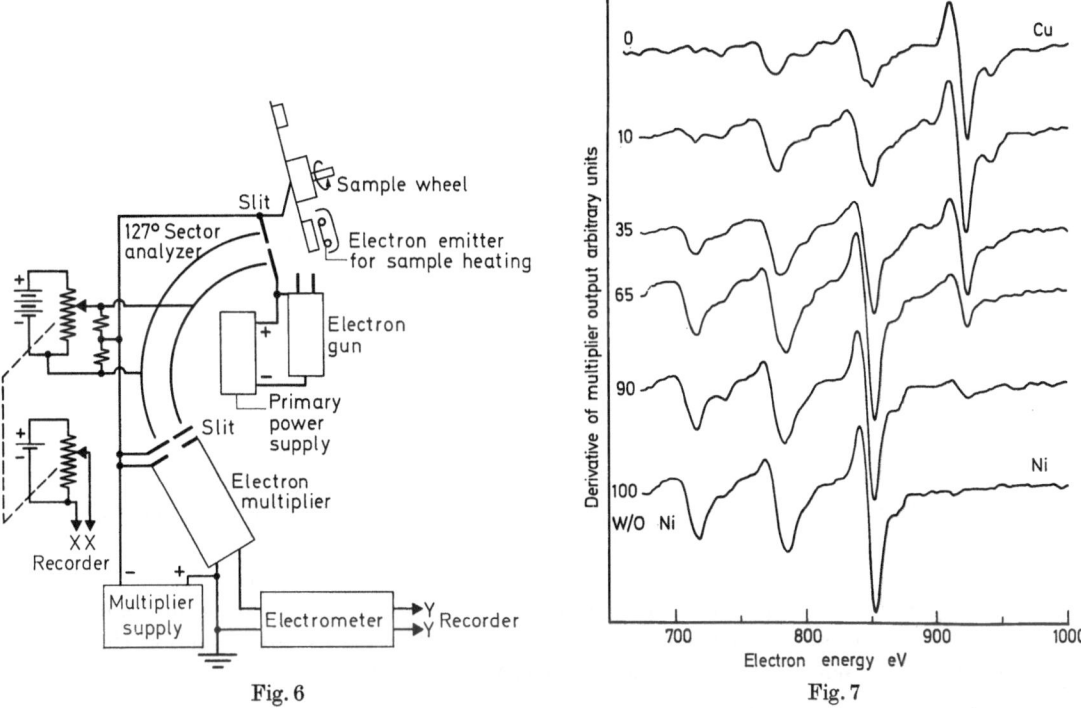

Fig. 6
Fig. 7

Fig. 6. An Auger electron detection system used by HARRIS [49] employing an electrostatic electron spectrometer

Fig. 7. Differentiated Auger electron spectra from alloys of copper and nickel. From HARRIS [49]

they come mainly from depths of only 10—100 Å below the surface. The Auger lines of greatest interest lie in the energy range of 50—1000 volts; the low limit being determined by difficulties due to background from secondary electrons and difficulties in obtaining adequate energy resolution. The high energy limit is due to increased probability of radiative decay of the excited states of interest.

Early attempts [48] to observe Auger electrons in a conventional electron probe microanalyzer were unsuccessful because of specimen contamination, stray magnetic fields and high background due to scattered electrons. More recently, several laboratories have successfully used this technique in special instruments [49, 50] or in low energy electron diffraction (LEED) apparatus [51, 52]. In most cases, the incident beam has an energy of 500—2,000 volts and ultrahigh vacuum conditions are required. In some cases [52], it has been possible to detect 1/10 of a monolayer of absorbed atoms using an incident beam focussed to a diameter of the order of a few mm.

Although Auger electron spectroscopy has not yet been done with an electron probe of small diameter, it is useful to mention some of the instrumentation that have been used and the results obtained. An example of a system for Auger electron spectroscopy used by HARRIS [49] is shown in Fig. 6. An electrostatic spectrometer is used with phase sensitive demodulation techniques to obtain the derivative of the electron spectrum, thereby minimizing the contribution of the background to the recorded traces. Differentiation also enhances the Auger spectrum, since the line shape is highly assymmetrical because of energy losses during escape. An example of the spectra obtained from copper nickel alloys is shown in Fig. 7. Auger electron emission due to L excitation of Cu and Ni can be seen to vary as the composition is changed. With this apparatus, HARRIS has detected carbon in iron at a concentration of 40 parts per million.

An example of the type of spectrometer used in LEED apparatus is shown in Fig. 8. This figure was taken from the work of PALMBERG [50] and shows a retarding field spectrometer consisting of several concentric hemispherical grids and a hemispherical collector. In this case, a dc retarding field is applied between sample and the grid nearest the collector, and an ac voltage

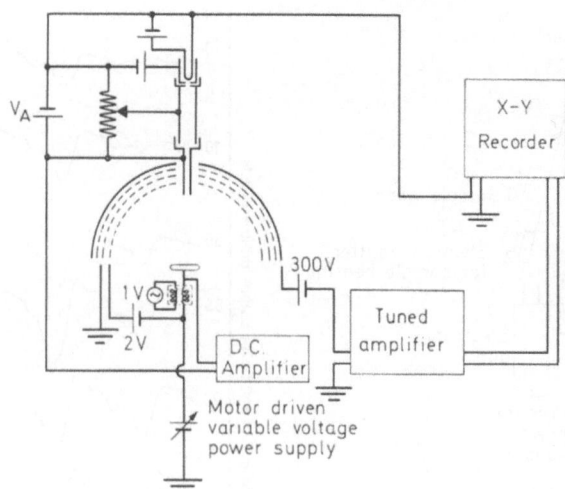

Fig. 8. A typical retarding field electron spectrometer for Auger electron spectrometry. From the work of
PALMBERG [50]

of small amplitude is superposed on the specimen. By detecting the signal at the frequency of
the modulation, normal $N(E)\,dE$ curves are obtained. Alternatively, by detecting the second
harmonic [52], it is possible to obtain curves of $[dN(E)/dE]\,dE$.

An advantage of the method of Auger electron spectrometry over X-ray analysis lies in the
high solid angles that can be used, particularly if the retarding field spectrometer is used. Moreover, it provides higher signals than those available in X-ray analysis when the energy levels
are very low, since low energy excited states tend to decay non-radiatively.

Ion Microprobe Mass Spectrometry

Local mass spectrometer analysis using ions sputtered from a solid surface has been done
recently using two different approaches. A selected area method was possible with the technique
developed by CASTAING and SLODZIAN [53] in which images were formed with ions of a particular
charge to mass ratio. The instrumentation for this approach was subsequently modified by
CASTAING et al. [54] by using an electrostatic mirror and a magnetic prism through which the
electrons pass twice. A commercial instrument based on this approach has been developed [55].

The microprobe approach has also been used for local mass spectrometry. In this technique,
shown schematically in Fig. 9, an ion probe is formed by using a duoplasmatron source and

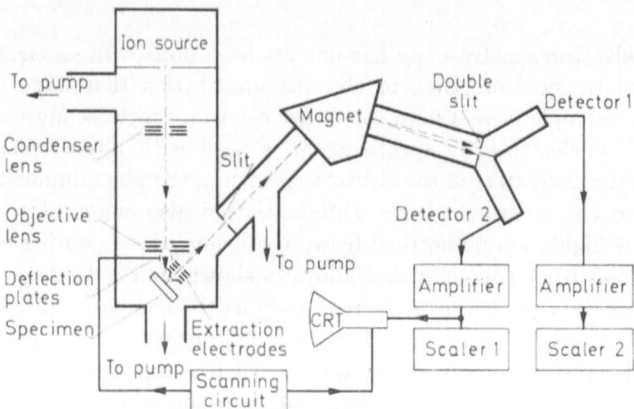

Fig. 9. Schematic diagram of an ion probe mass spectrometer. From LONG [56]. (Reprinted by permission of
the Institute of Physics and the Physical Society)

two electrostatic lenses. The ions sputtered from the specimen are extracted and pass through a mass spectrometer. The figure is taken from the work of Long [56]. The instrument described by Long is still under development but has already exhibited a resolution of the order of 1 micron with an ion probe of 20 kV and 4.10^{-9} amp. [57]. In another laboratory, a more elaborate ion probe microanalyzer has been completed by Liebl [58]. This instrument is the subject of two papers at this conference [59].

Summary and Conclusions

While instruments for microprobe analysis have changed only slightly over the past several years, it appears that more significant changes in the available instrumentation will take place in the near future. These changes will be due partly to the development of new components for electron probe microanalysis and partly to the development of new instruments that combine various analytical techniques. New electron beam instruments using new components or combining electron microscopy with electron probe microanalysis may differ greatly from the earlier versions.

The trend toward increased emphasis on the use of ancillary techniques in the electron probe is expected to continue. Incorporation of better vacuum conditions and methods for studying Auger electron emission appears to be inevitible. The development of completely new instruments for microprobe analysis, involving ion bombardment and ion emission, may provide new possibilities for attacking problems that cannot be dealt with by the more familiar methods of electron probe X-ray microanalysis.

References

1. Nixon, W. C.: This conference.
2. For a comprehensive bibliography covering work up to 1965, see Heinrich, K. F. J., The Electron Microprobe, ed. by T. McKinley et al., p. 841—974. New York: John Wiley & Sons 1965.
3. Liebmann, G.: Proc. Phys. Soc. (London) 68, 737 (1955).
4. Henke, B. L.: Optique des rayons X et microanalyse, p. 168—180. Paris: Hermann 1966.
5. Fernandes-Moran, H.: Electron Microscopy, vol. 1, p. 27. Tokyo: Marusen 1966.
6. Cosslett, V. E., and M. E. Haine: Proc. Int. Conf. Electron Microscopy, London 1954, p. 639 (1956).
7. Pattee, H. H., Jr.: X-ray Microscopy and Microanalysis, ed. by V. E. Cosslett et al., p. 367—373. New York: Academic Press 1957.
8. Wittry, D. B.: Thesis California Institute of Technology 1957.
9. Marton, L., R. A. Shrack, and R. B. Placious: X-ray microscopy and microanalysis, ed. by V. E. Cosslett et al., p. 287. New York: Academic Press 1957.
10. Crewe, A. V., D. N. Eggenberger, J. Wall, and L. M. Welter: Rev. Sci. Instr. 39, 576—583 (1968).
11. Broers, A. N.: J. Appl. Phys. 38, 1991—1992 (1967); erratta: J. Appl. Phys. 38, 3040 (1967).
12. Le Poole, J. B.: Proc. Third Regional Conf. on Electron Microscopy, Prague 1964, p. 439.
13. These proceedings, paper Nos. 12, 39, 41, 54, 64.
14. Davidson, E., and H. Neuhaus: First National Conf. on Electron Probe Microanalysis, College Park, Md. 1966, paper No. 9.
15. Nicholson, J.: Third National Conf. on Electron Microprobe Analysis, Chicago 1968, paper No. 30. — Nicholson, J. B., and M. F. Hassler: Advances in X-ray analysis, vol. 9, p. 420. New York: Plenum Press 1966.
16. Braybrook, R. F., A. Franks, F. J. Kirby, and K. Lindsey: Optique des Rayons X et Microanalyse, p. 477—479. Paris: Hermann 1966.
17. To be published.
18. Fitzgerald, R., K. Keil, and K. F. J. Heinrich: Science 159, 528—529 (1968).
19. Available for MS 46 DLC, Elmisonde.
20. Available for SEM 2 A, EMX-SM, XMA 5, JXA 3 A, JXA 5.
21. Available as for transmission except XMA 5.
22. Available for SEM 2 A, AMX, EMX.
23. Available for EMX (Shimadzu), JXA 3 A, JXA 5, Model 400 S.
24. Available for EMX-SM.
25. Nearly all models.
26. Used in MS 46 DLC, XMA 5, JXA 3 A, JXA 5.
27. Used in Microscan III, Microscan V.
28. Available in SEM 2 A, MS 46 DLC.
29. Neuhaus, H.: Third National Conf. on Electron Microprobe Analysis, Chicago 1968, paper No. 19.
30. A comprehensive review of Scanning Electron Probe Microanalysis was recently published by K. F. J. Heinrich, National Bureau of Standards Technical Note 278 (1967).

31. Ong, P. S.: Third National Conf. on Electron Microprobe Analysis, Chicago 1968, paper no. 23.
32. Hughes, K. A., D. V. Sulway, R. C. Wayte, and P. R. Thornton: J. Appl. Phys. **38**, 4922 (1967).
33. These proceedings, paper No. 100.
34. These proceedings, paper No. 41.
35. These proceedings, paper No. 39.
36. *K* Square Corporation, Pennwood and Lamar, Pittsburgh, Pa. 15221.
37. Duncumb, P.: Fifth International Conf. on Electron Microscopy, Philadelphia, paper KK 4. New York: Academic Press 1962.
38. Shippert, M. A., S. H. Moll, and R. E. Ogilvie: Anal. Chem. **39**, 867—876 (1967).
39. Browning, G. W., D. Cooknell, K. Heathcoat, I. P. Openshaw, J. L. Williams, and P. W. Wright: Second National Conf. on Electron Microprobe Analysis, Boston, Mass. 1967, paper No. 62.
40. Long, J. V. P.: Dept. of Mineralogy, Cambridge University, private communication.
41. Lifshin, E., and R. E. Hanneman: General Electric Research Report 66-C-250 (1966).
42. Dorfler, G.: Quantitative evaluation methods of alloy microstructures by microprobe methods, Seminar on Quantitative Microprobe Analysis, National Bureau of Standards; to be published as an NBS monograph.
43. Theisen, R.: Z. Metallk. **55**, 128—134 (1964).

44. Melford, D. A., and R. Widdington: Optique des Rayons X et Microanalyse, p. 497—505. Paris: Hermann 1966.
45. Tong, M., C. Conty, and R. Lewis: Third National Conf. on Electron Microprobe Analysis, Chicago, 1968, paper No. 21; see also these proceedings, paper No. 100.
46. These proceedings, paper No. 77.
47. These proceedings, paper No. 34.
48. Wittry, D. B.: Optique des Rayons X et Microanalyse, p. 168—180. Paris: Hermann 1966.
49. Harris, L. A.: J. Appl. Phys. **39**, 1419—1427, 1428—1431 (1968).
50. Palmberg, P. W.: J. Appl. Phys. **38**, 2137—2147 (1967).
51. Tharp, L. N., and E. J. Schreibner: J. Appl. Phys. 38, 2320—2330 (1967).
52. Weber, R. E., and W. T. Peria: J. Appl. Phys. **38**, 4355—4359 (1967).
53. Castaing, R., and G. Slodzian: Compt. Rend. **255**, 1893 (1962); — J. Microscopie 1, 395 (1962).
54. — Optique des Rayons X et Microanalyse, p. 48—63. Paris: Hermann 1966.
55. These proceedings, paper No. 88.
56. Long, J. V. P.: Brit. J. Appl. Phys. **16**, 1277 (1965).
57. Drummond, T. W., and J. V. P. Long: Nature **215**, 950—952 (1967).
58. Leibl, H.: J. Appl. Phys. 38, 5277—5283 (1967).
59. These proceedings, paper Nos. 89, 90.

A Precision Linear X-Ray Spectrometer

L. HAILES

S. T. L., London Road, Harlow, Essex, England

Introduction

Changes in soft X-ray spectra, due to chemical bonding, valency, co-ordination number, etc have been reported in the literature for many years. However, very little use of these effects have been made in electron probe microanalysis, rather, they have been regarded as a nuisance in quantitative analysis. A research programme to study the feasibility of using such measurements to increase the amount of information obtainable from the electron probe microanalyser has been set up at S.T.L. In particular, measurements have been made of the so-called Chemical Shift. It soon became evident that the repeatability of the standard spectrometers of our commercial microanalyser, although perfectly adequate for normal use, was not always sufficient for chemical shift measurements. This paper therefore describes a precision linear spectrometer (which is a direct replacement for the fixed centre bearing spectrometer of the Japan Electron Optics Laboratories JXA-3A), a digital control unit, a novel proportional counter design, and gives some data from a continuing investigation of large D-spacing crystals.

Mechanical Design of the Spectrometer

The basic linear geometry is shown in Fig. 1. In this configuration the Rowland circle centre is no longer fixed, and must rotate around the X-ray source. The detector path then becomes the complex 'four leaf rose' shape shown. The mechanical problem can be broken down into four parts:

1. Providing precise linear motion for the crystal.

2. Locating the Rowland circle centre (since the X-ray source is normally mechanically inaccessible).

3. Achieving an accurate 2:1 drive from the crystal arm to the detector arm.

4. Alignment of the crystal and the detector slit.

In the interests of precision it is generally desirable to make a design as simple as possible, with few moving parts. A detailed study of existing designs has been made, which has shown that there are many different ways of achieving this basic geometry, but many of them could not be used for a high precision spectrometer. The design finally adopted is shown in Fig. 2. The crystal holder is fixed to a carriage which moves on a linear bearing, driven by a double recirculating ball anti-backlash nut on a leadscrew. The leadscrew is 15 mm diameter with a 4 mm pitch, and is machined to an accuracy of 7 μm over 422 mm length. The Rowland circle centre locus is provided by a curved track, accurately machined relative to the linear bearing, and the pivot is spring loaded against the outer face of this track. The normal 2:1 bevel gearbox drive between the arms was rejected because of tooth to tooth pitch errors, and a new drive has been developed. It consists of a two-unit radius cam centred at the crystal axis, connected to a one-unit radius wheel at the counter arm pivot, by a steel tape tensioned by a Tensator spring. This system has been extensively tested on a full size prototype and was found to give satisfactory use over a period corresponding to two years typical E.P.M.A. operation, at which

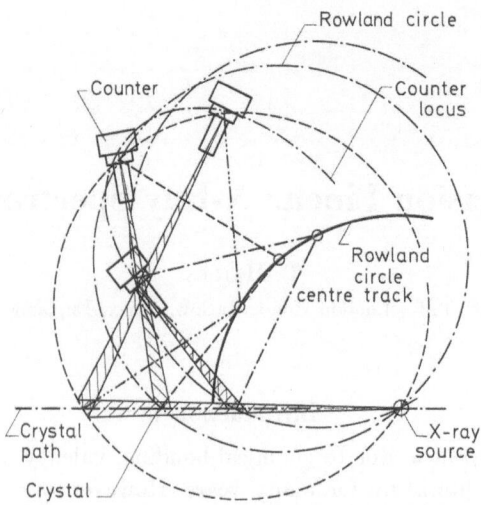

Fig. 1. Basic linear spectrometer geometry

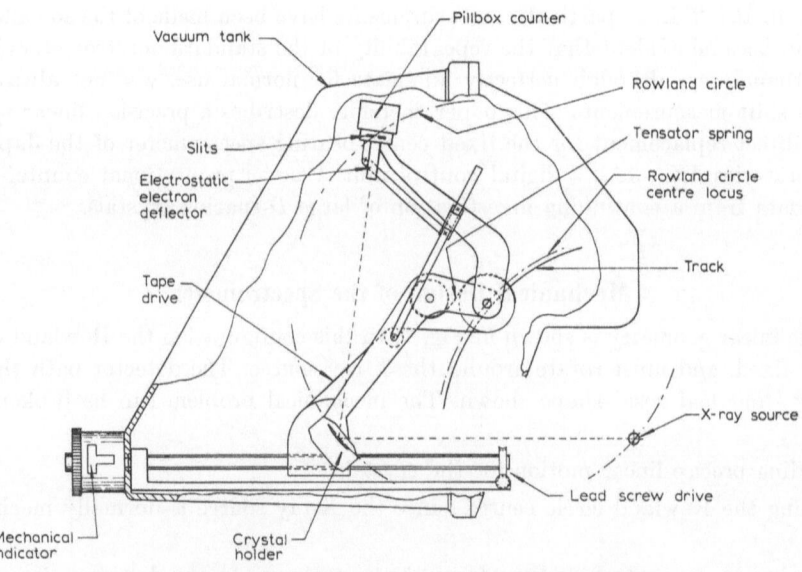

Fig. 2. Schematic diagram of S.T.L. spectrometer

time it is only necessary to replace the Tensator spring. The system is in fact limited by the fatigue life of the spring. The advantages of this drive are simplicity, low cost, very high accuracy and no backlash. The crystal to counter slit alignment is obtained by a conventional sliding link parallelogram.

Designing the spectrometer to fit to the existing specimen chamber casting imposed several restrictions. First, the take-off angle of 20° and the horizontal Rowland circle had to be retained. Secondly, the overall height was restricted if the minimum Bragg angle of 22° 2θ was to be achieved. It was found that by cranking the crystal and counter arms, the maximum 2θ could be increased from 80° to 145°. For high resolution, the Rowland circle radius of 250 mm and crystal dimensions of 29 × 8 mm were also retained.

An error calculation for this design gives the absolute accuracy for measurement of 2θ at the end of the leadscrew (by conversion from the linear source to crystal distance), to be ± 2 min/arc.

Fig. 3. Partially completed prototype spectrometer. (As shown at the Physics Exhibition, London 1968)

Fig. 3 shows a photograph of the partially completed prototype, illustrating the drive system.

The whole spectrometer is constructed on a subframe, which can be adjusted in three dimensions at right angles with respect to the X-ray source to facilitate final alignment. The vacuum tanks are made from aluminium alloy to reduce weight, and the working parts from stainless steel, to reduce the magnetic effects on the electron beam. The crystal and counter arms are hollow box sections. The three-crystal holder, (interchangeable in vacuum) and the electrostatic electron traps in front of the counter are the original J.E.O.L. equipment.

Spectrometer Control Unit

A simple digital control unit has been designed, mainly to ensure a high repeatability, but also to provide some automation facilities, and eventually the possibility of punched tape control. In the absence of space inside to measure the Bragg angle directly at the crystal, the following method has been devised to give a repeatability of ± 5 sec/arc from measurement outside the spectrometer. The functions of driving the spectrometer, and measuring the Bragg angle indirectly have been combined. At the electron optical column end, the leadscrew is driven via a 10:1 backlash-free worm gear, by a Slo-syn stepping motor. This motor has 200 steps per revolution and this gives approximately 2×10^5 steps over the full range of the spectrometer.

A five channel present control unit has been constructed using 'Digic' counting and logic modules (manufactured by Darang Electronics Ltd.).

The complete control unit operates in the following way: The motor driving pulses are produced in a variable frequency oscillator. These are then switched in the correct sequence to drive the motor by a Motor Translator Unit. The passage of these pulses can be interrupted by an electronic gate, so stopping the spectrometer immediately. The electronic gate is controlled by a six decade bidirectional indicating counter, (which counts the pulses from the V.F.O.) and one of five six-digit decade switches. When the number in the indicating counter coincides with the number preset in the decade switches, the electronic gate closes. Five sets of decade switches are provided, enabling up to five Bragg angles to be preset, but this can be increased by simple addition of further switches.

This is of course an accumulating method, and it is therefore necessary to start from a fiduciary zero, and rely on the logic to hold knowledge of the position of the spectrometer. For this reason the logic units have a separate power supply, which will continue to operate in the event of a mains failure.

Apart from presetting the decades, the operator only has to decide the direction of movement of the spectrometer, according to whether the number preset is greater or smaller than the number already in the indicating counter. Background offsets can be obtained simply by changing one digit in the decade switch.

In addition to stopping the spectrometer at the correct position, a pulse also starts a scaler in the X-ray counting channel (Nuclear Enterprises "International" Series), and on pressing the start button to obtain another wavelength, this scaler is automatically reset to zero. Printout facilities can also be utilised.

In the spectrum scanning mode, a pulse is available from each of the indicating decades to operate a marker pen on a recorder. The speed of the scan is determined by the setting of the V.F.O. control.

The Counter

The use of coaxial anode cylindrical proportional counters is now almost universal in conventional electron probe microanalysers. In this instrument a new type of proportional counter, known as the "Pillbox" [1] is being used. It is unusual in that the anode wire is perpendicular

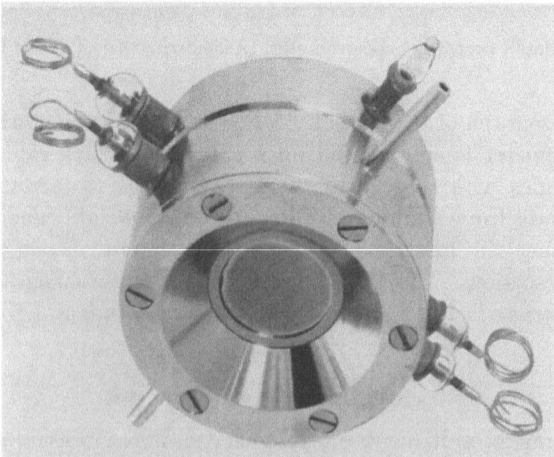

Fig. 4. Double Pillbox proportional counter

to the axis of the cylinder. This geometry means that it is very simple to construct a compact double proportional counter, consisting of a thin windowed flow counter in front of a sealed Xenon filled counter, the two being separated by a 250 μm thick beryllium window. This arrangement is shown in Fig. 4. The counter body diameter is 50 mm, and the depth 30 mm. The high voltage is simply switched from one anode to the other to change counters. The resolution of the flow counter using P.10 gas and an H.T. of 1,700 V has been measured as 17% for $Fe\,K_\alpha$, and varies by less than 1% over a window area of 25 mm in diameter. This property of constant resolution over a large window area, coupled with small overall size gives the Pillbox counter significant advantages over the conventional cylindrical type.

Window materials for use in the very soft X-ray region have been extensively investigated. In addition to the method of "floating" windows (described by Henke [2]) a method has been devised giving smoother films which are more reproducible. The apparatus consists of a 3 inch square optically flat glass plate mounted at right angles to the rotation axis on the end of a 10,000 rpm motor with a variable speed control. If a solution of window material is poured onto the plate while it is rotating at high speed, a very thin uniform film is obtained. This can then be mechanically stripped, or floated off, after hardening, and mounted on a window mesh. The film thickness can be controlled both by the solution concentration and the speed of rotation. Formvar, Collodion and Polycarbonate windows of approximately 0.1 μm thickness (judged from interference colours) have been made and tested. So far the polycarbonate windows are giving the best lifetimes in operation.

Crystals

Since peak shifts become measureable at about 5 Å, the crystals which are most useful are K.A.P. (potassium acid phthallate) and lead stearate. However any crystal which has a larger $2D$ spacing than K.A.P. (26 Å) is potentially interesting. The Chlorite group of minerals, are reported to have $2D$ values of approximately 28 Å, and in certain cases may exceed this. Chlorites are common minerals, related in composition and structure to the micas, but contain no alkalies. The 001 basal plane is the natural cleavage plane, but they are not so easy to cleave as mica, and the crystals obtained are not very elastic. Unfortunately large perfect single crystals, suitable for diffraction work are difficult to obtain. There are three main varieties: Clinochlore, Penninite and Prochlorite. CAUCHOIS [3] has recently obtained very good results with Prochlorite, and BAUN and FISCHER [4] have reported on the use of Clinochlore.

Samples of Clinochlore (from the Urals, U.S.S.R.), identified by powder diffraction as Leuchtenbergite, with $2D = 28.2$ Å (ASTM card 12.242) were cleaved and mounted in the spectrometer. The best results obtained gave a resolution slightly better than K.A.P., with a reflectivity of about 50% of K.A.P. for SiK_α.

Samples of Penninite (from Zermatt, Switzerland, and Ontario, Canada) with $2D = 28.6$ Å, (ASTM Card 10—183) gave much worse results. In the case of the Zermatt samples the Penninite was found to be intergrown with about 10% of Phlogopite. In the Ontario samples, the basal plane was rather wavy and it is thought that this gave rise to poor reflectivity because large areas of the surface were not at the Bragg angle. The resolution was of the same order as K.A.P.

Providing that good enough specimens of chlorite can be obtained, it is felt that this would be a very useful crystal.

Further evaluation of this group of crystals is continuing.

Acknowledgements. This work was carried out as part of the requirements for a collaborative Ph. D. with the University of Surrey. Thanks are extended to my joint supervisors: Dr. K. O. BATSFORD (S.T.L.) and Dr. D. S. ALLAM (Chemical Physics Dept. University of Surrey), to Mr. P. JONES (Darang Electronics) for help with the controller circuitry and to Mr. R. DANIELS (20th Century Electronics) for his advice on the Pillbox counter design.

References

1. SANFORD, P. W.: J. Sci. Instr. **43**, 908 (1966).
2. HENKE, B. L.: Advan. X-ray Analysis 8 (1965).
3. CAUCHOIS, Y.: X-Ray Optics Microanalysis 480 (1966).
4. BAUN, W. L., and D. W. FISCHER: Paper presented at Pittsburgh Spectroscopy Conf. 1968.

Iron-Free Lenses for Electron Probe Forming Systems

R. Bassett and T. Mulvey

Department of Physics, The University of Aston in Birmingham, England

Abstract

Recent developments in the techniques of electron probe micro-analysis have emphasized the need for improved lenses for the probe forming system. Iron-free magnetic lenses as proposed by Le Poole offer several interesting possibilities. The general properties of such lenses are discussed in the light of the authors' calculations and measurements, with a view to extending their field of application.

Introduction

From its earliest beginnings the electron probe micro-analyzer was closely associated with optical metallography. Consequently a focussed electron beam or probe of about a quarter of a micrometre in diameter was adequate. Indeed with solid specimens no great improvement in X-ray resolution could be expected with smaller probes since the electrons penetrate several micrometres into the sample, and this rather than probe size generally determines the limit of resolution of the technique. The aberrations of conventional lenses are adequate for probes of this size, but do not seem capable of substantial improvement without serious compromise in the other essential features of the instrument. However, the development of techniques for preparing thin samples of metallurgical, minerological, and even biological material makes a reappraisal of the design of the probe forming lens highly desirable.

In electron microscopy a similar situation has prevailed for many years in respect of the illuminating or probe forming system which has the task of controlling both the angle of the illuminating pencil and the area of specimen irradiated. A minimum electron probe size of some 2 micrometres at the specimen has usually proved adequate for most microscopical purposes. The aberrations of the final condenser in an electron microscope are usually large, but this is unimportant since the angle of illumination is small, typically $10^{-3}-10^{-4}$ radian, compared with the 10^{-2} radian needed in an electron probe analyser. X-ray micro-analysis in commercial electron microscopes is therefore usually handicapped by the poor performance of the probe forming system, so that the excellent conditions for viewing transmission specimens cannot be fully exploited. For this reason it is also useful to reconsider the design of the probe-forming lenses of the electron microscope.

The Design of Probe Forming Lenses

The theoretical and practical considerations in the design of electron probe-forming systems have recently been reviewed (Mulvey, 1966). The problem is how to obtain the maximum possible probe current at the specimen compatible with the minimum restriction on the optimum design of the other essential components such as X-ray spectrometers, secondary electron detectors, optical microscopes etc. The maximum current I_p that can be obtained in an electron probe is given by:

$$I_p = \text{const } \beta \, d^{\frac{8}{3}}/C_s^{\frac{2}{3}}$$

where β is the current density/per unit solid angle, d is the diameter of the probe and C_s is the spherical aberration coefficient of the probe forming lens.

In principle, and to a large extent in practice, the brightness β of electron guns relying on thermionic emission can attain the theoretical Langmuir limit. In practice, no difficulties should arise at cathode temperatures below 2,700° K; at more elevated temperatures space charge limits effects may be observed with conventional electron guns, but these can be eliminated by suitable design and choice of materials.

More recently (CREWE et al., 1968) much higher brightnesses have been achieved in demountable probe systems by the use of electron field emission guns, but this poses problems in maintaining a suitably high vacuum (10^{-10} Torr) in the system.

Although it is true that an increase in brightness can be used to offset the harmful effect of spherical aberration, it is always worthwhile to reduce the lens aberrations to as small a value as possible.

It would appear that conventional iron-clad lenses have reached a practical limit. The principal requirement, of course, is that the working distance, i.e. the distance between the focal plane and the last working surface of the lens must be as large as possible for a given value of the spherical aberration coefficient C_s. An alternative and possibly more useful criterion is that the solid angle subtended at the specimen by the lens body should be as small as possible for a given value of C_s. Presently available lenses usually achieve a value of C_s of some 4—8 cm for a working distance of 1 cm, but the windings and magnetic shroud are extremely bulky. LE POOLE (1964) proposed that solenoids without iron polepieces could profitably be used as lenses to avoid the above difficulties and succeeded in constructing a long (5 cm) solenoid of small internal diameter (2 mm) that was comparable in electron optical performance with the best available iron-clad lenses. A lens of this type was successfully incorporated into a single stage electron luminescence apparatus (LE POOLE et al., 1965) with a probe size of 30 micrometre. This was followed by a two stage miniature microanalyzer (TNO, 1967) with a spot size of 1 micrometre. A long "mini-lens" has also been inserted (LE POOLE, 1964) into the upper body of an electron microscope objective lens in order to produce a fine probe (1,000 A.U.) at the specimen. DUNCUMB and COOKE (1968) have described the use of a mini-lens of the Le Poole type, in EMMA, a combined electron microscope-microanalyzer.

In spite of all these developments little has been written about the electron optical properties of such lenses and how they compare with those of iron polepiece lenses. It is by no means obvious that the mini-lens in the form of a long solenoid is the best practical form of iron-free lens. It is clear that by reducing the diameter of the coil the heating effect of the exciting current will be reduced and that the lens will present less of a physical obstacle to an optical microscope or to X-rays emerging from the specimen. However, it is also known that the spherical aberration of a long solenoid increases rapidly as the ratio of length S to diameter D increases.

Comparison between Symmetrical Iron Polepiece Lenses and Simple Solenoid Lenses

Consider a solenoid (Fig. 1) of length S and constant diameter D. The angle θ_{Ext} (external clearance angle) indicates the angle not obstructed by the coil. According to the Biot-Savart Law the flux density B_z at a point P on the z axis is given by

$$B_z = \mu_0 \frac{NI}{2S} (\cos \alpha_1 - \cos \alpha_2) \tag{1}$$

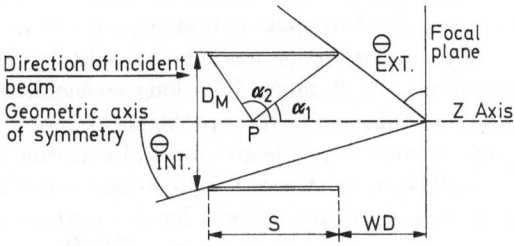

Fig. 1. Principal geometrical parameters of a magnetic solenoid lens

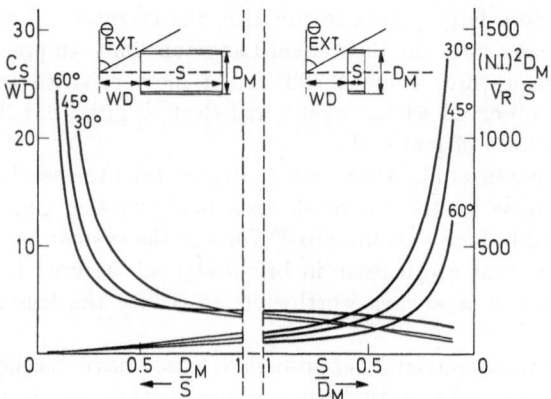

Fig. 2. Calculated focal properties and excitation of parallel-sided solenoid lenses, for various take-off angles

where NI is the excitation in ampere turns, and $\mu_0 = 4\pi \times 10^{-7}$ henrys/metre, and α_1 and α_2 are the angles that the lines joining the ends of the coil to P make respectively with the axis (measured in an anti-clockwise direction). The axial flux density distribution of Eq. (1) is a good approximation for solenoids in which the depth of the winding is small compared with its length. By tracing rays through such a field distribution, the focal properties and aberrations can be obtained. The focal properties of iron polepiece lenses are well-known from the work of van Ments and Le Poole (1947), Liebmann and Grad (1951), Lenz (1950), Dugas, Durandeau and Fert (1961) and others. Of particular importance here is the work of Durandeau (1957), who suggested that, provided $S/D > 0.2$, the axial field distribution of a polepiece lens is approximately the same as that of a solenoid as shown in Fig. 1 of the same length S but whose diameter is two thirds of that of the polepiece of the iron lens. This relation gives a qualitative explanation of the essential advantage of the solenoid lens, namely its smaller radial extension, since it is the axial field distribution that determines focal properties. In practice there is a further advantage, since in an iron lens the iron shroud even if conical in shape protrudes considerably in a radial direction. This is offset however by the fact that a solenoid lens generally requires a water cooling jacket which can often take up more space than the winding itself. Consequently a solenoid lens will produce an appreciable improvement over an iron lens only if certain conditions are satisfied. Two important factors are spherical aberration and coil loading.

To illustrate the foregoing remarks consider the solenoid lens shown in Fig. 1. Suppose that for a given take-off angle θ_{Ext} the lens excitation NI be adjusted so that the focal point occurs at a constant distance WD from one end of the lens as the length S of the coil is varied. Fig. 2 shows how the spherical aberration coefficient, plotted as C_S/WD and the excitation power dissipated in the coil, plotted as $\dfrac{(NI)^2 D_m}{V_r S}$ varies with coil geometry S/D (or D/S). Since S/D varies from zero to infinity, values of S/D greater than unity are plotted as D/S. The curves are plotted for three values of the external clearance (or take-off) angle θ_{Ext}, namely, 30°, 45°, and 60°.

The results shown in Fig. 2 indicate that there is no *optimum* design having either a minimum spherical aberration or minimum coil loading. At a given working distance and take-off angle, a solenoid lens can always be improved by making it shorter at the cost of increasing the power dissipated in the windings. A reduction in coil loading can only be obtained at the price of increasing the spherical aberration. Fig. 2 shows that long solenoids are associated with large values of spherical aberration but that coil heating problems are not severe. It is clear however, that although such lenses are capable of an electron-optical performance comparable with that of conventional lenses it is unlikely that they will lead to any substantial *improvement* in that performance. For this it is necessary to turn to short lenses i.e. those with S/D less than unity. Here the heating problem as judged by the parameter $\dfrac{(NI)^2 D_m}{V_r S}$ rises prohibitively as S/D is reduced below 0.75. In this region it is also noticeable that reducing the clearance angle e.g. from 60° to 30° effects a noticeable improvement in spherical aberration; this result is also found with iron polepiece lenses.

Conical Lenses

The foregoing results suggest that for a given take-off angle short conical coils might have some advantages. Fig. 3 shows a conical coil of length S and mean diameter $D_m = \dfrac{D_1 + D_2}{2}$, a cone semi-angle θ_c and a working distance WD as before. The axial flux density B_z at a point P is given by:

$$B_z = \mu_0 \frac{NI}{2S} \cos \theta_c \left[\cos(\theta_c - \alpha_1) - \cos(\theta_c - \alpha_2) \right. \tag{2}$$
$$\left. + \sin^2 \theta_c \ln \left\{ \frac{[1 - \cos(\theta_c + \alpha_2)]}{[1 - \cos(\theta_c + \alpha_1)]} \frac{D_2 \sin \alpha_1}{D_1 \sin \alpha_2} \right\} \right].$$

Eq. (2) reduces to the more familiar Eq. (1) if $\theta_c = 0$. Considerable changes in axial flux density distribution are now possible for a given value of S/D_m by altering the ratio D_1/D_2 of maximum to minimum coil diameter. It should be emphasized that Eq. (2) holds strictly for a winding of negligible thickness compared to the coil diameter, but in practice is an adequate representation of lenses in which the depth of winding is less than one tenth of the mean diameter of the coil.

Fig. 4 (full line) shows the axial flux density distribution calculated according to Eq. (2) for a coil of cone semi-angle $\theta = 40.5°$ and $D_2/D_1 = 12$ compared with that for a parallel sided coil with the same value of $S/D_m = 1.0$. As might be expected by analogy with asymmetric

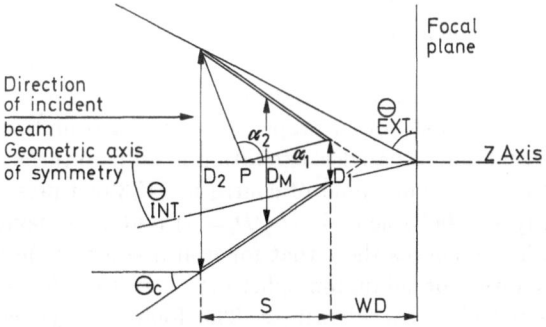

Fig. 3. Principal geometrical parameters of a conical magnetic lens

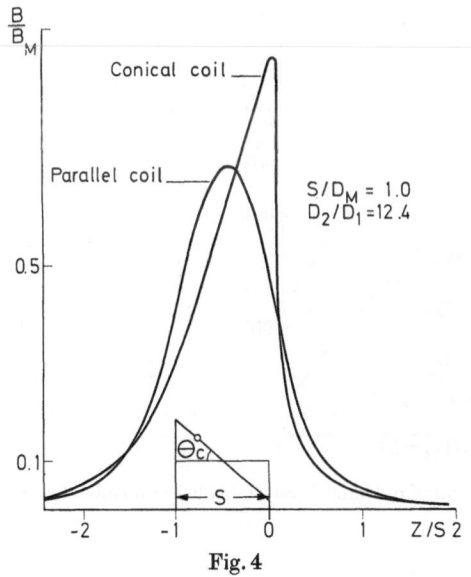

Fig. 4

Fig. 4. Calculated axial flux density distributions in parallel-sided and conical coils

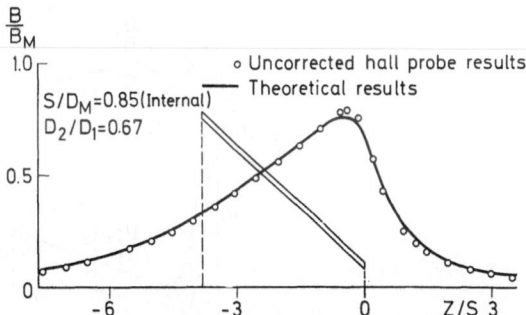

Fig. 5. Comparison of calculated and experimentally determined axial flux density distribution in a conical magnetic lens

15*

polepiece lenses, the conical coil has a greater maximum axial flux density compared with a parallel coil and the flux density falls off more rapidly towards the focal plane. Up to now however, no simple quantitative relationship has been found between the axial field distribution of asymmetrical polepiece lenses and conical coil lenses. The electron-optical properties of conical coils has therefore been calculated both from Eq. (2) and from experimentally measured axial field distributions. Fig. 5 shows the axial flux density (dashed curve) measured experimentally compared with that calculated from Eq. (2) (full line). The measuring system devised for this purpose made use of a Hall probe[1] of dimensions less than 1×1 mm. The deviations between experimental and calculated values may be attributed almost entirely to the finite size of the Hall detector. By taking design precautions to ensure a smooth set of experimental data by the use of a continuous X—Y plotter, it has proved possible to use the uncorrected Hall probe data to calculate focal properties without appreciable error. By using even smaller Hall probes which have recently become available a further improvement in accuracy will be possible.

Calculations with Conical Coils

The calculated properties of parallel-sided coils for a series of constant take-off angles shown in Fig. 2 were repeated for conical coils. The results led to essentially the same conclusions, namely, that any improvement in electron optical performance must be paid for in terms of increased power dissipation in the coil and are not reproduced here. In many applications a short conical coil is preferable to the corresponding solenoid as it presents less obstructions, for example, to a cone of X-rays. The cooling jacket can also be placed on the inside of the cone where it does not reduce the external take-off angle.

In order to give a more detailed comparison of parallel-sided and conical coils further optimisation is necessary. For a given working distance (WD) and lens geometry there will be one lens with minimum spherical aberration. Such optimized lenses are compared in Fig. 6. In this figure the working distance is unity and the other properties resulting from this optimisation procedure such as internal and external take-off angles, excitation in ampere turns at 30 kV are also shown.

At a kilovoltage V the excitation would of course be $(V/30)^{\frac{1}{2}}$ times as great. To avoid overcrowding the curves, only parallel solenoids $(D_2/D_1 = 1)$ and very asymmetrical conical lenses $(D_2/D_1 = 19)$ are shown. These curves show that for iron-free lenses there is no optimum design leading to minimum excitation or minimum spherical aberration. There is a practical limit to performance set by technological considerations. The following typical examples from Fig. 6 may allow some general conclusions to be drawn from the curves. For simplicity a working

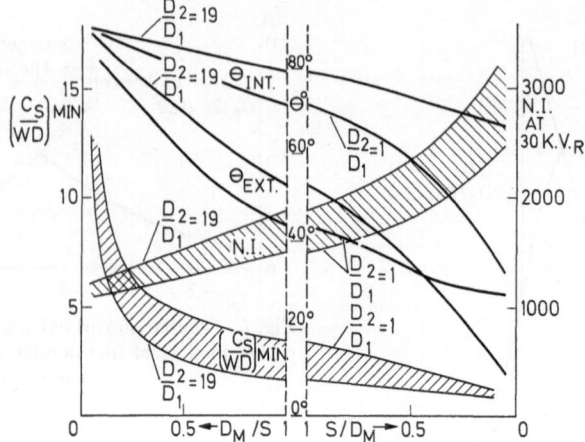

Fig. 6. Aberration (in terms of working distance) and excitation of optimized parallel-sided and conical magnetic lenses

1. Bell gaussmeter Model 120 and probe Model A-1201.

distance of 1 cm and operation at 30 kV is assumed. At large values of S/D, the ratio D_2/D_1 cannot depart greatly from unity for practical reasons. Thus for $S/D = 5$ and $D_2/D_1 = 1$, $C_S = 7.25$ cm and $NI = 1,150$ AT. The take-off angle read from the curves is 75° but as the water cooling jacket must be external to the coil this will reduce the angle to some 45—50°. Conical coils become practical for medium and small values of S/D. Unfortunately it does not seem technologically feasible to reduce S/D below 0.75. At this value of S/D, Fig. 6 gives $C_s = 1.45$ cm for $D_2/D_1 = 19$ and an excitation of 2,060 AT with $\theta_{\mathrm{Ext}} = 45°$. The main water jacket could be internal here but some reduction of angle, perhaps to 40° would probably occur in a practical design. Heating problems would be severe. A practical compromise might be a conical lens with $S/D_m = 1$ and $D_2/D_1 = 19$ with $C_s = 1.7$ cm and 1,870 AT at 30 kV and $\theta_{\mathrm{Ext}} = 53°$ (probably reduced to 50° in practice). These examples illustrate the fact that parallel-sided and conical lenses are not inherently suited to high external take-off angles.

If low take-off angles can be accepted, the smallest value of S/D_m compatible with lens overheating should be used. If a high take-off angle (60° or more) is required it is preferable to consider an internal take-off arrangement.

If heating effects could be overcome, the curves of Fig. 6 indicate that the ideal lens would be one in which S/D is extremely small. In this case $C_s/(WD)$ can be less than unity and a wide range of internal take-off angles becomes possible. It therefore appears worthwhile to investigate in more detail lens structures for which $S/D \rightarrow 0$.

Thin Helical (Pancake) Lenses

Consider a thin helix of internal diameter D_1 and external diameter D_2 as shown in Fig. 7. The axial flux density distribution B_z at a point P may be found from Eq. (2) by putting $\theta_c = \pi/2$.

This leads to:

$$B_z = \frac{\mu_0 N I}{2l} \left[\sin \alpha_1 - \sin \alpha_2 + \ln \left\{ \frac{1 + \sin \alpha_2}{1 + \sin \alpha_1} \right\} \frac{D_2 \sin \alpha_1}{D_1 \sin \alpha_2} \right] \cdots \qquad (3)$$

where $l = (D_2 - D_1)/2$. Note: $\cos \theta_c/S \rightarrow 1/l$ as $\theta_c \rightarrow \pi/2$.

Eq. (3) is strictly valid for a helix of vanishing axial extent but in practice is an adequate representation for lenses in which $S \leqq D_m/10$: With the arrangement shown in Fig. 7, two modes of operation are possible: (a) the usual one in which the electrons pass through the lens and come to a focus on the other side at a distance (WD) from the lens and (b) a mode in which the electrons are brought to a focus *before* they reach the lens structure. This results in a *negative* working distance $- (WD)$, an arrangement which gives a practically unrestricted range of take-off angles combined with low spherical aberration. In order to achieve this performance, a high but not prohibited excitation is required. As one half of the field plays no part in the imaging process under these conditions, it may be suppressed by placing a semi-infinite plane of ferromagnetic material behind the lens. This reduces the ampere-turn required by a factor of two. Such a lens may be referred to as a "half-lens". In practice a ferromagnetic plate of diameter equa

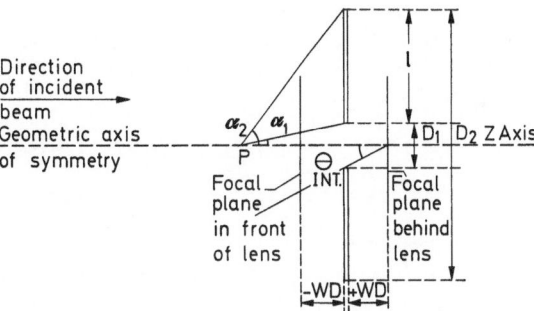

Fig. 7. Principal geometrical parameters of spiral coil (pancake) magnetic lens

to that of the maximum diameter of the lens is adequate. The focal properties of pancake lenses have been calculated and are illustrated by the following three examples. To facilitate comparison each lens has the same value of $C_s = 1.0$ cm. An operating voltage of 30 kV is also assumed.

Lens 1. $l/D_m = 0.5$, $D_1 = 2.38$ cm; $D_2 = 7.14$ cm, coil thickness $D_m/10 = 0.48$ cm. The working distance $WD = +0.79$ cm giving a *minimum* take-off angle of 43° for an excitation of 2,770 AT. The power dissipation is estimated at 200 watts. *Note.* The power dissipated depends on the thickness of the coil and is included as a general guide only.

Lens 2 (Half-lens). $l/D_m = 0.5$ as before, excitation 3,650 AT $D_1 = 6.06$ cm, $D_2 = 18.18$ cm $(WD) = -1.4$ cm, i.e. on the *same* side of the lens as the electron source. The power dissipation is estimated at about 250 watts in a coil of thickness $D_m/20 = 0.6$ cm backed by a ferromagnetic plate. If the electrons were brought to a focus at the near surface of the plate, this power would be reduced to 100 watts and the value of C_s would rise to 1.2 cm. This type of lens has the useful property that as the (negative) working distance is increased the spherical aberration is reduced.

Lens 3 (Half-lens). $l/D_m = 0.8$, $D_1 = 2.35$ cm, $D_2 = 20.1$ cm. $(WD) = -2.2$ cm at an excitation of 4,900 AT. Power dissipation is estimated as 300 watts in a coil of thickness $D_m/20 = 0.56$ cm.

Although the power to be dissipated in these lenses may appear at first sight to be high, it must be remembered that it is spread over a large surface area and the thin coils have a favourable geometry for cooling purposes. In addition they are easier to construct than conical coils. The combination of these features suggests that such lenses may find several interesting applications in electron probe X-ray microanalysis and other X-ray optical devices.

Acknowledgments. The above work was carried out in collaboration with the Division of Inorganic and Metallic Structures of the National Physical Laboratory with the support of the Ministry of Technology. The authors would like to thank Dr. A. Franks and R. Braybrook for many stimulating discussions.

References

Crewe, A. V., D. N. Eggenburger, J. Wall, and L. M. Welter: Rev. Sci. Instr. **39**, No. 4, 576—583 (1968).

Dugas, J., P. Durandeau et C. Fert: Rev. opt. **40**, 277—305 (1961).

Duncumb, P., and J. Cooke: These Proceedings 1968.

Durandeau, P.: Ph. D. Thesis No. 128, University of Toulouse 1957.

Lenz, F.: Z. Angew. Phys. **2**, 448—435 (1950).

Liebmann, G., and E. M. Grad: Proc. Phys. Soc. (London) B **64**, 956—971 (1951).

Le Poole, J. B.: 3rd. European Conf. Elec. Micros. Prague. Prague: Czech. Academy of Sciences 1964.

— A. B. Bok, and E. J. Boogerd: T.N.O. Nieuws **20**, 917 (1965).

Ments, M. van, and J. B. Le Poole: Appl. Sci. Res. Sect. B **1**, 3—17 (1947).

Mulvey, T.: Focussing of charged particles, vol. I, chapt. 2.6 ed. Septier. 469—494: Academic Press 1967.

TNO-TH (Delft) Jaarverslag, 21—23 (1967).

Elektronische Techniken
für die Elektronenstrahl-Mikroanalyse

H. Hoetzel und W. v. Loeben

Siemens AG, Karlsruhe, BRD

Die Elektronenstrahl-Mikroanalyse hat sich in den letzten Jahren sehr schnell von der zunächst wissenschaftlichen Anwendung zu einem Routine-Verfahren entwickelt, das aus den Laboratorien der Industrien und Institute nicht mehr wegzudenken ist.

Die Vielgestaltigkeit der Analysenprobleme verlangt universell einsetzbare Geräte, die neben dem heute selbstverständlichen Komfort vor allem übersichtlich angeordnete und somit bei der praktischen Arbeit schnell handhabbare Bedienungselemente enthalten.

Darüber hinaus erfordert eine schnelle und sichere Analyse Möglichkeiten, die verschiedenen Analysenparameter dem Meßproblem optimal anpassen zu können.

Ein wesentlicher Teil einer solchen Mikrosonde ist die Steuerelektronik und die Elektronik zur Verarbeitung der Meßsignale. Im folgenden wird auf die technischen Möglichkeiten und die Leistung einiger elektronischer Baugruppen eingegangen, die sowohl bei der Probendurchmusterung als auch bei der Analyse selbst wesentliche Erleichterungen bieten.

Der *Scan-Generator* steuert synchron den Elektronenstrahl der Sonde, der Bildteile sowie X-Y-Schreiber und Vielkanalanalysator. Er ermöglicht für die schnelle Durchmusterung von Proben eine periodische Abtastung der Probenoberfläche in 1, 3 und 10 sec. Für Bildschirmphotographie und die Messung der Konzentrationsverteilung mit Hilfe des Intensitätsdiskriminators ist eine unabhängige Einstellung von Bild und Zeilenzeit notwendig, wobei bei den Röntgenbildern Bildzeiten bis 100 min noch sinnvoll sind. Wichtig ist hierbei eine langsame Zeilenabtastung auch bei reduzierter Zeilenzahl, damit längs der Zeile eine gute Konturenschärfe erreicht wird. Bis 50 Zeilen/Bild ist die Flächenverteilung noch gut erkennbar.

Für schnelle halbquantitative und quantitative Analysen muß der „line-scan" in beliebiger Richtung und Lage über die Probe geführt werden (Abb. 1).

Probendrehung dafür zu Hilfe zu nehmen ist nicht sinnvoll, weil dann die X-Y-Zuordnung der Probe verlorengeht und die Durchmusterung erschwert wird.

Für die streng quantitative Analyse ist die Punktabtastung mit elektronischer Punktverschiebung notwendig. Eine wesentliche Hilfe für die Einstellung der Linien- bzw. der Punktabtastung auf die einzelnen Probendetails ist dabei der automatische Wechsel zwischen Rasterbild und Linien bzw. Punktabtastung.

Die *Bildausschnittechnik* ermöglicht besonders bei den intensitätsschwachen Röntgenbildern durch die Intensitätssteigerung an den interessierenden Bildpartien noch Details zu erkennen, die sonst nur über die Bildschirmphotographie erkannt werden könnten. In Verbindung mit allen obenerwähnten Scan-Techniken ist somit auch eine gezielte „Punkt- und Profilanalyse" dieser Detailbereiche möglich.

Die Aufzeichnung von Bildern der absorbierten Elektronen stellt hohe Anforderungen an die Verstärker, wenn bei kleinen Probenströmen unter 1 nA und schneller periodischer Abtastung noch eine gute Detailerkennbarkeit erzielt werden soll. Bei einer notwendigen Anstiegszeit unter 10 μsec für Rasterbilder von 1 sec läßt sich das Eingangsrauschen unter 10^{-10} A halten.

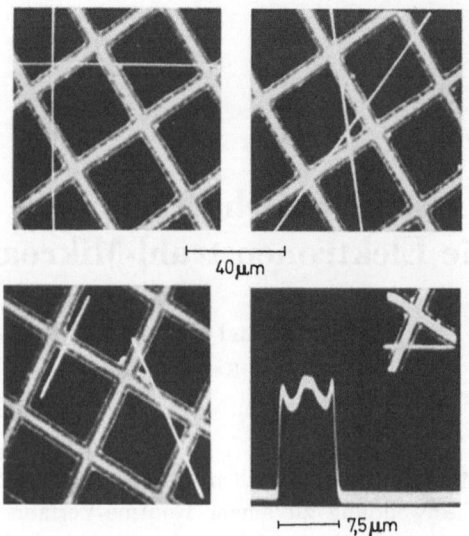

Abb. 1. Möglichkeiten der Linienabtastung

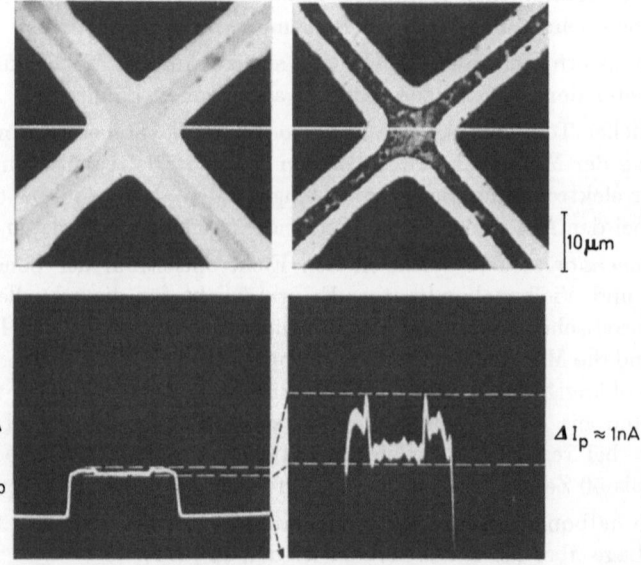

Abb. 2. Kontrastanhebung bei Gleichstrommeßkanälen

Abb. 2 links oben zeigt das Probenstrombild eines Testgitters. Die in der Mitte des Stegs aufgebrachte dünne Ag-Schicht hat nur eine geringe Probenstromänderung zur Folge, wie das Bild links unten zeigt. Durch Nullpunktsunterdrückung und Signalverstärkung (Bild rechts unten) läßt sich schon bei einer Probenstromänderung von 1 nA eine Weißschwarz-Zeichnung erreichen, wie Bild rechts oben zeigt. Für eine kontrastreiche Aufzeichnung von Bildpartien niedrigen Signalpegels ist eine Phasenumkehr der Signalspannung erforderlich.

Die Arbeitsweise des Impulsspektroskops, das für die nichtdispersive Röntgenanalyse uner-läßlich ist, zeigt die Abb. 3. Die von den Röntgendetektoren erhaltenen und verstärkten Dreiecks-Impulse werden durch Amplituden-Gleichrichtung in Rechteckimpulse mit konstanter Dauer (Dachzeit) umgewandelt.

Diese Spannung wird auf die x-Ablenkplatten einer Bildröhre gegeben. Die x-Auslenkung ist ein Maß für die Impulsamplitude und damit die Energie des diesen Impuls ausgelösten Röntgenquants.

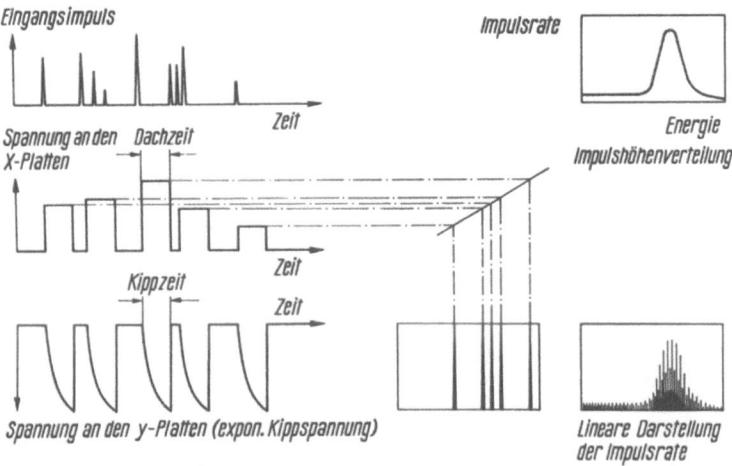

Abb. 3. Arbeitsweise des Impulsspektroskops

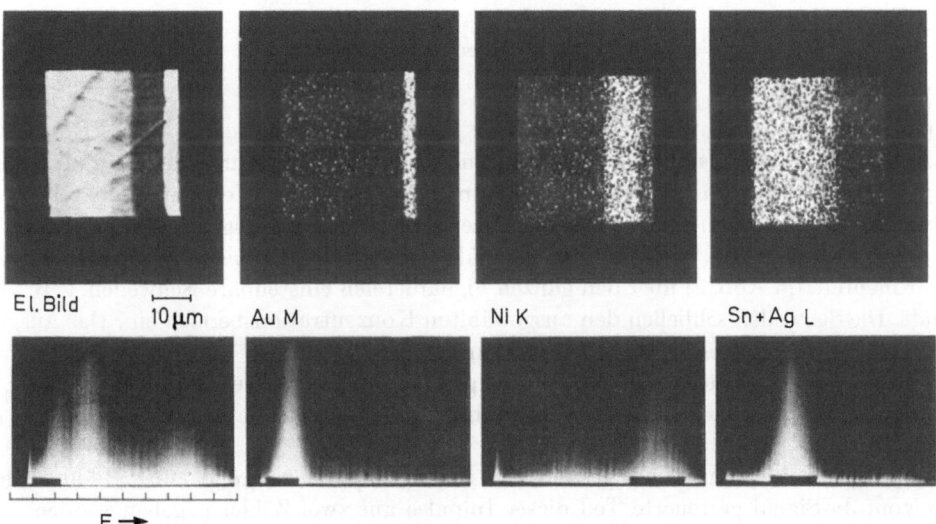

Abb. 4. Nichtdispersive Röntgenanalyse

Während der Dachzeit, in der der Elektronenstrahl in der ausgelenkten x-Position gehalten wird, erhalten die y-Platten eine exponentiell abfallende Kippspannung, wodurch der Elektronenstrahl auf dem Bildschirm von oben nach unten geführt wird. Dabei ist die Geschwindigkeit proportional und die Helligkeit umgekehrt proportional der Auslenkung y. Man erreicht durch diese Maßnahme eine lineare Intensitätsskala für das Impulsspektrum. Durch geeignete Schaltungsmaßnahmen wurde eine automatische Einblendung des verschiebbaren Diskriminatorkanals erreicht, der die Voraussetzung für den erfolgreichen Einsatz des Impulsspektroskops ist.

Abb. 4 zeigt den Einsatz des Impulsspektroskops für die nichtdispersive Röntgen-Analyse.

Das Bild der reflektierten Elektronen zeigt einen Teil eines Sandwiches, und zwar die Partie mit Gold (Au), Nickel (Ni), Zinn (Sn) und Silber (Ag). Unter diesem Bild sieht man die integrale Impulshöhenverteilung, wie sie während eines ganzen Bilddurchlaufes erhalten wird. Mit Hilfe des Diskriminators wählt man eine Spektral-Linie nach der anderen und zeichnet so jeweils das gewünschte Röntgenbild auf. Das Ergebnis sieht man in der oberen Zeile.

Die unteren Spektren wurden durch Punktanalysen in den jeweiligen Bereichen erzielt.

Der Einsatz des Impulsspektroskops in Verbindung mit einem Halbleiterdetektor mit Halbwertsbreiten unter 500 eV wird für die nichtdispersive Röntgenanalyse besonders bei niedrigen Konzentrationen bzw. erforderlichen sehr niedrigen Strahlströmen interessant werden.

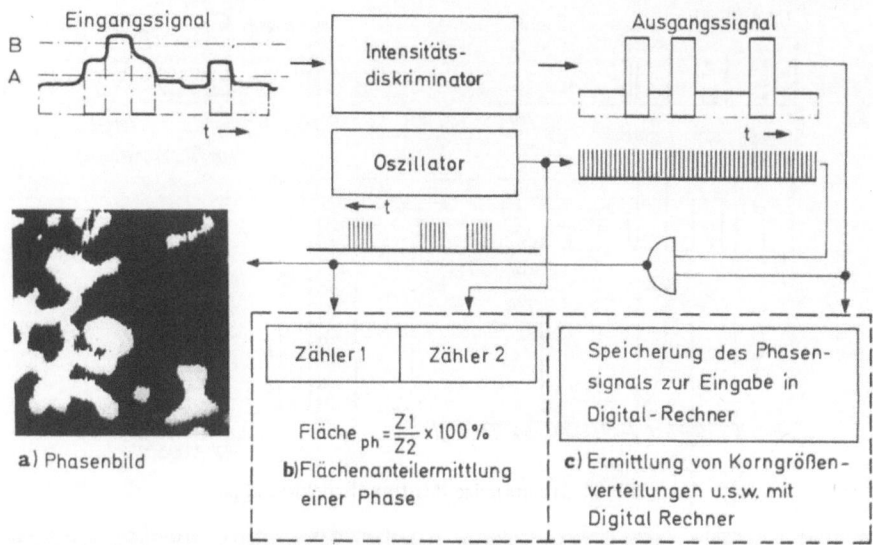

Abb. 5. Prinzip des Intensitätsdiskriminators

Abb. 5 zeigt schematisch die Arbeitsweise des *Intensitätsdiskriminators*. Dieses Gerät gehört heute schon zur Standardausrüstung von Mikrosonden und dient zur Aufzeichnung und zu quantitativen Messungen von Flächenanteilen bestimmter vorgewählter Konzentrationsverteilungen und ist damit Voraussetzung für Phasenanalysen [1]. Das Steuersignal (Ausgangssignal des Mittelwertmessers oder der Gleichstrommeßkanäle) wird einem Intensitätsdiskriminator zugeführt, der mehrere (in Abb. 2) über den ganzen Signalbereich einstellbare Schwellen, z.B. A und B, enthält. Die Schwellen schließen den ausgewählten Konzentrationsbereich ein. Das Ausgangssignal ist eine Ja-Nein-Aussage bezüglich des eingestellten Bereichs.

Dieses Signal wird mit einer periodischen Impulsfolge in der Weise verknüpft, daß nur während der Ja-Phase Oszillatorimpulse die Bildröhre aufhellen und damit das Phasenbild a erzeugen.

In ähnlicher Weise können die gesamten Oszillatorimpulse während eines Bilddurchlaufes und der vom Ja-Signal gesteuerte Teil dieser Impulse auf zwei Zähler gegeben werden.

Der Quotient der beiden aufgelaufenen Impulszahlen ergibt den prozentualen Flächenanteil b der Phase $A - B$.

Gibt man das Ausgangssignal auf eine entsprechende Steuereinheit c, dann läßt sich die Information auf Datenträger speichern. Man erhält so das Ja-Nein-Signal der Ausgangsspannung Zeile für Zeile gespeichert. Mit geeigneten Rechnerprogrammen lassen sich Flächenzuordnungen, Korngrößenverteilungen usw. berechnen.

Literatur

1. Dörfler, G., u. E. Plöckinger: Arch. Eisenhuettenw. 9 (1965).

Elektronenstrahl-Mikroanalyse mit der Elmisonde: Arbeitstechnik und Analysemöglichkeiten

U. Jecht und K. Tögel

Siemens AG, Karlsruhe, BRD

Auf dem Gebiet der Elektronenstrahl-Mikroanalyse sind die Grundlagen des Verfahrens und der Gerätetechnik längst zur Selbstverständlichkeit geworden. Im Mittelpunkt des Interesses steht heute vielmehr die schnelle und sichere Lösung der anfallenden Analysenprobleme. Dafür ist weniger der eigentliche Meßvorgang von Bedeutung — die reine Meßzeit ist vergleichbar gering —, sondern die schnelle und sichere Geräteeinstellung durch entsprechende Bedienungstechnik. Voraussetzung hierfür ist eine kompakte und leicht überschaubare Meßanlage (Abb. 1), bei der alle Bedienungselemente in Handreichweite liegen und bei der Fehlanalysen bei Einsatz nichtfachkundiger Kräfte durch einfache Bedienung weitgehend vermieden werden können.

Die Probenvorbereitung muß durch die Verwendung vielseitiger Probenträger auf das notwendige Mindestmaß beschränkt werden. Es stehen Probenhalterungen nicht nur für die in der Mineralogie und Metallurgie verwendeten Standardformen bis zu Probendurchmessern von 40 mm und für Objektträgerplatten der Mikroskopie, sondern auch für kleinere Handstücke mit unregelmäßigen Konturen zur Verfügung. Bei den großen Probendurchmessern ist eine Probendrehung sowie für Sonderprobleme eine Halterung für veränderlichen „take-off-Winkel" vorgesehen.

Eine schnelle Analyse fordert, daß die Höhenlage der Proben z.B. durch Anschlag gegen eine Fläche genau reproduzierbar ist. Nur so ist es möglich, bei Probenwechsel ohne zeitraubende

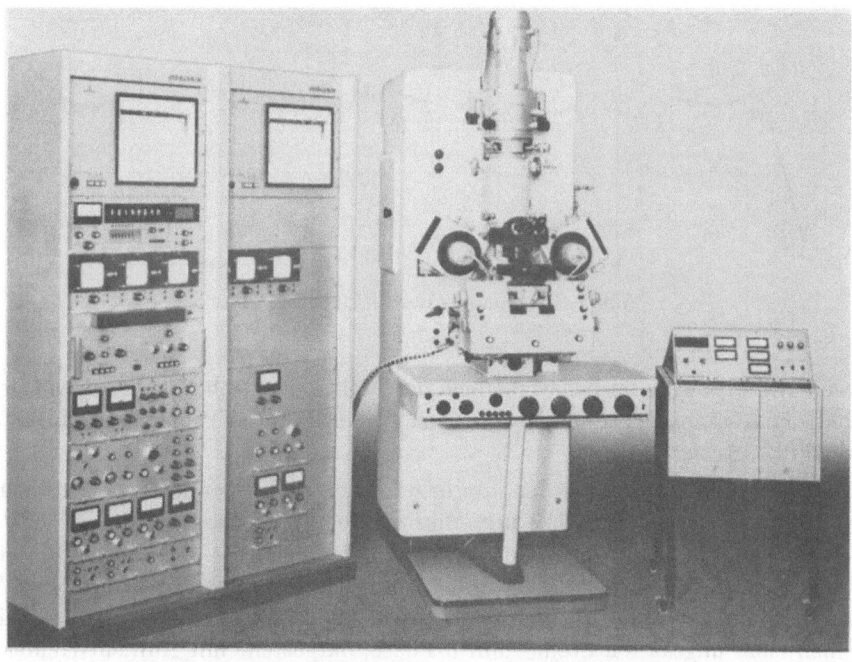

Abb. 1. Meßanlage für die Elektronenstrahl-Mikroanalyse

236   U. Jecht und K. Tögel:

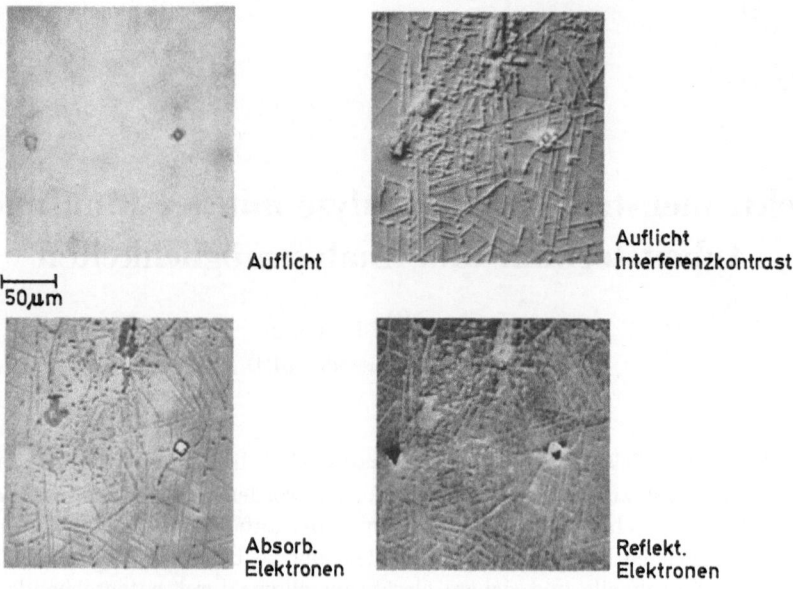

Abb. 2. Möglichkeiten der Probendarstellung

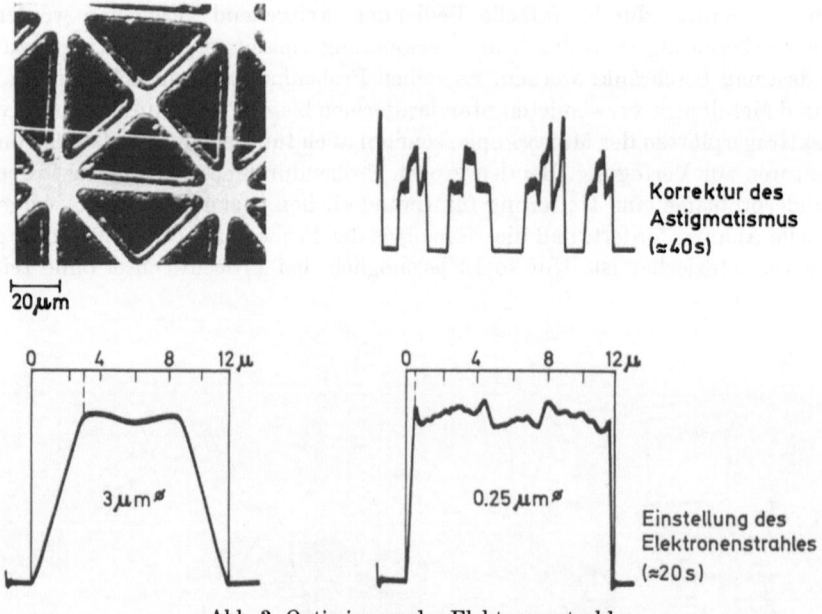

Abb. 3. Optimierung des Elektronenstrahles

Neujustierung sofort sowohl im Schärfebereich des Elektronenstrahles als auch auf dem Fokussierungskreis der Röntgenspektrometer zu liegen. Alle erwähnten Probenhalterungen sind nach diesem Prinzip konstruiert.

Zur Auswahl und Registrierung der interessierenden Probenbereiche sollten die bekannten Arbeitstechniken der Lichtmikroskopie und Mikrophotographie direkt an der „Sonde" zur Verfügung stehen. Das zur Darstellung von Strukturen erforderliche Anätzen der Probe führt bei der Elektronenstrahl-Mikroanalyse häufig zu Fehlanalysen. Das „Auflicht-Interferenz-Verfahren" nach Nomarski macht diese Strukturen auch ohne Ätzung gut erkennbar. Abb. 2 zeigt Mikrophotographien eines ungeätzten Stahles mit σ-Phase, dargestellt mit Auflicht-Hellfeld und mit dem „Auflicht-Interferenz-Verfahren", sowie Rasterbilder der absorbierten bzw. reflektierten

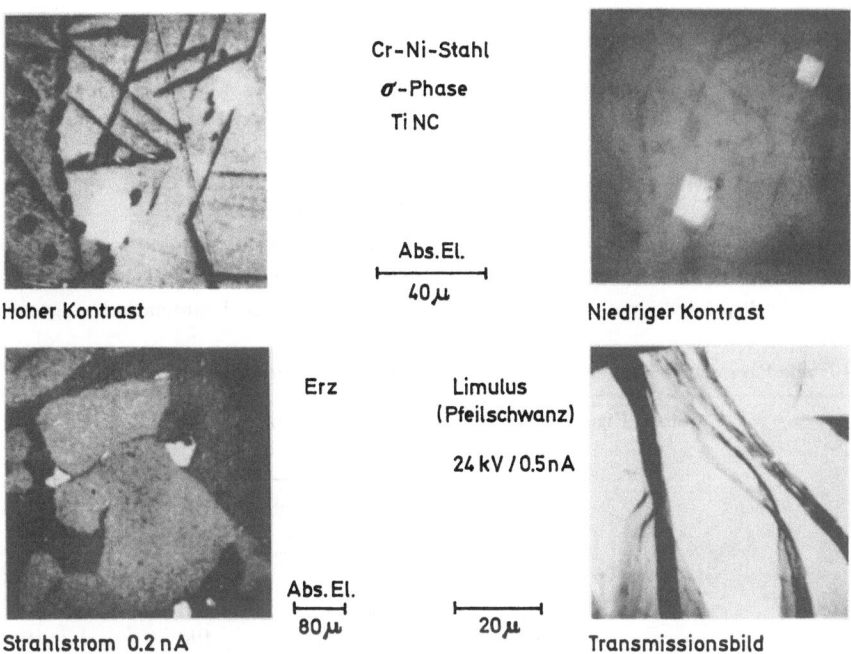

Cr–Ni–Stahl

σ-Phase

Ti NC

Abs.El.

40 μ

Hoher Kontrast

Niedriger Kontrast

Erz

Limulus
(Pfeilschwanz)

24 kV / 0.5 nA

Abs.El.

80 μ

20 μ

Strahlstrom 0.2 nA

Transmissionsbild

Abb. 4. Kontrasteinstellung; Transmissionstechnik

Elektronen. Die Orientierung auf den Präparaten wird wesentlich erleichtert, wenn mehrere Vergrößerungsstufen der Mikrophotographie mit denen der Rasterbilder übereinstimmen und wenn diese Vergrößerungen unabhängig von der gewählten Strahlspannung sind (Abb. 7).

Die Einstellung eines gewünschten Auflösungsvermögens durch entsprechend feinen und runden Elektronenstrahl ist für quantitative Messungen von Vorteil. Bei anderen Problemen wiederum muß eine hohe thermische Belastung der Probe durch zu feine Fokussierung unbedingt vermieden werden. Die Praxis erfordert also eine einfache, schnelle und genaue Astigmatismuskorrektur sowie die Möglichkeit, einen gewünschten Strahldurchmesser definiert einstellen zu können. Abb. 3 zeigt das Prinzip eines derartigen Verfahrens, bei dem die optimalen Einstellungen durch Beobachtung der Signalamplituden am Bildschirm in wenigen Sekunden vorgenommen werden können. Bei Linienabtastung eines speziellen Testgitters ist die gleichmäßige Steilheit der sechs Flanken ein Maß für eine gute Astigmatismuskorrektur. Andererseits kann unter hoher Vergrößerung bei bekannter Stegbreite der Durchmesser des Elektronenstrahles aus der Flankensteilheit direkt abgelesen werden.

Die Durchmusterung der Probe mit Hilfe der Elektronen-Rasterbilder dient in erster Linie der Orientierung des Elektronenstrahles für die Röntgenanalyse. Die Vielfalt der Analysenprobleme erfordert dazu die Möglichkeit, die Probenoberfläche wahlweise mit Hilfe der reflektierten, der absorbierten oder bei dünnen Präparaten auch der transmittierten Elektronen darzustellen. Bei empfindlichen Proben sind dabei kleinste Ströme unter 1 nA erforderlich. In Spezialfällen können auch Strahlspannungen bis 100 kV gewählt werden. Eine gute und genaue Orientierung ist jedoch nur möglich, wenn alle Probenbereiche unabhängig von den Ordnungszahlunterschieden mittels einer regelbaren Kontrasteinstellung in deutlich erkennbaren Grauabstufungen dargestellt werden können. Abb. 4 zeigt als Beispiel für hohen Kontrast die Darstellung der σ-Phase und für niedrigen Kontrast die Darstellung der Ti-Carbide im Stahl. Weiterhin wird ein Probenstrombild bei sehr niedrigem Probenstrom sowie das Transmissionsbild eines biologischen Präparates gezeigt.

Die Messung der charakteristischen Röntgenstrahlung bildet die Grundlage für die chemische Probenanalyse. Dem Anwender stehen dazu verschiedene Möglichkeiten zur Verfügung, angefangen von der intensitätsstarken, „nicht-dispersiven" Röntgenspektroskopie und -analyse über

Semifokussierendes Spektrometer	Vollfokussierendes Spektrometer
Robust; niedriger Preis; schnelle Handbedienung	Hohe Auflösung; großer Bedienungskomfort; Parameteranwahl elektrisch; programmierbar
Für häufig auftretende Elemente $Z \geqq 5$ 6 Kristalle + 2 Detektoren Winkelbereich bis 148°	Für gesamtes Spektrum $Z \geqq 5$ 4 Kristalle + 2 Detektoren Winkelbereich bis 148°
El. Rasterfeldgröße unkritisch	Fokussierungsbedingung automatisch gesteuert, daher el. Rasterfeld $\approx 800\ \mu m$
Det.-Stellung „nicht-dispersiv" für Kanaleinstellung und Analyse	Diskriminatorkanal automatisch eingestellt $1/b$ Verstärker
Einstellzeit für alle Parameter ≈ 45 sec	Einstellzeit für alle Parameter ≈ 35 sec

Abb. 5. Eigenschaften verschiedener Spektrometertypen

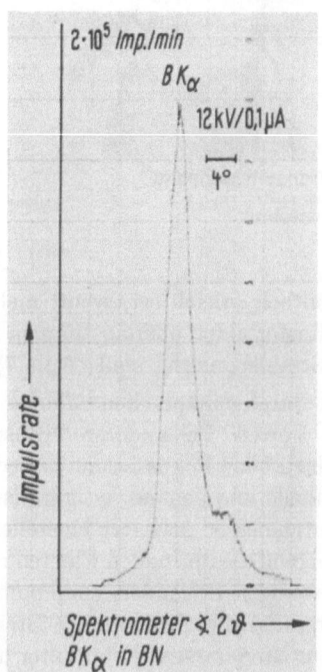

Linie	Probe	Strahl [kV]	Pr.-Strom [nA]	Intensität [I/sec]	P/B
BK_α	BN	12	1000	$2,5 \cdot 10^3$	75
CK_α	Graphit	12	100	$2,5 \cdot 10^3$	80
OK_α	SiO$_2$	12	100	$1,5 \cdot 10^2$	20
FK_α	LiF	12	100	$1,2 \cdot 10^3$	140
AlK_α	Al	24	100	$3 \cdot 10^4$	700
FeK_α	Fe	30	100	$5 \cdot 10^4$	366
CuK_α	Cu	30	100	$4,5 \cdot 10^4$	300

Abb. 6. Röntgenintensitäten, gemessen mit einem semifokussierenden Röntgenspektrometer

kompakte, schnell einstellbare semifokussierende Spektrometer bis zum automatisch steuerbaren vollfokussierenden Spektrometer mit höchstem Bedienungskomfort. Abb. 5 bringt einen Vergleich der wichtigsten Charakteristiken der beiden genannten Spektrometertypen.

Die Eigenschaften des semifokussierenden Spektrometers können nur dann sinnvoll für die Analyse genutzt werden, wenn auch bezüglich Intensität und Auflösungsvermögen befriedigende Ergebnisse erreicht werden. Abb. 6. gibt für einige Elemente die erreichbaren Intensitätswerte an.

Neben der Messung mit den Spektrometern sind weitere Techniken für die praktische Röntgenanalyse von Bedeutung. Abb. 7 gibt ein Beispiel für die schon erwähnte „nicht-dispersive" Analyse. Die Elemente Cr und Fe können durch zweckmäßige Einstellung des auf dem Impulsspektroskop gut sichtbaren Diskriminatorkanals noch getrennt werden, wie es die zugehörigen Röntgenbilder zeigen. Röntgenrasterbilder und Profilanalysen verschiedener Elemente in einer Probe können nur dann sicher miteinander verglichen werden, wenn der Abbildungsmaßstab der Elektronenabtastung bei Änderung der Strahlspannung erhalten bleibt (Abb. 7). Durch die Technik des „Ausschnittbildes" wird bei gleicher Belichtungszeit die Impulsdichte und damit die Konturenschärfe in den interessierenden Bereichen des Röntgenbildes wesentlich erhöht.

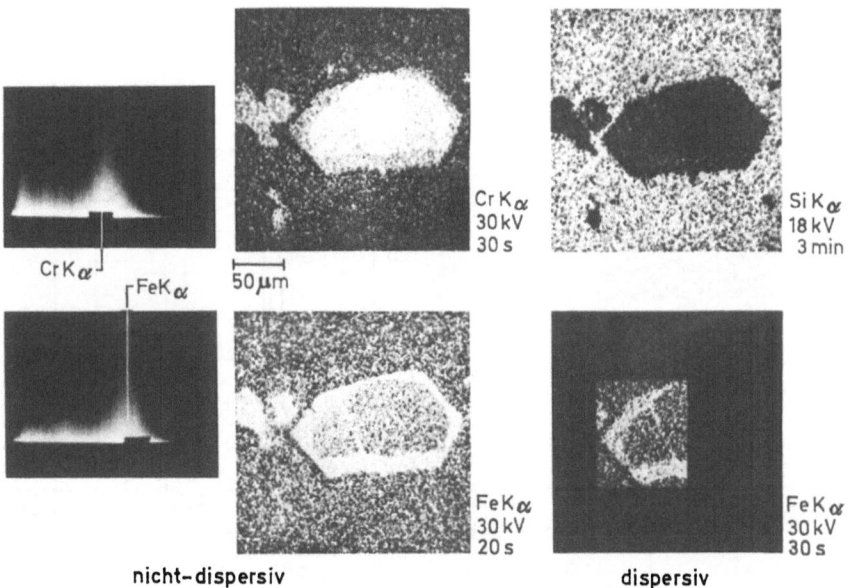

Abb. 7. Beispiel zur Röntgenspektralanalyse: Analyse eines Erzes

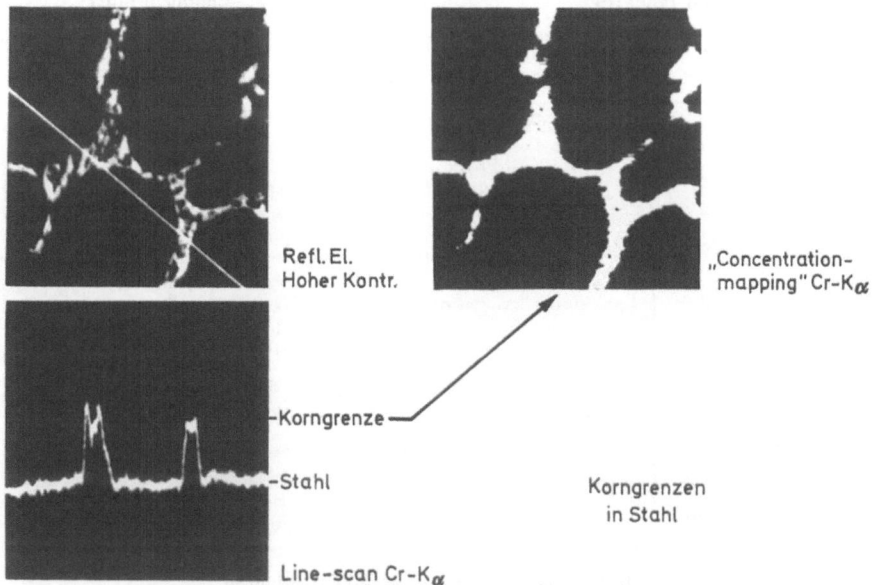

Abb. 8. Beispiel zur Röntgenspektralanalyse: Analyse von Korngrenzen in Stahl

Abb. 8 zeigt die Analyse von Korngrenzen in Stahl, und zwar das Elektronenbild in hohem Kontrast, darunter den Konzentrationsverlauf von Chrom über die angezeigte schräg verlaufende Bahn, sowie das Bild der Chromanreicherung, dargestellt mit Hilfe des Intensitätsdiskriminators (concentration mapping).

Zur sicheren Analyse der Elemente mit niedriger Ordnungszahl ist die Verminderung der Proben-Kontamination unerläßlich. Abb. 9 zeigt links die Wirksamkeit einer entsprechenden Vorrichtung am zeitlichen Verlauf der Kohlenstoff-Intensität, rechts daneben Spektrogramme der Linienverschiebung von $Si K_\beta$ in Silizium und Quarz auf Grund der chemischen Bindung (chemical shift), aufgenommen mit dem semifokussierenden Spektrometer.

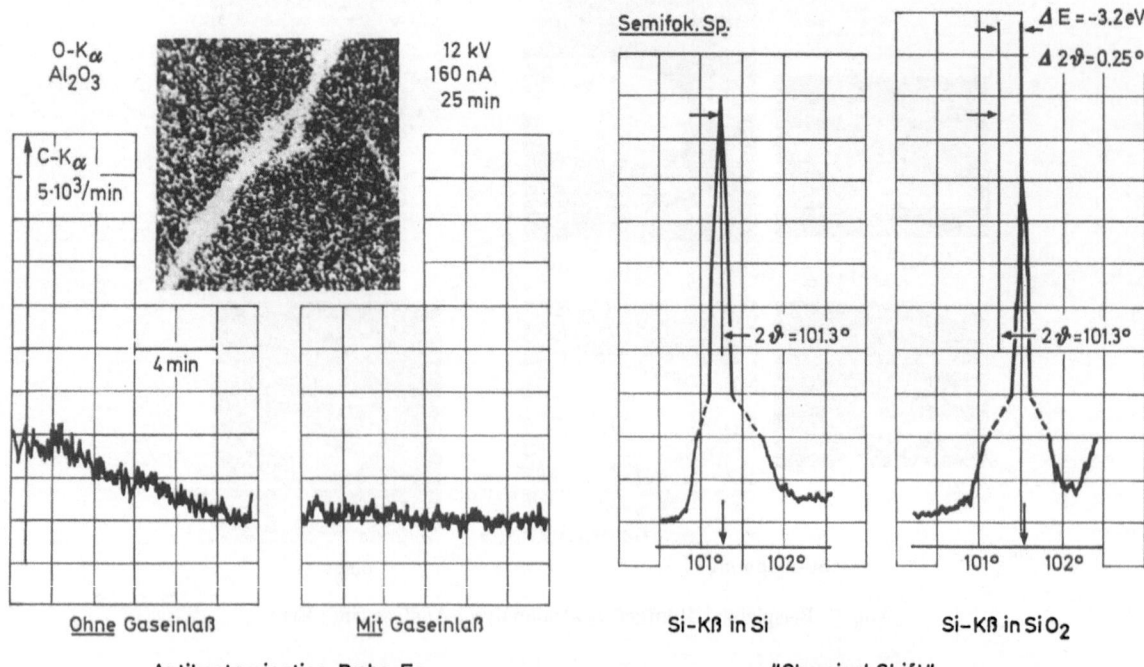

Abb. 9. Wirkung der Antikontaminationstechnik; Messung der „chemischen Verschiebung"

Gerät Information	Umrüstung ←	Elmisonde →	Zurüstung
	↓ Elektronenmikroskop 20 bis 120 kV	Mikroanalysator 4 bis 40 (100) kV	↓ Rastermikroskop 4 bis 20 (100) kV
„Auflösung"	$< 10\,\text{Å}$ (1 nm)	$\approx 10^4\,\text{Å}$ (1 m)	$> 100\,\text{Å}$ (10 nm)
Gestalt der Oberfläche	Lackabdruckverfahren	Rasterbilder Rückgestreute und Sekundärelektronen	Sekundärelektronen
Gestalt innerer Strukturen	Durchstrahlungs- mikroskopie	Rasterbilder in Transmission	
Kristallstruktur	Elektronenbeugung	Zusatzverfahren	
Chemische Zusammensetzung		Röntgen-Spektral- analyse	Röntgen-Spektroskopie, nichtdispersive Analyse

Abb. 10. Mikroanalytische Meßmöglichkeiten

Durch Vorwahl eines konstanten „Probenstrom-Integrals" können auch bei beliebigen Änderungen des Probenstromes während einer Messung oder bei zeitlich getrennten Messungen genau
reproduzierbare Röntgenintensitäten erhalten werden. Zur quantitativen Auswertung werden
alle Meßwerte übersichtlich registriert und zur Eingabe in ein Rechensystem, z.B. auf Loch-
streifen, gespeichert. Dabei wird neben den Impulszahlen der Röntgenkanäle auch das während
der Messung aufgelaufene Probenstrom-Integral als digitaler Meßwert mitgeführt.

Zum Gewinn weiterer Informationen ist es vorteilhaft, im gleichen Gerät ohne Umbau echte
hochauflösende Rasterelektronenmikroskopie oder — speziell in Forschungslaboratorien — nach
kurzer Umrüstungszeit Elektronenmikroskopie und Elektronenbeugung betreiben zu können.
Abb. 10 zeigt die Analysenmöglichkeiten bei Verwendung eines derartigen Universalgerätes.

A Microanalysis Attachment for the Elmiskop I

P. F. Chapman

Cavendish Laboratory, Cambridge, England

Abstract

The attachment consists of a modified airlock section, specimen stage and specimen holder. Normal microscope facilities are maintained, and an outlet provided for radiation generated within the specimen. A Le Poole lens incorporated in the airlock section allows the illuminating spot to be varied between 0.25 μ and 200 μ. This small illuminating spot has been used for both micro-diffraction studies, and microanalysis of normal transmission specimens.

Introduction

The microanalysis attachment consists of a replacement airlock section and specimen stage for the Siemens Elmiskop electron microscope. The essential feature of this attachment is that it provides an outlet for the radiation generated in the specimen without destroying any of the normal microscope facilities, and permits a small illuminating probe to be produced. This has been accomplished by using a simpler specimen changing device and a WARD [1] holder, leaving room for a LE POOLE [2] lens immediately above the specimen stage. The use of the inclined Ward holder means that the 13 mm objective polepiece has to be used.

As an electron microscope the attachment has a good performance. Specimens can be changed under vacuum and specimen movement is as normal. The inclined Ward holder accepts 3 mm grids, and provides tilt of ± 20° about two perpendicular, horizontal axes. The Le Poole lens contains its own stigmator and deflection coils and allows the electron beam to be focussed to

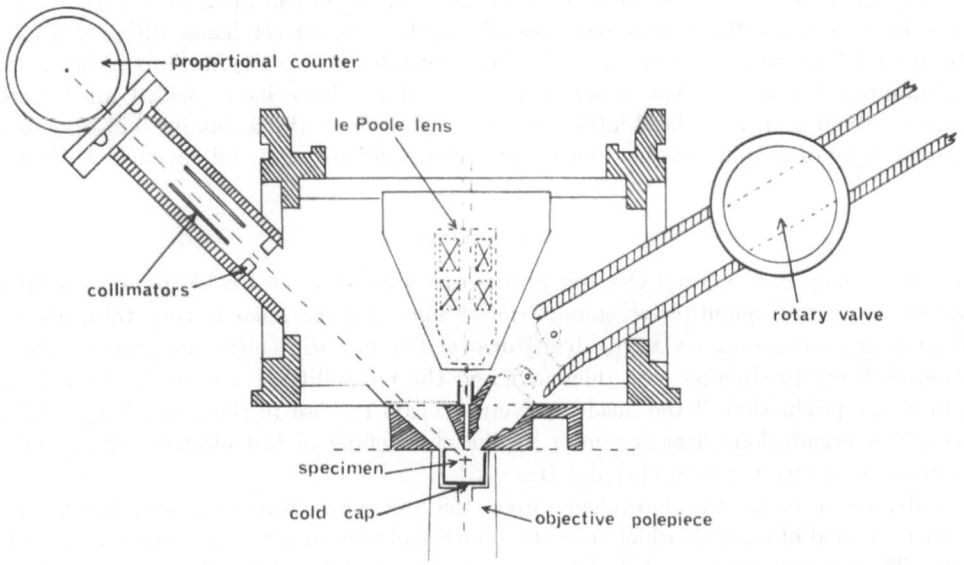

Fig. 1. Airlock section, showing the simple loading tube leaving room for the Le Poole lens

a probe 0.25 μ in diameter. The 13 mm polepiece causes a slight magnification of the diffraction pattern and has a quoted resolution of 15° A. The microscope is always used at 100 kV. The anti-contamination device is essential under focussed probe conditions, when it reduces the contamination rate from 100° A/sec to 3° A/sec.

Diffraction

In addition to allowing micro-analysis of small areas, the fine illuminating probe has been used to perform a series of accurate micro-diffraction experiments [3]. The inaccuracies of using a selecting aperture are avoided by selecting the areas with the focussed probe. Although the

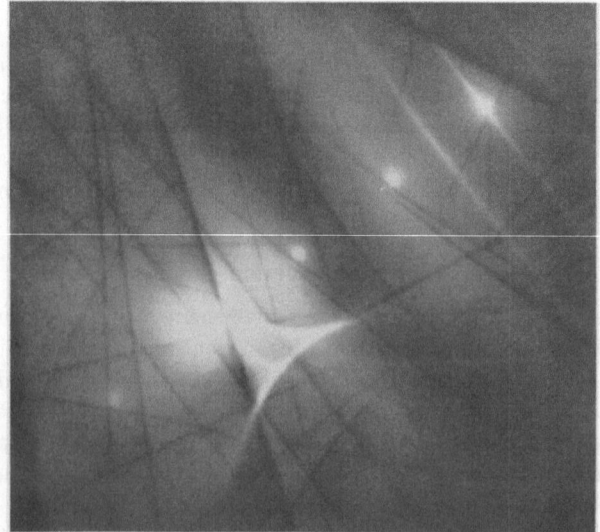

Fig. 2. Diffraction pattern showing the clarity of Kikuchi pattern

diffraction spots are large, due to the beam divergence, the Kikuchi patterns are very clear (see Fig. 2). This is because the Kikuchi patterns are only dependent upon the crystal perfection, and over the small area illuminated even strained crystals are perfect. These diffraction patterns can be used to determine relative and absolute crystallographic orientations to an accuracy of ± 2 minutes of a degree. The experiments referred to above have yielded much valuable information about highly localised lattice rotations. In general the technique will provide complementary crystallographic information to the electron micrograph and chemical analysis.

Microanalysis

The microanalysis of normal electron microscope specimens at 100 kV requires a different method of computing quantitative compositions. Since the specimen is very thin, absorption and fluorescence corrections are negligable. However two new difficulties are present; the linear variation of X-ray production with thickness, and the possibility of a variation by a factor of three in X-ray production, if the incident electron beam is close to the exact Bragg reflecting position. This orientation effect is caused by the channelling of fast electrons in crystals and has been examined by DUNCUMB [4] and HALL [5].

In order to overcome the dependence upon foil thickness, MARSHALL and HALL [6] have reported a method of analysis which uses the bremsstrahlung or white, radiation as a thickness corrector. The concentrations are deduced by experimentally determining the quantity Q, given by

$$Q = \frac{(\text{characteristic/white})_{\text{spec.}}}{(\text{characteristic/white})_{\text{standard}}}.$$

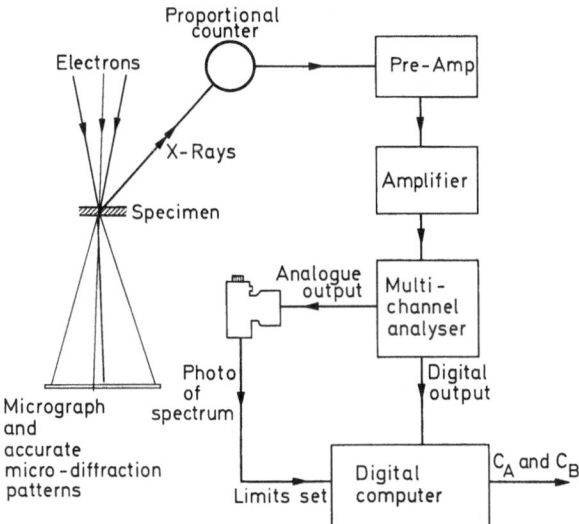

Fig. 3. The non-dispersive analysis method

However, since both the K shell radiation and the white radiations are excited by electrons passing close to the atomic nucleus, the variation with specimen orientation of the production of characteristic and white radiation will be somewhat similar, i.e. the ratio of characteristic/white will be independant of crystal orientation. The Marshall theory is an approximation (since it assumes a constant independent of Z in white radiation production), but has been experimentally tested [7] on non-crystalline specimens and found to have an accuracy of 8—10%.

Unless the probe forming system and beam current are highly stabilised, serious errors would arise in the ratios if the counts were measured sequentially. They are therefore measured simultaneously. Since there is only one radiation outlet available, a non-dispersive detector and multi-channel-analyser (MCA) are used to examine the X-rays. A single channel analyser can be used in conjunction with the deflection coils in the Le Poole lens to produce scanning X-ray images. Fig. 3 shows the quantitative analysis arrangement. The use of the multi-channel analyser has some advantages over other non-dispersive methods. Firstly the very quick method of examining the entire spectrum to decide the elements present and times required for an accurate analysis. Secondly it is applicable to systems containing any number of elements. Lastly it has the maximum possible efficiency, using every X-ray detected in the counter to compile the spectrum. The collection efficiency of the MCA is very important since with an irradiated volume of material of 10^{-13} cm³ the total X-ray count may be only 300 c/s. Another consequence of the low count rate is that to avoid pulse-height depression in the counter and dead-time corrections in the MCA, the total count rate for all spectra must be the same. Thus only the counter EHT and the total count rate have to be kept constant between recording different spectra for the analysis. This means that specimens and standards can be loaded into the microscope separately.

The MCA has an analogue output on a CRT and a digital output which is fed to a tape punch. After recording the specimen and standard spectra the tapes are fed into a digital computer, and the specimen spectrum is synthesised from the standard spectra. A weighted least squares method of fitting the spectra is used, the goodness of fit being an indication of the accuracy of the synthesis. A simple provision in the programme detects any systematic deviations between the synthesised and standard spectra. Since the MCA separates adjacent elements (about 5 channels in 150 channel mode), it is hoped that this last provision will detect the presence of elements present in the specimen but not provided in the synthesis.

This method of analysis, MCA, spectrum synthesis and the Marshall theory is expected to give an accuracy of about 10%. Fig. 4 is an example of the spectra obtained from small precipitates. The precipitate is Fe_2Ti, the third peak is a Cu peak and arises from the copper anti-

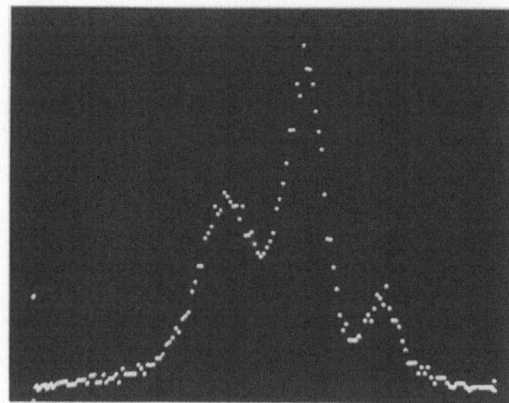

Fig. 4. Spectrum from Fe$_2$Ti precipitate, 0.5 μ in diam. and approx. 1,000° A thick

contamination cap in the forward scattering area. Coating this cap with Be will probably remove the copper peak (see Duncumb and Cooke [8]), it remains to be seen whether the bremsstrahlung caused by stray scattering causes inaccuracies in using the theory outlined above.

Conclusion

This attachment provides more information about the specimens normally studied in transmission electron microscopes. The additional modes of operation are:

1. Accurate selected area diffraction: This can be used to determine relative and absolute orientations to ± 2 mins°. of very small areas. It will also permit diffraction patterns to be obtained from smaller precipitates than with conventional techniques.

2. Scanning X-ray images. These will be valuable, in both biological and metallurgical systems, when correlated with the transmission micrographs.

3. Quantitative analysis. This will be possible along line scans, providing direct information on segregation effects, and also from areas down to 2,500° A in diameter.

The author is indebted to Dr. A. Howie for many useful discussions.

References

1. Ward, P. R.: J. Sci. Instr. **42**, 767 (1965).
2. Le Poole, J. B.: 3rd Eur. Conf. El. Mic., p. 329.
3. Chapman, P. F., and W. M. Stobbs: In publication.
4. Duncumb, P.: Phil. Mag. 7, 2101 (1962).
5. Hall, C. R.: Proc. Roy. Soc. (London), Ser. A **295**, 140 (1966).
6. Marshall, D. J., and T. A. Hall: IV. Congr. X-Ray and Micr., p. 374.
7. — Thesis. Cambridge.
8. Cooke, C. J., and P. Duncumb: IV. Int. Congr. X-Ray and Micr., p. 467.

Performance Analysis of a Combined Electron Microscope and Electron Probe Microanalyser "EMMA"

C. J. Cooke and P. Duncumb

Tube Investments Research Laboratories, Cambridge, England

Introduction

An instrument was constructed about six years ago as an electron-optical bench for studying the combination of a variety of electron-optical research techniques, principally transmission microscopy and microanalysis, in one instrument. The objects of the present paper are to examine the applications found for the instrument, to relate the scope of them to its design specification and to show how it is hoped to improve the performance in a new design.

The Present Instrument

Design features. The instrument has been described previously in the literature [1, 2] so that only a brief outline will be given here. Fig. 1 is a column cross-section showing the two-stage probe forming system and the three-stage electron microscope on either side of a large specimen chamber. This chamber is octagonal in plan so that there are sufficient radial faces around the specimen to mount two focussing crystal spectrometers, a non-dispersive X-ray channel consisting of a proportional counter and collimator, a scanning electron microscope type of secondary electron detector and the specimen stage itself. The kilovoltage range is 5—60 and the fundamental resolution limit in transmission is about 30 Å, set by the long focal length of the microscope objective lens necessitated by the specimen being outside the pole pieces. The probe forming system is capable of focussing the probe down to less than 1,000 Å in diameter. X-ray take-off is at 20° and the spectrometers are designed for high collection efficiency [2] rather than high resolution.

Applications. The applications divide broadly into two types categorised by location of areas for microanalysis by transmission or by scanning electron microscopy. On solid samples, operating the instrument as a conventional microanalyser, the high spectrometer sensitivity and small probe capability have proved very useful in the examination, for example, of sub-micron particles on the surface of polished metal samples. Scanning image formation with a sensitive secondary electron detector, as opposed to a primary backscattered detector, has enabled particles on very rough surfaces, such as fracture faces, to be located for qualitative microanalysis. The relative insensitivity of the spectrometers to specimen height variations, consequent upon the use of curved but not ground crystals subtending a large solid angle (0.02 steradians maximum) at the specimen, also contributes to this sort of investigation.

With electron-transparent specimens, the ease of selection and location under the probe of areas for analysis using a transmission image is a considerable advantage. This is especially true of biological specimens, which are already known by their thin section morphology in the electron microscope [3]. With carbon extraction replicas, the possibility of viewing small particles under a stationary probe during an analysis, taken together with the low X-ray background, has enabled chemical analysis of particles below 1,000 Å in size to be carried out, [4].

Two spectrometers have been found essential for accurate ratio measurements on small particles; the interpretation of measurements on thin specimens has already been discussed [4, 5]. Quantitative analysis employing a non-dispersive system only was found to be compli-

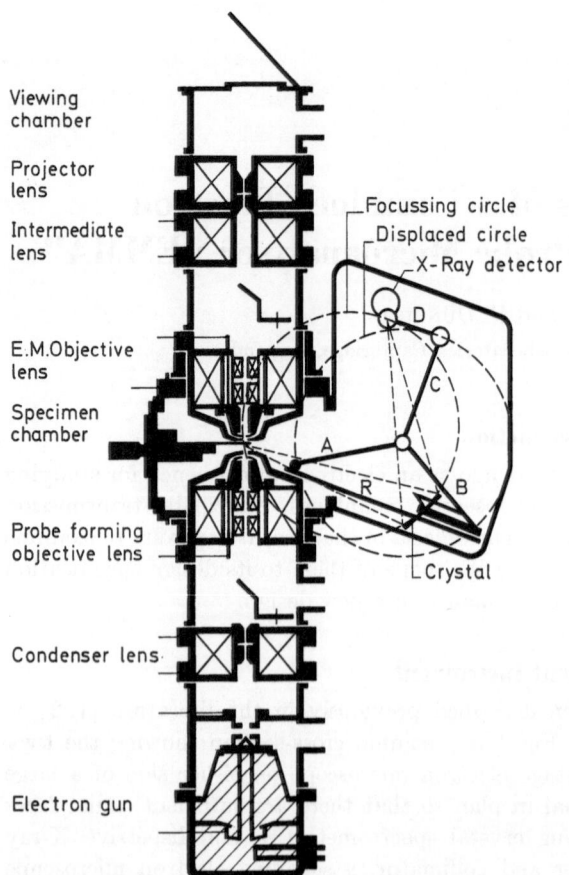

Fig. 1. Column cross section of the present instrument

cated by the proportional counters' inability to discriminate against X-rays generated by scattered electrons at metal surfaces near the specimen [6]. On the other hand, non-dispersive analysis gives a rapid qualitative guide to the major constituents of an unknown specimen.

On some specimens it has proved possible to get chemical and crystallographic information, the latter by electron diffraction, from the same particle [2]. More generally, on account of the 60 kV upper limit on accelerating voltage, diffraction patterns may only be obtained from particles to othin to analyse. The advantage of combining microscopy and microanalysis then rests on the operator being able to select at will the technique most suitable to a particular area under investigation. The application to metal foil specimens is severely restricted by the 60 kV limit, due to the difficulty of preparing sufficiently large areas of thin enough material.

Selected area diffraction can be carried out by focussing the probe on the appropriate area, the accuracy of selection then being as good as 0.1 μ — an order of magnitude lower than in normal electron microscopy [7]. The resolution in the pattern using this technique is, however, relatively poor due to the convergent beam conditions; its use in conjunction with the more conventional method, employing a selection aperture at the intermediate lens, would enable diffraction spots to be reliably assigned to the small area selected by the focussed probe.

The New Instrument

In order to overcome the limited beam energy and image resolution of the electron microscope in the present design, a new instrument is under construction. Fig. 2 shows a cross-section of specimen region. The specimen is placed inside a conventional type of electron microscope objective lens with X-ray take-off at 45° to the specimen surface. A "Le Poole" minilens [8] is employed as the probe forming lens. The microscope objective has a resolution parameter in transmission of better than 10 Å and the whole of the electron optical column, minilens included, is designed to operate up to 100 kV.

An X-ray take-off angle of 45° implies that with the same Rowland circle radius [6″] as in the present instrument, the locus of the Rowland circle centre is wholly outside the column for a Bragg angle range of 15—75°, thus permitting the use of a simpler mechanical linkage in the spectrometer than at present [2]. The counter arm drive, although developed independently, is of the form employed by Hailes [9]. The general performance is expected to be similar to that of the present spectrometer's except that, since four crystals interchangeable under vacuum are provided, a single spectrometer will cover the wavelength range 1—100 Å.

Scanning facilities, other than a fine manual adjustment of probe position are not required, as it is not envisaged that the machine will be used for solid specimens.

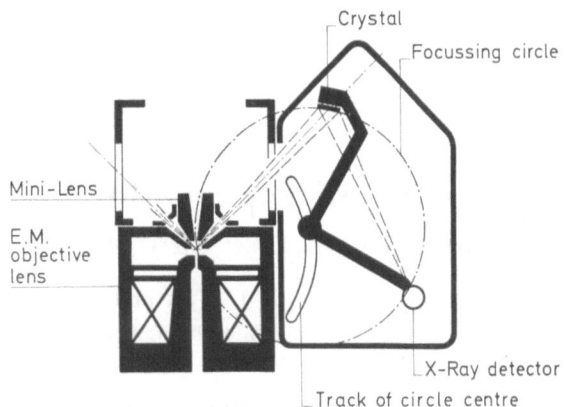

Fig. 2. Cross section of the specimen region of the new instrument

Conclusions

The performance of the present instrument demonstrates the extension of microanalysis to sub-micron particles on extraction replica specimens. This is made possible by the use of high efficiency X-ray spectrometers and by the availability of the transmission image to locate and maintain the 0.1 μm probe on individual particles. The microanalysis of biological sections is a second area of application which requires these facilities.

The main disadvantage is the limited kilovoltage range (maximum 60 kV), which rules out the examination of most metal foil specimens and severely handicaps the performance in electron diffraction. To overcome this, a new instrument exploits the possibility of integrating a mini-lens and high efficiency spectrometers into a conventional 100 kV electron microscope (for example, a G.E.C.-A.E.I., EM 8). Since the specimen is immersed in the objective lens, it is not practicable to analyse large solid specimens, but this disadvantage is out-weighed by the benefits of retaining the full performance of a high resolution electron microscope.

Acknowledgments. The authors would like to express their appreciation to Dr. L. A. FONTIJN of the Technical University, Delft, Holland, for information on mini-lenses and to staff at the G.E.C.-A.E.I. Laboratories, Harlow, England, for their help with the detailed conception of the new instrument. They are indebted to the Chairman of Tube Investments Limited for permission to publish this paper.

References

1. DUNCUMB, P.: Fifth Internat. Conf. on Electron Microscopy, Paper KK4. New York: Academic Press 1962.
2. — The electron microprobe, p. 490. New York: Wiley 1966.
3. HÖHLING, H. J.: Naturwissenschaften 6, 142—143 (1967).
4. DUNCUMB, P.: J. Microscopie (in press).
5. PHILIBERT, J., and R. TIXIER: Brit. J. Appl. Phys., Ser. II, 1, 6, 685 (1968).
6. COOKE, C. J., and P. DUNCUMB: Proc. Conf. on X-ray Optics and microanalysis, p. 467. Paris: Hermann 1967.
7. RIECKE, W. D.: Optik 5, 273 (1962).
8. LE POOLE, J. B.: Proc. 3rd Eur. Conf. on Electron Microscopy, Publ. House Czech. Ac. Sc. Prague, vol. A (1964).
9. HAILES, L.: This conference.

A Shielded X-Ray Microprobe for the Analysis of Radioactive Samples

V. G. Macres, O. Preston, N. C. Yew and R. Buchanan

Materials Analysis Company, Palo Alto, California, U.S.A.

Introduction

In the atomic energy field, the application of electron microprobe analysis is necessary in the development of materials for use in the nuclear reactor environment. However, the radioactive and/or toxic character of materials subjected to such an environment presents a unique problem to the microprobe instrument user. Until recently the only solution available has been to operate with very small specimens; however, the instrument described here changes this situation.

Instrument Description

A cut-away drawing of the instrument with the optionally available specimen transfer cask attached is shown in Fig. 1. A single fully focusing spectrometer is shown, however up to three can be used on the instrument, each with a 2θ coverage of 40° to 140°. With the complement of available crystals and detectors this allows the analysis of all elements from boron to uranium.

One of the most important aspects of the instrument is the manner in which the shielding is accommodated for contending with the penetrating gamma radiation of radioactive specimens. The main tungsten base shielding components are the column base and an intermediate chamber shielding ring. In addition, a tubular shield, which fits over the gun cable at the top of the column, is provided. The cone of radiation which escapes from the back of the spectrometer is blocked by Kennertium plugs which are held in the shielding wall located at the rear of the spectrometer. During operation, shielding equivalent to 4.3 inches of tungsten in all directions from the specimen is provided.

Specimens are introduced into the instrument by means of a specimen transfer slug which incorporates the specimen stage. A single specimen with dimensions of $1-1/4''$ diameter by $1''$ high can be accommodated. In addition, a removable ring segment is included which accommodates up to fifteen, single and multi-element standards and fluorescent target specimen. The simplicity of specimen loading is such that this operation can easily be carried out remotely using manipulators if necessary. The transfer slug includes the necessary shielding materials consistent with the total shielding provision of the column and the specimen cask.

The stage unit is mechanically connected by means of anti-backlash couplings to an external manipulator unit mounted on the front side of the column base. All specimen motions including focusing, translation and rotation can be carried out manually during instrument operation. Motor-driven specimen traverse motion is also included.

The traverse control provides for a total translation motion of $7/8''$. This control in combination with the 360° rotation control provides for complete coverage of any point within a $1-1/2''$ diameter circle. A specimen height adjustment which allows the specimen to be raised or lowered $0.02''$ is provided for focusing.

Results

Initial results of measurements performed at G. E. Vallecitos by Dr. Rosenbaum and Mr. W. Cummings on the first production Model 450 indicate that the maximum instrument contact dose rate with a 9 curie Co^{60} specimen is 60 mr/hr at a point below the first condenser

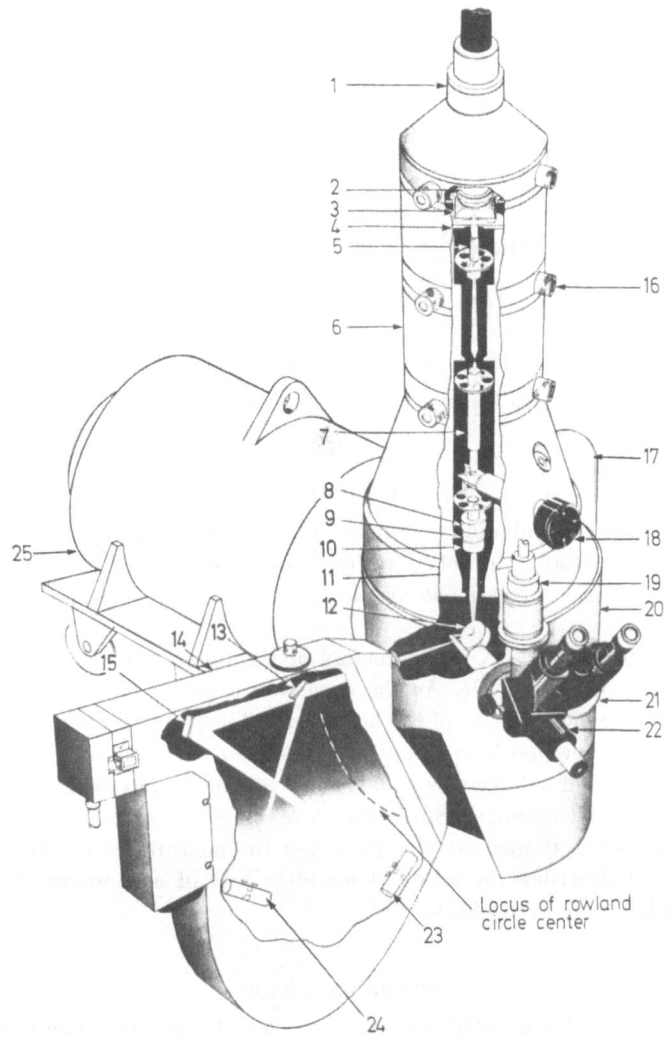

Fig. 1. Model 450 schematic. *1* Cathode *Z* axis transiator, *2* cathode assembly, *3* grid cup, *4* anode plate, *5* upper magnetic beam shield, *6* auxilliary lens, *7* lower magnetic beam shield, *8* beam centering coils, *9* stigmator, *10* beam scanning deflection assembly, *11* objective lens, *12* specimen stage, *13* analyzing crystal shown at intermediate 2 theta angle, *14* X-ray spectrometer (one of the three shown), *15* analyzing crystal shown at high 2 theta angle, *16* column alignment transiators, *17* vacuum connector, *18* aperture selector, *19* illuminator, *20* column base, *21* specimen stage manipulator, *22* microscope, *23* proportional counter shown at high 2 theta angle, *24* proportional counter shown at intermediate 2 theta angle, *25* specimen transfer cask

lens, the average readings being about 20 mr/hr over the whole instrument. A reading of 140 mr/hr was recorded at the junction of the column and cask during transfer of the specimen from the cask into the instrument.

Results of peak to background measurements made on a stainless steel specimen in the presence of a 9 curie Co^{60} radiation source indicate that the second order K_α peaks of Cr, Fe, and Ni are little different to those measured in the absence of a radiation source. The crystal used for these measurements was PET. These results are summarized in the table.

Table. *Comparison of peak to background ratios of elements in a stainless steel specimen measured in the Model 400 microprobe, and measured in the Model 450 in the presence of a 9 curie Co^{60} radiation source*

Line (second order)	Peak/background ratio	
	Model 450	Model 400
Cr K_α	$41.6/_1$	$57.5/_1$
Fe K_α	$73.1/_1$	$80.6/_1$
Ni K_α	$11.4/_1$	$13.7/_1$

Analysis of Radioactive Materials
by Means of an Electron Microprobe
Demonstrated on UO₂—Mo and UO₂-Zircalloy Fuel Cermets*

G. Giacchetti[1] and J. Ränsch

Euratom, European Institute for Transuranium Elements, Karlsruhe, BRD

Introduction

Microprobe analysis is a valuable complement to conventional methods for the analysis of irradiated fuel. It is especially suitable for the study of compatibility between fuel and cladding, diffusion and corrosion phenomena and to investigate the build-up of fission products and their distribution.

Difficulties arise because the X-ray detectors of the microprobe are strongly influenced by the ionizing radiation of the sample. As the dose rate depends on the volume of the sample, it would be feasible to reduce the size of the sample until its residual activity had become negligibly small. However, the microprobe is meant to supplement metallography; hence the equipment must be capable of handling all samples produced in metallography and do so without any subsequent metallographical treatment. Since the samples generally occurring have a maximum radioactivity around 10 Ci, it was our aim to design the microprobe to this activity.

The present paper describes the adopted modifications of a commercial microanalyser and the first results which have been obtained.

Preliminary Tests

For a general study of the problems we first carried out experiments with an available CAMECA microprobe[2]. For this purpose, samples of different but low activities were introduced into the unchanged equipment to assess the kind and intensity of disturbing effects as a function of energy and intensity of the gamma radiation.

After initial experiments with ^{60}Co sources we used an irradiated UO₂—CeO₂ cermet with a burn up of about 3,000 Mwd/ton and a cooling time of six months. The gamma dose was 100 mR/h at 30 cm distance, determined by means of a Jordan meter shielded with 20 mm of plexiglass. The gamma spectrum of this sample recorded with an NaI detector extended up to 1.5 MeV.

The radiation measured by the detectors of the microprobe, which we will call background, was measured first with two spectrometer settings, i.e. at Bragg angles of $\theta = 10°$ and 54°, respectively, which corresponds to the beginning and the end of the spectrometer ranges. The detectors used were the available proportional counters two of which are closed and two of which are designed as ArCH₄ flow counters.

It turned out that the pulse spectrum registered by the detectors showed a distinct peak, independent of the spectrometer setting, which was influenced by the sample detector distance, in intensity only, but not in position. The spectra furnished by the four detectors were recorded point by point by means of a single channel discriminator (Fig. 1) and afterwards compared

---

* This work has been performed within the framework of the association between the European Atomic Community and the Gesellschaft für Kernforschung mbH., Karlsruhe, on Fast Breeder Reactors.

1. Euratom delegated to the Fast Breeder Reactor project.
2. CAMECA, 103 Bd. St.Denis, Courbevoie (France).

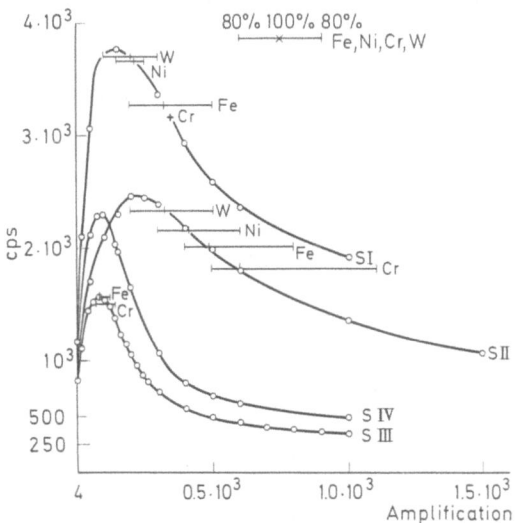

Fig. 1. Fluorescence spectra furnished by the microprobe counters

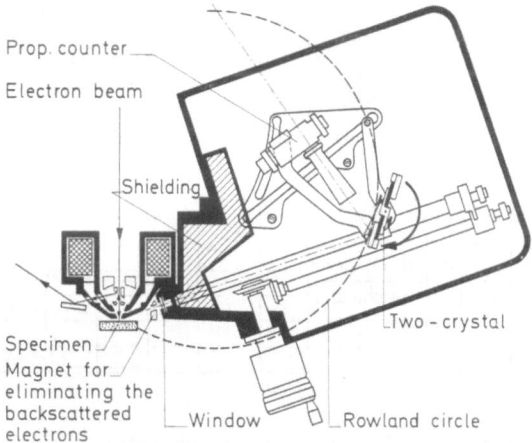

Fig. 2. Arrangement of the shielded spectrometer

with the line peaks of various materials. The peaks of the Cr-, Ni-, Fe-K_α lines all were in the range of the background peak. The fact that the counting tube material — 18/10 grade steel — consisted of these components permitted the conclusion that these were fluorescence effects of the walls of the counting tube due to gamma radiation. There are various possibilities of reducing this gamma radiation. As, for reason of weight, direct shielding of the counter housing is only possible to a limited extent, and since the double reflexion technique applied in fine structure systems is impossible because of the low X-ray intensities, the arrangement of a very strong shielding between the radioactive sample and the spectrometer was deemed to be the most effective measure.

Therefore we decided in favour of this solution and installed a lead shield at first in the right hand one of the opposed spectrometer chambers (Fig. 2).

Modification of the Equipment

Encapsulation of the sample in a shielding material was not used on purpose, because of the high fluorescence. The spectrometer shielding has replaceable collimators which are adapted to the X-ray optic and the angles of aperture of these collimators are arranged to suit the respective crystals (Table 1).

Table 1. *Wave length range in the first order (shielded microprobe)*

	Crystal						
	LiF	SiO₂	PET	ADP	KAP	LMª	LSDᵇ
Plane	100	10Ī1					
Grating space $2d$ [Å]	4.026	6.686	8.75	10.648	26.6	80	100.5
$K = R/d$ ᶜ	124.19	74.78	57.143	46.96	18.8	6	5
Spectrometer	I	I	II and III	II	III	IV	IV
λ-min. [Å]	1.29	2.14	2.8	3.41	8.5	26.7	32
λ-max. [Å]	3.22	5.35	7.0	8.53	21.3	66.5	80

ª LM = lead myristat.

ᵇ LSD = lead stearate decanoate.

ᶜ Sample-crystal distance $= L\,[\mathrm{mm}] = K \cdot \lambda\,[\mathrm{Å}]$, with $K = \dfrac{R\,[\mathrm{mm}]}{d\,[\mathrm{Å}]}$.

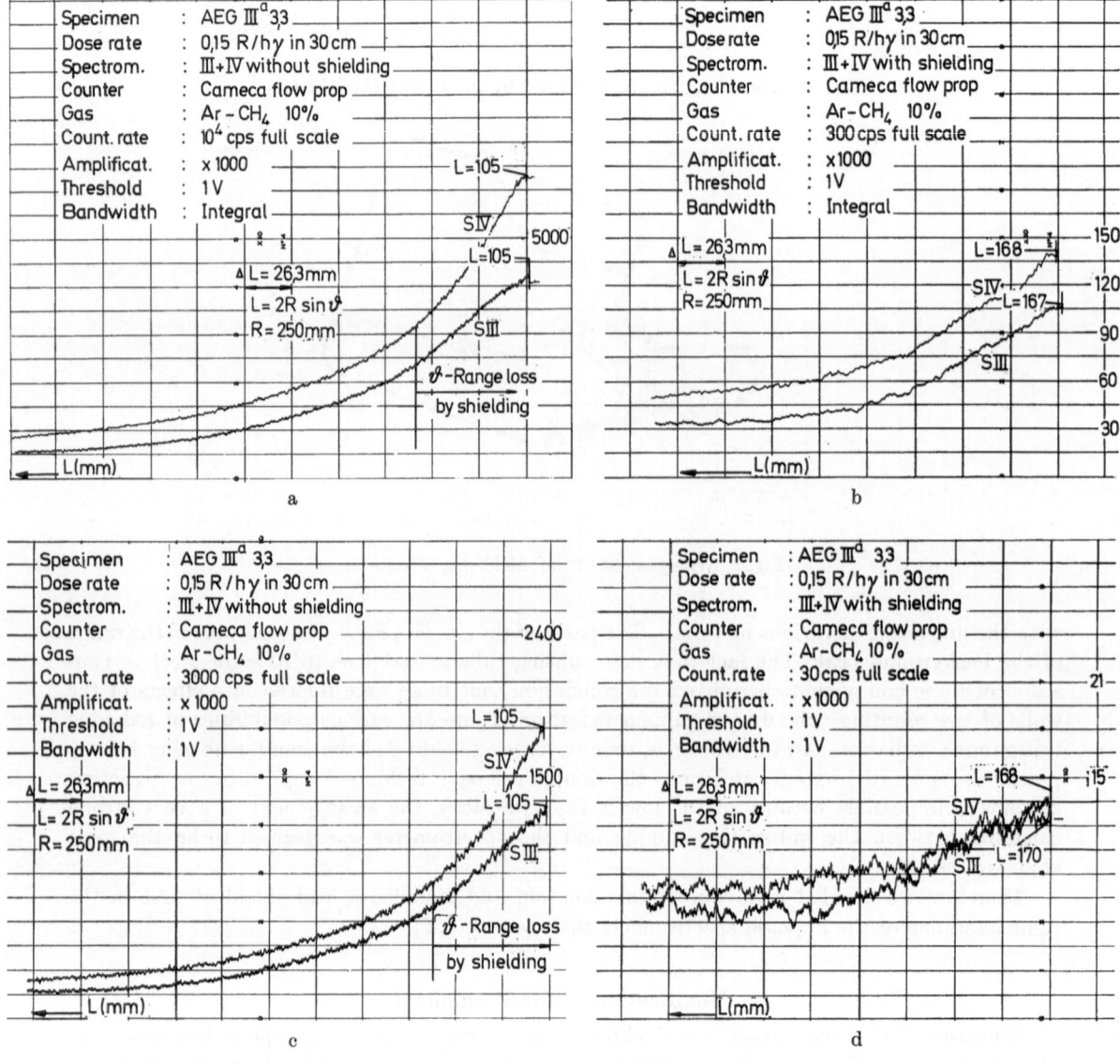

Fig. 3a—d. Spectrometer scanning: influence of the shielding and of the pulse-hight discrimination on the background. a Without shielding, integral. b With shielding, integral. c Without shielding, with discrimination. d With shielding, with discrimination

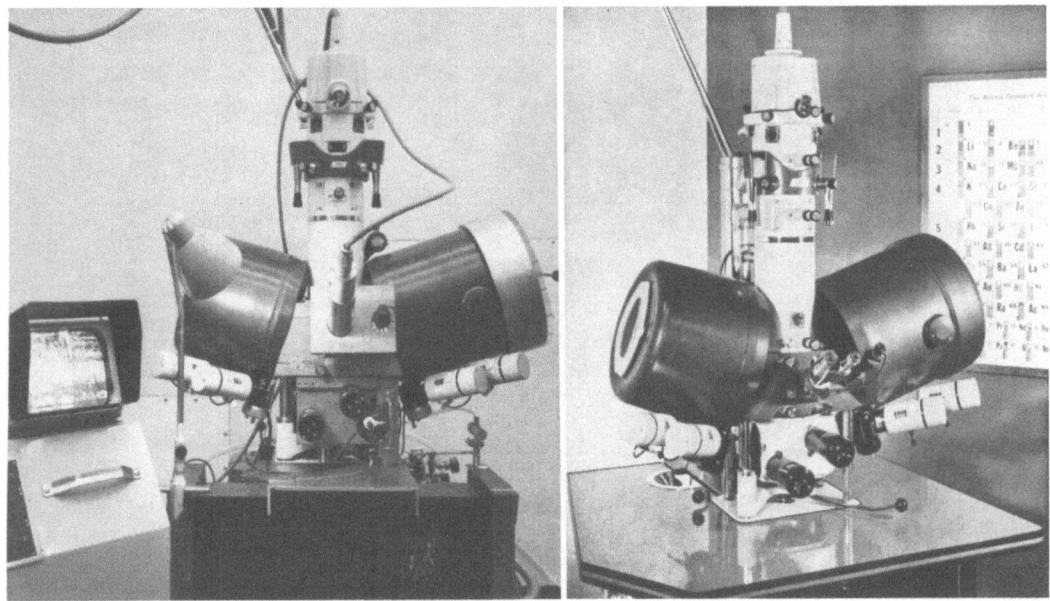

Fig. 4. The microprobe after and before the modification

In this way the X-radiation intensity is certainly not reduced while an optimum shielding effect is obtained.

Depending upon the position of the spectrometer we attain effective shielding thicknesses between 50 and 150 mm. The attenuation of the background is evident from the diagrams (Fig. 3) which we recorded with a sample that had a gamma dose of 150 mR/h at 30 cm distance. The direct influence of the shielding will be clear from the comparison of the two integrally recorded diagrams a) and b). Electronic discrimination further reduced this integral background, as is seen from diagram c). We see that the attenuation factor of the shielding only is about 25, while that of the electronics is not more than an additional factor of 10 with the threshold at 1 V and a bandwidth of 1 V and an amplification of 1,000.

Using an electronic readout from the Berthold Company[3], we have been able to significantly improve this result under rather more rigid conditions. A linear scanning (Fig. 10) was performed on a sample thirty times stronger, i.e. with a dose of 5,000 mR/h at a distance of 30 cm. Integrally and without shielding the interference level would reach some 40,000 pulses/sec. However, here the residual level is 4 pulses/sec, which corresponds to an attenuation factor of about 10,000 times. We backed the counter wall by 3 mm of Pb sheet on the side facing the sample. Thus, an attenuation of 70 is obtained by the direct shielding, and an attenuation of 140 is obtained through the electronic system.

However, even this result can be improved still further by mechanical and electronic measures (i.e. use of tungsten shield, development of a new counter, etc.).

To keep the weight of the equipment within reasonable limits and for easy handling we designed the biological shielding of the equipment (Fig. 4), so that samples up to a dose of 200 mR/h at 30 cm can be investigated without difficulty. For higher gamma dose rates of up to 100,000 mR/h at 30 cm a lead box with removable walls of 10 cm thickness was built into which the microprobe could be introduced (Fig. 5).

For focusing the electron beam and observing the sample in "hot operation" of the probe, a TV camera (875 lines) is used instead of the microscope eyepiece.

The sample is transported from the hot cells to the microprobe in a mobile lead container equipped with an electrically controlled telescopic lifting device.

---

3. Lab. Prof. Dr. BERTHOLD, 7547 Wildbad (West Germany).

Fig. 5. Lead box

The samples are taken from the sample chamber of the container into that of the probe by a simple direct grip; the observation into the lead-box is made through a lead glass window and by a second TV camera.

Since the hazard of contamination by radioactive particles, which may be released by the sample during transport and investigation, is also high all samples were decontaminated. They were cleared for analysis only after the last five swab tests of a series of consecutive swab tests showed loose contamination of less than 20 cpm for alpha and 1,000 cpm for beta radiation.

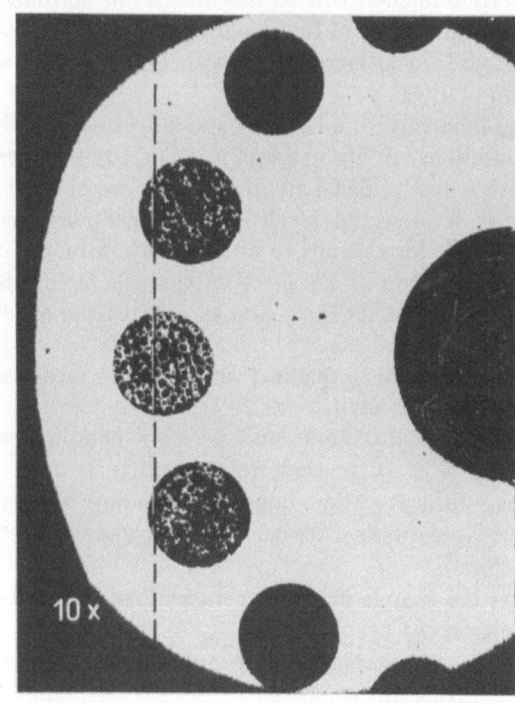
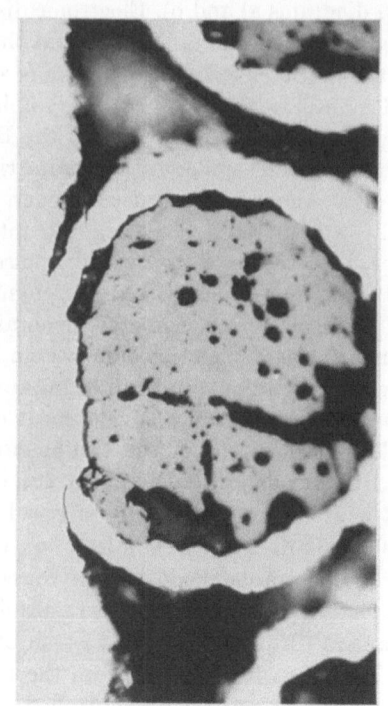

Fig. 6a and b. Emitter. a Cross section ($\times 10$ reduced to $^3/_4$ of its original size). b Analysed UO_2 coated particle ($\times 500$ reduced to $^3/_4$ of its original size)

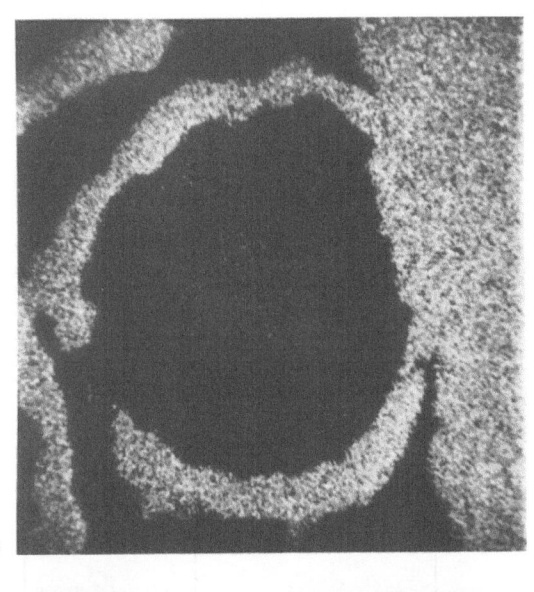

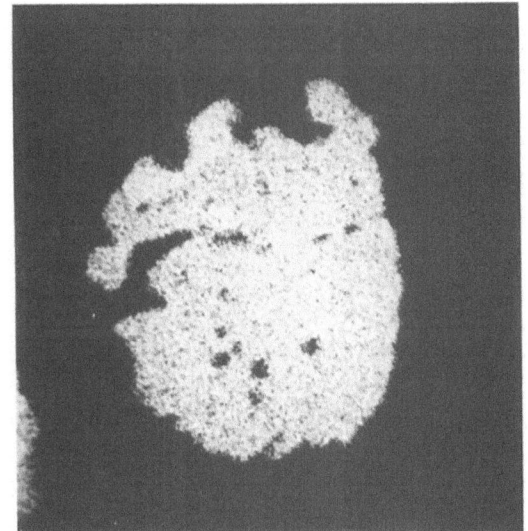

a   b

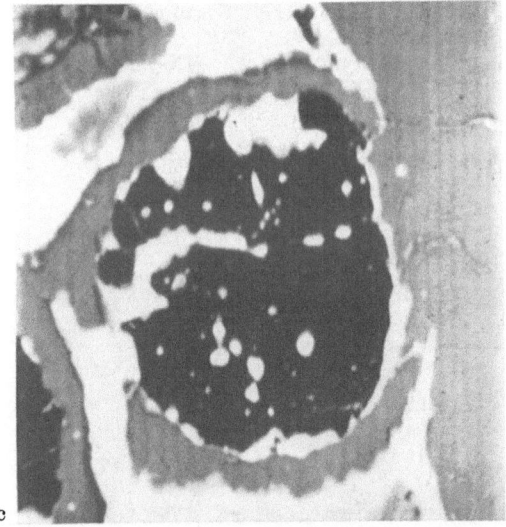

c

Fig. 7 a—e. Scanning pictures of the UO_2 coated par-
ticle. a $Mo\,L_\alpha$; 200×200 μ (expos. 10 min). b $U\,M_\alpha$;
200×200 μ (expos. 10 min). c Specimen current;
200×200 μ. d $Mo\,L_\alpha$; 100×100 μ (expos. 90 min).
e Specimen current; 100×100 μ

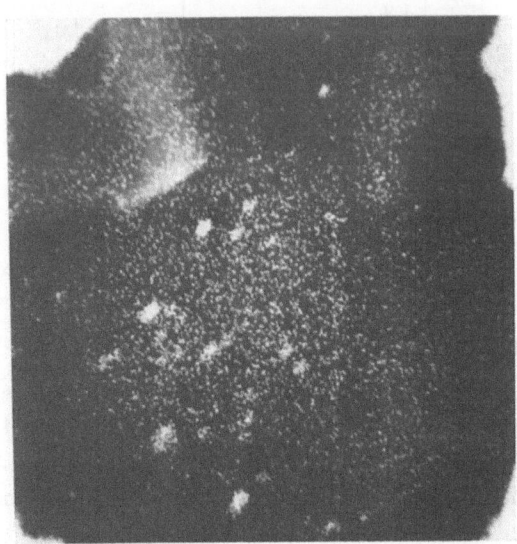

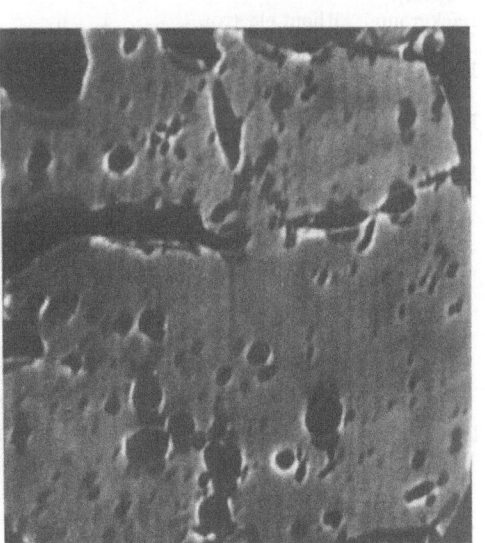

d   e

Table 2

Microanalyser Type		CAMECA old type MS 85		
Author		Ref. [1]	Ref. [2]	Ref. [3]
Sample: Material		U natur.	UC$_2$	UO$_2$
Enrichment		0.7% ^{235}U	—	—
Irradiation time		24 h	—	—
Th. neutron flux (n/cm^{2}sec)		10^{13}	—	—
Burnup		—	2—20% fiss./At. ^{235}U	4.6%
Cooling time		—	—	—
Dose rate $\beta + \gamma$ (mR/h)		—	—	220 in 30 cm
Dose rate γ (mR/h)		—	10^3 in 5 cm^a	12 in 30 cm
Microprobe: Spectrometer shield (cm)		2 + 1^b Pb	3 + 1^b Pb	none
Counter	1.	sealed prop. counter	prop. counter	NaI crystal
Counter	2.	flow prop. counter	prop. counter	NaI crystal
Crystal	1.	SiO$_2$ 1340 (22° 45′)	Mica (10° 40′)	Mica
Crystal	2.	SiO$_2$ 10Ī1 (7° 49′)	gypsum (14° 10′)	SiO$_2$ 10Ī1
Electron beam intensity (nA)	1.	1,200	—	—
Electron beam intensity (nA)	2.	1,200	—	—
Accelerating voltage (kV)	1.	—	13	17
Accelerating voltage (kV)	2.	—	13	17
Results: Line intensity (c/sec)	1.	220 U $L_{\alpha 1}$	82 U M_β	—
Line intensity (c/sec)	2.	11,206 U $L_{\alpha 1}$	84 U M_β	—
Peak/background (with electron beam)	1.	—	19 U M_β	—
Peak/background (with electron beam)	2.	—	47 U M_β	—
Background without electron beam (c/sec)	1.	6	1.9	—
Background without electron beam (c/sec)	2.	312	0.28	—
Detection threshold (wt.%)	1.	—	—	—
Detection threshold (wt.%)	2.	—	0.3 U in C	—

^{a} It is not indicated if γ or $\beta + \gamma$ radiation is concerned.

^{b} 3 cm brass (= 1 cm Pb) is the thickness of the column wall between sample and spectrometer.

^{c} 200 nA on Nb (absorbed current).

Experimental Results

After conversion of the microprobe, alpha, beta and gamma emitting metallographic specimens from projects of the Gesellschaft für Kernforschung Karlsruhe and the European Institute of Transuranium Elements were investigated for testing purposes.

A survey of the literature [1—10] and a comparison with our results is done in Table 2.

Table 2

AEI SEM 2		ARL EMX	NORELCO AMR/3	CAMECA MS 46 D	
Ref. [5]	Ref. [6]	Ref. [7]	Ref. [8]	this work	this work
UO_2	$(U_{0.85}Pu_{0.15})O_2$	$ThO_2 + 5.3\%$ wt. UO_2	Nb — 3% wt. Ti wire in irr. fuel	UO_2 with Mo coated	UO_2 + Zr-2 cladding
—	—	93% $^{235}U/U$	—	33% ^{235}U	2.3—2.7%
—	—	8 months	—	1,200 h	3 years
—	—	—	—	—	—
4.6%	8%	$1.6 \cdot 10^{20}$ fiss./cc	—	15,000 Mwd/t	15,000 Mwd/t
—	—	—	—	17 months	2 years
—	$15 \cdot 10^3$ in 30 cm	80 in 30 cm	—	100 in 30 cm	$15 \cdot 10^3$ in 30 cm
1 in 30 cm	300 in 30 cm	6 in 30 cm	$12 \cdot 10^3$ in 30 cm	10 in 30 cm	$5 \cdot 10^3$ in 30 cm
on the detector	on the detector	none	7.5 Pb	5—12 Pb	5—12 Pb
flow prop. counter	—	sealed prop. counter	—	flow prop. counter [d]	flow prop. counter [d]
flow prop. counter	—	—	—	flow prop. counter [d]	flow prop. counter [d]
Mica	Mica	—	—	SiO_2 1010	SiO_2 1010
LiF	—	—	—	SiO_2 1010	SiO_2 1010
—	—	—	300 [c] (estimated)	200	300
—	—	—	—	200	300
—	—	30	30	20	25
—	—	—	—	20	25
—	—	U $L_{\alpha 1}$	—	438 U $M_{\alpha 1}$ on UO_2	327 U $M_{\alpha 1}$ on UO_2
—	—	—	—	627 Mo $L_{\alpha 1}$	554 Zr $L_{\alpha 1}$ on Zr-2
—	—	—	—	($\pm 36'$) 80 on UO_2	(± 0) 40 on Zr-2
—	—	—	—	($\pm 36'$) 145 on Mo	
—	—	—	—	($\pm 36'$) 330 on UO_2	(± 0) 100 on UO_2
—	—	—	—	($\pm 36'$) 165 on Mo	
—	—	—	—	0,12	4
0.5 $\pm$ 0.05 Zr in UO_2	0.6 Mo in UO_2	see Ref. [7], Fig. 12	—	0,12	4
—	—	—	—	1,200 ppm U in Mo [e]	2,700 ppm U in Zr-2 [f]
—	—	—	—	640 ppm Mo in UO_2 [e]	1300 ppm Zr in UO_2 [f]

[d] The counter has been shielded by 3 mm lead.

[e] Counting time 100 sec.

[f] Counting time 200 sec.

Three examples will be described:

1. Within the development of thermionic energy converters (emitters) an emitter cross section (Fig. 6a) was analyzed on behalf of Brown, Boveri & Cie.[4] [11].

The right hand segment was cut out. For analysis a point was selected where the cladding of a UO_2 particle had grown together with the emitter wall as a consequence of re-crystallization

---

4. BBC AG, 6800 Mannheim 1, West Germany.

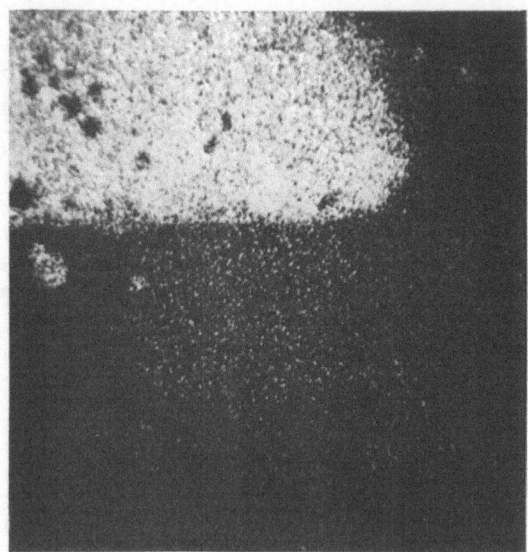

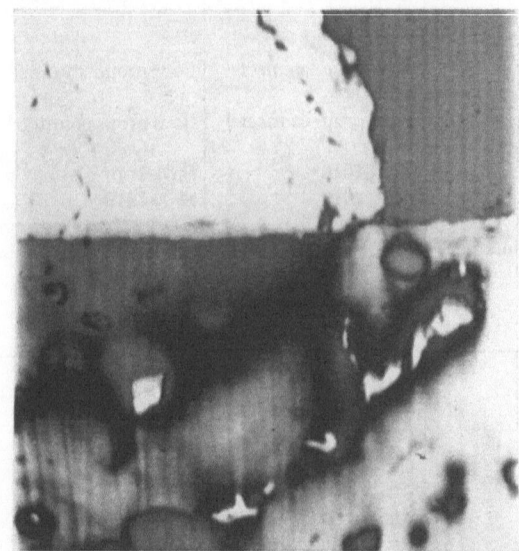

Fig. 8a—c. Scanning pictures of the UO_2—Zr-2 contact zone. a ZrL_α; 150×150 μ (expos. 10 min). b UM_α; 150×150 μ (expos. 10 min). c Specimen current; 150×150 μ

during the irradiation experiment (Fig. 6b). The scanned images of this point are seen in the Fig. 7.

A quantitative analysis was unable to detect any diffusion of fuel into the molybdenum cladding above the detection threshold of 1,200 ppm [12].

On the other hand, at a detection threshold of 640 ppm it turned out that there was a small Mo concentration of about 1,700 ppm in the UO_2 phase. Since equilibrium conditions were closely attained during the experiment (1,200 h at 2,000° C) it must be concluded that not more than 0.17 wt.% of Mo can be dissolved in UO_2. We believe however that no solubility at all exists and that the detected metal traces originated from the polishing of the specimen.

Thus the monotectic phase diagram of uranium dioxide — molybdenum [13—16] seems to be confirmed and supplemented in the areas of boundary concentration.

2. For the second example we selected the semi-quantitative analysis of the center rod of a boiling water rod bundle the burn-up of which was near 15,000 MWd/t. This was a sectional sample the γ dose rate of which was 150 mR/h at 30 cm distance.

Linear scanning across the edge of the fuel resulted in a peripheral Pu enrichment of about 1.5 wt.% of some $70-100 \mu$ width.

At a point of contact between the fuel and the cladding we discovered a zone of UO_2 of some 5μ width which contained Zr. Fig. 8 shows $Zr\,L_{\alpha 1}$ and $U\,M_{\alpha 1}$ images and the corresponding specimen current picture of the $150 \times 150 \mu$ section in this point. The beam intensity was 200 nA at 20 kV.

3. For the investigation of the corrosive layer of the cladding material (zircalloy) on the coolant side of these fuel rods an external rod of the same bundle was cut radially. This sample

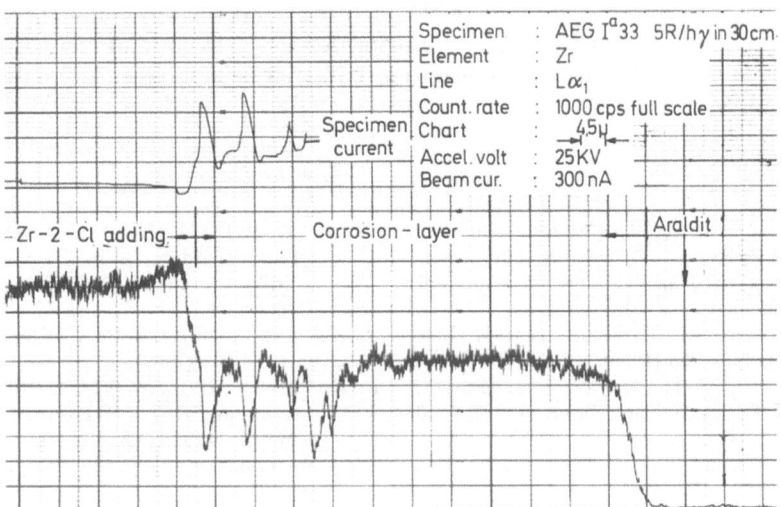

Fig. 9. Zr — linear scanning on the corrosion layer in zircalloy (cooling side)

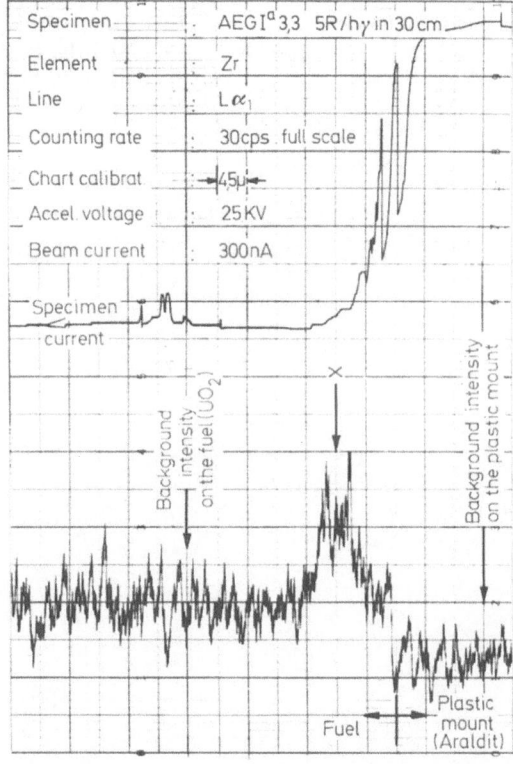

Fig. 10. Zr — linear scanning on the UO_2 boundary

had a gamma dose rate of 5,000 mR/h at 30 cm. Various linear scannings were made across the corrosive layer which was up to 200 μ wide. A Zr-intensity profile of the layer is shown in the Fig. 9.

On the basis of point analyses the corrosion layer was identified to be ZrO_2, as expected. A perpendicular Zr linear scanning above the fuel border indicated a small Zr concentration of a few μ width at the edge of the fuel (Fig. 10).

Conclusions

The experiments confirmed the possibility of using a microprobe for investigating active samples and gave the basic knowledge to construct for the Transuranium Institute (Euratom) a second shielded CAMECA microanalyser of an industrial type foreseen to analyse specimens with a γ-dose up to 100,000 mR/h at 30 cm.

References

1. Theisen, R., and J. P. Lecocq: Proceedings of the 10th Colloquium Spectroscopicum Internationale (1963), p. 383—389. Washington: Spartan Books.
2. Quataert, D., and H. W. Schleicher: J. Nucl. Mater. 19, 221—233 (1966).
3. Bradbury, B. T., J. T. Demant, P. M. Martin, and D. M. Poole: J. Nucl. Mater. 17, 227—236 (1965).
4. Jeffery, B. M., and G. K. Williamson: J. Sci. Instr. 41, 172—173 (1964).
5. — J. Nucl. Mater. 22, 33—40 (1967).
6. Bramman, J. I., R. M. Sharpe, D. Thom, and G. Yates: J. Nucl. Mater. 25, 201—215 (1968).
7. Padden, T. R., R. Burton, and C. E. Campbell: WAPD-TM-644 (5/67).
8. Scotti, V. G., J. M. Johnson, and R. T. Cunningham: Advanced in X-ray analysis, vol. 9, p. 314—322. Proceedings of Denver Conference (8/65).
9. Zelezny, W. F., and R. A. Moen: IDO 17218, p. 127—132 (1966).
10. Rosenbaum, H. S., W. V. Cummings, T. A. Lauritzen, and R. T. Peterson: GEAP-5344.
11. BBC-Report 1201/S: Forschung über Direktumwandlung in elektrische Energie.
12. Täffner, K., and R. Theisen: EUR 1819 e., p. 27—36 (1964).
13. Gebhardt, E., and G. Ondracek: J. Nucl. Mater. 13, 220—228 (1964).
14. Dayton, R. W., R. F. Dickerson et al.: BMI-1607, Sect. D-1 (1962).
15. — — BMI-1644, Sect. D-1 (1963).
16. — — BMI-1659, Sect. D-1 (1964).

The TPD Electron Probe X-Ray Micro Analyzer

L. A. FONTIJN, A. B. BOK and J. G. KORNET

Institute of Applied Physics TNO-TH, Delft, Netherlands

Summary

The article describes an electron probe X-ray micro-analyzer specially designed for mineralogical investigations and constructed for the Department of Mining Engineering of the Technological University, Delft. A miniature magnetic lens is used in the electron optics making it possible to employ a standard Leitz polarization microscope. The specimen can be rotated around the microscope axis and translated in two orthogonal directions.

Introduction

The principle of the method is already widely known: an accurately focused electron beam is directed at a particular point on the surface of a sample whose chemical composition it is desired to ascertain. Owing to the small diameter of the electron spot, which is in the order of 1 μm, and the slight depth of electron penetration into the specimen, which is likewise in the order of 1 μm, a very small volume of material is irradiated by the electron beam. The X-ray spectrum includes the characteristic radiations of the various elements at the points of impact of the electron probe. Spectrographic analysis of this X-ray spectrum permits the concentrations of these elements to be determined. The spot nature of the analysis is a refinement on chemical analysis by X-ray fluorescence and a more direct method of obtaining quantitative measurements.

In the microprobe, the required size and shape of the iron shrouded objective lens always limit the freedom of design for both light and X-ray optics. The various philosophies behind these limitations have resulted in a wide variety of microprobes. The TPD microprobe uses a

Fig. 1. General view of the TPD microprobe, with at the left: vacuum system, at the right: lens currents regulation

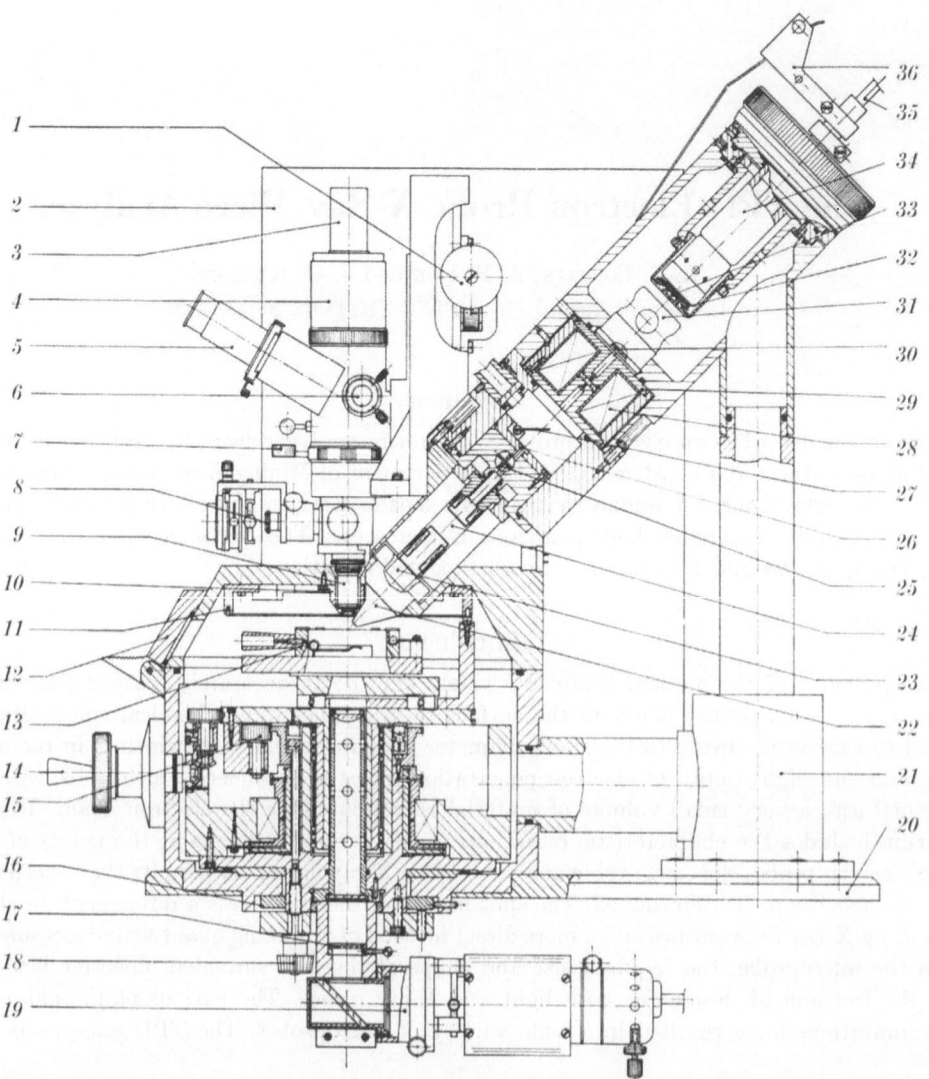

Fig. 2. Cross-section of the TPD-microprobe. *1* Microscope height control; *2* Philips X-ray spectrometer with mica cristal; take-off angle 30°; *3* adapter for camera; *4* Leitz polarization microscope; *5* binocular viewing head; *6* Bertrand lens; *7* analyzer; *8* polarizer; *9* objective lens; *10* vacuum seal for microscope; *11* contact ring for absorbed electron signal; *12* vacuum chamber door; *13* specimen holder; *14* knob of specimen rotation around the microscope axis; *15* part of the 2 orthogonal *x-y* traverses; *16* centring adjustment; *17* polarizer; *18* specimen height control knob; *19* transmitted light attachment; *20* table top; *21* vacuum connection to oil diffusion pump; *22* miniature magnetic lens with coil for defocusing correction; *23* stigmator and correction deflection coils; *24* scanning deflection coils; *25* beam limiting aperture holder; *26* aperture; *27* fluorescent screen; *28* condenser lens; *29* pole piece centring; *30* aperture; *31* anode; *32* Wehnelt cylinder; *33* filament; *34* electron gun; *35* high-tension cable; *36* gun-earth contact

miniature magnetic lens as the objective lens. The electron optics are scaled down, providing space for a good polarization microscope (Leitz), which is of particular importance for work on mineralogy. The light and electron optical system axes are separated, the electron optical axis being inclined at 45° to the horizontal specimen plane. This inclined position of the electron optics in theory permits any take-off angle of the X-rays to the specimen plane. In the TPD microprobe two Philips Norelco spectrometers allow a take-off angle of 30° owing to their being placed upside down.

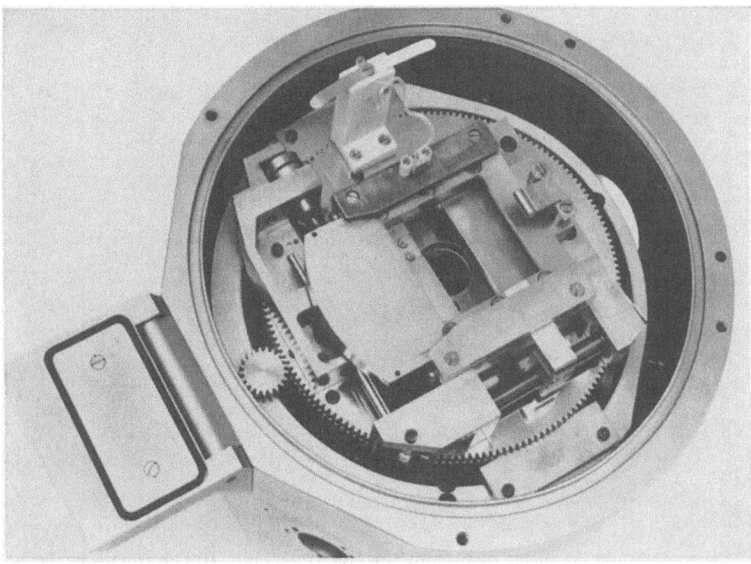

Fig. 3. The specimen movements system. Rotation at every point of the specimen (not shown) is possible. Translation in two orthogonal directions of the specimen on the turn-table

The main design features of the TPD microprobe are:

1. The use of a standard Leitz microscope mounted entirely outside the vacuum, permitting observation of the specimen during X-ray analysis. Resolution is 1 μm and polarization facilities are included.

2. The electron optical axis is inclined at 45° to the horizontal specimen plane, enabling the electron beam to clear the light objective and maintaining a short working distance in the miniature magnetic lens. Defocusing due to the skew illumination is corrected.

3. The X-ray take-off ranges from 0° ... 60° and may be extended to 90° if the light optics are omitted. Owing to the smallness of the miniature lens, the spectrometers and detectors may be mounted at an azimuthal angle of 300° around the electron probe.

4. The specimen holder fits on a cross-table mounted on a turn-table so that the specimen can be translated in x, y directions and rotated around the turn-table axis, which coincides with the microscope axis.

Besides a complete microprobe system built for the Department of Mining Engineering of the Technological University, Delft, the same electron and light optics as used in the apparatus have been tested with a Philips Norelco Probe at the Analytical Chemical Laboratory of the State University, Utrecht.

Description of Microprobe System

The microprobe consists of four main parts: electron optics, light optics, X-ray optics and specimen movements system. These four main parts will be described first and then the further equipment with reference to photographs and drawings.

Electron Optics

The electron optical axis, inclined at 45° to the horizontal specimen plane, is determined by the fixed positions of the anode and the miniature magnetic lens. Otherwise the electron optics are conventional and consist of the following parts:

a) A triode-type electron gun is used to accelerate the generated electrons, with a standard filament 0.125 mm in diameter. All electron gun adjustments are possible without breaking the vacuum, including:

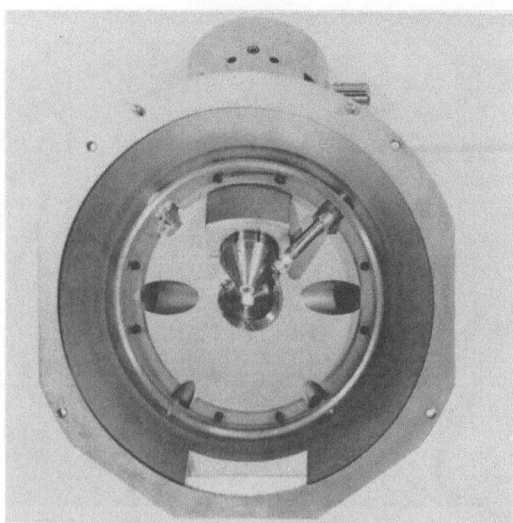

Fig. 4. Top plate of the specimen chamber seen from the specimen side. In the centre is the holder with a glass plate acting as a vacuum seal for the microscope. Opposite the opening for specimen changing, a part of the miniature objective lens can be seen. Two large openings for passing the generated X-rays to both Philips X-ray spectrometers. An electron backscatter detector and a feed-through of the specimen current are positioned in two of the four remaining openings

1. adjustment of filament height relative to the Wehnelt cylinder;
2. centring of filament in the Wehnelt opening;
3. centring of Wehnelt-cylinder and filament relatively to the fixed anode.

The electron gun H.T. supply is variable between 4 and 50 kV, while the Wehnelt tension can be changed with resistors in the H.T. supply. In order to obtain maximum gun performance, measurements of brightness were made. Special attention was paid to vacuum conditions in the gun chamber for reasons of filament life.

b) The condenser lens has an iron circuit in order, together with the objective lens, to obtain adequate geometrical demagnification (max. $250 \times$) of the gun cross-over. To avoid deflection defects by a change in lens excitation owing to misalignment of the pole pieces, the upper pole piece can be centred.

The beam-limiting apertures are positioned between the condensor and objective lens in a rotatable holder and are easy to change.

The scanning coils, a double deflection system, are mounted outside the vauum system around a brass tube between the aperture holder and the miniature lens. The stigmator and correction deflection coils are just above the objective lens.

c) The objective lens is a miniature magnetic lens constructed for operation up to 50 kV. It was tested on an electron optical bench. As regards its spherical aberration coefficient, close agreement was obtained between calculated and measured values. At ten times demagnification the value was $C_s/f^3 = 3.5 \times 10^{-3} \, \text{mm}^{-2}$ ($\pm 15\%$). C_s is the spherical aberration coefficient and f the focal length. The C_s value in the instrument is 40 mm. A first order stigmator, placed above the lens, is necessary to correct astigmatism.

When the specimen is scanned by the electron beam, defocusing of the electron spot due to skew illumination is corrected with an extra layer of windings around the main coil of the miniature lens. As the scanning coils are mounted outside the vacuum, they can easily be rotated so that defocusing occurs in one scanning direction.

In order to obtain the right variation of lens excitation to correct defocusing, the deflection coils for the direction in the plane of the electron optical axis and the normal on the specimen plane are connected in series with the extra windings on the miniature lens. This gives a good first order approximation for correcting defocusing. By induction in the copper jacket of the miniature lens, a frame frequency up to 10 Hz is usable for defocusing correction.

The Light Optics

are a Leitz refractive polarization microscope. This is completely outside the vacuum system, a glass plate forming a vacuum seal between the objective lens and the specimen. The microscope axis is perpendicular to the specimen surface. The specimen can be observed during X-ray analysis by transmitted or reflected light from the probe side. By using the different objective lenses and eyepieces, magnification can be chosen in the range from 50 to 500 ×. The field of view is at least 600 μm. The microscope assembly includes a binocular viewing head, camera tube, polarized light attachments and facilities for using illumination filters. The objective lenses are the same as used in the Leitz universal turn-table microscope. For our purposes these have the advantage of a large free object distance of about 15 mm.

X-Ray Optics

An X-ray take-off angle of 30° is obtained in this apparatus with a Philips Norelco spectrometer placed upside down. Normally, this spectrometer allows for an X-ray take-off angle of 15°. The maximum take-off angle will be about 60° owing to the dimensions of the objective lens in the polarization microscope. Left and right Philips Norelco spectrometers are used.

Specimen Movements System

To obtain full benefit from the polarization microscope and the fullest possibility of scanning the specimen with the electron probe, the microprobe specimen table permits the same movements as with a normal polarization microscope. They include manual rotation of the specimen around the light optical axis and two orthogonal traverses on the turn-table. The space available for specimens and standards is $50 \times 50 \times 15$ mm³. Specimen holders can be provided for different sample sizes. The specimen movements system is constructed so that if a perfectly flat glass plate is mounted in the specimen holder, the vertical displacement of the specimen over its whole area is less than 5 μm; the specimen remains within the focal depth of the light microscope objective lens and, with a stationary electron beam, within the focal depth of the miniature lens. The axis of the turn-table can be centred in the specimen chamber. The position of the specimen can be varied vertically within 0.6 mm. These adjustments are necessary to ensure that the fixed electron optical axis and the rotation axis intersect on the specimen plane. The microscope axis is afterwards centred on the axis of the turn-table of the specimen movements system. To hold the specimen at a fixed height, the microscope can be fixed in its lowest position.

Construction of this ideal specimen movements system with the cross-table on the turn-table has been made possible by using concentric toothed wheels and two differentials for both orthogonal traverses.

The vacuum system is automatic for convenience in operation. Electro-pneumatic valves are used to avoid strong magnetic fields in the vicinity of the electron optical column.

Standard Philips electronic measuring panels are used for measuring the generated X-rays, in combination with the Philips Norelco spectrometers. A modified Norelco beamscanner is used for displaying electron and X-ray images.

A Brandenburg E.H.T. power supply is used and the very stable lens power supply is our own design.

For meaningful light-element analysis with a suitable crystal, carbon contamination in the vicinity of the electron spot is reduced by an air-inlet anticontamination device. Fully focusing X-ray spectrometers can be used by readapting the electron optics and the specimen chamber top-plate. An instrument combining an electron probe X-ray analyzer and a scanning electron

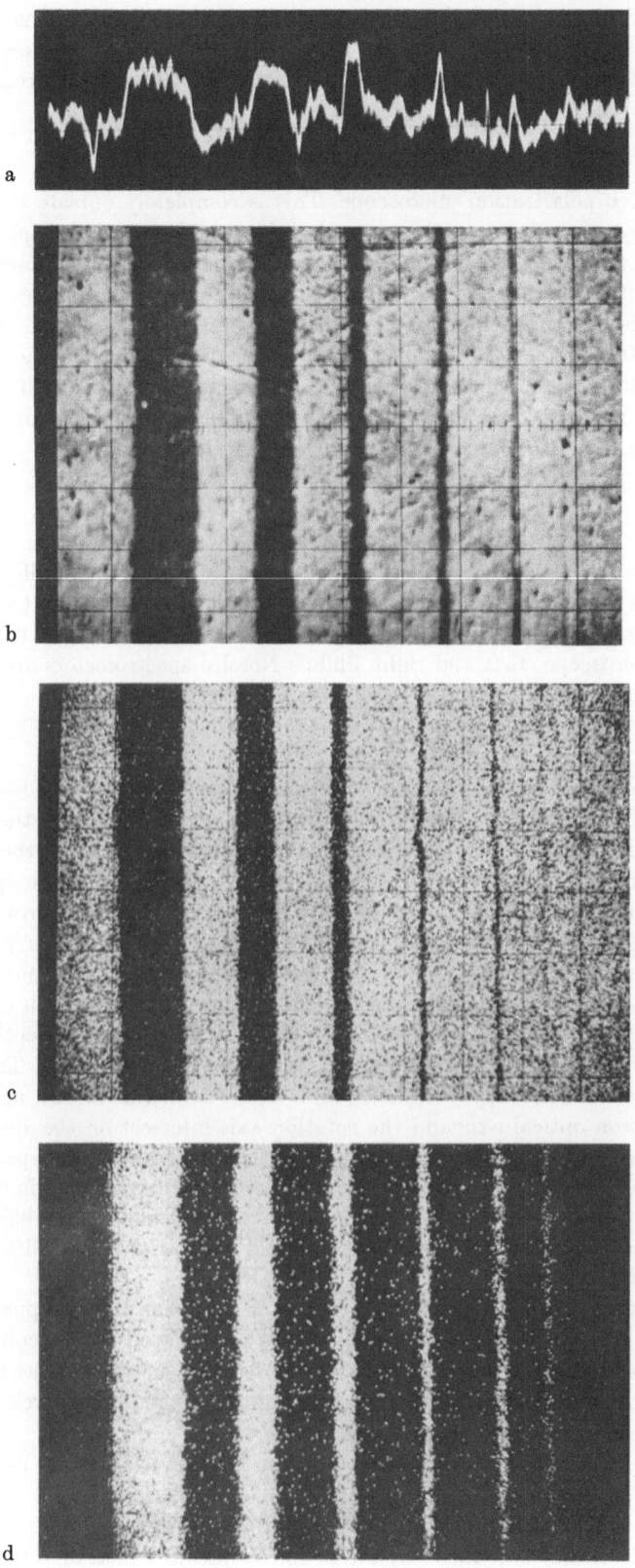

Fig. 5a—d. Results with a Cu—Ni sandwich; 19 kV; 0.1 μA sample current a line scan sample current. b Absorbed electron image. c CuK_α image. d NiK_α image

microscope is possible by using a stereo-scan accessory. This accessory consists of an extra condenser lens, which can easily be fitted in the electron optics for adequate geometrical de-magnification and a sensitive wide band detector for measuring the small number of secondary electrons.

The Theory of Electron Scattering and X-Ray Generation

Because almost all theoretical considerations and experiments relate to a normal incident electron beam, a relation must be found between the skew and the normal incident electron beam position with quantitative analysis. It should be noted that the distribution of the X-rays will be rotationally symmetrical about the electron spot in a homogeneous matrix, but the distribution of the backscattered electrons will then be non-symmetrical.

Three important effects modifying X-ray generation must be considered: absorption of X-rays by the specimen, fluorescence contributed by the continuous and characteristic X-rays, and the atomic number effect, which accounts for the fraction of electrons backscattered and the depth of electron penetration. In the inclined electron beam position, the total fraction of electrons backscattered is increased, causing a smaller production of X-rays in the specimen. But as electron penetration below the specimen surface is slighter, this results in less absorption of X-rays. In these conditions, the total X-ray intensity measured may increase.

Absorption correction of X-ray peak intensities can be obtained with the Duncumb and Shields modification of Philibert's absorption equation:

$$\frac{1}{f(\chi)} = \left(1 + \frac{\chi}{\sigma}\right) \cdot \left[1 + \left(\frac{h}{1+h}\right)\frac{\chi}{\sigma}\right]$$

in which

$$\sigma = \frac{2.39 \times 10^5}{V^{1.5} - V k^{1.5}}$$

and

$$h = \frac{1.2\,A}{Z^2}$$

V = beam acceleration voltage,
Vk = critical excitation potential,
A = atomic number,
Z = atomic weight.

χ for the inclined electron beam on the specimen can be found with a formula based on the assumption that the distribution of excitation in the direction of the incident beam is the same as that with depth for normal beam incidence

$$\chi = \frac{\mu}{\varrho}\,\mathrm{cosec}\,\theta \cdot \sin\alpha,$$

θ = take-off angle,
α = angle between electron beam and specimen surface,
$\frac{\mu}{\varrho}$ = mass-absorption coefficient.

One can also apply the formula proposed by BISHOP, obtained from the results of Monte-Carlo calculations and checked by using GREEN's data for the variation of $f(\chi)$ curves with incident angle

$$\chi = (\mu/\varrho)\,\mathrm{cosec}\,\theta\,(1 - 0.5\cos^2\alpha).$$

Under our conditions $\alpha = 30°$ and $\theta = 45°$, the difference between these two formulas for χ being only a small percentage. Another small advantage with the inclined electron beam position is the analysis of elements closer to the specimen surface owing to the smaller vertical penetration depth of the electrons. The fluorescence contribution as regards the direct X-ray emission generated in the electron spot will decrease slightly owing to the shorter penetration below the surface resulting in less fluorescence radiation excited by the direct emission towards the specimen surface. The fluorescence emission is generated on average at a depth six times greater than the direct emission. This means that for quantitative analysis the fluorescence correction procedure can be the same for both inclined and normal electron beam positions.

The inclination of the electron beam has a slight effect on the atomic number correction, wich is due to the variation of electron deceleration and backscatter properties of the elements. This is determined by the fraction of electrons backscattered from the specimen and the appropriate electron energy distribution for this inclined electron beam position at the average mass concentration of the specimen. Experimental and theoretical investigations of electron backscattering are necessary for proper atomic number correction.

For quantitative analysis with an inclined electron beam of 45° it will be possible to proceed as with a normal incident electron beam to determine the emission function of the characteristic X-rays with depth.

In this way the absorption correction can be found with the known techniques of variation of X-ray take-off angle, tracer method, specimen inclination or with Monte-Carlo calculations. Another possibility of quantitative analysis is to use standards whose composition is close to that of the specimen.

Although the correction procedures for the normal incident electron beam position are not yet reliable enough for general use, transformation of the correction procedures in the normal case to the inclined electron beam position, as described above, can be used in the first instance.

The first results obtained with the electron and light optics are given. They relate to a scan of the electron beam over a Cu—Ni sandwich. There are six tracks of Ni with thicknesses of 12, 6.8, 3.5, 2.0, 1.4 and 0.7 µm respectively. The 5th order Ni K_α and Cu K_α radiation in a Philips spectrometer with a take-off angle of 30° was used for detection.

References

1. Le Poole, J. B.: Miniature lens. Third European Regional Conference on Electron Microscopy, Prague 1964, p. 439.
2. Duncumb, P., and P. K. Shields: Calculation of fluorescence. Third Internat. Symposium, Stanford, A.P. 1963, p. 329—340.
3. Bishop, H. E.: The absorption and atomic number corrections. Brit. J. Appl. Phys., Ser. II, 1, 673—684 (1968).
4. Springer, G.: Investigations into the atomic number effect. Neues Jahrb. Mineral. Monatsh. 1967, 304—317.

New Spectrometers
and Accessories for the Electron Microprobe

J. B. Nicholson and H. Neuhaus

Applied Research Laboratories, Sunland, California, U.S.A.

M. F. Hasler

Hasler Research Center, Applied Research Laboratories, Goleta, California, U.S.A.

Abstract

Several new types of spectrometers have been developed for use with the EMX which will become available within the next year.

Two compact semi-scanning curved crystal monochromators of limited range have been designed to be mounted on either side of the microscope of the EMX spectrometer. One is designed to utilize the ARL standard 4″ radius curved crystals, the other the 11″ radius curved crystals. By choosing a crystal of the proper material and radius of curvature, most X-ray wavelengths of interest can be reached with one or the other of these two monochromators. Thus up to five curved crystal monochromators become available for the EMX.

The second is a grazing incidence grating spectrometer utilizing a gold grating with low blaze angle developed specially for work in the long wavelength X-ray region. By operating this scanning grating instrument at increased input angles for increased wavelengths, optimum performance can be achieved for first order diffraction, while discriminating against second and higher orders. Good performance from 20—170 A has been achieved with efficiencies from two to six times as great as obtainable from crystals or pseudo-crystals working in this range. The use of gold replica gratings provides good service life for this type of spectrometer when operated in the low contamination environment of the EMX spectrometer.

A new accessory which further reduces contamination in the microprobe has been developed. This consists of a means of cooling the large lower plate of the objective lens to −40° C by the use of a Freon refrigeration system. This accessory reduces the contamination rate of the EMX by a large factor. By controlling the temperature of the objective lens to very close limits, this markedly improves the stability of the electron beam position on the sample.

We wish to report three developments applicable to the EMX electron microprobe which we believe will be of general interest.

In electron microprobes it is conventional practice to provide more than one X-ray spectrometer. By this means each spectrometer may be provided with the particular geometry and diffracting means that is optimized for a particular part of the X-ray spectrum, and of course, the time required for a given analysis is reduced if data can be gathered through two or more channels simultaneously. Simultaneous measurement in multiple channels may in fact be imperative in ultra-microanalysis where it is desired to investigate the spatial correlation of two or more elements in a small region such as a particle or inclusion which may be less than a micron in size.

When there are as many as four or five spectrometers it is rarely necessary for all of them to be of the scanning type. More often there will be one or two elements that are of relatively permanent interest, and these elements can be measured on spectrometers that are at least

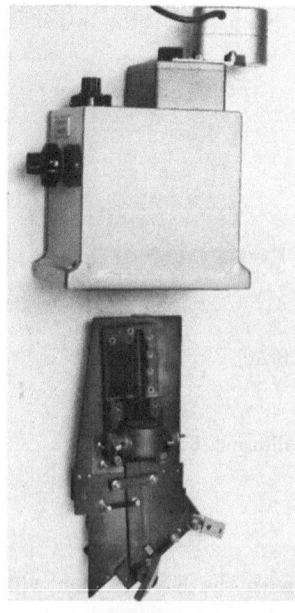 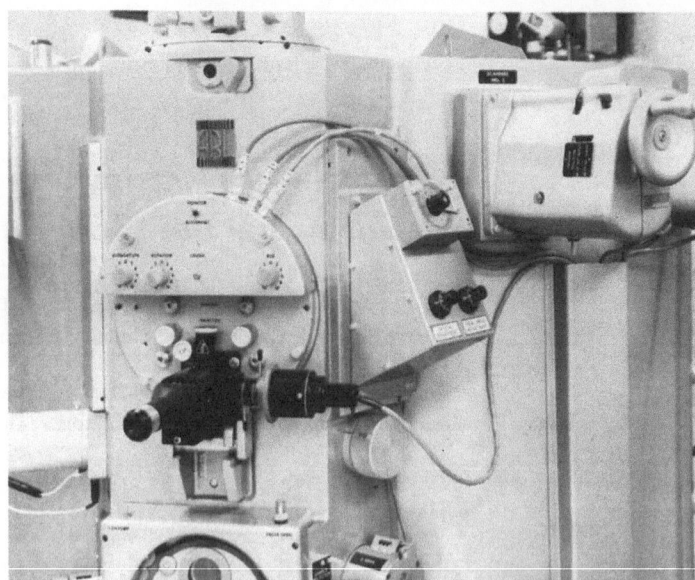

Fig. 1   Fig. 2

Fig. 1. Semi-scanning spectrometer laid on a table

Fig. 2. EMX with semi-scanning spectrometer installed

relatively fixed, scanning spectrometers being used to measure elements in which interest may vary from one problem to another.

Fig. 1 shows a semi-scanning spectrometer which has been developed by our associates Shimadzu Seisakoshu Ltd., and adapted to the EMX electron microprobe. It may be provided in either a 4″ or an 11″ radius and with a variety of crystals. Either one or two of these may be provided, mounted as shown in Fig. 2. In adapting these to the EMX, the usual high takeoff angle of 52.5° has been retained. The scanning range is ± 1.5 Angstroms from the mean position, sufficient not only to permit peaking on a particular line in the event of chemical wavelength shifts, but to permit measurement of line-to-background ratios directly, by measuring the count rate first at the peak of the line and then at the background level immediately adjacent. Using lead stearate decanoate pseudo crystals in the 11″ radius, with 0.1 μa sample current at 20 kV on an Al_2O_3 sample, OK_α count rates are typically 800 to 2,000 counts per second and line-to-background ratios between 75:1 and 85:1.

Many metals exhibit optical indices of refraction in the X-ray region that are slightly less than unity, and it should therefore be possible to achieve total reflectance of an X-ray beam from a metal surface at angles of incidence sufficiently near grazing. In fact, reflectances well in excess of 90% have been observed, as is well known [1]. Thus it is possible to use a ruled metal grating in X-ray spectroscopy and, at low angles of incidence, such gratings should in principle be more efficient than the more conventional crystals or pseudo-crystals. They have also the intrinsic advantage over pseudo crystals that they can be ruled with any desired groove density and so can function with good efficiency in the long-wavelength portions of the X-ray spectrum where suitable pseudo crystals do not exist. Such gratings have in fact been described [2, 3].

A *blazed* ruled grating may have the advantage, over either a pseudo crystal or a ruled grating of the Siegbahn type, that a large portion of the diffracted energy can be concentrated in a given order of interference. This property yields still further improvement in efficiency and reduction in interferences due to overlapping orders, especially due to specular reflection in the zero order which can be a problem with gratings of the Siegbahn type.

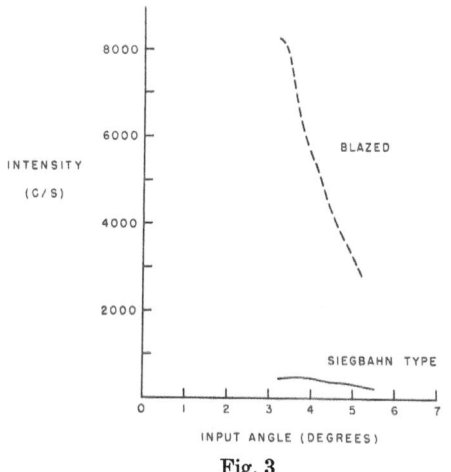

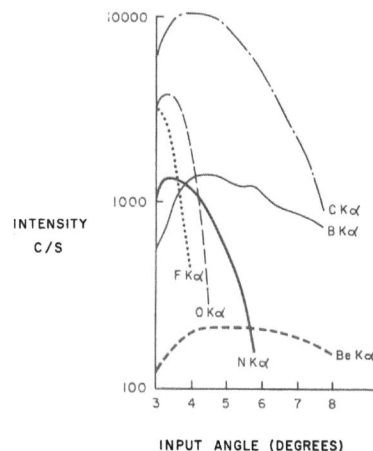

Fig. 3

Fig. 4. Grating efficiency 600 gr./mm, 1° blaze I M aluminized replica

We have reported on a blazed ruled grating in which efficiencies were higher than in a lead stearate decanoate pseudo crystal by about a factor of twenty [4]. This grating was coated with aluminum and the service life as limited by corrosion was unacceptably short. Further work in cooperation with the Grating Research Laboratory of BAUSCH and LOMB has resulted in a gold-coated replica blazed grating whose service life is completely satisfactory and whose performance is shown in the next figures.

Fig. 3 shows the relative intensities achieved with a blazed grating and with a Siegbahn ruling, also prepared in the Grating Research Laboratory of BAUSCH and LOMB, in which all variables except the form of the ruling were made the same as nearly as possible. The advantage of the blazed grating is quite apparent.

Fig. 4 shows the efficiency of the blazed grating as a function of input angle and wavelength. Here the advantage of the blazed grating in diffracting most of the incoming energy into a given order of interference is most apparent. For example, in detecting carbon (44.7 A) in the presence of second-order oxygen (47.4 A), setting the input angle to five degrees causes the grating to discriminate strongly in favor of the former.

Fig. 5 shows the first and second-order oxygen K_α lines from Al_2O_3. Note that carbon K_α and second-order oxygen K_α are completely resolved. The slide also shows the count rates achieved on first-and-second-order oxygen K_α and on background.

Fig. 6 shows first-order and second-order carbon K_α. It is particularly noteworthy that the background is quite low and almost independent of wavelength, which is not often the case in crystal spectrometers.

The use of a blazed grating in place of a pseudo crystal introduces some problems in alignment since the grooves must be accurately orthogonal to the plane of the Rowland circle if optimum resolution is to be obtained, and there is no analogous requirement with a crystal. Fig. 7 shows the resolution that can be obtained when alignment is correct. The figure shows the Si L group. The subsidiary figure shows the same group as obtained by LUKIRSKII in a vacuum grating spectrometer with X-ray line source and primary slit. It will be seen that the resolutions are quite comparable.

Fig. 8 shows an analytical working curve for a solid solution of carbon in iron. The limit of detectability is 120 parts per million, and a new development which I shall now describe reduces this limit to 50 parts per million.

A basic requirement in the electron microprobe is that the sample surface must be free of contaminants and must remain free during analysis. Even though the sample may be cleaned adequately during preparation, contaminants may still arise from condensation, on the sample surface, of gases and vapors that are residual in the vacuum system or which may be evolved from e.g. plastic compounds used to mount the sample. Such materials, present at a partial

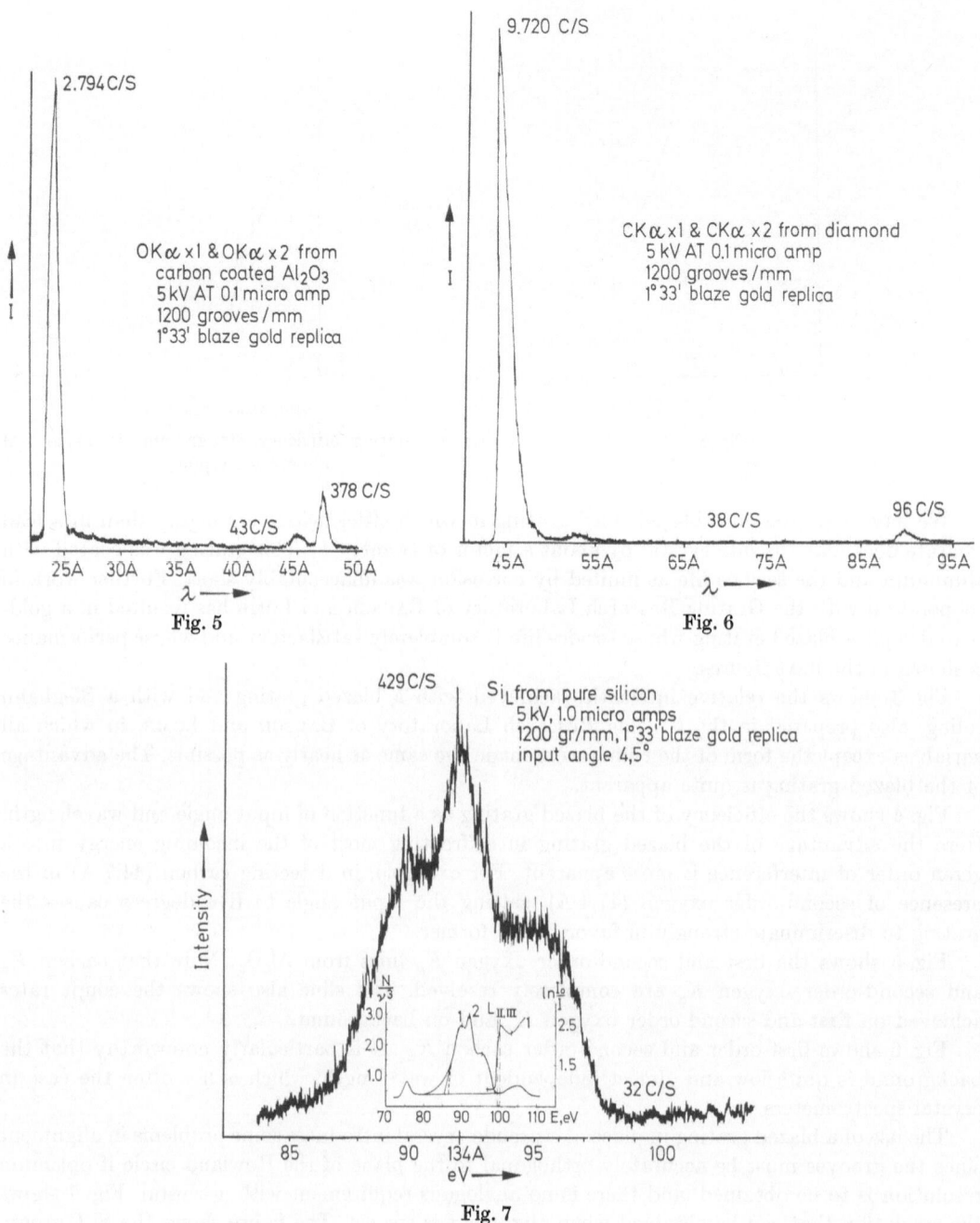

2.794 C/S

OKα x1 & OKα x2 from
carbon coated Al₂O₃
5 kV AT 0.1 micro amp
1200 grooves/mm
1°33' blaze gold replica

I

43 C/S

378 C/S

25A   30A   35A   40A   45A   50A

λ ⟶

Fig. 5

9.720 C/S

CKα x1 & CKα x2 from diamond
5 kV AT 0.1 micro amp
1200 grooves/mm
1°33' blaze gold replica

I

38 C/S

96 C/S

45A   55A   65A   75A   85A   95A

λ ⟶

Fig. 6

429 C/S

Si$_L$ from pure silicon
5 kV, 1.0 micro amps
1200 gr/mm, 1°33' blaze gold replica
input angle 4,5°

Intensity ⟶

32 C/S

85   90   134A   95   100

eV ⟶

Fig. 7

pressure of 10^{-6} Torr, may condense at a rate approaching one monolayer per second and, in the presence of a bombarding electron beam, may be pyrolyzed to result in a deposit on the sample surface which is thick enough to interfere seriously with a desired analysis.

Ultra high vacuum techniques involving prolonged high temperature baking, etc. are unfeasible in most cases. However if it is appreciated that such condensed layers are in dynamic equilibrium with their surroundings it becomes clear that the equilibrium thickness of such condensed layers may be drastically reduced by reducing the probability that a molecule which may evaporate from such a condensed layer can return to it. This can be accomplished by providing a chilled surface near the sample and such a procedure has been used effectively in electron microscopes for some years.

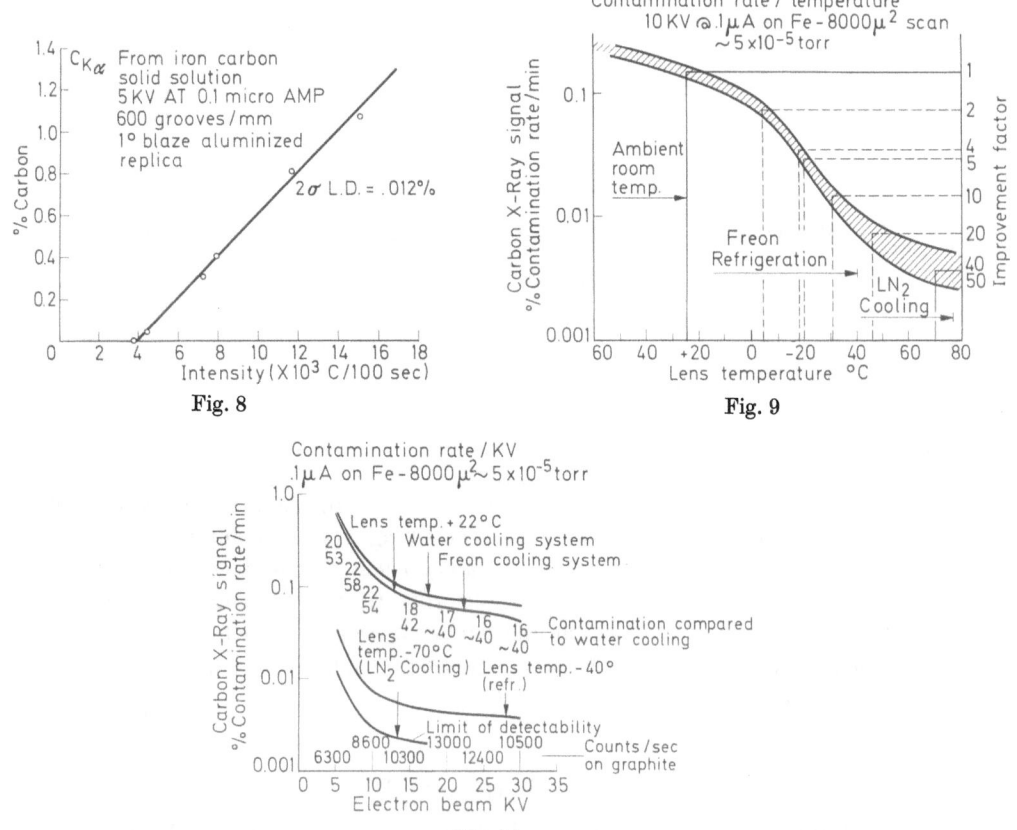

Fig. 8

Fig. 9

Fig. 10

For maximum effectiveness such a chilled surface should subtend the largest possible solid angle at the sample surface and should be maintained at such a temperature that any material which condenses on it has a negligible probability of reevaporating.

In the EMX the magnetic objective subtends a large solid angle at the sample, so that these requirements can be met simply by chilling the magnetic objective lens itself. Since the lens is already provided with means for water cooling, adaptation to cooling by mechanical refrigeration to $-40°$ C or by liquid nitrogen to even lower temperatures is relatively straightforward.

Fig. 9 shows the improvement in sample contamination rate which results from cooling the objective lens. It will be seen that such cooling may reduce the contamination rate, as measured by the carbon X-ray signal, by a factor of twenty or more. Indeed, Fig. 10 shows that with electron beams having more than 15 kilovolts of energy, cooling below $-70°$ C reduces the contamination rate below the limit of detectability.

Since the system cannot be vented when the lens is at very low temperatures, the time required to raise the lens to room temperature and to cool it again is of interest. Approximately 20 minutes are required to warm the lens before venting, and ten to fifteen minutes are required to chill the lens prior to running a new sample.

The three developments just described — the semi-scanning spectrometer, the blazed grating spectrometer, and the refrigerated objective lens — are believed to be significant, and broadly useful, in the field of electron microprobe analysis.

References

1. Lukirskii, A. P., E. P. Savinov, O. A. Ershov, and U. F. Shepelev: Opt. Spectry. (USSR) **16** (2), 314 (1964).
2. Holliday, J.: J. Appl. Phys. **33**, 3259 (1962).
3. Nicholson, J. B., and D. Wittry: Advances in X-ray analysis VII.: Plenum Press 1964.
4. —, and M. F. Hasler: Advances in X-ray analysis IX.: Plenum Press 1966.

Geräte und Methoden
zur quantitativen Gefügecharakterisierung
mit der Mikrosonde

G. DÖRFLER und G. REICH

Analytisches Institut der Universität Wien, Österreich

Die Elektronenstrahl-Mikroanalyse erschöpft sich nicht in der quantitativen Punktanalyse oder der halbquantitativen Registrierung der Verteilung von Elementen entlang von Linien oder auf Flächen. Wie in der Literatur [1—5] ausführlich beschrieben wurde, ist es möglich, mit der Mikrosonde und einigen Zusatzgeräten auch quantitative Informationen über den Anteil, die Größe und die Verteilung von Phasen und Konzentrationsbereichen zu erhalten.

Der vorliegende Bericht soll einige neuere Entwicklungen auf diesem Gebiet und ihre Anwendungen in Metallurgie und Mineralogie aufzeigen.

Stand der Technik

Durch die Entwicklung der Scanningtechnik in der Elektronenstrahl-Mikroanalyse wurde eine der wesentlichsten Voraussetzungen für die Durchführung einer Linearanalyse mit der Mikrosonde geschaffen. Das „concentration-mapping" von HEINRICH [6] und das „expanded contrast"-Verfahren von MELFORD [7] sowie einige ähnliche Verfahren erweiterten das Anwendungsgebiet der Scanningbilder, gestatten jedoch noch keine quantitative Aussage über Mengenanteile, Größe und Verteilung von Gefügekomponenten. Erste Versuche in dieser Richtung wurden von THEISEN [8] unternommen. Dieser bestückte einen Linienschreiber mit einem einfachen Mikrokontakt, um die Zahl der Teilchen zu zählen, bei denen die Röntgenimpulsrate eine gewisse Höhe übersteigt. Mittels Röntgenimpulsmessungen an Teilchen, Matrix und Standard konnten Gewichtsanteil und mittlere Größe dieser Phase ermittelt werden.

Unabhängig voneinander entwickelten MELFORD und WHIDDINGTON [1], RAMSEY und WEINSTEIN [9] sowie DÖRFLER und PLÖCKINGER [2] Geräte, die, auf dem Prinzip der Linearanalyse basierend, die einzelnen Gefügekenngrößen zu bestimmen gestatten. Während MELFORD et al. [1]

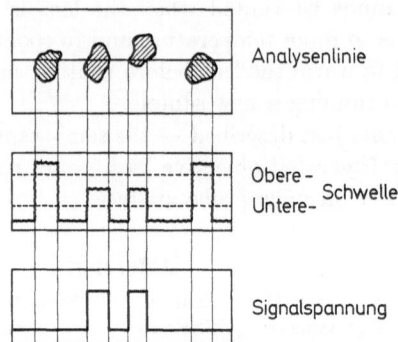

Abb. 1. Arbeitsprinzip des Phasenintegrators

nur eine qualitative Erkennung der Phasen einführen (Anwesenheit oder Abwesenheit eines oder mehrerer Elemente), wird bei den anderen Autoren [2, 9] die Impulsrate ausgewählter Elemente zur Charakterisierung herangezogen. Das Prinzip sei nochmals — obwohl bereits des öfteren publiziert — in Abb. 1 gezeigt. Wie man sieht, wird nur dann die Messung der Abtastlänge durchgeführt, wenn die Konzentration des entsprechenden Elementes in vorher festgelegte Grenzen (Schwellwerte) fällt.

Die apparativen Grundlagen, zu erwartende Fehlergrenzen und eine Reihe von Anwendungen wurden kürzlich in einer umfangreichen Arbeit zusammengestellt [5].

Neue apparative Entwicklungen

Nachdem im vorhin beschriebenen Phasenintegrator bereits die Bestimmung von Volumenanteilen, mittleren Korngrößen, spezifischen Kornoberflächen und mittleren freien Weglängen möglich war, erschien es von besonderem Interesse, auch die Korngrößenverteilung und die Anzahl der Teilchen pro Volumeneinheit ermitteln zu können. Wie kürzlich beschrieben [10], ist das einerseits durch eine Zusatzelektronik und die Auswertung im Computer möglich, andererseits kann — sofern dem Labor ein Vielkanalanalysator zur Verfügung steht — die Bestimmung auch an der Mikrosonde durchgeführt werden. Letztere Möglichkeit und ihre Realisierung soll nun eingehend besprochen werden.

Die räumliche Korngrößenverteilung (KGV) kann auf Grund statistischer Überlegungen aus der Schnittlängenverteilung ermittelt werden [11]. Diese Schnittlängen liefert uns der Phasenintegrator in der Form von Zeiträumen, die der Elektronenstrahl benötigt, um ein Korn zu passieren. Es geht nun darum, diese Schnittlängen entsprechend ihrer Größe zu klassieren und diese Häufigkeitsverteilung entsprechend zu speichern. Aus diesem Meßergebnis kann dann die KGV errechnet werden.

Die Zusatzelektronik zum Vielkanalanalysator (VKA) ist nicht allzu aufwendig. Sie bewirkt, daß entsprechend der Schnittlänge eine fortlaufende Adressierung erfolgt, so daß gleiche Längen immer im gleichen Kanal registriert werden. Die hierfür entwickelte Elektronik läßt für die Klassierung zwei Möglichkeiten zu: einerseits eine lineare Klassenteilung, wobei jedem Kanal ein bestimmter Längenbereich zugeordnet ist, andererseits die für die Umrechnung äußerst günstige [11] geometrische Klassenteilung, bei der die Längenintervalle zu hohen Kanälen hin immer breiter werden. Letztere Methode bietet den Vorteil, daß praktisch alle Gefüge mit ein und derselben Klassenteilung gemessen werden können. Beide Varianten haben gemeinsam, daß sehr große Schnittlängen, die nicht in die Klassen passen, im letzten Kanal registriert werden.

Abb. 2 gibt die gesamte Vorrichtung schematisch wieder. Dabei ist dargestellt, daß nicht nur die KGV einer Phase, sondern von 4 Phasen gleichzeitig registriert werden kann. Üblicherweise kann in VKA-Geräten das Magnetkerngedächtnis in eine Reihe von Subgruppen gespalten werden, beispielsweise bei einem 512-Kanal-Analysator in 4×128 Kanäle. Somit können — falls ein geeigneter Bereichswähler vorgesehen wird — die KGV von 4 verschiedenen Phasen während einer Messung registriert werden. Außerdem ist noch eine logische Schaltung eingebaut, die überprüft, ob der gemessene Längenabschnitt nicht am Zeilenanfang oder Zeilenende liegt, und somit zu kurz gemessen würde. Ist dies der Fall, dann wird diese Schnittlänge nicht registriert.

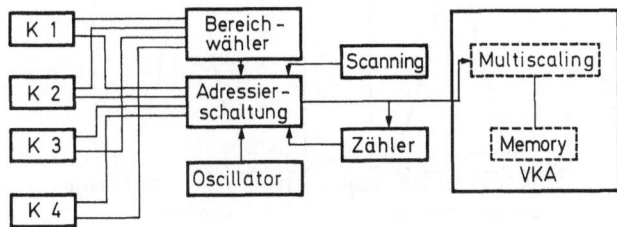

Abb. 2. Blockbild der Einspeicherung von Korngrößenverteilungen in einen Vielkanalanalysator (= VKA)

18*

Anwendungsbeispiele

Vorerst soll die Methode an zwei einfachen Beispielen demonstriert werden. Für die Überprüfung der Anlage wurde sphärisches Fe-Pulver in Kunststoff eingebettet und angeschliffen. Die Verteilung sollte, wie Exner [11] zeigte, logarithmisch normal sein und im Gaußschen Häufigkeitsnetz eine Gerade ergeben. An der Mikrosonde wurde die FeK_α-Strahlung gewählt und die Schnittlängenverteilung registriert. Das Ergebnis dieser KGV-Analyse ist in Abb. 3 dargestellt. Man sieht, daß die Bedingungen einer logarithmisch normalen KGV sehr gut erfüllt sind. In einem weiteren Versuch wurde die KGV von Kugelgraphit in Sphäroguß ermittelt. Als Signale der Mikrosonde dienten bei einigen Messungen die CK_α-Strahlung, bei anderen der Probenstrom. Beide Messungen zeigten idente Ergebnisse, jedoch gestattet der Probenstrom größere Analysengeschwindigkeiten. Abb. 4 zeigt die Ergebnisse, die bei einer Meßdauer von 1 Std erzielt wurden.

Besonders interessant erscheint die Analyse von KGV in komplex aufgebauten Festkörpern. So wurde eine Reihe von Steinmeteoriten untersucht [12], die das Mineral Troilit (FeS) enthielten. Wie die Schnittlängenverteilung zeigte, konnten hierbei eindeutig zwei Peaks registriert werden, die auf zwei verschiedene Verteilungen desselben Minerals im selben Meteoriten hinwiesen (Abb. 4). Nähere Untersuchungen ergaben, daß der bei kleineren Korngrößen liegende Peak der Sulfidverteilung in den sog. Chondren dieser Meteoriten entsprach, während der zweite Peak die Verteilung in der Matrix wiedergibt. Dies konnte durch Messungen, die ausschließlich in Chondren durchgeführt wurden, eindeutig belegt werden (Abb. 5).

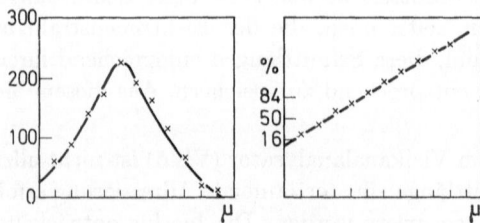

Abb. 3. Korngrößenverteilung von sphärischem Eisenpulver in Kunststoff. Links: Differentielle Verteilung; rechts: Summenhäufigkeitskurve

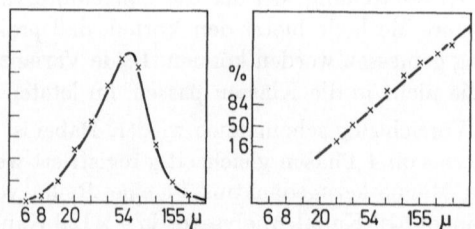

Abb. 4. Korngrößenverteilung von Kugelgraphit in Gußeisen. Rechts: Differentielle Verteilung; links: Summenhäufigkeitskurve

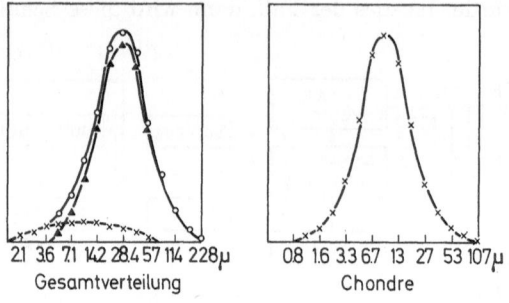

Abb. 5. Korngrößenverteilung von Troilit (FeS) in einem Steinmeteoriten

Es scheint, daß es hiermit möglich ist, zwei verschiedene Spezies desselben Minerals, die offensichtlich chemisch identisch sind, sich jedoch unter verschiedenen Bedingungen gebildet haben, zu trennen und somit sowohl ihre KGV als auch ihre relativen Anteile zu ermitteln.

Schlußfolgerungen

Die Bestimmung von Korngrößenverteilungen mittels Mikrosonde und Phasenintegrator wird durch den Einsatz eines Vielkanalanalysators sehr erleichtert. Da die zu messenden Phasen durch ihre chemische Zusammensetzung charakterisiert werden, können sehr viele metallurgische und mineralogische Proben untersucht werden, die bei Anwendung anderer Methoden erst schwierigen Ätz- oder Anfärbemethoden unterworfen werden müßten. Wie das Beispiel der Messung der Korngrößenverteilung an Meteoriten zeigt, eröffnen sich durch diese automatische Meßmethode neue Möglichkeiten, Keimbildungs- und Wachstumsvorgänge verschiedener Phasen quantitativ zu verfolgen.

Literatur

1. MELFORD, D. A., and R. WHIDDINGTON: X-ray optics and microanalysis, vol. IV, p. 497. Paris 1966.
2. DÖRFLER, G., u. E. PLÖCKINGER: Arch. Eisenhuettenwes. 36, 649 (1965).
3. — — X-ray optics and microanalysis, vol. IV, p. 506. Paris 1966.
4. — Z. Anal. Chem. 221, 375 (1966).
5. — Natl. Bur. Std. 1967. Publication in print.
6. HEINRICH, K. F. J.: Advan. X-Ray Anal. 6, 291 (1962).
7. MELFORD, D. A.: J. Inst. Metals 50, 217 (1962).
8. THEISEN, R.: Z. Metallk. 55, 128 (1964).
9. RAMSEY, J., and P. WEINSTEIN: The electron microprobe, p. 715. 1966.
10. DÖRFLER, G.: Mikrochim. Acta, Suppl. III (1968).
11. EXNER, E.: Z. Metallk. 57, 755 (1966).
12. DÖRFLER, G., u. H. G. HIESBÖCK: Int. Conference on Meteorite Research, IAEA Wien, Report SM 109/2.

Der Vielkanalanalysator als Zusatzgerät zur Mikrosonde für die Spurenanalyse und die Analyse mit geringen Strömen

G. Dörfler

Analytisches Institut der Universität Wien

E. Plöckinger

Forschungsdirektion der Gebr. Böhler u. Co. AG, Edelstahlwerke, Kapfenberg, Österreich

Der Elektronenstrahl-Mikroanalyse sind in gewissen Bereichen apparative Grenzen gesetzt. Oftmals reicht die Stabilität der Mikrosondenelektronik — von der Kathodenemission über die Linsenstabilisierung bis zu den Registriereinheiten — nicht aus, um über Stunden hinaus Daten zu sammeln, die uns über die Verteilung von Spurenelementen Aufschluß geben könnten. Andererseits ist in manchen Bereichen der Mineralogie und in vielen biologischen Fragen kein annehmbarer Kompromiß zwischen der noch für die Probe maximal erträglichen Intensität des Elektronenstrahles und der minimal zu fordernden Impulsrate zu finden. Das Wort „zerstörungsfrei" ist in diesen Fällen kaum mehr angebracht.

Die vorliegende Arbeit befaßt sich mit den Möglichkeiten, die sich zur Lösung der erwähnten Probleme durch den Einsatz apparativer Hilfsmittel ergeben.

Der Vielkanalanalysator als Zusatzgerät zur Mikrosonde

Die Anwendung des Vielkanalanalysators (VKA) bei der Elektronenstrahl-Mikroanalyse wurde bereits in mehrfacher Hinsicht in der Literatur beschrieben. Birks [1] und Heinrich [2] setzten den VKA sowohl in seiner konventionellen Anwendungsart, der Impulshöhenanalyse, als auch zur digitalisierten Einspeicherung von gesamten Scanningflächen ein. In ersterer Betriebsart registriert der VKA die Röntgenimpulse eines nichtdispersiven Röntgendetektors und speichert sie entsprechend ihrer Energie in sein Magnetkerngedächtnis ein. Als Resultat erhält man ein Energiespektrum, das in der Peaklage die anwesenden Elemente, in der Fläche unter den Peaks deren Konzentration angibt. In Verbindung mit höchstauflösenden Si-Halbleiterdetektoren lassen sich so qualitative und halbquantitative Analysen für sämtliche Elemente in ca. 1 sec durchführen [3].

Zur Einspeicherung ganzer Flächen in den VKA wird jedem Punkt der Fläche ein Kanal des VKA-Gedächtnisses zugeteilt. Dann kann, wie Birks [1] und Heinrich [2] gezeigt haben, schrittweise von Punkt zu Punkt die Röntgenintensität eines bestimmten Elementes eingespeichert werden. Dieses Ergebnis kann sowohl digital ausgedruckt [1] als auch „dreidimensional" dargestellt werden [2].

Zur Spurenanalyse wurde der Vielkanal-Analysator erstmals von Fergason [4, 5] eingesetzt. Fergason zeigte, daß man auf sehr lange Meßzeiten bei zu vernachlässigender Drift kommen kann, wenn der Elektronenstrahl zwischen Probe und Standard hin- und herspringt und jeweils auf Probe und Standard kurze Intervalle mißt. Der VKA arbeitet hierbei mit Impulshöhenanalyse, so daß keine Lokalisierung von Meßpunkten auf Probe und Standard angestrebt wird.

Die Registrierung der Verteilung niedrigster Konzentrationen
mit dem Vielkanal-Analysator

In der vorliegenden Arbeit wurde angestrebt, die lokale Verteilung von Spurenelementen zu registrieren. Auch hierfür bietet sich der VKA geradezu an. Bei den Experimenten wurde von folgenden Gesichtspunkten ausgegangen:

Die Lage der Meßpunkte entlang einer Linie kann im VKA als Kanalzahl eingespeichert werden. Ein bestimmter Meßpunkt ist damit eindeutig in seiner Lage sowohl auf der Probe als auch im VKA definiert. Als zweiter Gesichtspunkt diente eine Anregung aus der Meßtechnik

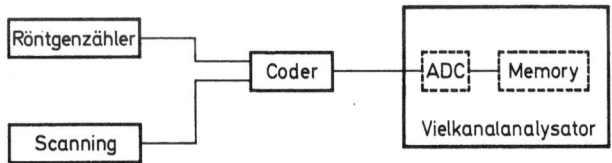

Abb. 1. Blockbild der Einspeicherung einer Analysenlinie der Mikrosonde in einen Vielkanalanalysator

in Biologie, Medizin und Astrophysik, das „signal-averaging". Bei diesem Verfahren werden sehr schwache, vom Rauschen verdeckte Signale dadurch klar isoliert, daß sie sehr oft und schnell in den VKA eingespeichert werden. Da immer im gleichen Zeitpunkt des periodischen Signalablaufes die Einspeicherung begonnen wird, überdeckt sich das Signal im VKA lagerichtig. Bei einer derartigen Behandlung analoger Signale steigt die Signalintensität des Rauschens jedoch nur mit der Wurzel der Anzahl der Abtastungen. Bei 100 Abtastungen ergibt sich also eine Signal/Rausch-Verbesserung von 10:1.

Da es sich bei der Konzentration entlang einer Linie um ein ebensolches periodisches Signal handelt, wurden auch hierfür ähnliche Überlegungen angestellt. Im Grunde genommen, handelt es sich hierbei nur um eine entsprechende Verlängerung der Meßzeit pro Punkt entlang der Linie, wie dies bei Spurenanalysen von THEISEN [6] oder GOLDSTEIN [7] gezeigt wurde. Der entscheidende Unterschied liegt jedoch auch hier wiederum in der Tatsache, daß sich diese lange Meßzeit aus einer Vielzahl kleiner Zeitinkremente zusammensetzt. Da während der gesamten Meßzeit die abzutastende Linie 50, 100 oder mehrere Male durchlaufen wird, üben Driftprobleme praktisch keinen Einfluß aus. Die Meßzeit kann erheblich ausgedehnt werden.

Abb. 1 zeigt das Blockbild der gesamten Anordnung. Die beiden, von der Mikrosonde her dem VKA angebotenen Signale sind die auftretenden Röntgenimpulse (Konzentration) und die für den Linescan erforderliche Ablenkspannung (Position).

Beide Parameter werden in einer für diese Problemstellung entwickelten Codierschaltung kombiniert und nachfolgend in den VKA eingespeist. Dieser speichert dann lagerichtig die Impulsanzahlen pro Linienelement ein.

Anwendungsbeispiele

Der Anwendungsbereich dieses Verfahrens ist natürlich weitgehend unbegrenzt. Über metallurgische Anwendungen wird in Kürze berichtet werden.

Im vorliegenden Bericht soll die Verteilung des Phosphors in fester Lösung in einem Eisenmeteoriten gezeigt werden. Wie bereits früher [8] gezeigt werden konnte, enthält die α-Phase des Meteoriten Steinbach 0,075% P, die γ-Phase je nach dem Abstand des Meßpunktes von der Phasengrenze 0,015—0,025% P. Die Konzentrationen wurden durch Punktmessungen ermittelt.

Im vorliegenden Fall wurde über eine ca. 30 μm breite γ-Lamelle des Steinbachmeteoriten ein 100 μm langer Linescan gelegt. Die Abtastgeschwindigkeit betrug 1 sec, die gesamte Meßzeit 30 min. Der Linescan wurde also während der Meßzeit 1800mal durchfahren. Abb. 2 zeigt die sich aus dem Gedächtnisinhalt ergebende Verteilung des Phosphors und dazu die Ni-Verteilung,

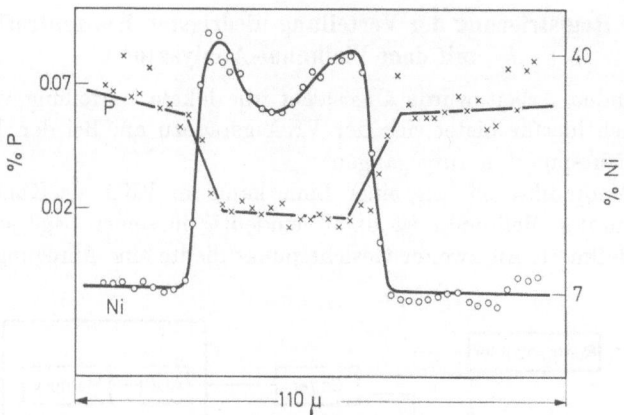

Abb. 2. Nachweis der Phosphorverteilung in einem Eisenmeteoriten; Liniendiagramm mit sehr niedrigem
Konzentrationsniveau

die während 3 min unter denselben Bedingungen aufgezeichnet wurde. Klar ersichtlich ist, daß
trotz der oftmaligen Abtastung der Linie (180mal bei Ni) die Phasengrenzen außerordentlich
scharf wiedergegeben werden.

Wie aus Abb. 2 hervorgeht, können selbst für die kurze Meßdauer von 30 min noch kleinere
Konzentrationsdifferenzen als die hier gezeigten 0,05 % unterschieden werden. Selbstverständlich
hängen Meßzeit und Empfindlichkeit der Methode von den zur Verfügung stehenden Impulsraten ab.

Die Messung mit niedrigen Strömen

Manche Minerale und ein Großteil der biologischen Präparate werden durch hohe Stromdichten des Elektronenstrahles zerstört oder verändert. Die Messung bei niederen Stromdichten
scheitert meist an den im Rauschen verschwindenden Impulsraten. In sehr vielen Fällen wirkt
sich auch eine langdauernde Belastung mit niederen Strömen sehr nachteilig aus.

Eine bedeutende Verbesserung kann die Verwendung der in Abb. 1 beschriebenen Anordnung
bringen. Hierbei wird die Probe entlang einer Linie bei sehr geringen Strömen rasch abgetastet.
Die einzelnen Phasen werden dadurch nur mit äußerst geringen Strombelastungen pro Zeit-
einheit belastet, was die Gefahr des Aufbaues von starken elektrischen Feldern oder thermischen
Belastungen weitgehend herabsetzt.

Daß auch bei extrem geringen Strömen noch eindeutige Konzentrationsprofile erhalten werden
können, zeigt Abb. 3. Hier wurde eine Linie über eine Al-reiche Schlacke in Stahl gelegt. Der

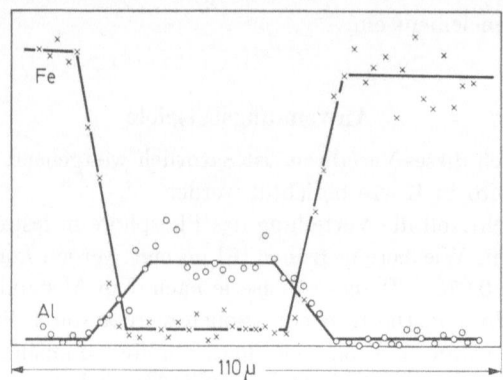

Abb. 3. Konzentrationsdiagramm über einen nichtmetallischen Einschluß unter Verwendung extrem niedriger
Strahlströme ($4 \cdot 10^{-10}$ A)

Probenstrom betrug 400 pA (0,0004 μA), die Abtastgeschwindigkeit 1 sec/Linie und die gesamte Aufnahmedauer 10 min.

Wie wichtig dieses Verfahren bei der Untersuchung von komplexen Schlacken sein kann, zeigt das Beispiel eines bariumhaltigen Schlackeneinschlusses. Tastet man einen solchen Einschluß mit einer normalen Vorschubgeschwindigkeit (10 μ/min) ab, so treten erhebliche Umwandlungserscheinungen auf, die am Absorberbild feststellbar sind. Verwendet man die iterative Schnellabtastung mit Strömen von 1 nA (0,001 μA) und weniger, so treten keine feststellbaren Beschädigungen des Einschlusses auf, und die quantitative Analyse kann wunschgemäß durchgeführt werden.

Schlußfolgerungen

Wie die gezeigten Beispiele demonstriert haben mögen, bietet die Kombination des Vielkanal-Analysators mit der Mikrosonde eine Reihe von Analysenmöglichkeiten, die bisher gar nicht oder nur mit großen Schwierigkeiten zur Verfügung standen. Die Ermittlung der Verteilung von Elementen in niedersten Konzentrationen, insbesondere auch der leichten Elemente, sowie die Analyse von sehr empfindlichem Material stellen heute Gebiete der Elektronenstrahlmikroanalyse dar, die noch wenig bearbeitet wurden, jedoch noch außerordentlich wertvolle Aussagen liefern können.

Wie in einer weiteren Arbeit [8] berichtet wird, läßt sich der VKA auch noch für andere Problemstellungen in Verbindung mit der Mikrosonde erfolgversprechend verwenden, so daß dieser als eines der erfolgversprechendsten Zusatzgeräte zur Mikrosonde erscheint.

Literatur

1. BIRKS, L. S., and A. P. BATT: Anal. Chem. **35**, 778 (1963).
2. HEINRICH, K. F. J.: NBS Technical Note 401 (1966).
3. FITZGERALD, R., K. KEIL, and K. F. J. HEINRICH: Science **159**, 528 (1968).
4. FERGASON, L. A.: Advan. X-Ray Anal. **9**, 265 (1966).
5. — Anal. Chem. **38**, 1955 (1966).
6. TÄFFNER, K., u. R. THEISEN: Euratom Rep. Eur. 1819, e, S. 27 (1964).
7. GOLDSTEIN, J. I.: J. Geophys. Res. (im Druck) (1967).
8. DÖRFLER, G., u. G. REICH: Dieses Symposium.

Développement d'accessoires adaptables au microanalyseur à sonde électronique pour l'étude des inclusions

C. CONTY, J.-M. ROUBEROL et M. TONG

CAMECA, Courbevoie, France

L'analyse élémentaire des inclusions constitue un domaine important des applications du microanalyseur à sonde électronique.

De nouveaux accessoires adaptables à la Microsonde CAMECA MS 46 viennent d'être développés pour faciliter ou compléter l'analyse des inclusions ou, d'une façon générale, des précipités ou des phases dans un alliage. Ce sont plus particulièrement:

L'Analyseur Métallographique Quantitatif,

le Microscope Electronique

et les «électrons secondaires».

L'Analyseur Métallographique Quantitatif

Certains auteurs [1—3] ont proposé des systèmes électroniques permettant de déterminer le diamètre moyen ou la concentration en volume de macro inclusions contenues dans une matrice.

Nous avons réalisé un accessoire dénommé Analyseur Métallographique Quantitatif, basé sur le principe suivant (fig. 1):

L'échantillon est exploré au moyen du système de balayage de la Microsonde. La présence de la phase analysée est détectée par un système logique, les informations analytiques fournies par la Microsonde répondant simultanément à un certain nombre de conditions fixées à l'avance. Ces conditions, choisies pour déterminer sans ambiguïté la phase considérée, peuvent être de deux natures: Intensité comprise entre deux limites, soit d'une raie X, soit d'un courant électronique (absorbé, rétrodiffusé ou secondaire). Trois signaux peuvent être utilisés simultanément pour caractériser une ou plusieurs phases.

La détermination de la concentration en volume se fait à partir de la mesure de concentration en surface. Cette dernière étant elle-même mesurée par le rapport du temps pendant lequel la sonde électronique se trouve sur la phase considérée au temps total du balayage. Pour ces mesures de temps, on utilise un *générateur d'impulsions pilotes* (à 50 Hz ou 60 Hz) dont on compte les impulsions, d'une part pendant toute la durée du balayage et d'autre part seulement lorsque la sonde électronique est sur la phase analysée.

La détermination du diamètre moyen et du nombre de précipités par unité de volume se fait à partir de la concentration en volume, de la longueur d'une ligne balayée, du nombre de lignes balayées et du nombre de traversées de la phase analysée par la sonde électronique, ces deux derniers nombres étant enregistrés par l'Analyseur Métallographique au cours de l'analyse.

En faisant l'hypothèse très simplifiée que les inclusions sont des sphères identiques dont la distribution en volume est erratique et isotrope, les formules suivantes (fig. 2) permettent de calculer les différents paramètres caractérisant ces inclusions en fonction des données fournies par l'Analyseur Métallographique Quantitatif.

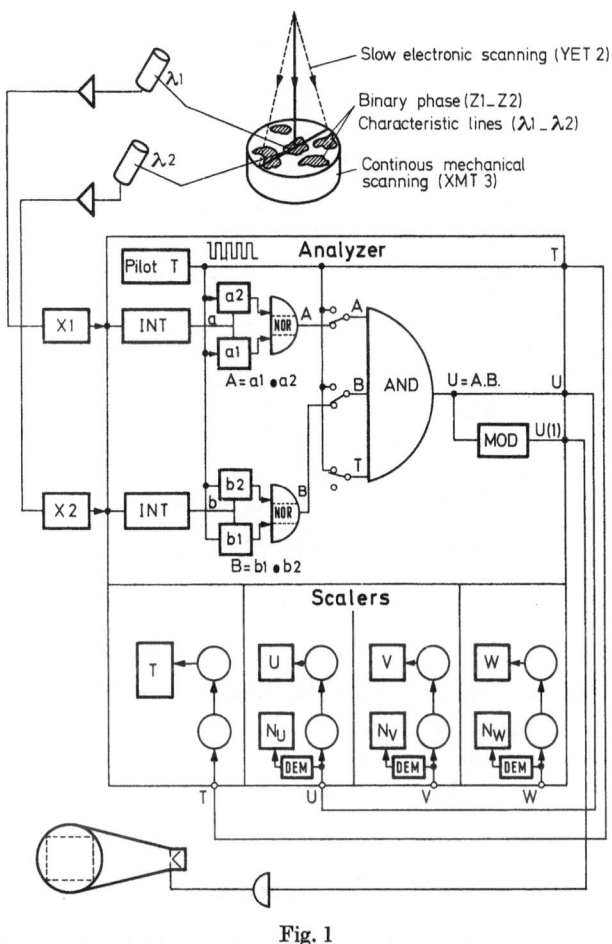

Fig. 1

Formules	Ecarts probables	
	$l_m/\Delta < 1$	$l_m/\Delta > 1$
$C_v = U/T$ $l_m = C_v N_L/N_U L$ $d_0 = \frac{3}{2} l_m$	$\left.\begin{array}{c} \\ \\ \\ \end{array}\right\}$ $1/(N_U)^{\frac{1}{2}}$	$(l_m/\Delta)^{\frac{1}{2}} \cdot 1/(N_U)^{\frac{1}{2}}$
$n_0 = 6 C_v/\pi d_0^3$	$2/(N_U)^{\frac{1}{2}}$	$(d_0/\Delta)^{\frac{1}{2}} \cdot 2/(N_U)^{\frac{1}{2}}$

Avec: L longueur d'une ligne balayée; N_L nombre de lignes par image; Δ distance (moyenne) entre deux lignes successives; U nombre d'impulsions totalisées par le compteur «U» durant le balayage; T nombre d'impulsions enregistrées par le compteur «T» durant le même balayage; N_U nombre de traversées; d_0 diamètre moyen des précipités; l_m valeur moyenne de la longueur d'une intersection de la phase analysée par le balayage ligne; C_v concentration en volume de la phase analysée; n_0 nombre de précipités de la phase analysée par unité de volume.

Fig. 2

Les applications de l'Analyseur Métallographique Quantitatif ne sont pas limitées à la mesure des concentrations de phases bien définies. Des problèmes de nature voisine peuvent également être résolus: on peut citer, par exemple, la mesure de la concentration en volume d'un constituant dont les concentrations élémentaires sont comprises entre certaines limites. On peut alors représenter, sur un oscilloscope, une image des domaines compris entre ces limites (fig. 3).

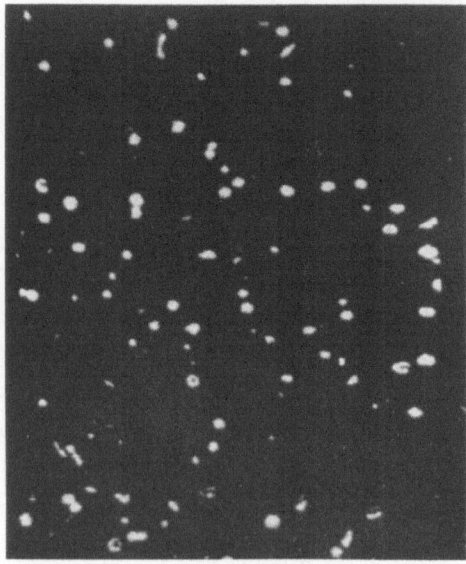

Fig. 3

L'étude des inclusions à l'aide de l'Analyseur Métallographique Quantitatif n'est possible que lorsque la surface de l'échantillon est parfaitement polie. Mais dans le cas d'étude de précipités dans les faciès intergranulaires, ceux-ci seront étudiés «in situ» à l'aide d'image utilisant l'émission électronique secondaire et par réplique au microscope électronique.

Electrons secondaires

Les images obtenues par émission électronique secondaire grâce à leur grande profondeur de champ et à la résolution voisine de 1000 Å permettent d'obtenir une reproduction fidèle de la topographie de surface très rugueuse. On a ainsi la possibilité d'observer les inclusions dans le relief d'un échantillon massif. Par exemple, on peut aisément observer (grossissement 500) le faciès de rupture d'une fonte à graphite sphéroïdal légèrement hypoeutectique (fig. 4). On note que les nodules de graphite sont situés au centre des cupules du faciès.

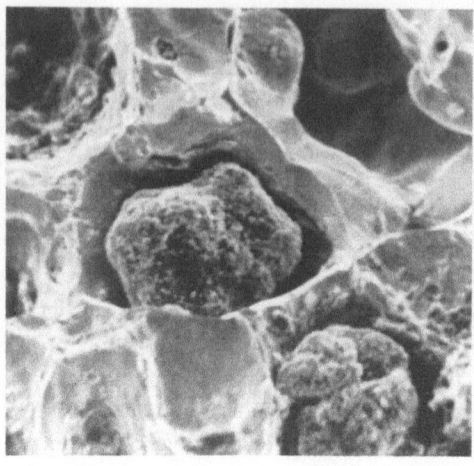

Fig. 4

Microscope Electronique

Pour l'analyse des inclusions de très petite taille, il est intéressant de les extraire sur réplique et, dans ce cas, l'utilisation d'un appareil combiné Microsonde — Microscope Electronique est particulièrement avantageux [4].

Nous avons donc développé un Microscope Electronique adaptable à la Microsonde. Ce Microscope est à deux lentilles: un objectif et un projecteur. Son pouvoir de résolution est de 50 Å et le grossissement de 10000. En défocalisant la sonde, on éclaire l'objet comme avec un double condenseur de microscope classique; au contraire, en focalisant la sonde, on réalise la visée sur un précipité que l'on veut analyser. La fig. 5 illustre les possibilités de cet appareil.

L'échantillon est une réplique d'une fracture intergranulaire d'un acier inoxydable. On observe deux types de carbures de composition $M_{23}C_6$, les gros carbures ont été analysés comme étant des carbures de Niobium et les petits carbures comme des carbures de chrome (fig. 6).

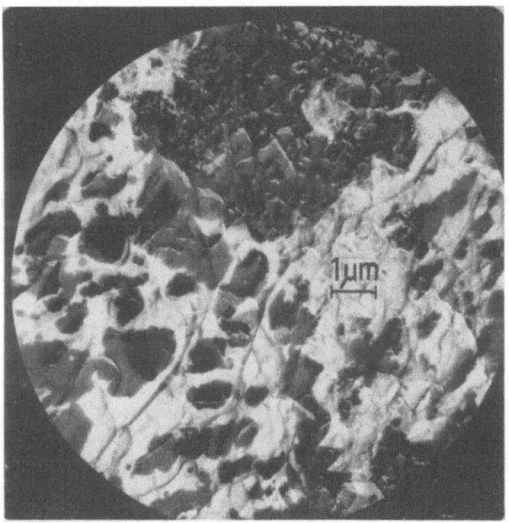

Fig. 5

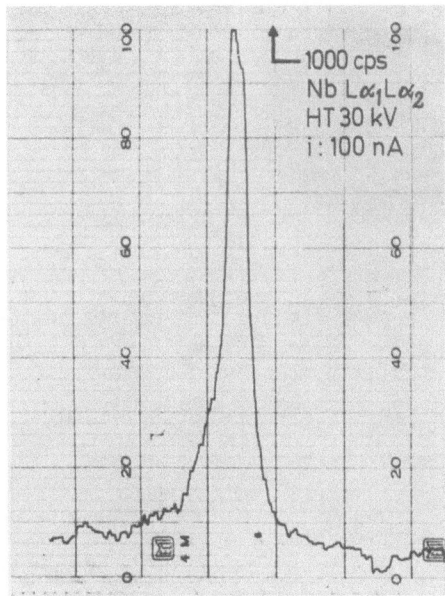
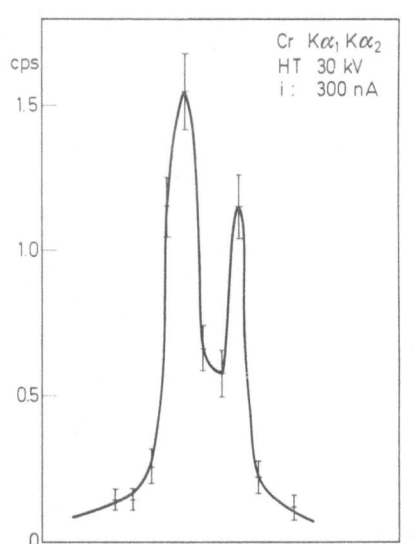

Fig. 6

L'épaisseur des échantillons couramment examinés et analysés à l'aide de l'appareil combiné et voisin de 500 à 1000 Å. Dans le cas où ces inclusions sont très petites, la quantité de matière excitée par le faisceau incident devient très faible. Finalement, la dimension des plus petites inclusions que l'on peut analyser est d'abord limitée par le niveau du signal X caractéristique émis par l'inclusion.

La photographie de la rupture intergranulaire que nous avons montrée a permis de voir que les fins carbures de chrome ($< 1\ \mu$) ont pu être analysés.

Cette méthode permet donc l'analyse des micro inclusions, mais elle nécessite l'emploi des techniques de réplique.

Ainsi équipé, le Microanalyseur est probablement l'un des instruments le mieux adapté à l'étude des inclusions par le nombre et la diversité des informations que l'on peut recueillir, dimension moyenne, concentration en volume, observation «in situ» et analyse chimique des macro et micro inclusions.

Tous ces accessoires sont évidemment compatibles, ce qui permet d'obtenir le maximum d'information dans un minimum de temps.

D'autre part, ils sont tous très facilement adaptables sur Microsonde standard.

Bibliographie

1. Theisen: Mém. Sci. Rev. Met. **60**, 1/8, No 3.
2. Melford: Optique des rayons X et microanalyse, p. 497—505. Hermann, Paris 1966.
3. Dörfler, G., et E. Plöckinger: Optique des rayons X et microanalyse, p. 506—512. Hermann, Paris 1966.
4. Duncumb, P.: Technical Report, No 182. Tube Investments Research Laboratories.

Sélecteur de fréquences X pour le rayonnement synchrotron. Etude de la réflexion spéculaire entre 6 et 14 Å

R. Barchewitz, M. Montel et C. Bonnelle

Laboratoire de Chimie Physique de la Faculté des Sciences de Paris, France

La réflexion sélective de Bragg entraîne, pour la plupart des systèmes réticulaires utilisés en spectroscopie X, la superposition de rayonnement réfléchi en ordres supérieurs sur le spectre étudié.

Lorsqu'on utilise comme source un tube à rayons X classique, le moyen le plus banal pour éliminer le rayonnement de longueur d'onde $\lambda/2$, $\lambda/3$ etc.... consiste à maintenir la tension d'excitation juste en-dessous du seuil correspondant à la longueur d'onde $\lambda/2$. Ceci limite l'intensité émise par le tube, car celle-ci est une fonction croissante de la tension d'excitation à la fois pour le rayonnement de freinage et pour les émissions caractéristiques jusqu'à plusieurs fois la valeur seuil.

On sait que l'utilisation du rayonnement continu émis par des électrons de très haute énergie en rotation dans un accélérateur circulaire présente un avantage considérable par rapport aux sources classiques du fait de la très grande intensité de ce rayonnement. Mais là encore, lorsqu'on se place dans des conditions d'intensité optimales, la proportion globale de rayonnement de courtes longueurs d'onde peut être importante par rapport au rayonnement étudié et gêner ainsi considérablement la mesure des distributions d'intensité en fonction de la longueur d'onde. En effet, la distribution spectrale de ce rayonnement s'étale vers les grandes énergies assez loin du maximum d'intensité. Ainsi, pour le synchrotron de Frascati, à l'énergie maximale de 1,1 Gev, la distribution spectrale présente un maximum vers 8 Å et s'étend de manière observable jusqu'à 1 Å environ [1—3].

Une méthode originale permettant de réduire ou d'éliminer l'intensité du rayonnement de longueur d'onde inférieure à la longueur d'onde étudiée, basée sur le principe de la réflexion totale, a été utilisée pour la première fois dans le domaine des rayons X par l'un de nous en 1956 [4]. On sait en effet que dans le domaine des rayons X, la partie réelle de l'indice de réfraction des milieux matériels est inférieure à l'unité (elle en diffère d'une quantité δ très petite, appelée décrément). Le pouvoir réflecteur d'une surface optique ne prend une valeur importante que lorsque l'angle d'attaque (complément de l'angle d'incidence) est inférieur à un angle critique $u_c \simeq \sqrt{2\delta}$, d'autant plus petit que la longueur d'onde est plus faible.

Nous avons exploité cette propriété en étudiant et en réalisant un dispositif qui, mis en place auprès du synchrotron de Frascati, permet l'élimination des courtes longueurs d'onde du spectre.

La mise au point d'un tel dispositif nous a amenés à effectuer tout d'abord une analyse du pouvoir réflecteur de différents miroirs en fonction de l'angle d'attaque du rayonnement, afin de choisir des substances réfléchissantes à «pouvoir de coupure» suffisamment net pour permettre la réalisation du sélecteur, c'est à dire présentant une chute rapide du pouvoir réflecteur pour une faible variation de l'angle d'attaque.

Mesures de pouvoirs réflecteurs

Cette étude a été faite à l'aide d'un tube à rayons X classique pour la longueur d'onde du doublet $K\alpha$ de l'aluminium à 8,34 Å.

Il a été utilisé pour ces expériences un spectrographe à cristal courbé fonctionnant sous vide [5] sur lequel a été adapté un montage porte-miroir déjà décrit [6].

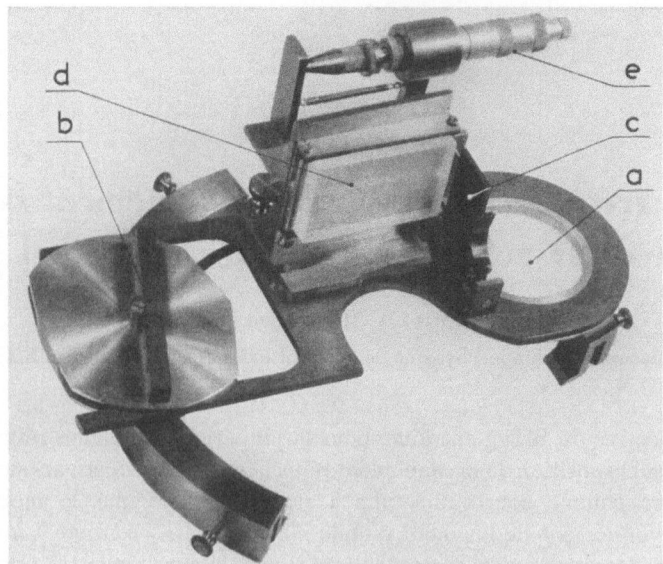

Fig. 1. *a* Emplacement du porte-cristal, *b* support du détecteur, *c* fente, *d* miroir, *e* vis micrométrique

Fig. 2

La fig. 1 représente une photographie du montage, la fig. 2 montre sa mise en place sur le spectrographe.

Grâce à ce dispositif, nous avons pu observer simultanément sur le même film les traces des faisceaux non réfléchi et réfléchi par le miroir, ce qui permet d'obtenir une précision de l'ordre de 5% sur la mesure des pouvoirs réflecteurs.

L'étude du pouvoir réflecteur R en fonction de l'angle d'attaque a été faite pour différents verres et dépôts métalliques [7]. Les résultats obtenus pour certains d'entre eux sont présentés fig. 3. La courbe expérimentale est tracée en trait épais. La courbe théorique qui s'ajuste le mieux à cette courbe expérimentale est tracée en trait fin. Elle est calculée dans chaque cas à l'aide de l'expression:

$$R = \frac{\left[\sqrt{2}\,X - \sqrt{\sqrt{(X^2-1)^2 + Y^2} + (X^2-1)}\,\right]^2 + [\sqrt{(X^2-1)^2 + Y^2} - (X^2-1)]}{\left[\sqrt{2}\,X + \sqrt{\sqrt{(X^2-1)^2 + Y^2} + (X^2-1)}\,\right]^2 + [\sqrt{(X^2-1)^2 + Y^2} - (X^2-1)]}$$

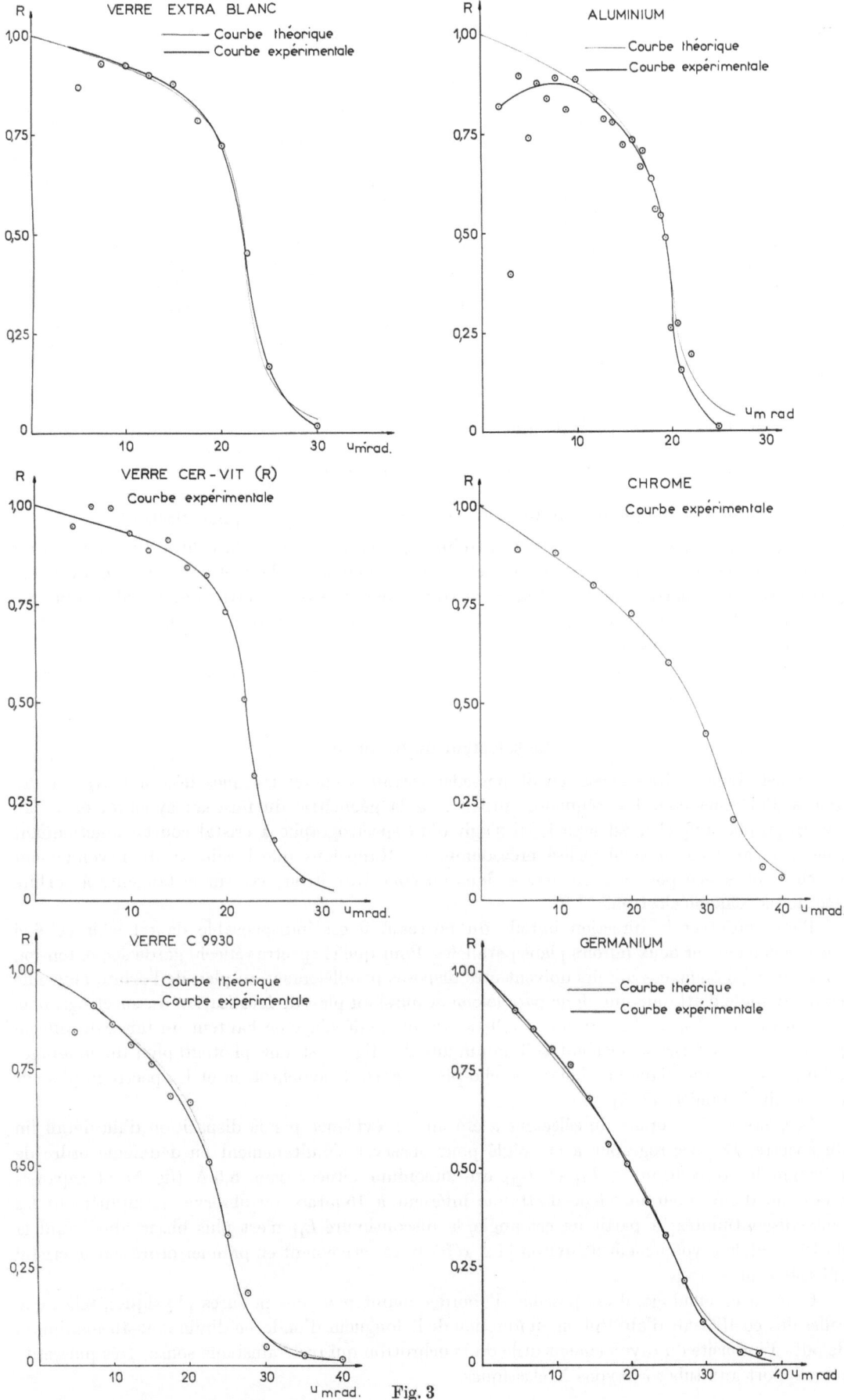

Fig. 3

Tableau

Substances	u_c mrad exp.	$\delta \times 10^4$ exp.	$\delta \times 10^4$ K-M	$\delta \times 10^4$ D-L	μ cm^{-1} exp.	μ cm^{-1} Jönsson
Verre extra blanc	$22{,}_3 \pm 1$	$2{,}_8$	2,19	2,34	3000	3035
Verre boro B 1664	$22{,}_5 \pm 1$	$2{,}_3$	2,25	2,34	4000	2895
Verre BK 7	$22{,}_5 \pm 1$	$2{,}_3$			3800	
Verre C 9930	$24{,}_7 \pm 1$	$3{,}_5$	2,58	3,38	7400	6500
Verre cer-vit (R)	$22{,}_3 \pm 1$	$2{,}_8$			2800	
Aluminium[a]	$19{,}_7 \pm 1$	$1{,}_4$	1,48	2,44	3200	900
Chrome[a]	$32{,}_2 \pm 1$	$5{,}_8$	5,77	6,24	17000	15900
Germanium[a]	$26{,}_5 \pm 1$	$3{,}_1$	3,38	4,43	18000	22250

[a] Dépôt métallique sur support en verre.

établie à partir des formules de Fresnel, où

$$X = \frac{u}{u_c} \quad \text{et} \quad Y = \frac{\beta}{\delta} \quad \text{avec} \quad \beta = \frac{\mu \lambda}{4\pi}$$

β est le coefficient d'extinction, μ le coefficient linéaire d'absorption photoélectrique.

A partir des valeurs de X et de Y introduites dans le calcul de la courbe théorique, il est possible de déduire les valeurs de l'angle critique u_c, donc aussi de δ, et de μ. Nos résultats expérimentaux sont portés dans le tableau comparativement aux résultats théoriques calculés pour δ, d'une part dans l'approximation de Drude-Lorentz (δ_{D-L}), d'autre part dans l'approximation de Kallmann et Mark (δ_{K-M}) et pour μ par la méthode semi-empirique de Jönsson.

Le verre extra-blanc qui possède un bon pouvoir de coupure a été retenu pour équiper le sélecteur de fréquences.

Le sélecteur de fréquences

Le sélecteur de fréquences devait posséder certaines caractéristiques liées à la valeur des angles d'attaque pour les fréquences utilisées, à la géométrie du faisceau synchrotron et au spectrographe auquel il est associé; il s'agit d'un spectrographe à cristal courbé fonctionnant sous vide analogue à celui utilisé précédemment. Rappelons que l'émission du rayonnement synchrotron a lieu pour ces fréquences dans un cône très étroit, axé sur la tangente à l'orbite décrite par chaque électron.

Pour conserver la direction initiale du faisceau, il est indispensable de réfléchir celui-ci successivement sur deux miroirs plans parallèles. Pour que le spectre réfléchi garde son extension en longueur d'onde, ces miroirs doivent être disposés parallèlement au plan de l'orbite moyenne pour un angle d'attaque nul, donc parallèlement aussi au plan de focalisation du spectrographe.

Nous avons disposé les miroirs de telle sorte que le décalage en hauteur du faisceau réfléchi par rapport au faisceau incident soit minimum. La fig. 4 est une photographie du montage; celui-ci est enfermé dans une boîte étanche placée entre le synchrotron et le spectrographe sur le conduit de lumière [2, 8].

Le pouvoir de coupure du sélecteur a été mis en évidence par la disparition d'un détail fin du spectre. Le spectrographe a été réglé pour observer simultanément en deuxième ordre de réflexion les discontinuités L_{II} et L_{III} du zirconium situées vers 5,5 Å (fig. 5) et séparées d'environ 0,2 Å. Pour un angle d'attaque inférieur à 15 mrad, on observe simultanément les deux discontinuités; à partir de cet angle, la discontinuité L_{II} n'est plus observable; au-delà de 16 mrad, le rayonnement d'environ 11 Å réfléchi sélectivement en premier ordre par le cristal subsiste seul.

Grâce à ce montage, il est possible d'aborder maintenant des mesures physiques, telles que celles des coefficients d'absorption en fonction de la longueur d'onde, en diminuant au maximum de 50% l'intensité du rayonnement utile du synchrotron qui reste ainsi une source très puissante par rapport aux tubes à rayons X classiques.

Fig. 4

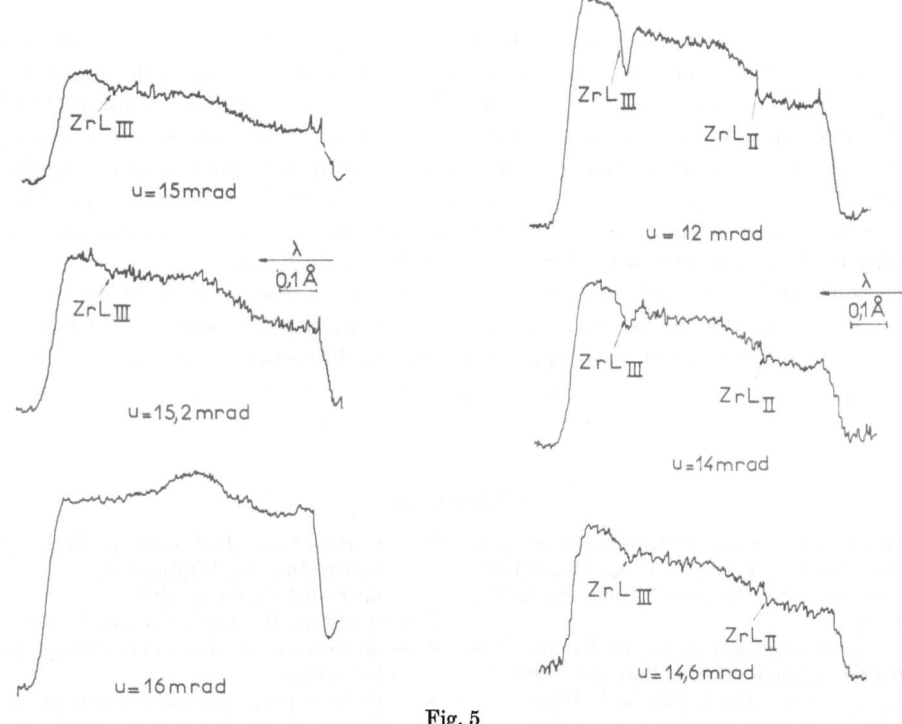

Fig. 5

La détermination de la distribution spectrale du rayonnement nécessite, quant à elle, de connaître la variation du pouvoir réflecteur des miroirs en fonction de la longueur d'onde. Mais il faudrait alors disposer d'un second sélecteur qui, placé avant le premier, permettrait d'effectuer

19*

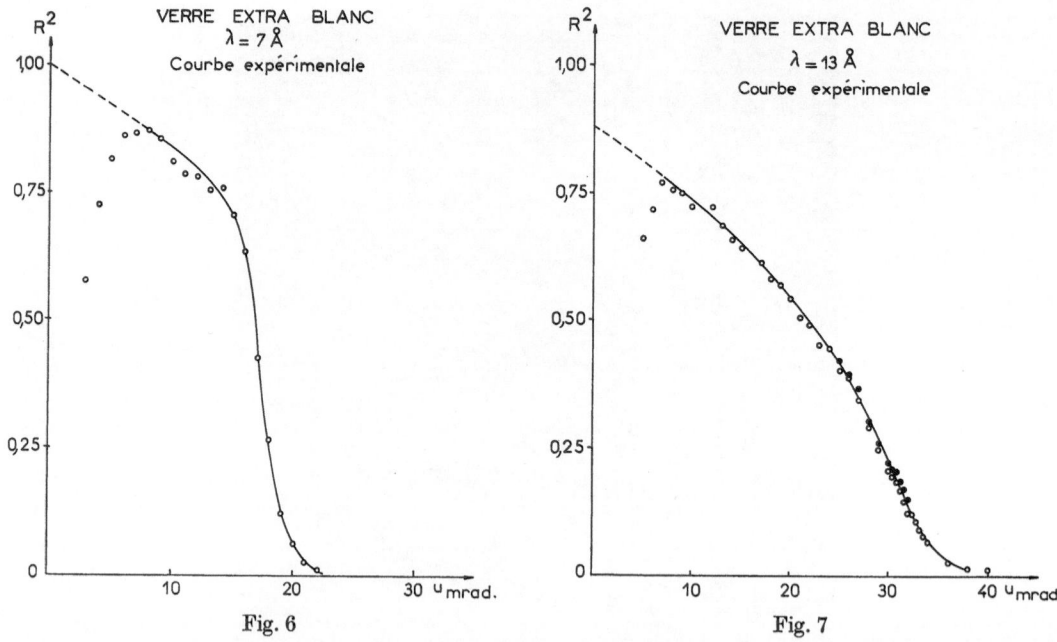

Fig. 6   Fig. 7

des mesures en rayonnement monochromatique; la réalisation de ce second montage est en cours. Nous avons effectué des mesures préalables de pouvoir réflecteur en abaissant l'énergie des électrons jusqu'à une valeur telle que l'intensité du rayonnement d'ordre supérieur reste faible; mais l'intensité globale se trouve alors fortement diminuée. De plus, la proportion de rayonnement de courtes longueurs d'onde n'est pas toujours négligeable; ainsi la variation du pouvoir réflecteur à 7 Å (cf. fig. 6) montre une structure vers 10 mrad due à la présence du rayonnement de longueur d'onde moitié. La fig. 7 représente la courbe de variation du pouvoir réflecteur à 13 Å mais obtenue plusieurs mois après la précédente courbe; la courbe extrapolée à un angle d'attaque nul correspond à un pouvoir réflecteur nettement inférieur à celui obtenu à 7 Å. Il est possible que cette diminution du pouvoir réflecteur soit due à la transformation physico-chimique du verre sous l'action des rayonnements secondaires très intenses au voisinage du synchrotron. Les expériences qui ont donné lieu aux figs. 6 et 7 ont été effectuées en collaboration avec le Laboratoire de Physique de l'Institut Supérieur de Santé de Rome.

Enfin nous mentionnerons une étude entreprise dans le cadre de cette collaboration qui porte sur la modification de la distribution spectrale du rayonnement au voisinage des discontinuités d'absorption des éléments présents dans les différents miroirs utilisés [9].

Les auteurs tiennent à exprimer leur vive gratitude à Mademoiselle Y. Cauchois, Professeur à la Faculté des Sciences de Paris, pour les conseils qu'elle leur a donnés au cours de ce travail.

Bibliographie

1. Cauchois, Y., et Y. Heno: Cheminement des particules chargées. Paris: Gauthier-Villars 1964.
2. — C. Bonnelle et G. Missoni: Compt. Rend. **257**, 409 (1963).
3. — — — Communication au 49ème Congrès National de Physique Italien, Bari, nov. 1963.
4. Montel, M.: Compt. Rend. 242, 2335 (1956).
5. Cauchois, Y.: J. Phys. Radium 6, 89 (1945).
6. — C. Bonnelle, P. Jaegle et M. Montel: IV° Congr. Int. sur l'Optique des rayons X et la microanalyse, Orsay 1965.
7. Barchewitz, R.: Thèse 3ème cycle, Paris 1968.
8. — M. Montel et C. Bonnelle: Compt. Rend. **264**, 363 (1967).
9. — C. Bonnelle, M. Cremonese et G. Onori: Compt. Rend. 268 B, 151, (1969).

Microanalysis in the Transmission Electron Microscope by Selected Area Electron Spectrometry

D. B. Wittry[1], R. P. Ferrier and V. E. Cosslett
Cavendish Laboratory, University of Cambridge, England

Abstract

Energy losses of transmitted electrons have been investigated as a basis for microanalysis using a simple homogeneous-field magnetic prism spectrometer below the camera chamber of an AEI EM 6 B. Adequate energy resolution and high efficiency are obtained by using the exit pupil of the electron microscope as an entrance slit for the spectrometer and an area-selecting diaphragm as the aperture stop. Chromatic aberration is minimized by changing the accelerating voltage; phase sensitive demodulation techniques are also used in studies involving inner-shell excitation. Some preliminary results are reported and the factors important in optimizing the signal to background ratio in microanalysis by selected area electron spectrometry are discussed.

Introduction

The possibility of using characteristic energy losses of electrons transmitted through thin films as a basis for microanalysis was first suggested by HILLIER [1] in 1943. This led to the development [2] of a "microanalyzer" using an electron beam focussed to a small probe with a magnetic spectrometer and photographic plate to record the energy loss spectra. At that time, this method of microanalysis attracted little interest because of its limited scope and because of the special instrumentation required. However, with the wealth of information that has subsequently been accumulated on characteristic energy losses [3—6] and with the recent development of instruments [7—12] that combine electron microscopy with energy selection or energy analysis, there has been renewed interest in the possibilities of microanalysis based on characteristic energy losses. As a result, the use of energy loss spectra for microanalysis has been discussed recently by several authors [13—15].

As in other methods of microanalysis, electron spectrometry can provide information on selected regions of a specimen either by focussing the incident beam into a probe or by imaging the transmitted (or emitted) beam and using appropriate apertures to receive signals corresponding to a selected area of the specimen. In the present work, we have used the selected area approach for several reasons, namely: (1) the transmitted electrons can be easily focussed to provide an image of the specimen, (2) electron spectroscopy of a selected area of the image can be done in conventional transmission electron microscopes with minimal modifications, and (3) the electron current density available in a given region of the specimen can be higher for selected area techniques than for probe techniques.

For the selected area technique, the study of large energy losses is usually restricted by chromatic aberration of the focussing system since electrons that have suffered large energy loss will not arrive at the proper image point. This limitation was recognized by HILLIER and

---

1. *Permanent address:* Department of Materials Science, University of Southern California, Los Angeles, California 90007, U.S.A.

BAKER [2] and was the principal reason for their choice of a probe method instead of a selected area method. The role of chromatic aberration in selected area electron spectroscopy was also considered by WATANABE [14]. He shows that chromatic aberration of electron microscopes using customary aperture sizes imposes a lower limit on the size of the region that can be studied using electrons of a given energy loss; for example the circle of confusion due to chromatic aberration is 400 Å for an energy loss of 400 eV at a beam voltage of 50 kV.

In the present work, the effect of chromatic aberration is minimized by using the electron spectrometer at a fixed setting and changing the voltage of the electron beam. Thus, larger energy losses can be studied, or alternatively, larger angular apertures can be used without limiting the spatial resolution by chromatic aberration.

When chromatic aberration is eliminated in this way, the smallest selected area that can be examined is limited by spherical aberration. In this case, the selected area technique can be a significant advantage over the probe technique only if adequate signal levels can be obtained without the use of large apertures. As will be shown, this appears to be true for specimens of low atomic number.

While previous authors have considered the use of small energy losses as a basis for micro-analysis it is apparent that this approach involves certain difficulties. The small energy losses are not always characteristic of the elements present in the specimen, but instead depend on the number of electrons per atom that can participate in collective excitation effects or on the details of the energy band structure. Moreover, the small energy losses usually have a large ratio of half-width to the energy loss and sometimes exhibit a significant shift in the peak position or half-width with chemical bonding effects. For these reasons, the use of small energy losses for microanalysis will probably be limited to special cases in which the small characteristic losses can be easily related to the chemical composition.

The use of larger energy losses (characteristic of inner shell excitation) provides a basis for microanalysis that is less ambiguous. However, as other investigators have noted [14] there is a significant reduction in the signals available with increasing energy loss. In addition, the use of large energy losses presents the following disadvantages; (a) the energy loss spectra near an inner-shell excitation potential has the form of an edge with fine structure similar to the structure observed near X-ray absorption edges, and (b) the ratio of intensities above and below the edge depends on many parameters such as film thickness, atomic number of the specimen, accelerating voltage, critical excitation potential and angular aperture for electron collection.

In the present paper, we evaluate the feasibility of microanalysis based on X-ray energy losses using selected area electron spectrometry and show that for elements of low atomic number, suitable specimen thickness and sufficiently high accelerating voltages, this approach to microanalysis compares favorably with other unambiguous methods of microanalysis.

Instrumentation

The electron spectrometer system was mounted below the image plane of an AEI EM 6 B transmission electron microscope as shown in Fig. 1. Since the details of the system will be published elsewhere [16], only a brief description will be given here. As shown in Fig. 1 the spectrometer consists of a magnetic prism, stigmator, exit slit and scintillation counter. The exit pupil of the electron microscope was used as the entrance slit of the spectrometer. By using a virtual rather than a real entrance slit, the addition of the spectrometer does not interfere with the normal operation of the electron microscope. The energy resolution in this case depends on the size of the objective aperture and the magnification used. However, this is not a serious disadvantage since the magnification is usually >5,000 when small selected areas are used.

An aperture near the image plane defines the selected area and also serves as an aperture stop for the spectrometer. While the size of this aperture can be changed, in practice it proved more convenient to use a fixed aperture and to vary the size of the selected area by changing

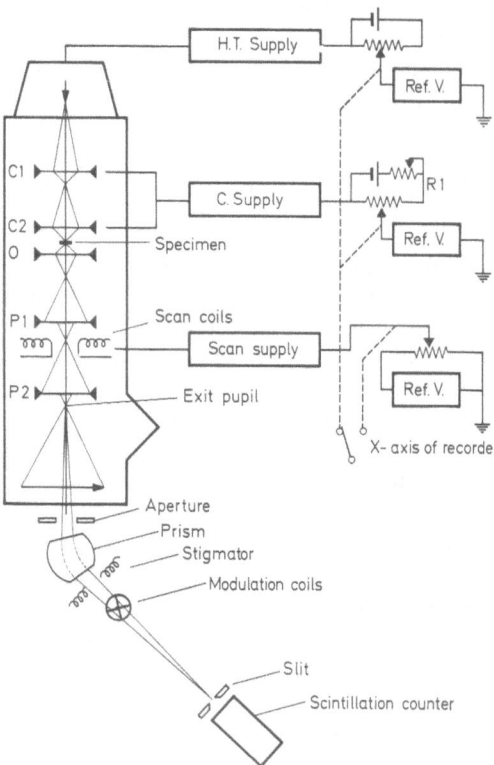

Fig. 1. Schematic diagram of the spectrometer system

the magnification. The magnetic prism was designed to provide stigmatic focussing at equal source and image distances. In this case, a deflection of the beam through an angle of 45 degrees in the spectrometer provides the necessary resolution without requiring a magnetic prism that is too large and bulky and without requiring exit slits that are too small. For example, the theoretical resolution is about 5 parts in 10^5 using a 15 micron slit. The resolution actually obtained is about twice this value, being limited mainly by ripple of the H.T. supply, stray fields, and difficulties in achieving perfect alignment. It appears that a resolution of 1 part in 10^5 could be achieved if these difficulties can be surmounted.

Beam voltages of 15, 30, and 60 kV were used in these experiments. For obtaining energy loss spectra, the beam voltage was scanned by changing the reference voltage of the H.T. supply, keeping the electron spectrometer at a fixed setting. In addition to minimizing chromatic aberration this also avoids difficulties from hysteresis of the iron in the spectrometer. However, there is a change in the intensity of illumination as the beam voltage is changed. In order to compensate for this, the reference voltage of the condenser lens supply is altered by a potentiometer mechanically coupled to the potentiometer used to change the voltage of the beam. By empirical adjustment of R 1 in Fig. 1 the current density at the specimen can be made nearly constant as the accelerating voltage is changed over a range of several hundred volts.

In studies of X-ray energy losses that are superposed on a high background, it is often useful to be able to obtain a differentiated energy loss spectrum. This can be done by periodically deflecting the electron beam from the prism across the exit slit and using phase sensitive demodulation with periodic integration of the pulse output of the scintillation counter. This is the purpose of the deflection coils shown in Fig. 1.

Variations in the energy loss spectra with position in an image or in a diffraction pattern can be investigated by using a scanning system. The scanning coils are located above the final projection lens (see Fig. 1) so that the magnetic field required is small and the distortion is minimal. However, with the spectrometer system used the amplitude and direction of the scan should be chosen so that the exit pupil of the microscope does not move significantly with respect to the virtual entrance slit of the spectrometer.

Microanalysis by X-Ray Energy Losses

The results obtained indicate [17] that large characteristic electron energy losses are more difficult to observe if the specimen thickness is not small compared to the mean free path for plasmon excitation. In these cases, the X-ray energy loss electrons are superposed on a high background due to multiple plasmon losses and there is also a loss of intensity at the X-ray loss edge due to the high probability that an electron undergoing an X-ray energy loss may also produce one or more plasmons. Thus, relatively thin films and large accelerating voltages seem to be more favorable for observation of X-ray energy losses than thick films or small voltages.

There is also some indication that electrons that contribute to the high side of the X-ray loss edge have a narrower angular distribution than the background electrons. Thus, it appears that small angular apertures may be more favorable in some cases than large angular apertures.

These conclusions are based on observations of the X-ray energy losses of K excitation in carbon, beryllium, and oxygen in films of low atomic number; it is not yet known whether similar conclusions will hold for L or M excitation of heavier elements. For the latter cases, the signal/background ratio is expected to be lower and it may be difficult to achieve comparable sensitivities.

An example of the use of X-ray energy loss spectra for microanalysis was provided by studies of Be foils thinned electrochemically by the "window" technique. After polishing as thin as practicable, most of the specimen was still too thick for meaningful energy loss measurements but several regions of high transparency were found. It was suspected that these regions were BeO instead of Be because of the polycrystalline diffraction pattern and because of the nature of the energy loss spectra at small energy losses. Energy loss spectra were obtained on a small part of one of these regions using a magnification of 10,000 so that the selected area corresponded to a disc of 0.1 micron in diameter.

The differentiated energy loss spectra from the selected area showed losses corresponding to K levels of Be, C, and O. In the normal or undifferentiated spectra, the Be K excitation appeared as a barely discernable hump on the background, but the differentiated spectra shows the Be K loss more clearly. The structure corresponding to C K excitation was attributed to contamination and the structure corresponding to O K excitation was interpreted as indicating that the specimen was BeO and not Be.

The undifferentiated spectrum near the O K excitation potential is shown in Fig. 2. In these measurements, the illumination was focussed to a spot of about 3 μm in diameter using a condenser aperture of 250 μm. By illuminating a region of the specimen much larger than the selected area, larger beam current densities are attainable since the current in an electron probe with a thermionic source varies as the 8/3 power of the diameter.

For an objective aperture of 400 microns, the intensities are considerably higher, but the signal to background ratio is lower than the values obtained with a 50 micron objective aperture. With the larger aperture, spherical aberration of the objective lens limited the spatial resolution, while for the smaller aperture, the spatial resolution of the analysis was determined by the size of the selected area diaphragm. The estimated film thickness was about 300 Å and the selected area was 1,000 Å; hence the volume was only about 2.4×10^{-4} μm^3. Assuming the specimen to be BeO, the amount detected in this case corresponds to about 4×10^{-16} gram. From counting statistics, it appears that 6×10^{-18} gram of oxygen could be detected in measurement times of the order of 100 seconds. Thus it appears that this method of microanalysis can provide significantly lower detection limits than that presently obtained in electron probe X-ray microanalysis for elements of low atomic number.

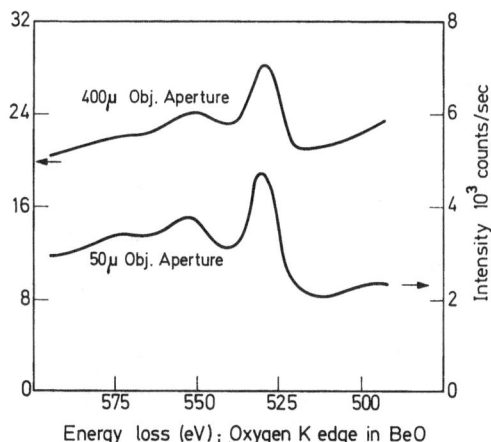

Fig. 2. The oxygen K edge from BeO obtained using a selected area aperture of 1 mm, a magnification of 10,000 and a primary beam voltage of 60 kV. The traces were recorded with a periodic integration system using a period of $\frac{1}{2}$ second

Summary and Conclusions

The use of selected-area electron spectroscopy can be a valuable ancillary technique in the conventional transmission electron microscope. It is possible to perform selected-area electron spectroscopy by the simple addition of a magnetic prism spectrometer using the exit pupil of the electron microscope as the entrance window of the spectrometer. The energy resolution attained in the present work was limited by the stability of the high voltage supply and by stray alternating fields; with improvement in these factors, it should be possible to achieve a resolution of 1 part in 10^5.

Characteristic losses due to inner shell excitations can be used for unambiguous qualitative analysis. The possibility of observing these losses can be increased by using thin specimens, differential recording techniques, high voltages, and suitable angle-limiting apertures. The angular distribution of the characteristic high energy losses can be investigated by using the electron miscroscope in the diffraction mode. Further work on the angular distribution of characteristic energy losses corresponding to inner shell ionization is of theoretical as well as practical importance.

It is not expected that the use of selected-area electron spectroscopy will compete with X-ray microanalysis for analysis of specimens containing predominantly heavier elements. However, for very thin specimens and for light elements, selected-area electron spectrometry has many advantages over the combined electron microscope-microanalyser approach (EMMA) because of the low fluorescence yield of elements of low atomic number and because of the difficulties in detecting long wavelength X-rays with high efficiency.

If the specimen contamination can be better controlled and if conventional thermionic electron sources are used, it appears that selected area techniques may be superior to electron probe techniques in microanalysis using electron energy loss spectra.

Acknowledgements. The authors would like to thank K. A. RIDAL of the English Electric Steel Corp., Ltd., for providing the special iron used in the construction of the spectrometer and T. W. STUBBINGS and R. J. RANDOLF for their skillful and rapid construction of the various parts of the spectrometer. N. BETT (Cavendish Laboratory) and J. McCOY (University of Southern California) contributed advice on the construction of the electronic recording system. Many others in the electron microscope group at the Cavendish Laboratory also aided by suggestions and technical assistance, particularly J. BELL and D. NEWLING. One of the authors (D.B.W.) gratefully acknowledges a grant from the Guggenheim foundation which made it possible to undertake this work during a sabbatical leave from the University of Southern California. Additional support for this work was also provided by the United States Air Force Office of Scientific Research, Office of Aerospace Research, grant number AF-AFOSR-68-1414.

References

1. Hillier, J.: Phys. Rev. **64**, 318—319 (1943).
2. —, and R. F. Baker: J. Appl. Phys. **15**, 663—675 (1944).
3. Marton, L., L. B. Leder, and H. Mendlowitz: Advances in electronics and electron physics, vol. 7, p. 183. New York: Academic Press 1955.
4. Pines, D.: Solid state physics. In: Advances in solid state physics, vol. 1, p. 432—450. New York: Academic Press 1955.
5. Klemperer, O., and J. P. G. Shepherd: Advances in physics. Phil. Mag., Suppl. **12**, 355—390 (1963).
6. Raether, H.: Solid state excitations by electrons (Plasma oscillations and single electron transitions). Springer Tracts in Modern Physics, vol. 38, 1965.
7. Castaing, R., et L. Henry: Compt. Rend. **255**, 76—78 (1962). — J. Microscopie **3**, 133—152 (1964).
8. Watanabe, H., and R. Uyeda: J. Phys. Soc. Japan **17**, 568—570 (1962). — Japan. J. Appl. Phys. **3**, 480 (1964).
9. Cundy, S. L., A. J. F. Metherell, and M. J. Whelan: J. Sci. Instr. **43**, 712—741 (1966).
10. Crewe, A. V.: Science **154**, 729 (1966).
11. Ichinokawa, T., and Y. Kamiya: Electron microscopy, vol. 1, p. 89—90. Tokyo: Maruzen Co., Ltd. 1966.
12. Curtis, G. H.: Thesis. University of Cambridge. 1968.
13. Castaing, R., A. El Hili et L. Henry: Optique des rayons X et microanalyse, p. 178—182. Paris: Hermann 1966.
14. Watanabe, H.: Optique des rayons X et microanalyse, p. 73—76. Paris: Hermann 1966.
15. Cundy, S. L., A. J. F. Metherell, and M. J. Whelan: Phil. Mag. **17**, 141—147 (1968).
16. Wittry, D. B.: To be published.
17. — R. P. Ferrier, and V. E. Cosslett: To be published.

Microanalysis by Electron Energy Analysis with a Cylindrical Magnetic Lens

T. Ichinokawa and H. Tochigi

Department of Applied Physics, Waseda University, Tokyo, Japan
Akashi Seisakusho, Ltd., Marunouchi, Tokyo, Japan

Introduction

The energy analysis of electrons is important not only for the study of absorption and inelastic scattering of electrons in thin films, but also for the exact explanation of electron micrographs and diffraction patterns. In some cases, the energy analysis of electron micrographs can be applied to the identification of material of a specimen part of the electron microscopic dimension by taking advantage of the characteristic energy loss spectra.

Up to the present, there are several kinds of high resolution electron energy analyzers:
1. Semi-circular magnetic deflector (Ruthemann [1] and Rang [2]).
2. Gegenfeld-type electron filter (Boersch [3, 4]).
3. Electro-static cylindrical lens (Möllenstedt [5]).
4. Wien filter (Boersch [6] and Legler [7]).
5. Electron mirror + Magnetic deflector (Castaing [8]).

Recently, however, the present author has found that a cylindrical magnetic lens is also available to the electron energy analysis when it is energized up to the lens strength of $k \simeq \sqrt{3}$. Thus, a simple cylindrical magnetic lens was designed and used in the place of the intermediate lens of a usual electron microscope.

Some results of the energy analysis of electron micrographs and electron diffraction patterns are shown in this paper. Furthermore, the energy selecting microscope of scanning type was tried by using the energy analyzer of this type.

Instrument

Fig. 1 is a schematic diagram of magnetic pole-pieces of the analyzer. The analyzer pole-pieces were inserted into the magnetic bore of the intermediate lens and energized through a magnetic yoke of the intermediate lens coil of 22,000 turns by a stabilized d.c. source. The pole-pieces were separated into two parts (right and left). The distance between the right and left was 4 mm and gap was 10 mm. The out-side diameter was 22 mm the same as that of intermediate lens pole-pieces. X, Y, and Z-axes are defined as shown in Fig. 1. Then the cylindrical lens has only lens action in X-direction, while it has not any action in Y-direction. The field limiting aperture of the intermediate lens was also replaced by a fine slit as shown in Fig. 1.

Characteristics of the Cylindrical Magnetic Analyzer

As increasing the analyzer current, the first and second focal lines of the electron source are formed on the screen by the analyzer lens at lens strengths of F_1 and F_2, and magnification of X-direction becomes infinity at G_∞. In the present experiment, the lens currents for F_1, G_∞, and F_2 were 8.5 mA, 70 mA and 105 mA, respectively, at 50 kV and values of $IN/\sqrt{V^*}$ were 1, 6.9 and 10.4 respectively.

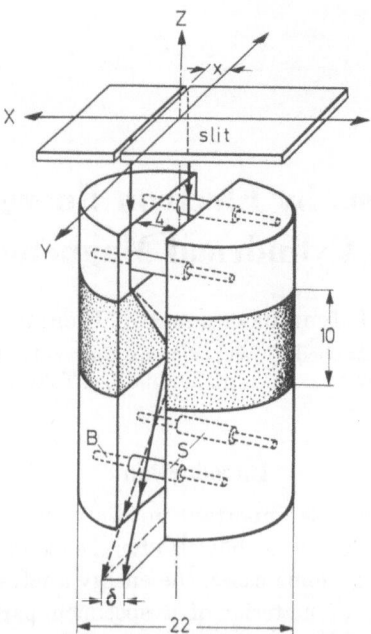

Fig. 1. Schematic diagram of magnetic pole-pieces used in this experiment

The chromatic aberration constant for off-axial rays of the analyzer lens was obtained by adding a small change of analyzer current, and using following equations

$$\delta = C_{ch} \cdot x \cdot \frac{\Delta V}{V},$$

$$\frac{\Delta V}{V} = -2 \frac{\Delta I}{I}$$

where δ is an amount of the image shift reduced to the object plane of the analyzer, C_{ch} is a chromatic aberration constant for off-axial rays, and x is a distance from the optical axis. The chromatic aberration constant becomes 0 at the lens strength of G_∞ and very large near F_2. Therefore, the analyzer lens should be used at the lens strength close to F_2. Fig. 2 shows the

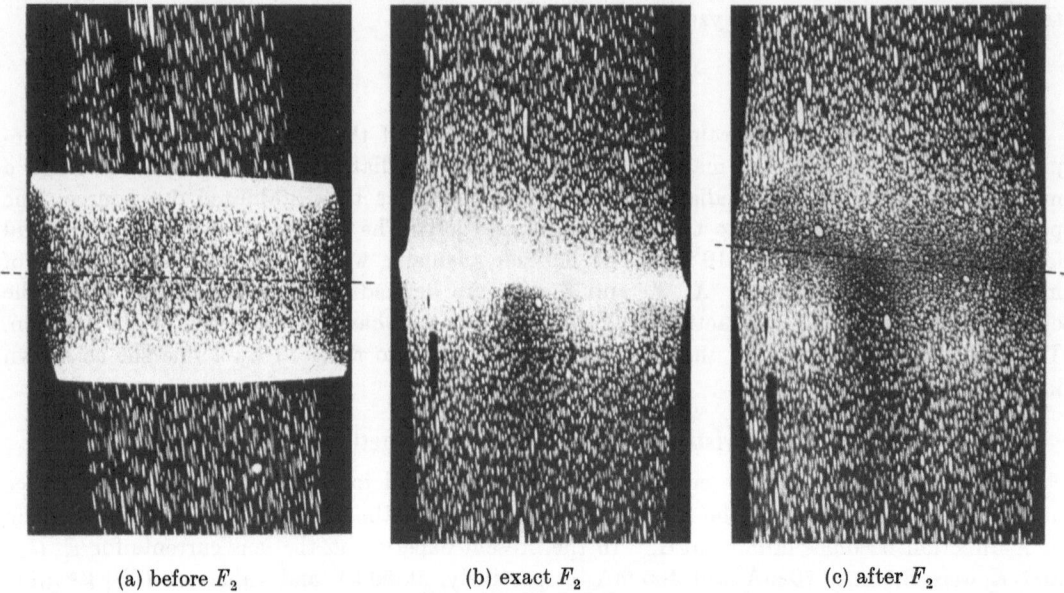

(a) before F_2 (b) exact F_2 (c) after F_2

Fig. 2. Images formed by the analyzer lens close by F_2

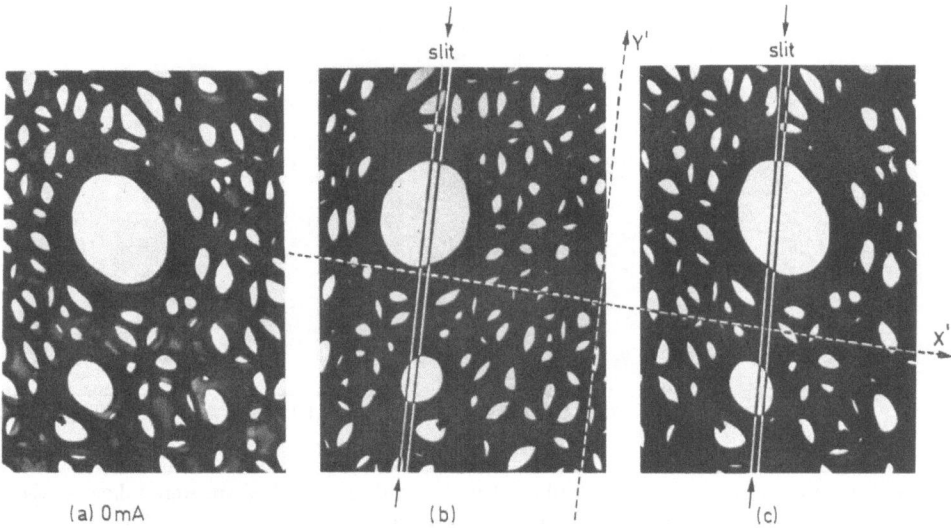

Fig. 3. Same areas of several tens micron from the optical axis in Fig. 2a and c were magnified by the projector lens as shown in b and c, respectively. a was taken at 0 mA

images formed by the analyzer lens close by F_2. Specimen is a microgrid. Fig. 2a is a micrograph taken by the lens strength weaker than F_2. Two horizontal lines in a are caustic lines, and the optical axis of the analyzer is indicated by a dotted line. Fig. 2b is taken at exact F_2, and c over F_2. The same areas of several tens micron from the optical axis were magnified by the projector lens. The images of those areas have equal magnifications in X and Y-directions. Fig. 3b shows a magnified image of Fig. 2a at the place close to a caustic line. Fig. 3c is a similar

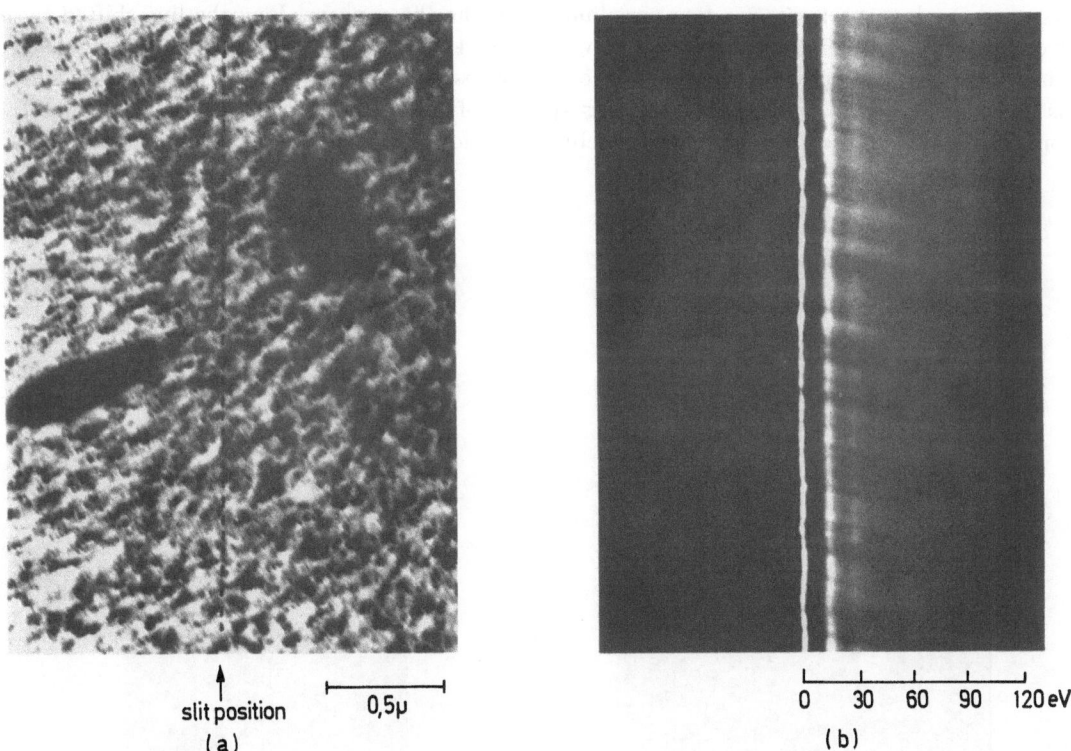

Fig. 4. a Electron micrograph of an evaporated Al film of 600 Å thick. b Energy spectrum along the line indicated in a

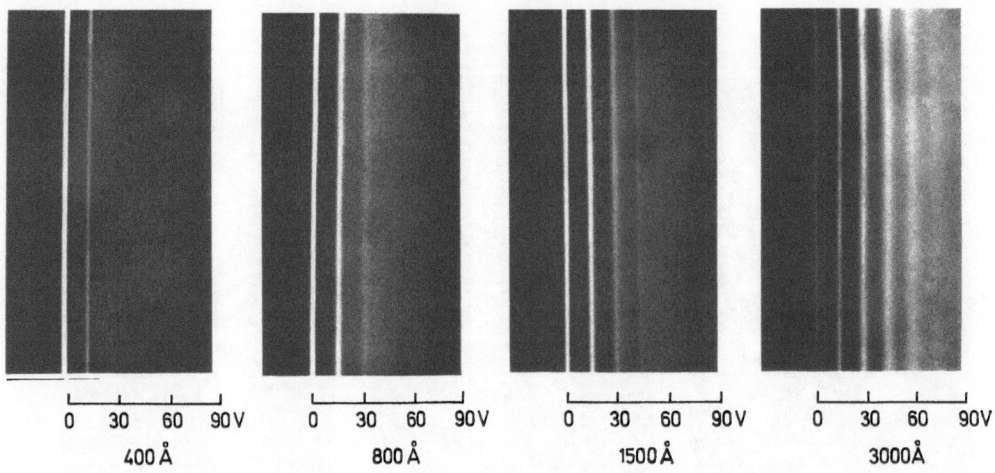

Fig. 5. Energy loss spectra of evaporated Al films of various thicknesses. Spectra were taken by the caustic method. Accelerating voltage was 80 kV

image of Fig. 2c at the place near the optical axis. Fig. 3a was taken at the analyzer current of 0 mA. Fig. 3c is alike the image of Fig. 3a, because c is a doubly inverted image of a with respect to the optical axis. In this figure, the optical axis is Y-direction, and analyzer lens acts to the X-direction as shown in Fig. 3.

Energy Analysis of Electron Micrographs

If we put the narrow slit on the images of Fig. 4b and c at a distance x from the optical axis, energy loss spectrum of a fine part limited by the slit appears along the line shifted by δ from the no-loss one. An example is shown in Fig. 4 for an evaporated Al film about 600 Å thick. Accelerating voltage was 50 kV. The energy loss spectrum along the line indicated in a is shown in b. Fig. 5 shows the energy loss spectra of Al films of various thicknesses taken by putting the slit on the caustic line, namely, by a caustic method. Accelerating voltage was 80 kV.

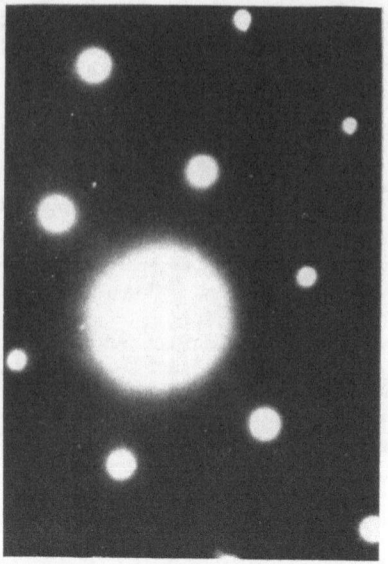
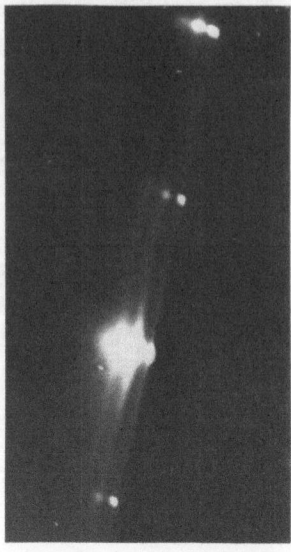

Fig. 6. Energy loss spectrum of a micro-diffraction pattern of molybdenite crystal film

Energy Analysis of Electron Diffraction Patterns

The energy analysis of electron diffraction patterns was made by inserting an analyzer lens between intermediate lens and projector, and by focusing the micro-diffraction pattern on the slit plane. Fig. 6 shows an example of the energy loss spectrum of a micro-diffraction pattern of a molybdenite film.

Energy Selecting Microscope

The entire image of energy selecting microscope was obtained by inserting scanning deflectors into the electron optical column as shown in Fig. 7. The principle of the energy selecting microscope has been reported by WATANABE and UYEDA [9]. Fig. 8 shows images taken by the specimen of a single crystal molybdenite film. a is an ordinary image, b a scanning image

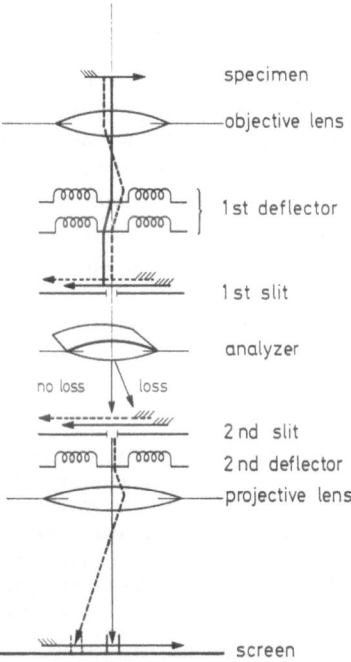

Fig. 7. Electron optical system of the energy selecting microscope

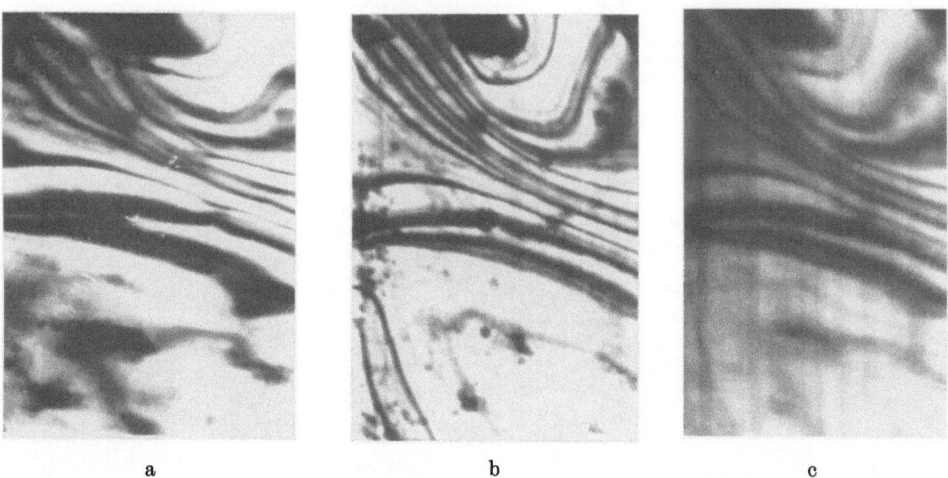

Fig. 8a—c. Energy selecting microscope images of a single crystal molybdenite film; a is an ordinary image, b no-loss image, and c loss image

formed by no-loss electrons, and c loss-electrons. Exposure times were 30 seconds in b, and 1.5 minuts in c. Bend extinction contour can be observed even in the loss image. This fact shows that inelastically scattered electrons produce a diffraction contrast. Furthermore, it should be noted that no-loss image has a higher contrast than that of the ordinary image, and ordinary image is largely affected by loss electrons. The energy resolution of energy selecting microscope is determined by an energy dispersion of the analyzer and a width of the 2nd slit, and was about 1 eV at 80 kV in this experiment.

Examples to identify the material of micro-specimens are not demonstrated here, but such an experiment is possible by this method for some specimens as performed by Castaing et al. [10].

Discussions

The highest resolution which has been attained by this analyzer is 0.5 eV at 50 kV, i.e. the width of the slit image was 0.2 mm and δ was 4 mm on the screen for $\Delta V = 10$ eV at 50 kV. Since the magnification of the projector lens was about 50 times, the dispersion of the analyzer itself was 0.1 mm/10 eV at the object plane of the projector. This value seems to be smaller than that of the Möllenstedt type. But, the spherical aberration of the magnetic lens is also smaller than that of the electro-static lens, hence the energy resolution seems to be comparable with that lens if the accelerating voltage would be kept to constant.

The stability of the accelerating voltage is a serious problem in the magnetic analyzer, but merits of the present one are (1) no change of instrumental arrangement is necessary, (2) it is very easy to handle, and (3) it can easily be used at any high voltage.

Acknowledgement. The authors wish to express their appreciations to Prof. R. Uyeda of Nagoya University for his encouragement.

References

1. Ruthemann, G.: Naturwissenschaften **29**, 648 (1941).
2. Rang, W.: Optik **3**, 233 (1948).
3. Boersch, H.: Z. Physik **134**, 156 (1953).
4. — Z. Physik **139**, 115 (1954).
5. Möllenstedt, G.: Optik **5**, 499 (1949).
6. Boersch, H., J. Geiger u. W. Stickel: Z. Physik **180**, 415 (1964).
7. Legler, W.: Z. Physik **171**, 434 (1963).
8. Castaing, R., et L. Henry: Compt. Rend. **255**, 76 (1962).
9. Watanabe, H., and R. Uyeda: J. Phys. Soc. Japan **17**, 569 (1962).
10. Castaing, R., A. El Hili et L. Henry: Compt. Rend. **261**, 3399 (1965).

Effect of Electron Source
to Energy Resolution in Electron Velocity Analysis —
Interpretation of Boersch Effect

T. ICHINOKAWA

Department of Applied Physics, Waseda University, Tokyo, Japan

Introduction

In 1954, BOERSCH [1] reported that a width of the emission spectrum of an electron beam increases with the beam current or the current density. Those experiments were carried out by a retarding potential analyzer at a beam energy of 30 kV. Later, similar experiments were repeated by several workers with using energy analyzers of various type [2—4]. All those workers agreed that the velocity distribution was broadened so that the effective cathode temperature might be as high as 3×10^4 °K, i.e. at least ten times the real temperature.

On the other hand, theoretical explanations of this effect were also proposed by several workers taking into account of the space charge oscillations in the electron beam, coupling mechanisms of the electron cynclotron resonance [5], plasma interactions between electrons and residual ions [6], and so on [7]. However, those explanations were not satisfactory to interpret the effect.

Recently, however, BECK and MALONEY [8] measured energy distributions of the emission spectra by using an energy analyzer of electro-static einzel lens for low energy electrons about 20 volts. Cathode temperatures deduced from the emission spectra agreed with temperatures measured by an optical pyrometer. They concluded that measurements showed no sign of any broadening of emission spectrum. However, in their paper, they did not refer at all to the origin of the Boersch effect. Therefore the origin of the Boersch effect has been yet a matter of controversy.

Since the Boersch effect gives an important effect to the energy resolution of electron velocity analysis and the position resolution of high resolution electron microscopy, the experiment to make sure of its origin have been tried by using a velocity analyzer of a cylindrical magnetic type.

Experiments

Fig. 1 shows the arrangement of the electron optical system. The electron gun and double condenser lenses are placed on the specimen chamber whose upper surface is a sphere with radius 155.6 mm and center located on a specimen plane, hence the illumination beam can be inclined without a movement of the irradiation spot on the specimen plane. Three sizes of 2nd condenser aperture, 0.1, 0.05, and 0.03 mm $\varnothing$, were used. Grid bias was changed by varying the self-bias resistors. Three kinds of cathodes, usual hair pin cathode of tungsten wire, pointed cathode, and oxide cathode were used. The brightness B of electron sources were measured and are shown in the table for each beam current of three cathodes.

A feed back circuit of the high voltage stabilizer is shown in Fig. 2. Total fluctuation of the accelerating voltage was less than 0.12 volts at 50 kV. On the other hand, the variations of lens currents were less than 2×10^{-6}. The total drift of the accelerating voltage and lens current was measured by observing the shift of the line spectrum for 10 minuts, and attained 0.03 volts/min. Thus, a total fluctuation became less than 0.15 volts/min.

Table. *Brightness of three cathodes for each beam current*

	Beam current (in μA)	Brightness (Amp/sterad·cm²)
Hair pin cathode	60	3.9×10^5
	20	6.0×10^4
	4	7.6×10^3
Pointed cathode	10	8.0×10^5
Oxide cathode	8	4.0×10^4

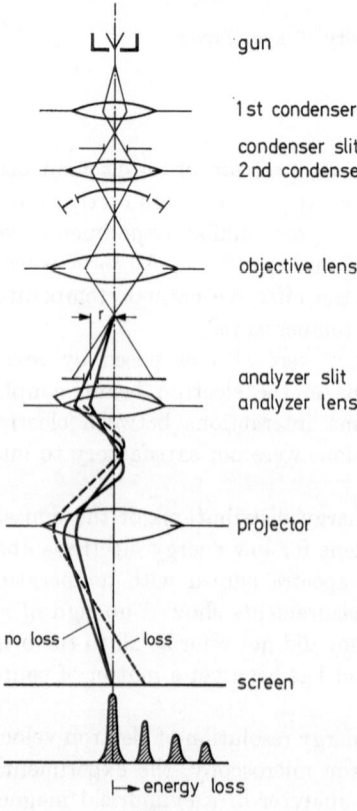

gun

1st condenser

condenser slit
2nd condenser

objective lens

analyzer slit
analyzer lens

projector

no loss   loss

screen

energy loss

Fig. 1

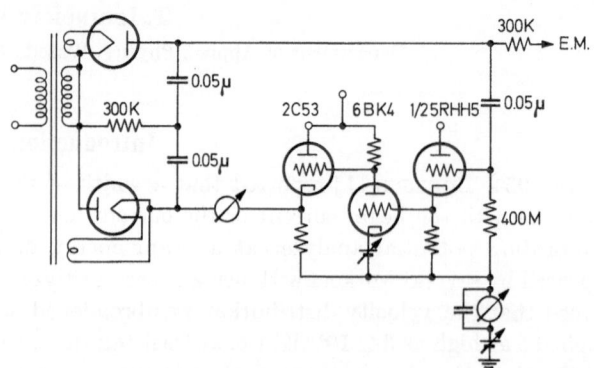

Fig. 2. The feed back circuit of the high voltage stabilizer.
(By Onoguchi)

Fig. 1. Optical arrangement of electron velocity analysis

Experimental Results

The energy width of the emission spectrum was measured by the caustic method. An example is shown in Fig. 3a. The distance between both lines corresponds to 10 eV. The half widths of the emission spectra were measured from the line profiles of the microdensitometer curves as shown in Fig. 3b. The beam current of hair pin cathode was changed by the Wehnelt bias or relative position of cathodes against the Wehnelt cap. In the present arrangement, the maximum current density of an electron probe was made at the back focal plane of the 1st condenser and was about 10^{-1} A/mm² for 60 μA of the hair pin cathode.

Fig. 4 shows a relation between line shift S and amount of the change of the accelerating voltage, ΔE. The relation is almost linear. The energy dispersion, $dS/d(\Delta E)$, was 0.4 mm/volt at the screen.

Similar experiments were made by varying the inclination angle α from the optical axis. α was changed mechanically by displacing the illumination system with screws along the spherical surface on the specimen chamber. A relation between S and α is linear as shown in Fig. 5. The

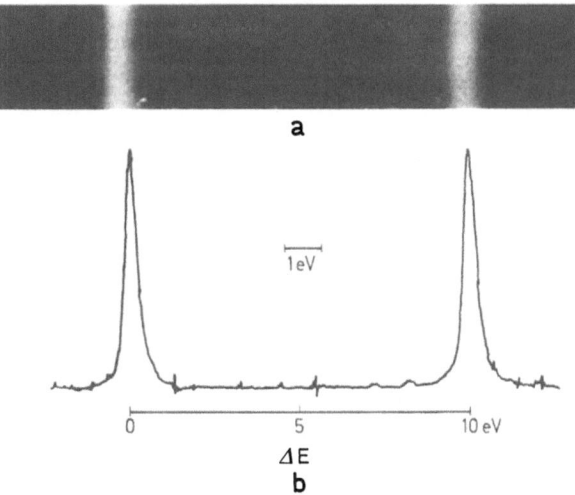

Fig. 3. a Line spectra of the incident beam. The distance between both lines corresponds to 10 eV.
b Microdensitometer curve of the line spectra of a

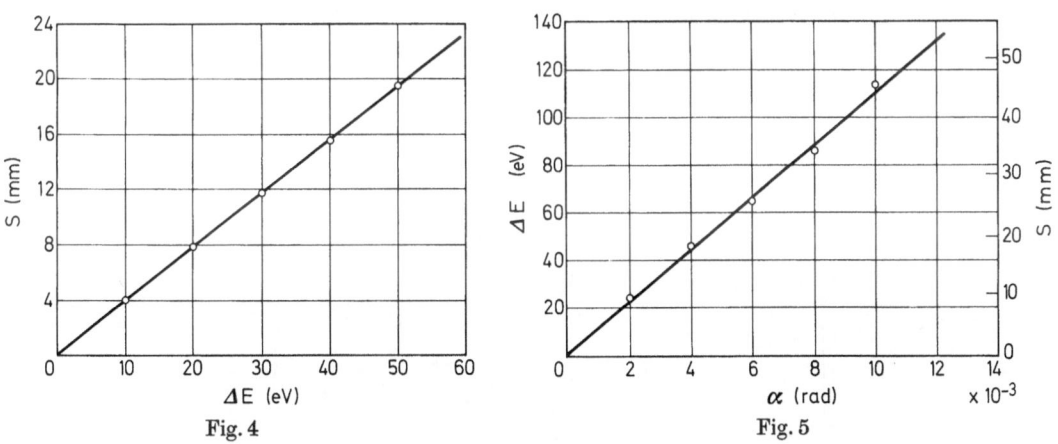

Fig. 4. A relation between line shift S and variation of the accelerating voltage $\varDelta E$

Fig. 5. A relation between line shift S and inclination angle α. Inclination angle was measured from the optical
axis of the objective lens

sensitivity of the line shift due to the inclination angle, $d(\varDelta E)/d\alpha$, was 1×10^4 volts/rad. This
value shows that a semi-divergent angle of 1×10^{-4} rad in the incident beam gives rise to an
energy spread of 1 eV.

Fig. 6 shows the variation of the line width by varying the divergent angle of the incident
beam with the 2nd condenser. 1st condenser current was fixed at 0.85 A. The value of 0.51 A
in the 2nd condenser current corresponds to a lens excitation by which the image of electron source
is focused onto the specimen plane. One group of curves shows relations between line width
and 2nd condenser current when the condenser aperture was 0.1 mm $\varnothing$. Another group corresponds to 0.03 mm $\varnothing$. Every curve decreases with the increase of the condenser current. A
converged value of these curves becomes 0.5 ± 0.05 eV for tungsten cathodes, and 0.4 ± 0.1 eV
for the oxide cathode.

Fig. 7 shows relations between line width and brightness. The measurements of these curves
were made under conditions that 1st condenser current was fixed at 0.85 A and 2nd condenser
focused the electron source onto the specimen. From this figure, it is easily found that line widths

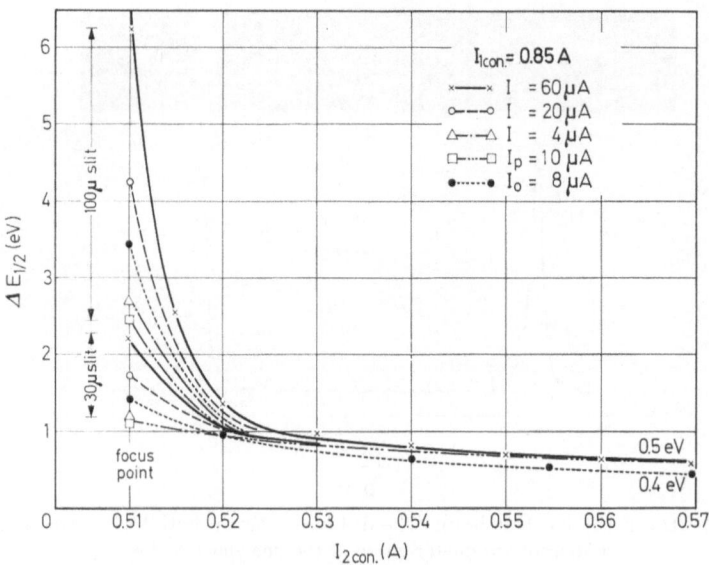

Fig. 6. The variation of half width of the emission spectrum against the 2nd condenser current as a parameter of condenser aperture size. 1st condenser current was fixed at 0.85 A

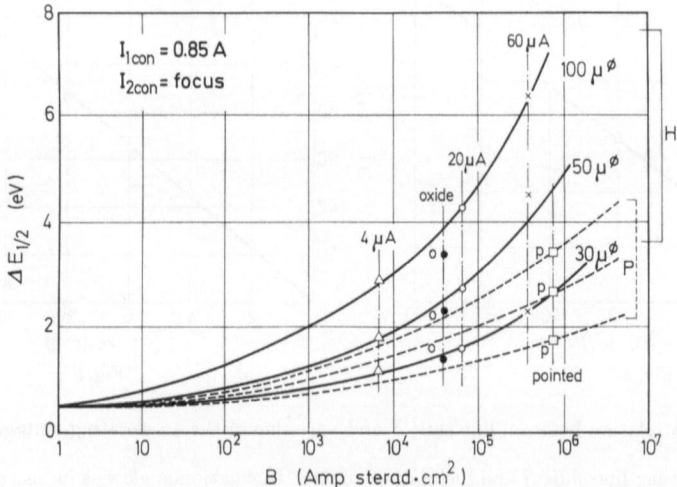

Fig. 7. The variation of half width against the brightness for each beam current of three cathodes. 2nd condenser apertures of 100, 50, and 30 μ ⌀ were used

of the pointed cathode are smaller than those of the hair pin cathode, and every curve decreases with the brightness. The reasons are caused by the facts that the spot size of the electron sources increases with the brightness and that of the pointed cathode is smaller than those of the hair pin cathode. Since the divergent angle of the illumination beam depends on the spot size of the electron source, the divergent angle of every cathodes increases with the brightness. Thus, if the infinitesimal diameter of the condenser aperture would be used, the divergent angle of the illumination beam should be small enough and the half width of the emission spectrum should become always 0.5 eV.

Discussions

From the experiments, we have found that the half width of the emission spectrum becomes 0.5 ± 0.05 eV for tungsten cathodes and 0.4 ± 0.1 eV for oxide cathode unrelated to the beam current, if the divergent angle of the incident beam is made to be as small as possible. Whereas

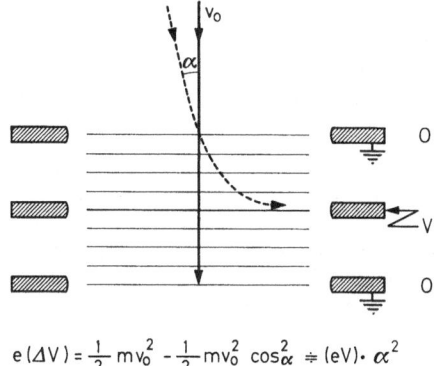

$$e\,(\Delta V) = \tfrac{1}{2}\,mv_0^2 - \tfrac{1}{2}\,mv_0^2\,\cos^2\!\alpha \;\doteqdot\;(eV)\cdot\,\alpha^2$$

Fig. 8. Schematic illustration of energy broadening due to the divergent angle on the retarding potential analyzer of einzel lens

the half width of the emission spectrum can be deduced from the cathode temperature by assuming the energy distribution Maxwellian. The half width of $\Delta E_{\frac{1}{2}}$ is given by

$$\Delta E_{\frac{1}{2}} = 2.45\,kT$$

where k is the Boltzmann constant and T absolute temperature of cathode. For $T = 2600\ ^\circ\mathrm{K}$ in tungsten cathode and 1500 °K in the oxide cathode, the equation gives $\Delta E_{\frac{1}{2}} = 0.55$ eV and 0.32 eV, respectively. These values agree with experiment in the range of the experimental errors even in taking into accont of the electric fluctuations.

In the present analyzer, the sensitivity of line broadening due to the divergent angle is considerably large. However, such effect may be more or less exist in any electron velocity analyzer. For instance, we consider the retarding potential filter of an electro-static einzel lens shown in Fig. 8. When the potential V of filter is put on a same potential as the accelerating voltage, only electrons normal to the equi-potential surface can pass through the potential maximum, but the inclined electrons from the normal direction are reflected before the maximum. On account of the simple calculation, electrons inclined by α are possible passing thorough the potential maximum if it is decreased by $\Delta V = V \cdot \alpha^2$. On the other hand, a practical value of $d\,(\Delta E)/d\alpha$ was experimentally obtained by OHNO and ABE [9] for a retarding analyzer of the einzel lens, and it was 2.5×10^3 volts/rad. This value means that a semi-divergent angle of 4×10^{-4} rad gives rise to an energy broadening of about 1 eV at 30 kV.

From the present experiment, properties of various cathodes will be discussed from a point of view of line broadening. From Fig. 7, it is found that the pointed cathode has a brightness about ten times larger than that of the hair pin cathode in the same divergent angle. Such a result consists in the experiment made by HIBI [10] with the electron interferometer of bi-prism from a point of view of the beam coherency.

Conclusions

BOERSCH considered that the increasing of the line width, ΔW, due to the beam current is intrinsically caused by the energy broadening due to the current density, i.e. $\Delta W = f(\Delta E)$ and $\Delta E = f'(J)$, where J is a current density. However, the present author has concluded that the energy broadening depending on the current density is caused by the change of the divergent angle of the incident beam, i.e. $\Delta W = F(\Delta E, \alpha)$. Therefore it can be said that the Boersch effect is not a physical significant problem but an effect concerning the experiment.

Acknowledgement. The present author wishes express his sincere thanks to Prof. R. UYEDA of Nagoya University for his discussions.

References

1. Boersch, H.: Z. Physik **139**, 115 (1954).
2. Dietrich, W.: Z. Physik **152**, 306 (1958).
3. Hartwig, D., u. K. Ulmer: Z. Physik **173**, 294 (1962).
4. Simpson, J. A., and C. E. Kuyatt: J. Appl. Phys. **37**, 3805 (1966).
5. Miller, M. H., and W. G. Dow: J. Appl. Phys. **32**, 274 (1964).
6. Lenz, F.: Proceedings of the Internat. Conference on Electron Microscopy, Berlin, 1958 (Berlin-Göttingen-Heidelberg: Springer 1960), p. 39.

7. Ulmer, K., u. B. Zimmermann: Z. Physik **182**, 194 (1964).
8. Beck, A. H., and C. E. Maloney: Brit. J. Appl. Phys. **18**, 845 (1967).
9. Ohno, T., and R. Abe: Mem. Fac. Eng., Nagoya Univ. **19**, 218; (1967); Private communication.
10. Hibi, T., and S. Takahashi: J. Electronmicroscopy (Tokyo) **12**, 129 (1963); **13**, 94 (1964).

Microanalyseur par émission ionique secondaire

J.-M. Rouberol, J. Guernet, P. Deschamps, J.-P. Dagnot et J.-M. Guyon de la Berge

CAMECA, Courbevoie 92, France

Le Microanalyseur par émission inonique secondaire que nous présentons aujourd'hui a été développé par CAMECA en utilisant le principe de double filtrage de la seconde version du Microanalyseur de Castaing et Slodzian et comporte d'importants perfectionnements par rapport aux appareils expérimentaux d'origine.

Principe. Le principe général du Microanalyseur Ionique ayant déjà été exposé par R. Castaing [1] au 4ème Congrès d'Orsay, nous rappellerons seulement que cet appareil permet d'obtenir, en utilisant à la fois les méthodes de la spectrométrie de masse et de la microscopie à émission ionique, la carte de distribution en surface des constituants d'un échantillon; la limite de résolution atteinte en pratique étant de l'ordre du micron.

Le spectromètre de masse est d'un type particulier, ayant des propriétés de focalisation dans les deux sections: radiale et transversale. Il est du type à double filtrage: le filtrage en quantité de mouvement étant réalisé par un secteur magnétique et le filtrage en énergie par un miroir électrostatique.

La fig. 1 montre la section horizontale générale du Microanalyseur. Nous décrirons seulement le mode de fonctionnement du convertisseur. Ce convertisseur d'image est du type de celui proposé par Möllenstedt et étudié par Castaing et Jouffrey [2]. Il peut être utilisé suivant quatre modes: Observation, enregistrement de l'image, mesure du courant ionique global et local.

Le faisceau d'électrons étant dévié par un aimant permanent, l'image électronique est envoyée, soit sur un écran fluorescent, soit par une rotation de 135° de l'aimant, sur le film d'une caméra placée à l'intérieur du convertisseur (fig. 2).

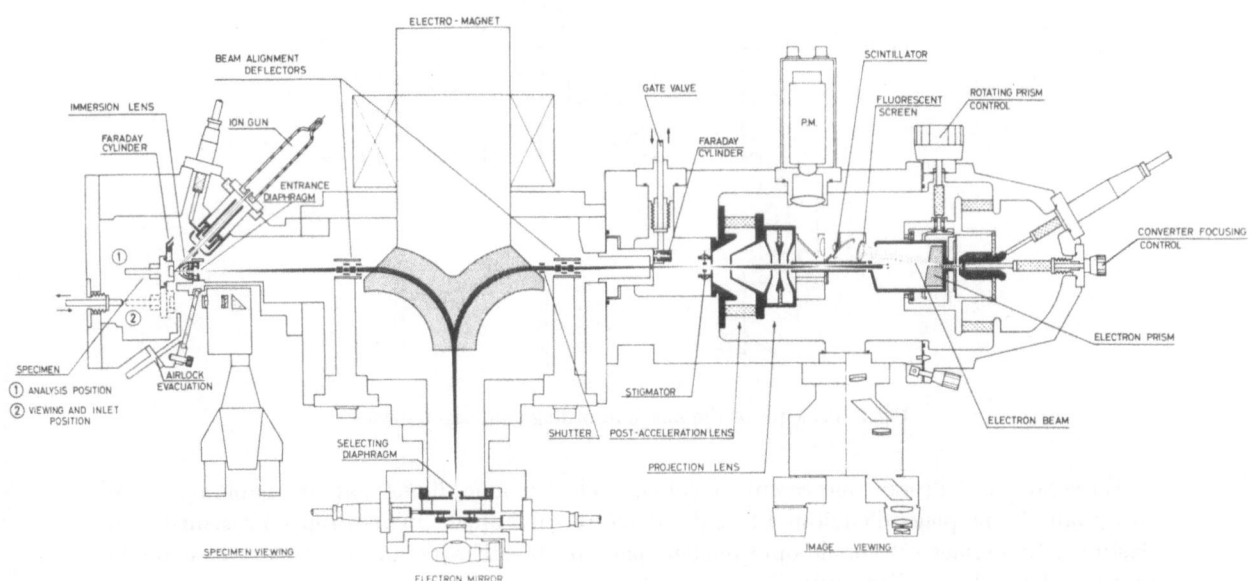

Fig. 1. Ion microanalyzer CAMECA

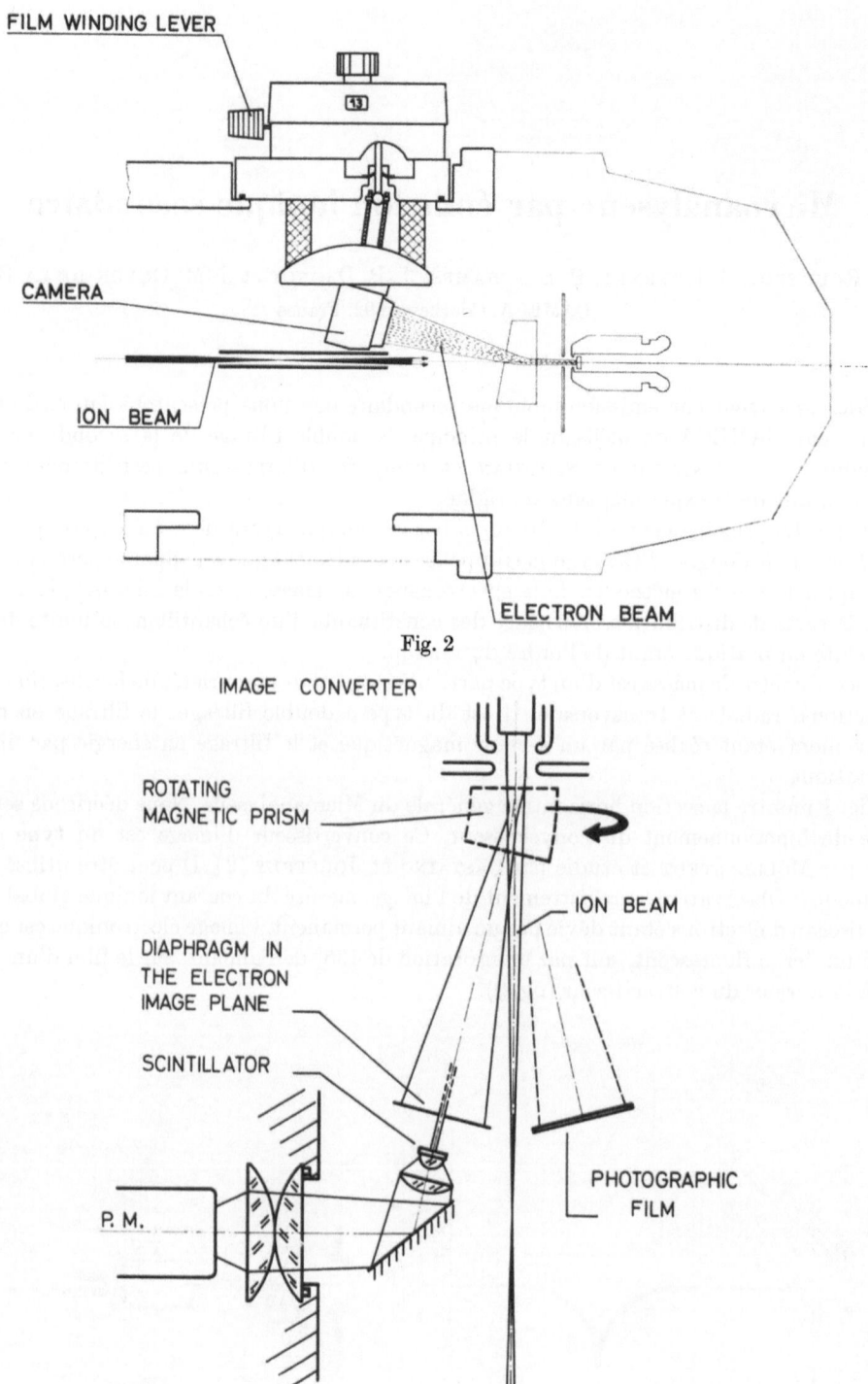

Fig. 2

Fig. 3. Principle of the image recording and local analysis

En escamotant l'écran fluorescent, le faisceau électronique, limité par un diaphragme correspondant à une plage d'environ 125 μ de diamètre sur l'objet, vient frapper un scintillateur plastique. Ce dernier est couplé optiquement par un objectif à grande ouverture à un photomultiplicateur placé à l'extérieur de l'appareil. On réalise ainsi la mesure du courant ionique global (fig. 3). Dans le cas de l'analyse ionique locale, un diaphragme placé au niveau de l'image donnée par le convertisseur, limite à un fin pinceau le flux électronique reçu par le scintillateur. Ce mode de fontionnement permet alors une analyse par spectrométrie de masse à l'échelle de quelques microns carré.

Performances et Caracteristiques

Rappelons tour d'abord le phénomène de renforcement de l'émission ionique secondaire par effet chimique [3] dont il faut tenir compte en choisissant les conditions de bombardement primaire :

Lorsqu'on bombarde un échantillon ayant fixé en surface de l'oxygène, il y a renforcement de l'émission ionique secondaire pour beaucoup d'éléments ; on suppose que l'oxygène forme avec ces éléments des composés ioniques dont le rendement d'ionisation est supérieur à celui de l'élément pur. Ainsi l'émission chimique intervient lorsque la surface de l'objet se recouvre, par adsorption, d'oxygène provenant de l'atmosphère résiduelle de la chambre-objet et on peut observer facilement ce phénomène avec l'aluminium, par exemple, dont l'émission en début de bombardement ou dans un mauvais vide peut être dix fois plus intense que dans les conditions normales.

Suivant les problèmes analytiques, on doit pouvoir utiliser, à volonté, l'émission chimique ou l'émission cinétique. Dans le premier cas, on peut employer des ions oxygène. Dans le second cas, on utilise des ions de gaz neutre tel que l'argon, en maintenant l'objet dans un vide aussi bon que possible et sous une densité ionique primaire suffisamment élevée pour rendre négligeable l'effet de l'oxygène de l'atmosphère résiduelle.

En pratique, ces conditions sont réalisées avec un vide voisin de 5×10^{-8} Torr dans la chambre-objet et une densité de bombardement de 200 $\mu A/mm^2$ obtenue au moyen d'un faisceau primaire d'environ 0,3 mm de diamètre et transportant une intensité de l'ordre de 15 μA.

A propos de ces conditions de bombardement primaire, il faut noter que les Microanalyseur ionique présente, sur la Sonde Ionique donnant des images par balayage [4], les avantages suivants :

La «luminosité» obtenue est bien supérieure puisque, pour obtenir la même densité moyenne bombardement ionique primaire, il faudrait que la sonde de 1 μm de diamètre ait la même intensité que le faisceau de 300 μm de diamètre.

Dans le microanalyseur, tous les points de la plage imagée sont soumis simultanément au bombardement, ce qui permet de garder propre la surface de l'échantillon et de rendre négligeable l'émission chimique.

Le système de double filtrage a permis d'obtenir un pouvoir de résolution suffisant pour analyser tous les éléments du tableau périodique. La fig. 4 donne, à titre d'exemple, un enregistrement du spectre du Tungstène obtenu en utilisant des diaphragmes d'entrée et de sélection de 0,2 mm et 0,4 mm et une limite supérieure en énergie de 2 volts : Le pouvoir de résolution «à 10 % de vallée» est égal à 350.

Le grandissement de l'image électronique, par rapport à l'objet, est compris entre 75 et 150, dont un facteur 10 est dû au convertisseur lui-même, le diamètre de la région imagée étant compris entre 350 μm et 150 μm environ.

La limite de résolution, liée principalement au rendement ionique, est en général de l'ordre de un micron dans la partie de l'image présentant les aberrations minimales. La fig. 5 montre un test de résolution obtenu au moyen d'éléments «lumineux» (Al et Ca) où la limite de résolution est inférieure au micron sur tout le champ (190 μm).

Deux facteurs contribuent à la sensibilité : la transmission du spectromètre et la sensibilité du récepteur incorporé au convertisseur.

Le spectromètre a un facteur de transmission élevé puisque les diaphragmes utilisés, à la place des fentes classiques, sont conjugués stigmatiquement et que, d'autre part, une fraction très importante des ions extraits de l'objet sont collectés par la lentille à immersion et passent par son diaphragme de cross-over. Pour fixer les ordres de grandeur, indiquons qu'avec le champ

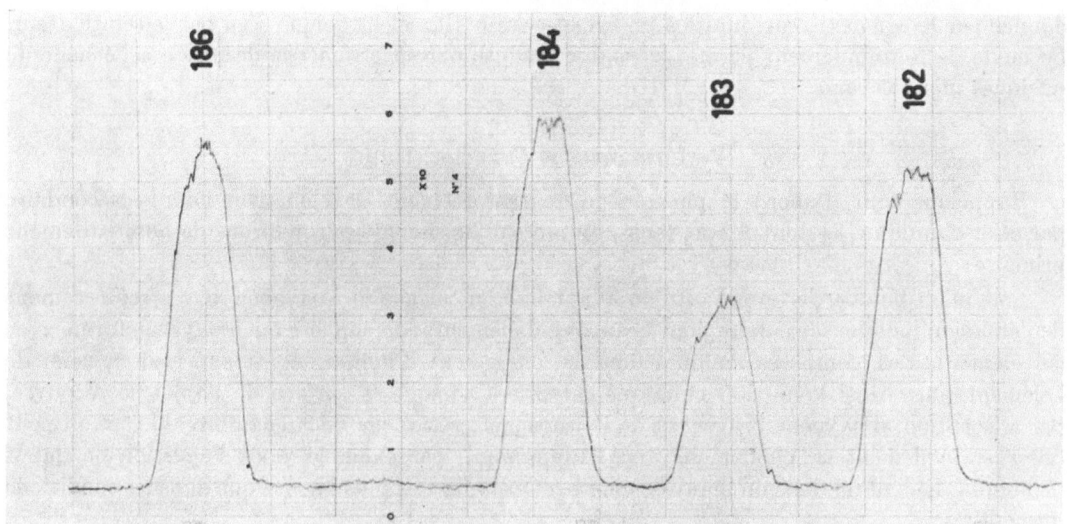

Fig. 4. Test of resolution. Tungsten spectrum. Entrance diaphragm 0.2 mm. Selecting diaphragm 0.4 mm. Accelerating voltage 4.5 kV. Energy threshold 2 V. Resolution (10% valley) $M/\Delta M = 350$

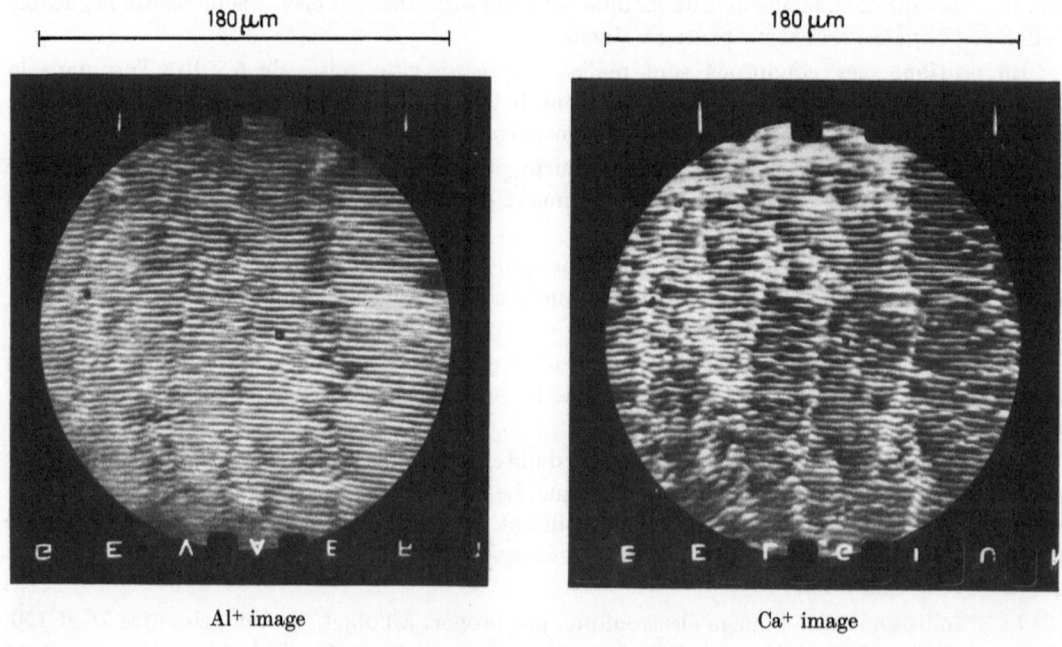

Al⁺ image   Ca⁺ image

Fig. 5. Test of spatial resolution, Al—Ca alloy

extracteur de 1 kV/mm normalement utilisé et un diaphragme de cross-over de 0,2 mm, l'angle limite de collection (pour des ions émis au centre du champ) est, par rapport à l'axe, de 20° pour les ions de 1 eV et de 9° pour ceux de 5 eV d'énergie initiale. Avec un diaphragme de 0,4 mm, ces valeurs sont respectivement de 45° et de 17°.

Le détecteur d'ions incorporé au convertisseur utilise la conversion ions — électrons de haute énergie comme celui décrit par Daly [5], mais présente sur ce dernier l'avantage de pouvoir détecter aussi bien les ions négatifs que les ions positifs. L'optique de concentration du flux lumineux émis par le scintillateur permet d'utiliser un photo-multiplicateur à photocathode de petit diamètre présentant un faible bruit de fond. Comme le détecteur de Daly, notre détecteur,

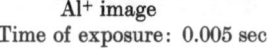

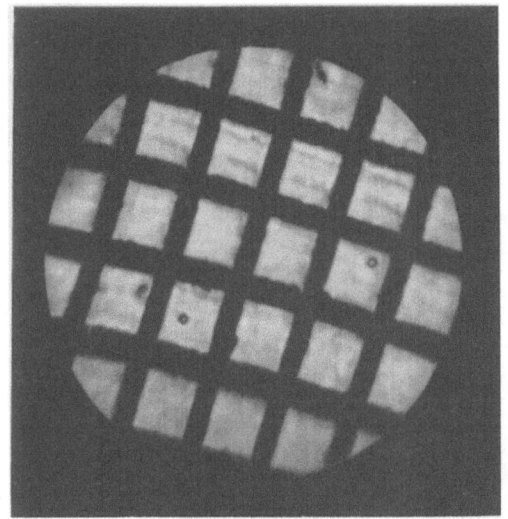
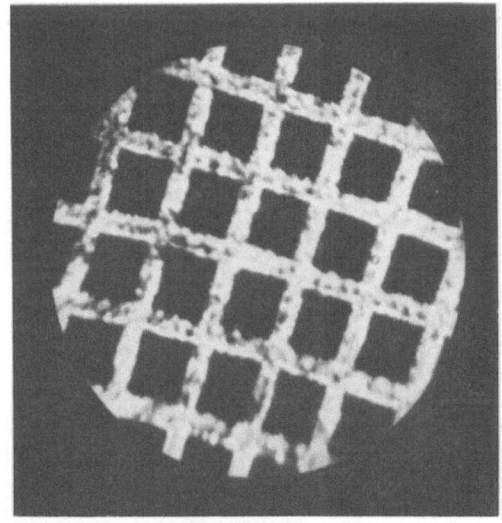

Al⁺ image   ⁶³Cu⁺ image
Time of exposure: 0.005 sec   Time of exposure: 1 sec

Fig. 6. Copper grid pressed on an aluminium block (primary ions: Ar⁺)

associé à un sélecteur d'impulsion, permet de détecter des courants correspondant à quelques ions par seconde. Pour les mesures courantes n'exigeant pas une telle sensibilité, on emploie un électronique d'amplification en courant continu dont les fluctuations de bruit de fond, traduites en courant ionique sont d'environ 10^{-16} A crête à crête (pour une constante de temps de une seconde).

Comme les images électroniques sont enregistrées directement par un film placé sous vide, les temps de pose nécessaires en pratique sont remarquablement courts et compris entre quelques millisecondes et quelques dizaines de secondes.

Sur la fig. 6, par exemple, l'image par les ions Al⁺ a été obtenue en quelques millisecondes seulement. Il faut noter toutefois que, dans ce cas, un faible pouvoir de résolution en masse est suffisant et que 50% environ des ions émis peuvent être utilisés pour former l'image.

On peut estimer le nombre de couches atomiques qu'il faut pulvériser pour obtenir une telle image, à partir des hypothèses suivantes:

le rendement ionique de l'aluminium est d'environ 0,1%,

pour isoler Al⁺, un faible pouvoir de résolution est suffisant et 50% environ des ions émis peuvent être utilisés pour former l'image,

un ion Al⁺ produit, en moyenne, l'émission de deux électrons dans le convertisseur,

enfin, le film doit recevoir environ 1,5 électron par micron-carré pour être noirci avec une densité égale à un.

Un calcul simple montre alors que l'image est obtenue par la pulvérisation d'une fraction de couche atomique.

Bien entendu, on ne peut en déduire que le pouvoir de résolution en profondeur de l'analyse est de l'ordre de la couche atomique, car il faut tenir compte de la pénétration des ions primaires dans le cible. Le pouvoir de résolution en profondeur n'a pas encore pu être déterminé par des expériences précises mais on suppose actuellement qu'il est compris entre 50 Å et 100 Å.

Notons aussi que l'image possède une définition très supérieure à la limite de résolution de l'appareil: l'image a été obtenue au moyen de $5 \cdot 10^8$ électrons alors qu'au point de vue résolution, elle ne comporte que quelques $5 \cdot 10^4$ éléments.

Applications

L'analyse isotopique est possible et, dans ce cas, les résultats sont quantitatifs. Comme exemple, nous citerons l'étude de l'autodiffusion. La fig. 7 montre la diffusion de l'^{18}O dans un oxyde d'uranium $U^{16}O_2$ stoechiométrique. Nous citerons aussi l'étude des segrégations isotopiques en minéralogie; comme exemple, sur la fig. 8, on peut observer une inclusion de plomb radiogénique dans une Pechblende.

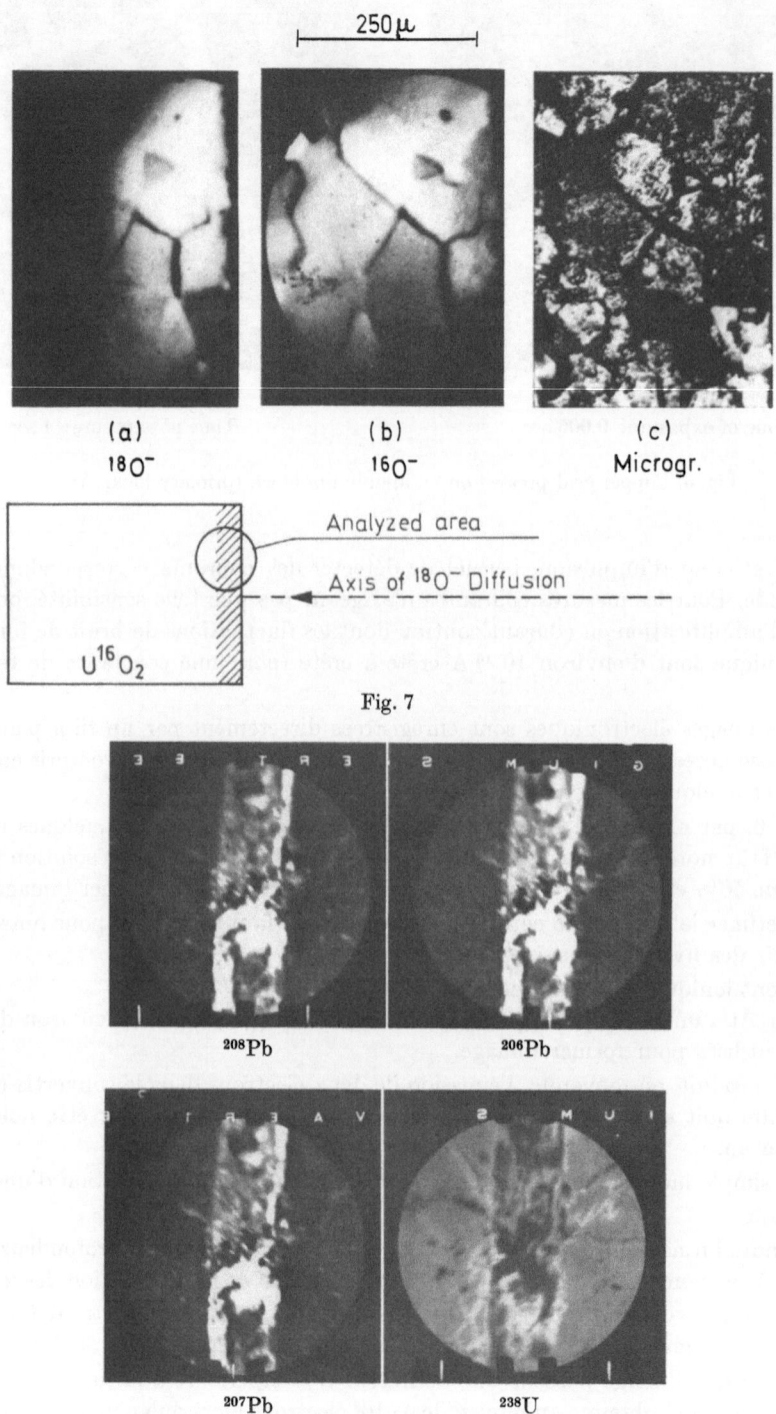

Fig. 7

Fig. 8. Inclusion de plomb radiogénique (Pechblende)

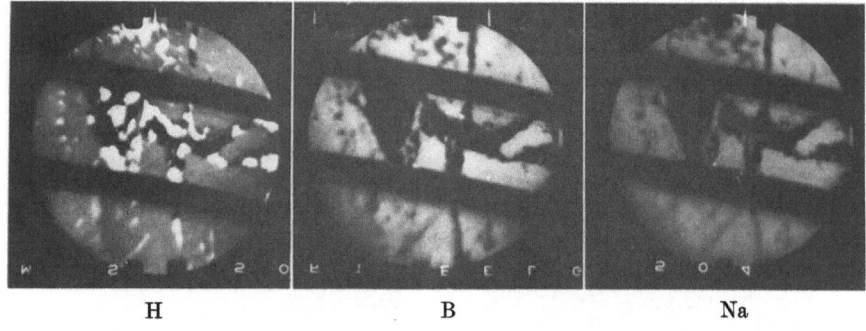

H   B   Na

Fig. 9. Biotite (granite)

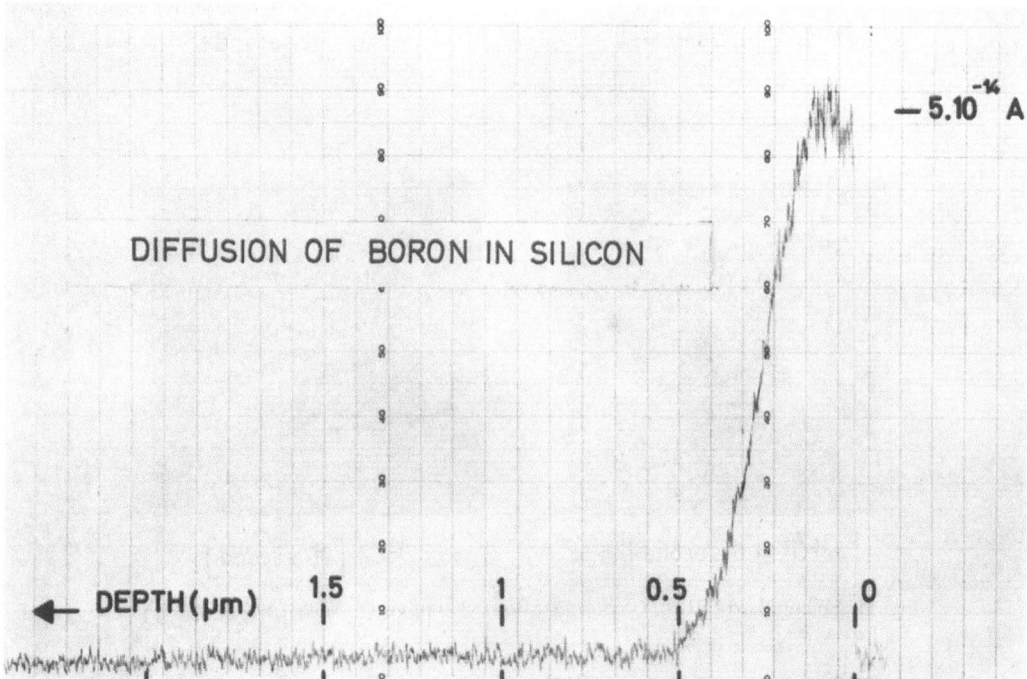

Fig. 10. Primary ions: 0^+. Surface concentration $= 7 \cdot 10^{19}$ At/cm³. Sputtering rate: 50 Å/sec

Il n'existe pas de limitation de nombre atomique: les éléments les plus légers sont accessibles. La fig. 9 montre la répartition de l'hydrogène, du bore et du sodium dans un cristal de «Biotite» (constituant d'un granit naturel).

Mais c'est surtout par son pouvoir de résolution en profondeur que l'analyse ionique est très supérieure à l'analyse par sonde électronique: on peut donc envisager d'étudier des phénomènes de surface ou de tracer des courbes de diffusion sur de très faibles profondeurs en érodant, couche après couche, la surface de l'échantillon. La fig. 10 est un exemple simple de résultat obtenu par cette méthode: il s'agit de la courbe de diffusion du Bore dans un monocristal de silicium.

Autre avantage important: il n'existe pas, en analyse ionique, de phénomène équivalent à celui du fond continu de rayons X qui, en analyse par sonde électronique, limite la sensibilité de la méthode. L'analyse de très faibles teneurs peut être envisagée et même donner lieu, si les éléments sont coalescés, à des images fortement contrastées (fig. 11).

Enfin, le microanalyseur ionique donne instantanément de images, ce qui est un avantage pratique non négligeable: l'échantillon est exploré aussi rapidement qu'avec un microscope métallographique ordinaire, alors qu'en analyse à la microsonde électronique, le système de balayage oblige à limiter la vitesse de déplacement de l'objet en cours d'examen.

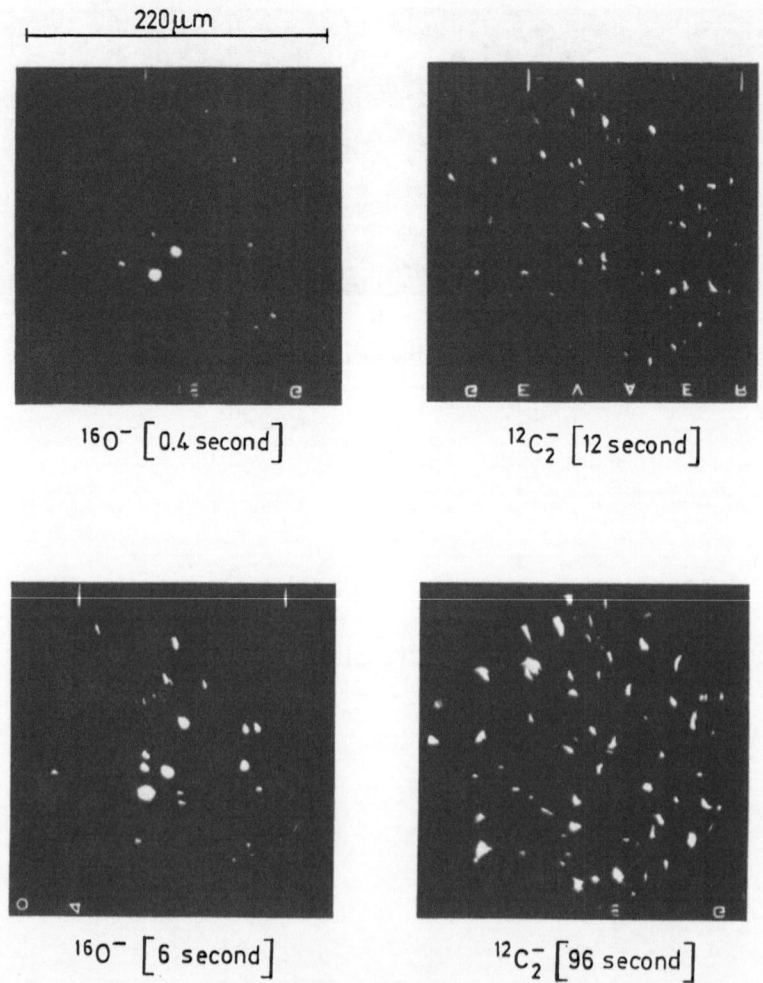

Fig. 11. Fe standard (NBS). C concentration: 260 ppm. O concentration: 480 ppm

Bibliographie

1. Castaing, R.: Optique de rayons X et micro-analyse, p. 48—63. Paris: Hermann 1966.
2. Jouffrey, B.: Diplôme d'etudes supérieures, 1960, Faculté des Sciences de Paris.
3. Slodzian, G., et J. F. Hennequin: Compt. Rend., Sér. B, **263**, 1246 (1966).
4. Liebl, H.: J. Appl. Phys. No. 38, 5277 (1967).
5. Daly, N. R.: Rev. Sci. Instr. **31**, 264 (1960).

Microanalysis with a Proton-Probe

D. M. Poole and J. L. Shaw

Atomic Energy Research Establishment, Metallurgy Division, Harwell, England

Introduction

The generation of Bremsstrahlung background in electron probe microanalysis limits peak/background figures on pure elements to no more than about 1000/1 and is a factor which limits the analytical sensitivity to perhaps 10's or 100's of parts per million in ordinary circumstances.

Proton excitation of characteristic X-rays, however, is virtually free of any accompanying Bremsstahlung background and both Khan [1] and Sterk [2] have discussed the use of broad proton beams of about 100 keV in the analysis of very thin layers on an extended specimen surface; in each case relatively soft radiations — e.g. O-K, Al-K, Cu-L — were detected directly in a proportional counter and high sensitivity was achieved. Khan reported a peak/background figure of 10,000/1 for aluminium-K from a thick target examined in his equipment.

Such work pointed to the possibility of using a *micro*-beam of protons of perhaps 10 μm diameter, in association with a crystal spectrometer, to provide an X-ray microanalysis method which would be virtually free of background and hence have a better sensitivity than the electron-probe. Extensive information exists in the literature [3] on the values of X-ray cross sections and thick target X-ray yields for protons of widely varying energies; examples of some X-ray yields versus energy are shown schematically in Fig. 1 where the corresponding curve for electron excitation of Cu-K is also given. The X-ray yields are strongly dependent on proton energy but it can be seen that protons of about $1\frac{1}{2}$ MeV are roughly equivalent to electrons of about 20 keV. The published data indicated that it should be possible to achieve reasonable counting

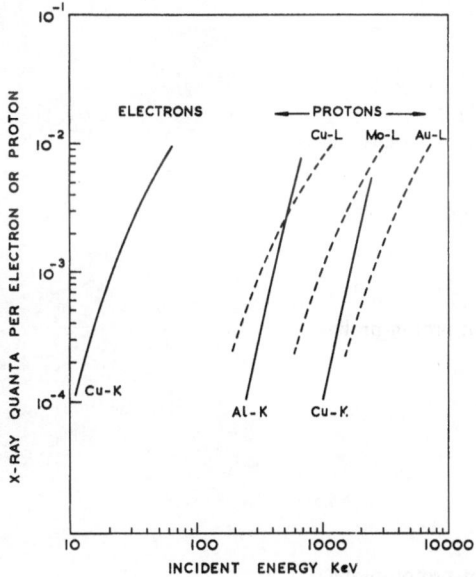

Fig. 1. Variation of X-ray yields with proton energy for various elements

rates in a crystal spectrometer by using a beam of protons to excite X-rays in the wavelength range 12 to 1.5 Å (including the K radiations from the elements Na to Cu); the proton energies required would be from $\frac{1}{2}$ to 2 MeV at a current of 0.01 μA or more. Furthermore it appeared that peak/background ratios should be very large since Bremsstrahlung intensities, calculated from the expression given by ADLER [4], were considerably less than 10^{-6} of the peak intensities; analytical sensitivities better than 1 ppm seemed a very real possibility. In order to test these predictions a crystal spectrometer apparatus has been assembled and is described here along with some preliminary results obtained on pure metals and alloys.

Experimental

The principle of the equipment is shown schematically in Fig. 2. The proton beam is produced in a Van der Graff machine and is converging towards a focus on the specimen along a path 45° to the specimen surface; in the present equipment a final aperture is necessary to limit the spot size to about 100 μm diameter, but improvements being made to the focussing system should soon enable this aperture to be eliminated and a beam diameter of about 10 μm should be achieved. The X-rays generated pass at 45° take-off angle into the crystal spectrometer. The specimen is positioned at the intersection of the proton and X-ray axes by reference to the fixed optical system viewing the specimen at normal incidence.

The present equipment is shown in Fig. 3. The proton beam enters axially down the tube on the right and on the left is the Materials Analysis Company linear spectrometer. The simple optical system is seen in the central position and employs only a single, low power objective. Specimen handling in the X, Y and Z directions is incorporated in the specimen chamber beneath the large circular plate.

The equipment was aligned initially using an electron beam from a microanalyser column and then transferred to the end of the accelerator flight tube for the proton experiments.

Measurements have been made of peak count rates from pure metals using proton currents in the range 0.01 to 0.1 μA and a range of proton energies up to $2\frac{1}{2}$ MeV. Backgrounds have been determined either by detuning the spectrometer and leaving all other factors constant, or by substituting another pure metal whilst the spectrometer remains tuned to the first element for which the peak figures were obtained.

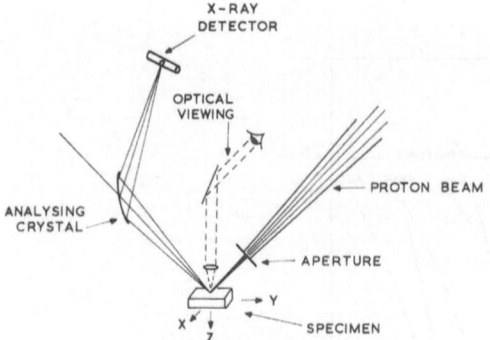

Fig. 2. Schematic diagram of proton-probe
microanalyser

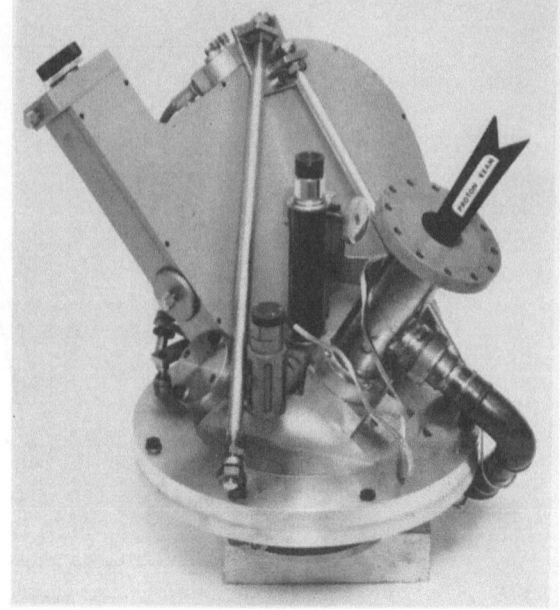

Fig. 3. Photograph of present proton-probe
microanalysis equipment

Fig. 3

Table 1. *Interim performance figures*

Radiation	Beam	Crystal	Peak c/sec	Background c/sec	Peak/background
Al-K_α	2 MeV protons	KAP	9,200	0.75	12,250/1
	$2\frac{1}{2}$ MeV protons	KAP	35,000	2.4	13,500/1
	(20 keV electrons	KAP	27,000	20	1,350/1)[a]
Si-K_α	$2\frac{1}{2}$ MeV protons	KAP	41,560	1.0	41,560/1
	$2\frac{1}{2}$ MeV protons	PET	17,000	0.83	20,700/1
Cu-K_α	2 MeV protons	PET	8,250	1.57	5,250/1
	$2\frac{1}{2}$ MeV protons	PET	20,000	4.0	5,000/1
	(20 keV electrons	LiF	26,000	57	450/1)[a]

Proton current $= 0.02\ \mu A$. Electron current $= 0.1\ \mu A$.

[a] Commercial EPMA performance.

Some of the results obtained at a typical proton current of 0.02 μA are shown in Table 1, which also includes electron excitation figures for comparison. It should be noted that all these figures are provisional and should shortly be improved upon.

In the case of Aluminium-K excitation at $2\frac{1}{2}$ MeV we observe a peak count rate similar to that achieved with a 0.1 μA electron probe at 20 KeV, but with a background rate which is lower by a factor 10 than that observed in the electron case.

A similar comparison can be made in the case of Copper-K where again the peak/background ratio shows a 10 times improvement over electron excitation.

The best peak/background ratio so far achieved was in the case of Silicon-K for which a value greater than 40,000/1 was measured.

These figures are very encouraging, and as they stand at present they represent a considerable improvement in sensitivity over electron excitation. For example with 40,000 c/sec peak and 1 c/sec background a conventional detection sensitivity calculation indicates that a 100 second analysis should detect 5 ppm. Furthermore there is no significant increase in X-ray background on going to high atomic number samples as there is with electron excitation; this means that the analytical sensitivity in, say, uranium will not be greatly less than that in aluminium — a major difference from the electron case.

However it is certain that these figures can be improved upon. Firstly the peak count rates achieved are lower than expected for both electron and proton excitation, and replacement of the KAP crystal and better spectrometer alignment should improve the aluminium count rate for example by a factor between 2 and 5 times.

Secondly backgrounds are higher than predicted due to the presence of γ-rays flooding the region of the detector. These γ's originate a) within the building which contains three accelerators and an operating nuclear pile, b) from the aperture material which at present stops a major proportion of the proton beam and emits γ's from nuclear reactions, and c) from the specimen itself. The γ's from the aperture will be lost when the aperture is dispensed with, and the extent to which the remainder generate a background count can be greatly reduced by cutting down the counter volume or by the use of anti-coincidence methods. A trial anticoincidence counter assembly , based on a design in the book by WATT and RAMSDEN [5], has shown a reduction of 15 × in background count rate for a given γ-flux; a final version, now under construction, to fit in the confined space within the M.A.C. spectrometer is expected to give at least a 10 × reduction in background over that observed with the normal M.A.C. proportional counter so far used in these experiments. It is important when designing new detectors for this purpose that the resolution be maintained as good as possible, since any broadening of the electronic "window" in the pulse height analyser, made necessary by deterioration in resolution, results in an increase in the background count rate.

Table 2. *Anticipated improvements in performance*

Item	Factor	
	peak counts	background counts
Spectrometer efficiency	×3	—
Elimination of aperture	—	×0.3
Small volume, anti-coincidence counter	—	×0.1

Expected peak @ 2 MeV	$= 9{,}000 \times 3$
	$= 27{,}000$ c/sec/0.02 μA
Expected background	$= 0.75 \times 0.3 \times 0.1$
	$\simeq 0.02$ c/sec/0.02 μA
∴ Expected peak/background	$= 540{,}000/1$ — neglecting Cosmic rays
	$\simeq 250{,}000/1$ — including the Cosmic ray contribution

Combining all the expected improvements the future performance of the equipment can be predicted, as in Table 2. Here is shown a 3 fold improvement in peak, combined with a 30 fold reduction in background, resulting in a peak/background ratio for Aluminium at 2 MeV of 500,000/1 which corresponds to a detection limit very much less than 1 ppm. However this figure neglects the contribution of cosmic rays to the detector background; anti-coincidence methods have enabled this contribution to be reduced to about 1 c/min and this should be added to the background figure. When this is done the peak/background ratio becomes 250,000/1 which still corresponds to a detection limit considerably below 1 ppm.

As well as detection sensitivity, the spatial resolution of a micro-method is of importance. Here it is clear that proton-probes will not be able to compete directly with electron-probes. Firstly the production of proton probes of sufficiently high current for our purposes and diameters below 10 μm will be difficult if not impossible, and for the present 10 μm is regarded as the likely minimum value.

Secondly the effective penetration of protons of the energy required can be very considerable. Using a very simple model consisting of a straight line proton trajectory into the target and considering a series of slices each corresponding to a small discrete energy loss, it is possible to estimate the total X-ray yield emerging from a target after absorption, and also to estimate the thickness of the layer from which the bulk of the X-rays come. As an example we show in Table 3 the total yield and depth for 90% of the total, for the two cases of Aluminium-K and Copper-K. For the crystal spectrometer equipment to show a reasonable count rate it is necessary for the yield figure to be in excess of 10^{-3} X-rays per proton; this means that the depth resolution is about 10 μm. Of course it may be possible in certain cases to sacrifice peak counts, operate at low beam energy and achieve a very much more superficial analysis with correspondingly reduced sensitivity. In the general case however, it appears that both the depth and the diameter of the analysed volume will be about 10 μm.

Table 3. *Total X-ray yield and effective emission depth versus proton energy*

Proton energy keV	Al-K_α		Cu-K_α	
	yield[a]	depth[b]	yield[a]	depth[b]
2,500	4.8×10^{-2}	15	3.0×10^{-3}	10
2,000	3.7×10^{-2}	14	1.3×10^{-3}	8
1,500	2.6×10^{-2}	11	3.9×10^{-4}	6
1,000	1.4×10^{-2}	7	7.5×10^{-5}	3
500	1.7×10^{-3}	3	4.1×10^{-6}	2

[a] Yield = X-ray quanta emitted at 45° per incident proton.
[b] Depth = depth in microns for 90% of total X-ray emission.

Table 4. *Proton-probe analysis of aluminium in copper with partially developed equipment*
($2\frac{1}{2}$ MeV protons, 0.025 µA)

Sample	Peak counts/sec	Peak minus background c/sec	Intensity ratio	2σ detection limit
Al	40,400			
Pure Cu	3.05[a]			5×10^{-6}
a) Cu—50 ppm Al	3.30[a]	0.25	6×10^{-6}	
b) Cu—500 ppm Al	4.72[a]	1.67	41×10^{-6}	

[a] Average of three 100 second counts.

$$\text{Absorption correction factor} = \frac{\text{calculated yield for Al in Al}}{\text{calculated yield for Al in Cu}} = 1.19/0.06 = 20.$$

∴ Concentration of Al: sample a) $\simeq$ 120 ppm; sample b) $\simeq$ 820 ppm.

Despite the equipment being in only a partially developed condition, it has been applied to a low level analysis problem, namely aluminium in copper at the nominal levels of 50 and 500 ppm. These levels are apparently higher than those which have been discussed previously but the high absorption coefficient for Aluminium-K in copper makes the system a particularly difficult one for X-ray analysis and the two alloys are equivalent to perhaps 5 and 50 ppm in a system where absorption effects are low.

The results of proton-probe analysis are shown in Table 4. The presence of aluminium in the 50 ppm alloy is shown as a difference between 3.3 c/sec on the alloy and 3.05 c/sec on a pure copper standard — the spectrometer and electronics being tuned to aluminium in each case. This difference corresponds to 6 parts in 10^6, whilst the standard deviation on the accumulated counts on the copper standard correspond to about $2\frac{1}{2}$ parts in 10^6, the observed difference is thus slightly over the 2σ detection limit of 5 parts in 10^6 and it is probable that aluminium is being detected at this level but with a large uncertainty in the actual amount present. This conclusion was confirmed by other runs carried out under different conditions.

In the case of electron-probe microanalysis the corresponding figures, for a similar peak on aluminium were: pure copper 32.5 c/sec, and alloy a) 32.75 c/sec. The difference here was totally insignificant compared to the standard deviation on the background and aluminium could not be detected.

Returning to the analysis by the proton-beam one notes that the apparent 6 parts in 10^6 requires correction for X-ray absorption. A rough estimate of the ratio of the yield for aluminium in aluminium to that for aluminium in copper has been made in the way described previously; this ratio is 20 and has been applied here as an absorption correction factor to yield an analysis of about 120 ppm for the 50 ppm nominal alloy. In a similar way the data for the 500 ppm alloy has been corrected to give a figure of about 800 ppm. Clearly this is not a good check on the procedure as the amounts involved are so near the detection limit, but it does indicate that the method should have a future for quantitative analysis of trace elements at the ppm level.

Conclusion

In summary, then, the method has a resolution of about 10 µm and a sensitivity presently about 10 ppm with excellent prospects for improvement to less than 1 ppm when certain modifications are carried out.

In addition to having a sensitivity somewhat better than that of the electron-probe, the proton-probe X-ray method forms a perfect partner for the proton excited nuclear γ emission analytical method already well established [6, 7]. In this method a proton beam is used to generate γ's by nuclear reactions with the light elements such as oxygen and fluorine; identification and counting of the γ's enables the excited nuclei to be identified and their concentrations and spatial distributions to be studied. The γ-emission method is best suited to the analysis of the light elements up to about atomic number 9 (fluorine), whilst the X-ray emission equipment here described can be used for all elements of higher atomic number.

Thus, addition to this apparatus of a suitable scintillation counter for γ-detection will enable high sensitivity analysis at or below 1 ppm, to be carried out for all the elements at the same point on a sample without the necessity for transfer from one equipment to another.

In conclusion the authors wish to thank their colleagues in the Nuclear Physics Division who provide and operate the proton beam facility which has been used in this work.

References

1. KHAN, J. M., D. L. POTTER, and R. D. WORLEY: J. Appl. Phys. 37 (2), 564 (1966).
2. STERK, A. A., W. P. SAYLOR, and C. L. MARKS: ORNL-IIC-10, vol. 2, p. 587.
3. See for example the series of papers in Physical Review by J. M. KHAN, D. L. POTTER, and R. D. WORLEY including 133, (3 A), 890 (1964); 134 (2 A), 316 (1964); 139 (6 A), 1735 (1965); 145 (1), 23 (1966).
4. ADLER, K.: Rev. Mod. Phys. 28, 432 (1956).
5. WATT, D. E., and E. RAMSDEN: High sensitivity counting techniques. Oxford: Pergamon 1964.
6. COLEMAN, R. F., and T. B. PIERCE: Analyst 92, 6 (1967).
7. PIERCE, T. B., P. F. PECK, and D. R. A. CUFF: Analyst 92, 143 (1967).

Microanalyse d'une surface solide par iono-luminescence

J. P. Meriaux, R. Goutte et C. Guillaud

Lab. de Phys. des Solides, Dépt. de Phys., I.N.S.A., Villeurbanne, France

Introduction

La surface d'un corps conducteur ou d'un isolant, bombardée par un faisceau d'ions positifs, émet des atomes neutres à l'état excité [1, 2]. La désexcitation de ces particules entraine une émission de lumière dont le spectre optique est équivalent au spectre d'arc de ces atomes pulvérisés. Il est donc possible de réaliser une analyse optique de la surface d'un échantillon en isolant une longueur d'onde correspondant à une raie d'émission d'un élément de la cible.

Étude du phénomène d'iono-luminescence

Nous avons bombardé des cibles métalliques au moyen d'un canon à ions solides classique. La lumière émise est analysée par un monochromateur à réseau et son intensité est mesurée à l'aide d'un photomultiplicateur dont le signal de sortie peut être enregistré.

Cette étude montre que, l'énergie du faisceau incident étant constante, il y a proportionnalité entre le nombre d'ions incidents et le nombre d'atomes émis à l'état excité. De plus ce dernier augmente avec l'angle d'incidence du bombardement et avec la masse de l'ion utilisé.

Nous avons ensuite étudié la variation de l'intensité lumineuse, proportionnelle au nombre total d'atomes émis dans un état excité, en fonction de l'énergie des ions (fig. 1). On remarque que la courbe passe par un maximum, résultat analogue à celui obtenu dans le phénomène de pulvérisation cathodique.

Enfin une étude de l'émission caractéristique du cuivre et du nickel dans le cas d'alliages Cu—Ni nous permet de conclure que pour un constituant donné, l'intensité d'une raie d'émission est proportionnelle à la concentration de l'élément dans cet alliage (fig. 2).

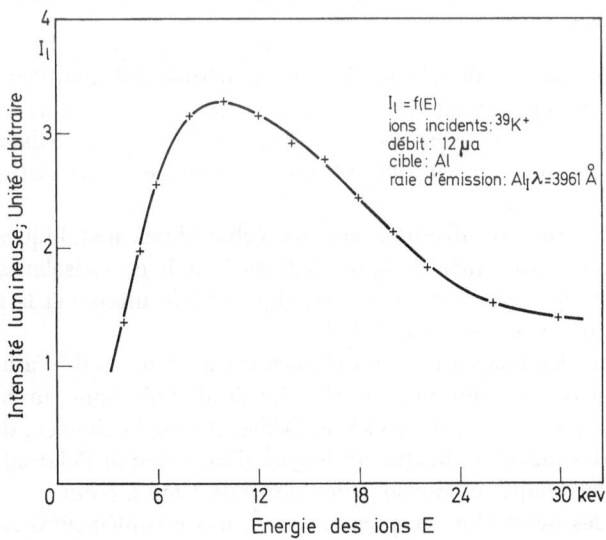

Fig. 1. Variation de l'intensité lumineuse en fonction de l'énergie des ions incidents

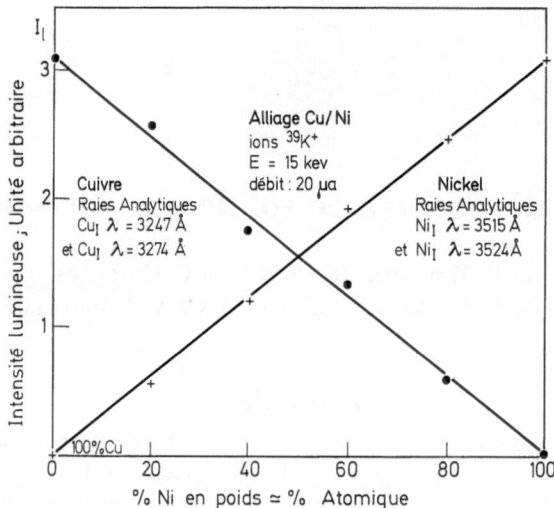

Fig. 2. Variation de l'intensité lumineuse en fonction de la concentration des constituants dans un alliage Cu—Ni

Application à la microanalyse

Ces résultats montrent que le phénomène d'iono-luminescence peut-être utilisé en micro-analyse. Deux dispositifs expérimentaux sont étudiés.

1. Utilisation de la microsonde ionique. Nous présentons tout d'abord un montage expérimental utilisant une microsonde ionique (fig. 3).

Une lentille réductrice forme sur la cible une image du «cross-over» d'un canon à ions potassium. Il y a émission de lumière au point d'impact. Un miroir concave forme l'image de ce point sur la photocathode d'un tube photomultiplicateur. Le filtre interférentiel correspondant à l'élément que l'on désire analyser est interposé dans le trajet du faisceau lumineux à l'entrée du photomultiplicateur. Un système déflecteur électrostatique permet d'utiliser la microsonde en microscopie à balayage [3].

Nous avons examiné une cible constituée par une vaporisation locale d'aluminium sur une surface de béryllium.

La fig. 4 montre d'une part l'image obtenue avec le filtre interférentiel centré sur la raie 3961 Å de l'aluminium, d'autre part l'image obtenue avec le filtre interférentiel centré sur la raie 3321 Å du béryllium.

2. Examen optique direct. Le deuxième dispositif expérimental représenté sur la fig. 5 permet d'obtenir directement une photographie de la surface qui nous renseigne sur les concentrations locales d'un élément. Un filtre interférentiel à bande étroite est interposé dans le trajet du faisceau lumineux à la sortie de l'objectif. On obtient donc une image caractéristique de l'élément que l'on désire analyser à la surface.

Après plusieurs expériences effectuées sur des échantillons métalliques nous avons réalisé par cette méthode des images caractéristiques de la surface de corps isolants.

Dans ce cas l'émission est aussi très caractéristique et très intense et la formation de l'image n'est pas influencée par les charges superficielles.

La fig. 6 représente les images caractéristiques du sodium et de l'aluminium dans le cas d'un échantillon constitué par une vaporisation locale d'aluminium sur une plaque de verre. Les ions potassium sont accélérés sous 15 kV et la densité sur la cible est de 150 µa/cm².

L'enregistrement des images s'effectue au moyen d'un appareil Polaroïd, le film a une sensibilité de 3000 Asa et les temps de pose sont de l'ordre de 1 mn à 5 mn.

Remarquons que les deux photographies ne sont pas complémentaires. Ceci est dû essentiellement à l'écart entre les durées de vie des transitions optiques correspondantes et aussi

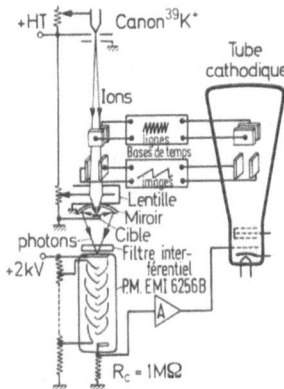

Fig. 3. Schéma de principe de la microsonde ionique utilisée en luminescence

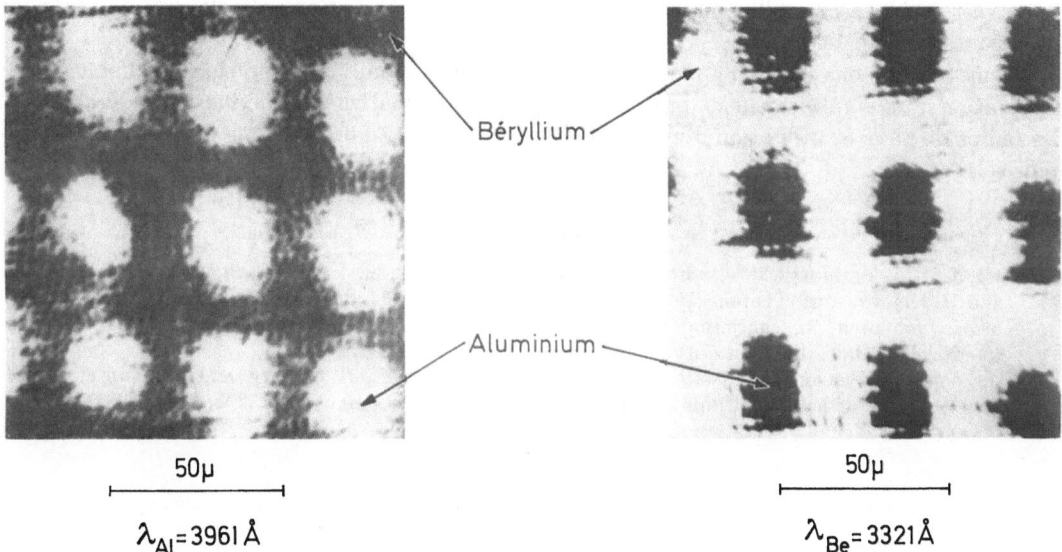

50μ   50μ

$\lambda_{Al} = 3961\,Å$ $\lambda_{Be} = 3321\,Å$

Fig. 4. Images d'une vaporisation d'aluminium sur du béryllium réalisées à la microsonde

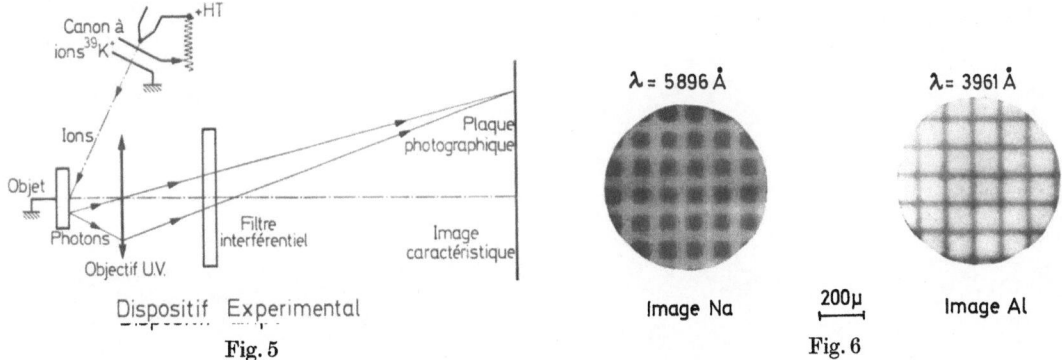

Dispositif Experimental   Image Na   200μ   Image Al

Fig. 5   Fig. 6

Fig. 5. Schéma du dispositif optique de microanalyse

Fig. 6. Images d'une vaporisation d'aluminium sur du verre réalisées avec l'analyseur optique

à la différence entre la dispersion de vitesse des atomes de sodium et des atomes d'aluminium. Le pouvoir séparateur des images dépend de ces deux variables.

Nous présentons enfin, lors du congrès, une série de micrographies donnant les distributions superficielles du calcium, de l'aluminium et du sodium à la surface d'une roche naturelle.

Conclusion

La luminescence sous impact ionique peut donc conduire à des applications interessantes dans le domaine de la microanalyse.

L'utilisation de la microsonde ionique permet d'effectuer une analyse ponctuelle de la surface d'une cible métallique et semble particulièrement interessante dans le cas de l'analyse d'éléments légers. On peut remplacer avantageusement dans ce système les filtres interférentiels par un monochromateur optique et la méthode peut être très précise et d'une grande sensibilité.

Remarquons qu'avec cet appareil la résolution est uniquement fonction des dimensions du spot. Au contraire, avec le dispositif expérimental que nous avons présenté ensuite, il n'est pas nécessairement possible d'atteindre le pouvoir séparateur optique car il existe une autre limitation qui est liée à la durée de vie des atomes à l'état excité et à leur dispersion de vitesse. Dans ce cas, des études systématiques doivent permettre d'effectuer un choix judicieux des différentes raies analytiques. Ce système est de réalisation bien plus simple que le précédent. Il présente de plus l'avantage de permettre une analyse directe de la surface de corps isolants. Sa mise en œuvre est rapide, l'échantillon massif ne subissant aucune préparation spéciale et l'interprétation des résultats est immédiate.

Nous envisageons enfin d'apporter à cet appareil quelques améliorations technologiques: adaptation d'un amplificateur de brillance, ce qui nous permettra de diminuer considérablement les temps de pose, et utilisation d'un interféromètre de Fabry-Pérot comme dispositif de filtrage plus sélectif.

Bibliographie

1. Fluit, J. M., L. Friedman, J. van Eck, C. Snoek, and J. Kistemaker: Photons and metastable atoms produced in sputtering experiments (5—20 keV) (Proc. Fifth Internat. Conference on ionisation phenomena in gazes, 1961, p. 131. Amsterdam: North Holland Publ. Co. 1962).

2. Kistemaker, J., and C. Snoek: Surface phenomena related with sputtering (Colloque international du C.N.R.S., No. 113, Paris, décembre 1962, p. 51).

3. Gabbay, M., R. Goutte, C. Guillaud et C. Monllor: Compt. Rend. **261**, 3325 (1965).

A High Resolution Electron Microscope
for Conventional Imaging and Scanning Mode of Operation

P. S. ONG

Philips Electronic Instruments, Mount Vernon, New York, U.S.A.

In instruments such as an Electron Microprobe and an Electron Scanning Microscope, one wants to have an electron spot of high current density and with a very small dimension. This is obtained by using an appropriate lens system to de-magnify an electron source. From an electron optical point of view, it would be desirable that the final lens or probe lens has a large reduction factor ($1/M \gg 10$). In such a case, the beam will be focused esentially in the focal plane of the lens.

The function of a conventional Electron Microscope objective lens is just the opposite. One wants to magnify a small area and image it with a large magnification on a screen. Here it is desirable that the objective lens has a large magnification ($M \gg 10$).

The ultimate resolution which can be obtained in both the Microprobe and the Microscope is determined among other things by the spherical aberration of the lens. Because spherical and chromatical aberration decreases with increasing lens strength, it would thus be desirable to use a very strong lens in both cases. A strong lens, however, has the focal plane in a strong magnetic field. In some cases, this may impose a restriction, for example, where the sample would influence the magnetic field symmetry. For the majority of cases however, the magnetic field does not limit its usefulness which has amply been proven by the use of existing Electron Microscopes.

There is a wide field of application in which a combined instrument for conventional viewing in transmission and spot analysis (either in stationary or scanning mode), will be of tremendous value. Instruments like the EMMA [1] and Probescope [2] were developed for this purpose. These two instruments were developed basically as an analytical instrument and the microscope part is used as an added feature to make observations easier. On the other hand, X-ray spectrometers have been built on existing electron microscopes to increase its usefulness. It would increase the usefulness even more if we could incorporate features of a scanning microscope in an Electron Microscope.

The following is a report of the results of some experiments carried out with a Philips EM 300 Electron Microscope to be used in the conventional mode of operation and as a scanning Electron Microscope. The approach to be described will theoretically be able to give a probe size equal to the electron microscope resolution. Whether the intensity is enough to be useful has still to be proven by future experiments.

Fig. 1 explains the principle of operation. On the left we have a strong probe-forming lens in which the sample is located between the pole pieces. On the right hand we see the same lens used as a microscope lens. We can now move the lower pole piece on the left or upperpole piece on the right, keeping the sample stationary and increasing the ampere turns of the lens to maintain the same focal length. The final arrangement is shown in the center figure. Now the sample is located midway the pole pieces. By doing so, the upper section is symmetrical to the lower section. The top lens which is a probe lens is now identical to the lower lens which is a microscope objective. A parallel beam of electrons is focused at the sample by the probe lens. The transmitted beam is imaged at infine by the microscope lens. By the use of a proper projector,

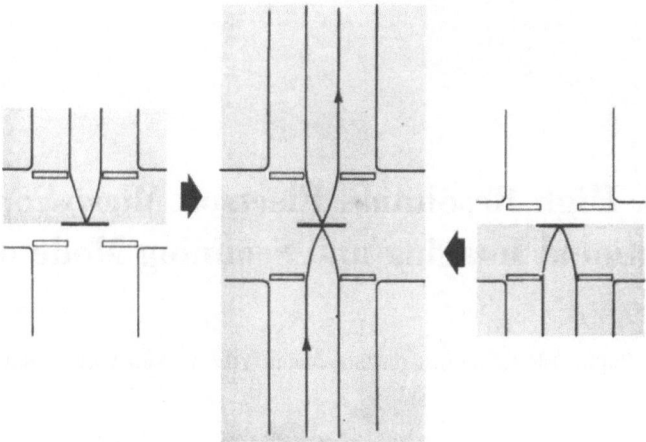

Fig. 1. Left microprobe lens, right microscope lens and center dual lens

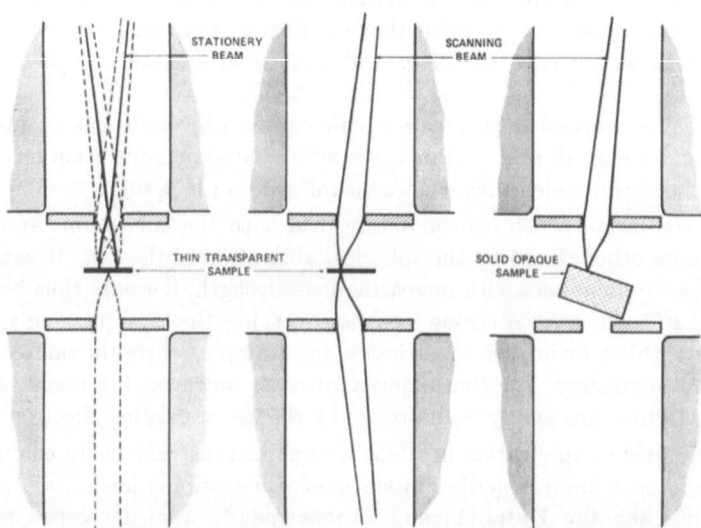

Fig. 2. Various modes of operation of the dual lens. Left conventional transmission microscope, center, scanning
mode in transmission and right scanning mode in "reflection"

the image is sharp on the viewing screen. By the use of a double condenser on top, the divergence
beam emerging from the electron gun can be rendered parallel before entering the lens.

Such a twin lens has been described by RUSKA [3] and used by RIECKE [4] as a combination
condenser objective lens and as a focusing aid. It has the advantage that both probe forming
lens and objective lens are strong lenses and thus can give inherent good resolution. The sample
is located at the focal planes of both lenses so that it is in focus for the microprobe as well as for
the microscope.

The field of view is controlled by the double condenser and can be used for normal viewing.
By adjusting the condenser, the field of illumination can be shrunk to the minimum spot size,
observable on the fluorescent screen. A set of deflection coils above the lens can move the spot
to a desired point of interest. By scanning the beam over a desired area, a scanning microscope
results.

Fig. 2 shows the lens in the various modes of operation. The left figure shows the conventional
microscope mode with a stationary wide field of illumination. The center figure shows the scanning
microscope mode of operation with a scanning beam and in transmission. The right hand figure

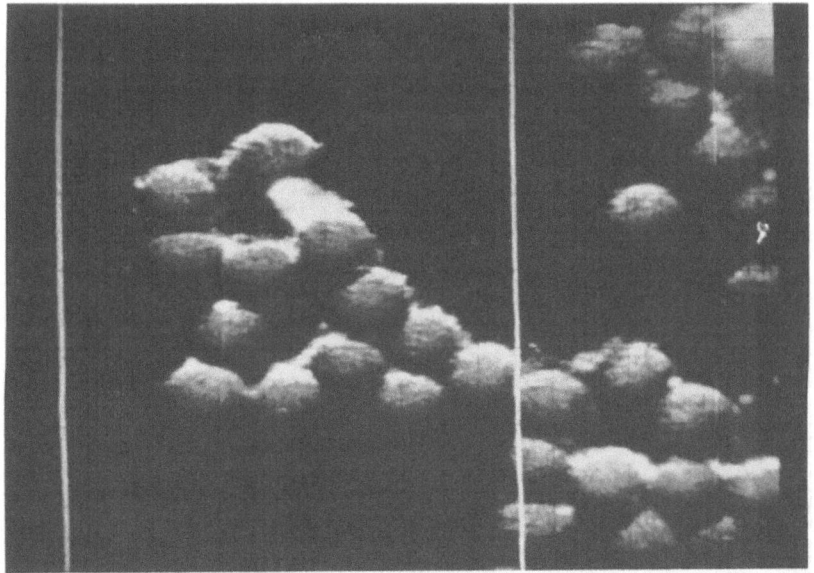

Fig. 3. 1/4 micron latex particles, scanning picture in transmission

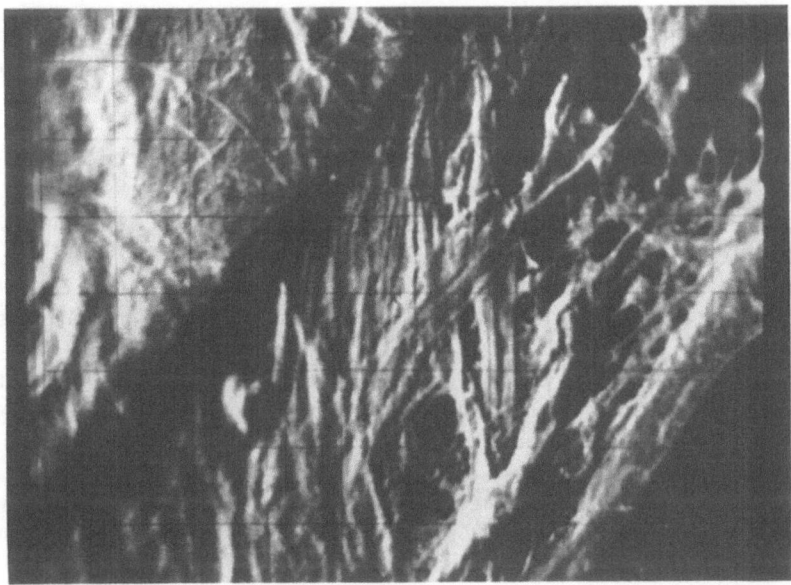

Fig. 4. Paper fibers, scanning in reflection field-of-view = 30 micron (horizontal)

gives the scanning mode with an opaque sample. The back scattered electrons are used here for image formation.

The field of application for opaque samples of the scanning microscope is well known and I will not elaborate on it here; but also, will a scanning microscope in transmission be of great value. Point analysis which reveals chemical components, crystal structure, electron density, electron scattering distribution, characteristic energy losses, etc., can be carried out. Methods for doing this type of work are described in literature. This combination instrument however, allows one to study not only the transmission pattern but also the topographic structure of a sample. We have used the Philips EM 300 Electron Microscope with the rotating tilting stage which has a gap width of 7 mm and a bore diameter of 2.5 mm. The sample holders has been

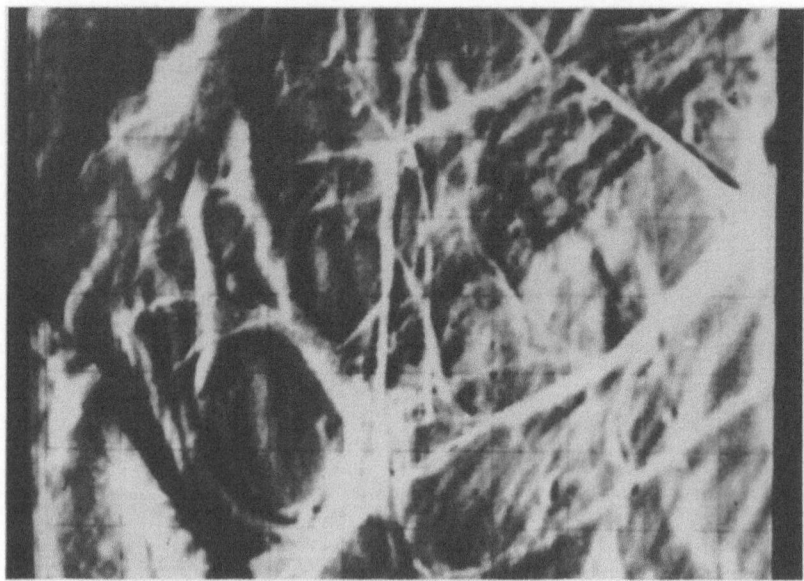

Fig. 5. Paper fibers, scanning in reflection field-of-view = 15 micron (horizontal)

modified to locate the sample in the center of the lens and the lens excitation increased. The beam wobbler coils were used as scan coils and operated by a NORELCO Beam Scanner. In the transmission mode of operation, a detector was located in the beam path to measure the transmitted beam.

In the back scattered mode of operation, the sample is tilted about 30° and the back scattered electrons picked up by an electron detector.

The following illustrations show some preliminary results of our experiments. Thes were obtained after only 4 weeks work and definitely not optimum.

Fig. 3 shows latex sphere, gold shadowed in transmission. Please note that these are scanning pictures and thus the lower lens does not contribute in any way to the sharpness — it only collects the transmitted electrons.

Fig. 4 and 5 are back scattered pictures of paper fibres. The sample was gold coated to make it electrically conductive and to enhance the back scattered coefficient yield. The results shown here are encouraging enough to continue this project.

Acknowledgement. I wish to thank Mr. Paul Shirra for his competent assistance in the operation of the EM 300 Electron Microscope and Mr. Al Sicignano for his patience and devotion for adapting the Beam Scanner to this project.

References

1. Duncumb, P.: 5th Internat. Conference Electron Microscopy, Philadelphia 1962.
2. Schippert, M. A., S. H. Moll, and R. E. Ogilvie: 2nd National Conference Electron Probe Microanalysis, Boston 1967.
3. Ruska, E.: Deutsche Patentschrift No. 875555 (29. 5. 1942).
4. Riecke, W. D.: 5th Internat. Conference Electron Microscopy, Philadelphia 1962.

Development of a Scanning Electron Mirror Microscope

D. M. KOFFMAN, S. H. MOLL, M. A. SCHIPPERT

Advanced Metals Research Corp., Burlington, Mass., U.S.A.

R. E. OGILVIE

Massachusetts Institute of Technology, Cambridge, Mass., U.S.A.

Abstract

The scanning electron mirror microscope can be employed for the study of topography as well as the quantitative determination of electric and magnetic field gradients.

The principle features of the instrument are described with emphasis on its application in solid state device evaluation. Data are presented to illustrate the principles of operation and feasibility of the instrument in material studies.

Introduction

In recent years the recognition of the possibility of focusing electron beams with magnetic or electrostatic lens systems has resulted in the development of a family of electron optical instruments which have been used with great success in the study of materials. A great deal of interest has been centered about the application of these instruments to the study and evaluation of sample surface topography, chemistry, magnetic or electric characteristics.

It has been observed that in many materials the impact of high energy electrons can produce a significant deterioration and in the case of solid state devices change the performance of the device. In addition, the electron beam will induce large circuit currents in the device which is not a normal operating condition. Consequently, the use of energetic electron beams must be considered, in a sense at least, a destructive test of the sample in question.

A research program has been sponsored at AMR by the National Aeronautics and Space Administration of the United States. The objective of this study will be to develop and design an electron optical instrument which will have the capability of examining integrated circuits under operating conditions for the determination of potential failure mechanisms. It is essential that the examination be harmless to the devices such that the technique may eventually be used as a screening method for integrated circuits.

To meet this requirement, a scanning electron mirror microscope is being developed which is based on a combination of pinciples which represent a new approach to the problem. In this instrument, a finely focused beam of electrons is directed at the specimen and scanned over an area in the form of a raster. In this respect, the instrument operates in a manner similar to a conventional Scanning Electron Microscope. The specimen surface, however, is maintained at a controlled potential such that as the electrons approach it, they are decelerated, stopped very close to the surface and then reaccelerated in the opposite direction. The specimen surface thus acts as an electron mirror. Precise control of the specimen potential permits the reflection point to be brought as close to the surface as desired but the electrons do not touch the surface so that there is not deterioration of the sample due to electron impact.

The fact that the method is nondestructive means that the test itself does not impair the specimen in any way and thus appears ideally suited for the inspection, testing and evaluation of integrated circuit devices, where extremely high reliability is the most important criterion.

D. M. KOFFMAN, S. H. MOLL, M. A. SCHIPPERT, and R. E. OGILVIE:

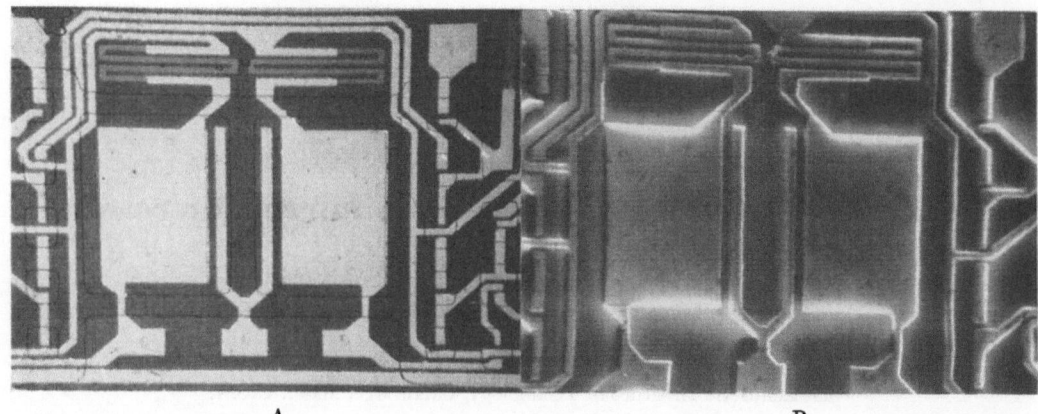

Fig. 1 A—D. Through electron mirror series mos integrated circuit. A normal mode (IOKV), B mirror mode
+ 50 volts, C mirror mode inverted signal + 50 volts, D mirror mode — 50 volts

The direction in which an electron is reflected in this process is very sensitive to the electric field gradients existing very close to the specimen surface, these gradients being the result of the currents flowing in the semiconductor device in its normal (or abnormal) operation, or formed by the contour of the specimen. A multiplicity of electron detectors determine the angular distribution of the reflected electrons and these detectors are coupled to the synchronized oscilloscope in such a way that an image or map of the surface potential gradient distribution is displayed. Sensitivity to potential gradients of different magnitudes may be varid over wide limits, the spatial resolution may be adjusted and various methods of image display may be employed to suit particular purposes.

Instrumentation

To determine the feasibility of this type of instrument, a prototype scanning electron mirror microscope was constructed. The electron optical system consists of two magnetic lenses and an experimental electron mirror.

A scan generator drives both the scanning coils in the column and in the display cathode ray tube. The signals derived from the output of the detector arrangement are amplified to modulate the brightness of the display tube.

The mirror potential can be adjusted over a wide voltage range to alter the position of the plane of electron reflection with respect to the sample plane.

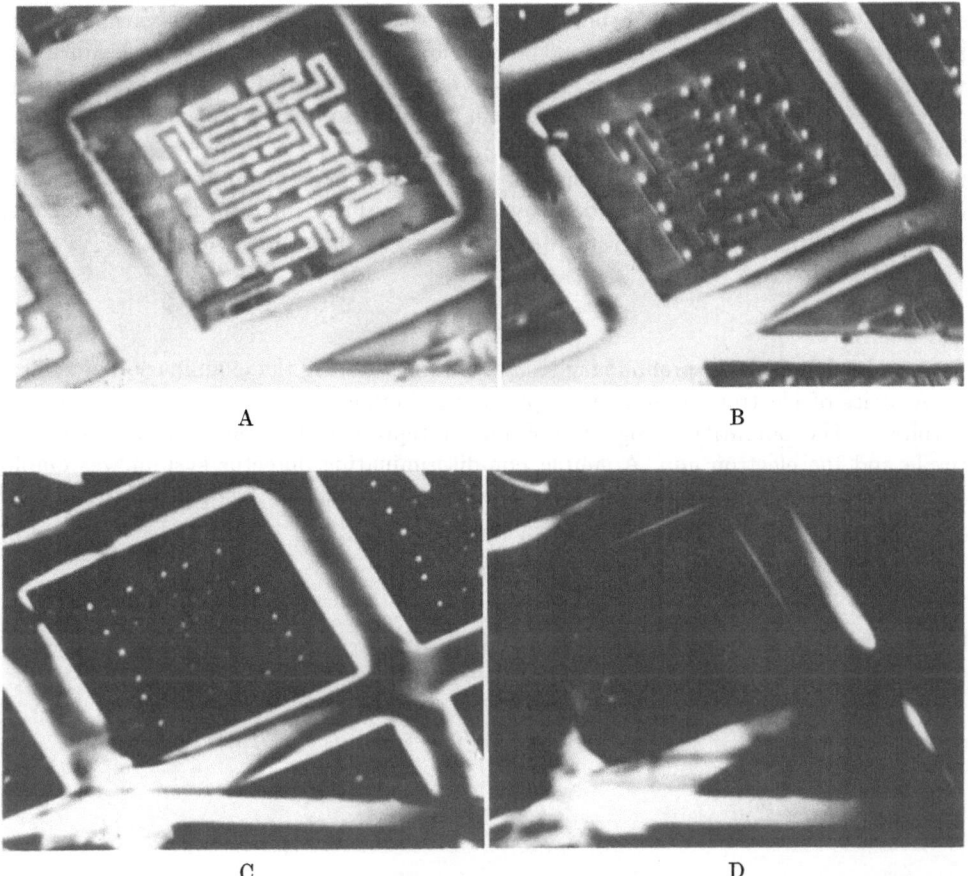

Fig. 2 A—D. Through electron mirror series-resistor integrated circuits. A mirror mode +50 volts, B 0 volts, C — 10 volts, D — 20 volts

To operate integrated circuit devices, power supplies are provided which are floated at mirror potential by employing a high voltage isolation transformer.

The specimen chamber which contains the prototype electron mirror consists of a polished sample holder floated at a controlled potential and provided with a tilt adjustment. The anode, located above this plate, is equipped with a wire mesh window to permit the reflected electron beam to reach the detector. The detector position may be adjusted during operation of the instrument. Two orthogonal micrometer translations are provided for precise location of the mirror under the electron beam.

If the incident electron beam is perpendicular to the electron mirror, it would return to the gun and could not be detected. A tilted mirror separates the reflected beam from the incident beam. Therefore, the reflected electron beam can be detected even in the absence of any field gradient on the specimen.

Experiments

In order to determine the influence of mirror tilt angle on the electron trajectory, electron reflection experiments were conducted employing a stationary electron beam. Excellent agreement was obtained with theoretical calculations based on the assumption of a parabolic electron trajectory as the beam undergoes reflection.

In order to determine the influence of sample electric field gradients on the electron trajectory, the following experiment was conducted. A sample was prepared by vapor deposition of Cu on a glass substrate, such that a 300 μ gap existed in the film. In order to develop a finite resistance across the gap, 50 Å of carbon was deposited on this assembly. A variable potential difference could be set up across the gap. The position of the reflected electron beam was then determined by translating the detector. With the detector located 4 cm above the sample the deflection of the reflected electron beam was 3 mm when the sample potential gradient was 40% of axial field gradient.

This experiment demonstrates that quantitative information can be obtained from the scanning electron mirror microscope. Suitable detector arrays would enable one to select and display discrete electric and magnetic field gradients which exist at the sample surface.

Results

Figs. 1 and 2 demonstrate preliminary results obtained during the examination of solid state devices. A series of electron photomicrographs were obtained by consecutively decreasing the mirror voltage. The potentials designated E mirror represent the voltage difference between the sample and the electron gun. A simple non-discriminating detector system was employed which permitted the collection of all electrons except those reflected nearly parallel to the incident beam.

Computer Controlled Scanning Electron Microscope

R. E. Ogilvie, D. Yankovich, R. E. Warren, J. H. Laning
Massachusetts Institute of Technology, Cambridge, Mass. U.S.A.

P. Reist and W. Burges
Harvard University, Cambridge, Mass., U.S.A.

Abstract

A study has been undertaken which will investigate the feasibility of developing a completely automatic, high speed particle counting and sizing device, utilizing a scanning electron microscope coupled with a digital computer. The computer will process the signal normally recorded on the cathode ray tube. From this information the computer will control the motion of the electron beam and the position of the specimen stage. The potential applications of this instrument appear to be almost limitless. Beside the routine counting tasks in air polution studies, this instrument is well suited for biological studies, for example, the counting of blood cells and chromosomes.

Introduction

At this time it seems most appropriate to refer to the German patent filed in 1927 by H. Stintzing [1] on a "Method and Device for Automatically Assessing, Measuring and Counting Particles of any Type, Shape and Size". In this patent the following statement is made "according to the invention the individual particles are assessed by directing onto them an energy impulse in the form of waves of rays or corpuscular radiation in beams smaller than the size of the individual particles. The attenuation or deflection of the beam is then recorded by devices responding to the energy impluses".

It is the aim of this paper to describe the use of a computer controlled scanning electron microscope which will fulfill the claims of the Stintzing patent.

One of the more difficult problems to be overcome before the system is completely automatic is the design of the interfacing gear between the scanning electron microscope and the computer. (At present conventional photographs taken with the SEM have been digitized by G. A. Moore [2] of NBS, for input to the computer.) In particular an A/D converter is required for digital conversion of position, signal intensity and magnification factor and a D/A converter is required to convert computer commands into control signals. An interface adapter, time controlled by the computer, is also required to gate digital signals in and out of the computer. This system is illustrated in block form in Fig. 1 a.

Procedure

Two methods appear promising for establishing the boundary of a visual image. The first of these involves classification of a point as a "boundary point" by application of a suitable Boolean function of the image intensity (0 or 1) at a particular point and its eight immediate neighbors. The resulting collection of points can then be approximated by standard curve-faring techniques with a sequence of smooth curvilinear line segments. The resulting boundary curve can then be further processed for recognition and sizing. The second method involves a point-by-point tracing of a sequence of straight lines connecting consecutive discrete boundary

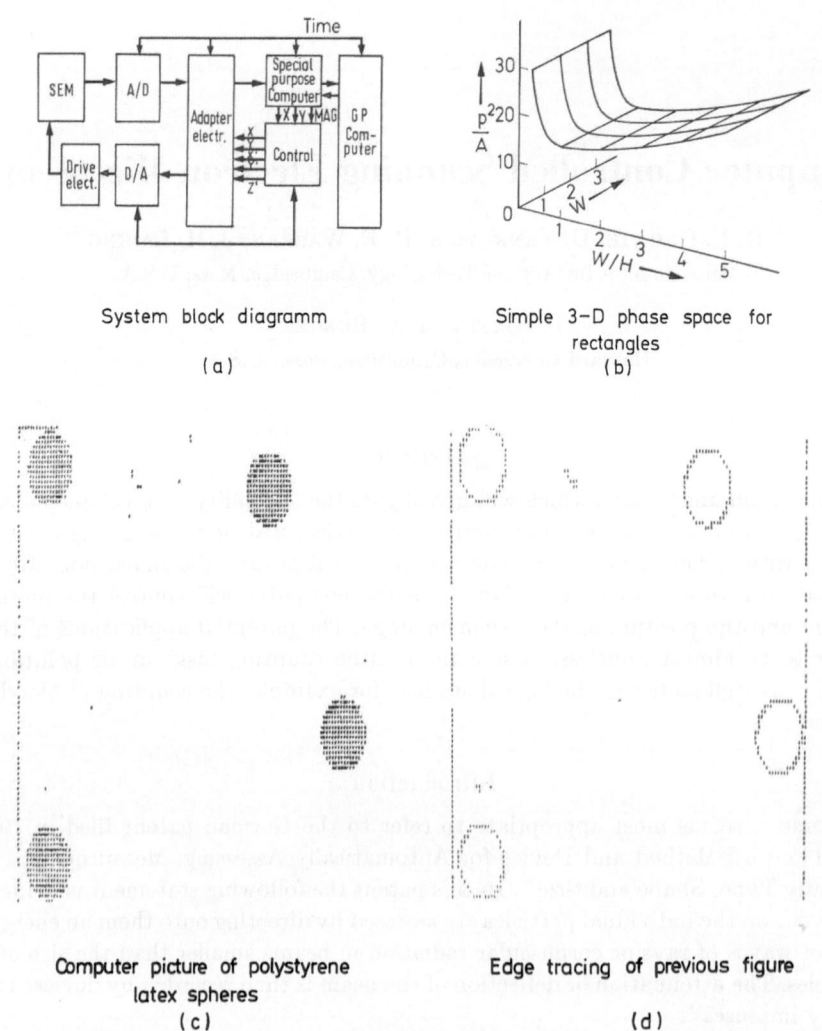

Fig. 1. a System Block Diagram, b simple 3-D phase space for rectangles, c computer picture of polystyrene latex spheres, d edge tracing of previous figure

points; the resulting broken line is then treated as the effective boundary. The algorithm by which this tracing is accomplished is as follows: if A and B are consecutive boundary points, the next boundary point C, is defined as the first neighbor at B, encountered in proceeding from A counter clockwise about B, which produces a value of 1. Points B and C are then used to define a new point D, and so on around the entire boundary.

Most, if not all, the data required for sizing and identification appears to be derivable from boundary information. Area and moments of inertia can be readily obtained as line integrals about the boundary using Stokes theorem. Perimeter and boundary roughness can be similarly calculated. All of these data can be generated step by step around the boundary without the need for storing the entire boundary. Fig. 1c illustrates the computer picture of polystyrene latex spheres. The tracing techniques of the same polystyrene spheres is illustrated in Fig. 1d.

Object Recognition

Object recognition is essentially the process of identifying a set of characteristics which are common to a class of objects. It is convenient to think of these characteristics as quantities which define points of coordinates in some abstract hyperspace. The process of identification

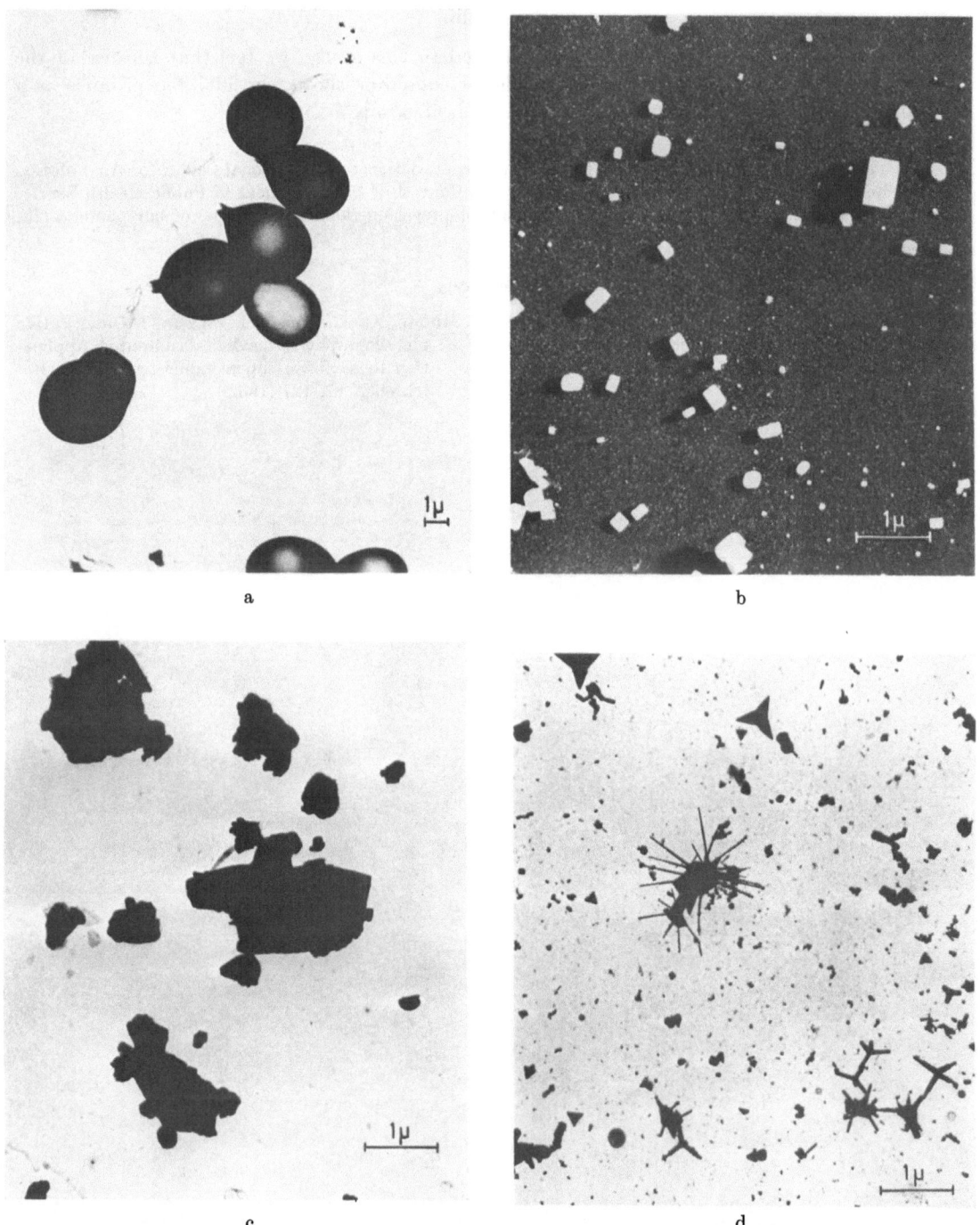

Fig. 2 a—d. Examples of air polution particles. a Puff ball spores, b sodium chloride, c silicon dioxide, d zinc oxide

reduces to: (1) selecting a set of suitable characteristics, and (2) deciding what sets of characteristic points fall within a particular region of similar objects. This area of common characteristics is defined by some hypersurface, which is usually called a "decision surface" for that class of objects. Such a surface for rectangular particles is illustrated in Fig. 1b where A is the area, P is the perimeter and W and H are the width and height of the rectangle respectively. To identify the types of particles illustrated in Fig. 2 will require a 5 to 10 dimensional phase space.

Conclusion

In spite of the limited amount of time devoted to this study, we feel that the use of the scanning electron microscope coupled to a digital computer shows considerable promise as a tool for high speed particle counting, sizing and classification.

Acknowledgements. The authors wish to express their appreciation to the National Center for Air Polution Control, Bureau of Disease Prevention and Invironmental Control, U.S. Department of Public Health Service for the support of this work, and to Dr. G. A. MOORE for preparing the digitized tapes of our photographs.

References

1. Dr. HUGO STINTZING, German patent No. 485155, Tube Investments Department of Technical Information, Translation No. 1666.

2. MOORE, G. A., and L. L. WYMAN: Quantitative metallography with a digital computer: Application to a Nb-Sn superconducting wire. J. Research A **67**, 127 (1962).

A New Scanning Electron Microscope

R. Buchanan, V. G. Macres, O. Preston and N. C. Yew

Materials Analysis Company, Palo Alto, California, U.S.A.

The design of the scanning electron microscope described here is based on the same modular concept as the Materials Analysis Company X-ray microprobe analyzers; and in fact, some parts are common to both instruments. The modularity of the instrument assures outstanding ease of disassembly and assembly for routine maintenance and cleaning, and allows the subsequent addition of accessories with a minimum of down time. A photo of the instrument is shown in Fig. 1.

The electron gun is of the well established self-biased triode type. Adjustment controls are provided for moving the filament laterally, relative to a fixed Wehnelt cylinder. In addition, a filament height control is provided which allows the filament/Wehnelt cylinder spacing to be adjusted during operation. This is very useful, as it allows the gun crossover brightness to be optimized over a wide range of operating potentials without having to resort to removing the gun.

The auxiliary lens has two focusing regions, one at either end of the bore. The lens is prealigned and uses a single energizing coil. Alignment controls are provided for moving the whole gun assembly relative to the auxiliary lens and for moving the gun/auxiliary lens combination relative to the objective or final probe forming lens.

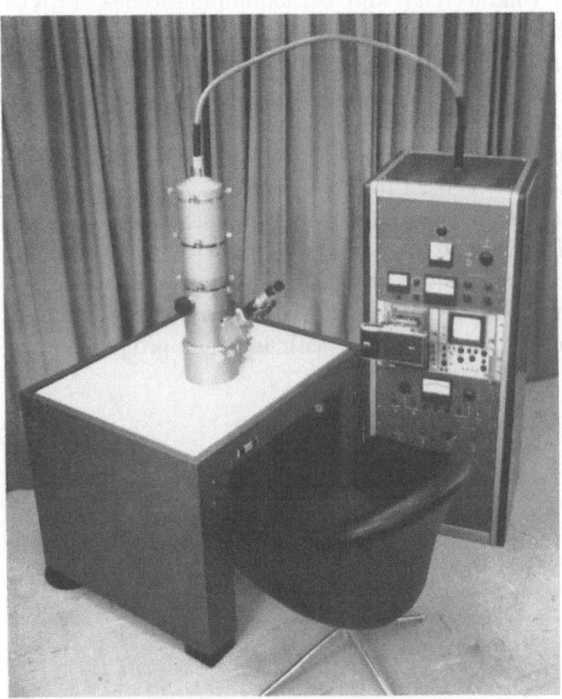

Fig. 1. Model 700 Scanning Electron Microscope

The design of the objective is unique in that a pulley arrangement is provided for moving the lens up and down during operation. In this way, the working distance is varied relative to a fixed specimen. This arrangement has the advantage of allowing components such as a light microscope and an optionally available X-ray spectrometer to be adjusted with reference to a fixed specimen. A further advantage is that it simplifies the specimen stage design. This is an important consideration, as a goniometer stage is necessarily mechanically complex and the elimination of a specimen height adjustment considerably simplifies the design and improves its reliability.

An aperture mechanism between the auxiliary and objective lens has provision for accommodating three easily interchangeable apertures. In addition, a small fluorescent screen which can be viewed through a window in the column wall, is provided as an aid for alignment.

The stage accommodates a single specimen up to $1''$ in diameter and $\frac{1}{2}''$ thick. The stage motions include 360° rotation, 90° tilt and $\pm\frac{1}{2}''x$ and y traverse. Limitations on the amount of tilt available are of course imposed when using large specimens at short working distances. The limitation exists because of mechanical interference between the specimen and objective lens.

A fully refracting binocular light microscope is provided for viewing the specimen during scanning. Provision is incorporated for moving the light microscope objective lens into position and removing it when it is required to operate with a short working distance. The microscope views the specimen at normal incidence when the specimen is tilted 45° to the electron optic axis. It is particularly useful for survey examination and for selecting areas for detailed examination in the various scanned imaging modes.

The complete column is mounted on a $1''$ thick steel plate and this is supported by an air vibration isolation system which has a resonant frequency of $\frac{1}{2}$ cps. An oil diffusion pump with liquid nitrogen cold trap is supported below the table and connected to the column via a flexible metal bellows. With these precautions, vibration problems are kept to a minimum, while the liquid nitrogen cold trap and a foreline oil vapor trap, which is provided, maintain specimen contamination at a very low level.

The scanned imaging modes available include secondary electron, backscattered electron, electron beam induced conductivity and cathodoluminescence. Provision is included for displaying the images in both intensity and deflection modulation. The videoamplifier incorporates a unique contrast control which allows the simultaneous display of detail in highlights and dark areas of the image. Unlike the more commonly used highlight suppression control which tends to sacrifice detail in the highlight as its intensity is suppressed.

Two separate systems are provided for the detection of backscattered and secondary electrons. A scintillator/photomultiplier is used for the secondary electrons, while a pair of surface barrier solid state detectors are used for the detection of backscattered electrons. As a result of the high electron collection efficiency and high inherent gain of the backscattered detection system, high resolution backscattered images of specimens free from severe shaddowing effects can be formed. A further feature of the system is that backscattered images of uncoated insulating specimens can be formed when operating with incident probe currents below about 100 picoamperes.

A Combined Scanning Electron Microscope/Electron Microprobe Analyzer

V. G. MACRES, O. PRESTON, N. C. YEW and R. BUCHANAN

Materials Analysis Company, Palo Alto, California, U.S.A.

The instrument described here is a development of the Materials Analysis Company Model 400 X-ray microprobe analyzer. It incorporates all the X-ray microanalysis capabilities of the earlier instrument, and in addition allows scanned electron images to be formed with up to 1,000 Å resolution.

A cutaway drawing of the instrument is shown in Fig. 1. The column comprises a triode electron gun, electromagnetic double condenser lens, and objective lens. The electron gun uses a conventional hairpin filament, autobiased Wehnelt cylinder and anode. An externally controlled filament/Wehnelt cylinder height adjustment is provided for optimizing gun performance at all operating potentials. The double condenser lens is unitized and has two lens regions and a common energizing coil. The estimated minimum focal lengths of each lens is about 0.33″, which provides a maximum total demagnification of about 100 X. The final demagnifying lens has a minimum focal length of about 1″ and this, in conjunction with the double condenser lens system, can produce an electron probe with a diameter of less than 0.1 µ containing up to 0.2 nanoamp. of electron current on a routine basis. The back bore of the objective lens houses the electromagnetic beam scanning coils, an electrostatic stigmator, and an additional set of electromagnetic coil which are used for fine positioning the probe on the specimen surface.

The objective lens also houses a purely refracting light microscope, which views the specimen surface at normal incidence through the objective lens bore. The magnification range of the microscope is variable up to 640 × and it has a resolution of 0.5 µ.

Up to three fully focusing Johansson-type crystal spectrometers can be used on the instrument. The crystals, which view the specimen at a constant effective X-ray take-off angle of 38.5°, cover a 2θ angular range of 20° to 145°. With the crystals available, namely LiF, PET, ADP, KAP, LOD, this allows the analysis of all elements in the periodic table down to boron. The mechanical design of the spectrometers is particularly simple, and this has been achieved at no sacrifice in performance e.g. the linearity is at least 0.02° at $2\theta = 20°$ and 0.1° at $2\theta = 145°$, while the reproducibility is better than 0.004° at $2\theta = 20°$ and 0.02° at $2\theta = 145°$.

A transistorized electron beam scanning system has recently been developed which provides a simple switch control for changing from one display mode to another, thus eliminating the need for patch plug connections. In addition to a wide range of scan speed controls a small area high speed (15 frames/second) scan control is provided for fine adjustment of the instrument. Also, a calibrated magnification control is provided which has automatic compensation to allow switching of accelerating potential without changing the magnification of the display. By designing and building the raster control circuitry, rather than using that in the oscilloscope, as was done previously, it has been possible to completely eliminate hum from the scanned image display.

Scanned images may be formed using selected monochromatic X-radiation, backscattered electrons, secondary electrons (including voltage contrast images), specimen current, electron beam induced conductivity, and cathodoluminescence. The images may be shown in either intensity modulation or deflection modulation. A Si surface barrier P—N junction is used to detect

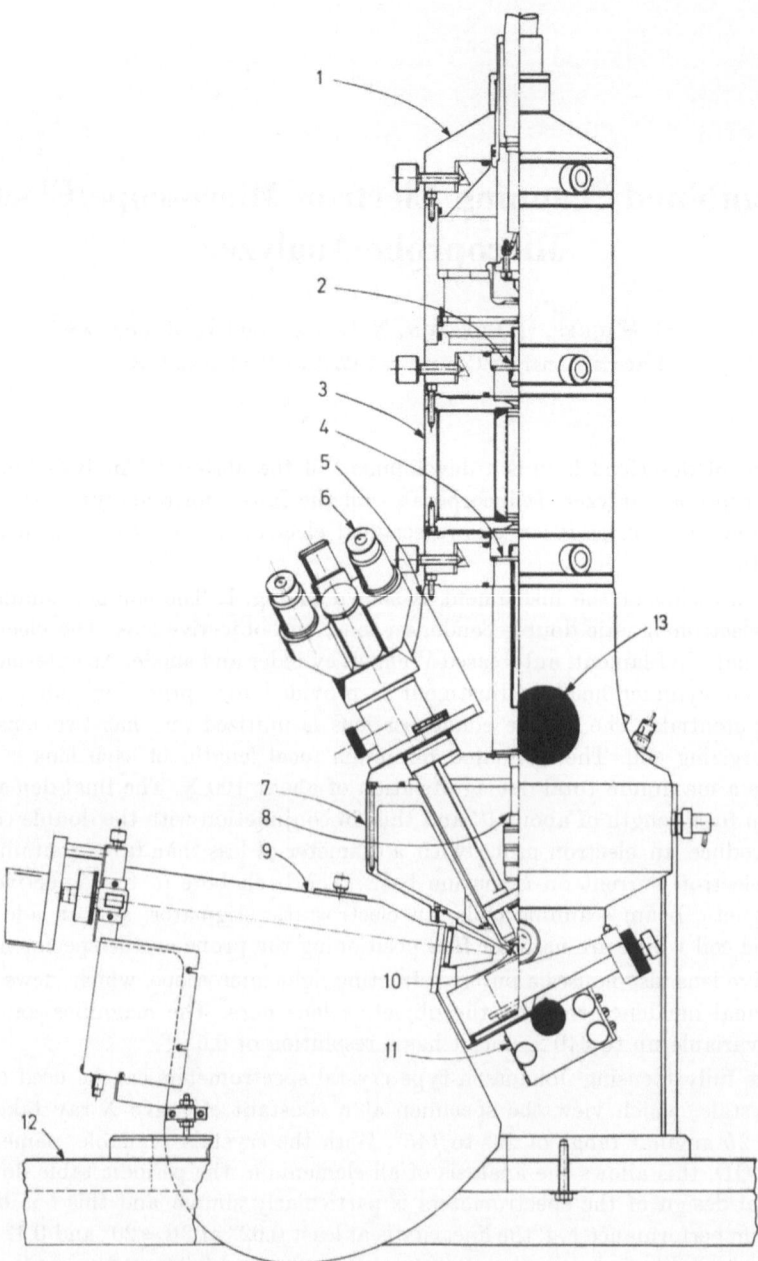

Fig. 1. Electron microprobe analyzer model 400 S. *1* Electron gun assembly, *2* upper shield, *3* auxiliary lens assembly, *4* intermediate shield, *5* intermediate chamber, *6* turret microscope assembly, *7* objective lens assembly, *8* suspender unit, *9* spectrometer, *10* static beam shield, *11* specimen stage assembly, *12* table top, *13* aperture control

the backscattered electrons. Due to the high collection efficiency and sensitivity of this system, it is possible to obtain high resolution images of uncoated insulating specimens at probe energies above 10 kV. The sensitivity of the system is such that images can be formed in about 20 seconds with detected backscattered electron currents as low as 20 picoamps. The secondary electron detector is of the conventional scintillator/electron multiplier type. The use of separate electron detection systems for backscattered and secondary electrons allows the detectors to be optimally located for their separate functions.

Photo Emission Electron Microscopy

L. WEGMANN

Electron and Ion Beam Department of the Balzers High Vacuum Ltd. Balzers, Fürstentum Liechtenstein

Recent investigations have confirmed, that the photo emission electron microscope represents a very powerful metallurgical electron microscope [1].

What the metallographer expects from an ideal *metallurgical microscope*, is direct information about the structure of a flat polished, but *unetched* metal surface. This information necessitates two types of contrast phenomena:

1. a substance differentiation, which enables to distinguish different phases (for example metals, intermetallic compounds) by different brightness or colour;

2. a face or orientation differentiation, which enables to distinguish grains of the same metallurgical phase but having different orientations.

Up to now, the light microscope represented the best approach to the ideal metallurgical microscope. However, this approach is still far away from the leading idea: generally the reflection and absorption properties of the different phases are too small for producing a sufficiently contrasted micrograph. Therefore a huge cataloque of individual preparation methods to reveal the surface structure by etching or to reinforce the contrast by oxidation or evaporation layers is necessary to achieve success with the metallurgical light microscope [2].

Many efforts have been undertaken to develop an electron microscopic method, which approches the ideal metallurgical microscope, with the hope to obtain at least the performance of the light microscope but with submicroscopical resolution. As yet, these efforts remained without any success. The X-ray and ion-beam microprobes which produce an excellent phase differentiation are not able to show an orientation contrast and their resolution is even worse than the resolution of the light microscope. A certain approach in delivering phase and orientation differentiation was realised with the scanning electron microscope as well as with the secondary emission electron microscope, but the results of both these methods are far away from satisfying the metallographer. Most of the work in submicroscopical metallography, therefore, is still dependent on topographical information in the transmission or scanning electron microscope, realized by structural etching preparations. (With the exception of the transmission of thin metallic films. This method however covers another region of information.)

According to recent investigations with the *photo emission* electron microscope it is now evident that this method represents a much better approach to the ideal metallurgical microscope than the light microscope. Photo emission microscopy bases on the emission of photoelectrons of a specimen surface irradiated by ultra-violet light. These photoelectrons are accelerated by the strong electrical field of a cathode lens. The image is formed through a three-stage electromagnetic electron microscope on a fluorescent screen exactly as in a transmission electron microscope.

We desist from a technical description of the photoemission electron microscope. The basic principles of the technique are summarized by G. MOELLENSTEDT and F. LENZ [3], modern photo emission microscopes are described by W. ENGEL [4], L. WEGMANN [1], and R. GRABER and M. GRIBI [5].

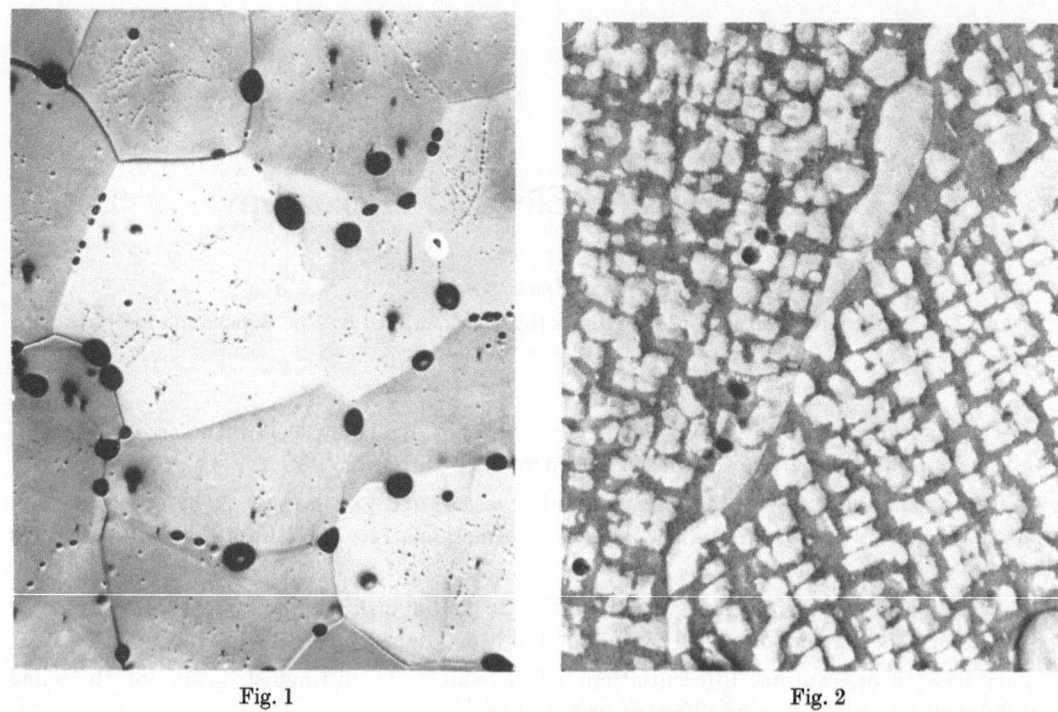

Fig. 1   Fig. 2

Fig. 1. Photo emission micrograph of precipitation of lead out of a nickel-silver matrix. The lead appears as black particles mainly in grain and subgrain-boundaries. The grains of the matrix show orientation contrast.
Magnification 1,200×

Fig. 2. Photo emission micrograph of nimonic. Phase differentiation between the γ and γ' phase.
Magnification 3,000×

The information in a photo emission micrograph is solely given by the distribution of electron density and this density is very strongly related to the emission density on the specimen surface. The emission density depends — very roughly said — on the work function of the different structural elements on the specimen surface.

Examination of a big number of metallurgical specimens within the photo emission electron microscope has shown that both two types of contrast, which are essential for a metallurgical microscope, exist in a very pronounced and very general manner:

1. Substance and phase differentiation is a quite general attribute of photo emission. Fig. 1 shows a very pronounced contrast between a precipitation of lead and a matrix of nickel-silver. This type of contrast is very sensitive to small differences between two phases. Fig. 2 shows the differentiation in brightness between the γ and γ' phase in nimonic and Fig. 3 shows a diffusion layer adjacent to a welding seam between steel and nickel. The sensitivity of the phase differentiation and the possibility to attain many graduations between black and white made it possible to differenciate easely up to five different phases within one micrograph. Another advance of this excellent phase differentiation is the omission of etching and with that of typical etching artefacts, as i. e. the broadening of perlite lamella. Fig. 4 shows a very fine perlite without lamella broadening. The best specimen preparation method for photo emission microscopy therefore is generally a very fine mechanical polishing. Fig. 4 represents also an example for the high performance of photo emission micrographs. The final resolution depends on different parameters and varies between 150 and 300 ÅU.

2. Orientation differentation likewise is a general attribute of photo emission. Fig. 1 shows a typical face contrast between the different grains of the matrix. In earlier publications [1, 5]

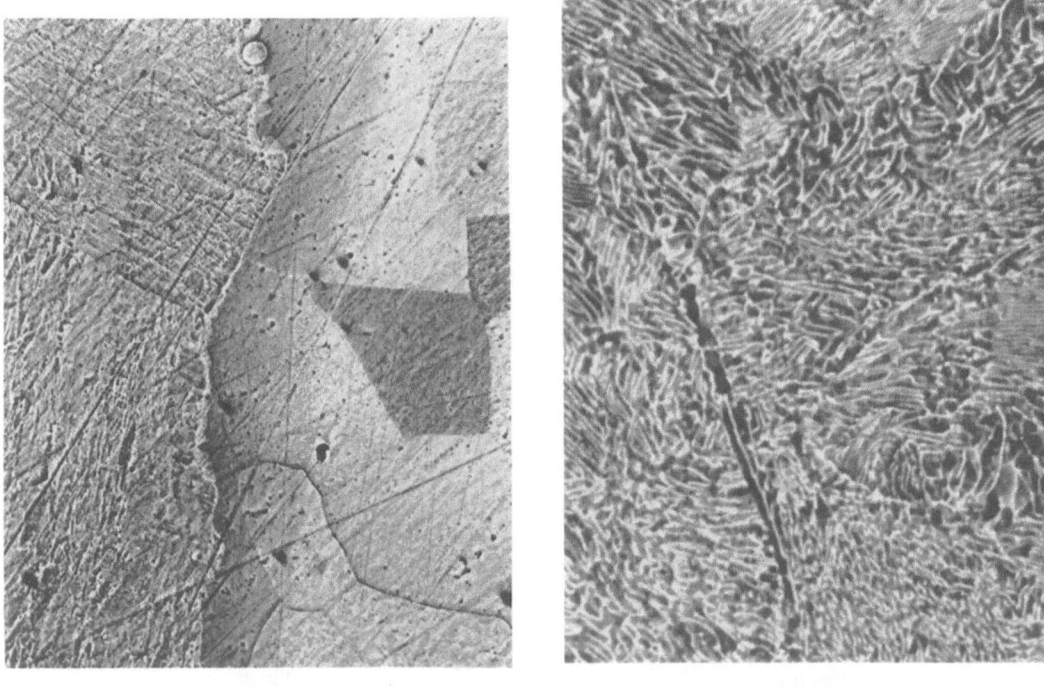

Fig. 3   Fig. 4

Fig. 3. Photo emission micrograph of a welding zone between steel (left hand side) and nickel (right hand side with orientation contrast). Diffusion zone from the welding seam about 10 μm into the nickel. Example of imaging a solid solution by concentration contrast. Magnification 1,500×

Fig. 4. Photo emission micrograph of a fine perlite (troostite). Magnification 13,000×

the particularly pronounced orientation contrast of twins was exemplified. Fig. 5 shows an example, where this effect is used to study the matrix deformation by cold work. One of the most important and striking attributes of this new technique is the fact that photoemission differentiates the grain orientation in metals having cubic lattices (Fig. 6). By far the most metals have a cubic lattice. As the reflection light microscope does not show orientation differentiation for cubic metals the photo emission has also in this respect considerable advantage. This is of high importance especially in all investigations of structural changes for example by temperature treatment. A direct and continuous pursuance of grain boundary migration, recrystallisation or twinning is only possible, if grain and twin-boundaries are observable by orientation differentiation (see Figs. 1, 3, 5 and 6). The same is valid for transformations and recovery effects. The photo emission microscope therefore represents and excellent *heating stage microscope*. An example of the recrystallisation of copper at medium temperature is given by ZAMINER [6]. Photo emission works well up to temperatures of 1,000° C; the continuation for the higher temperature region up to 2,000° C is made within the same microscope and with the same attributes of phase and face contrast by thermionic emission [7].

3. Topographic or relief contrast. Fig. 7 shows a photo emission micrograph of a cleavage fracture. This third type of contrast is partly a shadow contrast by the slightly oblique impact of the ultra violet light, partly an inclination contrast, the origin of which is very similar to the same effect in the scanning electron microscope [8]. This relief contrast makes photo emission in some respects useful for fracture studies; the field of application however is restricted by the depth of focus, which is about 50 times the resolution. This is excellent in comparison with the light microscope. At higher enlargements, the depth of focus of the photo emission micro-

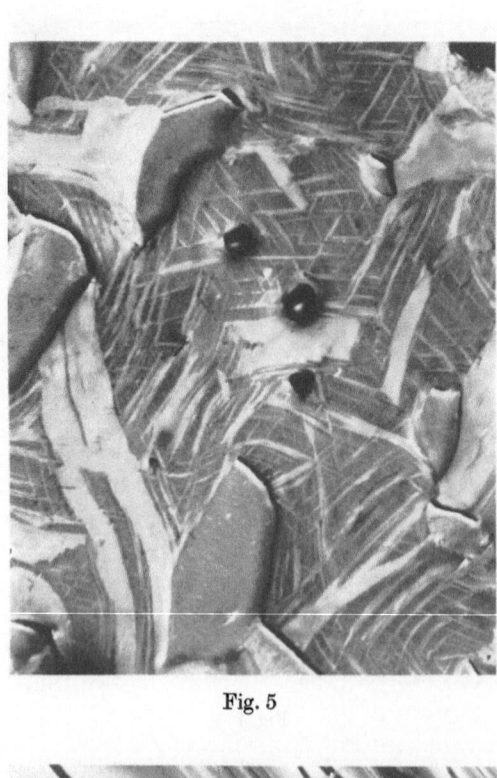

Fig. 5

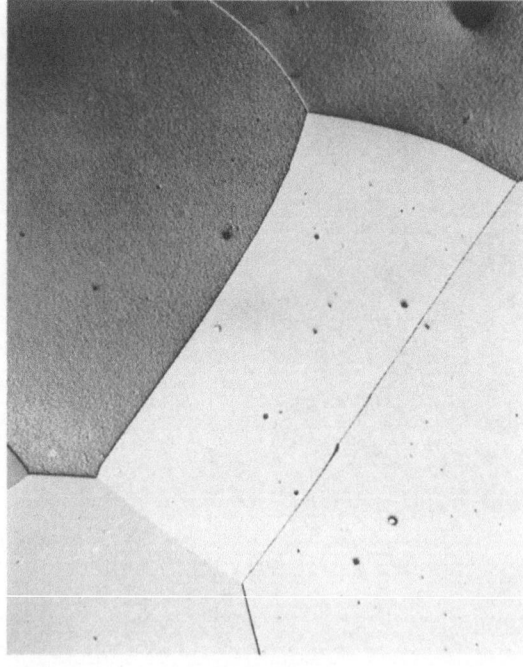

Fig. 6

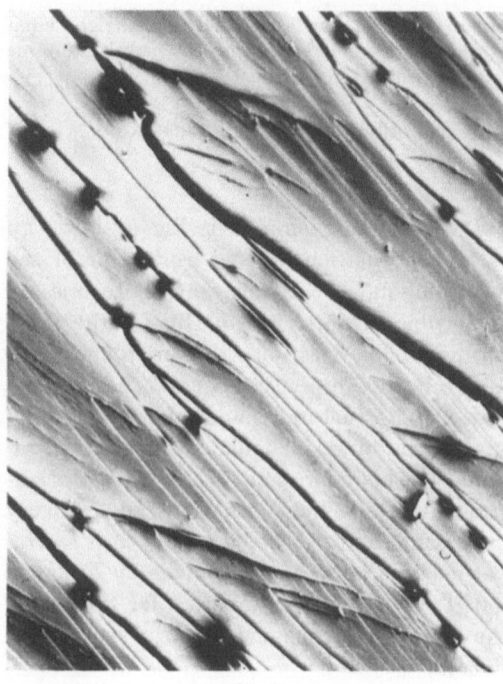

Fig. 7

Fig. 5. Photo emission micrograph of a cold worked brass with 6% deformation. β-grains "swimming" in the deformed α-structure. Deformation of the matrix visible by deformed twins. Magnification 1,700 $\times$

Fig. 6. Photo emission micrograph of tungsten (at room temperature). Typical example of face contrast. Magnification 700 $\times$

Fig. 7. Photo emission micrograph of a cleavage fracture of a zone melted single crystal of tungsten. Example of relief contrast and depth of focus. Magnification 1,400 $\times$

scope is even comparable with that of the scanning electron microscope; for small enlargements however the latter shows a much higher depth of focus.

These are the *empirical facts* of photo emission microscopy: very easy to express and very fascinating for the metallographer: a high resolution emission microscope including heating possibilities and producing an excellent phase differentiation and orientation contrast on pre-

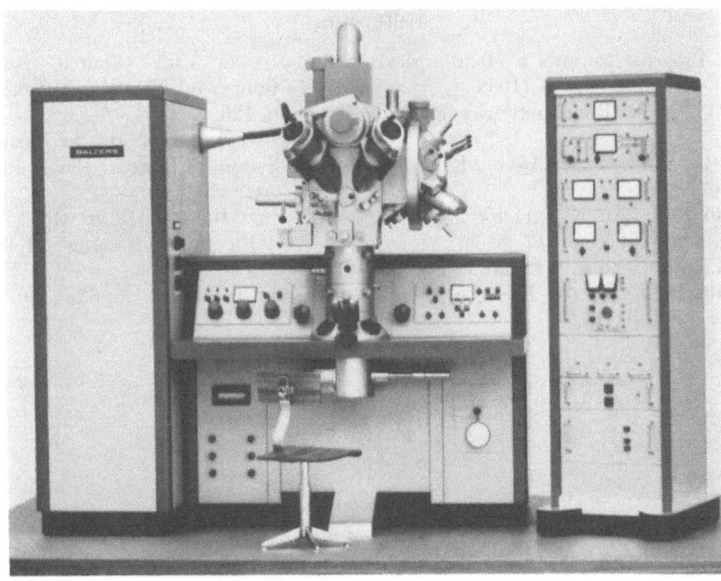

Fig. 8. Photo emission electron microscope Balzers metioscope KE 3. Description see [1] and [5]

sumably all metals. Much more difficult to understand is the *physical background* of these effects. One thing seems to be clear: The excellent contrast conditions in photo and thermionic emission are related to the fact that the energy offered to release the electrons is very close to the work function. This is not the case when using secondary emission by electron or ion impact, where it is not possible to introduce particles with a kinetic energy of some eV into the strong electric field between specimen and anode. A contrast sometimes observed in secondary emission is of another nature and related to the anisotropic scattering of the incident beam by the lattice of the specimen [9].

But what represents the work function of surface elements in a vacuum of 5×10^{-7} Torr?

The intensity of the emission spectrum of the utilised mercury-vapor high pressure lamp in combination with the quartz optics becomes practically zero at a wave length of 2750 ÅE. This $h\nu$-energy is slightly below the work function of pure Tungsten. Nevertheless Fig. 6 shows the photo emission micrograph of a tungsten surface at room temperature. This makes it clear that many of the fascinating contrast effects in photo emission microscopy are related to absorption layers on the surface of the specimen. However, it is not clear whether these absorption layers are so specific (and so reproducibly specific!) that they determine the work function, or if they simply reduce in a general manner the work function of the whole surface whereas the brightness of the structural element is determined by its nature and orientation, photo emission "seeing through" the absorption layer. There is some evidence that both of these effects are present, but the second preponderantly. There is still the effect of contrast reversal by changing the centre of the UV-spectrum demonstrated by ENGEL [4] and KOCH [10] which is difficult to explain without the first effect. Series of other effects are waiting to be physically investigated and explained, but it is an often repeated experience in microscopy, that their application was successful before the related theory of image generation and formation was established.

Photo emission microscopy represents an important analytical method for metallurgy. Photo emission enables metal microscopy with pronounced phase and orientation differentiation and therefore an analysis of shape, distribution and behaviour of components in a chemically known composition.

The question of application of photo emission microscopy for non conducting specimens cannot be answered at this time.

References

1. WEGMANN, L.: Progress towards a Metallurgical E. M. Prakt. Metallogr. **5**, 241 (1968).
2. PEPPERHOFF, W.: Arch. Eisenhuettenw. **36**, 941 (1965).
3. MÖLLENSTEDT, G., and F. LENZ: Advan. Electron. **18**, 251 (1963).
4. ENGEL, W.: Sixth Intern. Congr. for Electron Microscopy, Kyoto 1966, p. 217.
5. GRABER, R., M. GRIBI, and L. WEGMANN: Fourth European Regional Conf. on Electron Microscopy, Rome 1968, p. 111.
6. ZAMINER, CH.: Fourth European Regional Congr. on Electron Microscopy, Rome 1968, p. 123.
7. WEGMANN, L., and CH. ZAMINER: 6. Plansee-Seminar Reutte 1968, Pre-prints, vol. II, Art. 42.
8. SEILER, H., u. F. LENZ: Optik **27**, 438 (1968).
9. BAS, E., and S. ESCHER: Helv. Phys. Acta **40**, 353 (1968).
10. KOCH, W.: Optik **25**, 535 (1967).

Röntgenemissionsspektrometrie von Elementen niedriger Ordnungszahl ($Z < 15$)

W. HINK

Physikalisches Institut der Universität Würzburg, BRD

Abstract

A survey is given of recent work and instrumental developments in the spectrometry of ultrasoft X-rays. This includes the discussion of efficient sources as the proton-bombardment source and the synchrotron radiation source. Data of efficiency and resolution of the crystal-type spectrometer and the blazed-grating spectrometer for grazing incidence are compaired. The recent improvements of the detection devices, especially the proportional counter with low-noise semiconductor preamplifier, are discussed. Some applications are mentioned.

Einleitung

Röntgenspektrometrie von Elementen niedriger Ordnungszahl bedeutet Spektrometrie im Bereich der weichen und ultraweichen Röntgenstrahlung. So beträgt für das Element Phosphor mit der Ordnungszahl $Z = 15$ die Wellenlänge des langwelligsten Glieds der K-Serie (λ_{K_x}) 6 Å, das entspricht einer Photonenenergie von 2 keV. Das K-Band des Berylliums, des mit $Z = 4$ in seiner Ordnungszahl bisher niedrigsten der Mikroanalyse zugänglichen Elements, liegt bei 114 Å, entsprechend 109 eV. Verglichen mit anderen Bereichen des elektromagnetischen Spektrums ging die Entwicklung in dem Bereich der ultraweichen Röntgenstrahlung nur zögernd voran. Dieser Wellenlängenbereich von einigen Angström bis herauf zu einigen hundert Angström wird neuerdings auch XUV-Bereich genannt. Jedoch wuchs im letzten Jahrzehnt das Interesse am XUV-Bereich wegen der gesteigerten Aktivität auf den Gebieten der Plasmaphysik, der Raumforschung und nicht zuletzt der Mikroanalyse mit Röntgenstrahlen erheblich. Von den vielen die Spektrometrie ultraweicher Röntgenstrahlung berührenden Arbeiten auf diesen Gebieten seien insbesondere die zur Diagnostik der Hochtemperaturplasmen — wie sie in den Sternatmosphären vorhanden sind oder im Funken und Theta-Pinch im Labor erzeugt werden — und die zur Röntgenastronomie hervorgehoben.

Die Forschungsarbeiten auf diesen Gebieten erforderten die Verfeinerung bekannter und die Entwicklung neuer Techniken der Spektrometrie ultraweicher Röntgenstrahlung. Als Ergebnis dieser Aktivität stehen heute leistungsstarke Strahlungsquellen und mehrere Spektrometer- und Monochromatortypen, darunter ausgezeichnete kommerzielle, zur Verfügung. Besonders zu erwähnen ist in diesem Zusammenhang die zunehmende Nutzung der Synchrotronstrahlung als intensive Quelle weicher Röntgenstrahlen.

Auf der anderen Seite hat auch die nicht-dispersive Spektrometrie im langwelligen Bereich, also die Impulshöhenspektrometrie mit Proportionalzählrohr und Vielkanalanalysator, eine noch vor Jahren kaum vorstellbare Entwicklung erfahren. So ist es durch Einführung der Halbleiter-Elektronik gelungen, den Einfluß des elektronischen Rauschens auf die Impulshöhenspektren für Quantenenergien bis herab zu weniger als 100 eV fast vollständig zu eliminieren.

Neben den apparativen Entwicklungen haben in den letzten Jahren aber auch Grundlagenfragen der Erzeugung ultraweicher Röntgenstrahlen erhöhtes Interesse gefunden. Diese Fragen sind z. B. durch die Diskussion der Mikroanalyse der leichtesten Elemente stimuliert worden.

Ebenso kamen Impulse von der Astrophysik, die zu Voraussagen der Eigenschaften von Sternatmosphären auch auf genaue Daten der Wirkungsquerschnitte für den Elementarprozeß der
Erzeugung von charakteristischer Röntgenstrahlung und von Bremsstrahlung angewiesen ist.

Auf dem letzten Kongreß im September 1965 berichtete B. L. Henke [1] über die „Spektro-
skopie im 15—150 Å-Bereich der ultraweichen Röntgenstrahlung". Seitdem sind Instrumente
und Methoden dieses Wellenlängenbereichs in einem solchen Ausmaß weiterentwickelt worden,
daß ich mich beschränken muß auf Aspekte der XUV-Spektrometrie im Zusammenhang mit
der Mikroanalyse von Elementen niedriger Ordnungszahl.

Quellen ultraweicher Röntgenstrahlen

Quellen konventioneller Art, die auf Elektronenstoß oder auf Fluoreszenzanregung beruhen,
sind von mehreren Autoren beschrieben worden. Von diesen Quellen hat die von B. L. Henke [1]
entwickelte Röntgenröhre weite Verbreitung gefunden, weil sie durch ihre Geometrie die Kontamination der Anode durch verdampfendes Kathodenmaterial verhindert und große Anoden-
ströme zuläßt. Zunehmend gewinnt an Bedeutung die Methode der Erzeugung langwelliger
Röntgenstrahlung durch Protonenstoß. Während beim Stoß mit Elektronen dem Linienspektrum
der erzeugten charakteristischen Röntgenstrahlung ein Bremsspektrum überlagert ist, fehlt
ein solches beim Stoß mit schweren geladenen Teilchen fast vollständig, da die Intensität der
Bremsstrahlung umgekehrt proportional dem Quadrat der Masse M des stoßenden Teilchens
ist $\left(\sim \dfrac{1}{M^2}\right)$. Die durch Protonenstoß erzeugte Röntgenstrahlung ist also praktisch eine reine Linienstrahlung. Von A. Sterk u. Mitarb. [2—4] wird für ein massives Target eine Ausbeute von 10^{-2} bis
10^{-3} Photonen pro Proton in dem hier interessierenden Wellenlängenbereich angegeben, ein mit
der Ausbeute für Elektronenstoß vergleichbarer Wert [5]. Die Abhängigkeit der Ausbeute von
der Protonenenergie zeigt Abb. 1. Bei fester Protonenenergie wächst hiernach die Ausbeute mit
der Wellenlänge der erzeugten Strahlung an. Dies entnimmt man deutlicher aus Abb. 2, in
der die Abhängigkeit explizit dargestellt ist. Die Erzeugung von Röntgenstrahlung durch
Protonenstoß ist also wegen der starken Wellenlängenabhängigkeit der Ausbeute für ultraweiche
Strahlung besonders effektiv. Durch das Fehlen des Bremskontinuums lassen sich darüber hinaus
mit einer solchen Quelle besonders hohe Verhältnisse Linie zu Untergrund erzielen [4].

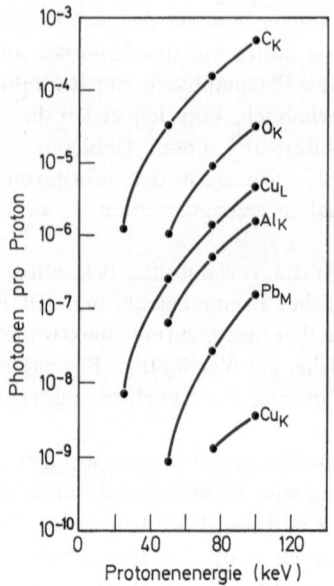

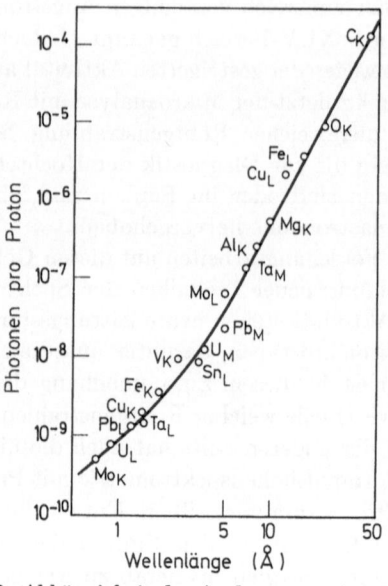

Abb. 1. Abhängigkeit der Ausbeute an
charakteristischer K-, L- und M-Röntgenstrahlung
von der Protonenenergie. (Nach Sterk
et al. [4])

Abb. 2. Abhängigkeit der Ausbeute an charakteristi-
scher K-, L- und M-Röntgenstrahlung von der
Wellenlänge bei fester Protonenenergie von 75 keV.
(Nach Sterk et al. [4])

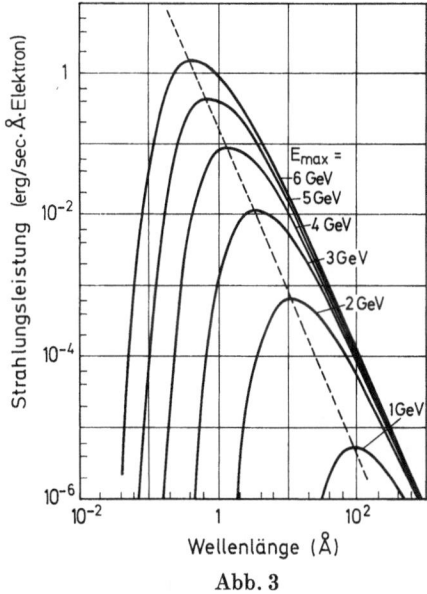

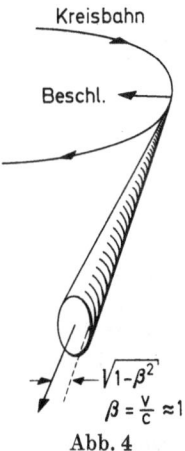

Abb. 3   Abb. 4

Abb. 3. Über die Beschleunigungszeit gemittelte Spektralverteilung der Synchrotronstrahlung der auf verschiedene Endenergien E_{max} beschleunigten Elektronen. (Nach HAENSEL und KUNZ [6])

Abb. 4. Qualitatives Strahlungsdiagramm der Synchrotronstrahlung

An Intensität übertrifft aber die Synchrotronstrahlung alle zur Zeit bekannten Quellen ultraweicher Röntgenstrahlen. Bei den heutzutage mit Beschleunigern erreichbaren Elektronenenergien erstreckt sich das Spektrum der Synchrotronstrahlung kontinuierlich vom Sichtbaren bis in das Röntgengebiet (Abb. 3). Nach der klassischen Elektrodynamik strahlt jede beschleunigte Ladung. Ein qualitatives Strahlungsdiagramm für Elektronen, die in einem Kreisbeschleuniger auf relativistische Energien beschleunigt sind, zeigt Abb. 4. Die Strahlung ist auf einen engen Kegel, der in Richtung der momentanen Elektronenbewegung weist, begrenzt. Die Winkelverteilung der Strahlung ist eine Funktion der Elektronenenergie und des Spektralbereichs. Für den Strahlungsanteil im XUV-Bereich eines 6 GeV-Synchrotrons liegt der Öffnungswinkel des Kegels bei 0,5 mrad [6].

Es ist möglich, die Synchrotronstrahlungsexperimente gleichzeitig mit den Hochenergieexperimenten laufen zu lassen. Die z.B. bei DESY verwendete Anordnung wird in Abb. 5 erläutert [6]. In dieser oder einer entsprechenden Anordnung werden an ungefähr einem Dutzend Beschleunigern Experimente mit der Synchrotronstrahlung durchgeführt. Ziel der ersten Experimente war, die Voraussagen der Theorie über die absolute spektrale Verteilung zu überprüfen. Die quantitative Überprüfung konnte allerdings nur im Sichtbaren und im kurzwelligen Röntgengebiet (Photonenenergien > 10 keV) durchgeführt werden, da in diesen Spektralbereichen geeichte Strahlungsempfänger existieren. Im Vakuum-UV und im weichen Röntgengebiet waren wegen des Fehlens solcher Empfänger nur qualitative Vergleiche möglich. Eine Berechnung der in diesem Wellenlängenbereich zu erwartenden absoluten Photonenflüsse wurde von TOMBOULIAN u. Mitarb. [7, 8] und von PARRAT [9] durchgeführt. Die Nachprüfung der Voraussagen erfordert ein absolut kalibriertes Spektrometer. Die mit der Realisierung eines solchen Spektrometers verbundenen Probleme sind jedoch bis heute noch nicht gelöst. So gelang es TOMBOULIAN und BEHRING [10] lediglich, den vorausgesagten relativen spektralen Verlauf zu bestätigen. Einen hoffnungsvollen Ansatz zur absoluten Kalibrierung stellt die Arbeit von CARUSO und NEUPERT [11] dar. Diese Autoren haben die Ausbeute der durch Elektronenstoß in einem massiven Kohlenstofftarget erzeugten K-Strahlung ($\lambda = 44$ Å) absolut gemessen. Der damit bekannte absolute Photonenfluß von einem Kohlenstofftarget diente ihnen zur Kalibrierung eines Gitterspektrometers für streifende Inzidenz bei dieser Wellenlänge.

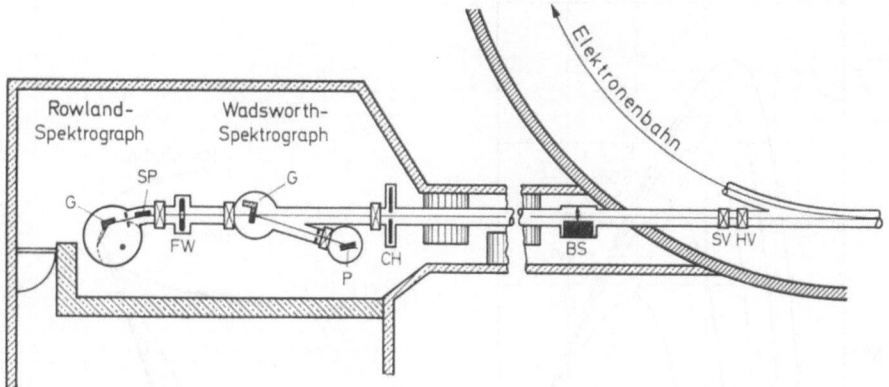

Abb. 5. Aufbau der Synchrotronstrahlungsexperimente beim DESY. *HV* Hauptventil, *SV* Schnellschlußventil, *BS* Strahlverschluß, *CH* Chopper, *FW* Filterwechsler. Der Wadsworth-Spektrograph enthält das aus dem Strahlengang schwenkbare Gitter *G* und eine Reflektorkammer mit dem Präparat *P*. Der Rowland-Spektrograph wird über einen Spiegel *SP* bei streifendem Einfall bestrahlt. (Nach HAENSEL und KUNZ [6])

Da die Experimente in den überprüfbaren Spektralbereichen Übereinstimmung mit der Theorie ergeben haben, erscheint es berechtigt, die Synchrotronstrahlung auch im XUV-Bereich als genau berechenbar anzusehen und sie als kontinuierliche Strahlungsquelle mit absolut berechenbarer spektraler Verteilung einzusetzen. Abschätzungen zeigen [6], daß z. B. die Synchrotronstrahlung eines 6 GeV-Beschleunigers von extremem Vakuum-UV herab bis zum Röntgengebiet ($1000 \text{ Å} > \lambda > 1 \text{ Å}$) im Vergleich zu anderen Kontinuums- oder Linienquellen, wie Funkenquellen, Glimmentladungen und Röntgenröhren, eine in Größenordnungen höhere Strahlungsstärke aufweist.

Spektrometer

Im augenblicklichen Entwicklungsstand finden praktisch nur zwei Typen von Spektrometern Anwendung zu spektralen Untersuchungen der ultraweichen Röntgenstrahlung, nämlich Kristallspektrometer, vorwiegend mit ebenen oder gekrümmten Pseudokristallen, die nach der Blodgett-Langmuir-Technik hergestellt werden, und Konkav-Gitterspektrometer für streifende Inzidenz mit Blazegitter. Nachdem die Eignung dieser beiden Typen schon vor Jahren sichergestellt war, zielte die Entwicklung auf die Verbesserung der beiden hauptsächlich interessierenden Parameter, Empfindlichkeit und spektrale Auflösung, ab. Besonderes Interesse erfuhr der Vergleich der Eigenschaften beider Spektrometertypen.

a) Kristalle und Pseudokristalle

Die Entwicklung der Kristallspektrometer war zunächst vorwiegend bestimmt durch die Suche nach Kristallen mit hinreichend großen $2d$-Werten. Kommerziell steht als Einkristall bisher nur der KAP-Kristall (Kaliumphtalat) mit dem $2d$-Wert von 26,6 Å zur Verfügung. Erfolgreiche Versuche zur Züchtung von Einkristallen mit größeren $2d$-Werten unternahmen RUDERMAN, NESS und LINDSAY [12]. Den Autoren gelang es, auf der Basis von organischen Estern Einkristalle der Größe 2 cm × 2 cm × 1 mm· zu züchten, die hinreichende mechanische Festigkeit besitzen, um auch in fokussierenden Spektrometern verwendet zu werden. Die erreichten $2d$-Werte sind in der von den Autoren angegebenen Tabelle aufgeführt (Abb. 6). Nach Meinung der Autoren sollte durch Variation der Kettenlänge der Moleküle jeder gewünschte $2d$-Wert zu realisieren sein.

Die Verwendung von Pseudokristallen, den geschichteten Fettsäurefilmen, gehört heute zu den Standardtechniken im XUV-Bereich. In vielen Laboratorien wurden Untersuchungen vorgenommen, um die Technik der Herstellung derartiger Schichtsysteme zu erforschen (zur Literatur vgl. EHLERT und MATTSON [13]). Heute stehen, teilweise sogar kommerziell, Vielschichten-

Verbindung	$2d$-Wert (Å)
Hexadecyl-Hydrogen-Maleat (HHM)	58
Octadecyl-Hydrogen-Maleat (OHM)	$63{,}5 \pm 0{,}05$
Behenyl-Hydrogen-Maleat (BHM)	74
Dioctadecyl-Terephthalat (OTO)	84
Dioctadecyl-Adipat (OAO)	$93{,}8 \pm 0{,}1$
Octadecyl-Hydrogen-Succinat (OHS)	$96{,}9 \pm 0{,}1$

Abb. 6. Netzebenenabstände einiger organischer Ester. (Nach RUDERMAN et al. [12])

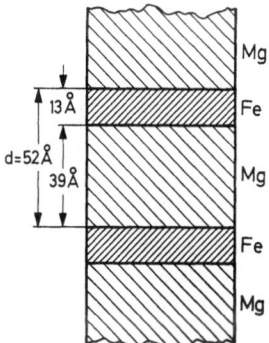

Abb. 7. Schematischer Aufbau eines mit der Aufdampftechnik gewonnenen Fe—Mg-Vielschichtensystems

kristalle mit $2d$-Werten von 70 Å über 100 Å (für die bekannten Barium- und Bleistearatkristalle) bis 165 Å zur Verfügung.

Einen aussichtsreichen Weg zur Herstellung von Vielschichtenkristallen anderer Art beschritten DINKLAGE und FRERICHS [14] und DINKLAGE [15]. Sie benutzten die Vakuumaufdampftechnik zum Aufbau eines Systems, bestehend aus etwa 100 Schichten eines stark streuenden Materials, nämlich Fe. Diese Schichten werden jeweils voneinander getrennt durch Schichten aus einem Stoff mit geringerem Schwächungskoeffizienten für Röntgenstrahlung (etwa Mg), die den gewünschten Abstand der streuenden Schichten bestimmen. Die Daten eines solchen Fe—Mg-Vielschichtensystems zeigt Abb. 7. Die von DINKLAGE durchgeführten Messungen mit Mg K- und O K-Strahlung zeigen, daß diese Systeme im Hinblick auf ihr Reflexionsvermögen (darunter soll verstanden werden: Verhältnis des in eine Ordnung gebeugten Strahlungsflusses zum einfallenden Strahlungsfluß) durchaus mit den besten Bleistearatkristallen zu vergleichen sind. Ihr Auflösungsvermögen ist allerdings um etwa den Faktor 3 geringer. Der Prozeß der Diffusion in den Schichtsystemen führt zu einer Verschmierung der Schichten und damit zur Herabsetzung des Reflexionsvermögens. Es ist zu hoffen, daß durch das genauere Studium dieses Prozesses Bedingungen zu seiner weitgehenden Unterdrückung gefunden werden.

b) Gitter mit Blazewinkel

Der verhältnismäßig geringe Aufwand bei der Herstellung und beim Einsatz der Pseudokristalle hat die Verwendung der früher einzig benutzten leicht geritzten Siegbahn-Gitter im XUV-Bereich stark reduziert. Durch die moderne Technik der Herstellung von Replika-Gittern mit Blazewinkel stehen jetzt aber Gitter zur Verfügung, die wegen ihres im Vergleich zu den Siegbahn-Gittern wesentlich erhöhten Reflexionsvermögens in Konkurrenz mit den Pseudokristallen treten. Beim Gitter mit Blazewinkel gelingt es nämlich durch geeignete Formgebung der Gitterfurchen, auf Kosten hauptsächlich der nullten Ordnung mehr Intensität in eine gewünschte Beugungsordnung zu bringen. Die ebenen Gitterfurchen bilden einen Winkel, Blazewinkel genannt, mit der Gitteroberfläche. Das Prinzip der Konzentration des Strahlungsflusses einer Wellenlänge in einer gewünschten Ordnung beruht einfach darauf, daß die regulär an den

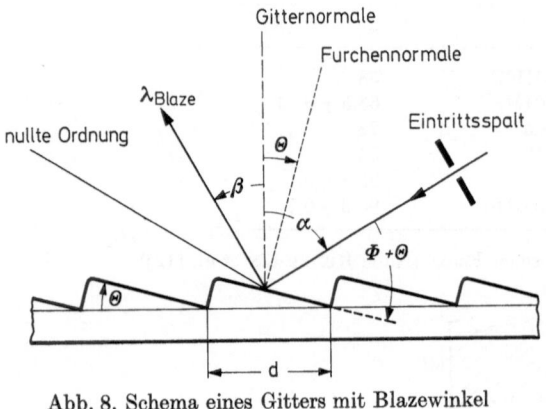

Abb. 8. Schema eines Gitters mit Blazewinkel

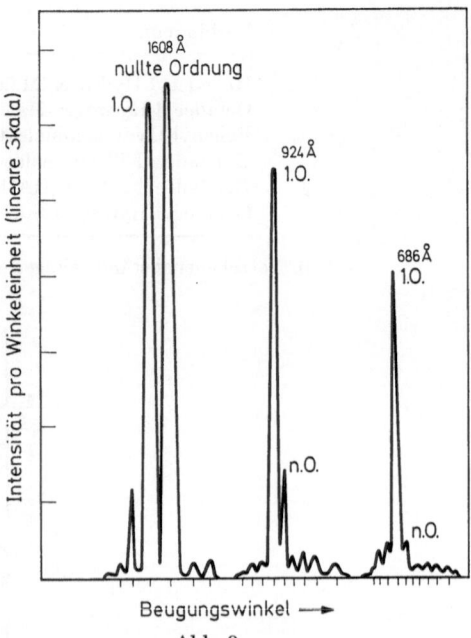

Abb. 9. Beugungsdiagramme für verschiedene Wellenlängen im Vakuum-UV, gewonnen mit einem Blazegitter. [Nach J. A. R. SAMSON, J. Opt. Soc. Am. **52**,
525 (1962)]

Abb. 9

ebenen Gitterfurchen reflektierte Strahlung in eine Richtung gelangt, die zusammenfällt mit
der Richtung, in welche die Strahlung der betrachteten Wellenlänge gebeugt wird. Unter Bezug
auf Abb. 8, in der Θ den Blazewinkel angibt, wird diese Bedingung durch Einführung des
Reflexionsgesetzes für die Furchenebenen

$$\alpha - \Theta = \beta + \Theta$$

in die Gittergleichung ausgedrückt. Man erhält die Wellenlänge λ_{Blaze}, für welche die maximale
gebeugte Intensität in Abhängigkeit vom Gittereinfallswinkel α beim Blazewinkel Θ auftritt, aus
der Beziehung:

$$m \cdot \lambda_{\text{Blaze}} = 2d \cdot \sin \Theta \cdot \cos (\alpha - \Theta)$$

(m: Beugungsordnung, d: Gitterkonstante).

In der Näherung für streifende Inzidenz ($\Phi = \pi/2 - \alpha$: Einfallsglanzwinkel) wird hieraus:

$$m \cdot \lambda_{\text{Blaze}} = 2d \cdot \Theta \cdot (\Phi + \Theta).$$

Abb. 9 zeigt im Vakuum-UV die Auswirkung des Blazewinkels auf die Intensitätsverteilung im
Beugungsdiagramm. Bei Annäherung der Wellenlänge an λ_{Blaze} (hier bei 686 Å) geht zunehmend
die Intensität aus der nullten in die erste Ordnung. Das Reflexionsvermögen des Gitters hat
für λ_{Blaze} ein Maximum. Im XUV-Bereich hat STERK [3] eine experimentelle Bestätigung des
erwarteten Maximums des Reflexionsvermögens als Funktion der Wellenlänge gegeben (Abb. 10).
Das von ihm gemessene breite Maximum überstreicht den Bereich von 100 bis 200 Å (Original-
gitter, Au-bedampft, 600 Striche/mm, 1 m Krümmungsradius). Der Maximalwert von 22%
für $\lambda \approx 150$ Å ist bemerkenswert hoch. Die Messungen von NICHOLSON und HASLER [16] mit
einem Al-bedampften Replika-Gitter sonst gleicher Daten sind besonders interessant, weil das
Reflexionsvermögen des Gitters verglichen wird mit dem eines Bleistearatkristalls (vgl. unten).

Damit die auf das Gitter mit Blazewinkel fallende Strahlung an den Furchenebenen mit
nennenswerter Intensität regulär reflektiert wird, muß der Einfallsglanzwinkel $\Phi + \Theta$ kleiner
sein als der wellenlängenabhängige Grenzwinkel ϑ_c der Totalreflexion. Das Reflexionsvermögen
einer ebenen Grenzfläche nimmt vom Wert 1 (Totalreflexion) allmählich ab, wenn der Einfallsglanzwinkel sich von kleineren Werten dem Grenzwinkel ϑ_c nähert. Bei ϑ_c selbst fällt das Reflexionsvermögen je nach dem Absorptionskoeffizienten des Mediums, an dem die Reflexion

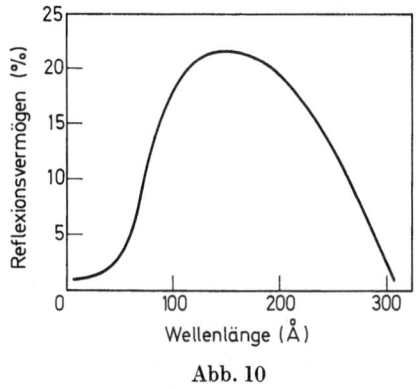

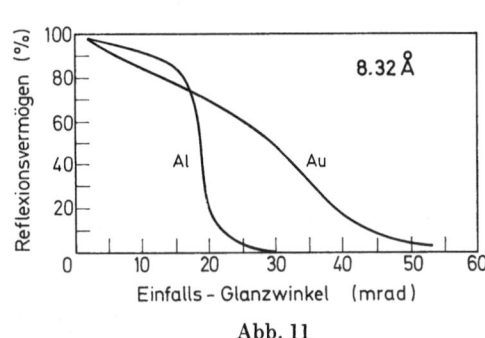

Abb. 10   Abb. 11

Abb. 10. Reflexionsvermögen eines Blazegitters. (Nach STERK [3])

Abb. 11. Reflexionsvermögen von Al und Au bei 8,32 Å in Abhängigkeit vom Einfallsglanzwinkel.
(Nach HENDRICK [17])

erfolgt, mehr oder weniger steil ab [17] (Abb. 11). ϑ_c bestimmt also die kürzeste noch mit nennenswerter Intensität totalreflektierte Wellenlänge. Da in einfacher Näherung ϑ_c proportional der Wurzel aus der Elektronendichte des Mediums ist, läßt sich durch Verwendung von Gittern mit Aufdampfschichten hoher Elektronendichte wie Platin und Gold [18] die Grenze für die Wellenlänge am weitesten hinausschieben. So gelang es GABRIEL et al. [19] mit einem platinierten Gitter Strahlung von 4,7 Å zu analysieren.

c) Empfindlichkeit und Auflösung

Unter der Empfindlichkeit einer Spektrometeranordnung wird das Verhältnis des in eine betrachtete Ordnung fallenden Strahlungsflusses zum einfallenden Strahlungsfluß bei gegebener Wellenlänge verstanden. Sieht man von der Geometrie ab, so wird die Empfindlichkeit durch das Reflexionsvermögen des Kristalls oder des Gitters bestimmt.

FRANS und DAVIDSON [20] sowie EHLERT und MATTSON [13] haben die Abhängigkeit des Reflexionsvermögens für die 1. Ordnung bei Seifenkristallen gemessen. Ein Beispiel ihrer Meßergebnisse zeigt Abb. 12. Für eine perfekte Struktur sollte das Reflexionsvermögen anwachsen wie $1 - e^{-a \cdot N}$, wobei N die Zahl der Doppelschichten bedeutet und a durch eine Funktion

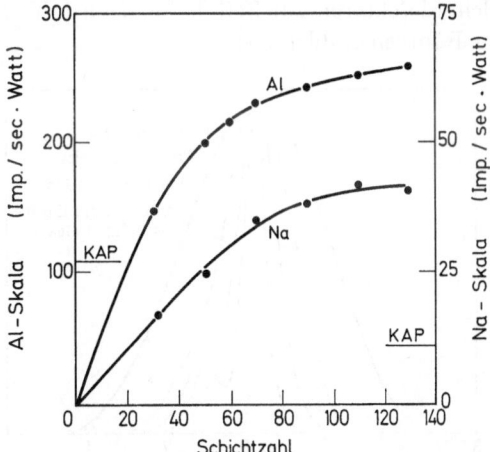

Abb. 12. Abhängigkeit der Intensität in der 1. Ordnung eines Pb-Stearatkristalls von der Zahl der monomolekularen Schichten für AlK_α-Strahlung ($\lambda_{Al\,K_\alpha} = 8,32$ Å, bezeichnet durch Al) und für NaK_α-Strahlung ($\lambda_{Na\,K_\alpha} = 11,88$ Å, bezeichnet durch Na). Die mit einem KAP-Kristall erzielten Werte sind mit KAP bezeichnet. (Nach FRANS at al. [20])

des Schwächungskoeffizienten der Schichten gegeben ist [13]. Mit etwa 20 bis 30 Doppelschichten wird bereits der Grenzwert des Reflexionsvermögens erreicht, der im Vergleich zum Reflexionsvermögen des KAP-Kristalls etwa um den Faktor 2,5 bzw. 4 höher liegt (vgl. Abb. 12). HENKE [1] gibt für diesen Faktor einen Wert von etwa 10 an. Das ebenfalls von N abhängige Auflösungsvermögen der Seifenkristalle erreicht bei der genannten Zahl von Doppelschichten auch sein Maximum, wie die Autoren zeigen konnten. Eine Verbesserung von Reflexionsvermögen und Auflösung mit wachsender Schichtenzahl berichtet DINKLAGE [15] auch für die Fe—Mg-Schichtsysteme.

NICHOLSON und HASLER [16] verglichen das Reflexionsvermögen eines Blazegitters mit dem eines Bleistearatkristalls. Für CK-Strahlung ($\lambda = 44,7$ Å) ergab sich beim Gitter ein Wert von 6% beim Seifenkristall hingegen von 0,27%. Der Vorteil des größeren Reflexionsvermögens des Gitters wird beim Seifenkristall keineswegs durch den im allgemeinen größeren nutzbaren Raumwinkel kompensiert. So beobachteten die genannten Autoren in allen Fällen des Vergleichs mit dem Gitter eine höhere Zählrate im Linienmaximum bei vergleichbarem Verhältnis Linie zu Untergrund. Blazegitter-Spektrometer haben also im XUV-Bereich eine durchaus vergleichbare, wenn nicht höhere Empfindlichkeit als Spektrometer mit Seifenkristallen. Da das Blazegitter in diesem Wellenlängenbereich nach HENKE [1] darüber hinaus ein 2- bis 5fach höheres Auflösungsvermögen liefert als die Seifenkristalle, wird sich der Einsatz des Blazegitter-Spektrometers trotz des erheblich größeren apparativen Aufwandes lohnen.

Nachweistechnik

Der für Photonenenergien über 5 keV vorwiegend verwendete Szintillationsdetektor ist bei niedrigeren Energien ungeeignet, da die Amplitude der Rauschimpulse des Photomultipliers die der Signalimpulse übersteigt. Bei der Suche nach Szintillationskristallen mit großer Lichtausbeute fanden HOFSTADTER u. Mitarb. [21] zwar, daß CaI$_2$(Eu) eine zweimal größere Lichtausbeute als NaI(Tl) gibt. Mit einem solchen Kristall kann aber der Anwendungsbereich auch nur bis zu etwa 1 keV erweitert werden, selbst wenn die neuentwickelten Photomultiplier (vgl. AITKEN [22] und dort angegebene Literatur) mit einer Quantenausbeute der Photokathode von $\approx 30\%$ benutzt werden.

Die Fortschritte der letzten Jahre in der Entwicklung der Halbleiterdetektoren haben ermöglicht, lithiumgedriftete Silizium- und Germaniumdetektoren mit dünnen Fenstern [22] zum Nachweis weicher Röntgenstrahlen einzusetzen. Die Energieauflösung dieser Detektoren überschreitet die des Szintillationsdetektors und des Proportionalzählrohrs wesentlich (Abb. 13). Wegen der Absorption im Fenster des zur Kühlung notwendigen Kryostaten liegt aber die Nachweisgrenze der Halbleiterdetektoren zur Zeit bei einigen keV. Somit scheiden auch diese zum Nachweis ultraweicher Röntgenstrahlen aus.

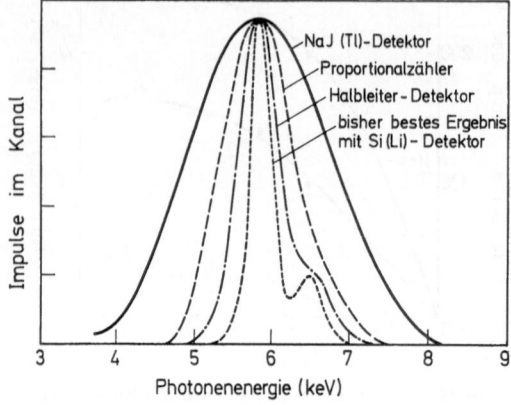

Abb. 13. Impulshöhenverteilungen für die MnK-Strahlung eines Fe 55-Präparats, gewonnen mit verschiedenen Detektoren. Mit dem Halbleiterdetektor wird die geringste Halbwertsbreite der Verteilung, also die beste Energieauflösung, erzielt. (Nach AITKEN [22])

Es bleiben der fensterlose Photomultiplier, das Geiger-Müller-Zählrohr und das Proportionalzählrohr. Diese drei Typen haben im XUV-Bereich einen vergleichbaren Quantenzählwirkungsgrad [1]. Der Vorteil, den der fensterlose Photomultiplier durch das Fehlen eines Fensters bietet, ist nach HENKE [1] jedoch nicht so wesentlich, da in einem Spektrometer im allgemeinen Filter, welche unerwünschte Strahlung eliminieren, verwendet werden. Demgegenüber besteht mit dem Proportionalzählrohr die Möglichkeit der Impulshöhendiskriminierung, durch welche der elektronische Rausch und Impulse unerwünschter Strahlung unterdrückt werden können. Die Impulshöhendiskriminierung ist herab bis zu Photonenenergien von 70 bis 100 eV durch die Verbesserung der Zählrohrauflösung und insbesondere durch die Entwicklung von rauscharmer Elektronik für die elektronische Verstärkung möglich geworden.

Die Faktoren, welche die Zählrohrauflösung wesentlich beeinflussen können, sind bekannt. Hierzu zählen elektronische Effekte, Druckschwankungen und Verunreinigungen des Füllgases, Gehäuse-Sauberkeit und insbesondere Homogenität des Anodendrahtes und Formgebung des Fensterbereichs. Die beste bisher experimentell erzielte Auflösung erreicht den nach der Theorie der Gasverstärkung von BYRNE [23] maximal zu erwartenden Wert [24, 25], wenn als Fano-Faktor der Wert 0,2 angenommen wird [26]. Da aber die Genauigkeit nicht abgeschätzt werden kann, mit der die Schwankungen der Gasverstärkung sowie der Fano-Faktor vorausgesagt werden, bleibt offen, ob eine weitere Verbesserung der Auflösung zu erzielen ist.

Den größeren Gewinn an Auflösung brachte jedoch die Einführung von rauscharmer Halbleiter-Elektronik. Höhe und Zahl der Rauschimpulse der dem Proportionalzählrohr nachgeschalteten Verstärkerelektronik legen nicht nur die untere energetische Nachweisgrenze fest, sondern sie liefern auch einen Anteil an der Streuung der beobachteten Impulshöhenverteilung. Da der Rauschanteil an der Auflösung mit abnehmender Photonenenergie relativ zunimmt, kommt seiner Eliminierung im Bereich der ultraweichen Röntgenstrahlung hohe Bedeutung zu.

An einem von AITKEN [22] gegebenen Beispiel seien die Verhältnisse etwas genauer erörtert. In einem Proportionalzählrohr mit der typischen Gasfüllung von 90% Argon und 10% Methan werden durch Röntgenstrahlen von 30 eV Photonenenergie die Elektronenlawinen von einzelnen Elektronen ausgelöst (der mittlere Energieverbrauch pro Ionenpaar beträgt für Argon ≈ 27 eV). Bei einer plausiblen Gasverstärkung von 1000 ist die auf den Vorverstärker gelangende Ladung die von etwa 1000 Elektronen mit einer Streuung σ von ≈ 800 Elektronen. Nach den Messungen der Streuung von Einelektronenlawinen [25] ist ein solcher Wert anzusetzen. Bis vor kurzem standen als rauscharme Vorverstärker röhrenbestückte ladungsempfindliche Typen zur Verfügung. Mit diesen erreicht man bei kleinen Zählrohrkapazitäten im praktischen Laborbetrieb ein Eingangsrauschen mit einer Streuung, die äquivalent etwa 600 Elektronen ist. Heute gibt es kommerziell bei Zimmertemperatur betriebene Halbleitervorverstärker mit Feldeffekttransistoren, mit denen ein Rauschäquivalent bis herab zu 150 Elektronen erreicht wird [22]. Der Anteil des elektronischen Rausches an der Linienbreite ist damit für 30 eV Photonenenergie auf einige Prozent reduziert. Dies bedeutet aber, daß praktisch im gesamten XUV-Bereich die Impulshöhenspektren frei vom Einfluß des elektronischen Rausches sind.

Anwendungen

Das Ziel der Spektrometrie ultraweicher Röntgenstrahlung ist die Ermittlung der spektralen Intensitätsverteilung, bei Spektrallinien speziell der Linienlage, der Linienform und der Linienintensität. Aus diesen Daten können für verschiedene Bereiche der Physik wesentliche Informationen gewonnen werden. So z. B. gibt die Form einer Emissionsbande, die durch Elektronenübergänge aus dem Valenzband auf eine Leerstelle eines inneren Niveaus entsteht, nach entsprechenden Korrekturen die Dichte der Energiezustände im Valenzband wieder. Da die Valenzelektronen am stärksten durch die chemische Bindung beeinflußt werden, liefert die Untersuchung der Emissionsbanden Aussagen über die chemische Bindung (Valenzbandstruktur-Analyse). Hochtemperaturplasmen, wie sie in den Sternatmosphären vorhanden sind oder im Funken und Theta-Pinch im Labor erzeugt werden, emittieren elektromagnetische Strahlung im Vakuum-UV und im XUV-Bereich. Erhöht man die Elektronentemperatur eines Plasmas, so verschiebt sich das Maximum seines Emissionsspektrums zu kürzeren Wellenlängen. Bei Temperaturen in der Größenordnung von 10^6 bis 10^7 °K liegt dieses Maximum im weichen Röntgengebiet. In einem solchen Hochtemperaturplasma sind Stoßprozesse im 100 bis 1000 eV-Energiebereich möglich, die zur Emission von weicher und ultraweicher Röntgenstrahlung führen. Neben Bremsstrahlung tritt auch Rekombinationsstrahlung beim Einfang freier Elektronen durch Ionen und Linienstrahlung der im Plasma vorhandenen Elemente auf. Aus der absoluten Spektralverteilung der Strahlung können z. B. die Strahlungsverluste und die Geschwindigkeitsverteilung der freien Elektronen ermittelt werden. Ist ein heißes, dünnes Plasma über ein kaltes, dichtes geschichtet — dies ist bei der Sonne der Fall, die Korona ist der Photosphäre überlagert —, so können durch Absolutmessung der Strahlung die lokale Temperatur und die Dichte des heißen dünnen Plasmas bestimmt werden. Eine solche Messung ist wegen des starken Eigenleuchtens der kalten Schicht im Sichtbaren und im nahen Ultraviolett nicht möglich.

Zu Voraussagen der Eigenschaften von Hochtemperaturplasmen müssen die Wirkungsquerschnitte für den Elementarprozeß der Erzeugung von charakteristischer Strahlung und von Bremsstrahlung bekannt sein. Weiterhin werden die experimentellen Daten benötigt, um die Gültigkeit der verschiedenen erforderlichen Näherungen in den Theorien dieser Elementarprozesse zu überprüfen. Die Arbeiten zu diesen Grundlagenfragen stehen aber erst an ihrem Beginn.

Aus der Linienlage und der Linienintensität kann auf die quantitative Zusammensetzung eines durch Röntgenemission angeregten Stoffes geschlossen werden. Die Analyse der leichten Elemente bis herab zum Beryllium ($\lambda_{\mathrm{Be}K} = 114$ Å) bildet einen weitgehend ausgearbeiteten Anwendungsbereich der Spektrometrie ultraweicher Röntgenstrahlung. Auf die jüngsten Ergebnisse der Analyse leichter Elemente braucht hier nicht eingegangen zu werden, da in den folgenden Vorträgen darüber berichtet wird.

Dieser Bericht gibt einen Überblick der Röntgenemissionsspektrometrie von Elementen niedriger Ordnungszahl. Wenn er vorwiegend instrumentelle Probleme behandelte, so erklärt sich dies daraus, daß die Spektrometrie ultraweicher Röntgenstrahlen erst in den vergangenen Jahren ihre Entwicklung erfahren hat. Es ist zu hoffen, daß diese Technik in der Zukunft viele nutzvolle Anwendungen erfährt.

Literatur

1. Henke, B. L.: X-ray optics and microanalysis (ed. R. Castaing, P. Deschamps, and J. Philibert), p. 440. Paris: Hermann 1966.
2. Sterk, A. A.: Advances in X-ray analysis, vol. 8, p. 189. New York: Plenum Press 1965.
3. — Advances in X-ray analysis, vol. 9, p. 410. New York: Plenum Press 1966.
4. — C. L. Marks, and W. P. Saylor: Advances in X-ray analysis, vol. 10, p. 399. New York: Plenum Press 1967.
5. Green, M.: X-ray optics and X-ray microanalysis (ed. H. H. Pattee, V. E. Cosslett, and A. Engström), p. 185. New York: Academic Press 1963.
6. Haensel, R., u. C. Kunz: Z. angew. Physik 23, 276 (1967).
7. Tomboulian, D. H., and P. L. Hartmann: Phys. Rev. 91, 1577 (1953).
8. —, and D. E. Bedo: J. Appl. Phys. 29, 804 (1958).
9. Parrat, L. G.: Rev. Sci. Instr. 30, 297 (1959).
10. Tomboulian, D. H., and W. E. Behring: Appl. Opt. 3, 501 (1964).
11. Caruso, A. J., and W. M. Neupert: Appl. Opt. 4, 247 (1965).
12. Ruderman, I. W., K. J. Ness, and J. C. Lindsay: Appl. Phys. Letters 7, 17 (1965).
13. Ehlert, R. C., and R. A. Mattson: Advances in X-ray analysis, vol. 10, p. 389. New York: Plenum Press 1967.
14. Dinklage, J., and R. Frerichs: J. Appl. Phys. 34, 2633 (1963).
15. — J. Appl. Phys. 38, 3781 (1967).
16. Nicholson, J. B., and M. F. Hasler: Advances in X-ray analysis, vol. 9, p. 420. New York: Plenum Press 1966.
17. Hendrick, R. W.: J. Opt. Soc. Am. 47, 165 (1957).
18. Landon, D. O.: Appl. Opt. 2, 450 (1963).
19. Gabriel, A. H., J. R. Swain, and W. A. Waller: J. Sci. Instr. 42, 94 (1965).
20. Frans, R. P., and F. D. Davidson: Rev. Sci. Instr. 36, 230 (1965).
21. Hofstadter, R., W. E. O'Dell, and C. T. Schmidt: Rev. Sci. Instr. 35, 246 (1964).
22. Aitken, D. W.: IEEE Trans. Nucl. Sci. NS-15, 10 (1968).
23. Byrne, J.: Proc. Roy. Soc. Edinburgh A 66, 33 (1962).
24. Charles, M. W., and B. A. Cooke: Nucl. Instr. Methods 61, 31 (1968).
25. Campbell, J. L., and K. W. D. Ledingham: Brit. J. Appl. Phys. 17, 769 (1966).
26. Alkhazov, G. D., A. V. Komar, and A. A. Vorab'ev: Nucl. Instr. Methods 48, 1 (1967).

The Detection of Light Elements

E. F. PRIESTLEY and H. K. PHELAN

Royal Armament Research and Development Establishment, Fort Halstead, England

The microanalyser built at the R.A.R.D.E uses a modified Duncumb lens with six re-entrant ports. Geometric and spatial restrictions make it difficult to design a fully-focussing spectrometer of high efficiency for use with a lens of this type. Focussing spectrometers of limited range have been constructed [1] and a multi-crystal instrument is being built using focussing geometry on the specimen side only. This spectrometer however, like the semi-focussing instrument it will replace, requires a counter with a large window in order to intercept the whole of the diffracted beam at the higher values of θ, and the difficulty of making thin windows of sufficient strength increases with their physical size.

Many workers have simplified the problem by operating their counters at less than atmospheric pressure, and this is an attractive solution if a second spectrograph or alternative counter is available for the detection of the shorter wavelengths.

In 1966 CULHANE, HERRING and SANFORD of University College, London and O'SHEA and PHILLIPS of 20th Century Electronics published a joint paper [2] in which they described a new type of proportional counter which from its shape has been appropriately named a "pill-box" counter, and it has a number of properties that make it particularly suitable for use in a microanalyser for the detection of light elements.

The general form of the pill-box counter has been described already in this conference by HAILES [3] and while our two versions differ in detail, their dimensions are identical and their performance is probably very similar. A measured 51% resolution for oxygen with the flow counter is probably not the best figure obtainable as no special precautions were taken to avoid contamination with water vapour.

The first advantage of the pill-box construction is the ease with which a tandem pair of flow and sealed-off counters can be accommodated in one unit by dividing it into two compartments separated by a beryllium window. The flow counter can then be operated at reduced pressure without overall loss of efficiency for shorter wavelengths.

Another benefit is the large window area available. A 2.5 cm diam. window can be fitted to a 5 cm diam. counter and no great variation of sensitivity is observed from point to point over the whole area. Loss of counts by recombination is sometimes observed in "orthodox" counters as a result of a pocketing effect when the window flexes under pressure, but no such effect has been observed with pill-box counters.

The circular window is advantageous since it gives a more even distribution of stress than a rectangular one and thus reduces the risk of failure. Also, the flat, circular face of the counter greatly facilitates the mounting of the window and its supporting grid. Fig. 1 shows a counter with and without a window in position. The window mounting seals on to the "0" ring, being clamped by a knurled ring not shown in the photograph.

Examination of fractured windows has suggested that catstrophic failure generally occurs as a result of breakdown of the supporting grid. We have therefore developed a method of making grids of 25 μm stainless steel wire which has a tensile strength of 120 tons/sq. in. Two parallel grids are wound under even tension at a spacing of about 0.24 mm and mounted at 90° rotation. A light spray with an aerosol resin solution cements the wires together and prevents their displacement under flexure. The nominal transmission is 80% but this is reduced to 75 to 77%

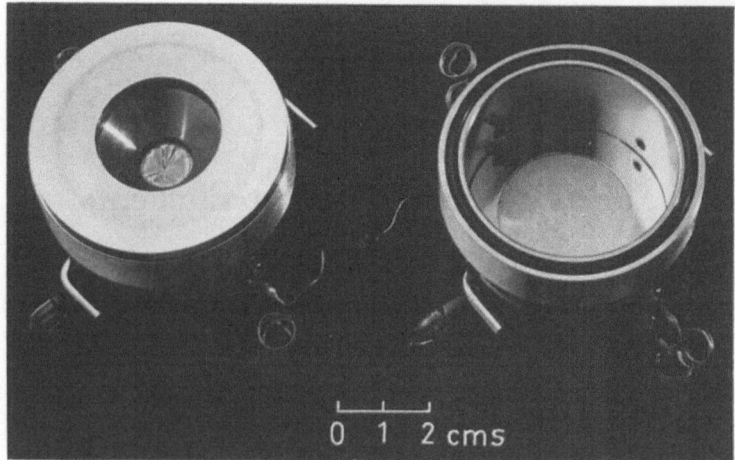

Fig. 1*. The Pillbox proportional counter, showing the flow section with, and without its window. The knurled ring holding the window assembly in position has been removed

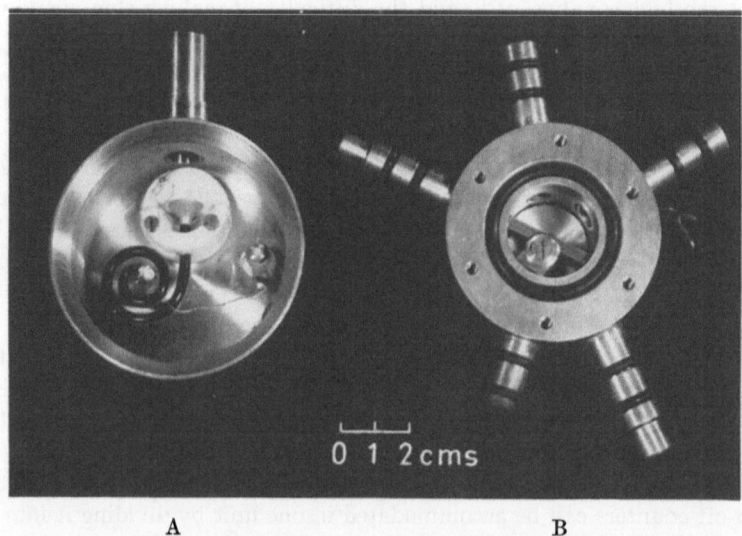

A   B

Fig. 2*. A. The Mullard channel multiplier in its vacuum housing, with lid removed. B. Miniature spectrometer with alternative ports for quick interchange of detectors covering the elements boron to fluorine

after spraying. No grid has ever broken under atmospheric pressure and no evidence of window puncture by the ribbed surface has been observed. During the testing of windows one grid has been used with as many as six successive plastic windows without deterioration.

The method of manufacture of these grids and the polycarbonate windows used with them was described at a recent Institute of Physics and Physical Society Conference in London and will not be elaborated here.

Experience has shown that window failure most often occurs near the periphery and a new type of window was developed to overcome this defect. The object was to obtain a disc shaped window, uniformly thin over a large central area but increasing in thickness at the outside edge. As in our previous experiments with window manufacture, polycarbonate was used in the form of a solution of Makrolon 3000 in dichlormethane, but the method should be applicable to any window material available in solution. A ring of PTFE is clamped to a glass plate and centred on a spinning table. A measured quantity of plastic solution is dropped in. Control of concentration, quantity of solution, spinning speed and drying conditions will produce films of predetermined characteristics provided that ambient temperature and humidity are constant. Films down to about 0.2 μm thickness, 25 mm in diameter have been made, with a large central area of constant thickness (as evidenced by the interference colours) and a periphery thick enough to allow easy handling with tweezers. This thicker edge is useful for sealing the window positively under a clamping ring.

Windows of 0.6 to 1 μm in thickness and wire grids have both been tested under normal conditions of operation for a long enough period to prove their reliability, withstanding daily pressure cycling from zero to full atmospheric pressure. Thinner windows have passed initial vacuum tests but have not been incorporated in the microanalyser.

An alternative and attractive way of solving the problem of light element detection is by the use of windowless counters, and some experiments have been commenced to assess the relative value of the method. Assuming 100% efficiency for a proportional counter in the long wavelength region and transmission efficiencies of about 80% for both window and grid a maximum overall efficiency of about 60% might be attained. A channel multiplier can have an efficiency of about 20% in this region [4], so that when used under comparable conditions the counting rate may be expected to be about one third of that obtained with a proportional counter.

In order to compare the performances of a number of counters under identical conditions the simplified spectrometer shown in Fig. 2 was constructed. Five tubular ports are provided and the detector housings can be plugged into any one of them. The spectrometer plugs into the microanalyser by means of any other of the ports and by the appropriate selection of analysing crystal and ports the spectrometer may be used for the detection of the elements boron to fluorine. Unused ports are vacuum sealed by plugs. A micrometer adjustment allows precise setting of the crystal angle. All experiments have been carried out so far using oxygen radiation from a beryl specimen. The analysing crystal was clinochlore, bent to 15 cm radius and lightly abraded. The figure also shows the channel multiplier in its housing, mounted so that the angle of incidence of radiation on the cathode can be varied through about 60°. The cathode is made about 2.5 kV negative and the output taken from across a load resistance inserted between the anode and earth. In order to reduce the risk of fracture during a series of measurements the proportional counter was fitted with a thick window of 1 μm polycarbonate, reducing the overall efficiency of the counter to an estimated 30% for oxygen. The efficiency of the spectrometer is low, since the focussing conditions are not critically fulfilled and the beam aperture is reduced so that it is less than that of the smallest detector used.

Under these conditions a pulse rate of about 400 counts/sec was observed with the proportional counter at a probe current of 1 μA at 10 kV. The corresponding count rate using the channel multiplier was about 100 counts/sec. Coating the cathode with magnesium fluoride has not yet been tried, but it is reported to increase the efficiency by a factor of at least two times [5], so that the counting rate would then be rather more than 200 counts/sec. This is not far short of the estimated figure.

Channel multipliers have no energy resolution and approximate to the Geiger counter in their characteristics. Pulse height is determined by operating conditions, the spread being less than with a multi-dynode counter. Pulse rate increases with applied voltage until a plateau is reached, usually at about 2 kV. Background is less than one pulse in ten seconds. They are stable in any vacuum high enough to prevent corona discharge, and at counting rates normal to microanalysis, no overload or ageing properties are likely to be troublesome. In view of their low efficiency their use may well be confined to very high vacuum equipment or to applications where window failure cannot be tolerated. Experiments continue and it is hoped to carry out tests on nitrogen radiation where the efficiency ratio should be better than with oxygen.

Some tests were also carried out using a Bendix magnetic electron multiplier, type M 306. Apart from the ease with which it can be cleaned it appears to offer no advantage over the channel type. It is heavier, larger, has a strong magnetic field and needs a more complex power supply. It also costs nearly five times as much.

It is proposed to include a channel multiplier in addition to the pill-box counter as an alternative detector in the spectrometer now being constructed and thus obtain a better assessment of its performance under normal conditions of microanalysis.

References

1. PRIESTLEY, E. F.: X-ray optics and X-ray microanalysis (H. H. PATTEE, V. E. COSSLETT, and
   A. ENGSTROM, eds.), p. 193. New York and
   London: Academic Press 1963.
2. CULHANE, J. L., J. HERRING, P. W. SANFORD, G.
   O'SHEAD, and R. D. PHILLIPS: J. Sci. Instr. **43**,
   908 (1966).

3. HAILES, L.: This Symposium (1968).
4. ADAMS, J., and B. W. MANLEY: Philips Tech. Rev.
   28, 156 (1967).
5. SMITH, D.: Private communication.

Quantitative Analysis
of Oxygen in Electron Probe Microanalysis

Toshio Shiraiwa and Nobukatsu Fujino

Central Research Laboratories, Sumitomo Metal Industries, Amagasaki, Japan

Introduction

In electron probe microanalysis of elements of low atomic number, the experimental data of quantitative analysis have not been enough because of the low intensity of ultra soft X-rays.

In the present report, oxygen of several oxides are analysed and the theoretical correction of them are studied.

In order to get the high sensitivity for OK_α for the purpose of the quantitative measurement, a new type spectrometer with 11″ lead stearate built-up film has been developed and attached to the ARL-Shimadzu type electron probe microanalyser, and more precise measurement can be done.

Instrumentation

The electron probe microanalyser is the ARL-Shimadzu type. The present authors have used a spectrometer with KAP analysing crystal for OK_α. This spectrometer has high resolving power, and it is useful for the measurement of peak profiles of OK_α in various oxides as shown in Fig. 1, but the sensitivity for OK_α is too weak to make quantitative analysis. Then the authors

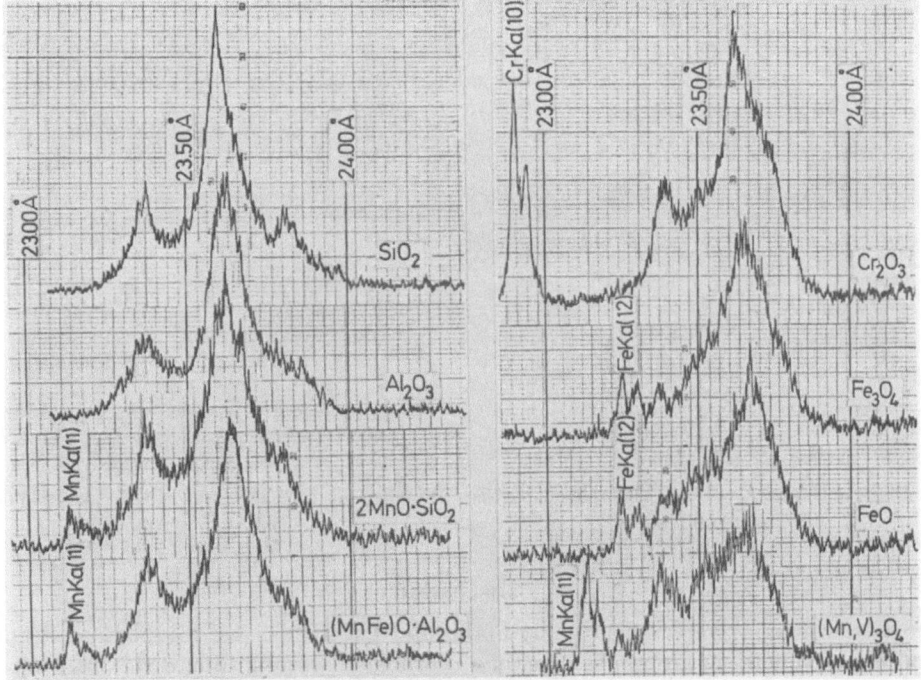

Fig. 1. Spectra of OK_α of various oxides obtained by KAP analysing crystal. Accelerating voltage 15 kV and sample current 0.1—0.3 μA

developed a new type spectrometer with 11″ Pb-stearate built-up film and attached to the electron probe microanalyser.

The crystal is moved on a stright line varying Bragg's angle so that the X-ray source point is on the Rowland circle and the detector is moved to receive the reflected beam of the reflection angle equal to the incident beam, but the distance between the analysing crystal and the detector is fixed. The reflected X-rays only focuss on the slit of detector at the wavelength of OK_α.

The general view of the oxygen spectrometers is shown in Fig. 1a (p. 65).

This oxygen spectrometer cannot resolve the three peaks which can be obtained by KAP crystal but can detect the total intensity of OK_α. Fig. 1b (p. 65) shows the peak profile of OK_α and AlK_α (III) from alumina and its sensitivity for OK_α about 1,200 cps per 0.1 μA and the P/B ratio is about 40 without pulse height analyser.

Experiment

Intensity of OK_α is examined for several oxides and the dependence of the accelerating voltage on the intensity is also determined. The sample used are BeO sintered beryllia, synthetic single crystal of MgO, natural sapphire (Al_2O_3), natural quartz (SiO_2), synthetic TiO, synthetic titania (TiO_2), single crystal glaxite in non-metallic inclusion of steel (Mn, Fe) $O \cdot Al_2O_3$, natural hematite (Fe_2O_3), natural magnetite (Fe_3O_4), synthetic single crystal of Cu_2O, natural zircon ($ZrSiO_4$), and synthetic zirconia (ZrO_2). The diameter of the electron probe is about 0.3—0.5 μ for the probe current of 0.03—0.1 μA. Intensity of OK_α is measured for ten seconds, and the sample current is fixed in the condition of 0.05 μA on alumina.

Result

Figs. 2 and 3 show the observed intensities of OK_α from several oxides using the fixed time method.

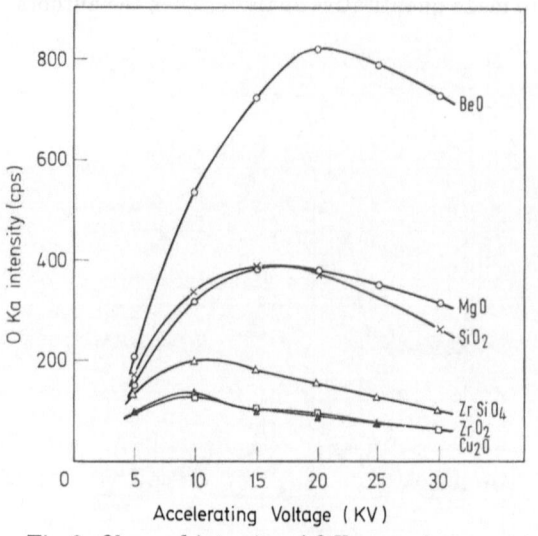

Fig. 2. Observed intensity of OK_α, sample current is 0.05 μA on alumina

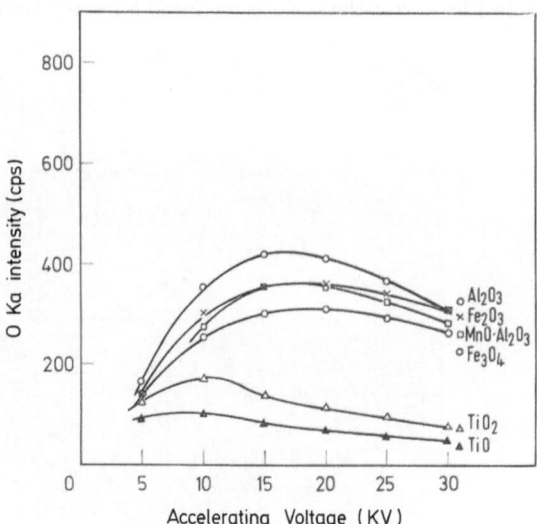

Fig. 3. Observed intensity of OK_α, sample current is 0.05 μA on alumina

Discussion

The atomic number correction given by Poole and Thomas [1] and the modified Philibert's absorption correction [2] are applied to the experimental results, where the mass absorption coefficients for OK_α are referred to Henke's data [3] and those of Ti, Mn, Fe and Cu are 22,500, 5,510, 6,030 and 7,755, respectively, which are the extrapolated values from Cooke's data [4].

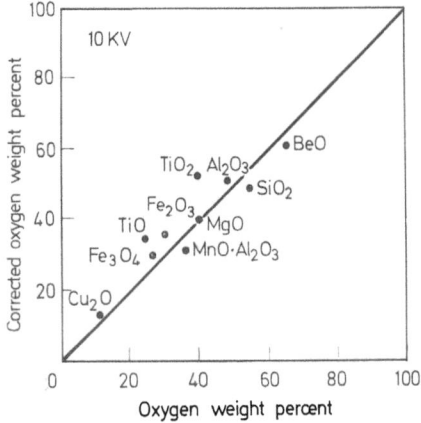

Fig. 4. Relationship between oxygen content and corrected experimental result. Accelerating voltage is 10 kV

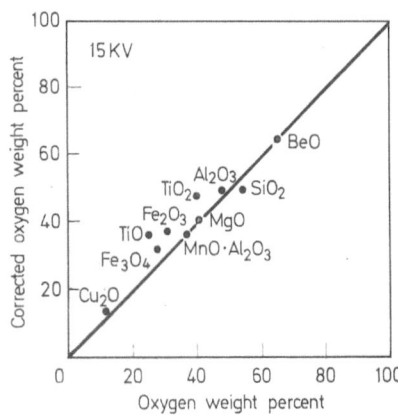

Fig. 5. Relationship between oxygen content and corrected experimental result. Accelerating voltage is 15 kV

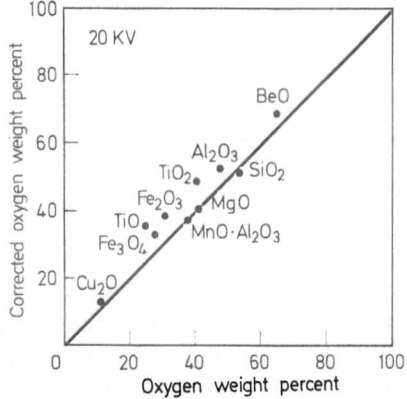

Fig. 6. Relationship between oxygen content and corrected experimental result. Accelerating voltage is 20 kV

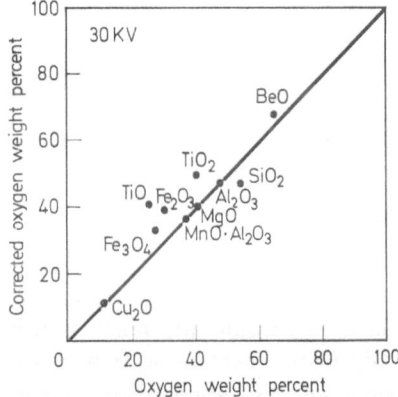

Fig. 7. Relationship between oxygen content and corrected experimental result. Accelerating voltage is 30 kV

The corrected results for 10 kV, 15 kV, 20 kV and 30 kV are shown in Figs. 4, 5, 6 and 7, respectively. The errors are a few percent except oxides of heavier elements. It is recognized that the OK_α intensity is relatively larger for the oxides of heavy element and for the oxide containing smaller amount of oxygen.

In the iron oxides, there are some differences between theory and experiment and so the iron oxides synthesized on pure iron are also examined but same results as the natural hematite and magnetite are obtained.

In Fig. 8, the experimental results at 15 kV and 30 kV versus χ are compared with the theoretical value by Philibert's correction fitted at MgO. It shows that the Philibert's corrections are roughly proper if χ values of iron oxides are about 4,000. But these values are very small compared with the extrapolated values from Cooke's data [4].

It has been known that Philibert's correction is successful for elements except those of low atomic number. In the present experiments, application of Philibert's correction is fairly well for oxygen analysis.

In analysis of the light elements, values of χ is large and the precise $\phi(\varrho\mathfrak{z})$ function at small value of $\varrho\mathfrak{z}$ is necessary.

Moreover, the experimental results in Fig. 1 obtained by KAP show that the spectrum of OK_α is affected by the chemical bonding and it suggests that the intensity of OK_α is also influenced by the chemical bonding because the $L_{\mathrm{II,\,III}}$ state of oxygen is a valence level.

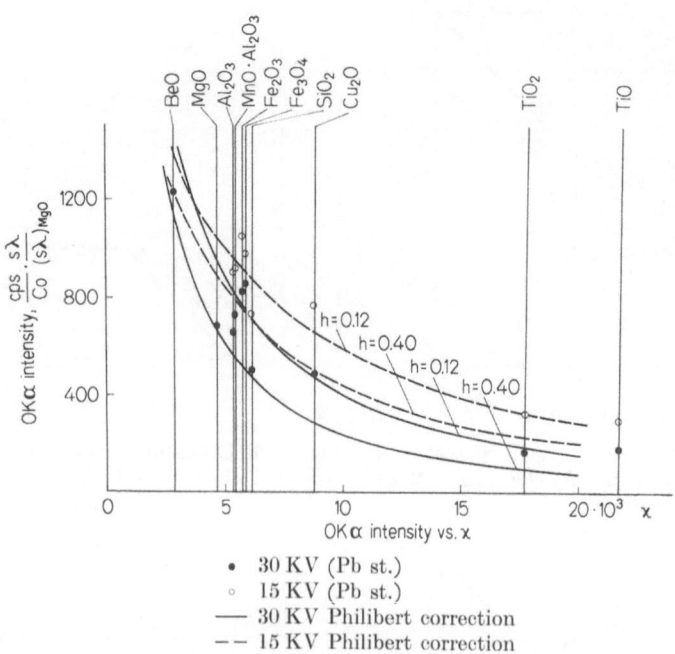

Fig. 8. Observed intensity and estimated one by Philibert's absorption correction versus χ

Conclusion

Oxygen in several oxides are analysed by an electron probe microanalyser with Pb-stearate crystal and KAP crystal and the atomic number correction and Philibert's absorption correction are applied. The results are fairly well and the reason of the disagreement is thought to be uncertainty of the mass absorption coefficient and the influences of the chemical bonding.

The authors express their heartfelt appreciation to Dr. M. Sumitomo, Director of Central Research Laboratories, Sumitomo Metal Industries, who has always encouraged them. They also wish to thank Dr. Y. Tachibana, Director of the Scientific Instruments Division, Shimadzu Seisakusho, Ltd., for his support and Drs. K. Tooyama and K. Kayashima who designed the oxygen spectrometer.

References

1. Poole, D. M., and P. M. Thomas: J. Inst. Metals 90, 228 (1961/62).
2. Philibert, J.: Proc. Symposium on X-ray Optics and X-ray Microanalysis, Stanford, p. 379. New York: Academic Press 1963.
3. Henke, B. L.: Advances in X-ray analysis, vol. 7, p. 460. New York: Plenum Press 1964.
4. Cooke, B. A., and E. A. Stewardson: Brit. J. Appl. Phys. 15, 1315 (1964).

Beryllium Determination in Electron Probe Microanalysis

S. Kimoto, H. Hashimoto and H. Uchiyama

Japan Electron Optics Lab. Co., Ltd., Tokyo, Japan

Today a microprobe can be used to analyze light elements down to boron with dispersive monochromators. In most cases the microprobe is provided with a lead stearate multilayer analyzer for the elements between fluorine and boron. As the $2d$ spacing of the stearate analyzer is about 100 Å and is shorter than the wavelength of beryllium K-radiation (113 Å), the beryllium cannot be detected.

Some other multilayer analyzers having longer $2d$ spacing than that of the stearate analyzer are made of lignoceric, cerotic or melissic acid. These analyzers have been applied to long wavelength X-ray spectroscopy. However, most applications of these analyzers have been made by the fluorescense X-ray method or by procedures other than electron probe microanalysis.

A large $2d$ spacing analyzer is built up on a mica base using the Langmuir-Blodgett technique. Lignocerate, cerotate and melissate analyzers are now available in our laboratory for the purpose of analyzing beryllium. Table 1 shows $2d$ spacing and detectable wavelength range and detectable element range of these analyzers. These $2d$ spacings are generally said to be 130 Å for lignocerate, 140 Å for cerotate and 160 Å for melissate. The actual spacings measured with the aid of 5th order aluminium K-radiation, however, are 125, 137 and 156 Å, respectively.

Since the upper limit of 2θ of the goniometer of the JXA-5 Microprobe used in this experiment is 130°, the detectable wavelengths for the analyzers are up to 113, 124 and 141 Å, respectively. Barium and lead are used as bivalent metals of the above fatty acid soap layer.

The focusing spectrometer has a Johann type analyzer and a 140 mm radius of Rowland circle. The take-off angle of X-rays is 40°. The dimensions of the effective area of the multilayer analyzer are 25 mm in length and 8 mm in width.

The gas flow proportional counter used is the same as a usual gas flow proportional counter in construction except for the nitrocellulose window approximately 0.2 μ in thickness. The flowing gas mixture is 90% argon and 10% methane. The aperture of the detector is 16 mm in length and 2 mm in width.

Fig. 1 shows the spectra of beryllium K-radiation from pure beryllium. In the case of the lead lignocerate analyzer, the detectable wavelength limitation is near the peak wavelength of the radiation; therefore, the perfect spectrumprofile cannot be obtained. The half widths of

Table 1. *Range of detectable wavelength and elements of available analyzers*

Analyzer	Chemical formula	Spacing ($2d$, Å)	Range of wavelength (Å)	Range of elements K	L	M
Myristate	M $(C_{14}H_{27}O_2)_2$ [a]	80 [b]	17— 73	$_5$B—$_9$F	$_{17}$Cl—$_{25}$Mn	
Stearate	M $(C_{18}H_{35}O_2)_2$	100	22— 91	$_5$B—$_8$O	$_{16}$S—$_{23}$V	
Lignocerate	M $(C_{24}H_{47}O_2)_2$	130	28—117	$_4$Be—$_7$N	$_{15}$P—$_{21}$Sc	
Cerotate	M $(C_{26}H_{51}O_2)_2$	140	31—126	$_4$Be—$_7$N	$_{14}$Si—$_{20}$Ca	
Melissate	M $(C_{30}H_{59}O_2)_2$	160	35—144	$_4$Be—$_6$C	$_{14}$Si—$_{19}$K	

[a] M corresponds to bivalent cation, as Ba, Pb, etc.

[b] Actual spacings measured in this experiment are 79, 98, 125, 137 and 156 Å for respective analyzer.

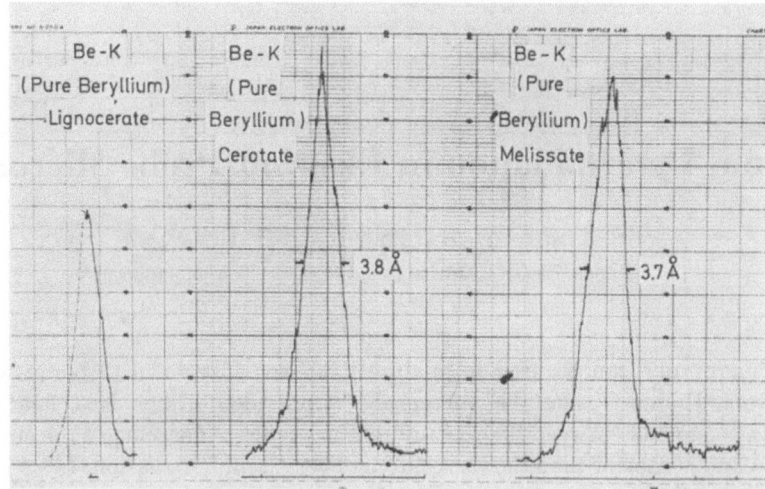

Fig. 1. Spectra of Be-*K* radiation from pure Be with lignocerate, cerotate and mellissate analyzers

these spectra are 3.8 Å for cerotate and 3.7 Å for melissate. These spectral profiles and half widths did not change even when a slit of 300 µ width was used in front of the detector aperture. This means that the resolution of the analyzer is restricted by the characteristics of the soap layers and not by the alignment of the analyzer or spectrometer.

The peak intensity and peak-to-background ratio of *K*-radiation spectra of carbon, boron and beryllium are indicated in Table 2. The sensitivity for beryllium is 0.16, 0.30 and 0.33 wt% for the lignocerate, cerotate and melissate analyzer respectively where the incident probe energy is 10 kV, specimen current is 1.0 µA and the counting time, 100 seconds.

For all analyzers: stearate, lignocerate, cerotate and melissate, the peak intensity for graphite is approximately three times that for boron. However, the ratio of the intensity for boron to that for beryllium is about 10 for lignocerate and about 40 for cerotate and melissate. The intensities measured for cerotate and melissate are lower than expected from the result of lignocerate. When using 300 µ slit, the peak intensity of each spectrum is reduced to approximately 1/3.

These analyzers were applied to beryllium—35% molybdenum, aluminum—13~15% beryllium, nickel—1~5% beryllium and copper—4% beryllium alloys. Fig. 2 shows the beryllium distribution in the aluminum—beryllium alloy. In this specimen, beryllium is concentrated at the grain boundary, and the concentration is nearly 100%. Fig. 3 shows the application to the beryllium—molybdenum alloy. The dark phase in the composition image contains 94% beryllium and the bright phase 67% beryllium. The probe energy and current in the above applications are 10 keV and 0.3—0.5 µA, respectively.

Table 2. *Comparison of peak intensity and peak-to-background ratio of K-radiation spectra of carbon, boron and beryllium*

	C-*K* (graphite)		B-*K* (pure boron)		Be-*K* (pure beryllium)	
	I_p (cps/µA)	I_p/I_b	I_p (cps/µA)	I_p/I_b	I_p (cps/µA)	I_p/I_b
Pb Stearate	3.5×10^4	70	1.5×10^4	50		
Ba Lignocerate	3.6×10^4	15	1.0×10^4	27	1.0×10^3	38
					(2.5×10^2)	(50)
Ba Cerotate	5.3×10^4	16	1.7×10^4	37	4.0×10^2	24
Ba Melissate	4.8×10^4	12	1.3×10^4	29	3.3×10^2	24

Accelerating voltage: 10 kV (5 kV).

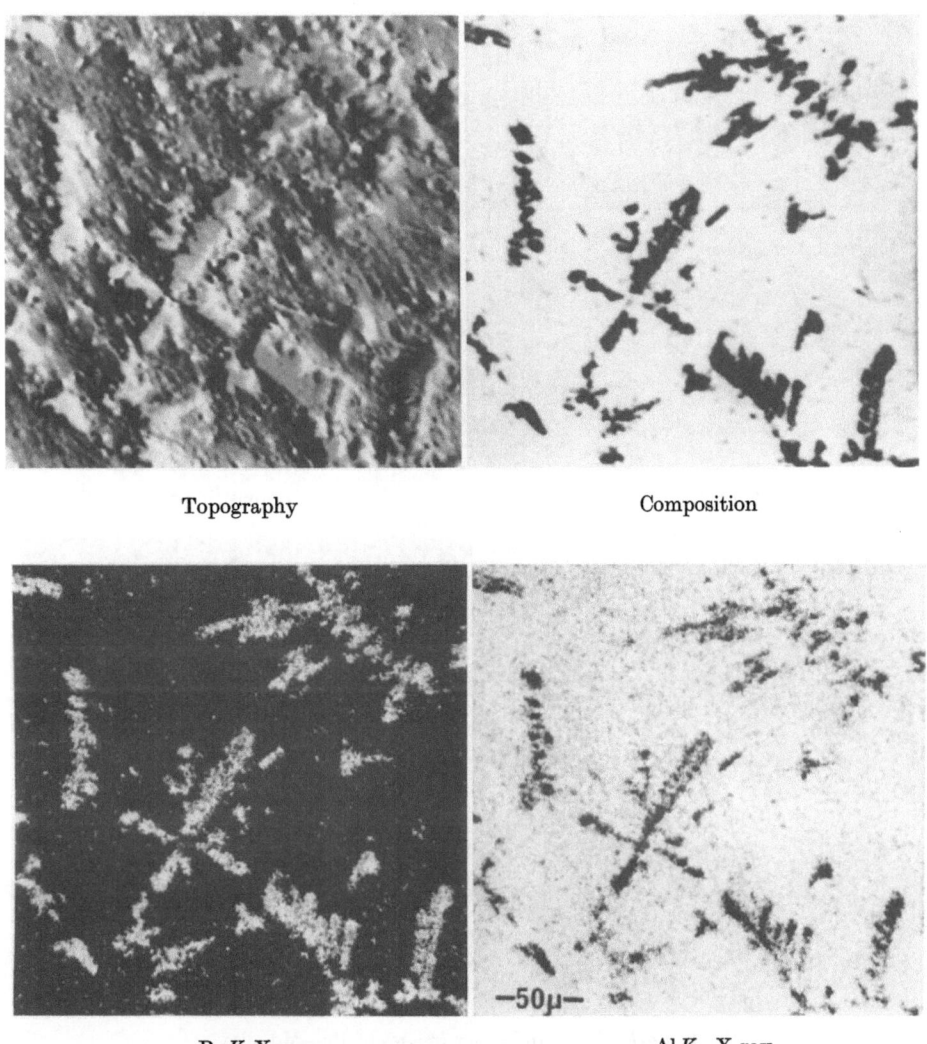

Topography   Composition

Be K X-ray   Al K_α X-ray

Fig. 2. Be-K X-ray and other images of Al—Be alloy

Practical sensitivity can be estimated from a result of the analysis of the copper —4% beryllium alloy in which the beryllium is distributed uniformly. With the lignocerate analyzer, the peak intensity of beryllium radiation is 964 counts for the conditions of 10 kV, 0.35 μA probe current and 100 second measuring time. The background intensity, which is measured using pure copper and setting the spectrometer in the same position as that for measuring above peak intensity, is 868 counts. In this case the sensitivity is 3.4%. This sensitivity is one tenth of that estimated from the data obtained using pure beryllium under the same operational conditions.

The results of the above experiment show that the multilayer soap film analyzers can be effectively used for analyzing beryllium in electron microprobe analysis without any instrumental modification. Under the same operational conditions as in measuring heavier elements, however, the sensitivity for beryllium is lower. Therefore, probe current for this purpose should be greater than in usual measurement. In this experiment, 0.1—0.5 μA probe current was used.

In order to improve the sensitivity for beryllium, and adequate number of layers of the analyzing crystal should be used. In this experiment, a 60 layer analyzer was used. However, it has not been confirmed as yet that this is the optimum number. The thickness of the windows of the detector restrict the intensity of beryllium X-rays critically. The contamination of the

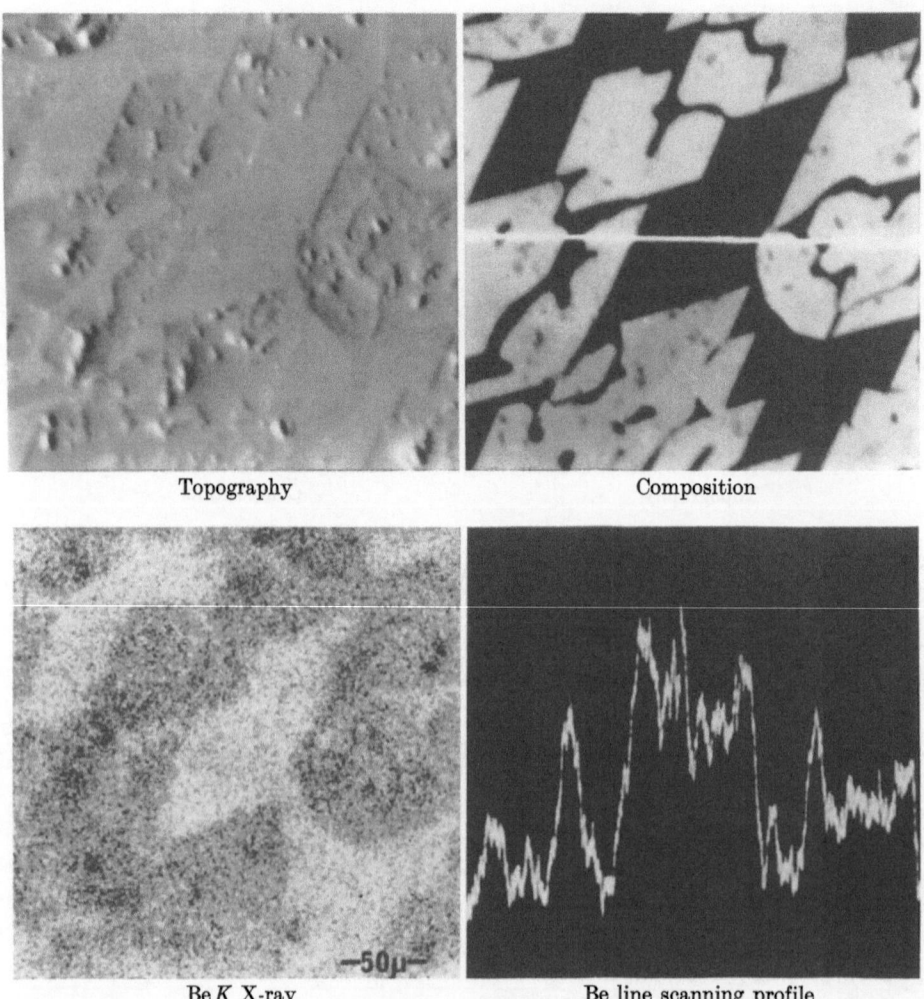

Topography   Composition

Be K X-ray   Be line scanning profile

Fig. 3. Be-K X-ray and other images of Be—Mo alloy

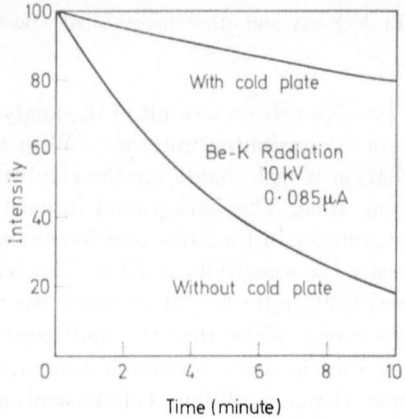

Fig. 4. Contamination effect for Be-K radiation

specimen influences the intensity of beryllium more than that of carbon or boron. Fig. 4 shows the effect of contamination on pure beryllium. The contamination reduces the relative intensity at the rate of 14% per minute. Using a cold finger anti-contamination device, the rate becomes 2% per minute.

Microanalyse des éléments très légers

Mme F. Pichoir

Ingénieur de recherches à l'Onera — 29, avenue de la Division Leclerc, 92 Chatillon, France

La microanalyse des éléments légers est le plus souvent réalisée à l'aide d'une méthode dispersive utilisant des cristaux artificiels. Toutefois la faible ouverture du faisceau de rayons X et sa forte absorption dans le cristal dispersif rendent cette méthode peu sensible lorsque, pour réduire les corrections d'absorption, les analyses sont effectuées avec une tension d'accélération des électrons de l'ordre de 5 kV.

Pour pallier ces inconvénients et augmenter la sensibilité des mesures, des analyses ont été effectuées à l'aide d'une méthode non dispersive utilisant une absorption différentielle (dans des filtres solides ou gazeux) [1]. Cette méthode permet de sélectionner des rayonnements dont les longueurs d'onde sont comprises entre les deux discontinuités d'absorption de deux éléments de numéros atomiques consécutifs. Il est alors possible de distinguer les émissions X de deux éléments légers voisins que la seule sélectivité des compteurs ne permet pas de séparer.

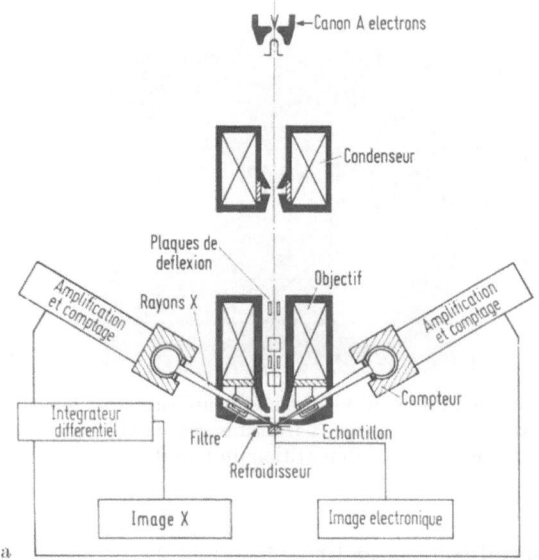
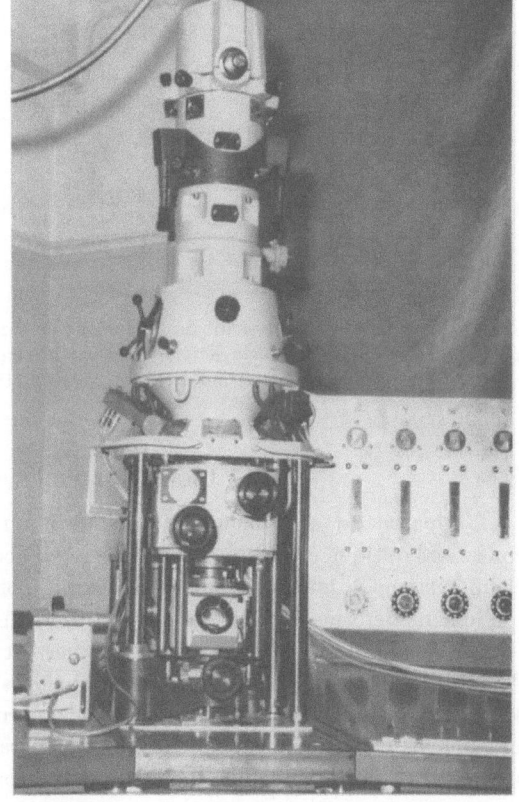

Fig. 1a et b. Microanalyseur à éléments légers. a Schéma simplifié, b Photographie du corps de l'appareil

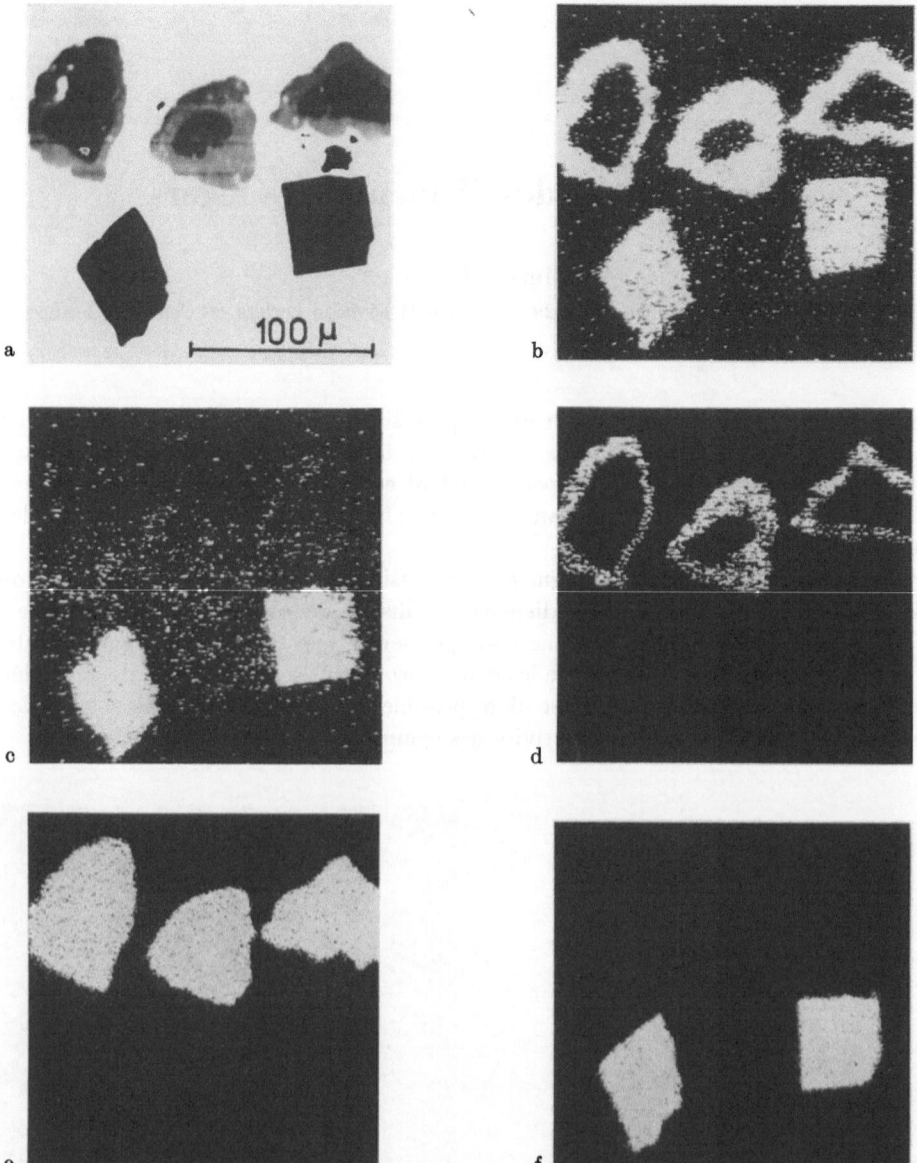

Fig. 2a—f. Aluminium contenant des inclusions de carbure de tantale et de chrome nitruré superficiellement.
a Image électronique, b Image X transmise par la voie azote seule, c Image X transmise par la voie méthane
seule, d Répartition de l'azote, e Répartition du chrome, f Répartition du tantale

Description de l'appareil (fig. 1a et b)

L'appareil est sensiblement identique à un microanalyseur classique, du moins en ce qui concerne l'optique électronique. La tension d'accélération des électrons est réglable de 0 à 10 kV et un système à double plaque de déflexion, placé avant l'objectif, permet le déplacement de la sonde sans déformation sur des plages inférieures ou égales à 300 microns de côté.

La visualisation de la surface à étudier est réalisée, non par un dispositif optique, mais à l'aide du courant d'échantillon. L'observation s'effectue avec un balayage rapide de 20 images/seconde dont la faible définition (100 lignes) permet cependant la recherche des plages à analyser. Les images électroniques sont ensuite photographiées à une vitesse plus lente (1 image en 60 secondes) ce qui permet une meilleure définition (1200 lignes).

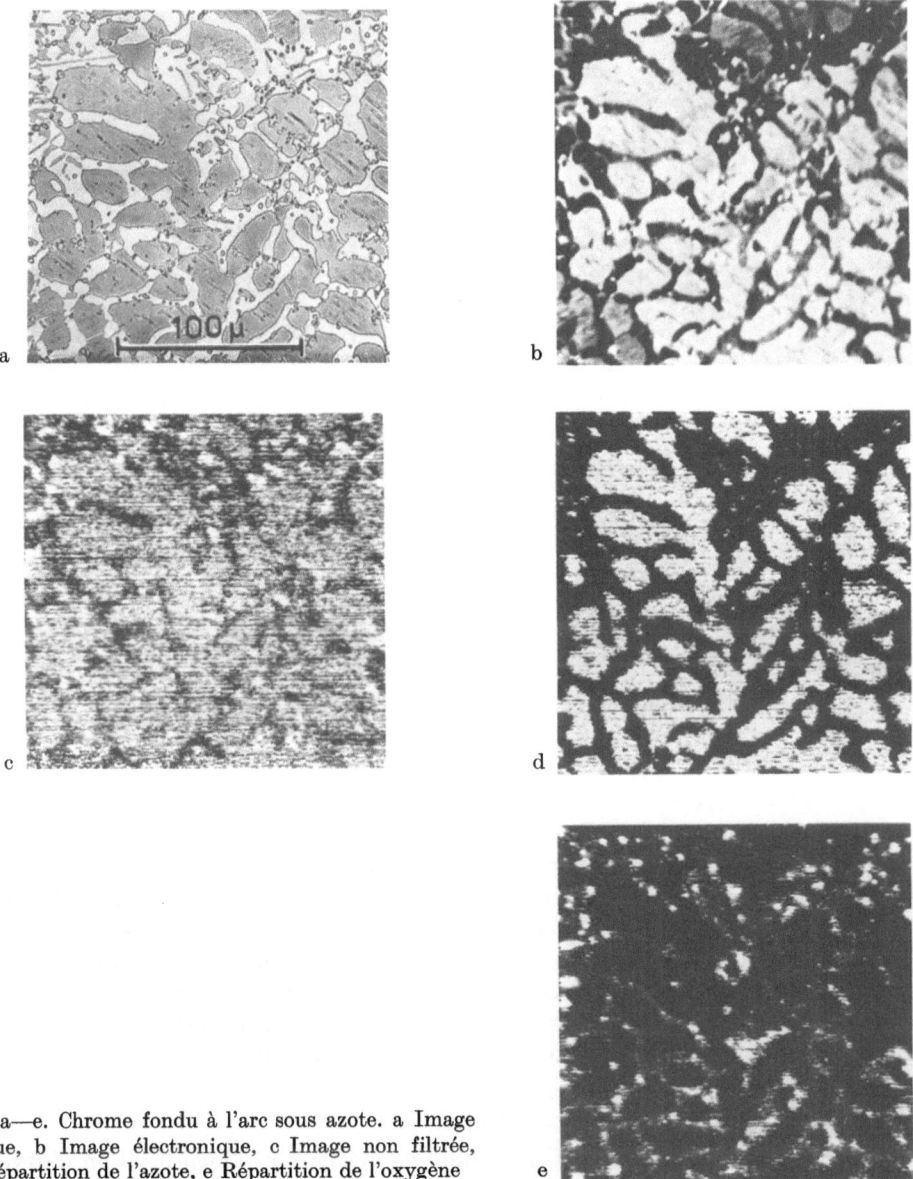

Fig. 3a—e. Chrome fondu à l'arc sous azote. a Image
optique, b Image électronique, c Image non filtrée,
d Répartition de l'azote, e Répartition de l'oxygène

Les échantillons sont introduits à travers un sas pour éviter une remise à l'air de l'appareil.
Ainsi le réglage de la pression à l'intérieur des filtres gazeux et des compteurs n'est pas modifié.
Les filtres dans lesquels les pressions sont inférieures à la pression atmosphérique, restent donc
équilibrés entre chaque mesure.

Deux faisceaux X de même ouverture, issus de l'échantillon, traversent les filtres situés
entre les pièces polaires de l'objectif et sont ensuite collectés par des compteurs proportionnels
à flux de méthane. Ces derniers sont placés à l'extérieur de l'appareil auquel ils sont réunis par
leur fenêtre d'entrée.

Le dosage de l'azote et de l'oxygène s'effectue à l'aide de filtres gazeux, celui du carbone,
du bore et du béryllium à l'aide de filtres solides.

Les filtres à circulation de gaz sont équilibrés en pression de telle sorte que les rayonnements
de longueur d'onde extérieure à la bande comprise entre leur discontinuité d'absorption soient
également transmis. Pour les filtres solides ce réglage s'effectue en variant leur inclinaison par
rapport au faisceau de rayons X.

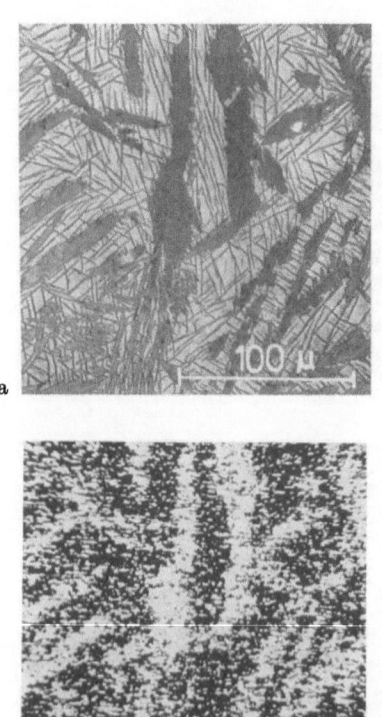

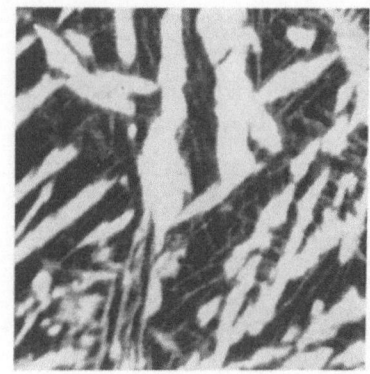

Fig. 4a—c. Niobium fondu à l'arc sous azote. a Image optique, b Image électronique, c Répartition de l'azote

Lorsque les deux sélecteurs d'amplitudes sont calés sur la raie analysée, les impulsions issues des deux chaînes peuvent être soit comptées pour des dosages ponctuels soit injectées sur un intégrateur différentiel. Ce dernier délivre un signal analogique qui module le wehnelt d'un oscilloscope et permet la formation des images de rayons X.

L'utilisation d'un intégrateur impose des vitesses de balayage relativement lentes: 5 secondes par ligne pour une fréquence de comptage de 500 coups/seconde et 15 secondes par ligne pour une fréquence de 150 coups/seconde, ce qui correspond respectivement à des images construites en 30 minutes avec une définition de 360 lignes et 1 heure avec 240 lignes.

Résultats

Les analyses portent essentiellement sur la mise en évidence et l'identification d'oxydes et de nitrures. Les carbures et les borures, aisément détectés par une analyse dispersive classique, présentent vis à vis de cette méthode un intérêt moindre, et n'ont pas été étudiés particulièrement.

Dosage de l'azote

Les deux filtres employés sont constitués l'un d'azote, l'autre de méthane. La transmission de la raie de l'azote dans l'ensemble des deux filtres est de 60%.

Le premier échantillon (fig. 2) est constitué par une matrice d'aluminium contenant des grains de carbure de tantale (TaC) et de chrome nitrurés superficiellement (Cr_2N).

La fig. 2b est construite avec les impulsions transmises par le filtre d'azote seul. Ce filtre transmet le rayonnement de l'azote et de façon moindre celui du carbone. Le compteur ne séparant pas les rayonnements K de ces deux éléments, on observe à la fois la couche nitrurée des grains de chrome et les pavés de carbure de tantale. Bien qu'en faible proportion (6%) et moins bien transmis que l'azote, le carbone, qui est peu absorbé par les fenêtres de collodion des filtres et des compteurs, y apparaît avec une intensité comparable à celle de l'azote dans Cr_2N (11,9%).

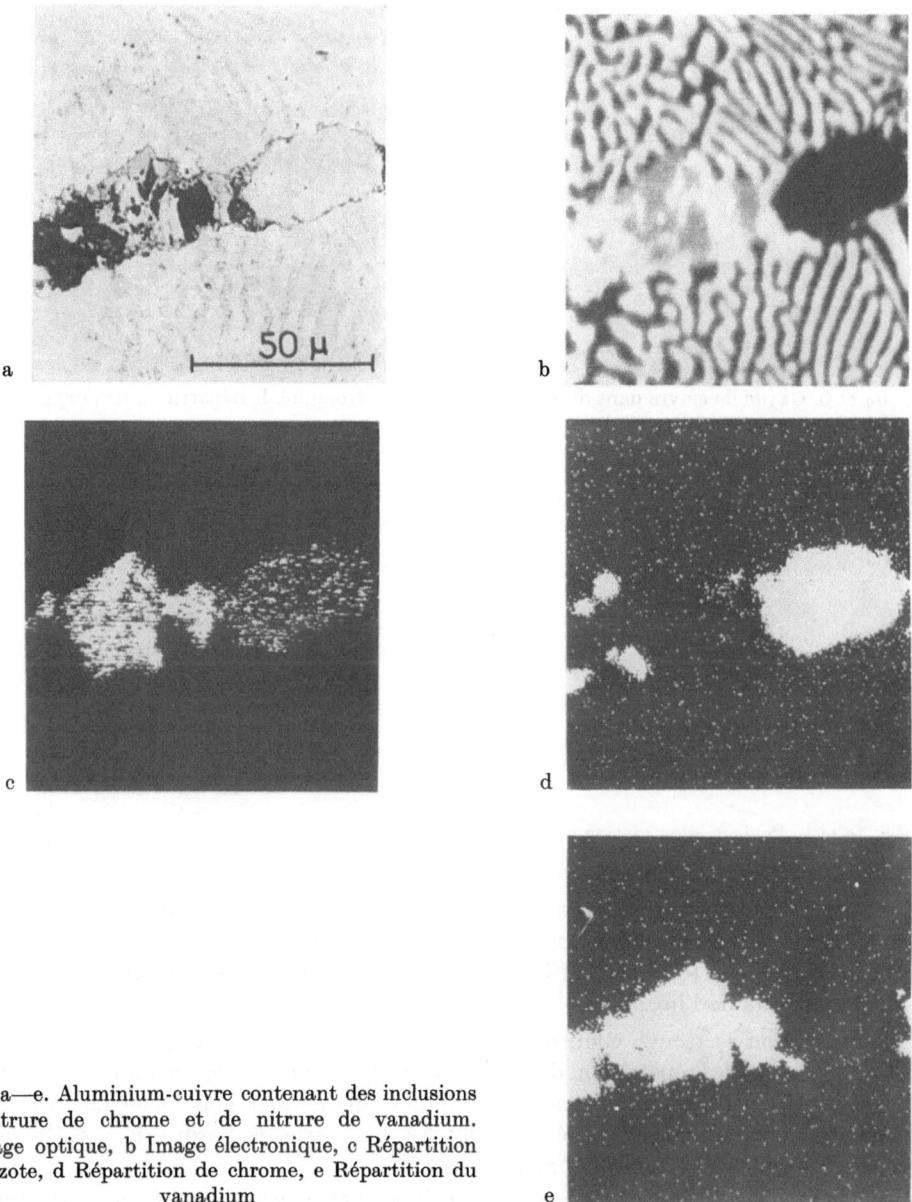

Fig. 5a—e. Aluminium-cuivre contenant des inclusions de nitrure de chrome et de nitrure de vanadium. a Image optique, b Image électronique, c Répartition de l'azote, d Répartition de chrome, e Répartition du vanadium

La fig. 2c correspond au rayonnement transmis par le filtre de méthane seul. La raie de l'azote étant fortement absorbée dans ce filtre, le nitrure n'est plus visible tandis que les carbures apparaissent avec une intensité égale à celle de la photographie précédente.

Après avoir effectué la différence, il ne reste plus sur l'image définitive (fig. 2d) que les zones de nitrure de chrome (Cr_2N).

L'échantillon suivant est un chrome fondu à l'arc sous azote et contenant des inclusions d'oxyde Cr_2O_3 (fig. 3).

La fig. 3c a été réalisée avec un seul compteur et sans filtre, le sélecteur d'amplitude étant centré sur la raie de l'azote (31 Å). On y distingue trois zones de luminosités différentes:

Les plus sombres proviennent des raies L du chrome qui se situent pour les plus intenses vers 21 Å; la sélection d'amplitude atténue ce rayonnement sans l'éliminer totalement.

Les zones plus claires sont constituées par la superposition des raies L du chrome et de la raie K de l'azote.

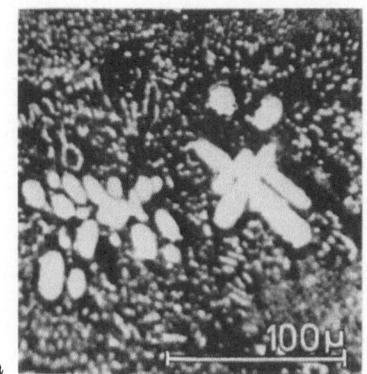 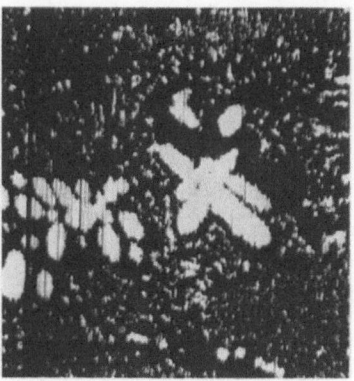

a   b

Fig. 6a et b. Oxyde de cuivre dans du cuivre. a Image électronique, b Répartition de l'oxygène

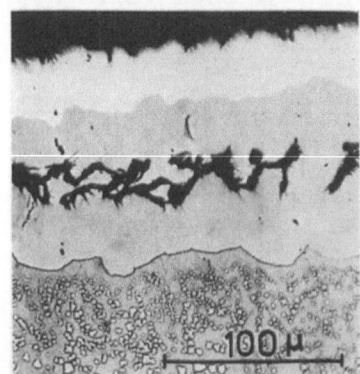 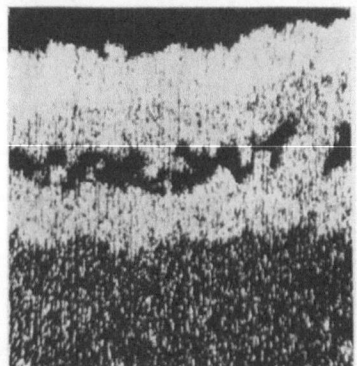

a   b

Fig. 7a et b. Pellicule d'oxydation du fer pur. a Image optique, b Répartition de l'oxygène

Les points brillants sont obtenus par la superposition des raies L du chrome et de la raie K de l'oxygène. Comme le rayonnement du chrome, celui de l'oxygène est atténué, mais la concentration de cet élément est toutefois suffisante pour que les inclusions d'oxyde constituent les points les plus brillants de l'image.

Cette interprétation se trouve confirmée par les images de rayons X suivantes:

La fig. 3d présente la répartition de l'azote (filtres d'azote et de méthane). Un dosage précis, effectué point par point par comptage d'impulsions permet non seulement d'identifier la phase Cr_2N mais de déterminer un écart à la stoechiométrie de l'ordre de $-2,5\%$ qui est confirmé par une étude de l'échantillon aux rayons X.

La fig. 3e présente la répartition de l'oxygène (filtres d'oxygène et d'azote). Sur cette dernière apparaissent nettement les inclusions de Cr_2O_3 ($31,6\%$ d'oxygène) et très faiblement les zones de chrome pur dont certaines raies L peu intenses sont comprises dans la bande $23,3\,\text{Å}-31\,\text{Å}$ correspondant aux discontinuités d'absorption des deux filtres.

Deux autres exemples sont encore présentés: l'un relatif à un échantillon de niobium fondu à l'arc sous azote (fig. 4) où l'analyse met en évidence la phase Nb_2N (7% d'azote), l'autre concernant un échantillon d'aluminium-cuivre contenant des inclusions de nitrure de chrome Cr_2N ($11,9\%$ d'azote) et de nitrure de vanadium ($21,5\%$ d'azote) (fig. 5).

Dosage de l'oxygène

Les filtres sont constitués d'oxygène et d'azote, la transmission de la raie de l'oxygène pour l'ensemble des deux filtres étant de 60%.

La fig. 6b présente la répartition de l'oxygène dans un échantillon de cuivre contenant des inclusions d'oxyde de cuivre Cu_2O ($11,2\%$ d'oxygène). On y distingue non seulement les grosses inclusions mais aussi la précipitation fine.

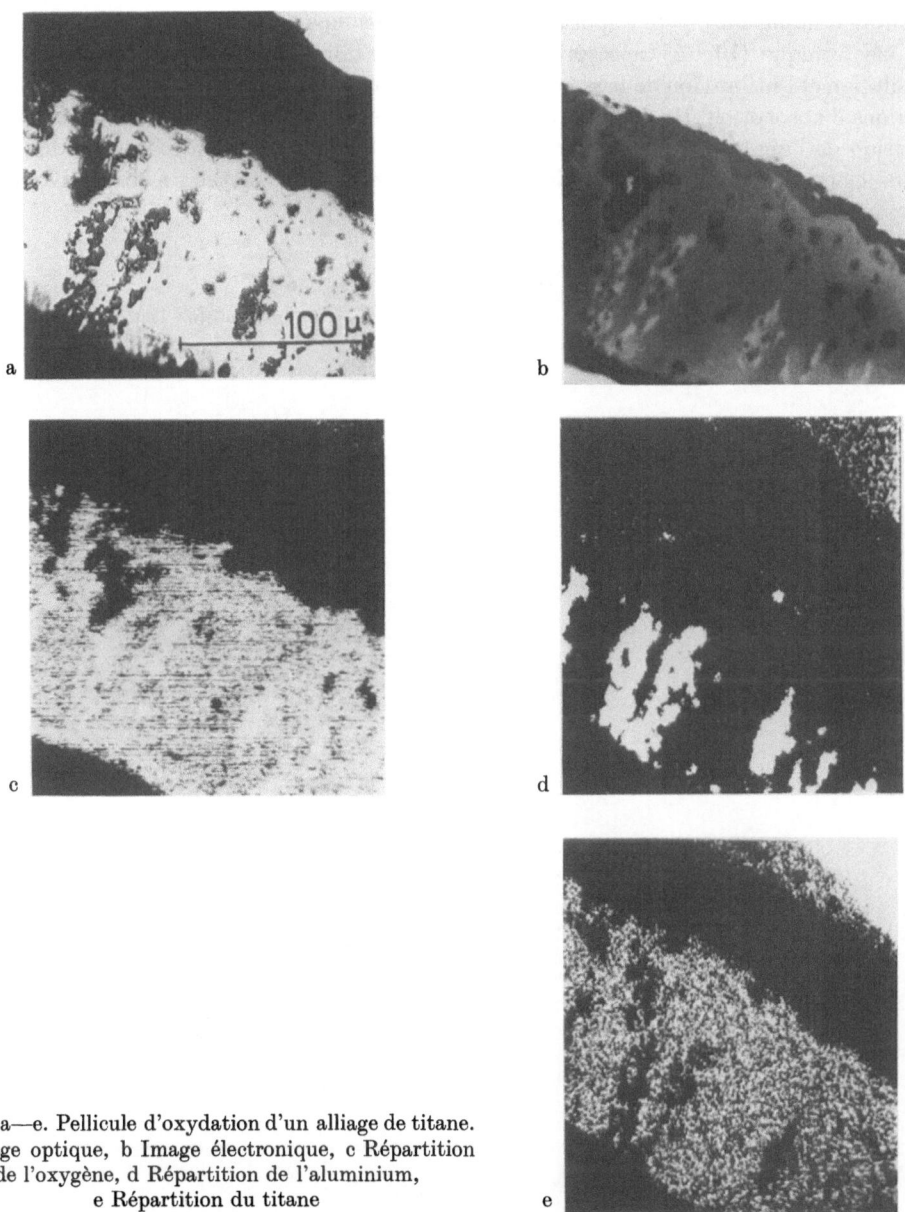

Fig. 8a—e. Pellicule d'oxydation d'un alliage de titane.
a Image optique, b Image électronique, c Répartition
de l'oxygène, d Répartition de l'aluminium,
e Répartition du titane

Sur la fig. 7b sont nettement visibles les trois oxydes d'une pellicule d'oxydation super-ficielle de fer pur: FeO (22,2% d'oxygène), Fe_3O_4 (27,6%), Fe_2O_3 (30%). Par contre la fine précipitation de Fe_3O_4 dans FeO n'apparaît pas nettement.

Enfin la fig. 8c a été réalisée sur une pellicule d'oxydation superficielle d'alliage de titane (TA6V, 6% Al, 4% V). Elle révèle la présence d'alumine au sein de l'oxyde TiO_2. Pour ce dernier, le dosage reste peu précis en raison de la très forte absorption du rayonnement de l'oxygène dans le titane, et de la présence des raies L les plus intenses de cet élément dans la bande comprise entre les deux discontinuités d'absorption.

Conclusion

Ces quelques exemples, relatifs à l'analyse de l'azote et de l'oxygène mettent en évidence les avantages que la méthode d'absorption différentielle, présente sur les méthodes dispersives lorsqu'il s'agit de doser les éléments très légers.

Le nombre d'impulsions reste important (3000 à 4000 coups/seconde), malgré la faible valeur du débit électronique (10^{-7} A) nécessaire pour la formation d'image de rayons X ayant une bonne résolution et l'utilisation de basse tension accélératrice (5 kV) indispensable pour minimiser les corrections d'absorption. Les images peuvent dès lors être obtenues en des temps raisonnables et la précision de l'analyse permet de détecter des écarts à la stoechiométrie de l'ordre de 1%.

Certains éléments comme l'azote difficilement accessible aux méthodes dispersive sont ici dosables avec précision.

Bibliographie

1. CASTAING, R., et F. PICHOIR: La recherche aérospatiale, No 108, Septembre-Octobre 1965.

Utilisation des compteurs à flux gazeux dans le domaine des rayons X ultra-mous pour l'étude des sources à émissions brèves

P. Dhez et P. Jaegle

Laboratoire de Chimie Physique de la Faculté des Sciences de Paris, 91-Orsay, France

Nous avons abordé avec un compteur proportionnel, l'étude de flux intenses et brefs de rayonnement électromagnétique pour des longueurs d'ondes de quelques dizaines à quelques centaines d'angströmns.

En effet, pour ce domaine, des sources de rayonnement pulsées, périodiques ou non, sont fréquemment utilisées. C'est le cas du rayonnement émis par les électrons accélérés dans les synchrotrons et de l'émission des plasmas de haute température obtenus par décharge électrique sous vide ou par impact d'un faisceau laser sur une cible. La durée d'émission d'une décharge électrique est de l'ordre de la milliseconde, celle d'un plasma laser de quelques dizaines de nanosecondes.

La mesure de flux intenses pendant des temps aussi courts ne peut s'effectuer par comptage de photons individuels. En effet, le temps de déplacement des électrons dans les compteurs proportionnels ou dans les photomultiplicateurs, et les caractéristiques des circuits électroniques associés, limitent le nombre d'évènements décelables par unité de temps. Nous avons donc utilisé la propriété principale du compteur proportionnel qui est de donner une réponse proportionnelle à l'énergie du photon reçu, en vérifiant que la proportionnalité de la réponse à l'énergie reçue était conservée si le nombre de photons incidents était important comme dans les émissions de sources pulsées, déduction faite du nombre de photons absorbés dans la fenêtre.

Le compteur utilisé est semblable a celui décrit par A. P. Lukirskii [1], il a été placé dans un spectrographe réalisé par l'un de nous [2]. Ce spectrographe, à réseaux tangents et à incidence variable, peut être équipé de deux réseaux successifs pour séparer les ordres d'interférences.

En faisant varier la valeur du $R\,C$ (R résistance de charge, C capacité de liaison totale du compteur) nous avons montré qu'il était possible, à une longueur d'onde choisie, soit d'obtenir un signal proportionnel au flux total émis pendant une décharge électrique, soit de suivre l'émission pendant chaque période de la décharge [3]. L'étude de la réponse du compteur en fonction du flux total a fait apparaître qu'il existait, pour une tension donnée appliquée au compteur, un flux maximum limite au-delà duquel sa réponse n'était plus proportionnelle [4]. Ceci est dû à la charge d'espace dans le compteur.

Nous avons aussi étudié l'émission, à différentes longueurs d'onde, du plasma obtenu par focalisation d'une impulsion laser sur une cible solide sous vide [5]. Dans le montage réalisé pour cette source le réseau est éclairé par une tranche de plasma d'environ 0,15 mm d'épaisseur et l'ensemble lentille-cible peut être déplacé devant la fente d'entrée du spectrographe le long d'un axe perpendiculaire à la surface de la cible. Le déplacement du compteur sous vide le long du cercle de Rowland permet de balayer le spectre. La fig. 1 montre le profil de la raie à 160, 07 Å de l'aluminium ionisé 3 fois, tracé point par point. La fig. 2 indique la localisation par rapport à la surface de la cible de l'émission de cette raie et du continuum de longueur d'onde voisine. Ces résultats sont obtenus en déplaçant l'ensemble lentille-cible devant la fente. La zone de plasma émettant l'intensité maximum du continuum est nettement séparée de la zone émettant

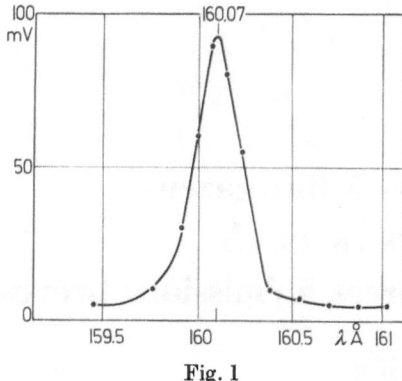

Fig. 1

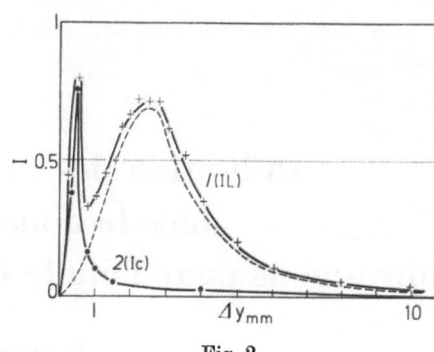

Fig. 2

Fig. 1. Profil de la raie d'Al IV

Fig. 2. Intensité émise par une tranche de plasma, de 0,15 mm de largeur, située à une distance Δy mm de la surface de la cible. Ic mesure faite à 160,5 Å; Il mesure faite à 160,1 Å. La courbe en pointillé obtenue par soustraction, donne l'allure de l'intensité qui peut être attribuée à la raie d'Al IV seule

l'intensité maximum de la raie. Des résultats analogues ont été obtenus pour plusieurs raies et confirmés photographiquement. Ceci permet d'attirer l'attention des utilisateurs de ce type de sources sur les précautions à prendre pour l'interprétation de leurs spectres. Signalons enfin que l'écart de temps entre l'émission laser et l'apparition de l'impulsion de compteur proportionnel peut être mesuré, pour une raie donnée, à différentes distances de la cible. Ces résultats permettent de faire des hypothèses sur certains aspects des mécanismes d'émission de ce type de source.

Les quelques exemples indiqués montrent de nouvelles possibilités d'utilisation des compteurs proportionnels pour les rayons X ultra-mous. Les caractéristiques des compteurs doivent naturellement être adaptées à ce domaine de rayonnement. En particulier la fenêtre doit être d'une épaisseur de l'ordre de 1000 Å, en collodion par exemple, ce qui permet d'utiliser une pression gazeuse de l'ordre d'une centaine de torrs. Par rapport à d'autres détecteurs, ces compteurs présentent certains avantages. Par exemple la loi de variation de leur réponse en fonction de l'énergie est simple. Sur le plan pratique leur fonctionnement ne nécessite pas de vide poussé et «propre». Enfin, ils sont insensibles au visible et s'ils sont utilisés avec une amplification gazeuse faible, leur bruit de fond est négligeable.

Bibliographie

1. Lukirskii, A. P., O. A. Ershow, and J. A. Brytov: Bull. Acad. Sci. USSR., Phys. Ser. English Transl. **27**, 798 (1963).
2. Jaegle, P.: Thèse Doctorat d'Etat, Paris 1965. — Communication à l'Internat. Symposium Roentgenspektren und chemische Bindung, publié par Wiss. Z. Leipzig (1966).
3. Dhez, P., et P. Jaegle: J. phys. appl. Paris **3**, 275 (1968).
4. — — Compt. Rend. **262**, 1432 (1966).
5. Cauchois, Y., P. Dhez, P. Jaegle, S. Leach, and M. Velghe: Scd. Int. Conf. on Vacuum, U.V. Physics Gattlinburg, Tennessee, U.S.A. 1968.

Microdiffraction

The Accuracy of the Orientation of Cubic Crystals from Back Reflection Kossel Patterns

N. SWINDELLS

Department of Metallurgy and Materials Science University of Liverpool, England

Introduction

The determination of crystallographic information by using Kossel patterns generated by an electron probe is now an established technique. Its use for the determination of the orientation of crystals 10 μm and larger has been the subject of several recent papers [1—4]. The advantages of the back reflection method using an electron probe microanalyser lie in the ease with which the scanning system can be used to select a crystal for examination and the simplified preparation of the specimens. The orientation of the microcrystal can be described by the Miller indices of the normal to the specimen surface at the point of impact of the electron probe and can be completely determined by the use of a reference line relating the position of the crystal to known directions on the film recording the pattern [2].

The camera is supplied as an attachment to the SEM 2 electron probe microanalyser and is shown in Fig. 1. The distance between specimen and film is 4 cm and the solid angle included by the film is approximately 90°. The resolution of the electron beam is 1 μm and full electron scanning facilities are retained. The camera is pre-aligned with the electron lens axis and then subsequent measurements on the film assume that the X-ray source and the centre of the pattern lie on the same axis. Since the centre of the film is removed, the centre of the pattern is established by measuring from the circumference, marked by the edge of the film cassette. The cassette also marks a reference direction on the film. The specimen surface should be maintained in a plane parallel to the film by the orthogonal specimen shifts which move on ball bearing tracks.

The purpose of this paper is to report the results of a limited examination of the accuracy of the assumptions given above.

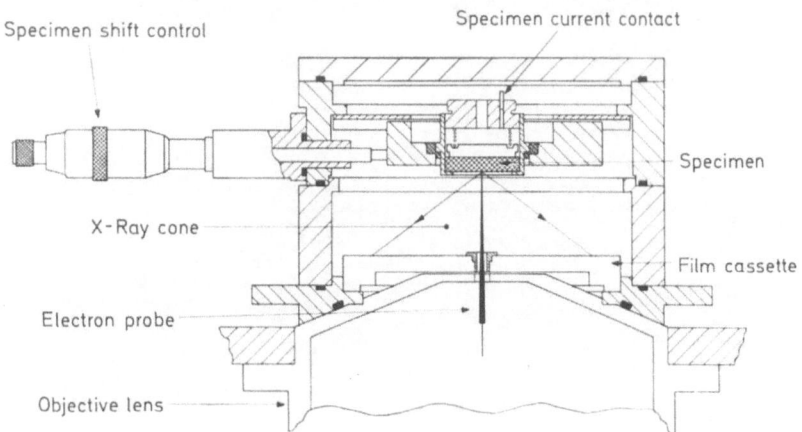

Fig. 1. Back reflection Kossel camera

Experimental

A Kossel pattern (Fig. 2) was obtained from a crystal of austenite and then measured by various devices developed for this purpose. Measurements were made directly on the film. The results were then compared with the Miller indices of the crystal surface pole obtained by the trace analysis method of Bevis, Hecksher and Crocker [5]. Three sets of slip traces, of {111} planes, were obtained by impressing the surface with a microhardness indenter.

The first measurements were made using the charts devised by Rowlands and Bevis [6]. These charts are rapid in use and also provide a convenient method of indexing some of the lines on the pattern. The results obtained are the positions of principal poles on a stereographic projection whose centre is the centre of the film. The indices of the centre are obtained by solving the equations for the known angles between the unknown and known indices.

The alternative method for achieving the same principal poles was first suggested by Peters and Ogilvie [1]. This requires that the polar co-ordinates of points on the Kossel lines be measured so that the conics can be transferred from the pattern to a stereographic projection. The centres of the circles on the stereographic projection are obtained by geometrical construction. The pole of the associated plane is then obtained from the angular diameter of the circle.

The polar co-ordinates are conveniently measured by aligning the reference marks of the film with corresponding marks on a rotatable jig marked in degrees. Over the jig a cross wire can be moved along a calibrated scale. The point selected for measurement is rotated to this axis, giving one co-ordinate, and its distance from the centre measured by moving the cross wire along the scale until it coincides with the point. Several jigs have been constructed on the principle, mentioned above. One, designed by P. Rowlands, is shown in Fig. 3. Constructed of perspex, it is a convenient device for obtaining the data for the method of Peters and Ogilvie. Distances from the centre can be measured to ± 0.1 mm.

More precise measurements, to ± 0.01 mm of the distance from the centre, can be achieved by using a measuring microscope over a rotatable jig holding the film. A jig for this purpose

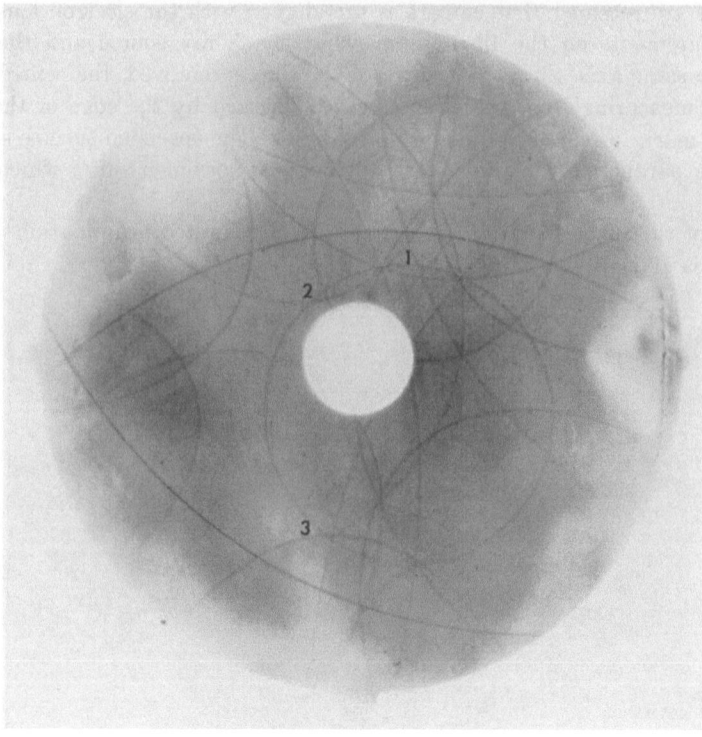

Fig. 2. Kossel pattern from austenite

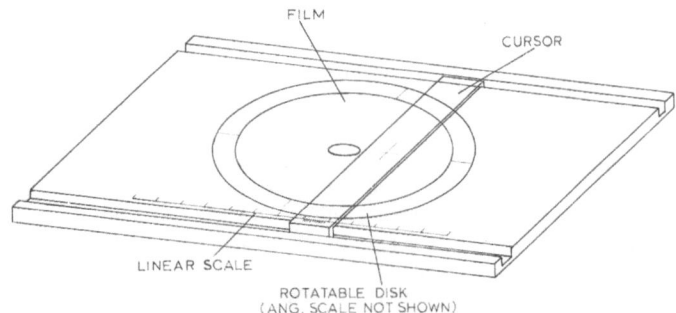

Fig. 3. Perspex measuring jig

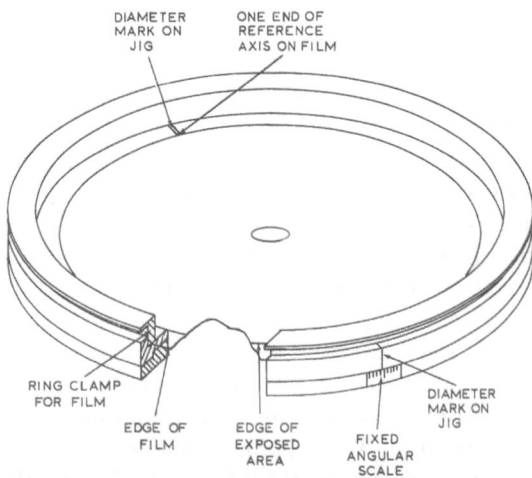

Fig. 4. Jig for use with measuring microscope

is shown in Fig. 4. The film is centred by locating the edge of the exposed region with the circumference of the jig. In this case the film position is observed by eye. HARRIS and FISHER [7] have developed a similar device but with the facility for centring the film by observation with the measuring microscope. Information from these devices is of sufficient precision for use with the analytical method of calculating the orientation described by BEVIS and SWINDELLS [2].

Including the HARRIS and FISHER device therefore, there are four ways of obtaining an orientation from a Kossel pattern; two of these are basically graphical. For the analytical method the principal difference between the two devices lies in the precision of the centring of the film.

Results

The results from the four methods are collected together in the table. In each case the original results have been normalised so that $\sqrt{h^2 + k^2 + l^2} = 1$. The pattern used for 2, 3, 4 and 5 was obtained with the scanning plates of the microanalyser earthed. In 6 the source was positioned in the centre of the image screen by the usual spot controls.

The graphical methods, 2 and 3, involved some selection of the data. The best checks on the accuracy of the plot are the angles between the poles. With the chart method, results that do not give consistent and correct angles can be easily re-checked or replaced by other planes. This approach is not so easy with method 3 and it appears to be more difficult to obtain three poles with the correct interplanar angles by this method. The successful result obtained from the charts was with three {110} poles whose interplanar angles were correct to $\frac{1}{2}°$. The third

Table

Method	Indices of surface normal $\times 10^5$	Deviation from trace result
1. Trace analysis	27053, 48887, 82935	—
2. Charts	25849, 49286, 80383	44'
3. Stereographic plot	23802, 46899, 85053	2°30'
4. Analytical[a]	25174, 49100, 83400	1°06'
5. Analytical[b]	25692, 48696, 83479	49'
6. Analytical[c]	25457, 48850, 83461	56'

[a] Without precise centring of the film, mean of nine results.
[b] With precise centring of the film, one result.
[c] Second film, measured as b, source positioned in centre of display screen.

result was the best that was achieved using the construction method but the interplanar angles were 2° in error for one of the poles used.

The problem of the selection of the poles is a considerable one when more than three poles are available. In general the results from different sets of any three from four will be different and normally one would have no indication of which result was the most accurate. It is only in the case reported here, where a standard result has been adopted, that a realistic selection can be made, and it should be remembered that the standard result suffers from the same problems and includes an error.

In the analytical method there is no selection of alternative results beyond the original selection of the intersections. The ones used in this determination are marked 1, 2 and 3 in Fig. 2. The original method of BEVIS and SWINDELLS used four intersections with an approximate value of the specimen-film distance. This approach requires that the positions of the intersections be measured without error, and clearly this is not possible. However it was found that the result was not very sensitive to the value of the specimen to film distance for the intersections used here. Therefore, a direct measurement of this parameter was made and used with the three intersections noted above.

The principal parameters used in the analytical method are the distances between the intersections and the centre. One would therefore expect to achieve the most reliable results when the effects of centring the film were reduced and result 5 appears to confirm this. The apparent error in 4 then appears to be due to the centring of the film because only the deviations from the standard have been calculated. Their positions with respect to the standard would vary.

A comparison between 5 and 6 shows that the effect of utilising the display image to centre the spot is lost in the combination of the other experimental errors because, although the same intersection positions on the two patterns differed by 0.1 mm approximately, the final deviations from the standard are very similar.

A larger investigation of the sensitivity of the analytical method to the errors on the parameters used is being carried out at the present time. In particular, the value of its use for the case of non-cubic crystals is being assessed.

Conclusions

The assumptions underlying the use of the back reflection Kossel patterns for the orientation of crystals appear to be justified and the result can be relied upon to an accuracy of 1° or better depending upon the method used for measurement. For the cubic crystal used in this limited experiment, the chart method of ROWLANDS and BEVIS appears to give as accurate a result as would normally be required. The analytical method requires precise positioning of the film for measurement before the same accuracy can be achieved.

Acknowledgements. This research is part of a larger project financed by the Science Research Council. The author is grateful to Dr. BEVIS and Mr. ROWLANDS of the University of Liverpool for the benefit of frequent discussions during the course of this work, and to Mr. P. CALVELEY for the reproduction of the Kossel pattern and other experimental assistance. I am pleased to record the generosity of Dr. DOROTHY FISHER in making the precise measuring jig available for measurements used in this paper.

References

1. PETERS, E. T., and R. E. OGILVIE: Trans. Met. Soc. AIME **233**, 89 (1965).
2. BEVIS, M., and N. SWINDELLS: Phys. stat. sol. **20**, 197 (1967).
3. ROWLANDS, P., E. O. FEARON, and M. BEVIS: The mechanism of phase transformations in crystalline solids, p. 194. Institute of metals, London, 1969.
4. SWINDELLS, N., and J. BURKE: The mechanism of phase transformations in crystalline solids, p. 92. Institute of metals, London, 1969.
5. BEVIS, M., F. HECKSHER, and A. G. CROCKER: Phys. stat. sol. **6**, 355 (1964).
6. ROWLANDS, P., and M. BEVIS: Phys. stat. sol. **26**, K 25 (1968).
7. HARRIS, N., and D. G. FISHER: This Conference.

Technique for Orientation Determinations
by Means of Kossel Diffraction in the Electron Microprobe

H. Hälbig, P. L. Ryder and W. Pitsch

Max-Planck-Institut für Eisenforschung, Düsseldorf, BRD

Abstract

A transmission Kossel camera for the AMX microprobe is described, which provides facilities for investigating any selected region of the specimen.

A technique for the determination of orientations from Kossel patterns is briefly summarized, and the application of the method is illustrated by means of an example. In combination with a computer programme the method provides a rapid means of orientation determination and is particularly useful in investigations where large numbers of orientations must be determined (e.g. preferred orientation studies).

1. Introduction

By means of the Kossel technique it is possible to determine the orientations or lattice parameters of small regions of a polycrystalline material. In order to be able to make full use of this technique it is desirable to be able to position the electron probe which generates the X-rays at any point on the specimen surface, so that the crystallographic parameters of any desired region may be determined. For this purpose a transmission camera has been designed for the AMX microprobe (manufactured by Applied Research Laboratories) and is described in this paper.

Various methods have been given in the literature [1—6], for determining crystal orientations from Kossel patterns. All these methods, with the exception of that due to Heise [3], in which a cylindrical film is employed, require a more or less exact knowledge of the distance of the specimen from the (plane) film and of the position of the "centre" of the diagram, i.e. the projection of the X-ray source onto the plane of the film. However, the specimen-to-film distance is difficult to measure exactly, especially in view of the vertical movement of the specimen necessary for focussing purposes. Further, it is not easy to determine the position of the centre. One possibility is to remove the specimen and allow the electron beam to fall momentarily onto the photographic film; the resulting spot on the film then indicates the position of the centre, provided that the electron beam is perpendicular to the film. However, it is possible that the electron beam may be deflected on inserting the specimen. Moreover, this method cannot be used with other geometrical arrangements, where the film is not normal to the beam. It is therefore useful to be able to determine orientations from Kossel diagrams without a knowledge of the position of the centre or the specimen-to-film distance. A method for such orientation determinations has been described elsewhere [7, 8]. A brief outline of the method (slightly modified) is given below and an example is used to illustrate its application.

2. Description of the Kossel Camera

For the AMX microprobe a transmission Kossel camera may easily be built without having to carry out structural modifications to the instrument. The specimen chamber, which contains the specimen holders and the mechanism for specimen movement, can be removed from the

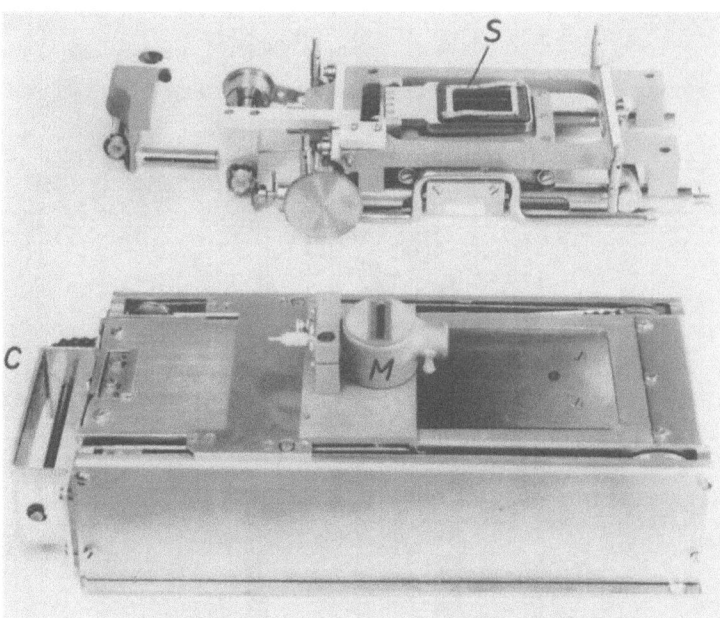

Fig. 1. Kossel camera removed from its housing, showing the specimen holder (S) with traverse mechanism and the film cassette changer (C) with exposure meter (M)

column for the purpose of changing the specimens. It was therefore decided to build a second specimen chamber and use this for the Kossel camera, which may then be fitted to the instrument simply by replacing the normal chamber.

In order that polycrystalline specimens may be investigated, the Kossel camera should satisfy the following requirements:

1. It must be possible to observe the microstructure while the specimen is under the electron beam.

2. The specimen must be moveable, so that any desired structural feature may be brought under the electron beam.

3. It should be possible to change the film a number of times under vacuum without moving the specimen or the electron beam, in order to avoid having to interrupt the investigation after every exposure.

4. An exposure meter is desirable, since the transmitted X-ray intensity is strongly dependent upon the specimen thickness, which may vary from point to point in a polycrystalline specimen.

For observing the microstructure the normal auxiliary equipment of the microprobe — generally the optical microscope — may be employed. For this purpose it is only necessary to ensure that the specimen is positioned at the correct level, with the surface perpendicular to the electron beam.

In order to provide specimen movement in three dimensions, a mechanism similar to that employed in the normal specimen chamber was constructed. However, no provision for rotating the specimen was included, and only one specimen stage, instead of the normal four interchangeable stages, was provided (see upper part of Fig. 1). This resulted in a considerable saving of space, so that only the top third of the chamber is occupied and the X-rays after traversing the specimen can pass freely into the lower part of the enclosure where the film is situated (see Fig. 2). The external specimen drive controls are also shown in Fig. 2. By means of these controls the specimen may be moved through a distance of 6 mm in the z (vertical) direction, in order to bring it exactly into the focal plane of the light microscope. In the horizontal plane the specimen may be moved 30 mm in the x (transverse) direction and the same distance in the y (longitudinal) direction, perpendicular to x and z. By means of counters driven by the specimen movement

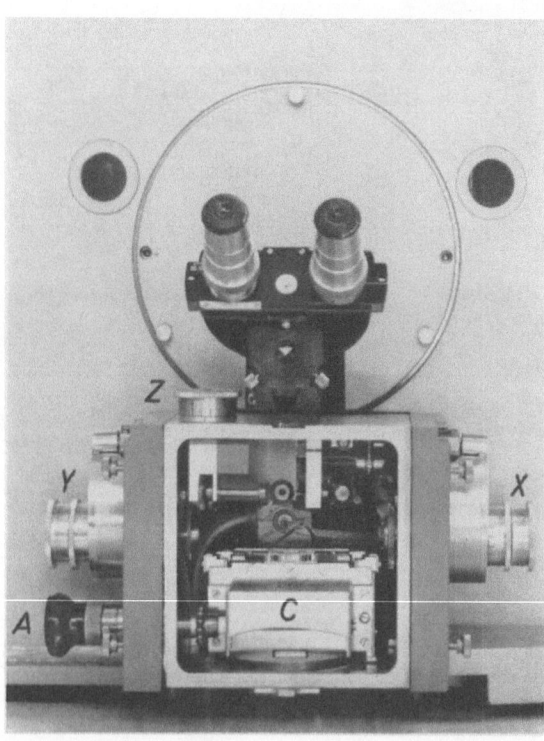

Fig. 2. Kossel camera in position on the electron microprobe, with the front cover removed, showing the controls for moving the specimen in the x, y and z directions (labelled X, Y and Z, respectively) and the film change knob (A)

and divisions on the control knobs, the specimen position coordinates may be read to an accuracy of 2 μm.

In order to satisfy the third requirement, the camera was fitted with a slightly modified version of the cassette changer used in the Siemens Elmiskop I electron microscope (see lower part of Fig. 1). The magazine holds 12 cassettes, which are normally used for photographic plates, but can just as easily carry the X-ray film used for the Kossel patterns. It is situated in the lower part of the Kossel camera and can be taken out after removal of the front cover (see Fig. 2). The opening through which the film is exposed is permanently closed with a thin sheet of plastic foil, which is transparent to the X-rays, but shields the film from scattered electrons. The film is changed by means of an external knob, which is also shown in Fig. 2.

A proportional counter of the same type as that used in the spectrometer of the microprobe serves as exposure meter, so that it can easily be connected to the existing counting equipment. The counter is mounted onto the film change mechanism in such a way that it is automatically moved under the specimen when a cassette is closed and moved away as the next cassette is opened. The detector preamplifier is situated outside the camera and connected to the detector through a socket in the rear wall of the chamber.

The specimen, which may have maximum dimensions of $25 \times 28 \times 0.5$ mm, is held in a metal frame (see Fig. 1), which has several openings, in order to be able to take smaller specimens if required, and also carries a polished thorium oxide specimen for adjusting the electron beam. The metal frame is put into an insulating holder and electrically connected to a socket in the rear wall of the Kossel camera for measurement of the specimen current. The specimen lies in a horizontal plane, parallel to the plane of the film and at a distance of approximately 8 cm from it.

A fine wire, fixed to the walls of the chamber and stretched transversely across the camera, projects a shadow onto the film and indicates the direction of the x-axis.

Normal commercial X-ray film with a format of 9×6.5 cm is used. For a specimen thickness of $30-100$ µm, a specimen current of 10^{-7} A and an accelerating voltage of 30 kV, the typical exposure times lie between 30 sec and 10 min for Fe-samples.

3. Determination of Orientations from Kossel Diagrams

The principles of the method described more fully in [7], for determining crystal orientations from Kossel diagrams will be briefly summarized here. The Kossel lines in the diagram must first be indexed. This can be done e.g. by comparing the pattern with a stereographic projection of all the Kossel cones for the particular crystal and X-ray wavelength [9], or by measuring the radii of curvature of the lines [7]. Next, the crystallographic indices of four vectors from the X-ray source Q (see Fig. 3) to four selected points (denoted by $P_1 \ldots P_4$) on the film must be determined. Suitable points are the points of intersection of Kossel lines or the intersections of the traces of known crystallographic planes (e.g. symmetry planes). From the four chosen points, a fifth point (P_5 in Fig. 3), the point of intersection of the lines $P_1 P_2$ and $P_3 P_4$, is found by construction, and the lengths $d_1 \ldots d_4$ of the lines $P_5 P_1 \ldots P_5 P_4$ are measured. Let r_j be the unit vector parallel to $Q P_j$ ($j = 1, 2, 3, 4, 5$). The determination of the vectors $r_1 \ldots r_4$ for the intersection of Kossel lines, symmetry elements etc., will be discussed below. Once they have been found, the vector r_5 is given by:

$$r_5 = (r_1 \times r_2) \times (r_3 \times r_4)/n \tag{1}$$

where n is a normalizing factor. Two vectors r_{12} and r_{34}, parallel to $P_1 P_2$ and $P_3 P_4$, respectively, which both lie in the plane of the diagram, may then be found by the relations [8]:

$$r_{12} = r_2 - \frac{d_1 \sin \varphi_2}{d_2 \sin \varphi_1} r_1 \tag{2}$$

and

$$r_{34} = r_4 - \frac{d_3 \sin \varphi_4}{d_4 \sin \varphi_3} r_3$$

where φ_j ($j = 1, 2, 3, 4$) is the angle between r_j and r_5 (see Fig. 3), given by:

$$\cos \varphi_j = r_j \cdot r_5 \quad (j = 1, 2, 3, 4).$$

The vector product $r_{34} \times r_{12}$ gives the vector normal to the plane of the diagram. The orientation of the crystal is thus completely determined, and may best be expressed by giving the positions of the crystallographic directions [100], [010], [001] with respect to three orthogonal reference axes x, y, z, where the z-axis is the normal to the film and the x-axis a reference direction (the shadow of a fine wire, see Section 2) in the plane of the diagram. For this purpose, the angle between the vector r_{12} (or r_{34}) and the reference direction (x-axis) must be measured.

The calculation of the indices of the intersection of two Kossel lines from the indices of the lines, the lattice parameters and the X-ray wavelength may be carried out as follows. In general, two Kossel lines intersect, if at all, in two points (see Fig. 4). Between these two points the Kossel lines form a "lens", which is either convex (Fig. 4a) or concave-convex (Fig. 4b). Let g_1, g_2 be the reciprocal lattice vectors corresponding to the two Kossel lines. These vectors are parallel to the respective cone axes and their magnitudes g_1, g_2 are the reciprocals of the reflecting plane spacings. We may arbitrarily define three unit vectors u_1, u_2, u_3 such that $u_1 // g_1$, $u_2 // g_2$, $u_3 // (g_1 \times g_2)$. The unit vectors r (the index $j = 1, 2, 3$ or 4 is omitted in the following for the sake of brevity) parallel to the intersection lines may be expressed as a linear combination of these three vectors:

$$r = \alpha_1 u_1 + \alpha_2 u_2 + \alpha_3 u_3. \tag{3}$$

By using Braggs' law it may be shown [8] that, for both intersections,

$$\alpha_1 = \frac{\lambda}{2 \sin^2 \phi} [g_1 - g_2 \cos \phi]$$

$$\alpha_2 = \frac{\lambda}{2 \sin^2 \phi} [g_2 - g_1 \cos \phi] \tag{4}$$

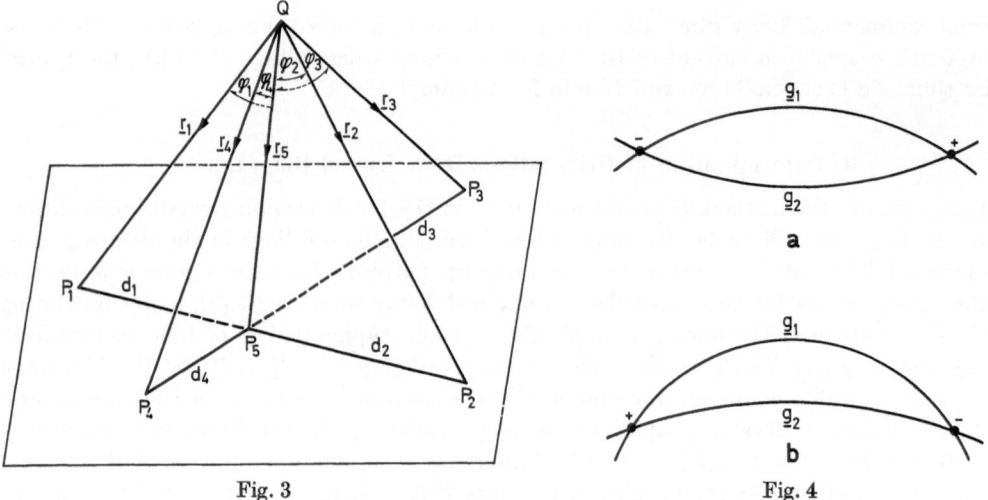

Fig. 3

Fig. 4

Fig. 3. Illustrating the measurements required to determine the orientation from a Kossel pattern when the specimen-to-film distance and the position of the centre are not precisely known

Fig. 4. Two types of lens-shaped intersections of Kossel lines. The lines are labelled with the reciprocal lattice vectors g_1, g_2 of the reflecting planes. For each intersection point, the sign of α_3 in Eq. (5) which must be used to obtain the correct vector r is indicated

where λ is the X-ray wavelength and ϕ the angle between g_1 and g_2. The third parameter is

$$\alpha_3 = \pm \sqrt{1 - \alpha_1^2 - \alpha_2^2 - 2\alpha_1\alpha_2 \cos\phi} \qquad (5)$$

where the plus sign refers to one intersection point and the minus sign to the other. The choice of the correct sign for each of the two types of Kossel intersection is shown in Fig. 4.

The determination of the orientation using the points of intersection of Kossel lines alone involves a somewhat laborious calculation, since the indices of the intersection lines are in general irrational. The calculation may often be considerably simplified, particularly in the case of a cubic crystal, by making use of the symmetry elements. The traces of the cubic symmetry planes of the forms {100} and {110} can usually be identified by inspection on the diagram. If two or more such traces are present, their points of intersection may be constructed. Since the common zone axis of any two symmetry planes has simple rational indices of the forms ⟨100⟩, ⟨110⟩ or ⟨111⟩, the calculation of the orientation is facilitated by making use of such intersections.

Symmetry planes are, however, not the only rational planes whose traces may be found in the diagram. It may easily be shown that the two points of intersection of the Kossel lines g_1, g_2 and the source Q lie on the plane s where

$$s = g_2^2\, g_1 - g_1^2\, g_2.$$

In the cubic system, s is rational; the line joining a pair of intersection points is therefore the trace of a plane with rational indices, which may be used in the same way as the symmetry planes to find the orientation. Even if these planes are (as in non-cubic systems) irrational, they have the advantage that, like the symmetry planes, their position is independent of the ratio of the X-ray wavelength to the lattice parameters of the crystal. A change in the wavelength, or an *isotropic* expansion or contraction of the crystal, causes a lens intersection to expand or shrink, but the two points of intersection remain on the same plane.

The positions of individual intersection points are, however, often very sensitive to changes in the lattice parameter, which must therefore be known very accurately if such intersections are used to determine the orientation. This is a potential source of error in orientation determinations, particularly in cases where, e.g. due to inhomogeneous composition, the lattice parameter varies from point to point in the specimen. Unfortunately, it is not always possible to find four suitable points from the intersections of symmetry planes or other fixed planes; hence the use of Kossel intersections cannot always be avoided. The method described above for determining the indices of such intersections and the orientation of the crystal is, however, suitable for computer programming, so that the lengthy calculations involved can be avoided.

Such a computer programme has been written for calculating orientations of cubic crystals from four Kossel-line intersections. The input data are:

1. the value of $\lambda/2a_0$, where a_0 is the lattice parameter of the crystal,

2. the Miller indices of the four pairs of Kossel lines corresponding to the four intersection points chosen, each pair taken in the order corresponding to the plus sign in Eq. (5),

3. the measured values of the lengths $d_1 \ldots d_4$ (see Fig. 3) and

4. the angle between the x-axis (reference direction) and the vector $\boldsymbol{r}_{12}$ ($P_1 P_2$).

The output consists of the angles between the three reference axes x, y, z and the three crystallographic axes [100], [010], [001], together with the three angles between the $x-y$ plane and the three planes containing the x-axis and each of the three crystallographic axes in turn. These data enable the orientation to be plotted on a stereographic projection, with the z-axis as pole, directly with the aid of a Wulff net, without having to rotate the net or perform constructions.

As described in [7] it is also possible to determine the position of the centre and the specimen-to-film distance from measurements of the Kossel diagram. This gives another possibility for simplifying the determination of orientations from a series of Kossel patterns if no computer facilities are available. If care is taken to keep the position of the electron beam and the specimen-to-film distance constant for such a series, they need only to be determined for one of the diagrams in the series; for the other patterns, any of the methods already described in the literature may be used to determine the orientation. It may be noted that the latter methods can also be simplified by making use of symmetry elements etc.

In the next section, the determination of crystal orientations from Kossel patterns, in the case where the specimen-to-film distance and the position of the centre are not known, is illustrated by means of an example.

4. Example of the Application of the Method

Fig. 5 shows a Kossel pattern obtained from the austenite phase of an Fe—31% Ni—0.66% P alloy ($a_0 = 3.584$ Å, $\lambda = 1.937$ Å), on which the four points used to determine the orientation are indicated. Of these points, P_1 is the point of intersection of four symmetry planes (two {100} planes and two {110}-planes) indicated by the dotted lines in Fig. 5, and is therefore the point

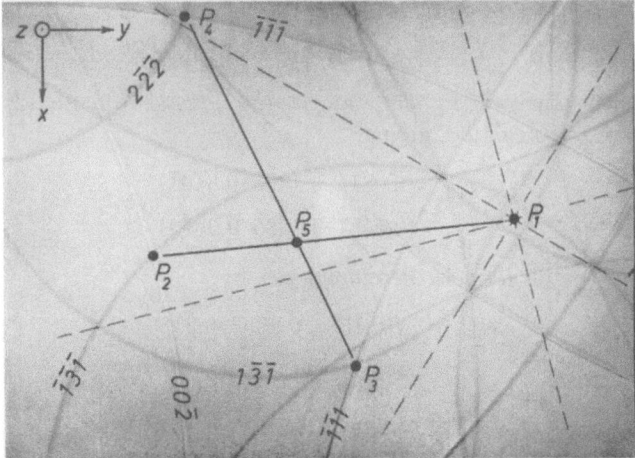

Fig. 5. Kossel diagram obtained from the austenite phase of an Fe—31% Ni—0.66% P alloy ($a = 3.584$ Å, $\lambda = 1.937$ Å), showing the indices of some of the Kossel lines, the traces of symmetric planes (dashed lines) and the points $P_1 \ldots P_5$ used in deriving the orientation

in which a $\langle 100 \rangle$ direction cuts the plane of the film. With the particular variant of the indexing chosen, P_1 has been assigned the indices $[0\bar{1}0]$. The points P_2, P_3, P_4 are defined by the intersections of pairs of Kossel lines and must be indexed by the method described above. As an example, consider the point P_3, which lies at the intersection of the Kossel lines $(13\bar{1})$ and $(\bar{1}11)$. The vector product is $[\bar{4}0\bar{4}]$. Substituting for u_1, u_2, u_3 in Eq. (3) we have:

$$r_3 = \frac{\alpha_1}{\sqrt{11}}\,[13\bar{1}] + \frac{\alpha_2}{\sqrt{3}}\,[\bar{1}11] + \frac{\alpha_3}{\sqrt{2}}\,[\bar{1}0\bar{1}].$$

The quantities g_1, g_2, ϕ are given by:

$$g_1 = \frac{\sqrt{11}}{a_0}, \qquad g_2 = \frac{\sqrt{3}}{a_0}, \qquad \cos\phi = \frac{1}{33}.$$

Hence, from Eqs. (4) and (5):

$$\frac{\alpha_1}{\sqrt{11}} = \frac{\lambda}{2a_0}\cdot\frac{33}{32}\left\{1 - \frac{1}{11}\right\} = 0.2533,$$

$$\frac{\alpha_2}{\sqrt{3}} = \frac{\lambda}{2a_0}\cdot\frac{33}{32}\left\{1 - \frac{1}{3}\right\} = 0.1858,$$

$$\frac{\alpha_3}{\sqrt{2}} = \pm\,0.2194.$$

Comparing Figs. 4 and 5 it is seen that the plus sign must be used for the point P_3. Hence:

$$r_3 = 0.2533\,[13\bar{1}] + 0.1858\,[\bar{1}11] + 0.2194\,[\bar{1}0\bar{1}]$$
$$= [\overline{0.1519}\ \overline{0.9457}\ \overline{0.2869}].$$

Similarly, for the points P_2 and P_4, it can be shown that

$$r_2 = [0.0814\ \overline{0.8378}\ \overline{0.5404}]$$

and

$$r_4 = [0.4053\ \overline{0.3891}\ \overline{0.8267}].$$

From Eq. (1) the vector r_5 is then found to be

$$r_5 = [0.05077\ \overline{0.9401}\ \overline{0.3370}].$$

By forming the scalar products of r_5 with each of the four vectors $r_1 \ldots r_4$, the angles φ_j of Eq. (2) may be calculated. The sines of these angles are found to be:

$$\sin\varphi_1 = 0.3409; \quad \sin\varphi_2 = 0.2270; \quad \sin\varphi_3 = 0.2086; \quad \sin\varphi_4 = 0.3704.$$

The distances $d_1 \ldots d_4$, measured on an enlarged projection of the diagram, were found to be:

$$d_1 = 75.2\ \text{mm}; \quad d_2 = 48.4\ \text{mm}; \quad d_3 = 46.3\ \text{mm}; \quad d_4 = 81.4\ \text{mm}.$$

Substituting the above values for r_j, $\sin\varphi_j$ and d_j ($j = 1, 2, 3, 4$) in Eq. (2) the vectors parallel to $P_1 P_2$ and $P_3 P_4$ may be calculated, giving

$$r_{12} = [0.0814\ 0.1968\ \overline{0.5404}]$$
$$r_{34} = [0.5587\ 0.1285\ \overline{0.0993}].$$

The vector normal to the diagram is therefore given by:

$$r_{34} \times r_{12} /\!/ [\overline{0.1588}\ 0.9352\ 0.3167]$$

(expressed as a unit vector).

5. Summary and Discussion

With the help of the Kossel camera described in this paper it is possible to determine crystal orientations or lattice parameters of small selected regions (of the order of 10 μm in diameter). It has further been shown that orientations may be derived from Kossel patterns without an exact knowledge of the specimen-to-film distance or the position of the centre. A method for deriving such orientations has been derived in earlier papers [7, 8] and briefly summarized here.

The computations involved in the method have been programmed for a digital computer, thus providing a rapid and efficient method of determining large numbers of orientations. In the course of a study of preferred orientation in cold-rolled and recrystallized pure iron (not yet published), the orientations of 400 individual grains have been determined by the Kossel technique with the aid of this programme. The technique has also been used for habit plane determination by two-surface trace analysis in the martensite-austenite transformation in an Fe—32.5% Ni alloy [10], where it was necessary to measure the orientations of very small regions of the martensite phase.

Acknowledgement. The financial support of the Landesamt für Forschung des Landes Nordrhein-Westfalen is gratefully acknowledged.

References

1. BEVIS, M., and N. SWINDELS: Phys. stat. sol. **20**, 197 (1967).
2. BRÜMMER, O., G. DRÄGER, and W. SCHÜLKE: Exptl. Techn. Physik **14**, 73 (1966).
3. HEISE, B. H.: J. Appl. Phys. **33**, 697 (1962).
4. IMURA, T.: Bull. Naniwa Univ., Ser. A **2**, 51 (1954).
5. MORRIS, W. G.: J. Appl. Phys. **39**, 1813 (1968).
6. PETERS, E. T., and R. E. OGILVIE: Trans. AIME **233**, 89 (1965).
7. RYDER, P. L., H. HÄLBIG, and W. PITSCH: Mikrochim. Acta, Suppl. **2**, 123 (1967).
8. HÄLBIG, H., P. L. RYDER, and W. PITSCH: Mikrochim. Acta, Suppl. **3**, 201 (1968).
9. LONSDALE, K.: Phil. Trans. Roy. Soc. London, Ser. A **240**, 219 (1948).
10. HÄLBIG, H., H. KESSLER, and W. PITSCH: Acta Met. **15**, 894 (1967).

A Method for the Precise Determination of Lattice Spacings from Kossel Patterns from Selected Grains in Steel Samples

D. G. FISHER and N. HARRIS

GEC-AEI Engineering Ltd. Research Laboratory, Manchester, England

Introduction

By using the probe-forming system and electron imaging facility of an electron probe X-ray microanalyser in association with an X-ray divergent beam camera it has been possible to obtain sharp Kossel patterns by back reflection from selected microregions in polycrystalline steels. Fig. 1 shows an example of such a pattern from a commercial grade of austenitic stainless steel with a grain size of about 10 μm.

The long term objective of the work is to use the technique in the study of stress in microregions in steel samples especially in the vicinity of incipient cracks or other discontinuities. However, since X-ray stress measurements are essentially measurements of small changes in lattice spacings the first stage of the work was to establish a technique for the precise determination of lattice spacings from patterns of this type.

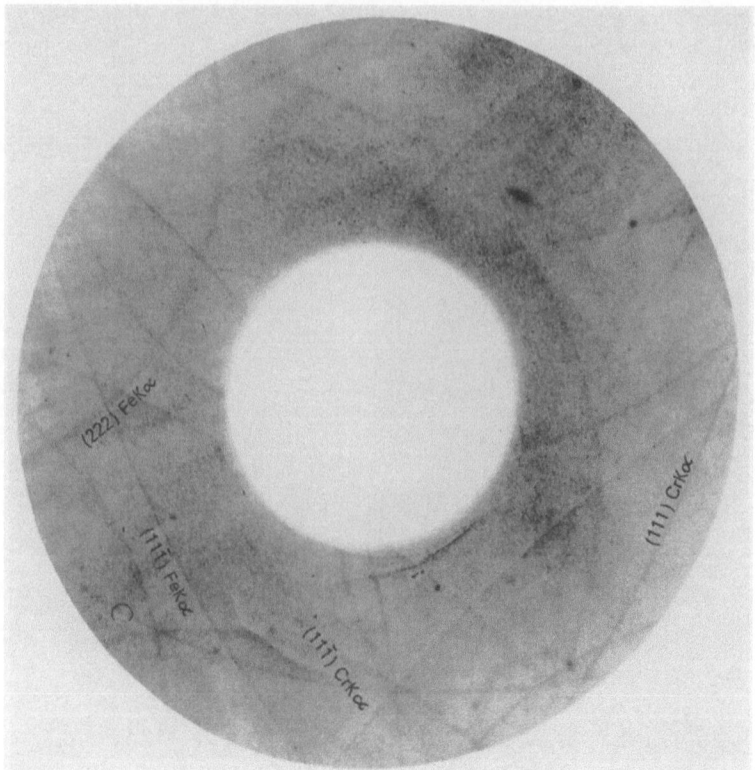

Fig. 1. Back reflection Kossel pattern from single grain of polycrystalline steel

The method now put forward, while designed primarily to meet the requirements of this project, is of general applicability and has several attractive features. It is simple in concept and does not require the intermediate use of stereographic projection.

Description of the Method

The method treats the Kossel pattern as a three-dimensional array of intersecting circular cones with a common vertex, and the interpretation depends on the fact that if three or more positions of a cone generator are specified the cone is defined and its axis and semivertex angle can be derived. It has features in common with the method used by ELLIS et al. [1] in the study of strain in single crystals and with methods reported more recently by SCHNEIDER [2] and MORRIS [3] but differs from each in several significant respects.

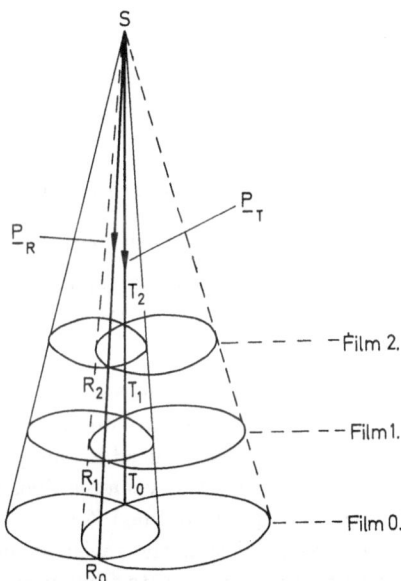

Fig. 2. Lines of intersection of two Kossel cones

$R_0\, R_1\, R_2$⎱ sets of corresponding intersections of Kossel curves
$T_0\, T_1\, T_2$⎰ defining common cone generator positions $\boldsymbol{P}_R$ and $\boldsymbol{P}_T$

In studying a particular cone the lines along which it intersects other cones are used as easily identified positions of its generator (Fig. 2). A series of points along each line is obtained by recording Kossel sections at different specimen to film distances and the best straight line through each set is obtained by a least squares method of fitting without assuming that the line passes through the X-ray source. When sufficient positions of the cone generator are obtained in this way the axis and semivertex angle are again fitted by a least squares method.

Although precision of measurement is improved by arranging for the cone studied to have its axis roughly normal to the plane of the photographic film the method is not restricted to such cases but can be applied to any cone which cuts the recording film whether in a closed ellipse or in a portion of an ellipse or hyperbola. It is this feature which renders it particularly useful in the study of grains in a polycrystalline matrix where only one face is available for examination and the range of crystal orientation which can be used is consequently limited.

In order to carry out the analytical treatment the positions of the points derived from the different Kossel sections must be expressed on a single three-dimensional reference system and for this purpose orthogonal axes are set up as indicated in Fig. 3. The origin of coordinates is placed in the film farthest from the specimen at a centre defined by three circumferential marks imprinted on the film by the cassette; X and Y directions lie in the film plane, and the Z axis

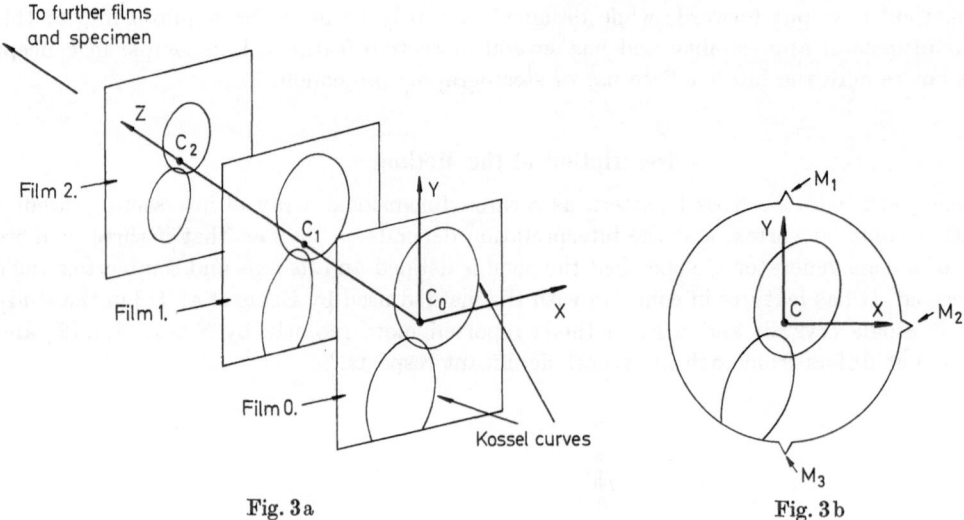

Fig. 3a. System of coordinate axes

X, Y, Z, orthogonal axes with origin at centre C_0 of film 0
X, Y,   in plane of film
Z normal to film and passing through centres C_1, C_2 — of films 1, 2 — —

Fig. 3b. Location of film centre and choice of axial directions $X\,Y$, imprinted on film by cassette top. C is centre of circle through points M_1, M_2, M_3. Axis Y is on line M_1, M_3

passes through the centres (defined as above) of successive films. In the experimental set-up X and Y coordinates are derived from measurements on each film in polar coordinates and Z values are known in terms of precision spacers introduced to change the specimen to film distances.

It should be noted that the position of the X-ray source does not enter into the calculations and there is no assumption that the arbitrarily defined Z axis coincides with the electronoptic axis of the camera. It is, however, essential that the registration of the pattern in relation to the axial system is not changed when the specimen to film distance is changed. This is ensured by locating pins in the film cassette and in the specimen holder.

Once the required points have been measured and defined in the specified reference system the analysis is carried out as follows:

The best straight line through a set of corresponding intersections $R_0 R_1 R_2 \ldots$ or $T_0 T_1 T_2 \ldots$ (Fig. 2) is obtained by a least squares method and its direction cosines a, b, c are derived. It is then expressed as a unit vector

$$\boldsymbol{P} = a\boldsymbol{X} + b\boldsymbol{Y} + c\boldsymbol{Z}. \tag{1}$$

The cone axis $\boldsymbol{N}$ may similarly be treated as a unit vector, and direction cosines u, v, w assigned to it, giving

$$\boldsymbol{N} = u\boldsymbol{X} + v\boldsymbol{Y} + w\boldsymbol{Z}. \tag{2}$$

The scalar product

$$\boldsymbol{P} \cdot \boldsymbol{N} = au + bv + cw = \cos\alpha \tag{3}$$

then defines the semivertex angle α of the cone.

In the relationship (3) the direction cosines u, v, w and the angle α are unknown, but since $u^2 + v^2 + w^2 \equiv 1$, three vectors $\boldsymbol{P}_n$ are sufficient to determine the cone. However, the precision of the results is enhanced by taking the maximum available number of vectors $\boldsymbol{P}_n$ and again deriving the best values of the coefficients by a least squares method. When the coefficients, u, v, w have been obtained a check of the method and an indication of the errors is obtained by calculating a value of $\cos\alpha$ as the scalar product $\boldsymbol{P}_n \cdot \boldsymbol{N}$ $(= ua_n + vb_n + wc_n)$ for each cone generator vector in turn, the figure finally used being the mean of these.

Finally, the interplanar spacing d of the plane producing the cone is given by

$$d = \frac{\lambda}{2 \cos \alpha}$$

where λ is the wavelength of the radiation, and the angle β between two sets of planes with normals N_1 and N_2 can be derived from the expression

$$\cos \beta = N_1 \cdot N_2.$$

A vector in the direction N but of length $1/d$ can be treated as a vector in the reciprocal lattice of the crystal and if enough such vectors are obtained the arbitrary coordinate system can be related to the conventional crystallographic axes and Miller indices assigned to the Kossel curves.

The least squares method of averaging is used twice in the above treatment and is regarded as an important feature of the method. Its use to find the best straight line through a set of experimental points is well known but the situation in this case is complex since the points are obtained in three-dimensional coordinates all of which are derived experimentally and are therefore subject to error. In our treatment some simplification was achieved by considering projections on to the axial planes. The second case where it is used to find the cone axis from a set of vectors P_n on the cone surface is less usual and the procedure is as follows:

The expression (3) above is re-written in the form

$$au + bv + cw - \cos \alpha = 0. \tag{3a}$$

Normal equations are then constructed by summation over the set of Eq. (3a)

$$\begin{aligned}
u \sum a_n^2 + v \sum a_n b_n + w \sum a_n c_n - \cos \alpha \sum a_n &= 0 \\
u \sum a_n b_n + v \sum b_n^2 + w \sum b_n c_n - \cos \alpha \sum b_n &= 0 \\
u \sum a_n c_n + v \sum b_n c_n + w \sum c_n^2 - \cos \alpha \sum c_n &= 0 \\
u \sum a_n + v \sum b_n + w \sum c_n - j \cos \alpha \qquad &= 0
\end{aligned} \tag{4}$$

where the summation runs from $n = 1$ to $n = j$ and j is the number of vectors P_n determined.

Since $u^2 + v^2 + w^2 \equiv 1$ only three of the Eq. (4) are required to determine the coefficients and the final one may be omitted or used for checking.

A computer program has been written in Algol language to handle the arithmetical operations involved in this method of analysis.

Experimental Trial of the Method

The method has been tested on a polycrystalline sample of Swedish iron with a grain size of about 20 μm.

Fig. 4 shows the central part of a pattern from a grain oriented with a (220) set of planes almost parallel to the photographic film. Six clear intersections occur on the (220) Kossel curve defining six positions of the cone generator which were used for the first tests. In these tests four different film positions were used to give four points on each cone generator but instrumental modifications have been introduced to increase the number to seven in future work and enhance the precision attainable.

Results obtained from four different but similarly oriented grains gave values for the (220) spacing in the range 1.0119 Å to 1.0122 Å (or a spread of 3 parts in 10,000) which is an encouraging result for a pattern of this type. Although the data are as yet insufficient for statistical analysis the figures appear to compare well the single crystal results of ELLIS [1] who quotes a standard deviation of 0.00029 Å on a spacing of 1.00085 Å and SCHNEIDER [2] who quotes a precision of ±0.0004 Å on a spacing of 1.0095 Å.

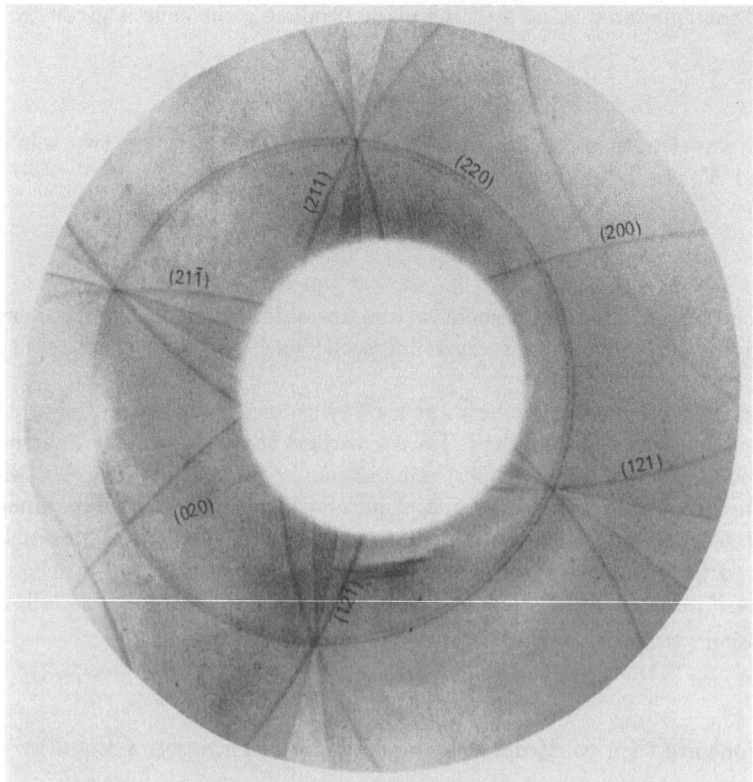

Fig. 4. Back reflection Kossel pattern from grain in Swedish iron

Results from less favourably oriented sets of planes in the same grain (e.g. (211) in Fig. 4) are predictably somewhat more variable and very sensitive to the number of cone generators measured but there are so far too few results on these to give reliable figures.

Acknowledgement. The authors are indebted to Professor R. S. Tebble and members of his staff in the Physics Department of Salford University for helpful discussions in relation to this work and they particularly wish to thank Dr. P. Lissberger of that Department for suggestions leading to the least squares method adopted in processing the results. They are also grateful for the assistance provided by the University's computer service and especially for the help in programming given by Miss J. Whittle.

Finally they wish to record their appreciation of several stimulating discussions with Dr. N. Swindells and Dr. M. Bevis of Liverpool University.

They thank the director of Research, GEC-AEI Engineering LTD., Trafford Park, Manchester for permission to publish this work.

References

1. Ellis, T., et al.: J. Appl. Phys. **35**, 3364 (1964).
2. Schneider, J. R.: Ph.D. Thesis, University of Cincinnata, 1965.
3. Morris, W. G.: J. Appl. Phys. **39**, 1813 (1968).

A Variable Orientation Specimen Holder
for Precision Microdiffraction in the Reflection Mode

R. D. CORNFORTH, D. G. FISHER and N. HARRIS

GEC-AEI Engineering Research Laboratory, Manchester, England

1. Introduction

A specimen holder incorporating several novel features has been constructed for back-reflection Kossel microdiffraction in an electron probe X-ray microanalyser. The device is designed to replace the standard reflection specimen holder in the AEI microfocus X-ray camera but could readily be modified to suit other systems.

The holder was developed as part of the project (1) for examination of strain in localised regions of polycrystalline samples and is designed to meet the specific requirements of this work, but should also be useful in many other applications of the Kossel technique.

The special facilities included are 1. means for holding the specimen under known strain during test, 2. means for tilting the specimen to bring the desired part of the Kossel pattern into the field of view, and 3. means for changing the camera length by a succession of known steps without introducing lateral or rotational movement of the specimen relative to the photographic film. A simple graticule can also be incorporated when required.

2. Description of the Holder

2.1. General Assembly

Fig. 1 shows a general view of the holder set up on the stage of the microfocus camera with stretching device and graticule in position. The individual features are illustrated in Figs. 2—4 and are described below.

Specimen

Stretching device

Graticule

Fig. 1. General view of holder set up on specimen stage with graticule and stretching device in position

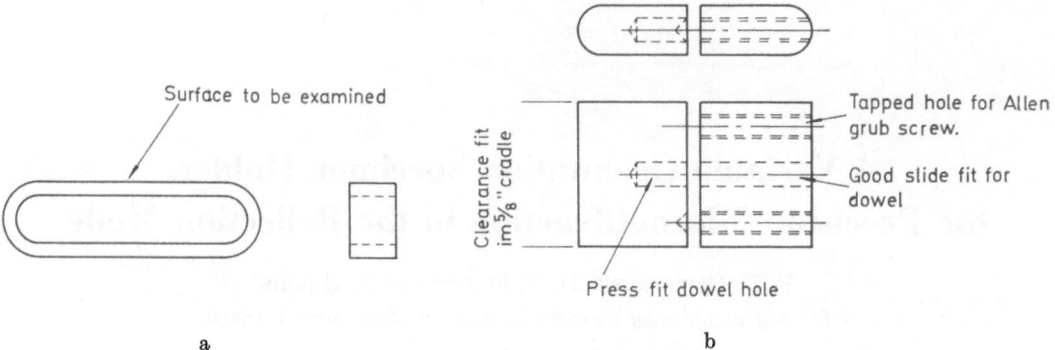

Fig. 2a and b. Specimen straining device. a Specimen in form of flat loop. b Stretching jig — designed to fit cradle of specimen holder

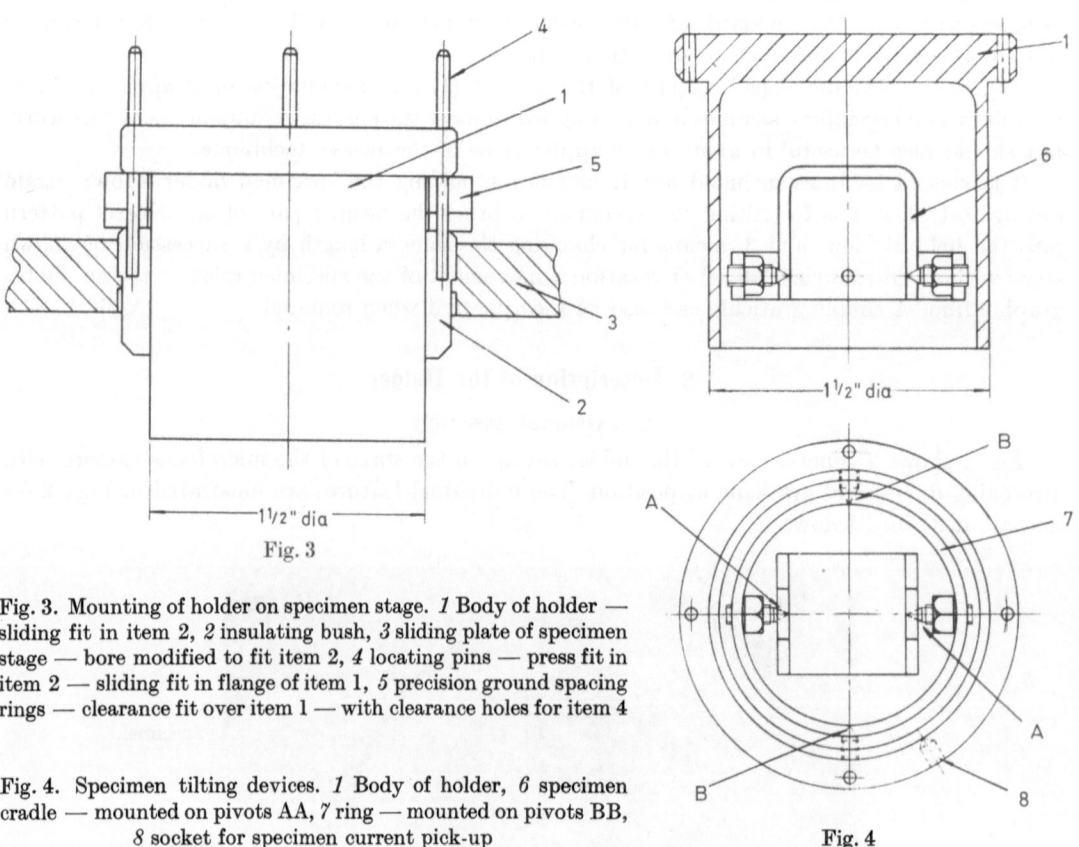

Fig. 3

Fig. 3. Mounting of holder on specimen stage. *1* Body of holder — sliding fit in item 2, *2* insulating bush, *3* sliding plate of specimen stage — bore modified to fit item 2, *4* locating pins — press fit in item 2 — sliding fit in flange of item 1, *5* precision ground spacing rings — clearance fit over item 1 — with clearance holes for item 4

Fig. 4. Specimen tilting devices. *1* Body of holder, *6* specimen cradle — mounted on pivots AA, *7* ring — mounted on pivots BB, *8* socket for specimen current pick-up

Fig. 4

2.2. Specimen Straining Device (Fig. 2)

This is an independent unit made to fit into the jaws of the holder in place of any other specimen. Its method of operation is clear from Fig. 2. It requires a special type of specimen machined in the form of a flat loop into which the jig fits snugly. Strain is introduced by forcing the two members of the jig apart with the aid of the screws shown. For measurement of the strain two reference lines are diamond scribed on the polished experimental surface of the specimen and their separation is measured with a travelling microscope; the desired strain is introduced before mounting the unit in the holder.

2.3. Mounting of Holder and Provision for Changing Camera Length (Fig. 3)

The holder is built into a cylindrical body, $1\frac{1}{2}$ inches (38 mm) in diameter which is a push fit in a bakelite bush in the sliding base plate of the specimen stage.

The baseplate, 3, is a standard part of the microfocus X-ray camera except that the bore of the lip on which the bakelite bush rests has been slightly enlarged. By this means an extra 2.5 mm in internal diameter of the body cylinder has been gained which is valuable in accommodating the tilting and stretching facilities.

Changes in camera length are achieved by introducing spacers, 5, between the bush and the flange on the body of the holder. These are precision ground to ± 0.0025 cms and provide for a range of camera lengths from 2.75 cms to 4.50 cms in 0.250 cm steps. Locating pins, 4, prevent accidental changes of azimuthal orientation but permit rotation in steps of 90° if required.

2.4. Tilting Device (Fig. 4)

This consists of a gimbals arrangement in which the specimen cradle, 6, is mounted on pivots AA in a ring 7 which is mounted in the holder body on pivots BB. The two pivot axes are at right angles to one another but lie in the same plane. With this arrangement tilts of up to $\pm 30°$ about either axis can be obtained. The cradle is provided with means for fixing the specimen so that its face contains the axis AA.

The desired tilt is preset with the aid of a protractor and the pivots are then tightened to clamp the specimen cradle firmly. In order to produce pure tilt without lateral movement the relevant microregion of the sample must lie at the point of intersection of the two pivot axes. However, in practice this setting needs only to be approximate because small lateral movement can be corrected by adjustment of the stage micrometers in association with the electron image of the specimen to reposition the desired feature. Any change in specimen height resulting from the tilt is not corrected by this method but this does not matter provided the multiple exposure method is used at each specimen tilt.

Fig. 5 shows a Kossel pattern from a selected grain in a sample of Swedish iron with the specimen tilted to bring a (211) reflection into the centre of the field. Fig. 4 of the preceding paper shows a pattern with a (220) reflection central.

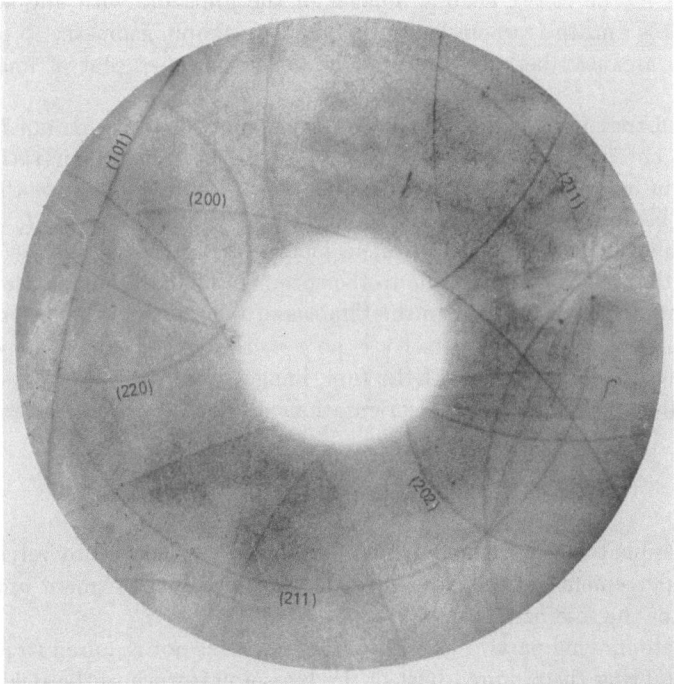

Fig. 5. Kossel pattern from Swedish iron tilted to put (211) reflection in centre of field

2.5. Graticule

The open end of the body cylinder provides a useful frame on which a graticule can be mounted when required.

A form of graticule which has been found useful consists of two parallel 0.001 inch wires offset with respect to the centre of the specimen. This has been used for three purposes: — determining camera length by a method independent of any Kossel pattern, determining the X-ray centre of the film, and checking the flatness of the film by inspection of the images of the wires.

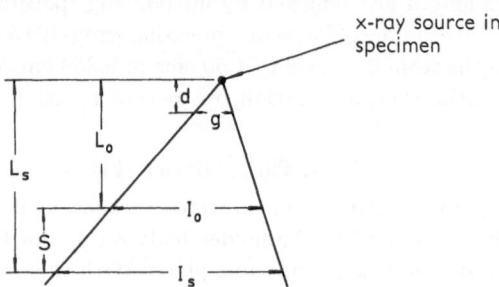

Fig. 6. Determination of distance between specimen and film (camera length). g = Reference distance in graticule. d = Distance of graticule from specimen. I_0 = Image of g without spacer (measured). I_S = Image of g with spacer (measured). S = Thickness of spacer (known)

$$\frac{I_S}{I_0} = \frac{L_S}{L_0} = \frac{L_0 + S}{L_0}$$

Hence $\quad L_0 = \frac{I_0 S}{I_S - I_0} \quad$ and $\quad L_S = \frac{I_S S}{I_S - I_0}$

$$\text{Sensitivity } \frac{\Delta I}{\Delta L} = \frac{g}{d}$$

For the purpose of measuring camera length the graticule is used in conjunction with one or more of the spacers referred to in Section 2.3 above. Measurement of the separation of corresponding points in the X-ray shadow images of the graticule with and without the spacer in position are made and the camera length is found by simple geometry as indicated in Fig. 6. If several spacers are used better values can be derived from a plot of image length against spacer thickness.

It will be noted that neither the reference distance g in the graticule nor its distance d from the X-ray source enter into the expression for camera length but their ratio g/d controls the sensitivity. Thus in using the method it is not necessary to know either of these quantities but their ratio should be kept as large as is consistent with recording the image I_s within the confines of the photographic film at the longest camera length used. It is perhaps worth noting that it follows from this that fixing the graticule to the specimen holder is preferable to the alternative which was first considered of fixing it to the film cassette because in the latter case the ratio g/d changes with camera length and sensitivity is poor at the longer camera lengths.

For determining the X-ray centre of the film, images were recorded and superimposed with the holder in its four possible azimuthal orientations and the centre of the resulting figure was found.

3. Discussion

It has been possible to keep the holder simple and easy to construct by relying to some extent on the precision movements already available in the microfocus camera and on the electron imaging facilities of the microanalyser.

It has no delicate moving parts. Since the tilting device is not required to provide a precision setting but only to bring the required part of the Kossel pattern into the field of view it can be fairly crude.

The use of precision spacers for introducing changes in camera length is an established technique and similar spacers are provided with the standard specimen holder. However, the arrangement described here which puts the spacers outside the holder leaves the inside of the holder clear for the introduction of the tilting device and also permits the use of a wider range of camera lengths. This, together with the introduction of locating pins to prevent misalignment greatly enhanced the value of the holder for the multiple exposure method adopted for the determination of lattice parameter.

Neither camera length nor specimen tilt can be changed while the equipment is under vacuum. These facilities would introduce considerable complexity of design, and in the case of specimen tilt would have little value since the Kossel pattern cannot be viewed in situ. The ability to change specimen height while under vacuum might be an advantage especially if multiple exposures on a single film were used, and the possibility for introducing this may be considered later.

Several other refinements of design and construction could be suggested, but this very simple laboratory-built device has been found to do all that is required of it so far.

Acknowledgement. The authors thank the director of Research, GEC-AEI Engineering Ltd., Trafford Park, Manchester, for permission to publish this work.

Reference

FISHER, D. G., and HARRIS, N.: Proceedings of Fifth Internat. Conference on X-ray Optics and Microanalysis.

Hochtemperatur-Kossel-Technik

H.-J. ULLRICH, S. DÄBRITZ und H. SCHREIBER

II. Institut für Experimentalphysik der Technischen Universität Dresden, DDR

1. Problemstellung

Mit der Entwicklung von Hochtemperaturzusätzen für Elektronenstrahl-Mikroanalysatoren konnte gezeigt werden, daß man chemische Konzentrationsbestimmungen auch im Hochtemperaturbereich vornehmen kann. Bisher war noch nicht geklärt, ob bei hohen Temperaturen Orientierungsbestimmungen, Symmetrieermittlungen und Gitterkonstantenmessungen in mikroskopischen Bereichen von der gleichen Probenstelle ausgeführt werden können. Bei der Untersuchung von Festkörpereigenschaften besitzt aber das Stadium von temperaturabhängigen Phänomenen gleichzeitig mit der chemischen Konzentrationsbestimmung besondere Bedeutung.

Zu ihnen gehören u. a.

1. Diffusionsvorgänge, d. h. Zusammenhang zwischen Konzentration und Gitterparameter sowie Orientierung der Diffusionsfront,

2. Phasenumwandlungen, die besonders bei der Untersuchung von Zustandsdiagrammen notwendig sind, wobei sich Symmetrie, Orientierung und Gitterparameter ändern können und deren Korrelation mit der ursprünglichen Phase oder der Matrix gesucht wird,

3. Bestimmung von Ausdehnungskoeffizienten und Fehlstellendichte in Abhängigkeit von der Konzentration.

Für derartige kombinierte Messungen erscheint die ursprüngliche Kossel-Methode besonders geeignet, weil mit der Anregung der Röntgeneigenstrahlung in einem einkristallinen oder polykristallinen Gebiet (Korndurchmesser größer als der Brennfleckdurchmesser) das Reflexsystem nahezu immer auftritt. Die kurzen Belichtungszeiten sind ebenfalls für die Untersuchung temperaturabhängiger Vorgänge ein wesentlicher Vorteil.

Die aufgezählten Gründe veranlaßten uns, Kossel-Hochtemperatur-Experimente auszuführen. Über die ersten Experimente soll im folgenden berichtet werden.

2. Experimente

Gegenüber der Zimmertemperatur sind im Hochtemperaturbereich insofern Schwierigkeiten zu erwarten, weil die Intensität der an sich schon schwachen Kossel-Reflexe mit wachsender Temperatur abnimmt. Besonders die für die Präzisionsbestimmung der Gitterkonstanten erforderlichen hochindizierten Reflexe werden stark geschwächt. Es war deshalb zunächst zu untersuchen, bis zu welchen Temperaturen auswertbare Kossel-Aufnahmen hergestellt werden können. Dabei konnten wir auf Erfahrungen zurückgreifen, die wir im Normaltemperaturbereich bei der Herstellung von kontrastreichen Kossel-Aufnahmen einer großen Zahl von Substanzen gesammelt hatten.

Als Probensubstanzen wurden Cu, Fe und Fe_3O_4 ausgewählt. Folgende Gründe waren dafür maßgebend: Cu bleibt bis zum Schmelzpunkt einphasig und liefert bei Zimmertemperatur besonders kontrastreiche Aufnahmen. An Fe sollte über Gitterkonstantenmessungen die magnetische Umwandlung erfaßt und nach Orientierungszusammenhängen bei der α-γ-Umwandlung gesucht werden. Fe_3O_4 gehört zu den Substanzen, die wegen ihres ungünstigen λ/a-Verhältnisses für Kossel-Untersuchungen weniger geeignet erscheinen. Es galt, den Gitterkonstantenverlauf im Curie-Temperaturbereich zu messen.

3. Ergebnisse

3.1. Cu. Bei Zimmertemperatur eignet sich der [011]-Pol am günstigsten zur Präzisionsbestimmung. Es wurde versucht, hier auch die Gitterkonstantenbestimmung im Hochtemperaturbereich vorzunehmen. Die 042-024-Koinzidenz wird gegen die Reflexe 133-1̄33 ausgespielt. Das Stereogramm (Abb. 1) verdeutlicht die Lage der Linien zueinander. Abb. 2 zeigt eine Tieftemperatur-Kossel-Aufnahme (−188,5° C) am Rhombendodekaederpol. In Abb. 3 überdecken sich die 042- und 024-$K_{\alpha 1}$-Linien im [011]-Pol bei 262° C. Ohne irgendwelche Linienabstände zu vermessen, folgt daraus sofort die Gitterkonstante $a = 3,63111$ Å (λ-Cu$K_{\alpha 1} = 1,54055$ Å, Brechung korrigiert). In Abb. 4 (363° C) bilden der 042- und 024-Reflex eine für die Gitterkonstantenmessung sehr empfindliche Linse.

Oberhalb von 750° C nimmt die Intensität der 042-Reflexe so stark ab, daß die Linien im Untergrund verschwinden und die 133-1̄33-Linien zur Auswertung benutzt wurden.

Oberhalb 900° C scheiterte eine Gitterkonstantenbestimmung am [011]-Pol. Für eine kontinuierliche Gitterkonstantenmessung vom Tieftemperaturbereich bis zum Schmelzpunkt wurden deshalb die Reflexe vom Typ 111 und 002 am Oktaederpol herangezogen.

3.2. Fe. Die Aufnahmen wurden am [011]-Pol angefertigt und die Begegnung der 112- und 022-Reflexe gegen den Durchmesser des 022-Kreises ausgespielt. Die Experimente ergaben, daß sich die genannte Auswertungsmöglichkeit auf den gesamten α-Bereich anwenden läßt. Bei

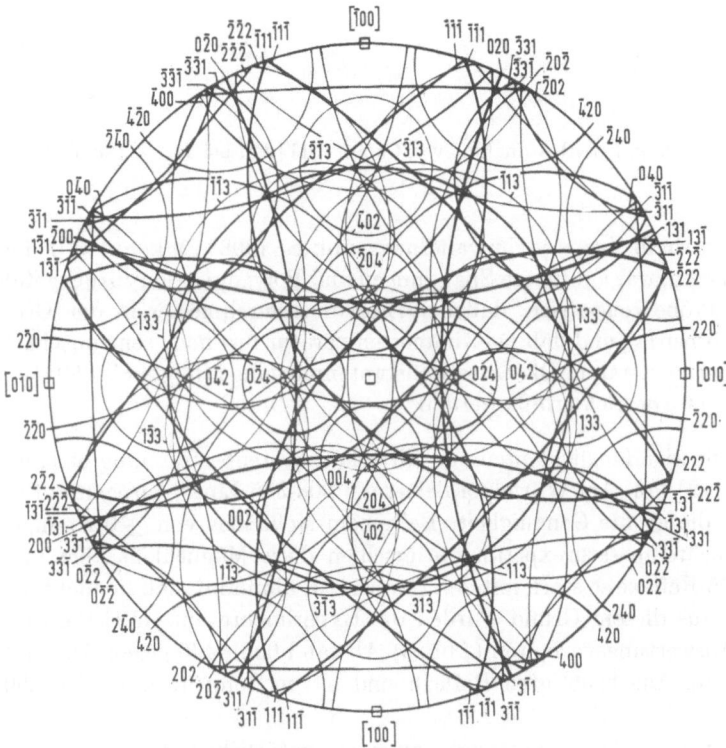

Abb. 1. Stereogramm der Kossel-Linien von Cu ($a = 3,61$ Å) mit Cu-K_α-Strahlung ($\lambda = 1,54$ Å) symmetrisch zum [001]-Pol

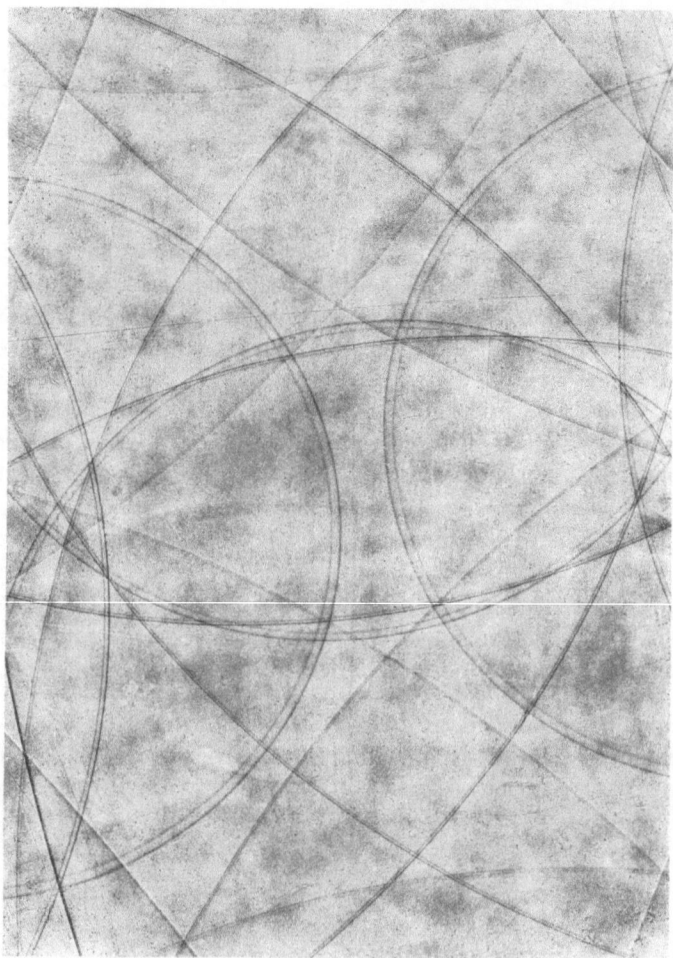

Abb. 2. Kossel-Aufnahme von Cu am [011]-Pol bei $T = -188,5°$ C

Temperaturen oberhalb des α-γ-Umwandlungspunktes (908° C) konnten keine Kossel-Linien nachgewiesen werden, was auf innere Spannungen und Polygonisation zurückgeführt wurde. Nach Abkühlung der Probe zeigte sich, daß der Brennfleckdurchmesser in der Größenordnung der neugebildeten Körner lag und daß bei Zimmertemperatur nur noch sehr unscharfe Linien infolge innerer Spannungen auftraten. Die Gitterkonstantenmessung erfolgte bis 850° C. Deutlich konnte der Curie-Temperaturbereich erfaßt werden.

3.3. Fe_3O_4. Günstige Möglichkeiten zur Gitterkonstantenbestimmung existieren bei Magnetit am [001]- und [112]-Pol. Am Würfelpol liefert die 008-355-Linse, bezogen auf den Durchmesser des 008-Kreises, die größte Genauigkeit. Bei einem Meßfehler von $\pm 0,1$ mm wird $\Delta a/a \approx 10^{-5}$ erreicht. Für Hochtemperaturexperimente erschien diese Möglichkeit aber ungeeignet, da die Intensität der 355-Reflexe so stark mit der Temperatur abnimmt, daß sie nicht mehr ausgemessen werden konnte. Aus diesem Grund wurden die Experimente am [112]-Pol durchgeführt. Hier boten sich die Auswertungsvarianten (Abb. 5) Abstand 008—440 gegen Abstand 404—044 bzw. gegen 008—220 an. Alle benötigten Reflexe sind bis zu Temperaturen von 850° C noch nachweisbar.

Es wurden Gitterkonstantenmessungen an einem natürlichen Magnetit-Kristall bis zu 850° C ausgeführt, wobei die Gitterkonstantenänderung die bekannte Anomalie innerhalb des Curie-Temperaturbereiches zeigte.

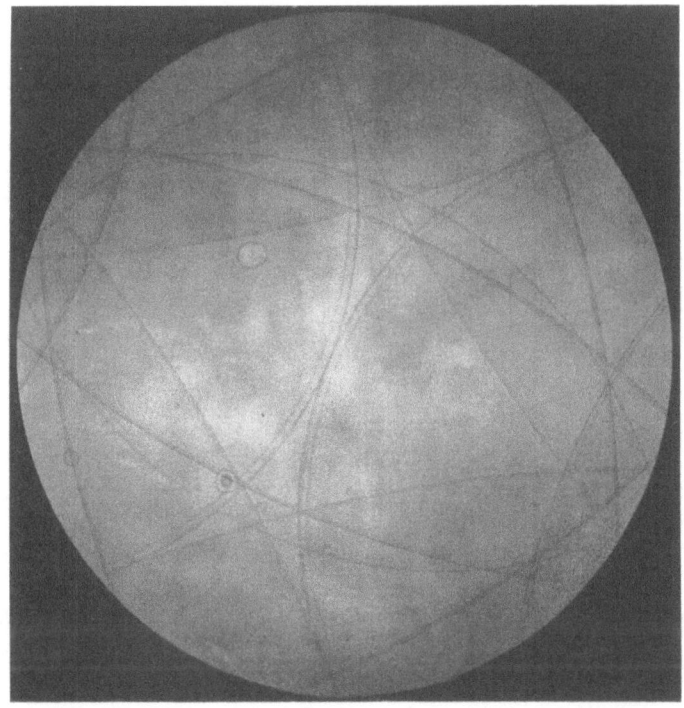

Abb. 3. Kossel-Aufnahme von Cu am [011]-Pol bei $T = 262°$ C

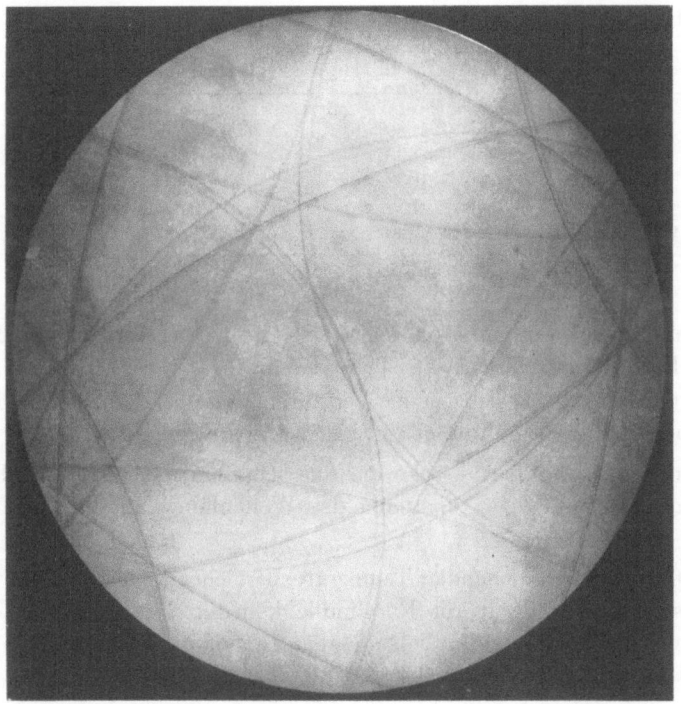

Abb. 4. Kossel-Aufnahme von Cu am [011]-Pol bei $T = 363°$ C

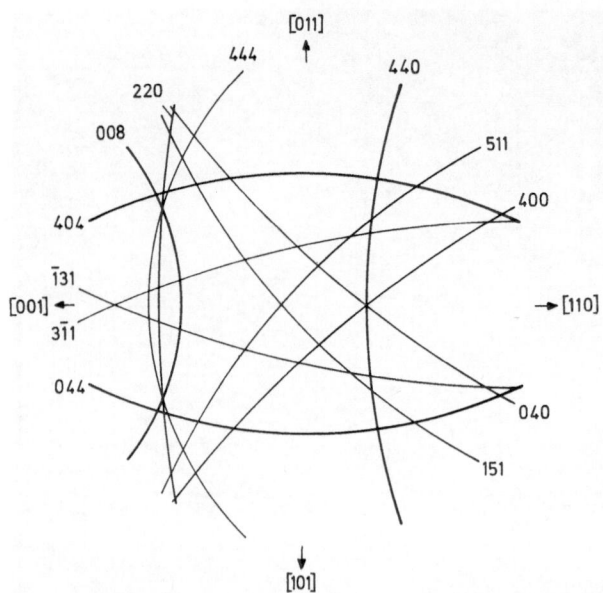

Abb. 5. Detail aus dem Stereogramm der Kossel-Linien von Fe_3O_4 ($a = 8,394$ Å) mit Fe-K_α-Strahlung ($\lambda = 1,937$ Å) symmetrisch zum [1$\bar{1}$2]-Pol

4. Diskussion

In der folgenden Tabelle sind die untersuchten Substanzen mit Temperaturbereich und relativer Genauigkeit für die Gitterkonstantenbestimmung zusammengefaßt.

Tabelle

Substanz	Temperaturbereich/°C	Relative Genauigkeit der a-Bestimmung (Einzelmessung)
Cu	-188 bis $+\ \ \ 80$	$5 \cdot 10^{-5}$
Cu	$+\ \ 20$ bis $+\ \ 800$	$5 \cdot 10^{-5} - 1 \cdot 10^{-4}$
	$+800$ bis $+1050$	$4 \cdot 10^{-4}$
Fe	$+\ \ 20$ bis $+\ \ 850$	$1 \cdot 10^{-4}$
Fe_3O_4	$+\ \ 20$ bis $+\ \ 850$	$3 \cdot 10^{-4}$

Wie daraus geschlußfolgert werden kann, sollte sich die Kossel-Technik für die Lösung der eingangs genannten Probleme erfolgreich einsetzen lassen.

Gegenüber der in der Literatur beschriebenen Verfahrensweise bei der Auswahl der günstigsten Reflexe zur Gitterkonstantenbestimmung gelten hier andere Kriterien:

1. Bei der wahren Kossel-Technik ist man auf die Strahlung der in der Probe enthaltenen Elemente angewiesen. Eine günstige Anpassung der Wellenlänge an die Gitterkonstante ist daher nicht möglich.

2. Infolge des großen zu untersuchenden Temperaturbereiches ändert sich die Topologie des vollständigen Reflexsystems und damit die Empfindlichkeit einzelner Schnitte.

Vorhandene Linsen werden unempfindlicher, andererseits können aber auch neue, für die Gitterkonstantenbestimmung empfindliche Linsen entstehen.

3. Durch die Intensitätsabnahme mit der Temperatur verschwinden bestimmte Reflexe im Untergrund.

4. Der in den Punkten 2 und 3 erwähnte Sachverhalt bedingt, daß sich die Genauigkeit der Gitterkonstantenbestimmung über große Temperaturbereiche ändert, wie aus der Tabelle zu entnehmen ist. Daraus folgt, daß Voruntersuchungen über die Reflexauswahl entscheiden müssen. In gewissem Maße kann dies auch schon theoretisch über die Intensität der Reflexe erfolgen; z.B. wurden die Intensitäten im Stereogramm unabhängig von der Lage der Kossel-Kegel zur Probenoberfläche durch verschiedene Strichstärken bereits berücksichtigt.

5. Bei Temperaturerhöhung entstehen sehr häufig Schnitte, die unmittelbar ohne Vermessung von Linien genaue Gitterkonstantenwerte liefern. Als Beispiel sei die Berührung der 042- und 024-Reflexe in Abb. 3 angeführt.

5. Zusammenfassung

Die Möglichkeiten zur Durchführung von Orientierungs- und Symmetriebestimmungen gehen über Präzisionsgitterkonstantenmessungen hinaus, da sie nicht an empfindlichste Reflexe geknüpft sind. Die Gitterkonstantenbestimmung niedriger Genauigkeit (besser als 10^{-3}) lassen sich für viele Substanzen unter Ausnutzung der intensitätsreichsten Reflexe, die meist bei niedrigen Bragg-Winkeln liegen, bis zum Schmelzpunkt durchführen.

Aus diesem Grunde ist die Hochtemperatur-Kossel-Technik ein wertvolles Hilfsmittel bei mikroanalytischen Untersuchungen.

Nouvelle méthode d'obtention des diagrammes de diffraction en rayonnement X divergent

J. Despujols et F. Jordi

Paris, France

Pour obtenir des diagrammes de rayons X en faisceau divergent, il est nécessaire de disposer d'une source de rayonnement X fine (quasi-ponctuelle) et monochromatique. Si la source est intérieure au cristal à examiner, les diagrammes résultants sont habituellement appelés: «diagrammes de Kossel»; si la source est extérieure au cristal (quoique très voisine de celui-ci), les diagrammes, semblables aux précédents, sont quelque fois appelés: «pseudo-Kossel».

La méthode présentée ici[1] permet d'obtenir, avec un appareillage très simple et peu coûteux, des diagrammes «pseudo-Kossel» par transmission, utilisables, par exemple, pour la détermination précise des paramètres cristallins. Nous en donnerons un exemple.

1. Principe et réalisation (fig. 1)

Un tube à rayons X puissant, du genre de ceux utilisés en spectrométrie de fluorescence X, irradie une cible; le rayonnement secondaire, utilisé ici, comprenant principalement les raies caractéristiques des éléments constituant la cible, peut être considéré comme composé uniquement d'un petit nombre de radiations monochromatiques.

La source est constituée par un diaphragme fin du type «pinhole», le faisceau étant limité par des protections métalliques en forme de boîte emprisonnant le radiateur secondaire.

Le cristal à examiner est placé contre le diaphragme. Le film photographique plan est disposé à la distance désirée (de l'ordre de quelques centimètres ou de quelques dizaines de centimètres) de l'échantillon.

L'aspect des diagrammes dépend en grande partie de l'homogénéité du faisceau divergent issu du diaphragme. Cette homogénéité est réaliséegrâce à la disposition particulière du radiateur secondaire par rapport à l'anticathode du tube à rayons X. Il est important aussi de bien choisir

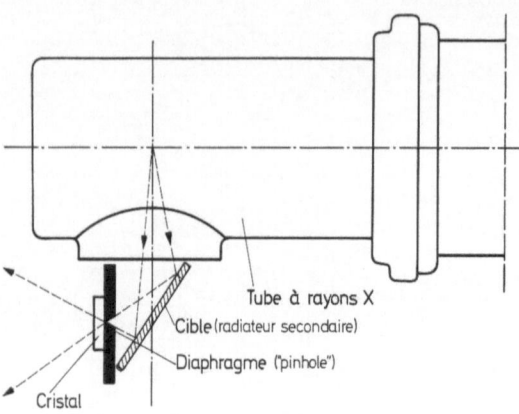

Fig. 1. Schéma général du dispositif

---

1. Cette méthode a été conçue et réalisée au Laboratoire de Chimie Physique sous la direction de Mademoiselle Y. Cauchois.

la nature du radiateur secondaire: poudre fine agglomérée ou plaque métallique exempte de texture. Les longueurs d'onde utilisées ont varié dans nos expériences de 0,5 A (rayonnement $K\alpha$ de l'argent) à 1,9 A (rayonnement $K\alpha$ du fer).

Le diamètre du diaphragme joue un rôle important. Il paraît séduisant à priori de le réduire le plus possible; mais s'il devient inférieur à quelques dizaines de microns, les temps de pose sont prohibitifs. Il faut d'autre part tenir compte de l'épaisseur et du manque de perfection des échantillons cristallins. Différents essais ont été effectués à l'aide d'une série de diaphragmes interchangeables, du type de ceux utilisés en microscopie électronique, dont les diamètres s'échelonnaient entre 30 et 300 microns. Les meilleurs résultats ont été obtenus pour des diamètres de 70, 100 et 200 microns.

2. Résultats et applications

A l'aide de ce montage, nous avons pu obtenir de beaux diagrammes pour des cristaux d'origines variées: halogénures alcalins, cristaux métalliques, etc (fig. 2). Les clichés obtenus sont assez nets pour permettre notamment la détermination précise de paramètres.

A titre d'exemple nous indiquons ici la manière dont nous avons pu mesurer le paramètre d'un cristal d'aluminium commercial recristallisé épais de 0,12 mm, à l'aide du rayonnement du cuivre; pour cela nous avons choisi, parmi les différentes méthodes de dépouillement possibles, une méthode dérivée de celle de HEISE, basée sur l'observation d'intersections de lignes de Kossel en forme de «lentilles» biconvexes. Pour des lignes correspondant à des indices de Miller élevés, les doublets $K\alpha$ sont résolus, et il est possible de mesurer les cordes L_1 et L_2 d'une «lentille» correspondant respectivement aux rayonnements $K\alpha_1$ et $K\alpha_2$ (fig. 3). Les calculs sont facilités si le film photographique est disposé perpendiculairement à une direction privilégiée, qui est par exemple la direction d'un rayon issu du diaphragme et tombant au centre de la lentille («cas de Heise»). Lec arré du rapport des cordes L_1/L_2 est alors une fonction simple du paramètre a (fig. 4).

Pour déterminer le paramètre de l'aluminium commercial étudié, nous avons utilisé la «lentille» due aux réflexions 115 et 204; il est possible aussi d'utiliser la réflexion 222 (à la place de 204) mais la précision de la mesure est moins bonne. Un cliché pris à la distance de 28 cm du cristal nous a donné: $(L_1/L_2)^2 = 12$, valeur correspondant, d'après la courbe de la fig. 4 à $a = 4,05038$ A. La température était de 25° C. La correction de l'effet dû à la réfraction dans le cristal n'a pas été faite; elle est probablement négligeable. La précision de la mesure dépend surtout de la précision sur la détermination des longueurs L_1 et L_2 et de la température. On peut l'estimer à environ 10^{-5} en valeur relative.

Le tube à rayons X utilisé était un tube Machlett AEG 50 T à anticathode de tungstène qui pourrait supporter près de 2 kW; il était alimenté par un générateur de faible puissance,

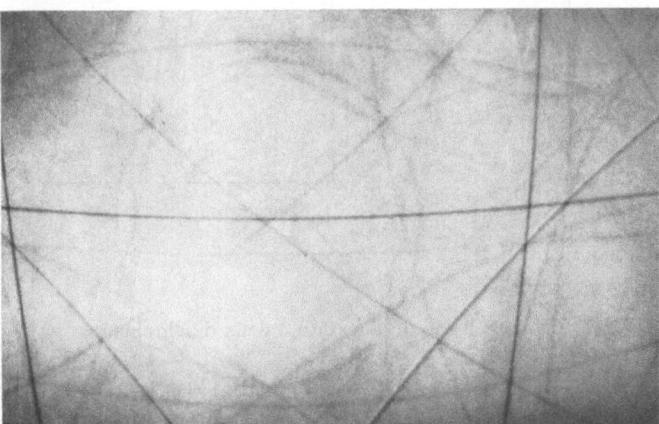

Fig. 2. Portion de diagramme obtenu pour un monocristal d'aluminium commercial épais de 0,12 mm
(rayonnement du cuivre)

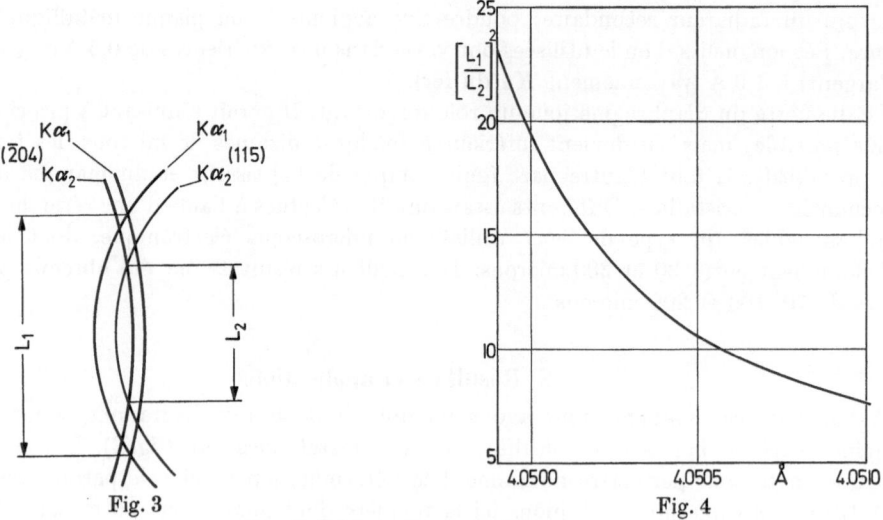

Fig. 3

Fig. 3. Schéma d'une «lentille»

Fig. 4. Variation de $(L_1/L_2)^2$ en fonction de a, pour le rayonnement $K\alpha$ du cuivre (les longueurs d'onde de Cu $K\alpha_1$ et de Cu $K\alpha_2$ sont prises dans la référence 2), dans le cas de la lentille de la fig. 2

débitant 15 mA sous 30 kV. Aussi les temps de pose étaient très longs, 60 heures dans le cas cité. L'on peut toutefois gagner un facteur 10 en adoptant un tube et surtout un générateur plus puissants.

Conclusion

Nous venons donc de montrer qu'à l'aide d'un montage très simple, il est possible d'obtenir des diagrammes suffisamment nets pour être exploitables. Certains avantages sont évidents:

il est facile de maintenir constante la température du cristal,

il est possible de faire entrer en jeu les raies caractéristiques de plusieurs éléments (le radiateur secondaire étant dans ce cas constitué par un mélange de poudres ou d'un alliage métallique).

Il est possible encore, grâce à deux diaphragmes disposés l'un à côté de l'autre (fig. 5) d'obtenir simultanément sur un même film, les diagrammes de deux cristaux ou de deux parties d'un même cristal, ceci à des fins de comparaison; les clichés sont encore parfaitement lisibles.

Les temps de pose ne sont pas prohibitifs si l'on dispose d'un générateur et d'un tube suffisamment puissants.

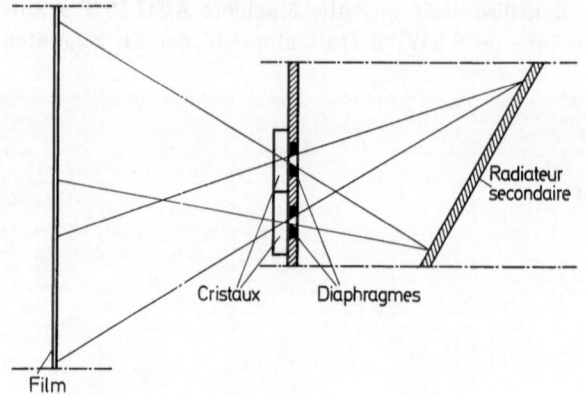

Fig. 5. Schéma du dispositif à deux diaphragmes

Références

1. Heise, B. H.: J. Appl. Phys. **33**, 938 (1962). — Gielen, P., H. Yakowitz, D. Ganow, and R. E. Ogilvie: J. Appl. Phys. **36**, 773 (1965).

2. Bearden, J. A.: X ray wavelengths. U. S. Atomic Energy Commission 1964.

Metallurgical and Mineralogical Applications

Advances in the Metallurgical Application
of Electron Probe Microanalysis

K. F. J. HEINRICH

Spectrochemical Analysis Section, National Bureau of Standards, Washington, D.C., U.S.A.

Electron probe microanalysis is widely used in the investigation of phenomena of general metallurgical interest as well as in the study of specific materials and processes. New devices for X-ray detection and automatic data collection, increase the usefulness of the microprobe for such applications. Besides the well-known methods for qualitative and quantitative electron probe analysis, the instrument can also be used advantageously for quantitative metallography and related techniques.

Introduction

Metals and alloys are good conductors of heat and electrical current. They suffer little or no structural damage under the impact of electron beams of low intensity. Compositional changes of the irradiated area, which frequently occur in other types of targets, are not observed in metallic targets. Alloys are thus eminently suitable specimens for electron probe microanalysis.

The technological importance of the microstructure of alloys is well established, as shown by the extensive work in classical metallography, transmission electron microscopy, and chemical microanalysis. The scanning electron beam has proven to be a powerful tool for the investigation of microstructure. The topographic resolution of the scanning electron microscope is superior to that of the light microscope. Furthermore, the analysis by means of X-ray spectrometry is more specific and quantitative in nature than the metallographic methods. For these reasons, electron probe microanalysis and, more recently, scanning electron microscopy, have been used widely and with great success in metallurgical research.

Several review articles provide insight into the impact of microprobe analysis on metallurgy [1—4]. Many recent references may be found in the reviews on X-ray emission and absorption by W. CAMPBELL and J. D. BROWN [5—7]. This literature does not fully reflect the growing importance of electron probe microanalysis applied to metallurgical problems. It indicates, however, certain significant trends.

The microprobe has been used successfully in the study of some general aspects of the physics of metals. The static aspects include the study of phase diagrams of systems in equilibrium (particularly composition ranges of intermediate phases and limits of solubility). The dynamics of diffusion, segregation, and reaction in the solid state are also the object of research (Fig. 1).

Another area of application deals with more specific problems concerning materials and processes. Typical examples are the identification of precipitates and inclusions in steels and other alloys [8]; the study of the structure of alloys as a function of treatment; studies of oxidation and corrosion, and of compatibility of working materials at high temperatures; the analysis of failures and defects, of coatings and weldings, etc.

The number of publications concerning basic aspects of metal physics is growing steadily, but not spectacularly. There is a tendency towards a more critical evaluation of the conditions and limitations of the measurement, and of the underlying theories. For instance, the limits of validity of the use of diffusion couples for establishing the equilibrium conditions are discussed in several papers. Possible pitfalls of this method were discussed by ADDA, BEYELER, KIRIANENKO,

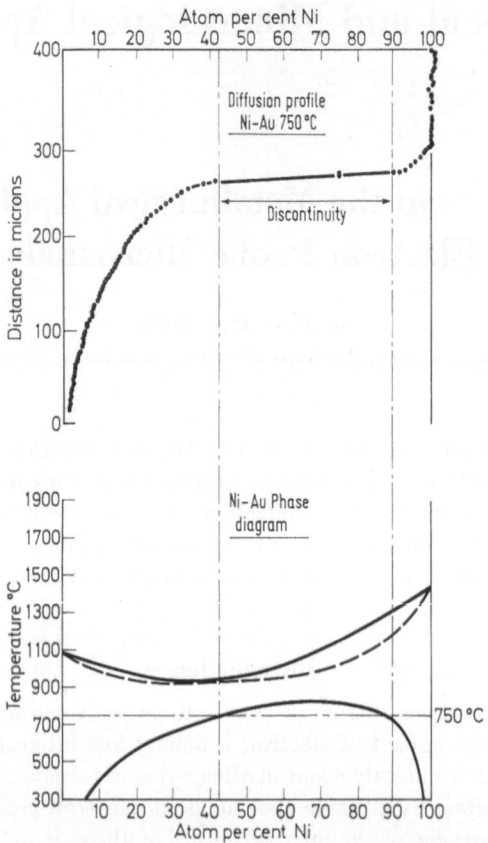

Fig. 1. Relation between the concentration profile of a binary diffusion couple and the corresponding phase
equilibrium diagram. Courtesy of E. Lifshin, General Electric Co., Schenectady, N.Y.

and MAURICE in 1961 [8]. Further studies are due to KIRKALDY [9], BROWN and POWELL [10],
and BOROVSKII and MARCHUKOVA [11]. The recent study of EIFERT et al. [12] suggests that
the precise location of phase boundaries in the equilibrium diagram should be obtained from
equilibrated specimens rather than from diffusion couples.

In these basic studies, more, and perhaps better work of the same nature as before is being
done. However, we have not observed significantly new areas of application, nor have the
instrumental techniques changed to any great extent. In particular, I have not seen any quantitative work with soft X-rays, or analysis on trace-levels of concentration.

Changes are more noticeable in the area of specific applications. The range of materials
being investigated is ever-widening, with particular emphasis on more exotic specimens, such
as semiconductor materials, microcircuitry, and devices used in reactor techniques. The number
of publications dedicated to the study of such specific problems has increased much faster than
that of communications concerned with more fundamental work. However, not all of the contributions in this category are easily spotted in literature reviews, since much of the material
is not published, or appears in company reports which are difficult to obtain. Furthermore, the
microprobe is frequently mentioned in passing in publications which center on the specific
problems to which microprobe analysis has been applied.

Techniques

Progress in the application of microprobe analysis is closely related to advances in the technique of obtaining information. At present two techniques are widely used: qualitative exploration by means of area scans — occasionally supported by semiquantitative line scans — and
quantitative analysis of individual points on the specimen by the photon counting procedure.

The area scanning technique provides summary information about the topographic distribution of the elements of interest. It is particularly useful in the investigation of materials and processes. This microscopic technique is usually combined with topographic image registration using specimen current, backscattered electrons, and secondary electrons. The possibility of investigating the distribution of elements of low atomic number, such as carbon, oxygen, nitrogen, and fluorine, has greatly extended the usefulness of the scanning technique. In most instances the simple technique of producing a clear dot on the image for each detected photon produces the desired information. However, more effective procedures using ratemeter signals have also been developed [13]. COSSLETT and DUNCUMB [14] showed in 1956 that the correlation of distributions of several elements can be illustrated by means of color composites in which each primary color represents one element. This method has been employed by several authors. YAKOWITZ and HEINRICH [15] developed a color copy technique by means of which such composite images can be prepared quickly and easily.

Scanning electron microscopy, besides its role as an adjunct to X-ray microanalysis, is quickly becoming an important metallographic tool on its own merits. Three-lens instruments capable of better resolution than the conventional electron probe are now employed in many

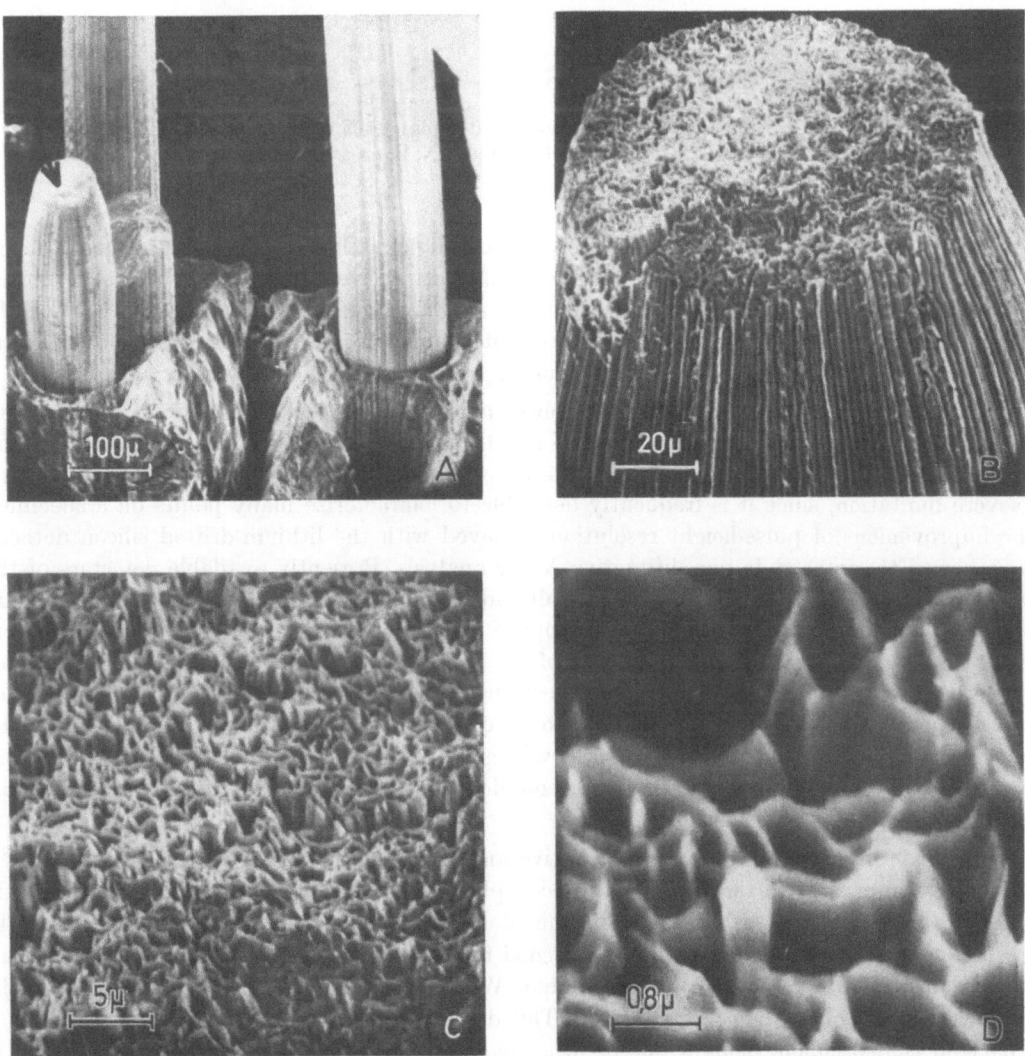

Fig. 2. Scanning electron micrographs of tungsten fibers imbedded in a silver matrix at different magnifications. Courtesy of O. JOHARI, ITT Research Institute, Chicago

Fig. 3. Iron platelets grown by halide decomposition. Center platelet is oxidized. Target current micrograph taken by W. G. MORRIS, General Electric Co., Schenectady, N.Y. on a modified Cambridge electron probe

laboratories. The excellent resolution in depth, and the possibility of observing surfaces without the need of preparing replicas render this technique attractive whenever the lateral resolution of the scanning microscope (approximately 100 Å) is sufficient (Fig. 2). The success of this technique has stimulated the efforts to improve its application in conventional microprobes by perfecting the scanning systems and introducing secondary electron detectors (Fig. 3).

The complete qualitative characterization of a microscopic region is of great practical importance. With present-day spectrometers of the diffractive type, all elements of atomic number five or higher can be detected. However, the time necessary for a complete wavelength scan is a severe limitation, since it is frequently desirable to characterize many points on a specimen. The improvement of pulse-height resolution achieved with the lithium-drifted silicon detector has renewed the interest in non-diffractive X-ray analysis. Presently available detectors of this type have an energy resolution of 500 eV full width at half maximum or better. This permits in most cases the detection of all components of a specimen present in concentrations of one percent or higher. Radiation at an energy of 1 keV can be separated from the detector noise. Hence, the silicon detector permits the detection of elements of atomic number 13 or higher (Fig. 4). In our laboratory we use this detector in combination with a flow-proportional detector in order to obtain a qualitative analysis covering all elements of atomic number higher than five, in two to three minutes. This detector combination was used frequently in the exploratory analysis of metallographic specimens.

Often it is necessary to obtain quantitative analysis of many points of a specimen. This is particularly important in the study of diffusion phenomena and of homogeneity. Although the duration of the analysis is basically determined by the time needed to collect a statistically satisfactory number of photons, the time needed for the analysis can be shortened if the beam translation and data collection are automated. We have developed for this purpose a device which we call the matrix generator (Fig. 5). This device displaces in steps the point of the beam impact on the specimen, along a line or in a raster configuration. The distance as well as the number of steps in both matrix dimensions can be varied over a wide range. At present, a maximum raster of 100×100 steps can be provided. At each point, a series of scalers accumulates

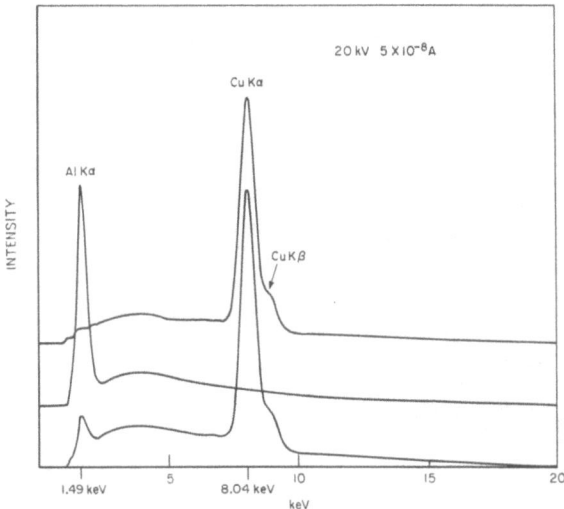

Fig. 4. Pulse-height spectrum of copper, aluminum, and a copper-aluminum alloy, obtained with a lithium-drifted silicon detector

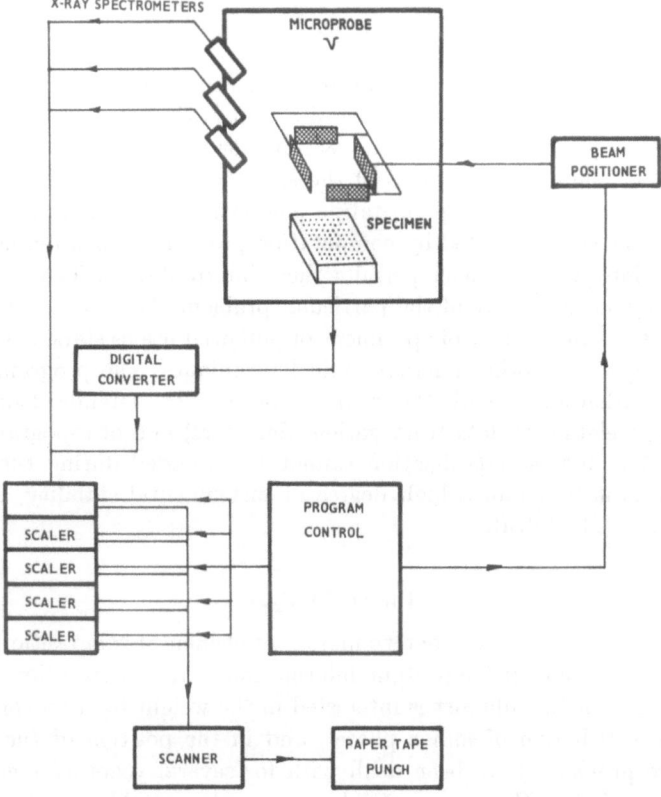

Fig. 5. Schematic of the matrix generator

X-ray signals, as well as digitalized specimen current and target current information. All information of interest is punched on a paper tape, and then the device advances the beam to the next position. The pre-set raster can be repeated several times if desired. At present, the displacement of the electron beam is achieved by electrostatic deflection. The useful range of deflection is limited by the defocusing of the spectrometers. This problem can be eliminated by

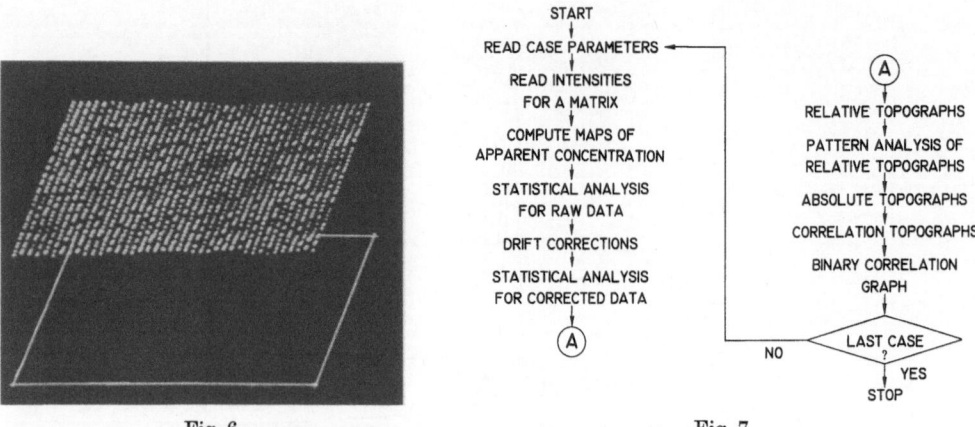

Fig. 6   Fig. 7

Fig. 6. Three-dimensional representation of X-ray counts from a slightly inhomogeneous alloy. Size of matrix: 40×40 points. The axes in the horizontal plane represent spatial dimensions on the specimen. Variations in the count-ratio are plotted in the vertical direction, and as variation of brighteners

Fig. 7. Flow sheet of a computer program for the characterization of the degree of homogeneity of alloys

using the non-diffracting detectors mentioned previously. Alternatively, one could provide mechanical movement of the specimen in one direction [16].

This matrix technique has been used successfully for over a year in the analysis of alloys. It provides a "total analysis" of an area of the specimen, combining some of the features of both qualitative area scanning and quantitative point analysis (Fig. 6). To be of practical use, the procedure must be complemented by computation programs which reduce the large number of data generated into a form which permits their interpretation. The form of computation may vary depending on the nature of the particular problem. At NBS we are mainly concerned with investigating the homogeneity of specimens of potential use as standard reference materials for microprobe analysis and other microanalytical techniques. The programs which we use at present give information about both the statistics of the data (standard deviation, frequency distribution, and correlation of data from various detectors) and of topographic distribution of signal levels (Fig. 7). Since standardization cannot be provided during the data taking on a matrix, it is important to attain a high degree of instrumental stability, and to compensate by computation for residual drifts.

Phase Analysis

Until now, we have discussed the determination of elemental composition of points or zones within the specimen. An equally important information is the distribution of phases within a multiphase specimen. The metallurgist is interested in the weight-fraction corresponding to each phase, in the size distribution of minor phases, and in the position of the phases relative to one another. These problems have been dealt with for several years by means of quantitative optical microscopy [17, 18]. However, optical micrographs suitable for this technique are difficult to obtain. The electron probe, which can identify phases by recognition of their elemental constituents, is an ideal tool for this purpose.

Applications of the electron probe to this problem include the investigation of inclusions in steel by Melford and Whittington [18]. The subject has been studied in depth by Dörfler [19, 20], who has thoroughly discussed the theory, constructed the read-out equipment for this technique, and demonstrated the application of the microprobe to phase analysis in numerous applications (see p. 274).

With the aid of appropriate computer programs, phase analysis can be performed using the matrix generator. Since in most cases the requirements for quantitative accuracy of the X-ray measurements can be lowered, matrices containing many points can be generated in a reasonable time interval. The possibility of replacing the special electronic components presently necessary by a simple change in programming further demonstrates the flexibility of the matrix technique.

Special Instruments

Although instrumentation is the subject of another review in this book, we should not omit the mention of the application of some instruments related to the microprobe to metallurgical problems. Instruments similar to Duncumb's Electron Microscope Microanalyzer [21] have been built in several laboratories, commercial instruments were adapted to include transmission electron microscopy, and combined instruments of the EMMA type are now commercially available. Such instruments have been shown to be useful in the analysis of small precipitates and other particles, of microtome sections of specimens, and of evaporated films.

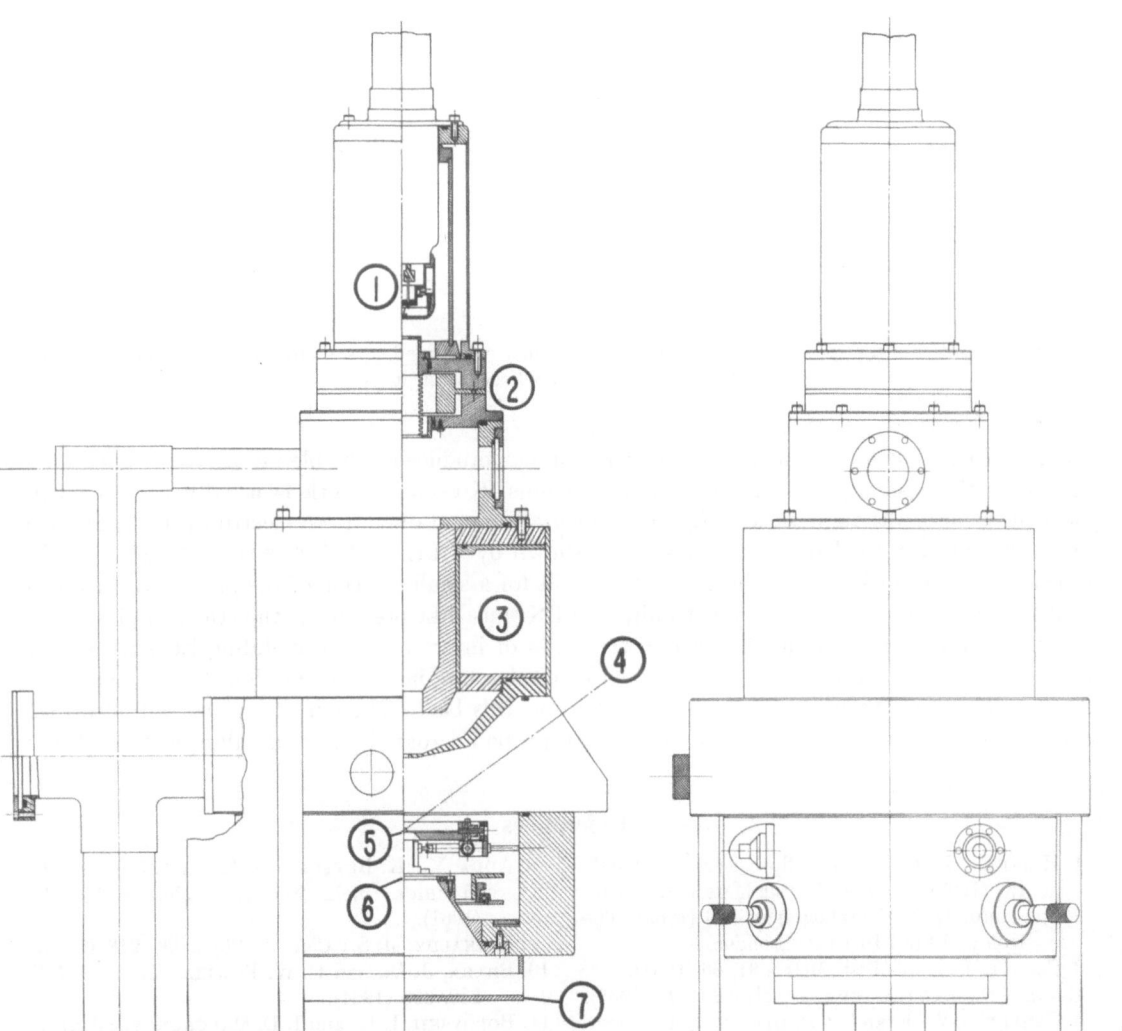

Fig. 8. Schematics of the Kossel line generator constructed by VIETH and YAKOWITZ [22]. *1* Electron gun, *2* Aperture centering device, *3* Electron lens, *4* Center line of optical microscope, *5* Specimen, *6* Mylar film separating the specimen chamber from the film chamber which is at atmospheric pressure, *7* Photographic film

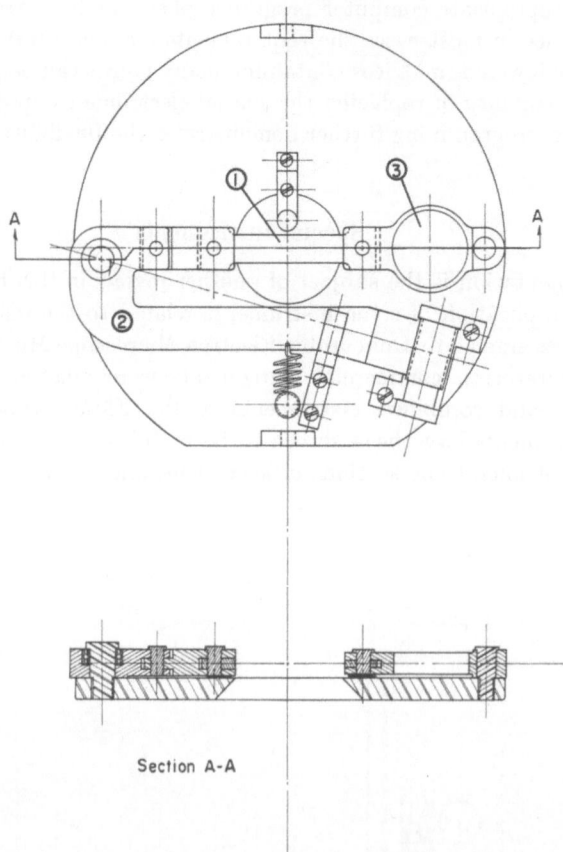

Section A-A

Fig. 9. Schematics of the specimen manipulator for the study of strained crystals by the Kossel line technique.
1 Specimen, *2* Lever, *3* Force transducer

The Kossel-line technique of X-ray diffraction has produced valuable information concerning precise lattice spacing data of microscopic domains. Kossel-line work is usually performed in electron probes, although most of the instrumentation contained in an electron probe performs no useful function for Kossel-line studies. As shown by VIETH and YAKOWITZ [22, 23], a highly efficient instrument for this technique can be built for a small fraction of the price of an electron probe microanalyzer. The instrument built at NBS is used at present for the study of elastically strained crystals (Figs. 8—9). Further instruments of interest for the metallurgist, such as ion probes with mass spectrometers and devices for the characterization of thin foils by characteristic deceleration of electrons are discussed elsewhere in this book. Therein the reader will also find excellent examples of the application of electron probe microanalysis to metallurgical problems.

References

1. HEINRICH, K. F. J.: ASTM Spec. Tech. Publ. **349**, 163 (1963), Am. Soc. Testing Materials, Philad.
2. MELFORD, D. A.: Metallography 206 (1963), The Iron and Steel Institute, London.
3. CARROL, K. C.: J. Inst. Metals **91**, 66 (1962).
4. MERZ, D., u. G. WASSERMANN: Metall **19**, 10 (1965).
5. CAMPBELL, W. J., and J. D. BROWN: Anal. Chem. **36**, 312 R (1964).
6. — —, and J. W. THATCHER: Anal. Chem. **38**, 416 R (1966).
7. — — Anal. Chem. **40**, 346 R (1968).
8. ADDA, Y., K. BEYELER, A. KIRIANENKO, and F. MAURICE: Mem. Sci. Rev. Met. 8 (9), 716 (1961).
9. KIRKALDY, J. S.: Can. J. Phys. **36**, 918 (1958).
10. BRAUN, J. D., and G. W. POWELL: Trans. AIME **230**, 698 (1964).
11. BOROVSKII, I. B., and I. D. MARCHUKOVA: J. Inst. Metals A. A. Baikov **1958**, 309.
12. EIFERT, J. R., D. A. CHATFIELD, G. W. POWELL, and J. W. SPRETNAK: Trans. AIME **242**, 66 (1968).

13. HEINRICH, K. F. J.: NBS Tech. Note **278** (1967), National Bureau of Standards, Washington, D.C.

14. COSSLETT, V. E., and P. DUNCUMB: Nature **177**, 1172 (1956).

15. YAKOWITZ, H., and K. F. J. HEINRICH: J. Res. Nat. Bur. Std. A **73**, 113 (1969).

16. PHILIBERT, J., u. E. WEINRYB: J. Microscopie **1**, 13 (1962).

17. MOORE, G. A., in: Automatic photointerpretation, p. 275. Washington, D. C.: Thompson 1967.

18. MELFORD, D. A., et K. R. WHITTINGTON, in: Optique des rayons X et microanalyse (R. CASTAING, P. DESCHAMPS, and J. PHILIBERT, eds.), p. 497. Paris: Hermann 1966.

19. DÖRFLER, G., et E. PLÖCKINGER, in: Optique des rayons X et microanalyse (R. CASTAING, P. DESCHAMPS, and J. PHILIBERT, eds.), p. 506. Paris: Hermann 1966.

20. — in: Quantitative electron probe microanalysis (K. F. J. HEINRICH, ed.). NBS Spec. Publ. **298**, 215 (1968), National Bureau of Standards, Washington, D.C.

21. DUNCUMB, P., in: The electron microprobe (T. D. MCKINLEY, K. F. J. HEINRICH, and D. B. WITTRY, eds.), p. 490. New York: Wiley 1966.

22. VIETH, D. L., and H. YAKOWITZ: J. Res. Nat. Bur. Std. C **71**, 313 (1967).

23. — — Rev. Sci. Instr. **39**, 1929 (1968).

Mineralogical Applications
of the Electron Probe Microanalyser

G. SPRINGER

Institut für Mineralogie und Lagerstättenlehre, Technische Hochschule Aachen, BRD

An analytical method with a spatial resolution of less than 1 μm, a quantitative precision of about 1% and the possibility — by means of scanning — of revealing concentration variations within an area of a few hundred microns, quite naturally arouses the interest of the mineralogist and it is therefore not surprising that the electron-probe microanalyser has become an indispensible research tool in many mineralogical laboratories.

A number of reviews have appeared that deal with mineralogical applications of the electron-probe, among them articles by LONG (1962), KEIL (1967) and CAMPBELL and BROWN (1968). Sections of books on the subject also contain information (ADLER, 1966; LONG, 1967). It is not the object of this review to recapitulate the cases in which microprobe analysis has been applied so far. It is rather intended to show what new aspects in mineralogy have appeared with the introduction of this technique and what kind of problems can now be solved that could scarcely be attacked before.

I. Identification of Minerals

The purpose for which the electron probe is probably most employed is mineral identification. The commonly used optical methods are not always conclusive, in particular with opaque minerals where fewer parameters can be measured than with transparent material. Other means such as X-ray diffraction or chemical analysis can in practice not be applied if the minerals are only a few microns in size and closely intergrown. In many of these cases the electronprobe makes the task of identification possible.

A number of minerals that have been left unidentified in the past can now be determined and many others that have been supposedly identified have turned out to be quite different. An illustration of this is provided by the mineral mackinawite. Mackinawite occurs in some ore-deposits usually in the form of minute grains and is characterised by its very strong optical anisotropy. It was long thought to be identical with valleriite which is also conspicuously anisotropic and which was believed to have the composition $Cu_2Fe_4S_7$. Later on microprobe analyses revealed that mackinawite and valleriite are entirely different substances, the former being an iron sulfide with minor contents of Ni and Co and very little Cu. The latter has turned out to consist of a complex layer structure with sheets of Cu-Fe-sulphide alternating with others consisting of Mg-Al-Fe-hydroxide.

More cases like this are described in the literature and it is thus not astonishing that more than 50 new minerals have been found in recent years with the microprobe crucially involved in their discovery. It seems worth noting in this context that in two investigations alone a total of 19 mineral phases have been identified that were unknown before (STUMPFL, 1961; JAMBOR, 1968). Appropriately enough a mineral has been named after Professor R. CASTAING in recognition of his fundamental work on the electron-probe microanalyser. This mineral, castaingite, is a copper-molybdenum-sulphide and has been described by SCHÜLLER and OTTEMANN (1963).

II. Accurate Composition of Minerals

A qualitative check is often adequate to identify a mineral, but the greatest value of an electron probe lies in the quantitative results that can be obtained. For chemical analysis a mineral has to be separated from the rock in which it is embedded and the danger of contamination is therefore very great. With the microprobe the material can be investigated at the polished surface of the rock sample and very little interference from the surrounding usually occurs. This has led to the realisation that many minerals need to be newly examined even though their composition has long been regarded as well established.

An example is provided by the mineral tetrahedrite. On the basis of previous chemical analyses the formula $(Cu,Ag,Zn,Fe,Hg,Co,Ni,Pb)_3(As,Sb,Bi,Te)S_{3-4}$ is often quoted in textbooks. The number of sulphur atoms in this molecule cannot accurately be determined from the available evidence but crystallographic studies have suggested the value 3.25. The table shows a selection from a large number of quantitative microanalyses of tetrahedrites. The investigated material came from deposits all over the world. The top row of figures in every analysis is the measured, uncorrected concentration and the bottom row the weight percentage calculated therefrom. These values are accurate to about 1—2%. The accuracy can to some extent be judged from the total which is very close to 100%. The limit of detection was taken at about 0.1% and, as can be seen, only Cu, Ag, Fe, Zn, As, Sb and S are present above that limit. Ni, Co and Pb could be detected in none of the samples and Hg, Bi and Te only in some rare cases. The formulae that have been derived from the analyses confirm that the index for S is 3.25 and not 3 or 4 as has been suggested in earlier work. A further fact not clearly recognised previously becomes obvious from the table. There is no unrestricted substitution between Cu, Ag, Fe and Zn; rather the atomic ratio $(Cu + Ag):(Zn + Fe)$ is constant and equals 5:1. The general tetrahedrite formula can therefore be written as $(Cu,Ag)_{2.50}(Fe,Zn)_{0.50}(As,Sb)S_{3.25}$. This formula applies to the vast majority of all tetrahedrites and there are few exceptions.

Similar investigations of the rock-forming minerals olivine, pyroxene and feldspar have been carried out by SMITH and coworkers (see SMITH, 1967). In several cases the electron probe produced results different from chemical analyses and the discrepancies were attributable to small inclusions of other minerals.

Solid solubility as it occurs in tetrahedrite can be observed in many minerals and for various reasons it can be important to determine the exact ratio of the end members. One such case is iron-bearing sphalerite, $(Zn,Fe)S$. A method which has been extensively employed in the past utilizes the fact that the position of an X-ray diffraction line of this material is dependent on its iron content. Fig. 1 adopted from a paper by BARTON and TOULMIN (1966) shows how this technique is less precise in comparison with electronprobe analysis although a precision of ± 0.0006 Å has been assumed for the diffraction measurements and one of merely $\pm 2.5\%$ for

Table. *Electron-probe analyses of tetrahedrites (20 kV, 75° X-ray take-off angle)*

Radiation Standard	CuK_α Cu	AgL_α Ag	ZnK_α Zn	FeK_α FeS_2	SbL_α Sb	AsK_α NiAs	SK_α FeS_2	Total
1. Meas'd %	41.5	—	8.3	—	2.7	17.2	24.2	
Weight %	41.7	—	8.3	—	3.3	18.1	27.7	99.1
2. Meas'd %	37.3	—	6.7	0.8	24.6	—	23.6	
Weight %	37.9	—	6.8	0.7	28.9	—	24.9	99.2
3. Meas'd %	37.2	1.0	5.5	2.2	22.9	1.5	23.4	
Weight %	37.9	1.2	5.5	1.9	27.3	1.5	25.2	100.5
4. Meas'd %	35.4	2.3	2.0	5.7	23.5	1.0	23.5	
Weight %	36.3	2.7	2.1	5.0	27.7	1.0	24.9	99.7

1. $Cu_{2.49}Zn_{0.48}(Sb_{0.90}As_{0.10})_{1.01}S_{3.27}$
2. $Cu_{2.51}(Zn_{0.89}Fe_{0.11})_{0.49}Sb_{1.00}S_{3.25}$
3. $(Cu_{0.98}Ag_{0.02})_{2.51}(Zn_{0.71}Fe_{0.29})_{0.49}(Sb_{0.92}As_{0.08})_{1.01}S_{3.24}$
4. $(Cu_{0.6}Ag_{0.04})_{2.49}(Zn_{0.26}Fe_{0.74})_{0.51}(Sb_{0.94}As_{0.06})_{1.01}S_{3.24}$

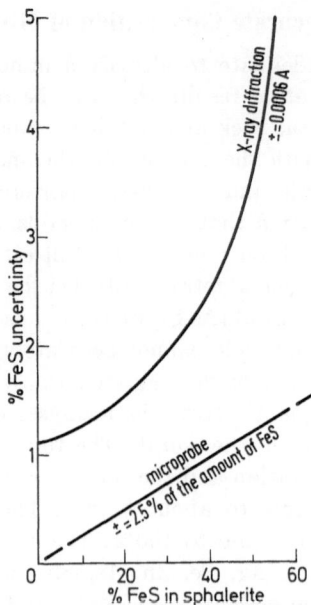

Fig. 1. Comparison of precision obtainable with X-ray diffraction and microprobe analysis for Fe determination in (Zn, Fe)S. (Courtesy of P. B. BARTON and P. TOULMIN)

the electronprobe. It should to be pointed out that the diffraction method can only be applied in the pure ZnS−FeS system unless special precautions are taken if further elements are present. For the electronprobe additional elements present no particular problem.

III. Variations in Composition

Many minerals are not homogeneous but reveal considerable variation in composition even in the same specimen. Such features have so far mostly been investigated optically with a microscope since the compositional changes frequently show up as differences in colour. The microprobe, of course, can tell not only that there are differences in composition but also which elements are concerned and to what extent.

This point is illustrated in Fig. 2 which shows reflected light and scanning images of germanite, a Cu-Fe-Ge-sulphide from Tsumeb in Southwest Africa. As can be seen, considerable colour and brightness differences appear on the photomicrograph. It was believed earlier that these differences are due to a variable ratio of the cations Cu, Ge and Fe. The scanning pictures reveal, however, that the presence of W and Mo is responsible for the zoning. These elements replace Fe in the lattice whilst Ge, Cu and S are virtually constant.

The optical appearance does not always reflect the inhomogeneity of a mineral and sometimes the X-ray signal is the only indication of concentration gradients. Fig. 3 shows scanning images of a tetrahedrite crystal. In the microscope practically no difference in reflectivity could be observed and the whole mineral exhibited the usual grey colour. As mentioned already, both Fe and Zn, and As and Sb, can substitute for each other in tetrahedrite and this process has taken place to a variable extent in different parts of the crystal shown.

Concentration differences can also be recorded quantitatively and interesting conclusions can be drawn from the results. Fig. 4 shows quantitative point analyses of the two dimorphous minerals enargite and luzonite. The former is orthorhombic, the latter tetragonal and both have the composition $Cu_3(As,Sb)S_4$. Arsenic and antimony can replace each other in the crystal lattice. To what extent this happens is shown separately for each of the structures in the figure. As can be seen not more than about 6% Sb can be dissolved by enargite whilst there is evidence for complete solid solubility between luzonite (Cu_3AsS_4) and

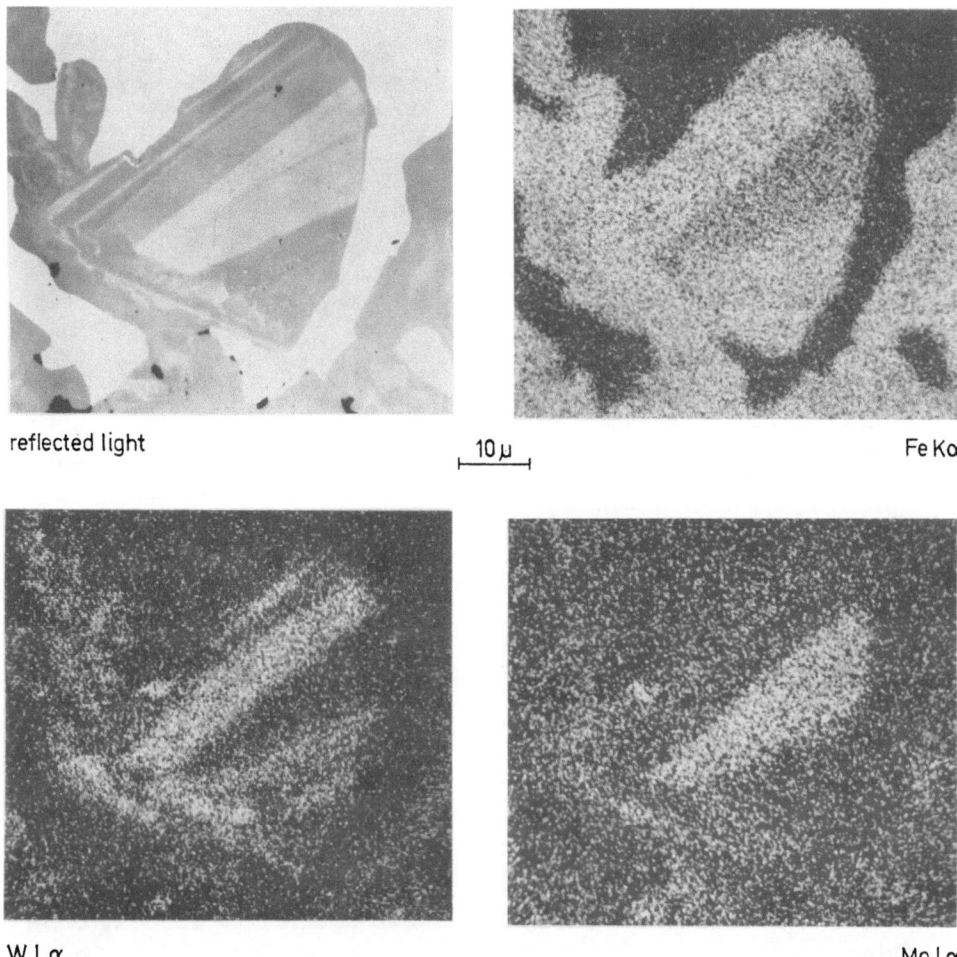

reflected light $\qquad$ 10 µ $\qquad$ Fe Kα

W Lα $\qquad$ Mo Lα

Fig. 2. Photomicrograph and scanning images of germanite from Tsumeb, S.W. Africa

its pure antimony analogue famatinite (Cu_3SbS_4). The amount of Sb present is sometimes very variable within the same sample over the range of a few millimeters thus indicating widely changing conditions of deposition.

IV. Application of Auxiliary Techniques

The applications of the electron probe mentioned so far are mainly based on its ability to provide qualitative and quantitative analyses by means of X-rays. There are, however, a few other ways in which it can profitably be used in mineralogy.

1. Mineralogical Modal Analysis. In various branches of geology the information is often required which fraction a certain mineral takes up in an ore or rock specimen. Normally the phase content is determined by point counting techniques under a microscope using optical properties for mineral identification. Obviously the chemical composition could also serve as a means to distinguish between the various minerals present and a microanalyser could then be employed instead of the microscope. A method which utilises this possibility has been suggested by KEIL (1965).

Since almost every instrument is fitted with a scanning system or a motor driven specimen carriage the counting procedure can be made automatic. This has been demonstrated success-

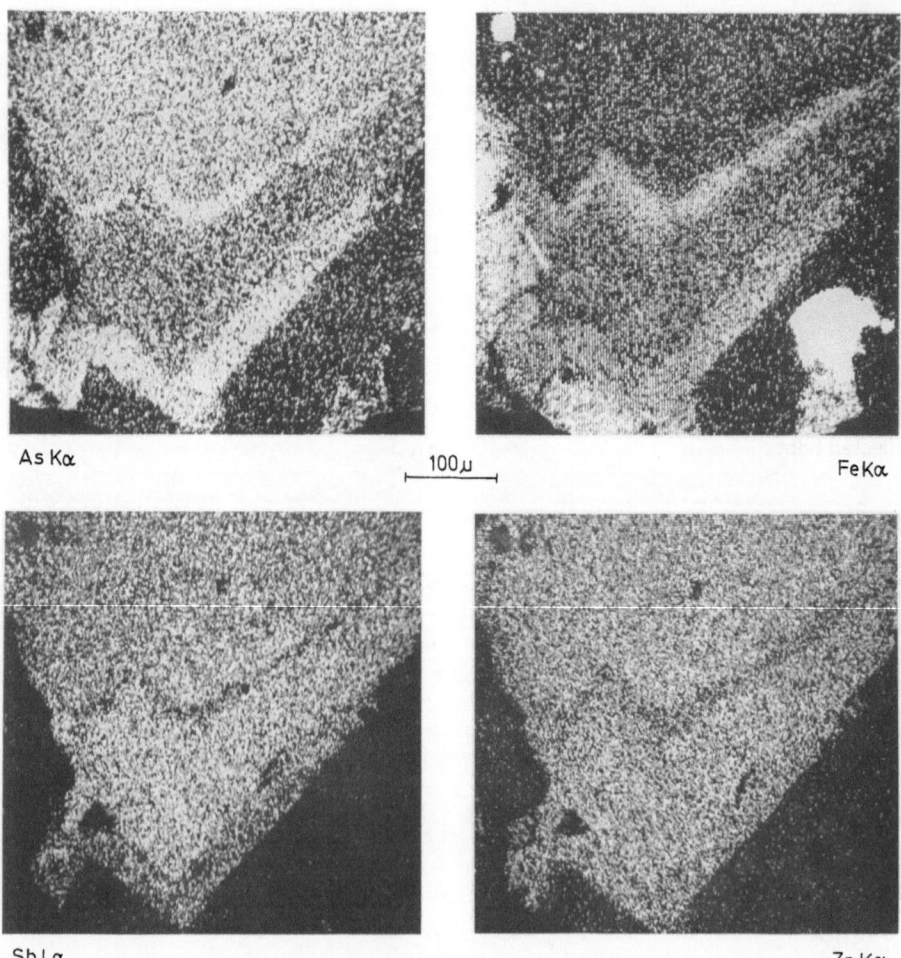

Fig. 3. Scanning pictures of zoned tetrahedrite from Guanaco, Chile

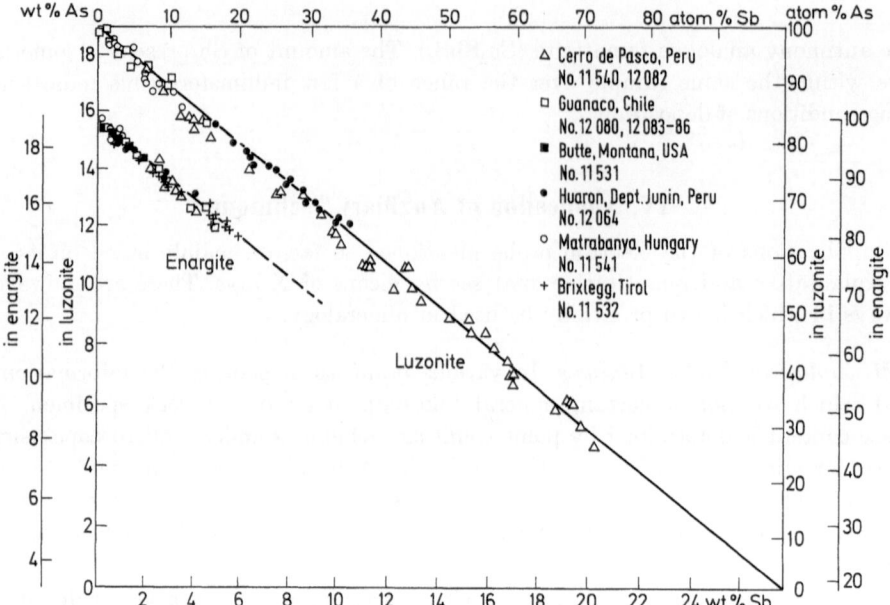

Fig. 4. Point determinations of As and Sb content of enargite (lower curve) and luzonite (upper curve)

fully in the field of metallurgy (MELFORD and WHITTINGTON, 1965; DÖRFLER and PLÖCKINGER, 1965). Similar developments for mineralogical specimens are under way (JONES and GAVRILOVIC, 1968).

2. Cathodoluminescence. The possibility of observing luminescence generated by the electron beam can be very useful in mineralogy. Cathodoluminescence is sometimes an intrinsic property of a mineral and it can then be employed for identification purposes. In other instances it is caused by small amounts of certain elements in otherwise non-luminescent material. The concentrations of these elements are mostly below the limits of detectibility for X-ray spectroscopy and several interesting applications arise from this. For instance, subsequent growth periods of a crystal can easily be recognised if the inclusion of trace elements was variable during these phases. Or fossil remains of life from earlier geological ages become visible even if the organisms had no hard parts. Only small concentration changes retained in the embedding rock can be sufficient to give distinct luminescence effects. Two articles have appeared on applications of cathodoluminescence (LONG and AGRELL, 1965; STENSTROM and SMITH, 1965) and these may be referred to for further examples.

3. Effects of Chemical Combination. Several authors have shown that soft X-ray spectroscopy can provide information on the chemical combination of an element in a compound (e.g. FISCHER and BAUN, 1965; HOLLIDAY, 1965). Such information can be obtained from the position of an X-ray line on the wavelength scale or from the relative intensities of the lines within one spectral series. The applicability of the electron probe to these studies has been demonstrated (BAUN and FISCHER, 1967; ANDERSEN, 1967). Mineralogists are often particularly interested in determining the valence state. However, it is not always easy to obtain this information since the X-ray spectrum is not only influenced by valency but also by factors like coordination number, bond character and the nature of the ligands. Furthermore there are experimental difficulties; the accuracy of the spectrometer is often not sufficient to measure precisely small differences in wavelength and absorption effects are not easy to assess when intensities of different X-ray lines are compared. Despite this, detailed studies of the low energy part of the X-ray spectrum have been made to solve certain mineralogical problems. Thus LOVERING and WIDDOWSON (1968) could demonstrate from the wavelength shift of the SK_α line whether sulphur in certain minerals of the scapolite, sodalite and mica groups occurs in the form of sulfate or sulfide. WHITE and GIBBS (1967) made a general study of the structural and chemical effects on the SiK_β line for a number of silicates. SMITH and ALBEE (1967) investigated the minerals garnet and piemontite. They were able to show that the presence of di- or tervalent iron or manganese is reflected by the intensity ratio of the L_α and L_β lines and using samples of known composition they could decide in which valence state these metals were in unknown material.

4. Crystallographic Information. The potentialities of the electron probe for the production of X-ray microdiffraction patterns have been realised very early (CASTAING, Thesis, Univ. Paris, 1951). The Kossel technique has been applied in metallurgy to determine lattice cell dimensions very precisely (HANNEMAN et al., 1962) and to study the plastic behaviour of aluminium (UMENO et al., 1965), but no specific use has been reported in the field of mineralogy. The restriction to a few specialised applications is probably due to difficulties in recording the patterns.

It has recently been shown that under certain circumstances the usual electron images of an electronprobe microanalyser can provide crystallographic information. The method has been adopted from scanning electron microscopy (COATES, 1967) and is based on the fact that in some directions a crystal is more transparent to electrons than in others (channelling). If a parallel electron beam is scanned over a crystal face the regions of higher or lower transmission can be observed on the cathode ray tube images formed by the backscattered or absorbed electrons. These orientation dependent channelling patterns can be obtained with certain types of microanalysers (HOLT et al., 1968) and have demonstrated for various minerals (JONES and GAVRILOVIC, 1968). Fig. 5 taken from the work of the latter authors shows a pattern of an artificially grown CdS single crystal. The information which this technique provides can be of use in studies of crystal orientation and defects.

Fig. 5. Channelling-pattern of (0001)-face of cadmium sulphide. (Courtesy of M. P. JONES and J. GAVRILOVIC)

Acknowledgments. This review was written while the author was staying in the Department of Mineral Technology, Imperial College, London during the tenure of a Royal Society fellowship. The financial support of the Royal Society and the hospitality of the department under Prof. M. G. FLEMING are gratefully acknowledged. Thanks are also due to Dr. J. V. P. LONG, University of Cambridge, for critically reading the manuscript and many helpful suggestions.

References

ADLER, I.: X-ray emission spectrography in geology. Amsterdam: Elsevier 1966.

ANDERSEN, C. A.: The quality of X-ray microanalysis in the ultra-soft X-ray region. Brit. J. Appl. Phys. **18**, 1033—1043 (1967).

BARTON, P. B., and P. TOULMIN: Phase relations involving sphalerite in the Fe-Zn-S system. Econ. Geol. **61**, 815—849 (1966).

BAUN, W. L., and D. W. FISCHER: Effect of alloying on aluminium K and copper L X-ray emission spectra in the aluminium-copper system. J. Appl. Phys. **38**, 2092—2096 (1967).

CAMPBELL, W. J., and J. D. BROWN: X-ray absorption and emission. Anal. Chemistry (Ann. Rev.) **40**, R 346—375 (1968).

COATES, D. G.: Kikuchi-like reflection patterns obtained with the scanning electron microscope. Phil. Mag. **16**, 1179—1184 (1967).

DÖRFLER, G., and E. PLÖCKINGER: A new apparatus for the determination of phase contents of metallic and non-metallic samples by electron-microprobe-analysis. Optique des rayons X et microanalyse (R. CASTAING et al., eds.). Paris: Hermann 1966.

FISCHER, D. W., and W. L. BAUN: The effect of chemical combination on some soft X-ray K and L emission spectra. Advances in X-ray analysis, vol. 9 (G. R. MALLET et al., eds.). New York: Plenum Press 1966.

HANNEMAN, R. E., R. E. OGILVIE, and A. MODRZE-JEWSKI: Kossel line studies of irradiated nickel crystals. J. Appl. Phys. **33**, 1429—1435 (1962).

HOLLIDAY, J. E.: Determination of electron distribution and bonding from soft X-ray emission spectroscopy. Advances in X-ray analysis, vol. 9 (G. R. MALLET et al., eds.). New York: Plenum Press 1966.

HOLT, D. B., J. GAVRILOVIC, and M. P. JONES: Scanning electron beam anomalous transmission patterns. J. Material Sci. **3**, 553—558 (1968).

JAMBOR, J. L.: New lead sulf-antimonides from Madoc, Ontario. Can. Mineralogist **9**, 7—24 (1967).

JONES, M. P., and J. GAVRILOVIC: The application of scanning electron beam anomalous transmission patterns in mineralogy. Mineral. Mag. **37**, 270—274 (1969).

— — An automatic searching unit for the quantitative location of rare phases by electronprobe X-ray microanalysis. Trans. Inst. Mining Met. London **77**, B 137—143 (1968).

KEIL, K.: Mineralogical modal analysis with the electron microprobe X-ray analyzer. Am. Mineralogist **50**, 2089—2092 (1965).

— The electron microprobe X-ray analyzer and its application in mineralogy. Fortschr. Mineral. **44**, 4—66 (1967).

LONG, J. V. P.: The application of the electronprobe microanalyser to metallurgy and mineralogy. X-ray Optics and X-ray Microanalysis (H. H. PATTEE et al., eds.). New York: Academic Press 1963.

—, and S. O. AGRELL: Cathode-luminescence of minerals in thin section. Mineral. Mag. **34**, 318—326 (1965).

LONG, J. V. P.: Electronprobe microanalysis. Physical methods in determinative mineralogy (J. ZUSSMAN, Ed.). London: Academic Press 1967.

LOVERING, J. R., and J. R. WIDDOWSON: Electron microprobe determination of sulfur coordination in minerals. Lithos 1, 264—267 (1968).

MELFORD, D. A., and K. R. WHITTINGTON: Application of the scanning microanalyser to inclusion counting and identification. Optique des rayons X et microanalyse (R. CASTAING, et al., eds.). Paris: Hermann 1966.

SCHÜLLER, A., u. J. OTTEMANN: Castaingit, ein neues mit Hilfe der Elektronenmikrosonde bestimmtes Mineral aus dem Mansfelder Rücken. Neues Jahrb. Mineral., Abhandl. 100, 317—321 (1963).

SMITH, D., and A. L. ALBEE: Petrology of a piemontite-bearing gneiss, San Gorgonio Pass, California. Contr. Mineral. and Petrol. 16, 189—203 (1967).

SMITH, J. V.: X-ray emission microanalysis of common rock-forming minerals. VI. Clinopyroxenes near the diopside-hedenbergite join. J. Geol. 74, 463—477 (1966).

STENSTROM, R. C., and J. V. SMITH: Electron-excited luminescence as a petrologic tool. J. Geol. 73, 627—635 (1965).

STUMPFL, E. F.: Some new platinoid-rich minerals identified with the electron microanalyser. Mineral. Mag. 32, 833—847 (1961).

UMENO, M., H. KAWABE, and G. SHINODA: Electronprobe microanalysis Kossel techniques applied to the plastic deformation of aluminium macrocrystals. Optique des rayons X et microanalyse (R. CASTAING et al., eds.). Paris: Hermann 1966.

WHITE, E. W., and G. V. GIBBS: Structural and chemical effects on the $SiK\beta$ X-ray line for silicates. Am. Mineralogist 52, 985—993 (1967).

Standards and Correction Procedures in Electron-Probe Analysis of Rock-Forming Minerals

T. R. SWEATMAN* and J. V. P. LONG

Department of Mineralogy and Petrology, University of Cambridge, England

Introduction

A large proportion of the substances which fall under the heading of rock-forming minerals consist of silicates and oxides incorporating elements from $Z = 11$ (Na) to $Z = 28$ (Ni) in addition to oxygen and hydrogen. The range of possible compositions is very large since such minerals often contain seven or eight component elements and the problem of selecting suitable standards for empirical calibration is considerable. Many workers in the field have advocated the use of analysed material covering a restricted range of composition near to that of the unknown (e.g. SMITH, 1965, 1966 a, b). More recently BENCE and ALBEE (1968) have described the use of a few simple primary standards and have derived empirically the correction factors by comparison with analysed minerals under a given set of measuring conditions. These authors state that "The uncertainties which shroud the calculation of individual matrix effects are so great that an empirical determination of correction parameters is clearly necessary".

The practice in this laboratory for some time has been to use a single oxide, silicate or metal standard for each element and this paper attempts to show that the use of these materials together with the currently available correction procedures gives an accuracy in the final analysis which compares well with that from other methods without resort to empirical calibration curves or factors.

Choice of Standards

For a material to qualify as a standard, it should, in our view, fulfil a specification including the following properties:

1. It should be readily available.

2. It should be homogeneous both on a sub-micron scale and over a sample of a few grammes.

3. The composition should be accurately known.

4. It should be stable under vacuum and electron bombardment and also in air.

5. Any differences which exist between the bonding in the standard and unknowns should not be such as to give rise to significant errors in measurements of intensity ratio.

Reference to the analyses of the rocks G_1 and W_1, (FLEISCHER, 1965; FLEISCHER and STEVENS, 1962) which illustrate the errors which may exist in chemical analysis of silicates, and to the microprobe analyses of olivines by SMITH and STENSTROM (1965) emphasise the difficulty of obtaining a wide range of minerals which fulfil even the first three conditions. Artificial glasses have been used but may exhibit inhomogeneity and, when alkalis are present, are frequently unstable under electron bombardment.

Effect of Wavelength Shift. Differences in bonding of the analysed element in standard and unknown are important for the lighter elements since they give rise to the well known "wavelength shifts" of the characteristic emission lines. The magnitude of the error in intensity ratio

---

* *Present address:* C.S.I.R.O., Division of soins, Glen Osmond, South Australia.

Table 1. *Wavelength shifts*a *relative to aluminium metal, expressed as minutes of arc (2θ), for representative minerals. Analyzing crystal — potassium acid phthalate (KAP)*

Mineral	Wavelength shift	Mineral	Wavelength shift
Anorthite	0.21	Corundum	0.59
Cordierite	0.32	Garnet	0.64
Hornblende	0.33	Epidote	0.79
Orthopyroxene	0.43		

a Measurements made on Cambridge Instrument Company Geoscan.

Table 2. *Percentage error due to wavelength shift in the measured intensity of $AlK_{\alpha 1,2}$ peak when corundum is used as a standard. Analyzing crystal — potassium acid phthalate*

Mineral	Percentage error	Mineral	Percentage error
Anorthite	0.3	Orthopyroxene	<0.1
Cordierite	0.2	Garnet	<0.1
Hornblende	0.2	Epidote	<0.1

which results is dependent in any particular case on the magnitude of the shift and on the resolution of the X-ray spectrometer. The error can be expressed in the form of perfectly general curves applicable to any instrument (SWEATMAN and LONG, 1969). The magnitude of the measured peak shift of the $AlK_{\alpha 1,2}$ line for representative minerals relative to pure aluminium is listed in Table 1. Also given (Table 2) is the error encountered when corundum is used as a standard for the determination of aluminium in these minerals with a spectrometer having a peak width at 95% maximum of 3 min for the $AlK_{\alpha 1,2}$ line. Anorthite is a case of an unknown having a large shift but the error is still below 0.5% if a KAP analysing crystal is used. In some instances where appreciable error is unavoidable it may be worth while to apply a correction factor, which for a particular instrument will be a constant provided that the spectrometer is accurately set to give the peak intensity from the standard.

Effect of Changes in Line Profile. The dependence of distribution of energy in the K band of the light elements on bonding is also a factor which may influence the analysis of these elements by means of chemically dissimilar standards. The magnitude of the possible error may be assessed by measuring the ratio of the K_α peak intensity to the integrated intensity under the whole of the K band. Measurements have been made on a wide range of minerals for the elements Mg, Al and Si; in no case has an discrepancy greater than the experimental error of 1.5% been observed except in the case of the pure metals. Here the deviation is appreciable and in the case of aluminium is 3% and for magnesium 5%.

Standards

Oxides and metals are an obvious choice for many of the elements but in the case of Na, K, and Ca, the chemical instability of these elements and their oxides forces the choice of a silicate. The results described in the latter part of this paper are based on the use of the following materials.

Sodium. Sodium chloride satisfies conditions 1—3 but fails on account of its instability under electron bombardment and possibly also on grounds of chemical bonding. The remaining possibilities are rather limited and the pyroxene jadeite $(NaAl(SiO_3)_2)$ appears to be the most satisfactory alternative. Its stability under electron bombardment is much greater than that of the feldspars, and relatively large, nearly stoichiometric specimens e.g. are obtainable. It may be noted, however, that the corresponding artificial glass is very unstable under electron bombardment.

Magnesium. Magnesium metal is unsatisfactory on several counts: it is easily oxidised, the wavelength shifts relative to Mg in silicates are large and the line profile error is considerable; moreover, it is difficult to polish. On the other hand, synthetic periclase (MgO) is readily obtainable[1] in the form of clear single crystal chips of high purity and has been adopted in all measurements.

Aluminium. Considerations very similar to those listed for magnesium apply. Natural corundum (Al_2O_3), checked for trace impurities by probe analysis has been used.

Silicon. Analyses of natural quartz (DEER, HOWIE and ZUSSMAN, 1963) show little evidence of contamination. The assumption of stoichiometry in this mineral almost certainly introduces a smaller error than the assumption of perfection in the chemical analysis of silica in a complex silicate. Again, the same considerations with regard to wavelength shift and line profile apply in the case of the pure element. We have been unable to confirm the reported instability of quartz (ADLER, 1963) under electron bombardment. The coefficient of variation obtained in ten successive measurements of the $Si K_\alpha$ intensity each lasting 10 seconds at a probe current of 0.5 µA and an accelerating voltage of 20 kV was less than 0.3% (comparable with the statistical error of the measurement). Even when the $1-2$ µm diameter beam was kept at the same point for periods of 1,000 seconds, the observed decrease of intensity of the order of 2% was almost certainly the result of the contamination of the bombarded area. On occasions when erratic behaviour of this and other materials has been observed the cause has invariably been due to an imperfect conducting film on the specimen surface.

Potassium. Choice is again restricted. Orthoclase has been used, although it is unsatisfactory on account of instability under bombardment.

Calcium. Rapid hydration prevents the use of the oxide. Wollastonite ($CaSiO_3$) is readily available, homogeneous and stable. Fluorite (CaF_2) and calcite ($CaCO_3$) both exhibit instability under electron bombardment.

Transition Elements. Among the first period transition elements, titanium and manganese are exceptional in that the metals do not provide ideal standards. Titanium metal often contains dissolved oxygen and it would appear preferable to use synthetic TiO_2. Manganese metal tarnishes rather readily when exposed to air although a satisfactory alternative is difficult to suggest. The closest approach to the ideal is probably rhodonite ($MnSiO_3$) (FREDRIKSSON and REID, 1967). In the measurements reported here the pure metal has been used for this element and for all other transition elements.

Measurement Technique. All specimens and standards have been prepared according to the procedure described by LONG (1967). Errors due to variations in conducting film thickness are discussed elsewhere (SWEATMAN and LONG, 1969) and it is believed that they are appreciably below 1% in the present series of measurements.

Two different instruments, a Cambridge Instrument Company Geoscan with a take-off angle of 75° and an instrument constructed in the Department with a take-off angle of 40° were used. In each case the voltmeter on the high voltage supply was calibrated by means of elements of known excitation potential and the error in the voltages quoted is of the order of ± 200 V. Probe current was held constant to ± 1 part in 500 during measurement (LONG and SWEATMAN, 1969).

Results

Fourteen different silicate minerals and oxides (Table 3) of well established chemical composition have been analysed at different accelerating voltages and at two different take-off angles, and when possible the bulk chemical analyses have been compared with determinations by X-ray fluorescence analysis. In all, 99 complete mineral analyses have been made. The results

---

1. W and C Spicer Ltd., St. Mary's Winchcomb Cheltenham, Gloucester, England.

Table 3. *Composition of minerals and oxides*

Mineral	Composition (wet chemical analysis) %
Synthetic TiO_2	Ti 59.95; O 40.05
Synthetic V_2O_5	V 56.02; O 43.98
Synthetic Cr_2O_3	Cr 68.43; O 31.57
Synthetic CoO	Co 78.65; O 21.35
St. John's Island Olivine	MgO 49.14; SiO_2 41.42; FeO 9.54
Marjalahti Olivine	MgO 48.08; SiO_2 40.24; MnO 0.28; FeO 11.53
Eagle Station Olivine	MgO 42.31; SiO_2 39.22; FeO 18.83
Sphene	Al_2O_3 0.38; SiO_2 30.02; CaO 27.51; TiO_2 39.08; MnO 0.05; Fe_2O_3 1.30; Nb_2O_5 0.44[a]; RE 0.37[a]; V_2O_3 0.06[a]; SrO 0.59[a]
Garnet (almandine)	MgO 5.81; Al_2O_3 22.04; SiO_2 37.85; CaO 2.22; TiO_2 0.04; MnO 1.16; FeO 31.33
Wollastonite	SiO_2 51.35; CaO 48.02; Fe_2O_3 0.43
Sillimanite	Al_2O_3 62.91; SiO_2 37.09 (stoichiometric composition; probe analysis shows 0.26 Fe_2O_3)
Kyanite	Al_2O_3 62.91; SiO_2 37.09 (stoichiometric composition)
Anorthite	Na_2O 0.49; MgO 0.06; Al_2O_3 35.31; SiO_2 43.75; CaO 19.39; TiO_2 0.03; Fe_2O_3 0.64
Bytownite	Na_2O 1.30; MgO 0.15; Al_2O_3 33.79; SiO_2 45.72; CaO 18.30; TiO_2 0.05; Fe_2O_3 0.64; K_2O 0.06

[a] Determined by X-ray fluorescence analysis.

Table 4. *Correction factors for some minerals for 40° angle of take-off and 15 kV accelerating potential*

Mineral	Element	Atomic number	Absorption	Fluorescence
Marjalahti Olivine	Mg	0.98	1.17	0.99
	Si	0.98	1.26	1.00
	Fe	1.21	0.99	1.00
Albite	Na	1.00	1.05	0.99
	Al	1.00	1.09	0.98
	Si	0.99	1.14	1.00
Hedenbergite	Si	0.95	1.14	0.99
	Ca	0.97	1.01	0.99
	Fe	1.17	1.00	1.00
Spinel	Mg	0.98	1.04	0.99
	Al	1.00	1.19	1.00
Almandine	Al	0.95	1.30	0.99
	Si	0.94	1.29	0.99
	Fe	1.15	0.99	1.00
Sphene	Si	0.95	1.10	0.99
	Ca	0.97	0.99	0.97
	Ti	1.13	1.04	1.00

presented here are based on the mean values for each mineral. The experimental data has been treated by applying an absorption correction of the type proposed by PHILIBERT (1963) and modified by DUNCUMB and SHIELDS (1964), a fluorescence correction according to REED (1965) and an atomic number correction using a variable mean ionization potential (J) (DUNCUMB and REED, 1967). Calculations have also been carried out with variants of these procedures, namely the use of the alternative value of σ in the absorption correction (HEINRICH, 1967) and the use of a fixed rather than a variable value for J.

The minerals analysed are listed in Table 3. Table 4 gives examples of the magnitude of the individual correction factors at 15 kV for a number of mineral compositions at a take-off angle

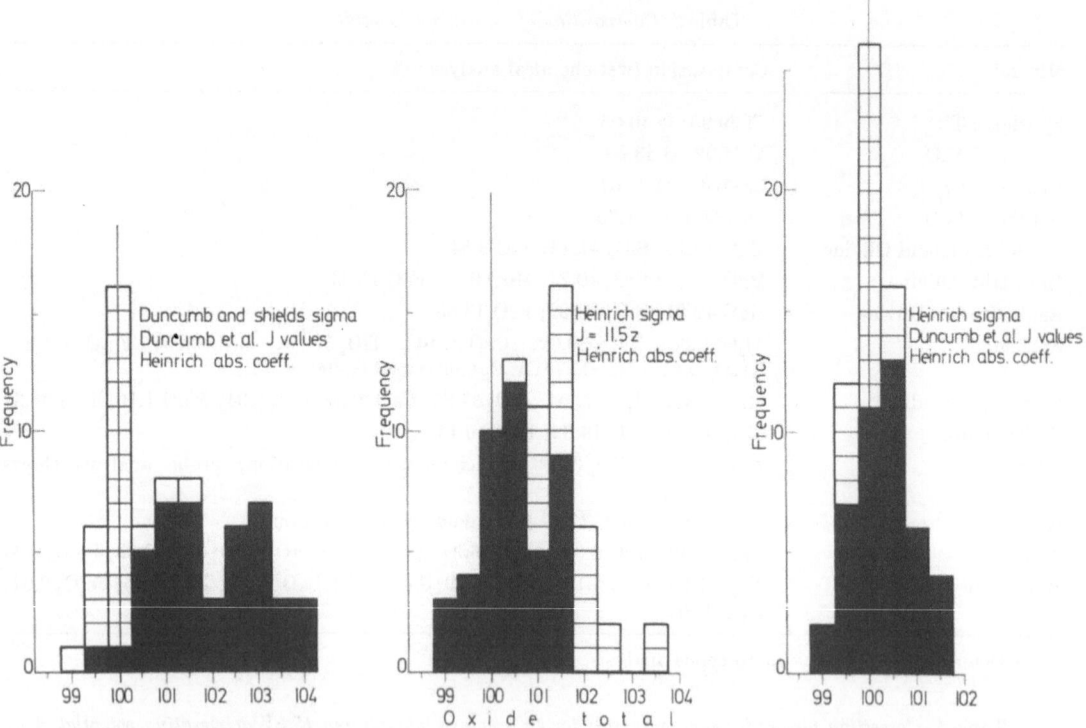

Fig. 1. Distribution of oxide totals obtained in the analysis of 14 silicates and oxides under different conditions of accelerating potential and take-off angle. Open squares are pure oxides

of 40°. The factor given is that by which the measured composition has to be multiplied in order to give the true composition. It is clear that in most cases the fluorescence factor is close to unity and that the atomic number factor dominates for the heavy elements when a metal standard is used, while for the light elements the absorption correction is the most important.

The distribution of the oxide totals about the expected value of 100% is one measure of the effectiveness of the total correction. The improvements which occurs as a result of using the HEINRICH value of σ and the fitted J values of DUNCUMB and REED is clear from Fig. 1. The analyses of the pure single crystal oxides of Ti, V, Cr and Co (the open squares in the histograms) provide a sensitive test of the effectiveness of the DUNCUMB and REED J values as opposed to the fixed $J = 11.5Z$. The heavy element analyses of the silicates do not provide such a good test for deciding between the alternative J's since the weighted mean of the J's for SiO_2 is fairly close to that given by $11.5Z$. On the other hand, these same results do provide a strong vindication of the atomic number correction in its present form since the factor is in some cases as high as 20% and yet gives agreement of chemical and probe data to within about 2%.

Similarly, the histograms showing the errors in the determination of the light elements (Fig. 2), where the correction is largely due to absorption, indicate the applicability of the voltage modified form of the Philibert equation to these elements. It is tempting to ascribe the bias in the silica and alumina results to an inverse error in the chemical analysis of these oxides (FAIRBAIRN and SCHAIRER, 1952). In view of the slight bias shown by the magnesia results however, it may be worth examining in more detail the sensitivity of these errors to changes in the mass absorption coefficients.

It must be emphasised that the distributions shown include the error in the chemical data. In a more detailed analysis of the correction procedures (SWEATMAN and LONG, 1969) which is in progress at the present time, it will be important, for critical tests of the probe data, to attempt to select particular chemical determinations where for various reasons the errors are almost certainly very small e.g. SiO_2 in wollastonite or FeO in St. John's Island olivine.

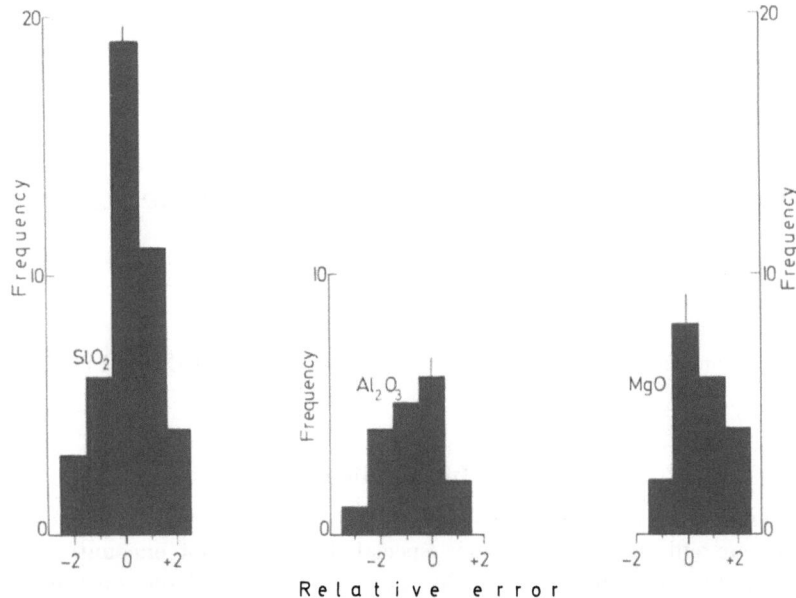

Fig. 2. Distribution of errors in analysis of silicon, aluminium and magnesium in 10 silicate minerals under different conditions of accelerating potential and take-off angle

In conclusion, we believe that the present evidence shows that the use of a single standard for each element and the application of individual matrix corrections gives results in silicate analysis which compare favourably with those obtained by empirical calibration methods.

Acknowledgements. One of us (T.R.S.) is indebted to the Division of Soils of the Australian Commonwealth Scientific and Industrial Research Organisation for the award of a Divisional Studentship.

References

BENCE, A. E., and A. L. ALBEE: J. Geol. **76**, 382 (1968).

DEER, W. A., R. A. HOWIE, and J. ZUSSMAN: Rock forming minerals, vol. 4. Longmans 1963.

DUNCUMB, P., and S. J. B. REED: Tube Investments Research Laboratories Technical Report No. 221 (1967).

FAIRBAIRN, H. W., and J. F. SCHAIRER: Am. Mineralogist **37**, 744 (1952).

FLEISCHER, M.: Geochim. Cosmochim. Acta **29**, 1263 (1965).

—, and R. E. STEVENS: Geochim. Cosmochim. Acta **26**, 525 (1962).

FREDRIKSSON, K., and A. M. REID, in: Researches in geochemistry (ed. P. H. ABELSON), p. 143. Wiley 1967.

HEINRICH, K. F. J.: In abstracts of the "Second National Conference on Electron Microprobe Analysis". Boston 1967.

LONG, J. V. P., in: Physical methods in determinative mineralogy (ed. J. ZUSSMAN), p. 215. Academic Press 1967.

—, and T. R. SWEATMAN: J. Sci. Instr. (in press) (1969).

PHILIBERT, J.: X-ray optics and X-ray microanalysis (ed. PATTEE, COSSLETT, and ENGSTRÖM), p. 379. Academic Press 1963.

REED, S. J. B.: Brit. J. Appl. Phys. **16**, 913 (1965).

SMITH, J. V.: J. Geol. **73**, 830 (1965).

— J. Geol. **74**, 1 (1966a).

— J. Geol. **74**, 463 (1966b).

—, and R. C. STENSTROM: Mineral. Mag. **34**, 436 (1965).

SWEATMAN, T. R., and J. V. P. LONG: J. Petrol. (in press) (1969).

Specimen Damage during Microprobe Analysis of Silicate Glasses

S. Scholes and F. C. F. Wilkinson

James A. Jobling and Co. Ltd., Advance Research Unit, Brancepeth Castle, Durham, England

Introduction

The use of an electron probe microanalyser for the examination of alkali — containing glasses and, to a lesser extent, minerals, presents special difficulties not encountered in conducting materials. Under the action of the beam, the glass apparently undergoes a composition change, as shewn by an approximately exponential decrease in intensity of the characteristic X-ray emission of the alkali metal present. This decrease in intensity has been reported by RIBBE and SMITH [1] in the analysis of sodium containing felspars, and by VARSHNEYA, COOPER and CABLE [2] for $K_2O-SrO-SiO_2$ glasses. These last named authors interpreted their findings in terms of a theory put forward by LINEWEAVER [3]. LINEWEAVER irradiated a variety of glasses by scanning a rectangular area of the glass surface $3'' \times \frac{3}{4}''$ with a 20 kV/150 μA beam of electrons for periods of time of up to 24 hours. He found that if the glass was subsequently heated to about 200° C in vacuo, up to 10% of the total oxygen content of the affected volume of the glass was driven off. This he equated with the non-bonding oxygen content of the glass. He proposed a stepwise movement of sodium ions deeper and deeper into the glass by the following process. Sodium ions near the surface migrate into the glass, attracted by electrons which have penetrated to greater depths. These ions are then neutralised by the emplaced electrons, only to be re-ionised by the arrival of further electrons. The ions then move a further step into the glass, again attracted by electrons at greater depths, and the cycle is repeated again and again. It was suggested that this movement of sodium ions was balanced by movement of oxygen ions in the reverse direction. The amount of outgassed oxygen was in good quantitative agreement with such a theory and the presence of a brown colour in the glass was also claimed to support the theory, the colour being thought to be due to a layer of neutral sodium atoms marking the maximum depth of penetration of the electrons.

BOROM and HANNEMAN [4] studied composition changes occurring in potassium silicate glasses, and in sodium silicate glasses containing up to 7 mole % FeO. They found that in glasses consisting only of alkali and silica the intensity of the characteristic X-ray emission of the alkali metal fell off slightly at first, then rose rapidly reaching, within about 15 seconds, a value three times that of the initial intensity. With increasing proportions of FeO in the glass this rise became less rapid, until at 7 mole % FeO the effect was similar to that reported by VARSHNEYA, COOPER and CABLE. BOROM and HANNEMAN suggested that the effect is more complex than the picture given by LINEWEAVER and that, in addition to his mechanism, vaporisation, surface diffusion, and attraction of alkali ions to the impingement area by the emplaced negative charge are all important. In those glasses containing higher proportions of FeO, BOROM and HANNEMAN obtained results which seem to suggest movement of Fe^{2+} ions as well.

In view of the possible restrictions such composition changes might place on the use of the microprobe analyser to study glasses, and also because of the increasing use of glasses as standards in the quantitative analysis of silicate minerals, it seemed desirable to study the nature of the damage created by the electron beam in glass specimens.

Experimental Details

A Cambridge Instruments "Stereoscan" scanning electron microscope has been used to examine the damage caused by a static beam of electrons in the microprobe. The glass on which this work has been carried out is an ordinary soda-lime window glass having the composition:

SiO_2	72.2 wt.%	Na_2O	13.0 wt.%
CaO	8.0 wt.%	K_2O	0.6 wt.%
MgO	3.7 wt.%	R_2O_3	1.5 wt.%

This glass was selected because it provided a readily accessible source of reasonably reproducible and homogeneous samples for study; because it had a high soda content; and because it was free from other complicating effects, such as phase-in-phase separation which was found to occur in Corning 7740 glass under the action of the beam. Samples were polished with diamond paste by normal petrographic techniques and coated with a layer of gold, calculated to be about 100 Å thick.

The electron probe microanalyser used was an A.E.I. SEM 2a, in which the electron beam strikes the surface at an angle of 30° to the normal. Accelerating voltages of from 10 to 50 kV have been used, at beam currents of from 0.04 to 4.0 μA. The size of the beam was kept as small as possible, and in some cases it has proved possible to confirm the calculated beam diameters (which range from about 0.3 to 2.0 μ) by measurement of the entry hole caused by the beam. After allowing the beam to act on the sample for periods of time varying between 1 second and 60 minutes, the gold coating was stripped with aqua regia, and the sample re-coated with a similar thickness of gold before examination in the Stereoscan. Checks were carried out to ensure that the aqua regia did not visibly modify the features of the glass surface in any way.

Results

Figs. 1—5 illustrate a typical sequence of damage caused by a 50 kV beam at successively higher beam currents. In Fig. 1 a blister some 33 μ across can be seen, surrounded by a ring of cracks. In Fig. 2 the blister is higher although its diameter has only increased very slightly, to 35 μ. In Fig. 3 the blister has become almost conical, and is beginning to open, shewing the interior. In Fig. 4, where the beam current was very much larger, a crater has developed, with smooth sides shewing evidence of melting, and possibly of vaporisation. The crater is about 100 μ across, and is surrounded by a roughly circular crack system, part of which is visible in the top left hand corner of the photograph. Also visible as white dots near the rim of the crater are numerous hollow spheres shewn in greater detail in Fig. 5. These cenospheres are of problematic origin: they may be due to "bloating" of molten or semi-molten glass by a volatile constituent; alternatively, they may have been formed by disproportionation of SiO vapour. In either case a high temperature must have been involved.

These same general features were also present at lower accelerating voltages except that, at the lowest voltages, no crater developed within the range of probe currents used.

The size of the blister or crater depends on the accelerating voltage and on the beam current, and only to a minor extent on the time for which the beam is allowed to impinge on the surface. If the blister (or crater) diameter is plotted against beam current a family of straight lines is obtained, whose slope increases with the accelerating voltage (Fig. 7). If these are extrapolated back to intersect the axis, the resulting "zero current" blister diameters are directly proportional to the accelerating voltage (Fig. 8). Similarly, if the blister diameter is plotted against accelerating voltage, a family of straight lines is obtained, passing through the origin (Fig. 9).

Examination of a number of sequences such as that shewn in part in Figs. 1—4 suggests that there is some critical value of the beam current for any given accelerating voltage which, if exceeded, will result in the formation of a crater. It has not proved possible to determine these critical values with sufficient accuracy to be able to decide whether this is simply a function of the energy of the beam. However, it does appear to be significant that whereas the craters

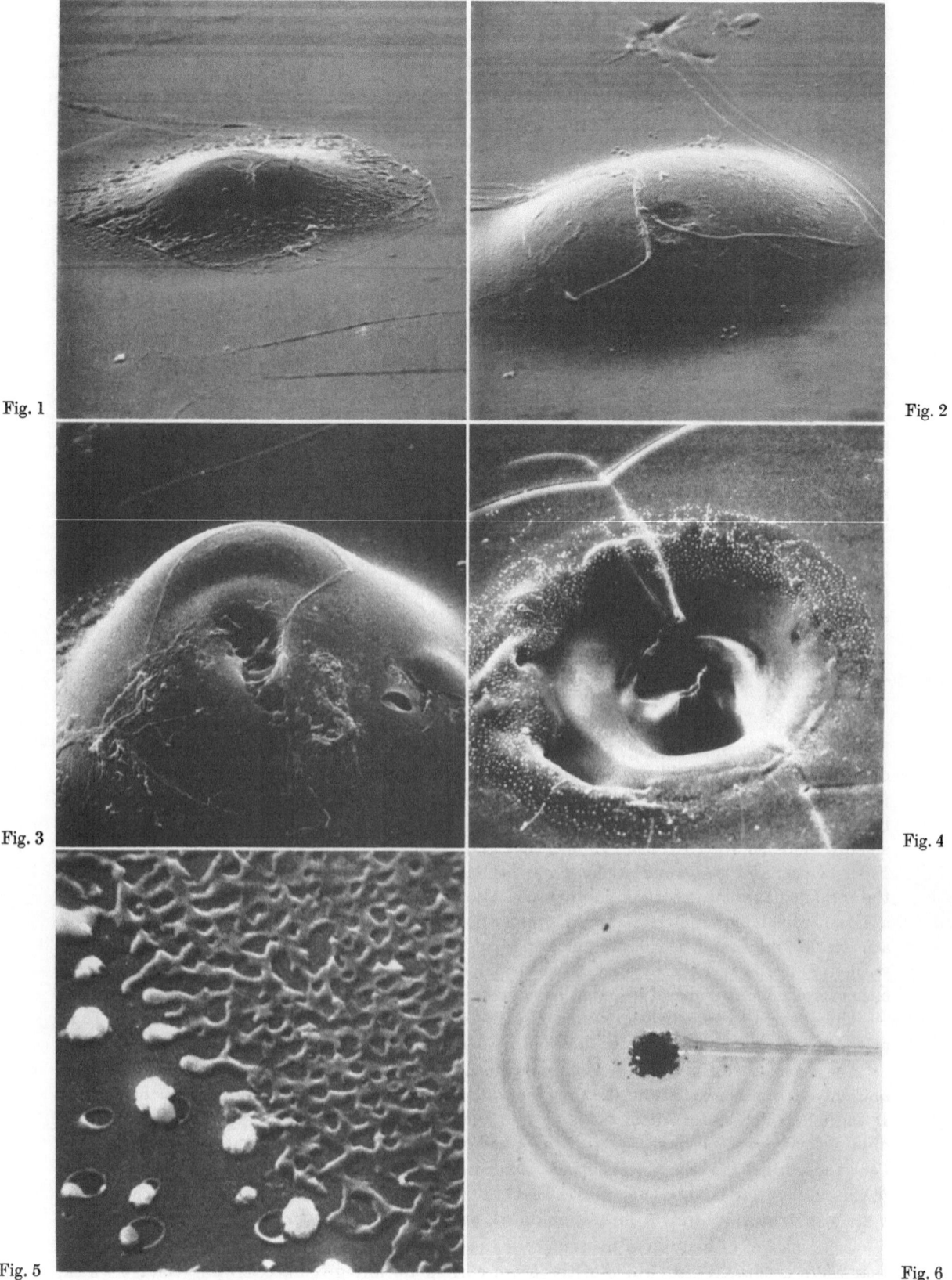

Figs. 1—6. Beam damage. Figs. 1—5 are scanning electron micrographs of damaged areas caused by a 50 kV beam at beam currents of 0.18, 0.6, 1.2, and 4.8 μA respectively. Fig. 5 is an enlarged portion of the area in Fig. 4, shewing the cenospheres, which are about 0.5 μ across. Fig. 6 is an optical micrograph, taken by reflected light, of the damaged area caused by a 30 kV/3.3 μA beam in 10 minutes. The diameter of the outermost fringe is about 400 μ

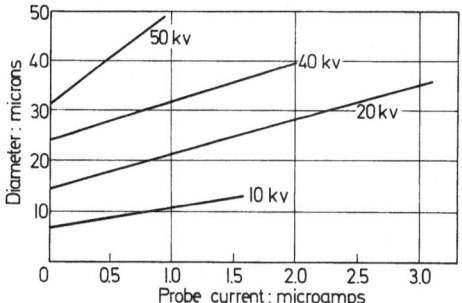

Fig. 7. Relationship between the blister or crater
diameter and the beam current

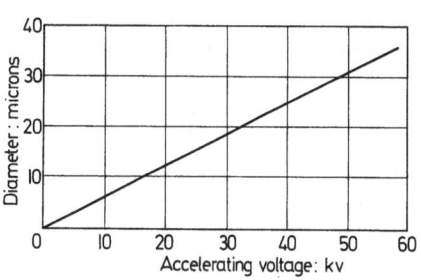

Fig. 8

Fig. 8. Relationship between the "zero current blister
diameter" and the accelerating voltage

Fig. 9. Relationship between the blister or crater
diameter and the accelerating voltage

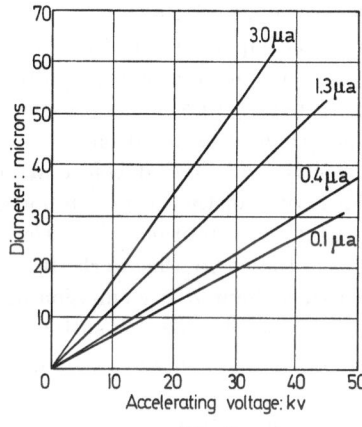

Fig. 9

have "splashy" rims indicative of melting, there is no evidence of melting at the blister stage:
if the blister is broken off, the underlying crater has a fissured surface with no suggestion of
it having been melted by the beam.

Fig. 6 shews an optical micrograph, taken by reflected light, of the damage caused by a
30 kV/3.3 µA beam in 10 minutes. Interference fringes indicate a refractive index change in a
region of the glass extending 200 µ from the centre of the crater, which is itself about 25 µ across.
By scanning across the area with the microprobe it was found that the entire area bounded by
the outermost fringe was completely devoid of sodium. Consideration of the radii and separation
of the fringes leads to the conclusion that the thickness of glass affected was less than 1 µ. No
concentration of sodium around the periphery was detected, nor had the sodium entered into
the gold layer. It was also possible to see a brownish colour throughout the area, which was
assumed to be metallic sodium in near-atomic dispersion within the glass.

The diameter of the sodium-depleted areas was found to be strongly dependent on beam
current, and on time though the time dependence proved to be far from simple, and it has not
yet proved possible to assign a simple function to describe the relationship.

The gold coating did not appear to play any part in protecting the specimen: at the point
of impingement the coating was removed virtually instantaneously. With uncoated specimens,
however, it was impossible to keep the beam still: it wandered erratically over the specimen
surface.

Discussion

Castaing has derived [5] an expression for the maximum temperature rise at the centre
of the beam

$$T_m = \frac{3w}{4\pi J k_{r_B}},$$

where W is the power in the beam (kV $\times \mu$A), $\overline{k}$ is the mean thermal conductivity of the specimen, and r_B the radius of the beam. Although, as mentioned above, the life of this layer appears to be transient, if the energy loss which would be experienced by the electrons in traversing the gold layer is taken into account, the formula predicts temperature rises of some $10^4 - 10^6$ °C for the sequence shewn in Figs. 1—4. The formula assumes that all the energy of the beam is converted into heat: its predictions do not appear to be in accord with the observed type of damage, and this suggests that a substantial part, if not the majority, of the energy is stored as charge within the glass. Some of this can leak away by movement of ions within the glass, but charge breakdown of the glass can occur. The appearance of the surface revealed by breaking of the blister was very suggestive of such breakdown.

The work so far has demonstrated that the effect is indeed complex, as Borom and Hanneman have suggested. The relative importance of the various mechanisms involved undoubtedly varies with composition, and the predominating charge effects postulated in the present work may well not be so important in other glasses. Of the other mechanisms suggested by Borom and Hanneman, some form of surface diffusion mechanism obviously plays a very considerable role in the glass investigated in the present work. It seems likely that this would be sensitive to the method of specimen preparations.

It is clearly evident that before attempting quantitative analysis of glasses, or using glasses as standards, it is of the utmost importance that their stability under the electron beam should be proved for the conditions of accelerating voltage and beam current to be used. In the case of the glass studied in this work, it was found that no detectable change in intensity of the characteristic sodium X-ray emission occurred over periods of up to 1 minute with a stationary 12 kV/0.05 μA beam: this was sufficient for quantitative analysis. If the beam was maintained on one spot for longer periods, an exponential decrease in intensity was observed.

Thanks are due to Mr. D. M. C. Thomas, Executive Director Advance Research, James A. Jobling & Co., Ltd., for permission to publish this paper.

References

1. Ribbe, P. H., and J. V. Smith: J. Geol. 74, 217—233 (1966).
2. Varshneya, A. K., A. R. Cooper, and M. Cable: J. Appl. Phys. 37, 2199 (1966).
3. Lineweaver, J. L.: J. Appl. Phys. 34, 1786—1791 (1963).
4. Borom, M. P., and R. E. Hanneman: J. Appl. Phys. 38, 2406—2407 (1967).
5. Castaing, R.: Application des sondes electroniques a une methode d'analyse ponctuelle chimique et cristallographique, thèse de doctorat. Publication O.N.E.R.A. No. 55 (1951).

Investigation of Ni-Zn Ferrite Formation by Electron Probe Microanalyser

D. Cerović, I. Momčilović and S. J. Kiss

Institute "Boris Kidrič", Vinča-Beograd, Yugoslavia

Summary

The reaction of the formation of stoichiometric Ni-Zn ferrite from Ni ferrite and Zn ferrite in the temperature range 1,100—1,300° C has been investigated. Change in the chemical composition along the reaction layer has been determined with an electron probe microanalyser. It has been found that the equimolecular ferrite $(NiO)_{0.5} \cdot (ZnO)_{0.5} \cdot Fe_2O_3$ forms on the surface of contact of two phases. Subsequent reaction forms a reaction layer with a continual change of the molar ratio Ni:Zn from 0 to ∞. The control process of the reaction is the diffusion of Zn into Ni-Zn ferrite with a minimum content of nickel.

1. Introduction

The ever increasing application and production of Ni-Zn ferrites as electromagnetic materials have imposed extensive investigations in this system. Since NiO, ZnO and Fe_2O_3 are used as the starting raw materials, most of the investigations have been devoted to the phase composition of this triple system and the solubility of the components in stoichiometric Ni-Zn ferrites [1—10]. The basic method used was the X-ray structural analysis. Many authors have obtained contradictory results for the solubility of the components in excess and the composition of the ferrites formed in dependence on the starting mixtures.

In order to understand better the reaction of Ni-Zn ferrite formation, we tried to divide the entire process into several elementary reactions. Our studies are based on the fact that in the triple system of real NiO, Fe_2O_3 and ZnO powders whose particle size is measurable (of the order of a micron) a triple point at which all three phases contact practically cannot be obtained. The triple point could only be imagined theoretically at the contact of the mathematical points of these components. From such considerations it may be concluded that, in practice, in investigating the reaction of the Ni-Zn ferrite formation from the starting oxides NiO, Fe_2O_3 and ZnO, in the beginning only two phases contact in all possible combinations ($NiO-Fe_2O_3$, $ZnO-Fe_2O_3$, ZnO—NiO). At higher temperatures a chemical reaction of the formation of a new two-component system develops on the surface of contact of the two phases. The thickness of the newly-formed phase increases until it contacts the other phase with which it reacts. Thus, the process of Ni-Zn ferrite formation can be decomposed into three elementary processes: Ni ferrite formation from NiO and Fe_2O_3, Zn ferrite formation from ZnO and Fe_2O_3 and Ni-Zn ferrite formation from Ni ferrite and Zn ferrite when contact is established between Ni ferrite and Zn ferrite. Studying the course of these elementary processes participating in the formation of Ni-Zn ferrite using an electron probe microanalyser, we have obtained data which considerably elucidate the complex reaction of Ni-Zn ferrite formation. In our previous works we studied the reaction of Ni ferrite [11] and Zn ferrite [12] formation, while in the present work we have followed the reaction of the formation of Ni-Zn ferrite from Ni ferrite and Zn ferrite.

2. Experimental Procedure

2.1. Preparation of Specimens. The reaction of the formation of Ni-Zn ferrite from Ni ferrite and Zn ferrite was investigated on specimens whose coarse grains of sintered Ni ferrite were surrounded by fine-grained Zn ferrite so it was possible to follow the course of reaction at the contact between these two phases. Nickel and zinc ferrite powders obtained by the method previously described [13] were used as the starting raw materials. Nickel ferrite powder was pressed in tablets at a pressure of 1 t/cm², sintered at 1,200 °C for two hours, pulverized and passed through a sieve to obtain fractions of 200—300 μm. Thus obtained Ni ferrite granules were mixed with $ZnFe_2O_4$ powder in a 20:80 weight ratio. After homogenization, the mass was pressed into tablets 9 mm in diameter and 10 mm thick. The tablets were air-dried and fired at 300° C for 24 hours to remove the PVA used as the binder. Microscopic examination showed that specimens thus prepared are composed of course grains of Ni ferrite, 200—300 μm in diameter, surrounded by Zn ferrite powder. The specimens were fired at 1,100—1,300° C for 1—7 hours and then the composition of the reaction layer formed on contact between the Ni and Zn ferrites investigated.

2.2. Investigation of the Composition of the Reaction Layer. The course of reaction in nickel-zinc ferrite formation was followed using an SEM-2 electron probe microanalyser. The possibility of local analysis was employed to determine the composition of the reaction layer at different points along a certain line which extended normally to the layer. The measurements were made at intervals of 4—5 μm. Pure elements were used as the standards. Results were corrected for absorption according to Colby's method [14, 15]. Evaluation of the corrections for the atomic number and fluorescence has shown that they can be neglected in this system.

The measurement used K_α lines of zinc and nickel excited by an accelerating voltage of 25 kV. A LiF crystal and a gas flow proportional counter were used.

3. Results and Discussion

The distribution of Ni and Zn in all the specimens investigated was photographed. For example Fig. 1 shows the electron image of the specimen fired at 1,250° C for 5 hours. The distribution of all the three elements along the line intersecting the reaction layer on the electron image (Fig. 1) is shown in Fig. 2. This figure shows that Fe concentration along the cross section is constant while Ni and Zn concentrations vary along the line observed. The relative concentrations of Ni and Zn were photographed (Fig. 3) along the same line but at larger magnification.

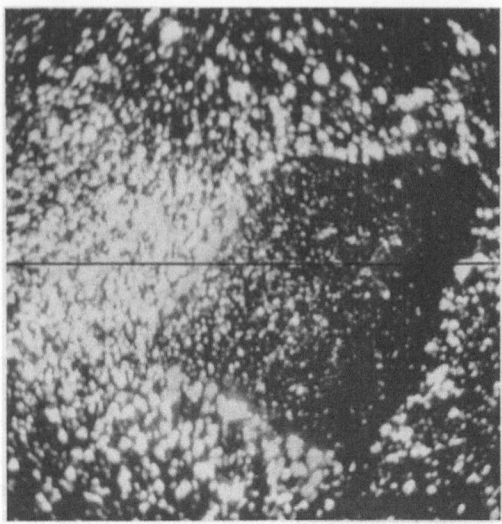

Fig. 1. Electron picture with the line intersecting the reaction layer. ×220

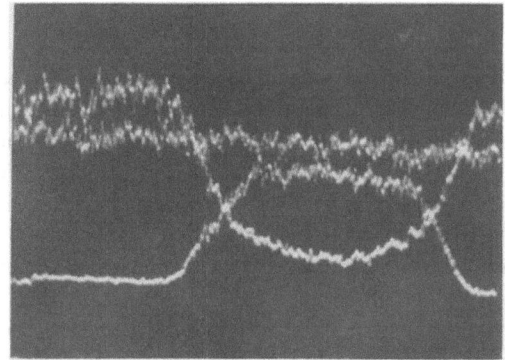

Fig. 2. Concentration variation of Ni, Zn and Fe along the line of Fig. 1

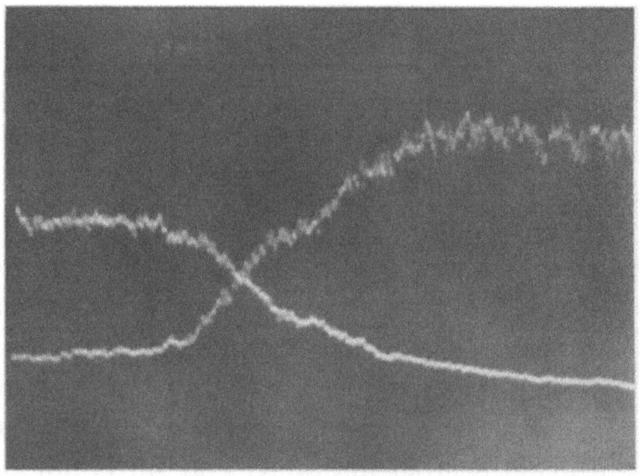

Fig. 3. Variation of concentration of Ni and Zn along the line of Fig. 1. ×660

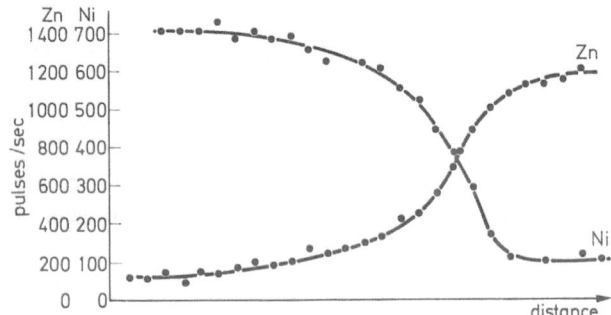

Fig. 4. Change of intensity of $Zn K_\alpha$ and $Ni K_\alpha$ along the reaction layer

In order to obtain absolute concentrations of these elements we made quantitative measurements of Ni and Zn concentrations along the line investigated at $4-5$ μm intervals. The curves plotted in Fig. 4 were obtained from the intensities determined in pulses/sec of the characteristic X-radiation of $Ni K_\alpha$ and $Zn K_\alpha$ as a function of the distance. The values obtained for the first approximation were corrected for absorption. The distribution of all the three elements on the specimens fired at other temperatures for certain times was the same, except for the reaction layer thickness which increased with increasing temperature and longer firing time. In the unreacted phase of Zn ferrite, the Ni content was equal to zero. Along the reaction layer, the Ni concentration gradually increased from 0 to 24 wt.% in unreacted Ni ferrite. In all the specimens this gradual transition of concentration from 0 to 24 wt.% has a point of inflection determined by intersecting the curve with a common tangent on both parts, which corresponds to a concentration of approximately 12 wt.% Ni. The change of Zn concentration along the reaction layer, in the opposite direction, varies from zero in unreacted Ni ferrite to about 27 wt.% in unreacted Zn ferrite. In all specimens the point of inflection on this curve corresponds to a concentration of 13.4 wt.% of Zn. Analysis of such results can give some conclusions about the course of reaction between Ni and Zn ferrite. Nickel and zinc concentrations at the points of inflection correspond exactly to the concentrations of Ni and Zn in Ni-Zn ferrite of the composition $(ZnO)_{0.5} \cdot (NiO)_{0.5} \cdot Fe_2O_3$. Since in the beginning of reaction on the boundary surface Ni ferrite — Zn ferrite the molar ratio Ni/Zn is equal to unity, according to PISKAREV [10] Ni-Zn ferrite of the composition $(NiO)_{0.5} \cdot (ZnO)_{0.5} \cdot Fe_2O_3$, i.e., equimolecular stoichiometric Ni-Zn ferrite, is formed by chemical reaction. The reaction further proceeds by diffusion of Ni and Zn through the reaction layer and by chemical reactions on the boundary surfaces of the reaction layers with Ni ferrite on the one side and Zn ferrite on the other. Now the molar ratio Ni/Zn is no longer equal to unity, so that the chemical reaction will form stoichiometric ferrites which are richer in nickel towards Ni ferrite and richer in zinc towards Zn ferrite. This permanently variable Ni to Zn ratio on the boundary surfaces causes formation of ferrites increasingly rich in Ni on the one side and in Zn on the other. The result of such a course of reaction is the formation of a reaction layer of stoichiometric Ni-Zn ferrites whose molar ratio of Ni to Zn varies from 0 to ∞. The boundary values correspond to pure Ni ferrite and Zn ferrite. Nickel and zinc diffuse through the reaction layer at a concentration gradient which causes the diffusion rate variation along the reaction layer. As the diffusion coefficient increases with the concentration gradient, the diffusion coefficient of Ni through the reaction layer increases from Ni ferrite to Zn ferrite, while the diffusion coefficient of Zn increases from Zn ferrite towards Ni ferrite. Accordingly, the smallest value of the diffusion coefficient of Ni and Zn approaches the diffusion coefficients of these elements in one-component ferrites, i.e. Ni in Ni ferrite and Zn in Zn ferrite.

Because of the lack of available data for diffusion of Ni in Ni ferrite the direct comparison of diffusion coefficients of Ni and Zn is not possible. Some conclusion may be drown from the comparison of the known diffusion coefficient of Zn in Zn ferrite [16], which is expressed by

$$D = 10^3 \cdot e^{-\frac{86,000}{RT}}$$

and the diffusion coefficient of Ni in Ni chromate [17], which is given by

$$D = 1.5 \times 10^{-3} \cdot e^{-\frac{61,400}{RT}}.$$

Ni chromate has also a spinel structure, and Cr is very close to Fe (atomic numbers 24 and 26; the valency is the same). Therefore it may be expected that the diffusion of Zn in stoichiometric Ni-Zn ferrite with maximum content of Zn is the control process of Ni-Zn ferrite formation and that the activation energy of this process is the activation energy of the diffusion of Zn in Zn ferrite, i.e. 86 kcal/mol.

Conclusion

In the reaction of formation of stoichiometric Ni-Zn ferrite from Ni ferrite and Zn ferrite, an equimolecular stoichiometric Ni-Zn ferrite layer is formed on the surface of contact. Further reaction causes the formation of a reaction layer with continual change of the Ni and Zn content going from Ni ferrite to Zn ferrite. The rate of diffusion of Ni and Zn through the reaction layer permanently changes due to the variation in the composition of the reaction layer. The diffusion of Zn is the control process of this reaction.

Acknowledgment. We wish to thank B. Djurič for his assistance in the microprobe examinations and for many valuable discussions.

References

1. Krause, O., u. W. Thiel: Ber. Deut. Keram. Ges. 15, 3 (1934).
2. Van Arkel, A., E. Verwey u. M. van Bruggen: Rec. tran. chem. 55, 331 (1936).
3. Gerasimov, Ja., T. I. Bulgakova u. Ju. Simonov: Ž. obš. himii 18, 154 (1948).
4. Toropov, H. A., i A. I. Borisenko: Dokl. Akad. Nauk SSSR 62, 6 (1948); 71, 1 (1950); 76, 1 (1951).
5. — — i E. A. Poraj Košic: Dokl. Akad. Nauk SSSR 66, 5 (1949).
6. —, u. E. Ž. Frajdenfeld: Ž. tehn. fiz. 23, 9 (1953).
7. Feirweather, A., F. F. Roberts, and A. J. Welch: Rept. Progr. Phys. 15, 142 (1952).
8. Erastova, A. P.: Dissertacija, Leningr. in-t točnoj mehaniki i optiki, 1955.
9. — Naučno tehn. sb. NII MRTP, ONTI. 3, 40 (1955).
10. Piskarev, K. A.: Izv. Akad. Nauk SSSR, Ser. Fiz. 23, 289 (1959).
11. Cerović, D., I. Momčilović, and S. Kiss: Study of solid state reaction in the NiO-Fe₂O₃ system (in press).
12. — Lj. Petrović, M. Radulović u. S. Kiss: Proc. XII Jugoslav Conf., ETAN, Rijeka (1968).
13. Kiss, S.: Development of ferrites. Nucl. Sci. Inst. "B. Kidrič", Vinča IBK-534 (1966).
14. Colby, J. W.: NLCO 944 (1965).
15. —, and Niedermeyer: NLCO 914 (1964).
16. Lindner, R.: Z. Naturforsch. 10a, 1027 (1955).
17. —, u. A. Akerström: Z. Physik. Chem. 18, 303 (1958).

An Electron Microprobe Analysis of Cu_{2-x} Layers Chemiplated on Single Crystals and Thin Films of CdS

M. Fabbricotti*, A. v. Aerschodt*, J. J. Loferski**, K. K. Reinhartz*,

A. P. v. Rosenstiel*** and A. P. Voskamp***

Abstract

The cupreous sulphide barrier layer in cadmium sulphide solar cells was investigated using an electron microprobe. In typical high efficiency thin film cells the cupreous sulphide was relatively homogeneous and its average thickness was approximately 1,100 Å. In low efficiency cells the cupreous sulphide was very inhomogeneous. In model experiments with thick cupreous sulphide layers on cadmium sulphide single crystals surface defects and crystallographic orientation had a strong effect on the cupreous sulphide growth rate and pattern.

1. Introduction

During the last decade, considerable work has been performed on the development of the thin film cadmium sulphide solar cell as a potentially less expensive and lighter alternative for silicon single crystal solar cells, both for space and terrestrial solar generators.

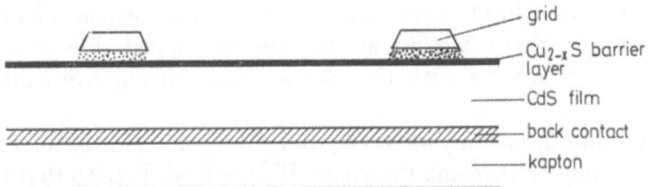

Fig. 1. Section of a $Cu_{2-x}S/CdS$ thin-film solar cell

The cadmium sulphide cell (Fig. 1) basically consists of an n-type semiconducting layer of polycrystalline cadmium sulphide, 10 to 20 μ thick, which is covered by a p-type copper sulphide barrier layer, less than 1 μ thick. The copper sulphide layer is normally formed by dipping the cadmium sulphide film into a hot aqueous solution of cupreous chloride. The substitution reaction of Cd^{++} by Cu^+ occurs according to the equation:

$$CdS + (2-x)Cu^+ \rightarrow Cu_{2-x}S + Cd^{++}.$$

The performance of the cadmium sulphide cell appears to be strongly dependent on the properties of the copper sulphide layer but, in spite of a considerable effort, very little is known about this layer.

---

* European Space Research Organisation, European Space Research and Technology Centre Noordwijk, Holland.

** Metaalinstituut, T.N.O., Delft, Holland.

*** On leave from Brown University 1967—1968.

As a part of a study of the operating mechanism of the cadmium sulphide solar cell, it was interesting to know the thickness and the homogeneity of the copper sulphide layer and the mechanism of the formation of this layer. This paper presents the results of an electron microprobe study of these parameters. Thickness and homogeneity were measured on actual thin film solar cells. The rate and growth mechanism of the copper sulphide were studied in model experiments on cadmium sulphide single crystals.

2. Thickness Calibration Curves for Cu_2S

One of the main parameters necessary to analyse the spectral response of the thin film solar cell is the thickness of its barrier layer.

So far, the thickness was determined by removing the copper sulphide layer partially with a cyanide solution and by measuring the thickness by means of interferometric methods. This method is inaccurate because of the roughness of the surface of evaporated CdS films and it only gives an average value for the thickness. Furthermore, it is destructive.

Measuring the thickness by means of the microprobe is very attractive since it is rapid, non-destructive and gives the thickness distribution with a resolution of 1 to $2\,\mu$. Calibration curves are required for accurate thickness measurements. In order to obtain these curves, relative $Cu\,K_\alpha$ intensities were determined from Cu layers evaporated on single crystals of CdS.

From these experimental data, a calibration curve for thin layers of Cu_2S on CdS was deduced.

2.1. Calibration Curve for Copper Films on Cadmium Sulphide

The standards were prepared by coating flat, polished CdS single crystals with evaporated Cu layers of different thicknesses. The thicknesses of the Cu layers measured by interferometry were between 125 and 4,200 Å.

After ascertaining by line scanning with the $Cu\,K_\alpha$, $Cd\,L_\alpha$ and $S\,K_\alpha$ characteristic radiations that the evaporated layers were uniform and homogeneous, point counting was performed at 10, 15, 20, 25 and 35 kV with a constant beam current of 10 nA. This very low beam current was chosen in order to avoid local overheating of the sample, which might cause a diffusion of copper in the substrate. Static measurements done on dipped thin films have shown that the $Cu\,K_\alpha$ intensity was decreasing with time for a beam current exceeding 90 nA at an accelerating voltage of 20 kV.

The measured absolute intensities for $Cd\,L_\alpha$ and $Cu\,K_\alpha$ are plotted in Fig. 2a and Fig. 2b as a function of accelerating voltage and thickness. We see from Fig. 2b that the $Cu\,K_\alpha$ intensity

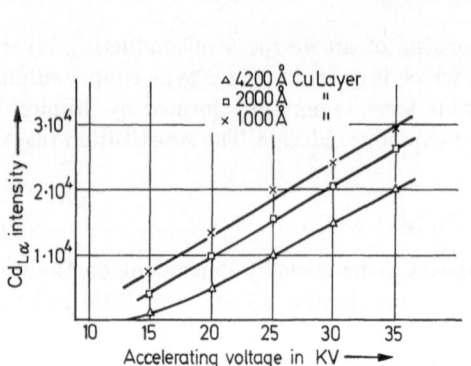

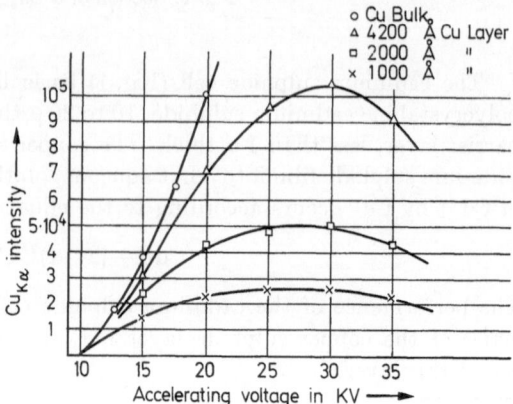

Fig. 2a. $Cd\,L_\alpha$ intensity as function of accelerating voltage from CdS single crystals covered with evaporated layers of Cu

Fig. 2b. $Cu\,K_\alpha$ intensity as function of accelerating voltage for Cu layers evaporated on CdS single crystals

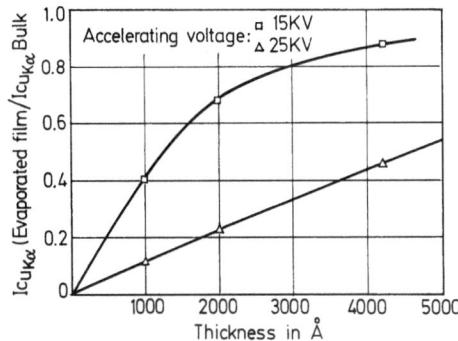

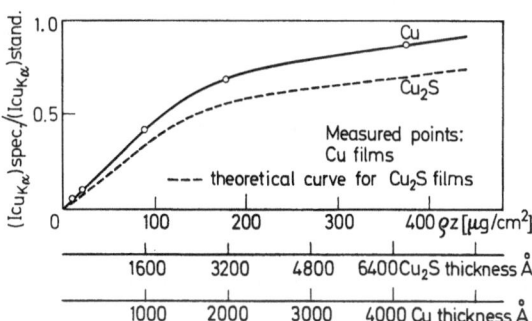

Fig. 3a. Relative CuK_α intensity versus thickness of Cu layers evaporated on CdS single crystals

Fig. 3b. CuK_α intensities for Cu and Cu$_2$S films on CdS single crystals as a function of measured film mass ϱz, at 15 kV

for the various layers reaches a maximum between 25 and 35 kV. This behaviour can be explained qualitatively by saying that, starting from the excitation voltage V_0, the intensity at constant current first increases with $(V - V_0)^n$ like in a bulk specimen. At high voltages, the intensity will increase less rapidly. With increasing energy the peak of the energy dissipation will lie deeper into the material. Most of the electrons will pass through the layer without exciting the characteristic CuK_α line. On the other hand, the CdL_α intensity increases linearly with a voltage above approximately 15 kV, as predicted by classical theory where $I \div (V - V_0)$ for $V > 3V_0$.

Relative intensities versus thickness are plotted on Fig. 3a for copper for accelerating voltages of 15 and 25 kV. At 15 kV, the relative CuK_α intensity varies linearly up to about 1,000 Å whereas, for 25 kV, it is linear to approximately 3,000 Å. It can also be seen that, for Cu layers which do not exceed 1,000 Å, an accelerating voltage of 15 kV offers a higher sensitivity than an accelerating voltage of 25 kV. The accelerating voltage of 15 kV was selected throughout the main part of our study.

2.2. Calibration Curve for Cu$_2$S Films on Cadmium Sulphide

For a thin film of a pure element on a substrate, the characteristic relative intensity can be described conveniently by the general equation [1]:

$$\frac{I_{FS}}{I_{bulk}} = \alpha(1 + A\eta)\left[\varrho z + \left(\frac{B}{2} - \frac{x}{2}\right)(\varrho z)^2 + \cdots\right]$$

which simplifies for small ϱz to:

$$\frac{I_{FS}}{I_{bulk}} = \alpha(1 + A\eta)\varrho z \tag{1}$$

where: I_{FS} = the measured intensity for the film on the substrate
I_{bulk} = the measured intensity for an infinitely thick film
α = the normalisation factor
$1 + A\eta$ = the substrate correction
ϱz = the film mass thickness (µg/cm²).

If the film is a compound and W the weight percent of the element in the compound, then the relation (1) has to be modified to:

$$\frac{I_{FS}}{I_{bulk}} = W\alpha(1 + A\eta)\varrho z. \tag{2}$$

This means that for sufficiently thin films and high accelerating voltages, the intensity of the characteristic emission of an element in a compound, is, in a first approximation, proportional to the mass per unit area of this element.

Eqs. (1) and (2) allow to compare the $Cu\,K_\alpha$ intensity of thin films of Cu and Cu_2S respectively on a CdS substrate. The ratio of the intensities according to Eqs. (1) and (2) will be:

$$\frac{I_{Cu K\alpha}^{Cu_2S}}{I_{Cu K\alpha}^{Cu}} = \frac{W_{Cu}^{Cu_2S} \cdot \varrho_{Cu_2S} \cdot z_{Cu_2S}}{\varrho_{Cu} \cdot z_{Cu}}. \tag{3}$$

The substrate correction factor can be eliminated for films of similar ϱz evaporated on the same substrate.

From relation (3) follows that, for two films of Cu and Cu_2S of the same mass per unit area, evaporated on CdS, the $Cu\,K_\alpha$ intensity ratio is equal to:

$$W_{Cu}^{Cu_2S} = 0.8.$$

The computed thickness calibration curve for Cu_2S is given in Fig. 3b where the relative $Cu\,K_\alpha$ intensities have been plotted as a function of ϱz. In the linear region of the graph which corresponds to the range where Eq. (1) is valid, the ratio of the $Cu\,K_\alpha$ intensities in Cu_2S and Cu films of equal ϱz is ± 0.8. For infinitely thick Cu_2S films, this ratio should also be $W_{Cu} = 0.8$ but, experimentally, 0.76 was measured. In the intermediate region, the $Cu\,K_\alpha$ intensity will be comprised between these two values.

3. Thickness and Variation in Thickness of the Barrier Layer on a CdS Film Solar Cell

The thickness determination of a thin copper sulphide layer on an evaporated polycrystalline cadmium sulphide film is complicated by the presence of crystallite boundaries and the roughness of the surface which can influence the measured X-rays intensities. Furthermore, an accurate determination of the thickness requires the knowledge of the stoichiometry of the compound forming the barrier layer.

3.1. Effect of Surface Roughness and Stoichiometry

In order to determine the influence of the surface roughness on the measured characteristic emission, line scans were made on 500 Å thick Cu layers evaporated on CdS films. Variations of the $Cu\,K_\alpha$ intensity were of the order of 7%. The effect of the surface roughness was smaller than expected. Electron micrographs and secondary electron images shown in Figs. 4 and 5 indicate that the surface roughness varies between 0.1 and 1 μ.

The stoichiometry of the compound forming the barrier layer has not yet been determined with certainty. In thick $Cu_{2-x}S$ layer (several microns), the composition of the sulphide has been reported to be Cu_2S [2, 3], but for thin layers it might be close to $Cu_{1.8}S$.

In thin films of unknown thickness, it is difficult to determine the composition with the microprobe. However, from Eq. (3), the error of the thickness measurement can be estimated when the compound in the barrier layer is $Cu_{1.8}S$ instead of Cu_2S.

The $Cu\,K_\alpha$ intensity ratio for two thin films of Cu_2S and $Cu_{1.8}S$ of equal thickness is:

$$\frac{W_{Cu}^{Cu_2S} \cdot \varrho_{Cu_2S}}{W_{Cu}^{Cu_{1.8}S} \cdot \varrho_{Cu_{1.8}S}} \cong 1.035.$$

The maximum error in thickness determination due to the unknown composition will be 3.5% since the intensities are proportional to the thickness. The total uncertainty for thickness measurement of the barrier layer on a polycrystalline film will not exceed 10%.

Fig. 4. Electron micrographs of the surface of CdS thin films evaporated under different conditions, Pt/C replica.
×8,000

29*

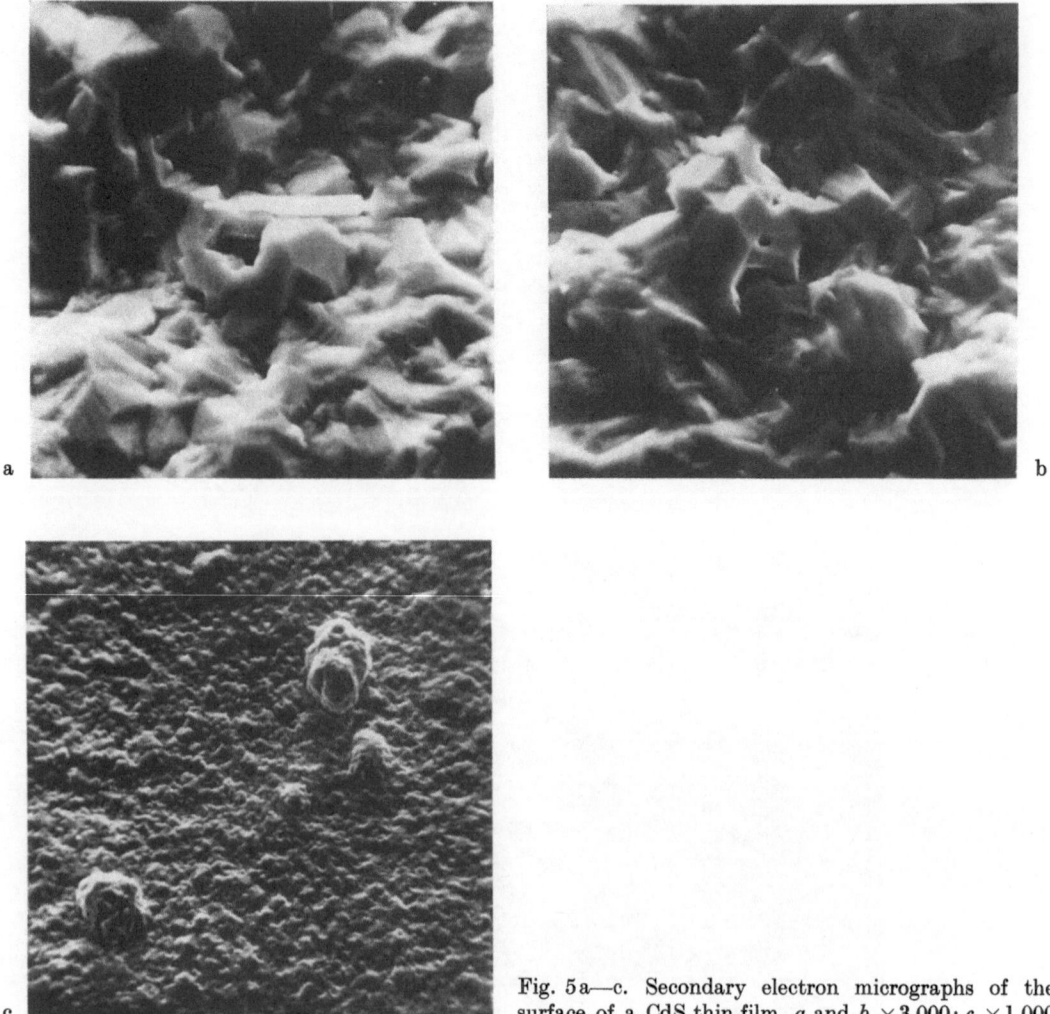

Fig. 5a—c. Secondary electron micrographs of the surface of a CdS thin film. *a* and *b* ×3,000; *c* ×1,000

3.2. Thickness of the Polycrystalline Copper Sulphide Layer

In order to determine the thickness of the copper sulphide layers in actual solar cells, line scans of the $Cu K_\alpha$ and the $Cd L_\alpha$ radiations were recorded simultaneously at very low scanning speeds of 1 and 2.5 μ/min. The smallest possible electron beam diameter of 1 to 2 μ was used in order to obtain a good spatial resolution.

The copper intensity distribution showed a strong correlation with the performance of the solar cells. Fig. 6 shows the copper and cadmium distribution in a typical good cell. The average copper intensity corresponds to a Cu_2S thickness of 1,100 Å according to the calibration curve in Fig. 3b. The intensity variations do not exceed ±10%. Since the cadmium intensities changed in the opposite direction, the $Cu K_\alpha$ intensity variations are not only due to the surface roughness or inclination of the crystallites faces but could indicate a variation of the copper sulphide thickness in the cell on a microscopic scale.

"Bad" cells, however, have a much larger variation of the copper intensity as shown on Fig. 7. Here the intensity variations have a range of several hundred percent. The peaks in the $Cu K_\alpha$ curve are probably due to the very deep diffusion of copper along crystallites boundaries since the distance between peaks corresponds to the average crystallite size of 1 to 5 μ in the evaporated cadmium sulphide films. This assumption is consistant with the observed relative

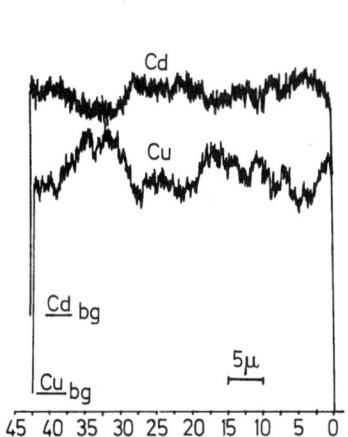

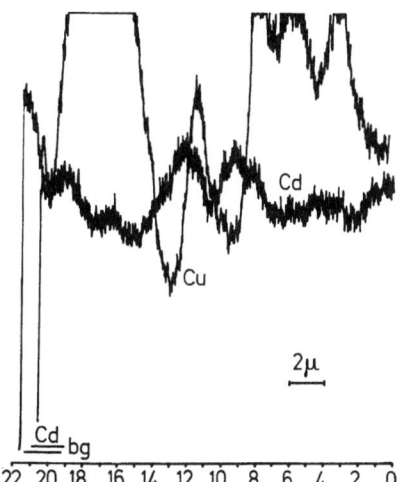

Fig. 6. Variation in intensity of the CuK_α and CdL_α radiation versus position in a good efficiency solar cell

Fig. 7. Variation in intensity of the CuK_α and CdL_α radiation versus position in a poor efficiency solar cell

change of the CuK_α and CdL_α curves in Fig. 7. If the area of the high copper concentration is small compared to the beam diameter, as in the case of crystallites boundary diffusion, the CuK_α intensity should change much more than the CdL_α intensity.

4. Analysis of Dipped Single Crystals of CdS

Because of their well-defined crystallographic structure as compared to thin films, single crystals of CdS are better suited than evaporated layers to study the growth rate and the perfection of the barrier layer and to measure the variations of Cu, Cd and Cl concentrations with depth in $Cu_{2-x}S$. We have studied therefore the formation of the barrier layer on CdS single crystals with the microprobe. Most of the results can be applied to thin films which are composed of small crystals.

4.1. Copper Distribution on the CdS Surface

In order to study the distribution of the copper concentration in the Cu-Cd exchange reaction, polished single crystals of cadmium sulphide were dipped in a cupreous chloride solution at 90° C for times ranging from 30 seconds to several hours.

The samples were first mechanically polished with diamond or tin oxide, then chemically by immersion in a solution of 10 g of potassium dichromate for 100 cc of orthophosphoric acid heated at 190° [4].

As an alternative, we also used the so-called "rotating disc method", described by M. V. SULLIVAN and W. R. BRACHT [5]. A 30% aqueous solution of hydrochloric acid was used.

On samples still containing mechanical damage, the electron microprobe showed that, even after a short dip, the copper distribution on the surface was very unhomogeneous. After a three minutes reaction, the average CuK_α intensities corresponded to a thickness of 0.12 μ but, in small areas, it increased to 1 μ. The X-ray images in Fig. 8a show that the enhanced diffusion of copper started from surface scratches caused by the mechanical polishing.

Samples without visible mechanical damage had a relatively homogeneous copper distribution over the whole surface. In such samples the effect of the crystal orientation on the plating reaction was studied using crystal wafers cut normal to the c-axis and the a-axis, respectively.

Diffusion proceeded at a considerably higher rate along the c-axis than along the a-axis. When the C-face was plated the cupreous sulphide layer was approximately 30 μ thick after 1 hour and the whole surface was covered by randomly oriented cracks as shown in Fig. 8b.

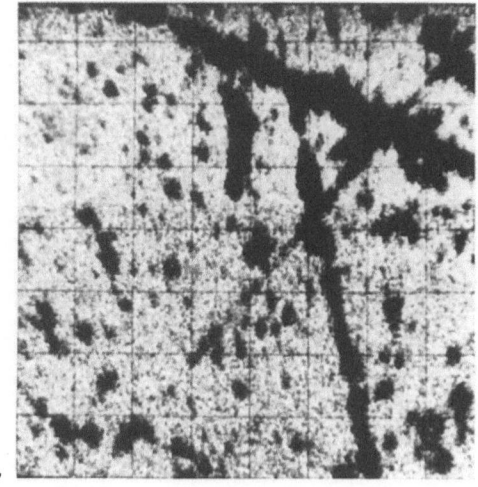

Backs. el. image

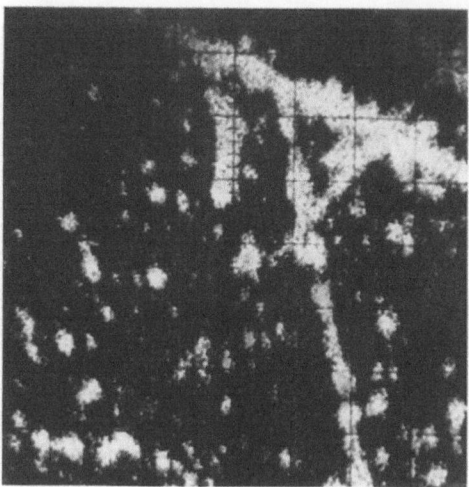

Cu K_α image

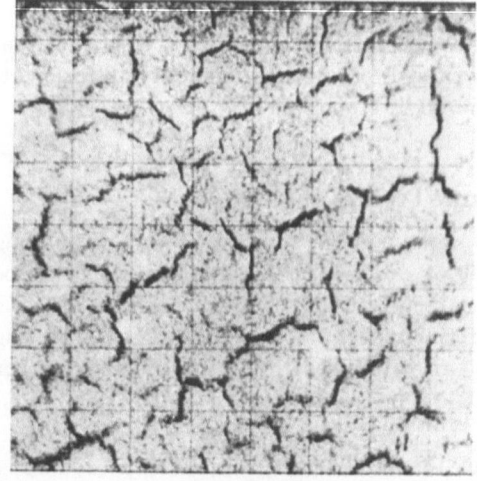

Backs. el. image

Fig. 8a. Rapid diffusion of copper at scratches

Fig. 8b. Cracks in the $Cu_{2-x}S$ layers chemiplated on CdS single crystal C oriented

On the A-face only some isolated spots with cracks appeared after 1 hour plating and these spots were separated by several millimeters.

After $2\frac{1}{2}$ hours plating, many more cracks appeared on the surface. The electron and the X-ray images (Fig. 9) show that these cracks appeared in areas where the copper concentration is very high. In the copper rich areas, the $Cu K_\alpha$ intensities corresponded to the bulk value for Cu_2S whereas the average $Cu K_\alpha$ intensity corresponded to a thickness of $0.28\ \mu$.

On the A-faces most cracks are not randomly oriented but are aligned parallel. Since Singer and Faeth [3] observed that cupreous sulphide grows epitaxially on single crystal cadmium sulphide the cracks apparently occur along one crystallographic plane in the cupreous sulphide (Fig. 10).

Cracks were also observed by Shiozawa et al. [2] in chemiplated layers. The molar volume of chalocite Cu_2S is 9% less than the molar volume of CdS. This creates high stress at the interface of the two layers. With increasing thickness of the copper sulphide layer the stress causes this layer to crack.

After very long plating times, the crystals break up into small pieces. Breaking occurs in areas where the copper sulphide has penetrated the whole crystal.

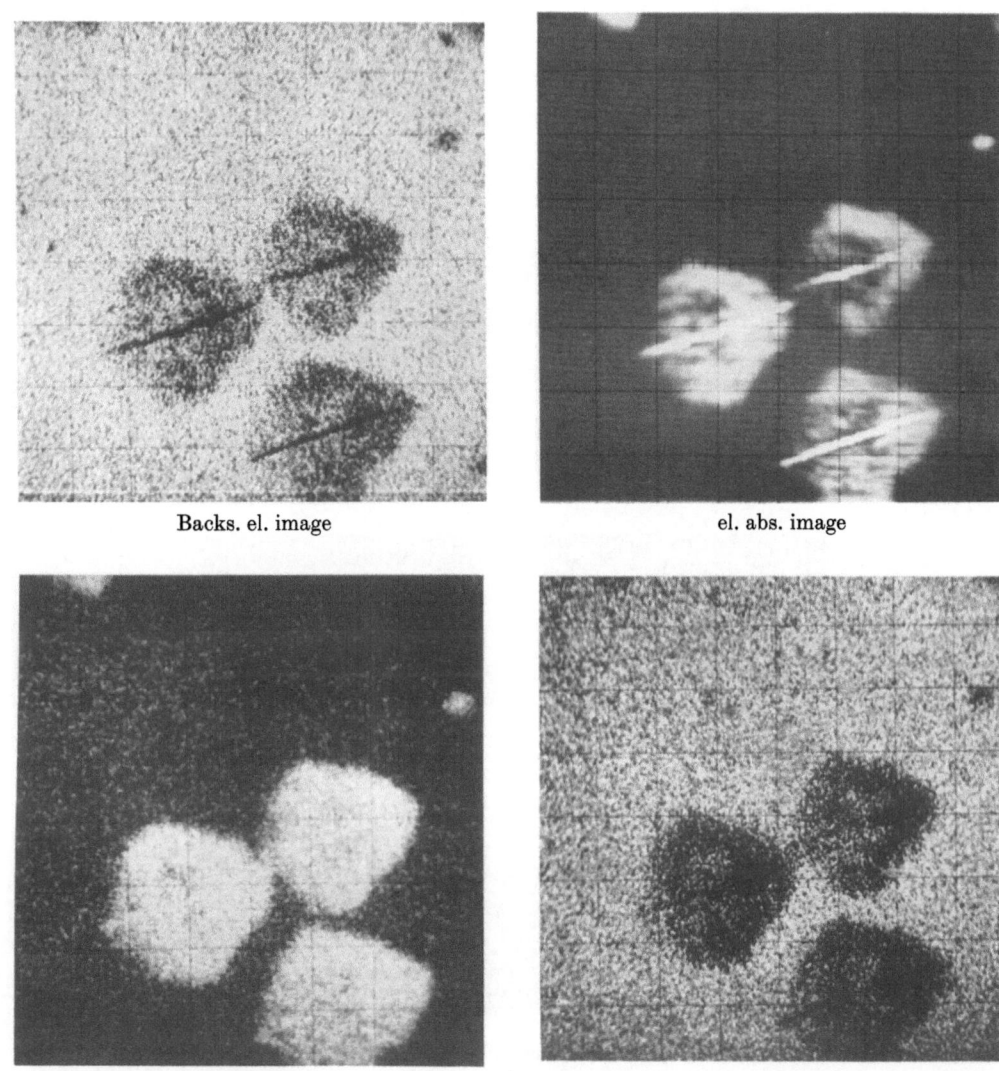

Backs. el. image   el. abs. image

$Cu\,K_\alpha$ image $Cd\,L_\alpha$ image

Fig. 9. Electron and X-ray images of areas containing a crack. The CdS single crystal is A oriented

4.2. Variation in Cu, Cd, Cl Concentration with Depth

Thin single crystals of CdS cut normally to the c-axis were polished and chemiplated in a CuCl solution for 0.5, 1.2 and 5 hours. They were then cut perpendicularly to the surface, the section polished and scanned in the microprobe normal to the interface $Cu_{2-x}S/CdS$. The speed was $5\,\mu/min$. The smallest electron beam diameter available was used at an accelerating voltage of $15\,kV$.

The copper concentration was constant in the $Cu_{2-x}S$. It dropped rapidly by several orders of magnitude when reaching the interface $Cu_{2-x}S/CdS$ and continued to decrease slowly until the background intensity was reached about 15 to 25 μ away from the $Cu_{2-x}S/CdS$ interface. This indicates that a Cu doped CdS region exists below the $Cu_{2-x}S$ layer. The content in copper of the region is low. The concentration in cadmium varied in a similar manner when scanned starting from the CdS side. Around the interface and in the $Cu_{2-x}S$ layer, chlorine was observed.

All $Cu_{2-x}S$ layers contained cracks perpendicular to the interface. At the sites of these cracks, the layer penetrated deeper in the CdS.

The mean thickness of the copper sulphide layer increased linearly with plating time in the experimental range between 0.5 and 2 hours. The reaction rate was approximately $30\,\mu/hour$.

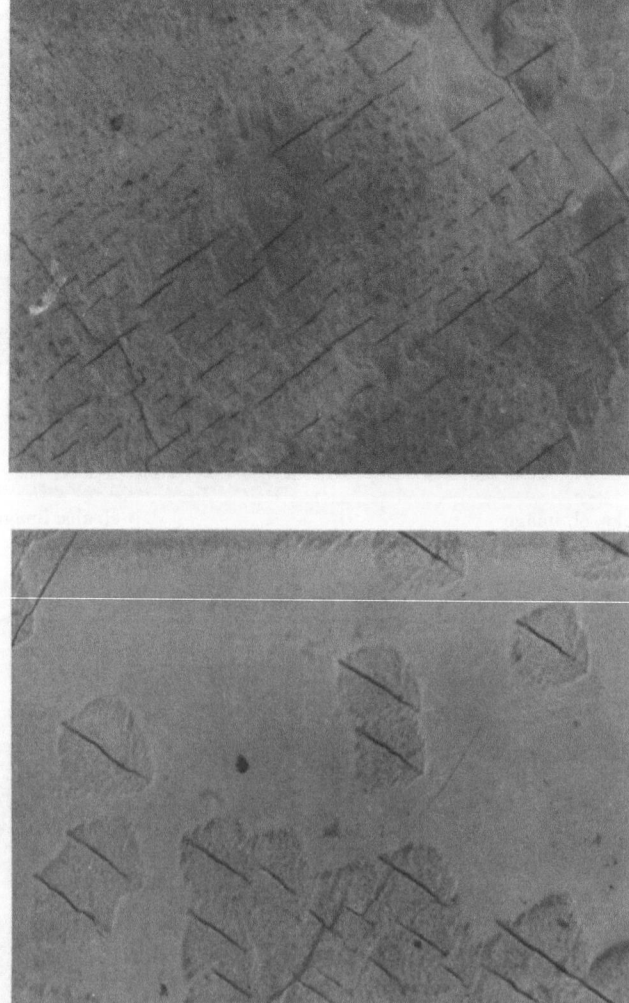

Fig. 10. Aligned cracks in the $Cu_{2-x}S$ layer chemiplated on CdS single crystals A oriented. $\times 400$

5. Summary

With an electron microprobe the relative $Cu K_\alpha$ X-ray intensities were measured from Cu films of known optical thickness evaporated unto polished CdS single crystals. From these data microprobe calibration curves for the thickness determination of $Cu_{2-x}S$ layers on cadmium sulphide were deduced. The effect of surface roughness and of deviations from stoichiometry in the cupreous sulphide was investigated. It has been estimated that these parameters introduce an error not exceeding ± 10 percent when the thickness of chemiplated cupreous sulphide layers in polycrystalline cadmium sulphide solar cells is measured.

In high efficiency cadmium sulphide solar cells the mean thickness of the cupreous sulphide layer was approximately 1,100 Å and its variations did not exceed ± 10 percent. In low efficiency cells the mean thickness of the cupreous sulphide was considerably lower and the copper distribution was very inhomogeneous. High peaks in the measured copper intensities were probably due to prefered diffusion along grain boundaries.

The investigation of chemiplated cupreous sulphide layers on CdS single crystals showed that the crystallographic perfection of the surface strongly influenced the growth rate of the cupreous sulphide. The growth rate was much higher starting from surface defects than on defect free surfaces. On well polished surfaces an effect of crystallographic orientation on the growth rate and the growth pattern could be observed. On C-faces approximately $30\,\mu$ $Cu_{2-x}S$ were formed after 1 hour chemiplating and the surface was completely covered by cracks.

On A-faces less than $0.3\,\mu$ $Cu_{2-x}S$ were formed after 1 hour chemiplating and only isolated spots with higher copper concentration were observed. In these spots also cracks appeared which were parallel oriented.

The analysis of cross sections of chemiplated CdS crystals with $Cu_{2-x}S$ layers of 50 to $100\,\mu$ thickness showed that the copper concentration was constant in the bulk of the $Cu_{2-x}S$. A slightly Cu doped region extended into the cadmium sulphide and a slight Cd doped region existed in the cupreous sulphide near the interface. Some chlorine was also found at the $Cu_{2-x}S$-CdS interface and in the $Cu_{2-x}S$.

References

1. HUTCHINS, A. G.: The electron microprobe (ed. T. D. KINLEY, K. F. G. HEINRICH, and D. B. WITTRY), p. 390—404. New York: John Wiley 1966.
2. SHIOZAWA, L. R. et al.: Fifth quarterly progress report — 1st June, 1967 to 1st August, 1967. Contract AF 33 (615)-5224.
3. SINGER, J., and A. FAETH: Appl. Phys. Letters 4, 130 (1967).
4. MAEDA, H., K. DOKEN, and H. FUKUI: Jap. J. Appl. Phys. 6, 652 (1967).
5. SULLIVAN, M. V., and W. R. BRACHT: J. Electrochem. Soc. 3 vol. 114, 295 (1967).

Application of X-Ray Microanalyser to Cast Iron "Relation between Few Elements of Nodular Graphite Cast Iron and Nodular Graphite"

H. Okamoto*, H. Soejima** and S. Kozawa**

1. Introduction

Nodular graphite cast iron is broadly used and well known as one of the important materials for industrial use because of its excellent mechanical properties. The purposes of this study are; to analyse the changes of segregation of Mg and remarkable spheroidizing harmful elements such as S and Ti, and to examine the relation between nodular graphite and these elements, when the amount of added Mg in the nodular graphite cast iron is changed, by using X-ray microanalyzer.

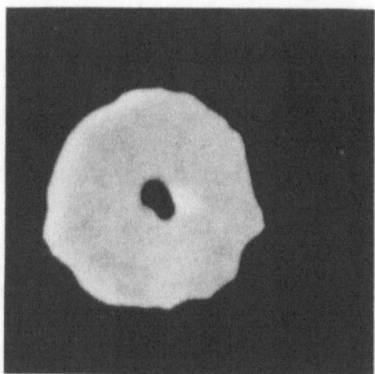

Sample Current

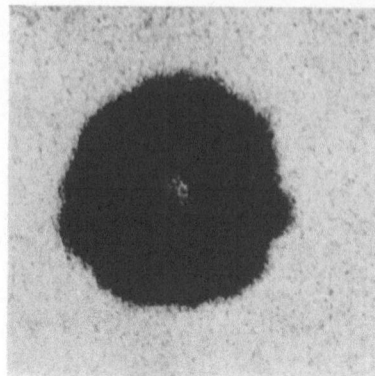

$Si K_\alpha$

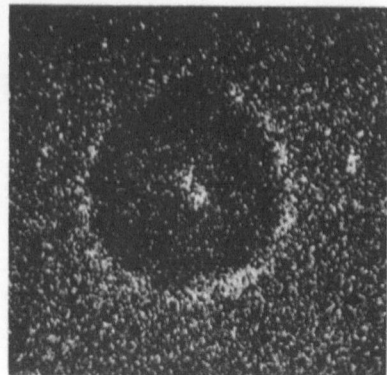

$Mg K_\alpha$

Fig. 1. Sample Current, $Si K_\alpha$ and $Mg K_\alpha$ images at the nodular graphite carbon of sample 1-3

  * Tokyo Shibaura Electric Co., Kawasaki, Japan.
 ** Shimadzu Seisakusho Ltd., Kyoto, Japan.

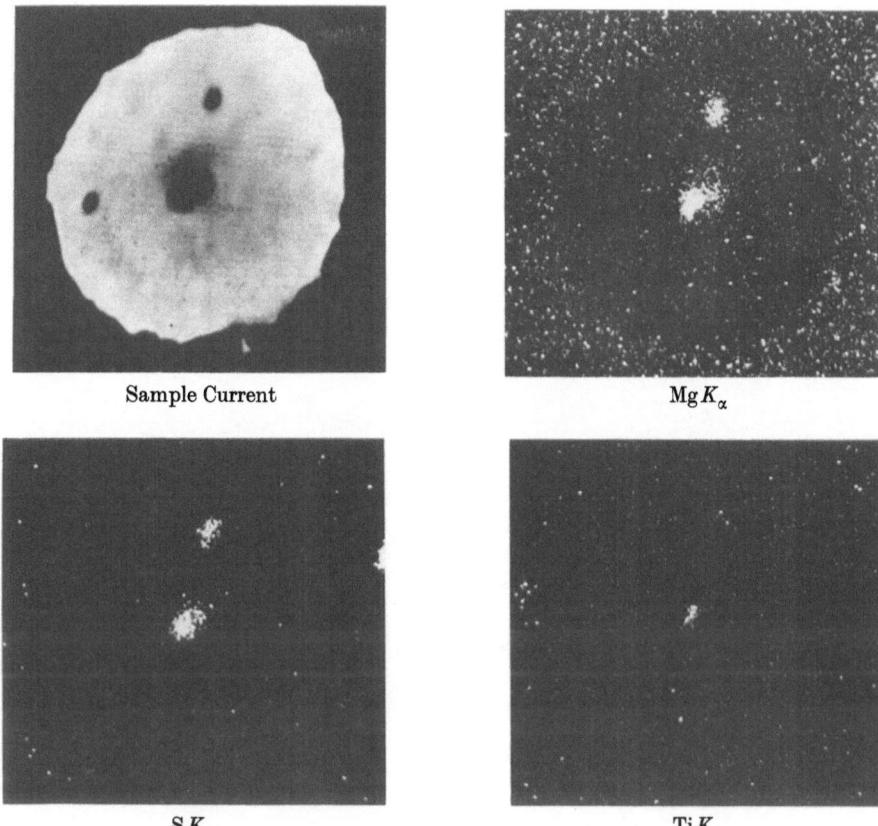

Fig. 2. Sample Current, MgK_α, SK_α and TiK_α images at another nodular graphite carbon of sample 1-3

OKUMOTO and others [1] reported that there are two kinds of nodular graphite, one around which Ca and Ce swarm, the other around which no element swarms.

I. YE. LEV [2] and others carried out the analysis of nodular graphite cast iron made by adding a large quantity of Ce to increase the detecting sensibility and found in 1965 that Ce is concentrated on the center of eutectic colony and also in the double nodular graphite.

R. J. WARRICK and others analysed the nodular graphite cast iron which contains Rare Earth and Mg in 1966, and found that Rare Earth, Mg and S are located in the center of nodular graphite, and then estimated that the nodular graphite is created by the nucleus of sulfide.

2. Experiment

The chemical components of the samples are shown in the table. Samples 1-3 and 1-6 are those with different quantities of Mg made by the high frequency melting method; 2-1 through 3-16 are samples with different quantities of Mg by the electro-slag method; and sample 4-1 is meechanite nodular graphite cast iron.

Table. *Chemical components of samples*

	C	Si	Mn	P	S	Mg	Ti	Ce
1-3	3.58	2.99	0.34	0.032	0.045	0.190		
1-6	3.01	3.06	0.35	0.044	0.032	0.055		
2-1	3.86	3.19	0.65	0.041	0.033	0.083		
2-2	3.72	3.30	1.32	0.037	0.042	0.077		
3-12	3.27	2.29	0.79	0.089	0.028	0.018	0.067	
3-16	3.37	2.40	0.59	0.038	0.010	0.388	0.047	
4-1	3.17	3.20	0.52	0.078	0.020	0.010		0.054

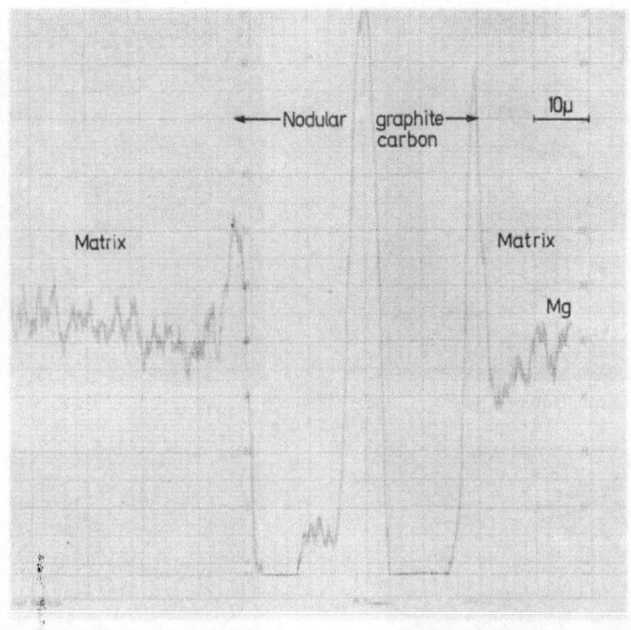

Fig. 3. Mg K_α line profile across the nodular graphite carbon shown in Fig. 1

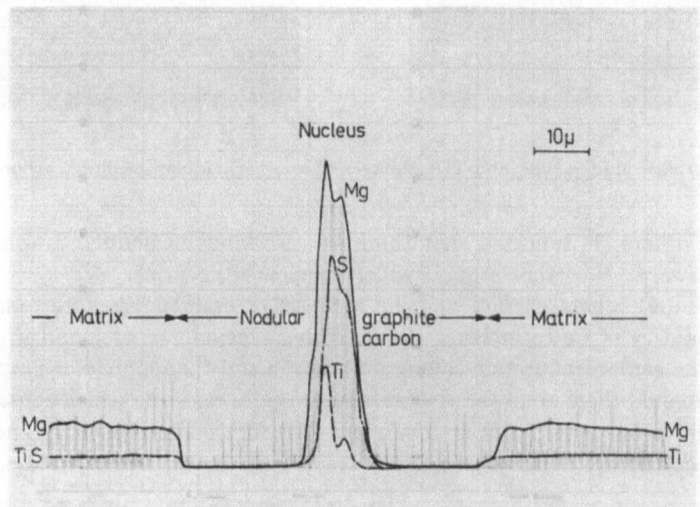

Fig. 4. Mg K_α, S K_α and Ti K_α line profile across the nodular graphite carbon shown in Fig. 2

The samples were subjected to dry polish up to No. 600 emery, then polish with diamond paste of 15, 5 and 1 μ successively, and then ultrasonic cleaning in chlorosen. X-ray micro-analyser, EMX Model 1, was employed. The measurement was done under the following conditions; accelerating voltage 15 kV, sample current 0.12 μA, X-ray spot diameter 1.2 μ, crystal 4″ ADP and KAP, detector Ne Exatron. The typical data of experiments are shown in Figs. 1 ∼ 6.

3. Conclusion

1. Mg does not always exist as a nucleus but a substance containing Mg happens to become nucleus and grows into a nodular graphite. There is no direct relation between Mg and nodular graphite. The similar phenomena can be seen in the case of Ce treatment for cast iron.

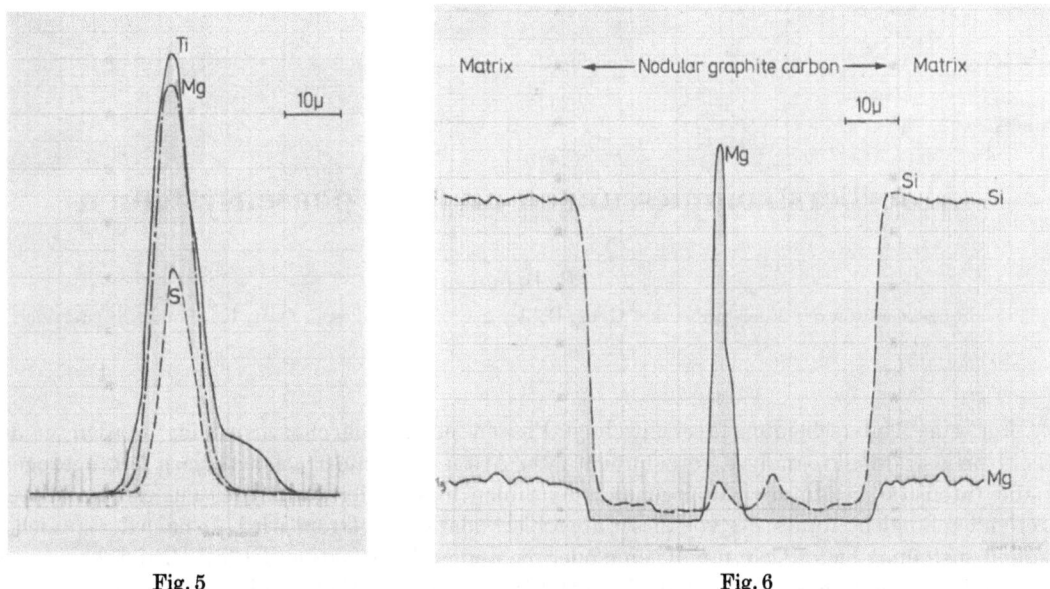

Fig. 5   Fig. 6

Fig. 5. MgK_α, SK_α and TiK_α line profile across a precipitate in sample 1-6

Fig. 6. MgK_α and SiK_α line profile across the nodular graphite carbon of sample 2-2

2. Mg seems to exist in the state of sulfide or carbide and to cover the sulfide or carbide of Ti. By this, Ti can not work as the nucleus of graphite.

3. Mg which swarms around nodular graphite has no direct relation with the growth of nodular graphite.

4. Numerous spheroidal matrix irons are contained in nodular graphite. However, any of them can not be considered as the nucleus of the nodular graphite. It has not yet been clear in whether state they are contained, in molten state or in solidified state.

References

1. OKUMOTO, IIJIMA: Journal of Japan Foundrymens Society **13** (**12**), 728 (1963).
2. LEV, I. YE., BELAY, G. YE.: Fiz. Metal. Metalloved. **20** (**2**), 236 (1965).
3. WARRICK, R. J.: AFS Cast Metals Res. J., Sept. 97 (1966).

Absorberstrommessungen an Mehrphasensystemen

R. BLÖCH

Forschungsanstalten der Edelstahlwerke, Gebr. Böhler & Co., Aktiengesellschaft, Kapfenberg, Österreich

Für eine Unterscheidung der einzelnen Phasen eines Mehrphasensystems werden in der Elektronenstrahlmikroanalyse vornehmlich zwei Arten von Informationsträgern herangezogen: 1. die Intensität bestimmter Röntgenemissionslinien und 2. der Anteil des auf die Probe auftreffenden Sondenstromes, der von dieser absorbiert oder rückgestreut wird. Je nach der speziellen Aufgabenstellung bietet sich nun die eine oder die andere Informationsart auf Grund ihrer spezifischen Eigenschaften als optimal an. Nachstehend seien kurz die Charakteristika von Röntgen- und Absorberstromsignalen einander gegenübergestellt.

Die Röntgensignale enthalten in Form der Wellenlänge und Impulshäufigkeit elementspezifische Konzentrationsinformationen. Durch geeignete Wahl der zur Phasencharakterisierung verwendeten Röntgenlinie können große relative Intensitätsänderungen zwischen den verschiedenen Phasen eines komplexen Systems erhalten werden. Auf Grund der Quantennatur der Röntgenstrahlen treten jedoch große statistische Schwankungen der Momentanintensität um den zeitlichen Mittelwert in Abhängigkeit von der Impulsrate (Imp./sec) und der Zeitkonstante des Meßsystems (rate meter) auf. Für eine sichere Differenzierung unterschiedlicher Phasen mittels Rasterverfahren, wie sie in der quantitativen Metallographie angewendet werden, ergeben sich daher relativ niedrige Abtastgeschwindigkeiten. Die optimale Auflösung liegt in kompakten Proben bei etwa 3 μ.

Im Gegensatz zu den Röntgenstrahlen enthält der Absorber- und Rückstreustrom, wenn man von den Sekundärelektronen absieht, Informationen über eine mittlere Ordnungszahl, die nur in binären Systemen mit Elementen hinreichend unterschiedlicher Ordnungszahlen Rückschlüsse auf die Zusammensetzung gestattet. Bei Drei- und Mehrstoffsystemen kann wohl einer Zusammensetzung ein bestimmter Absorberstrom zugeordnet werden, umgekehrt kann jedoch aus dem Absorberstrom nicht die Zusammensetzung abgeleitet werden. Wegen der Kleinheit der elektrischen Ladung eines Elektrons kann der Strom praktisch als Kontinuuminformation aufgefaßt werden und ist mit keinen größeren statistischen Schwankungen behaftet; so entspricht beispielsweise ein Sondenstrom von 100 nA einem Elektronenfluß von $6 \cdot 10^{11}$ Elektronen pro Sekunde, die bei diesem Strom auf Reinelementen erzielbaren Röntgenintensitäten hingegen liegen meist in der Größenordnung von 10^2-10^4 Impulse pro Sekunde. Auf Grund des Kontinuumcharakters des Stromes kann mit hohen Abtastgeschwindigkeiten gearbeitet werden. Die maximal erzielbaren relativen Stromänderungen beim Übergang von einer Phase auf eine andere sind meist klein im Vergleich zu den entsprechenden Änderungen der Röntgenintensitäten. Das Auflösungsvermögen des Elektronenstrahles kann in den Mikrosonden unter 1 μ und in Elektronenrastermikroskopen sogar unter 0,1 μ betragen.

Wegen der hohen erzielbaren Abtastgeschwindigkeiten erscheint die Verwendung des Absorber- bzw. Rückstreustromes in der quantitativen Metallographie besonders attraktiv. Auf diese Möglichkeiten wurde bereits verschiedentlich u. a. von HEINRICH [1], DÖRFLER, PLÖCKINGER und SWOBODA [2—4] hingewiesen.

Zahlreiche Autoren [5—9] befaßten sich mit der absoluten Messung des Absorberstromanteils in Abhängigkeit von Ordnungszahl, Beschleunigungsspannung, Probenvorspannung und

Probengeometrie. In vorliegender Arbeit sollte an Hand einiger Beispiele die Möglichkeit der Differenzierung verschiedener Phasen in Mehrphasensystemen auf Fe- und Ni-Basis mittels des Absorberstromes untersucht werden. Oxydische Schlackenphasen wurden in die Untersuchungen nicht mit einbezogen, da darüber bereits an anderer Stelle [10] berichtet wurde.

Versuchsdurchführung und Ergebnisse

Zur Untersuchung gelangten folgende Legierungen:

Tabelle

Nr.	Legierung	Zustand	Phasen
1	X 10 CrNiTi 18 9	verformt und abgelöscht	Matrix, Ti (N, C)
2	X 12 CrNiS 18 8	verformt und abgelöscht	Matrix, (Mn, Cr)S
3	X 210 Cr 12	gegossen und gehärtet	Matrix, $(Fe, Cr)_7C_3$
4	C 110 W 1	verformt und weichgeglüht	Matrix, $(Fe)_3C$
5	S 10-4-3-10	verformt und überhitzt gehärtet	Matrix, $(V, W, Mo)C$, $(W, Mo, Fe, Co)_6C$
6	18% Cr, 80% Ni, 2% B	gegossen und bei 1200° C geglüht	Matrix, $(Ni, Cr)_3B$, CrB

Aus den angeführten Legierungen wurden Probenplättchen von 16×24 mm und 6 mm Dicke herausgearbeitet und in der Mitte mit 4 Löchern von 1,8 mm Durchmesser versehen, die zur Aufnahme von 3 Referenzstandards aus Cr, Fe, Ni oder Ti, Cr, Fe und zur Messung des Faradaystromes dienten. Bei den beiden Legierungen X 210 Cr 12 und S 10-4-3-10, die sich nicht bohren ließen, wurden zylindrische Proben von ca. 6 mm Durchmesser in Plättchen aus austenitischem Stahl eingepreßt und in diese austenitischen Plättchen wieder die vier Bohrungen angebracht (Abb. 1). Alle Proben wurden geschliffen und mit Diamantpaste bis zu einer Körnung von 1 μ poliert. Die Proben 3—6 wurden zur Entwicklung des Gefüges leicht geätzt.

Die Absorberstrommessungen wurden an einer CAMECA-Mikrosonde MS 85 bei einer Beschleunigungsspannung von 20 kV und einem Sondenstrom von ca. 100 nA durchgeführt. Die Probenströme wurden als Spannungsabfall über 1 MΩ Wiederstand mittels eines KNICK-Verstärkers 7 s und einem nachgeschalteten RIKADENKI-Kompensationsschreiber (max. Empfindlichkeit 10 mV/250 mm Vollausschlag; Linearität besser 0,5%) gemessen, dessen Nullpunkt mittels einer Eichspannungsquelle HONEYWELL-„Potentiometer 2705" unterdrückt werden konnte.

In Vorversuchen ergab sich, daß über den in dieser Arbeit in Frage kommenden relativ kleinen Bereich der Ordnungszahl Z die Steilheit der Kurve $(1 - \eta) = f(Z)$, wobei $1 - \eta$ den Absorberstromanteil darstellt, durch Probenvorspannungen zwischen -90 V und $+90$ V praktisch nicht

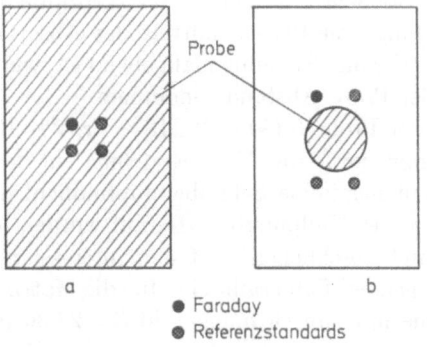

Abb. 1. Probengeometrie für Absorberstrommessungen

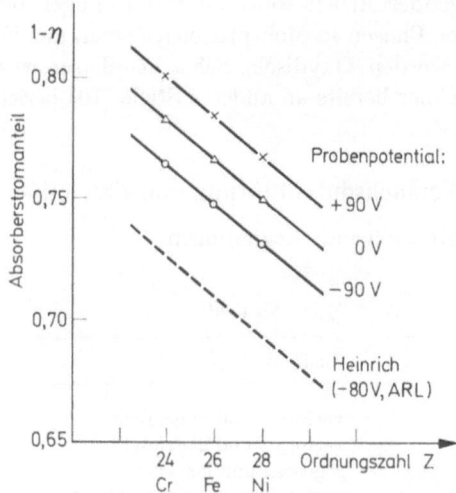

Abb. 2. Einfluß des Probenpotentials auf den Absorberstrom

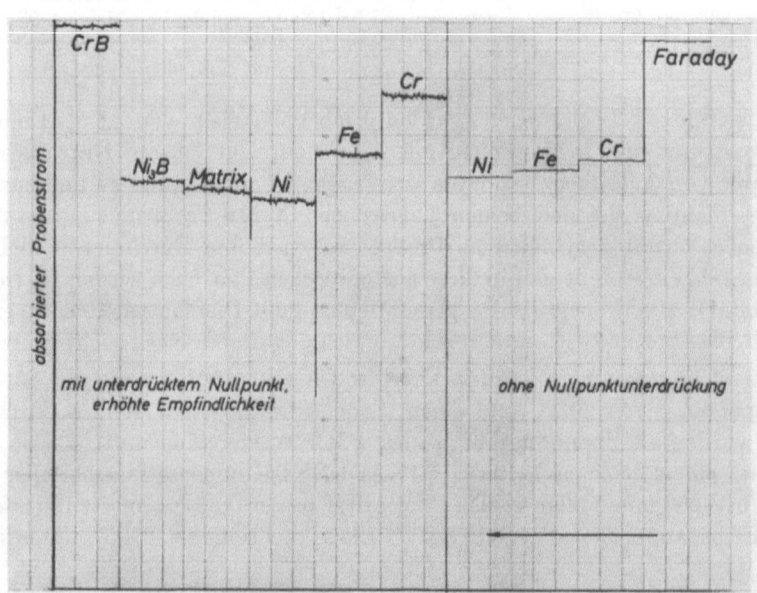

Abb. 3. Meßfolge an einer Legierung mit 80% Ni, 18% Cr und 2% B

beeinflußt wird und mit der Steilheit der Kurve von HEINRICH [11] übereinstimmt (Abb. 2). Da nun für eine Unterscheidung von Phasen mittels des Absorberstromes nicht die absolute Lage der Kurve, sondern vielmehr ihre Steigung maßgebend ist, wurde bei allen weiteren Messungen von einer Vorspannung der Probe Abstand genommen.

In Abb. 3 wird an Hand der Probe 6 (80% Ni, 18% Cr, 2% B) der Meßvorgang illustriert. Zuerst wurde der Faradaystrom sowie die Absorberströme auf den Referenzstandards (Cr, Fe, Ni) gemessen, anschließend auf maximale Schreiberempfindlichkeit umgeschaltet unter gleichzeitiger starker Unterdrückung des Nullpunktes. Hierauf wurden nochmals die Absorberströme auf den Referenzstandards und abschließend auf den interessierenden Phasen (Matrix, Ni_3B, CrB) aufgenommen. Im vorliegenden Fall ergibt sich für die Matrix mit einer Zusammensetzung von 16% Cr und 84% Ni eine mittlere Ordnungszahl $\bar{Z} = 27,36$ (errechnet aus den Gewichtsprozenten der Elemente) und der niedrigste Absorberstrom. Einen nur unwesentlich höheren Absorberstrom weist das Borid Ni_3B mit rund 7% Cr und $\bar{Z} = 26,52$ auf. Stark kontrastiert

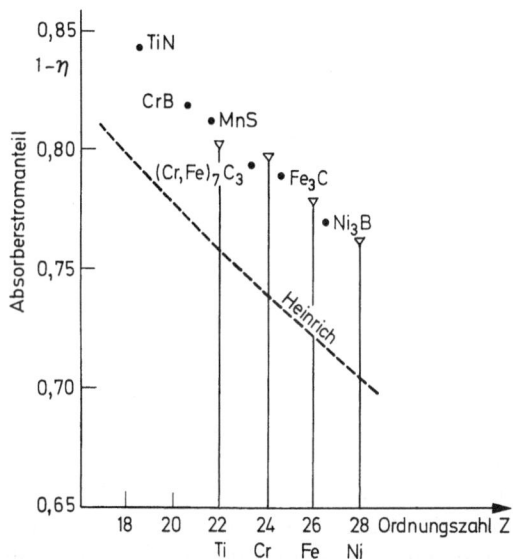

Abb. 4. Absorberstromanteile als Funktion der mittleren Ordnungszahl

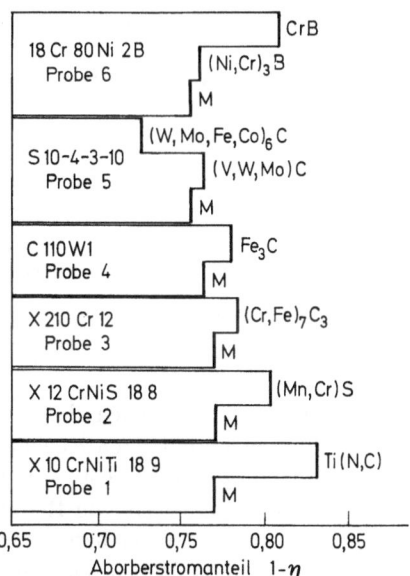

Abb. 5. Absorberstromanteile von Mehrphasensystemen

demgegenüber das Borid CrB mit 0,7 % Ni und $\bar{Z} = 20,74$ durch den stark erhöhten Absorberstrom.

Die Höhe des Absorberstromanteils für einige Phasen, wie sie in unserer Meßgeometrie erhalten wurden, sind in Abb. 4 über der mittleren Ordnungszahl aufgetragen. Vergleichsweise wurde auch die von HEINRICH für eine Beschleunigungsspannung von 20 kV und ein Probenpotential von −80 V an einer ARL-Mikrosonde erhaltene mittlere Kurve eingezeichnet. Die von uns gefundenen Werte liegen mit nur geringer Streuung auf einer Geraden, in Übereinstimmung mit dem in der Literatur [5] angegebenen linearen Teil der $\eta - Z$-Kurve für $Z \leq 30$.

Abb. 5 bringt einen schematischen Überblick über die unterschiedlichen Absorberstromanteile in den untersuchten Mehrphasensystemen. Bei den vier Zweiphasensystemen ist eine einwandfreie Unterscheidung zwischen Matrix und Zweitphase möglich. Bei dem bereits früher angeführten Dreiphasensystem der Probe 6 ist eine Unterscheidung zwischen Matrix und Ni_3B in der Praxis nicht mehr gesichert, wohingegen das Chromborid sehr deutlich absticht. Im Falle der Probe 5 liegen Karbide vom Typ MC und M_6C in relativ feiner Verteilung und teilweise stark verwachsen vor, so daß die angegebenen Absorberstromanteile optimale Werte bei langsamem Abfahren der Probe darstellen. Die Unterscheidung zwischen Matrix und MC ist in dieser Probe unsicher, hingegen wird M_6C auf Grund seines hohen Mo- und W-Gehaltes durch den stark erniedrigten Absorberstrom deutlich angezeigt.

Allgemeine Überlegungen

Wenngleich absorbierter Probenstrom und Rückstreustrom als komplementäre Größen grundsätzlich die gleiche Aussagekraft besitzen, bestehen doch in der praktischen Anwendung gewisse Unterschiede, die kurz besprochen werden sollen. Für die Unterscheidbarkeit verschiedener Phasen sind, abgesehen von deren Größe und Verteilung, in erster Linie die Unterschiede im Rückstreukoeffizienten η sowie die Stabilität des Sondenstromes I_0 von Bedeutung. Betrachtet man vergleichsweise die Gesamtänderungen des absorbierten Probenstromes I_p bzw. des Rückstreustromes I_R, so ergibt sich

$$\Delta I_p = \Delta \left[(1 - \eta) I_0\right] = -\Delta \eta \cdot I_0 + (1 - \eta) \cdot \Delta I_0, \tag{1a}$$

$$\Delta I_R = \Delta \left[\eta \cdot I_0\right] = \Delta \eta \cdot I_0 + \eta \cdot \Delta I_0. \tag{1b}$$

Das erste Glied auf der rechten Seite der Gleichungen ist das eigentliche Informationssignal, während das zweite Glied einen unerwünschten Driftterm darstellt. Bei längeren Meßzeiten, wie sie besonders in der quantitativen Metallographie mit Rücksicht auf eine gute räumliche Auflösung und auf die Auswertesysteme auftreten können, stellt die Drift ein ernstes Problem dar. Die Empfindlichkeit einer Anordnung gegenüber der Drift läßt sich durch das Verhältnis Q vom Nutzsignal zu Drift charakterisieren (analog signal — noise ratio)

für den absorbierten Strom

$$Q_p = \frac{\Delta \eta}{1-\eta} \bigg/ \frac{\Delta I_0}{I_0} \qquad\qquad (2\,\mathrm{a})$$

und für den Rückstreustrom

$$Q_R = \frac{\Delta \eta}{\eta} \bigg/ \frac{\Delta I_0}{I_0}. \qquad\qquad (2\,\mathrm{b})$$

Bei unseren Messungen lagen die η-Werte etwa im Bereiche zwischen 0,2—0,3. Rechnet man nun nun mit einem mittleren Rückstreukoeffizienten von $\eta = 0,25$, ergibt sich für den rückgestreuten Strom ein dreimal günstigeres Verhältnis von Nutzsignal zu Drift. In einer erst kürzlich bekanntgewordenen Arbeit von RIDAL und BEADLE [12] wird ein Auswertesystem (Vickers Automatic Inclusion Classifier) beschrieben, das sich des Rückstreustromes als Informationssignal bedient. Bei dem Rückstreustromverfahren besteht ferner die Möglichkeit, mit Hilfe eines sog. Stereomonitors Stromänderungen, die durch die Topographie der Oberfläche entstehen, von jenen zu unterscheiden, die durch Ordnungszahlunterschiede bedingt sind. Der gegenüber dem absorbierten Probenstrom niedrigere Wert des Rückstreustromes dürfte beim derzeitigen Stand der Verstärkertechnik keinen ernstlichen Nachteil darstellen.

Zusammenfassung

Bei Absorberstrommessungen an stehenden oder langsam bewegten mehrphasigen Proben konnten Phasen, die im Absorberstromteil $1-\eta$ um 0,01 differieren, noch eindeutig unterschieden werden. Dies entspricht im linearen Bereich der $\eta - Z$-Kurve etwa einem Ordnungszahlunterschied von $\Delta Z \cong 1,5$. In einer Gegenüberstellung von Absorberstrom- und Rückstreustrom wird auf die potentielle Überlegenheit des Rückstreustromsignals hingewiesen. Inwieweit diese Überlegenheit des Rückstreustromes bei praktischen Messungen in Erscheinung tritt und wie sich die Unterscheidbarkeitsschwellen für hohe Abtastgeschwindigkeiten ändern, bedarf weiterer Untersuchungen.

Literatur

1. HEINRICH, K. F. J.: NBS Tech. Note 401, S. 45 (1966).
2. DÖRFLER, G., and E. PLÖCKINGER: 4th Int. Congr. on X-Ray Optics and Microanalysis (XRMO), 1965, p. 506. Paris: Hermann 1966.
3. — — u. K. SWOBODA: Berg- Huettenmaenn. Monatsh. 111, 438 (1966).
4. SWOBODA, K., R. MITSCHE u. H. MALISSA: Radex Rundschau 1966, H. 4, 233.
5. PHILIBERT, J., and E. WEINRYB: XRMO 1962, p. 451. New York: Academic Press 1963.
6. POOLE, D. M., and P. M. THOMAS: J. Inst. Metals 90, 228 (1962).
7. HEINRICH, K. F. J.: Advances in X-ray analysis, p. 325. New York: Plenum Press 1963.
8. BISHOP, H. E.: XRMO 1965, p. 153.
9. WITTRY, D. B.: XRMO, p. 168.
10. BLÖCH, R.: Radex Rundschau 1967, H. 3/4, 785.
11. HEINRICH, K. F. J.: XRMO 1965, p. 159.
12. RIDAL, K. A., and R. BEADLE: English Steel Corporation R 39/67-N. 16, January 1968.

Die Anwendung der Mikrosonde
bei der Aufstellung von Mehrstoffsystemen

G. LANGENSCHEID

Hoesch AG Hüttenwerke, Dortmund, BRD

F. K. NAUMANN

Max-Planck-Institut für Eisenforschung, Düsseldorf, BRD

1. Einführung

Eine der häufig angewendeten Methoden, Mehrstoffsysteme für den festen Zustand zu untersuchen, ist die Aufstellung eines solchen Systems mit Hilfe von Temperaturschnitten. Dabei werden aus den Komponenten Legierungen verschiedener Zusammensetzungen hergestellt, bei der gewünschten Temperatur bis zur Einstellung des Gleichgewichtes geglüht und dann von dieser Temperatur zwecks Einfrierung des Zustandes abgeschreckt. Für die Untersuchung dieser Legierungen gibt es eine Reihe von Verfahren, von denen die metallographischen und röntgenographischen die wichtigsten und gebräuchlichsten sind. Als neues Verfahren bei der Aufstellung eines Mehrstoffsystems bietet sich jedoch die Mikrosondenuntersuchung an. Es ist das Ziel der vorliegenden Arbeit, darzulegen, daß sich die Aufstellung eines solchen Systems mit Hilfe der Mikrosonde einfacher und mit geringerem experimentellen Aufwand durchführen läßt als mit den erwähnten herkömmlichen Untersuchungsmethoden. Dies soll zunächst schematisch an dem Temperaturschnitt eines Dreistoffsystems erläutert werden (Abb. 1).

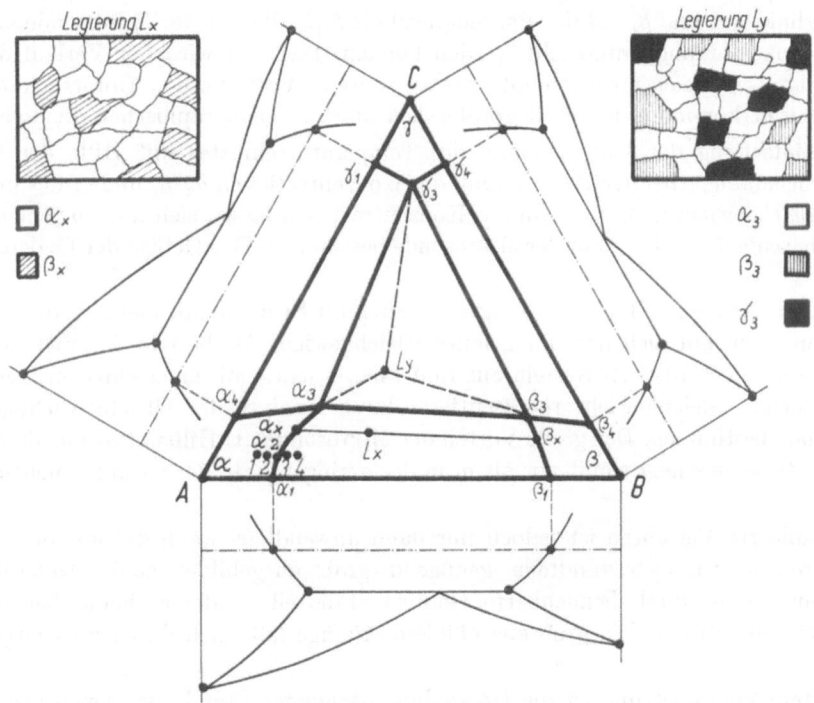

Abb. 1. Temperaturschnitt durch ein Dreistoffsystem (schematisch)

2. Erläuterung des Verfahrens

Die Komponenten A, B und C bilden mit ihren Nachbarkomponenten die Mischkristalle α, β und γ. Die Mischbarkeit soll nicht lückenlos sein. Im Temperaturschnitt existieren dann die Zweiphasengleichgewichte $\alpha+\beta$, $\alpha+\gamma$, $\beta+\gamma$ und das Dreiphasengleichgewicht $\alpha+\beta+\gamma$. Bekannt seien die Randsysteme und damit auch die für die Temperatur des gewünschten Temperaturschnittes maximalen Löslichkeiten der Phasen α, β und γ für ihre Nachbarphasen, d.h. bekannt sind die Grenzkonzentrationen α_1, α_4, β_1, β_4 usw. Unbekannt sei der Verlauf der Phasengrenzlinien $\alpha_1\alpha_3$, $\alpha_4\alpha_3$ usw. aus den Randsystemen ins ternäre System sowie die das univariante Dreiphasengleichgewicht kennzeichnenden Konzentrationen α_3, β_3 und γ_3.

Den Verlauf der Linie $\alpha_1\alpha_3$ findet man dadurch, daß man einen oder mehrere Punkte auf dieser Linie bestimmt. Bei den *herkömmlichen metallographischen oder röntgenographischen* Untersuchungsverfahren stellt man sich dazu eine Reihe von Legierungen, z.B. 1 bis 4, mit unterschiedlichen A- und B-Gehalten bei zweckmäßigerweise konstantem C-Gehalt her, und zwar in dem Konzentrationsbereich, in dem man die Zusammensetzung von α_2, das auf der Linie $\alpha_1\alpha_3$ liegen soll, erwartet. Es gilt nun, die B-reichste Legierung (2) zu finden, die gerade noch aus homogenem α-Mischkristall besteht, und die B-ärmste Legierung (3), die gerade schon als zweite Phase den β-Mischkristall enthält. Zwischen den Legierungen 2 und 3 liegt dann α_2. Will man α_2 genauer bestimmen, muß man weitere Legierungen mit Zusammensetzungen zwischen 2 und 3 untersuchen. Man findet α_2 auch durch Bestimmung der Gitterkonstanten der α-Mischkristalle in den Legierungen 1 bis 4. In der geschilderten Weise lassen sich auch die Punkte auf den Phasengrenzlinien $\alpha_3\alpha_4$, $\beta_1\beta_3$ usw. bestimmen. Man kann sich leicht vorstellen, welchen experimentellen Aufwand in Anbetracht der großen Zahl von Legierungen man treiben muß, um mit den genannten Verfahren die einigermaßen genaue Zusammensetzung der Punkte α_3, β_3 und γ_3 zu bestimmen.

Die Untersuchung eines solchen Temperaturschnittes mit Hilfe der *Mikrosonde* vereinfacht sich nun dadurch, daß man mit ihr die im Gefüge vorhandenen Gleichgewichtsphasen unmittelbar auf ihre chemische Zusammensetzung analysieren kann, wodurch sich die Zahl der zu untersuchenden Legierungen wesentlich verringern läßt. Denn eine beliebige Legierung L_x im Zweiphasenfeld $\alpha+\beta$ enthält als Gefügebestandteile (Abb. 1a) die Gleichgewichtsphasen α_x auf der Phasengrenzlinie $\alpha_1\alpha_3$ und β_x auf der Phasengrenzlinie $\beta_1\beta_3$, die mit der Mikrosonde auf ihre chemische Zusammensetzung untersucht werden können. Dadurch wird der Verlauf der Konode im Phasenfeld $\alpha+\beta$ gleich mitbestimmt, was als weiterer Vorteil dieser Untersuchungsmethode gegenüber den herkömmlichen metallographischen und röntgenographischen anzusehen ist.

Die Vereinfachung der Untersuchung des Temperaturschnittes mit Hilfe der Mikrosonde wird noch augenfälliger bei der Bestimmung der Konzentrationen α_3, β_3 und γ_3 des univarianten Dreiphasengleichgewichtes. Die genannten Konzentrationen lassen sich mit nur einer Legierung L_y im Dreiphasenfeld $\alpha+\beta+\gamma$ mit der Mikrosonde bestimmen. Das Gefüge der Legierung L_y zeigt Abb. 1b.

Das hier für das Dreistoffsystem Gesagte gilt auch für Systeme mit mehr als drei Komponenten. In jedem Fall läßt sich das univariante Gleichgewicht (es besteht im Vier-Stoff-System aus vier Phasen, im Fünf-Stoff-System aus fünf Phasen usw.) mit einer einzigen Legierung, die das entsprechende Gleichgewicht repräsentiert, durch Analyse der Gleichgewichtsphasen mit der Mikrosonde bestimmen. Der große Vorteil der Mikrosonde als Hilfsmittel für die Aufstellung eines Mehrstoffsystems liegt somit vor allem in der geringen Zahl der zu untersuchenden Legierungen.

Das geschilderte Verfahren ist jedoch nur dann anwendbar, wenn die mit der Mikrosonde zu analysierenden Gefügebestandteile genügend grob ausgebildet sind. Andernfalls kann das Analysenergebnis durch benachbarte Gefügebestandteile anderer chemischer Zusammensetzung verfälscht werden. Ein grob ausgebildetes Gefüge läßt sich durch langzeitiges Glühen erzielen.

Eine weitere Voraussetzung ist die Herstellung geeigneter Standards, denen der Vorzug zu geben ist, solange die vorhandenen Korrekturverfahren nicht völlig gesichert sind.

3. Untersuchung der Eisenecke des Systems Fe−P−C für 900° C

Die Untersuchung eines Temperaturschnittes mit der Mikrosonde soll nun am Beispiel des Systems Fe−P−C für 900° C erläutert werden. Das System ist bisher für diese Temperatur nicht eingehender untersucht worden [1].

Dazu wurden mehrere Fe−P−C-Legierungen unterschiedlicher P- und C-Gehalte hergestellt, bis zur Einstellung des Gleichgewichts, wozu etwa 100 h ausreichten, bei 900° C unter Schutzgas geglüht und anschließend in kalter Kochsalzlösung abgeschreckt[1]. Fünf dieser Legierungen, deren Gefüge die charakteristischen Phasengleichgewichte $\alpha + \gamma$, $\gamma +$ Phosphid (Fe_3P), $\alpha + \gamma +$ Phosphid und $\gamma +$ Phosphid $+$ Zementit (Fe_3C) aufweisen, wurden für die Mikrosondenuntersuchung ausgewählt (Abb. 2). Die γ-Phase liegt in allen untersuchten Legierungen als Martensit vor.

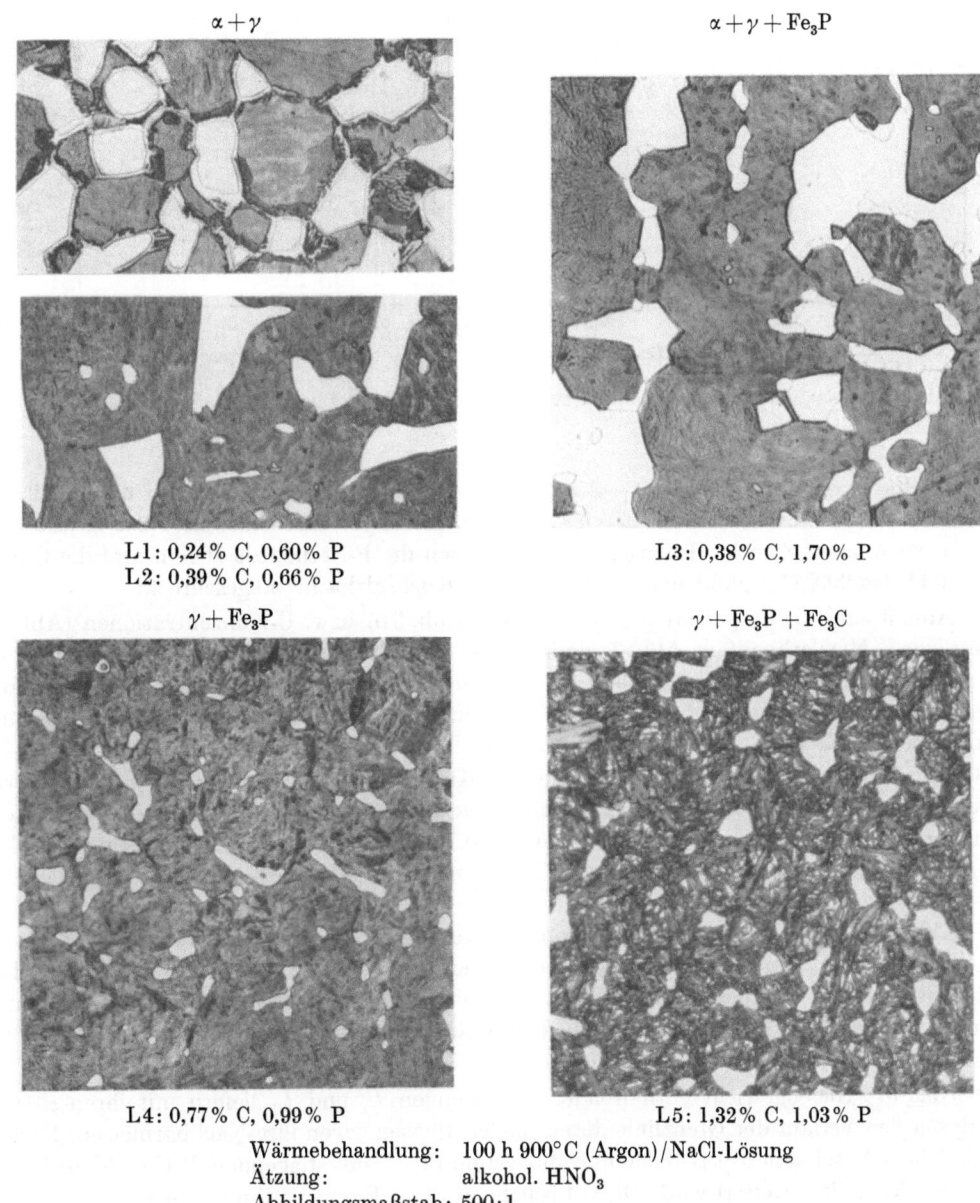

$\alpha + \gamma$ $\alpha + \gamma + Fe_3P$

L1: 0,24% C, 0,60% P
L2: 0,39% C, 0,66% P   L3: 0,38% C, 1,70% P

$\gamma + Fe_3P$ $\gamma + Fe_3P + Fe_3C$

L4: 0,77% C, 0,99% P   L5: 1,32% C, 1,03% P

Wärmebehandlung: 100 h 900°C (Argon)/NaCl-Lösung
Ätzung: alkohol. HNO_3
Abbildungsmaßstab: 500:1

Abb. 2. Gefüge von Legierungen des Systems Eisen-Phosphor-Kohlenstoff

_____

1. Von der von J. I. Godstein und R. E. Ogilvie [2] angegebenen Möglichkeit, Phasengleichgewichte in Mehrstoffsystemen mit der Mikrosonde messen zu können, wenn die verschiedenen Phasen noch nicht vollständig ihre dem Gleichgewicht entsprechenden Zusammensetzungen haben, soll hier kein Gebrauch gemacht werden.

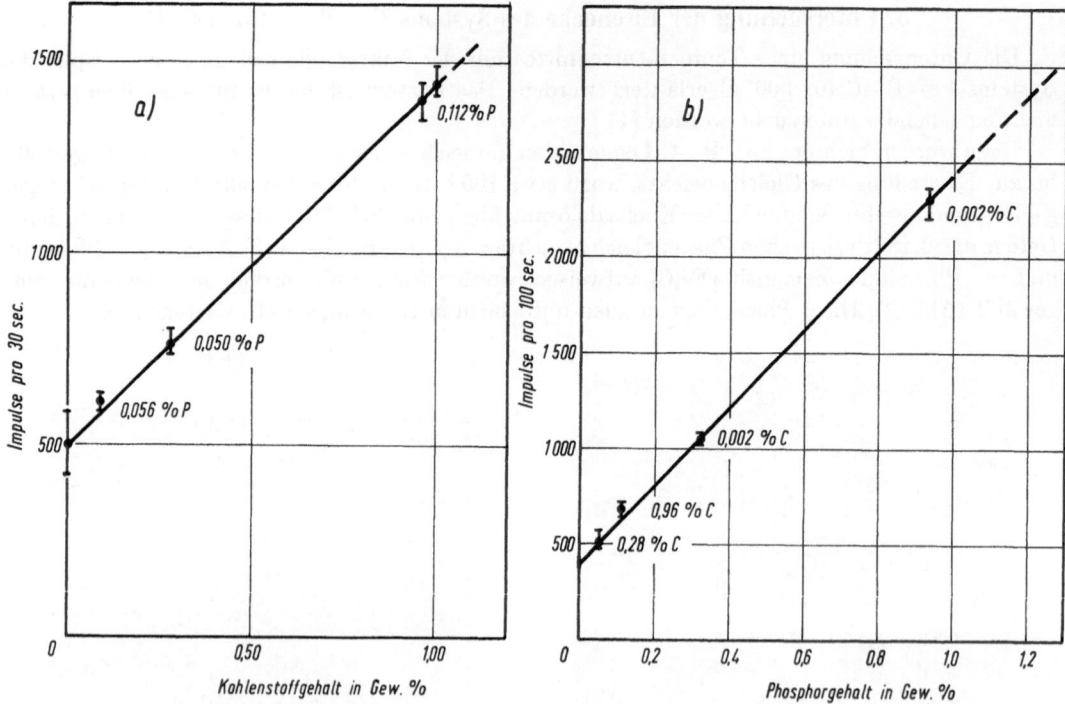

Abb. 3. Abhängigkeit der Impulsraten vom Kohlenstoff- bzw. Phosphorgehalt der Standards

Als P-Standards wurden vier Fe—P-Legierungen mit 0,05 bis 0,93% P und bis zu 0,96% C hergestellt. Als C-Standards wurden vier Fe—C-Legierungen mit 0,09 bis 1,0% C und max. 0,12% P verwendet. Zwecks Homogenisierung wurden die P-Standards 1000 h und die C-Standards 100 h bei 900° C geglüht und anschließend in Kochsalzlösung abgeschreckt.

In Abb. 3 sind die Impulsraten gegen die P- (Abb. 3 b) bzw. C-Konzentrationen (Abb. 3 a) aufgetragen[2]. Man erkennt in Abb. 3, daß bei den P-Standards auch die C-haltigen, bei den C-Standards auch die P-haltigen Proben auf der gezeichneten Geraden liegen, so daß die P- bzw. C-Bestimmung durch die entsprechenden Fremdelemente in den vorliegenden Gehalten nicht merklich beeinflußt wird.

Bei der C-Bestimmung wurde ein Blei-Stearat-Kristall verwendet, die Hochspannung betrug 10 kV, der Probenstrom 0,300 μA. Die P-Bestimmung wurde mit einem KAP-Kristall, einer Hochspannung von 20 kV und einem Probenstrom von 0,1 μA vorgenommen. An jedem untersuchten Gefügebestandteil einer Probe wurden fünf Messungen vorgenommen und die Ergebnisse gemittelt.

In Abb. 4 ist das Ergebnis der Mikrosondenuntersuchung im 900° C-Temperaturschnitt des Systems Fe—P—C dargestellt. Die bekannten Randsysteme Fe—P [3] und Fe—C [4] sind für den interessierenden Temperaturbereich mit eingezeichnet. Die Meßpunkte sind mit den Streubereichen für die C- und P-Analyse versehen. Der Kreuzungspunkt stellt jeweils den Mittelwert der Ergebnisse dar.

Die das α-γ-Gleichgewicht enthaltenden Legierungen L_1 und L_2 liefern mit ihren α- bzw. γ-Analysen den Verlauf der Grenzlinie dieser beiden Phasen gegen ihre Nachbarphasen. Es zeigt sich, daß der sehr schmale α-γ-Bereich im Randsystem Fe—C mit steigendem P-Gehalt auf Kosten des γ-Mischkristalls erweitert wird. Die C-Löslichkeit des α-Eisens wird mit steigendem P-Gehalt nicht wesentlich verändert. Die Konoden im α-γ-Feld erfahren mit steigendem P-Gehalt eine starke Drehung in dem Sinne, daß der Phosphorgehalt im α-Mischkristall stärker zunimmt als

2. Die Kohlenstoffbestimmung wurde freundlicherweise von Herrn Astruc mit der Cameca-Sonde durchgeführt, wofür wir ihm vielmals danken.

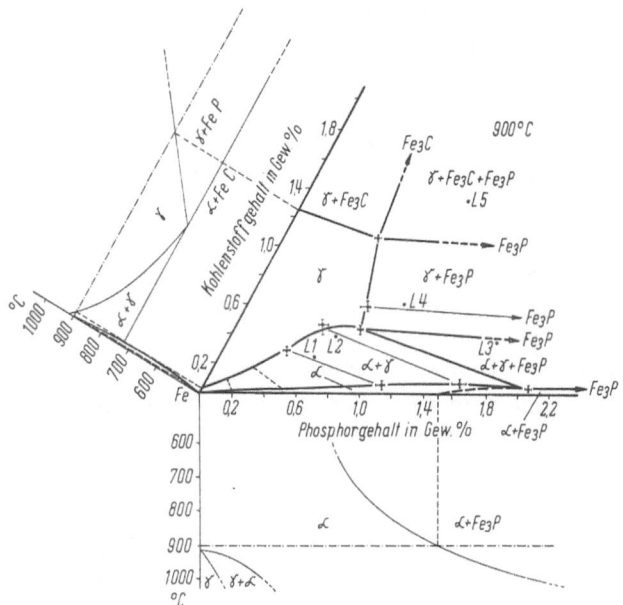

Abb. 4. Temperaturschnitt der Eisenecke des Systems Eisen-Phosphor-Kohlenstoff bei 900° C

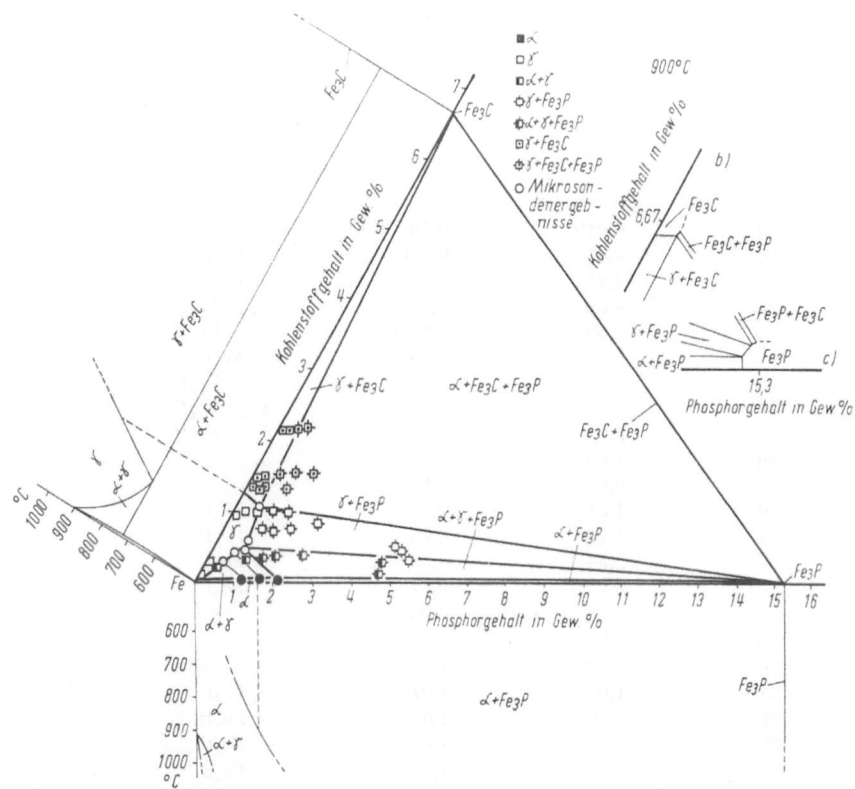

Abb. 5. Temperaturschnitt der Eisenecke des Systems Eisen-Phosphor-Kohlenstoff bei 900° C

im γ-Mischkristall. Dies wird durch die nicht experimentell ermittelten gestrichelten Konoden deutlich gemacht.

Legierung L_3 besteht aus den Gleichgewichtsphasen α, γ und Phosphid (Fe$_3$P). Die Analyse des α- und γ-Mischkristalls liefert zwei Eckpunkte dieses Dreiphasengleichgewichts.

Legierung L_4 enthält das Gleichgewicht γ + Phosphid. Mit der Zusammensetzung des γ-Mischkristalls hat man einen wichtigen Punkt auf der Linie gefunden, die das Feld des γ-Mischkristalls gegen das Phosphid abgrenzt.

An der dreiphasigen Legierung L_5 kann die Zusammensetzung des γ-Mischkristalls bestimmt werden, der sowohl mit Phosphid als auch Zementit im Gleichgewicht steht.

Mit Hilfe dieser fünf Legierungen läßt sich der 900° C-Temperaturschnitt der Eisenecke des Systems Fe—P—C in seinem wesentlichen Teil bestimmen. Abb. 5 zeigt diesen Schnitt vollständig bis zum Phosphid und Zementit. Die Ergebnisse der Mikrosondenuntersuchung sind durch Kreise besonders hervorgehoben. In diesen, allein auf Grund der Mikrosondenuntersuchung gezeichneten Temperaturschnitt sind eine größere Zahl weiterer Legierungen unter Kennzeichnung ihres Gefügezustandes eingezeichnet (siehe hierzu auch Tabelle). Alle Legierungen lassen sich sehr gut in die vorgegebenen Phasenfelder einordnen.

Die P- bzw. C-Löslichkeit des Zementits bzw. Phosphids wurde nicht untersucht. In Abb. 5b und c ist in stark vergrößertem Maßstab angedeutet, wie Zementit und Phosphid im Gleichgewicht mit ihren Nachbarphasen nach theoretischen Vorstellungen aussehen müssen.

Tabelle. *Zusammensetzung und Gefügezustand verschiedener, bei 900° C 100 h geglühter Eisen-Phosphor-Kohlenstoff-Legierungen*

Proben-Nr.	Gew.-% P	Gew.-% C	Gefügezustand
1	0,22	1,38	γ, Fe_3C
2	0,303	1,34	γ, Fe_3C
3	0,420	1,35	γ, Fe_3C
4	1,03	1,32	γ, Fe_3C, Fe_3P
5	0,056	0,09	γ
6	0,60	0,250	α, γ
7	0,662	0,39	α, γ
8	0,325	0,25	α, γ
9	0,049	0,28	γ
10	4,66	0,510	γ, Fe_3P
11	4,85	0,460	γ, Fe_3P
12	4,56	0,218	α, γ, Fe_3P
13	4,59	0,128	α, γ, Fe_3P
14	5,20	0,295	γ, Fe_3P
15	1,36	0,35	α, γ, Fe_3P
16	1,70	0,38	α, γ, Fe_3P
17	2,38	0,39	α, γ, Fe_3P
18	1,00	0,32	α, γ
19	0,64	0,66	γ
20	1,31	0,71	γ, Fe_3P
21	1,75	0,75	γ, Fe_3P
22	2,34	0,83	γ, Fe_3P
23	0,99	0,77	γ, Fe_3P
24	0,112	0,96	γ
25	0,31	1,00	γ
26	0,62	1,00	γ
27	1,05	1,01	γ, Fe_3P
28	1,46	1,00	γ, Fe_3P
29	0,76	1,03	γ, Fe_3P
30	0,11	1,49	γ, Fe_3C
31	0,32	1,51	γ, Fe_3C
32	0,66	1,52	γ, Fe_3P, Fe_3C
33	1,03	1,54	γ, Fe_3P, Fe_3C
34	1,56	1,53	γ, Fe_3P, Fe_3C
35	0,116	2,16	γ, Fe_3P
36	0,32	2,14	γ, Fe_3C
37	0,52	2,17	γ, Fe_3C, Fe_3P
38	0,71	2,19	γ, Fe_3C, Fe_3P

Zusammenfassung

Die Vorteile der Mikrosondenuntersuchung gegenüber den gebräuchlichen metallographischen und röntgenographischen Analysenverfahren bei der Aufstellung von Mehrstoffsystemen sind kurz zusammengefaßt folgende:

1. Man braucht relativ wenig Legierungen. Für die Untersuchung des univarianten Gleichgewichtes eines Temperaturschnitts benötigt man theoretisch eine einzige Legierung;

2. eine mehrphasige Legierung läßt sich relativ leicht herstellen, da die mehrphasigen Felder eines Mehrstoffsystems sich im allgemeinen über einen größeren Konzentrationsbereich erstrecken;

3. diese Methode liefert gleichzeitig den Verlauf der Konoden;

4. man braucht die im Gefüge vorliegenden Gleichgewichtsphasen nicht durch spezifische Ätzverfahren unterschiedlich anzuätzen oder anzufärben;

5. gegenüber neueren quantitativen metallographischen Verfahren hat diese Methode außerdem den Vorteil, daß man Strukturformeln von intermetallischen Verbindungen — sofern solche am Gleichgewicht beteiligt sind — nicht zu kennen braucht, da man diese Verbindungen unmittelbar auf ihre chemische Zusammensetzung untersuchen kann.

Literatur

1. VOGEL, R.: Arch. Eisenhuettenw. **3**, 369 (1929).
2. Optique des rayons X et microanalyse: IVe Congr. Internat. sur l'Optique des Rayons X et la Microanalyse, Orsay, Septembre 1965. Hermann, 115, Boulevard Saint Germain, Paris VI.
3. HANSEN, M.: Constitution of binary alloys. New York and London: McGraw-Hill Book Co., Inc. 1958.
4. Das Zustandsschaubild Eisen-Kohlenstoff und die Grundlagen der Wärmebehandlung der Eisen-Kohlenstoff-Legierungen. 4. Aufl. Bericht 180 des Werkstoffausschusses des Vereins Deutscher Eisenhüttenleute, Düsseldorf 1961.

Trace Elements in Ferro-Alloy Deoxidants and Their Influence on Non-Metallic Inclusion Compositions

A. G. FRANKLIN, G. RULE and R. WIDDOWSON*

British Iron and Steel Research Association, Sheffield, England

The use of the electron probe microanalyser to trace the sources of apparently exogeneous oxide inclusions in deoxidised steel melts is described. Correlation of inclusion analyses with melt oxygen contents and amounts of trace elements added with the ferro-alloys indicate that 98% or more of the inclusion population can be adequately explained by the purity of the alloys used. Ferro-silicon was found to contain calcium and aluminium and ferro-manganese and silico-manganese contained titanium.

Introduction

Many studies have been carried out on the sources of non-metallic inclusions in steel melts. Inclusions of two distinct types are recognised.

a) Indigeneous, resulting from precipitation of dissolved oxygen in the melt by the deoxidants added, and precipitation of sulphide during solidification.

b) Exogeneous, arising from external sources such as refractories and slags.

The steelmaker has a large degree of control over both inclusion types; over the former by selection of steelmaking technique and deoxidation practice, and over the latter by the use of adequate quality refractories, correct metal temperatures, teeming rates, etc.

Of the two inclusion types it is accepted that the exogeneous type on a weight for weight basis is often the more damaging since *macroscopic* as distinct from *microscopic* sizes are usually involved.

Considerable effort has therefore been applied to determine the magnitude of contamination of exogeneous inclusions and the influence of modified steelmaking practices. The most direct method consists of systematically labelling external sources such as refractory bricks, ramming compounds, furnace slags, etc., by means of suitable radioisotopes followed by inspection of the ingots or rolled products. Such trials have been criticised however because of doubts as to whether the tracers employed do behave in a truly similar way to the materials labelled. The results of such trials show very wide variations in indicated contamination from external sources, from approximately 20% of the total inclusion population to zero.

More recently the electron probe microanalyser has been extensively applied to quantitative inclusion analysis, the results of which may be used to identify the sources of inclusions [1]. Thus the presence of elements in inclusions peculiar to certain refractories or slags enable the probable sources of individual inclusions to be ascertained, e.g. the presence of potassium has been related to contamination by certain refractories, and lime and magnesia together indicate the presence of furnace or ladle slag [2]. This paper describes work at BISRA on the use of the microanalyser to determine inclusion sources with particular reference to the trace elements present in the deoxidants.

---

* Dr. FRANKLIN, Mr. RULE and Mr. WIDDOWSON are in the Steelmaking Division of BISRA, the Inter-Group Laboratories of the B.S.C. Mr. RULE has since joined New Zealand Steel.

Study of Deoxidation Techniques and Compositions of Inclusions Formed

An extensive study of deoxidation techniques is being carried out at BISRA which involves a detailed study of changes in inclusion and oxygen contents during melting, refining, casting and in the solidified ingot from 1.25 tonne basic arc melts of 0.2% carbon killed steel. A micro-analyser and automated inclusion counter are used to study inclusion compositions and distribution. Details of the techniques and principal results are to be reported elsewhere.

The ternary oxide system, $MnO-SiO_2-Al_2O_3$, in Fig. 1, shows the ratios of the three main components in the preferred inclusion compositions resulting from ferro-silicon plus ferro-manganese, and silico-manganese plus ferro-silicon deoxidation.

The relatively high Al_2O_3 content (20—40 wt.%) is noteworthy in view of the fact that no aluminium additions were made. In addition to the three principal oxides, which usually totalled 80 to 90%, small quantities of superficially exogeneous oxides were analysed. These were typically 1% TiO_2 after ferro-manganese additions and 3% CaO after ferro-silicon additions. Inclusions resulting from silico-manganese deoxidation were higher in TiO_2, averaging 4% TiO_2. Typical inclusion compositions microanalysed are shown in Table 1. Approximately 10% of the inclusions analysed were much richer in CaO than the mean quoted above, at between 10 and 20% CaO. Similar amounts of MgO were detected in only 2 or 3 inclusions out of several hundred analysed, indicating furnace or ladle slag as the main source of oxide in only 1 or 2% of the inclusions. Many of the inclusions were analysed for potassium, but none was detected. Table 2 gives mean "trace" element analyses for some hundreds of inclusions microanalysed in samples taken from the liquid steel after each of the two main deoxidation practices studied.

The content of the trace elements aluminium, titanium and calcium in the oxide inclusions was not immediately explainable in terms of furnace slag, refractories, or on the basis of nominal compositions of deoxidants. A more detailed examination of the ferro-alloys used was therefore carried out using the microanalyser.

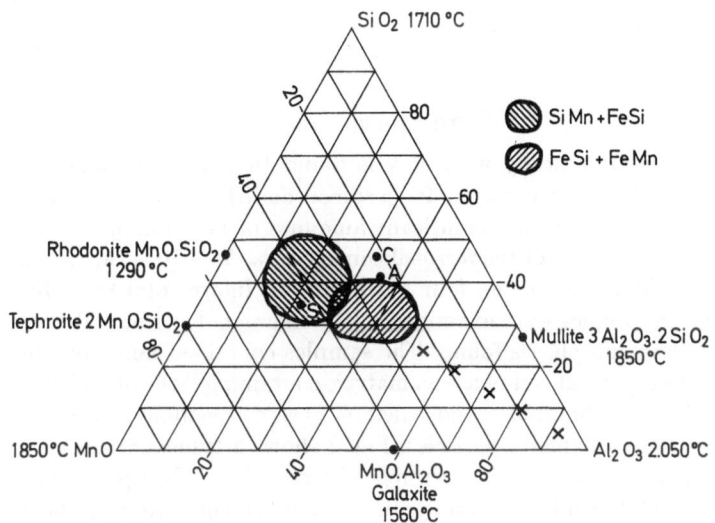

S Spessartite 1.195°C; A Mn Anortnite; C Mn Cordierite

Fig. 1. Oxide ternary for $MnO-SiO-Al_2O_3$ system

Table 1. *Typical inclusion analyses from samples taken from 1 tonne arc furnace, 0.2% C steel*

Time	Size (μm)	Wt.%					
		Al_2O_3	SiO_2	MnO	FeO	TiO_2	CaO
Melt out (0.6 C)	10	15	65	20	—	—	—
End of refining (0.2 C)	2	—	30	—	70	—	—
After FeSi + FeMn deoxidation	10	35	23	30	5	1.5	4.5
After SiMn + FeSi deoxidation	10	15	23	20	2	39	0.7

Table 2. *"Trace" elements in inclusions in liquid steel after deoxidation. Averages for over 100 inclusions from each of two practices*

0.4% Si + 0.8% Mn added to steel by	Average analyses, wt.%		
	Al_2O_3	TiO_2	CaO
a) FeSi + FeMn or	40	1.0	3.1
b) SiMn + FeSi	20	3.8	1.6

Microanalysis of Ferro-Alloys

Ferro-alloy samples were mounted in conducting plastic and given a normal metallographic polish. An A.E.I. SEM II microanalyser was used with a 0.100 micro-amp aperture current (corresponding to a 0.070 μA beam current, or 0.050 μA specimen current) and accelerating voltages ranging from 10—35 kV, depending on the particular phase being analysed. Comparison of X-ray counts for each of the elements Mg, Al, Si, Ti, Mn and Fe was made with pure metal standards. For sulphur an FeS standard was used and for calcium CaF_2. The apparent concentrations of the elements present in the samples were corrected for atomic number and absorption effects by the Philibert method and, where necessary, for fluorescence by the Castaing technique [3].

Errors in the quantitative analyses of the phases in the samples occurred because of experimental limitations, particularly in the counting statistics, and uncertainties in correction theory. The expected accuracy at the 95% confidence level was generally given by a relative error of 5% for each determination, e.g. $60 \pm 3\%$, $6.0 \pm 0.3\%$.

In addition to quantiative analysis, qualitative information was obtained from scanning electron and X-ray distribution images of areas between 100 and 200 μm square. The electron images, formed by the absorbed specimen current, gave information on atomic number (assuming a flat polish). The darker regions in the electron images are of higher mean atomic number than light regions [3].

Ferro-Silicon (75% Si)

Early in the deoxidation study work it was found that high alumina contents occurred in inclusions in the steel immediately after ferro-silicon additions. It is widely known in the steel industry that ferro-silicon contains significant amounts of aluminium, usually at the 1 to 2% level. The nominal composition of the ferro-silicon used was 74—76% Si, 1.5—2.0% Al, 0.6% Ca, balance iron. Aluminium analyses on four samples gave figures of 1.84, 1.76, 2.08 and 1.79%.

Three ferro-silicon samples were examined in the microanalyser and Table 3 lists the quantitative analyses of the main phases found. The samples contained numerous fissures and cracks. There appeared to be a pure silicon phase as matrix, containing laths of an iron-silicon-aluminium phase. Concentrations of aluminium and calcium were occasionally found associated with the latter phase, in 8—10 μm diameter particles. Fig. 2 shows a typical area and X-ray distribution photographs for the main elements in a sample of ferro-silicon. The pure silicon phase occupied approximately 75% of the microstructure, the remainder consisting of the Fe—Si—Al phase (51% Si, 42% Fe, 3 to 5% Al, 1% Mn) and the particles rich in Al and Ca. The literature on the iron-silicon phase diagram gives the following phases: — pure silicon (M.P. 1,430° C), FeSi (34% Si, 66% Fe, M.P. 1,410° C), zeta phase (53—57% Si, 47—43% Fe, M.P. 1,220° C). It is

Table 3. *Microanalyses of phases in ferro-silicon*

$\frac{3}{4}$ of microsection	100 wt.% Si
$\frac{1}{4}$ of microsection	51% Si, 42% Fe, 3 to 5% Al, 1% Mn (probably zeta phase)
Less than 1% of microsection	Al_2O_3 and CaO particles

Average analysis of ferro-silicon: 76% Si, 21% Fe, 1.8% Al, about 0.3% Ca.

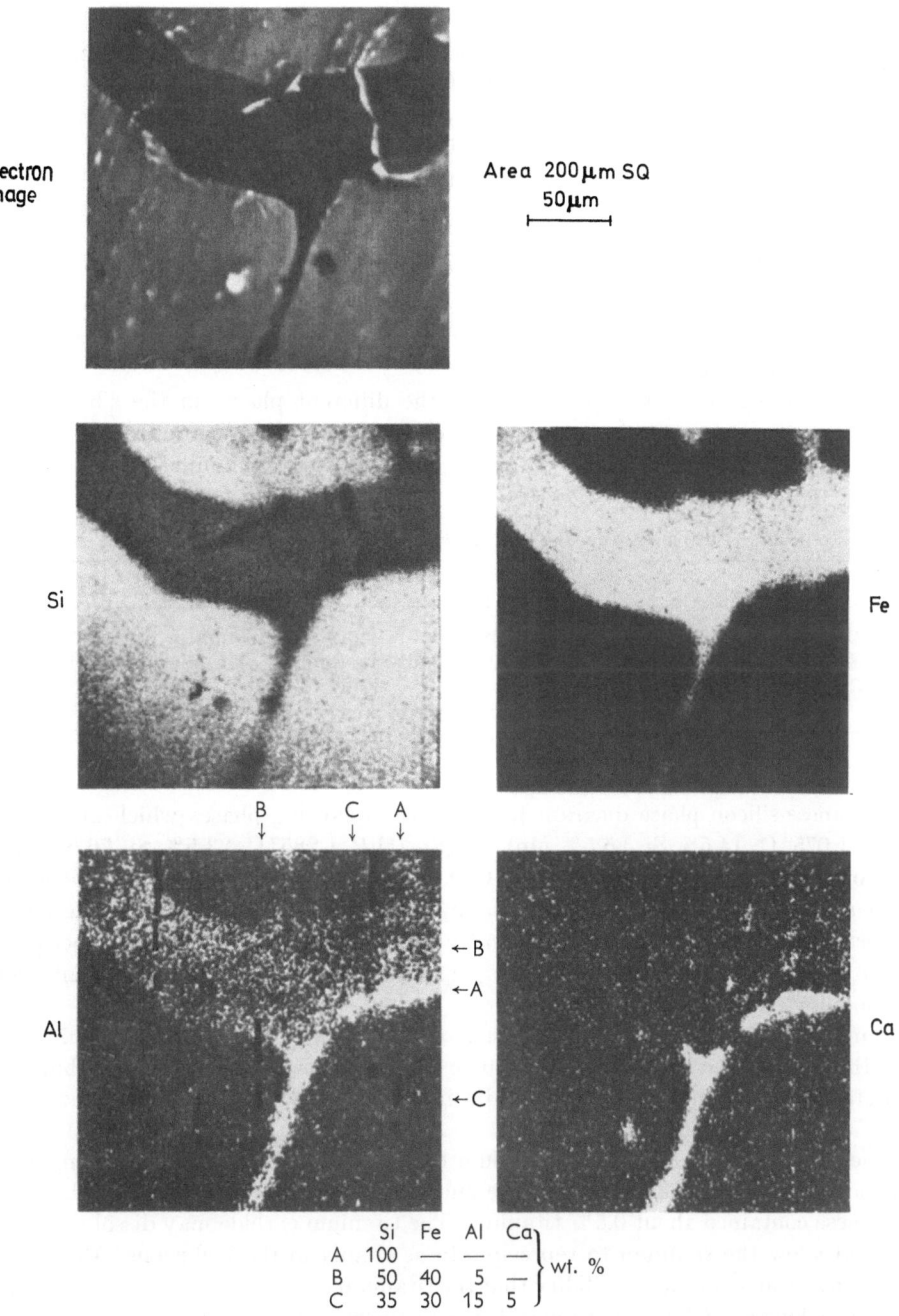

Fig. 2. Microanalysis of ferro-silicon used for deoxidation of 0.2 C steel

concluded that the iron rich phase in the ferro-silicon is the zeta phase, and the aluminium is largely present in this phase in solid solution, although there may be some alumina-lime inclusions scattered throughout both phases of the ferro-silicon.

Ferro-Manganese (80 % Mn, 6 % C)

The samples contained cracks and fissures and were predominantly single phase. Microanalysis of selected areas gave results of 74 to 81 % Mn, and 13 to 20 % Fe. The balance was assumed to be carbon. A trace of silicon was found. Overall contents of titanium, calcium and aluminium were all less than 0.1 %, the limit of detection.

It was thought however that the 1% TiO_2 present in inclusions originated in the ferro-manganese, although even if it did, the amount present was probably below the limit of detection. A typical chemical analysis [4] included 1% Si, 0.08% Ti.

Titanium, in the form of occasional large angular titanium carbide particles was positively identified by microanalysis after further examination of the ferro-manganese. The probable content, on the basis of the frequency of the particles, was about 0.05% Ti.

Silico-Manganese (70% Mn, 20% Si)

After additions of silico-manganese to the steel, the inclusions contained between 1 and 40% TiO_2. Table 1 records an analysis with rather an extreme TiO_2 content. The mean analysis of inclusions was about 4% TiO_2.

Table 4 gives the results of microanalysis of the different phases in the silico-manganese. Two main phases were present, together with frequent clusters of titanium carbide particles. A typical cluster is illustrated in Fig. 3. Titanium, calcium and aluminium were below the limits of detection.

Table 4. *Microanalyses of phases in silico-manganese*

	Wt.%
Main phase	70% Mn, 24% Si, 6% Fe (probably Mn_5Si_3)
Second phase	84% Mn, 10% Si, 6% Fe (probably Mn_3Si—Mn eutectic)
Clusters of particles	TiC

Average analysis of silico manganese: 72% Mn, 20% Si, 6% Fe, 2% C, 0.5% Ti.

The manganese-silicon phase diagram [5] gives the following phases which are of interest; Mn_3Si (M.P. 1,075° C, 14.5% Si, 85.5% Mn), Mn_5Si_3 (M.P. 1,285° C, 23.5% Si, 76.5% Mn), and an eutectic of Mn_3Si and Mn forms at 1,040° C at 12% Si, 88% Mn. If the iron and manganese were considered together in the alloy, then the high atomic number phase in the electron image probably corresponded to the eutectic between Mn_3Si and Mn. The low atomic number phase was probably Mn_5Si_3. As with the ferro-silicon, the phases have melting points which span the ranges given for the alloy (1,070 to 1,320° C).

The main features of interest were the clusters of 1 to 3 μm diameter particles which occurred frequently throughout the microstructure. Some of the particles were sulphide, but the vast majority were rich in titanium and carbon and contained no oxygen. They were clearly titanium carbide.

It is concluded that the clusters of titanium carbide particles in the silico-manganese lead to a significant amount of titanium being present in the deoxidation product in the steel. The silico-manganese contained about 0.5% titanium. The titanium carbide may dissolve in the steel, dissociate, and allow the titanium to reprecipitate as titania in the inclusions. Alternatively it is interesting to speculate on the possibility that the titanium carbide particles remain undissolved in the liquid steel long enough to act as nuclei for oxide inclusions.

Discussion

Table 5 lists the weights of fero-alloys added in two deoxidation procedures to give a nett addition of 0.4% Si and 0.8% Mn to the bath. Also listed are the amounts of the trace elements aluminium, calcium and titanium inadvertently added with the ferro-alloys. These are given in terms of parts per million, or ppm, (where 100 ppm = 0.01 wt.%) for convenience.

For ferro-silicon plus ferro-manganese deoxidation the powerful deoxidisers aluminium, calcium and titanium were added to the extent of 100 ppm Al, 15 ppm Ca and 4 ppm Ti. These were sufficient to form 190 ppm Al_2O_3, 20 ppm CaO and 7 ppm TiO_2, removing some 110 ppm of oxygen from the steel. These calculations assume 100% recovery of the deoxidisers, which

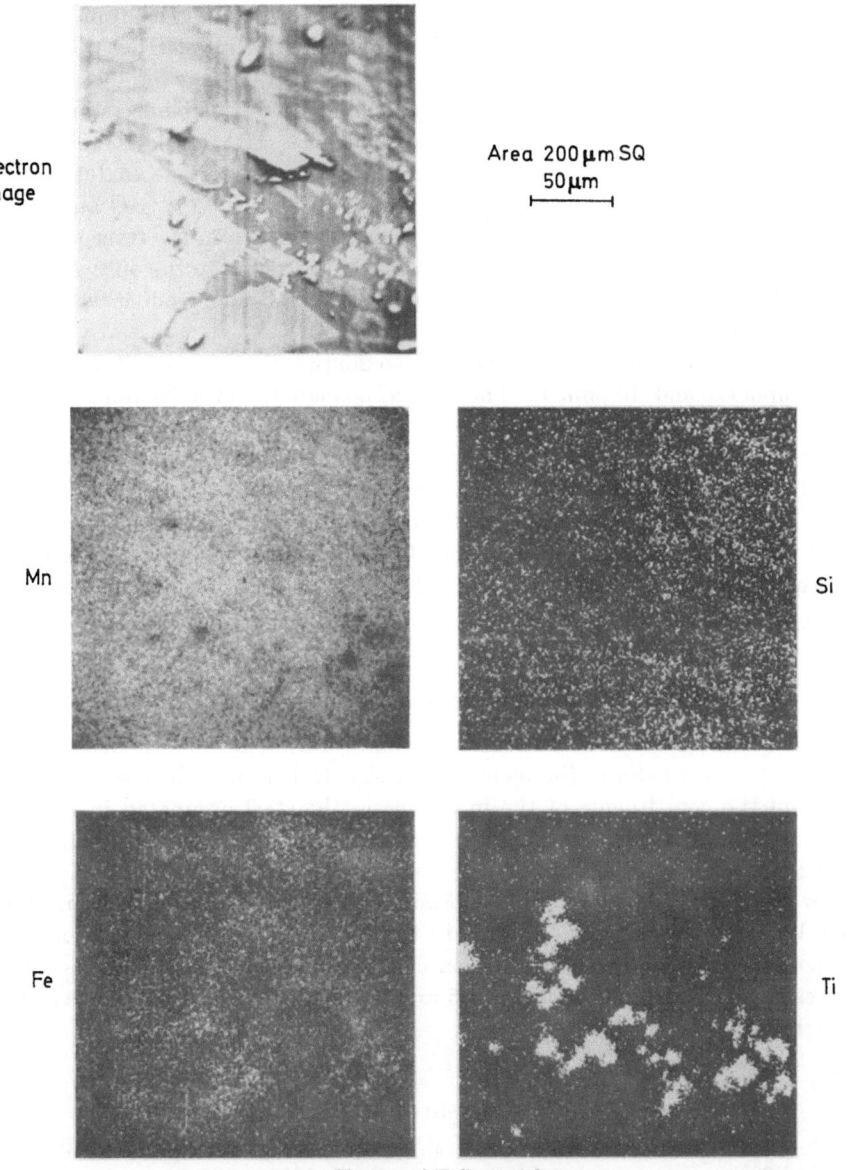

Fig. 3. Microanalysis of silico-manganese used for deoxidation of 0.2 C steel

Table 5. *Comparison of trace elements added in ferro-alloys with recovery as oxides in inclusions in 0.2 C steel. Weight of steel bath = 1,230 kg (2,700 lb.). Oxygen content before deoxidation = 0.025%, or 250 ppm. This was sufficient to form about 500 ppm oxide*

Deoxidation practice		Weight of ferro-alloys added to bath	Trace elements added, as proportion of bath (ppm)	Deoxidation product formed[a] (ppm)
FeSi +		6.8 kg (15 lb.)	100 Al	200 Al_2O_3
			15 Ca	15 CaO
	FeMn	11 kg (24 lb.)	4 Ti	5 TiO_2
SiMn +		12.4 kg (27 lb.)	40 Ti	20 TiO_2
	FeSi	3.4 kg (7½ lb.)	50 Al	100 Al_2O_3
			8 Ca	8 CaO

[a] The remainder in each case was about 150—200 ppm MnO; 150 ppm SiO_2.

was likely for aluminium and calcium, but some titanium could remain in solution in equilibrium with oxygen. Since there was about 250 ppm of oxygen in the refined bath, enough strong deoxidisers were added, as trace elements in the ferro alloys, to remove over 40% of the oxygen. When all the oxygen in the bath was subsequently combined with deoxidisers on solidification a deoxidation product including 38% Al_2O_3, 4% CaO and 1.5% TiO_2 would therefore have been expected. The remaining oxide would be SiO_2 and MnO. In fact the average analyses of some two hundred inclusions in the steel, recorded in Table 2, was 40% Al_2O_3, 3% CaO and 1% TiO_2, in excellent agreement with what might have been expected to arise from the trace elements in the ferro-silicon and ferro-manganese. The final column in Table 5 refers to the 40% Al_2O_3, 3% CaO and 1% TiO_2 as proportions of the estimated 500 ppm of deoxidation product which must have been formed in the liquid steel.

Similarly for silico-manganese plus ferro-silicon deoxidation the trace additions corresponded to 50 ppm Al, 8 ppm Ca and 40 ppm Ti. These were enough to form 95 ppm Al_2O_3, 10 ppm CaO and 70 ppm TiO_2, removing 75 ppm out of the 250 ppm oxygen in the bath. This was about a third of the total oxygen. The final deoxidation product could therefore be expected to contain 20% Al_2O_3, 2% CaO and 15% TiO_2. In fact the average composition of one hundred inclusions was 20% Al_2O_3, 1,6% CaO and 3.8% TiO_2, but 10 of the inclusions contained over 7% TiO_2. The recovery of aluminium and calcium was therefore very good, but less titania appeared in the inclusions than might have been expected. However, as mentioned previously, titanium is more likely to remain in solution in the steel.

Conclusions

A rough balance has been obtained between the weights of trace elements added by the deoxidants and their recovery as oxides in the inclusions in the steel. It was therefore unnecessary to assume that any of the constituents of the inclusions in the steel originated in refractories or furnace slag, since the ferro-alloys supplied all the oxide-forming elements needed for the formation of the inclusions in the steel.

The inclusion compositions were determined primarily by the powerful deoxidisers aluminium, titanium and calcium added as trace elements in the ferro-alloys. Any oxygen left over after the powerful deoxidisers had formed oxides then combined with silicon and manganese. Even small percentages of trace elements between 0.05 and 2% of the ferro-alloys, made a significant contribution to the inclusion composition.

References

1. Salmon Cox, P. H., and J. A. Charles: J.I.S.I, **201**, 863—872 (1963).
2. — — J.I.S.I. **203**, 493—499 (1965).
3. Franklin, A. G.: BISRA Open Report SM/B/108/67.
4. Bureau of Analysed Samples Ltd., Middlesborough, March, 1966.
5. Binary phase diagrams, Hansen, 1964.

Untersuchungen an supraleitenden Diffusionsschichten mit einer Elektronenstrahlmikrosonde

G. OTTO und H. WIZGALL

Institut für angewandte Physik der Universität Gießen, BRD

Abstract

Superconducting diffusion layers of the systems Nb—Sn, V—Si, and V—Ga containing intermetallic compounds of the A 15-structure have been investigated with a JEOL electron probe microanalyser of the type JXA-3A. The influence of different methods of diffusion as well as different conditions of preparation on the composition and thickness of the layers is shown. Diffusion wires of the system Nb—Sn prepared by vapour phase diffusion consist of only one phase with a composition being richer in niobium than the stoichiometric phase Nb_3Sn. The composition of the A 15-phase on corresponding samples of the system V—Si is very close to V_3Si. Diffusion layers of the most thoroughly investigated system V—Ga include the two phases V_6Ga_5 and V_3Ga besides an outer layer V—Ga—Si built up at higher temperatures. In addition the phase V_2Ga_5 is found in layers prepared by the dipping process. There is one sample with a diffusion temperature of 1,000° C showing all phases of the phase diagram by van VUCHT et al.

I. Einleitung

Der Aufbau der hier untersuchten supraleitenden Diffusionsschichten ist deshalb von besonderem Interesse, weil sie allesamt intermetallische Verbindungen vom A 15-Typ (β-Wolfram-Typ) mit den höchsten bisher gemessenen supraleitenden kritischen Daten enthalten. Es soll deshalb an den Zweikomponentensystemen Nb—Sn, V—Si und V—Ga in Abhängigkeit von den verschiedenen Diffusionsverfahren und -bedingungen der Anteil dieser Verbindungen an der Diffusionsschicht, ihre möglichen Abweichungen von der genau stöchiometrischen Zusammensetzung sowie die daneben auftretenden Phasen untersucht werden. Da die Dicke der Diffusionsschichten im Bereich zwischen 1 und 100 µm liegt, ist die Elektronenstrahlmikrosonde ein für derartige Messungen geeignetes Gerät.

II. Experimentelles und Methodisches

Die folgenden Untersuchungen sind mit einer Mikrosonde der Firma Jeol vom Typ JXA-3A durchgeführt worden. Der auf die Proben auftreffende Elektronenstrahl war stets auf 1 µm Durchmesser fokussiert. Die Anregungsspannung betrug für sämtliche Elemente 25 kV, wobei zum Nachweis von V, Si und Ga die K_a-, zum Nachweis von Nb sowie Sn die L_a-Röntgenlinie diente. Das Meßverfahren bestand in einer Linienabtastung der Diffusionsschichten bei feststehendem Elektronenstrahl und vorgeschobener Probe. Für die quantitative Auswertung sorgten Punktanalysen an einzelnen Stellen längs der abgetasteten Linie. Um die verschiedenen Messungen miteinander vergleichen zu können, betrug der Probenstrom in allen Fällen 2×10^{-7} A. Als Standards kamen die reinen Elemente zur Verwendung, die in der Diffusionsschicht enthalten sind. Die in Kunststoff eingebetteten, geschliffenen und polierten Proben sowie Eichsubstanzen waren mit einer ca. 500 Å dicken Aluminiumschicht bedampft, um eine Aufladung der verhältnismäßig kleinen Proben (Drähte mit 0,5 mm Durchmesser) zu vermeiden. Eine elektrolytische Verkupferung der Proben verhinderte wenigstens größtenteils das Ausbrechen der dünnen und sehr spröden Diffusionsschichten.

Tabelle. *Zusammensetzung und Dicke von Diffusionsschichten im System V—Ga, die bei verschiedener Diffusionstemperatur und einer Diffusionsdauer von 20 Std hergestellt sind (zum Vergleich Gew.-% Vanadium im V_3Ga 68,7; im V_6Ga_5 46,7)*

Probenart	Diffusionstemperatur (°C)	Diffusionszone	Gew.-% Vanadium, gemessen	Gew.-% Vanadium, korrigiert	Dicke der Diffusionsschicht (μm)
Dampfphasendrähte	700	I	44,6	46,8	1
	800	I	69,5—61,0	71,5—63,2	2
		II	46,0	48,3	6
	950	I	72,0—60,5	73,8—62,7	15
		II	46,3	48,7	15
		III	schwankend (V—Ga—Si)	—	10
	1100	I	64,8	67,0	20
		II	56,8	59,2	90
		III	schwankend (V—Ga—Si)	—	20
	1200	I	Mischkristall von 100 bis 85	Mischkristall von 100 bis 86	75
		II	67,2	69,2	12
		III	schwankend (V—Ga—Si)	—	25
Durchziehdrähte	700	I	46,2	48,5	13
	800	I	70,5—62,0	72,5—64,3	5
		II	45,1	47,5	15
	950	I	65,7	68,0	12
	1100	I	66,3	68,5	25
		II	schwankend (V—Ga—Si)	—	15
	1200	I	66,6	68,8	25
		II	schwankend (V—Ga—Si)	—	15

Die zu den Messungen verwendeten Proben der Systeme Nb—Sn und V—Si waren Dampfphasendrähte [1, 2]. Bei ihrer Herstellung diffundiert die niedrigschmelzende Komponente aus der Dampfphase in einen Niob- bzw. Vanadium-Draht. In dem am eingehendsten untersuchten System V—Ga gelangten außerdem Durchzieh- [3] und Tauchdrähte [1] zur Messung. Bei der Herstellung von ersteren wird vor dem Diffusionsprozeß ein Vanadium-Draht durch ein Gallium-Bad gezogen und dabei gleichmäßig mit Gallium benetzt. Bei letzteren erfolgt die Diffusion des Galliums aus der Schmelze mit anschließender Nachdiffusion in der Dampfatmosphäre.

Die gemessenen Röntgenstrahlintensitäten von Nb, Sn, V, Si und Ga (allgemein als Element A bezeichnet) müssen hinsichtlich des Untergrundes und der Totzeit des Zählers korrigiert werden; nach diesen Korrekturen ist das Intensitätsverhältnis von Legierung zu Standard in der Tabelle als „Gewichtsprozent A, gemessen" angegeben. Um die tatsächliche Konzentration des Elements A zu erhalten — in der Tabelle und in Abb. 1 als „Gew.-% A, korrigiert" angegeben —, erfolgt die Atomnummer-Korrektur nach dem Verfahren von Thomas [4], die Absorptionskorrektur nach dem von Duncumb und Shields [5] modifizierten Verfahren von Philibert [6], die Sekundärfluoreszenz-Korrektur für die K-Strahlung nach der graphischen Methode von Wittry [7], für die L-Strahlung nach dem Verfahren von Birks [8]. Die verwendeten Massenabsorptionskoeffizienten sind den Tabellen von Heinrich [9] entnommen.

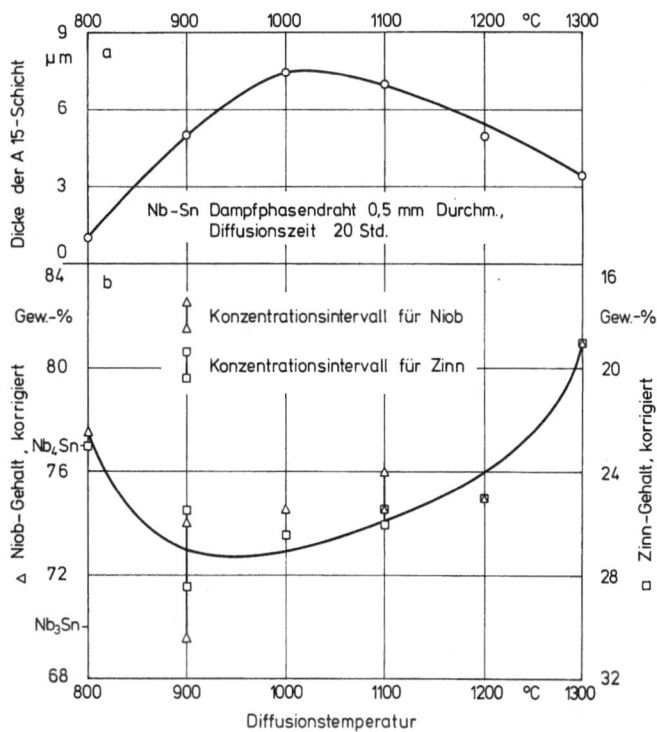

Abb. 1 a u. b. Dicke (a) und Zusammensetzung (b) der an Nb—Sn Dampfphasendrähten gemessenen Diffusionsschicht mit A 15-Struktur in Abhängigkeit von der Diffusionstemperatur

III. Meßergebnisse mit Auswertung und Diskussion

Sämtliche bisher durchgeführten Messungen an Dampfphasendrähten des Systems Nb—Sn zeigen, daß unabhängig von den Herstellungsbedingungen nur eine Phase gebildet wird, die niobreicher als die stöchiometrische Verbindung Nb_3Sn ist und nach Debye-Scherrer-Aufnahmen die A 15-Struktur besitzt. Aus Abb. 1 b ist die Zusammensetzung dieser Phase in Abhängigkeit von der Diffusionstemperatur bei konstanter Diffusionsdauer von 20 Std ersichtlich. Für die bei 900 und 1100° C hergestellten Proben ist ein Konzentrationsintervall angegeben, innerhalb dessen die Zusammensetzung der Phase variiert. Proben mit einer Diffusionstemperatur zwischen 900 und 1000° C kommen der stöchiometrischen Zusammensetzung Nb_3Sn am nächsten. Sie besitzen als Supraleiter die höchsten kritischen Daten. Wie aus Abb. 1a hervorgeht, nimmt die Dicke der Schicht mit A 15-Struktur bei Proben mit Diffusionstemperaturen über 1100° C aus bisher nicht geklärten Gründen ab.

RINDERER et al. [10] haben an Diffusionsproben und einer Sinterprobe festgestellt, daß die Phase mit A 15-Struktur die Zusammensetzung Nb_4Sn besitzt. Grundsätzlich bestätigen die hier durchgeführten Messungen den Niob-Reichtum der A 15-Phase, jedoch zeigen sie eine Abhängigkeit der exakten Zusammensetzung von der Diffusionstemperatur.

Linienabtastungen mit der Mikrosonde an Dampfphasendrähten des Systems V—Si waren nicht immer durchführbar, weil die Diffusionsschichten äußerst spröde und porös sind und leicht vom Vanadium-Kern abblättern. Sichergestellt ist, daß Proben, die bei 1000, 1100 und 1300° C während 20 Std geglüht sind, eine Diffusionsschicht enthalten, die um höchstens 3 Gewichtsprozente Vanadium von der Zusammensetzung V_3Si abweicht. Die entsprechenden Schichtdicken bei diesen Proben betragen ca. 5, 10 bzw. 30 μm.

Sehr eingehend sind Diffusionsschichten des Systems V—Ga untersucht worden. Die Tabelle zeigt, daß alle Dampfphasen- und Durchziehdrähte, die 20 Std lang zwischen 700 und 1200° C wärmebehandelt sind, neben einer bei höheren Temperaturen auftretenden siliziumhaltigen Zone nur Schichten enthalten, die der stöchiometrischen Zusammensetzung der Phasen V_6Ga_5 und

V_3Ga sehr nahe sind. Die Numerierung der Diffusionszonen erfolgt in der Reihenfolge vom Vanadium-Kern nach außen hin. Die galliumreiche Phase V_2Ga_5, wie sie Maier et al. [11] finden, tritt nicht auf. Die Ursache für den Unterschied dürfte in einem beschränkten Gallium-Angebot liegen, wobei das Gallium immer weiter in den Vanadium-Kern eindringt.

Die äußerste Schicht besteht bei Proben mit höheren Diffusionstemperaturen aus einem Dreikomponentensystem V—Ga—Si, deren Dicke und Silizium-Gehalt mit steigender Diffusionstemperatur zunimmt. Das eindiffundierte Silizium rührt von den Quarzampullen her, in denen der Diffusionsprozeß stattgefunden hat. Diese Beobachtung unterstützt die schon von van Vucht et al. [12] geäußerte Vermutung, daß das Ansteigen der Sprungtemperatur von V—Ga-Proben mit der Diffusionstemperatur bis in die Nähe der Sprungtemperatur von V_3Si durch diese siliziumhaltige Schicht V—Ga—Si bedingt ist. Die Sprungtemperatur von reinem V_3Ga liegt bei rund 14° C.

Durchziehdrähte, die vor dem Diffusionsprozeß mit einer 10 bis 20 μm dicken Kupferschicht überzogen sind, zeigen nicht die von Tachikawa und Tanaka [13] angegebene wesentliche Zunahme der Schichtdicke von V_3Ga. Auch keine der von diesen Autoren gefundenen galliumreicheren Phasen V_2Ga, V_3Ga_2, VGa und VGa_2 wird an den hier untersuchten Dampfphasen-, Durchzieh- und Tauchdrähten beobachtet.

Bei Tauchdrähten des Systems V—Ga mit den Herstellungsbedingungen 800 bzw. 900° C, einer Tauchzeit von 5 sowie einer Nachdiffusionszeit von 15 min können nur die beiden Phasen V_6Ga_5 (25 bzw. 35 μm Dicke) und V_2Ga_5 (jeweils 3 μm Dicke) beobachtet werden. Eine Verlängerung der Tauchzeit führt zu einer Vergrößerung der Schichtdicke. Im Gegensatz zu den Dampfphasen- und Durchziehdrähten wird durch das reichliche Gallium-Angebot die Phase V_2Ga_5 gebildet, während bei diesen niedrigen Diffusionstemperaturen eine mögliche Ausbildung von V_3Ga unterhalb der Beobachtungsgrenze liegt. Ein Tauchdraht mit den Herstellungsbedingungen: 1000° C Diffusionstemperatur, einer Tauchzeit von 1 und einer Nachdiffusionszeit von 15 min zeigt neben der äußersten Schicht V—Ga—Si (8 μm Dicke) sämtliche im Phasendiagramm von van Vucht et al. [12] vorkommenden Phasen: V_3Ga (1,5 μm Dicke), V_6Ga_5 (30 μm Dicke), V_6Ga_7 (18 μm Dicke) und V_2Ga_5 (17 μm Dicke).

Herrn Prof. Dr. E. Saur danken wir für die kritische Durchsicht des Manuskripts. Die Deutsche Forschungsgemeinschaft und die Fraunhofer-Gesellschaft haben diese Untersuchungen in dankenswerter Weise durch Bereitstellung von Personal- und Sachmitteln unterstützt.

Literatur

1. Rinderer, L., E. Saur u. J. Wurm: Z. Physik 174, 405 (1963).
2. Koch, D., G. Otto, and E. Saur: Phys. Letters 4, 292 (1963).
3. Bonnenberg-Koch, D., G. Otto u. E. Saur: Z. Physik (in Vorbereitung).
4. Thomas, P. M., and R. Theisen: AERE Report 4593 (1964, entnommen aus Quantit. electron microprobe analysis. Berlin-Heidelberg-New York: Springer 1965).
5. Duncumb, P., and P. K. Shields: The electron microprobe. New York: Academic Press 1966.
6. Philibert, J.: X-ray optics and X-ray microanalysis. New York: Academic Press 1963.
7. Wittry, D. B.: Diss., Calif. Inst. Technology (1957).
8. Birks, L. S.: Electron probe microanalysis. New York: Interscience Publ. 1963.
9. Heinrich, K. F. J.: The electron microprobe. New York: J. Wiley 1966.
10. Rinderer, L., J. Wurm, W. Züllig u. Z. de Beer: Z. Physik 179, 407 (1964).
11. Maier, R. G., Y. Uzel u. H. Kandler: Z. Naturforsch. 21a, 531 (1966).
12. Vucht, J. H. N. van, H. A. C. M. Bruning, H. C. Donkersloot, and A. H. Gomes de Mesquita: Philips Res. Rept. 19, 407 (1964).
13. Tachikawa, J., and Y. Tanaka: Japan. J. Appl. Phys. 6, 782 (1967).

Untersuchung von Diffusionsvorgängen an Mehrstoff-Gleitlagern mit der Elektronenstrahl-Mikrosonde

M. Semlitsch

Metallkundliches Labor der Gebrüder Sulzer, Aktiengesellschaft, Winterthur, Schweiz

1. Einleitung

In Lokomotiv-Dieselmotoren werden an Gleitlager besonders hohe Anforderungen in bezug auf Ermüdungsfestigkeit, Belastbarkeit, Benetzbarkeit, Verschleißwiderstand sowie Korrosionsbeständigkeit gestellt [1].

Im Betriebseinsatz erweisen sich hierfür Mehrstoff-Gleitlager mit Blei/Zinn- bzw. Blei/Indium-Laufschichten auf Lagerausgüssen aus Aluminiumlegierungen bzw. aus Bleibronze als brauchbar (Abb. 1).

Für die Korrosionsbeständigkeit gegenüber schwefelhaltigen Schmierölen ist ein Mindestgehalt von etwa 6 Gew.-% Zinn bzw. Indium in der Bleibasis-Laufschicht erforderlich.

Um die Abdiffusion dieser Elemente von der Laufschicht in den Lagerausguß zu verhindern, werden dazwischen dünne Nickelsperrschichten galvanisch abgeschieden.

Beim Überschreiten der an Gleitlagern herrschenden Betriebstemperatur von etwa 110° C zufolge lokal auftretender Überbelastung der Laufschicht (z. B. durch Kantenpressung) erfolgen

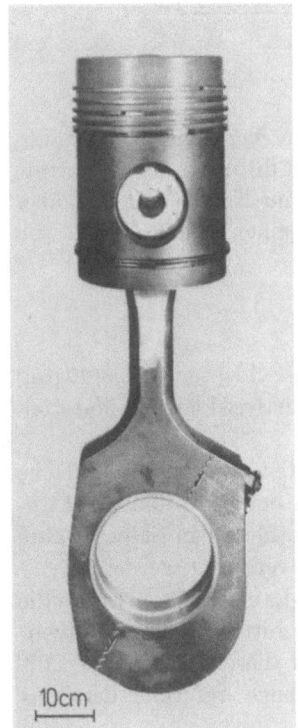

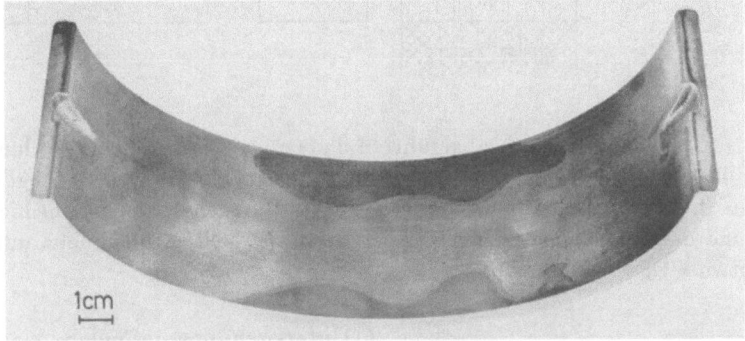

Abb. 2. Schwarze Korrosionsflecken auf Gleitlagerschale mit Blei/Indium-Laufschicht auf Bleibronzeausguß in Stahlstützschale nach 1000 Betriebsstunden

Abb. 1. Schubstange eines Dieselmotors mit Kolben (oben) und Lager (unten)

derartige Diffusionsvorgänge in verstärktem Maße. Durch die Bildung intermetallischer Verbindungen an den Übergangszonen der einzelnen Gleitlagerschichten tritt eine Verarmung der Laufschicht an Zinn bzw. Indium ein. Die Folge davon sind Korrosionsschäden in der Laufschichtoberfläche (Abb. 2), welche unter Umständen Betriebsstörungen verursachen könnten.

Zum besseren Verständnis dieser Diffusionsvorgänge erfolgten metallkundliche Untersuchungen an Querschnitten betriebsgeprüfter Gleitlager sowie an unterschiedlich wärmebehandelten Gleitlagerproben.

2. Untersuchte Mehrstoff-Gleitlager

Zur Untersuchung standen die bisher in der Praxis bewährten Mehrstoff-Gleitlager mit Sn- bzw. In-haltigen Bleibasis-Laufschichten und eingebauten Diffusionssperrschichten auf dem darunterliegenden Lagerausguß plus Stahlstützschale zur Verfügung (Tabelle 1).

Tabelle 1. *Querschnittsaufbau und Zusammensetzung von Mehrstoff-Gleitlagern für die Untersuchung von Diffusionsvorgängen*

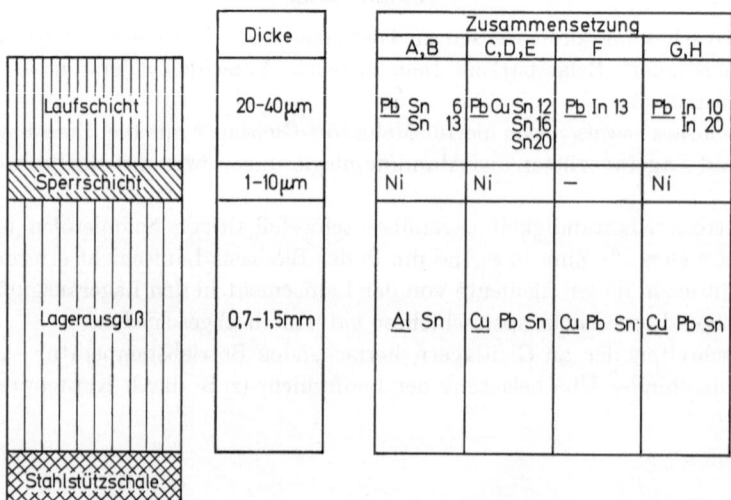

	Dicke	Zusammensetzung			
		A,B	C,D,E	F	G,H
Laufschicht	20–40 μm	<u>Pb</u> Sn 6 Sn 13	<u>Pb</u> Cu Sn 12 Sn 16 Sn20	<u>Pb</u> In 13	<u>Pb</u> In 10 In 20
Sperrschicht	1–10 μm	Ni	Ni	—	Ni
Lagerausguß	0,7–1,5 mm	<u>Al</u> Sn	<u>Cu</u> Pb Sn	<u>Cu</u> Pb Sn	<u>Cu</u> Pb Sn
Stahlstützschale					

Die der Reihe nach angeführten Typen *A* bis *H* stellen verschiedene Versuchsgleitlager dar, mit dem Ziel, eine bei erhöhten Temperaturen (110 bis 170° C) möglichst diffusionsarme Variante zu finden. Der anfängliche Zinn- bzw. Indiumgehalt der Laufschicht im Neuzustand soll über eine möglichst lange Betriebsdauer konstant bleiben und nicht unter einen Mindestgehalt von etwa 6 Gew.-% absinken.

3. Untersuchungsmethoden

Bei der Untersuchung der Mehrstoff-Gleitlager mit relativ dünnen Schichten (1 bis 40 μm) wurden hauptsächlich solche Analysenmethoden eingesetzt, welche die Analyse kleinster Materialbereiche erlauben.

Die Analyse des Zinn- bzw. Indiumgehaltes in der Laufschicht erfolgte an der Lageroberfläche entnommenen Spänen (Analysenmenge 1 bis 5 mg) mittels eines Atomabsorptions-Spektrophotometers [2] — Perkin-Elmer, Zweistrahlmodell 303 —, wobei das Metall in Salpetersäure gelöst und mit Wasser auf die jeweils geeignete Analysenkonzentration verdünnt wurde.

Zur zerstörungsfreien Bestimmung des Indiumgehaltes in Bleibasis-Laufschichten wurde die Methode der Gitterkonstantenermittlung des Pb/In-Mischkristalls mittels eines Röntgen-Zählrohrgoniometers, Philips (Cr—K_a Strahlung), eingesetzt [3]; bei diesem Verfahren läßt sich indirekt die in der Laufschichtoberfläche vorhandene Indiummenge aufgrund der Verminderung der Gitterkonstante von Reinblei bestimmen.

Die bei den verschiedenen Wärmebehandlungen gebildeten Diffusionszonen in den einzelnen Gleitlagertypen wurden nach galvanischer Verkupferung der Laufschichtoberfläche in Mikro-quer- bzw. Mikroschräg-Schliffen metallographisch untersucht. Zur Gefügeentwicklung eignete sich besonders das Aufdampfen von Interferenzschichten aus Zinkselenid [4], wobei ein hoher Farbkontrast eine deutliche Trennung der einzelnen das Gleitlager aufbauenden Schichten möglich machte.

Zur Bestimmung der chemischen Zusammensetzung in kleinsten Bereichen der Ausgangs-schichten und der neugebildeten Diffusionszonen wurde die Elektronenstrahl-Mikrosonde ein-gesetzt (Typ CAMECA MS 46 der Eidgenössischen Materialprüfungs- und Versuchsanstalt in Dübendorf). Bei dieser Untersuchungsmethode löst ein feinfokussiertes Elektronenstrahlbündel (25 kV Beschleunigungsspannung) auf dem Mikroschliff eine Röntgenfluoreszenzstrahlung aus, die man sowohl hinsichtlich Wellenlänge (Art des Elements) wie Intensität (Menge des Elementes) analysiert. Damit ist es möglich, von einem punktförmigen Bereich ($\varnothing$ 1—3 μm), längs einer Linie oder über eine quadratische Fläche die chemische Zusammensetzung zu ermitteln.

4. Diffusionsvorgänge in Gleitlagerschichten

4.1. Gleitlager mit Blei/Zinn-Laufschicht und Nickelsperrschicht auf Lagerausguß

Dieser Typ von Gleitlagern mit einem Lagerausguß aus Bleibronze auf einer Stahlstützschale repräsentiert eines der bis heute bestbewährten Hochleistungslager [5]. In letzter Zeit versuchten verschiedene Gleitlagerhersteller den Lagerausguß aus Aluminiumlegierungen herzustellen [6].

An zwei derartigen Lagern A und B mit folgendem Aufbau im Ausgangszustand (Tabelle 2) ergaben sich nach einer Wärmebehandlung 400 Std/170° C[1] interessante Veränderungen zufolge Abdiffusion von Zinn aus der Laufschicht in die Nickelsperrschicht.

Bei dem Vergleich der beiden Mikrogefüge vom Querschnitt des Gleitlagers A im Neuzustand und im wärmebehandelten Zustand zeigt sich nach der Wärmebehandlung deutlich die Bildung einer neuen Zone zwischen Laufschicht und Nickelsperrschicht; ein ähnliches Ergebnis wurde für das gleich wärmebehandelte Gleitlager B erhalten.

Tabelle 2. *Aufbau und Zusammensetzung zweier Mehrstoff-Gleitlager mit Pb/Sn-Laufschichten auf Aluminium-Lagerausguß*

Zonen	Schichtdicke		Zusammensetzung	
	A	B	A	B
Laufschicht	22 μm	40 μm	*Pb* Sn 6	*Pb* Sn 13
Diffusionsschicht (nach 400 Std/170° C)	5 μm	3 μm	Ni_3Sn_2	Ni_3Sn_4
Sperrschicht (Neuzustand)	10 μm	4 μm	Ni	Ni
Lagerausguß	1 mm	0,5 mm	*Al* Sn 6 Pb 18	*Al* Cu Mg Si

Untersuchungen mit der Mikrosonde[2] (Abb. 3) ergaben in der Grenzzone der Nickelsperr-schicht der beiden wärmebehandelten Gleitlager eine unterschiedlich starke Zinnanreicherung zufolge Zinndiffusion aus der Pb/Sn-Laufschicht. Bei den neugebildeten Diffusionszonen handelt es sich um intermetallische Ni/Sn-Verbindungen von der Zusammensetzung Ni_3Sn_2 beim Gleit-lager A (Abb. 3 b) und der zinnreicheren Verbindung Ni_3Sn_4 beim Gleitlager B (Abb. 3 c).

Im Falle des Gleitlagers B ist der Zinngehalt der Laufschicht im Neuzustand gegenüber dem des Gleitlagers A wesentlich höher und außerdem befand sich auf der Laufschichtoberfläche vor der Wärmebehandlung ein 2 μm dicker Zinnbelag; diese Tatsache des höheren Zinnangebotes an der Nickelsperrschicht erklärt die Bildung der an Zinn reicheren Ni/Sn-Phase.

1. Wärmebehandlung zum Beschleunigen der bei 110° C relativ langsam ablaufenden Diffusionsvorgänge.
2. Im Röntgen-Rasterbild entspricht zunehmende Schwärzung einem steigenden Gehalt des entsprechenden Elementes.

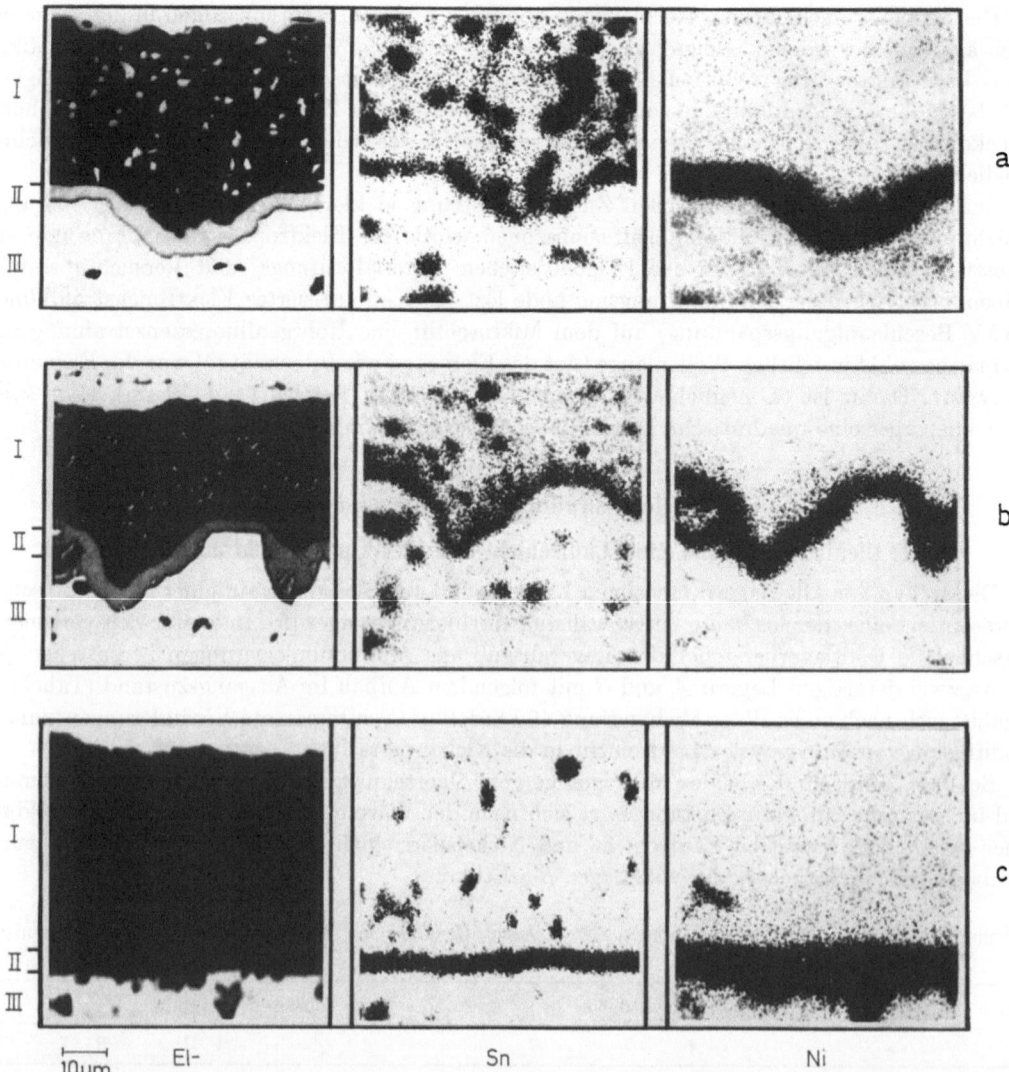

Abb. 3a—c. Querschnitte der Gleitlager *A* und *B* mit Blei/Zinn-Laufschichten (I) auf Aluminiumlagerausguß (III) und dazwischenliegender Nickelsperrschicht (II) sowie gebildeten Diffusionszonen; Mikrosonde-Rasterbilder der Rückstreuelektronen (*El-*) bzw. mit Verteilung der Elemente Sn, Ni. a Neuzustand (*A*); b, c 400 Std/170° C (*A*, *B*)

Tabelle 3. *Aufbau und Zusammensetzung dreier Mehrstoff-Gleitlager mit Pb/Sn/Cu-Laufschichten auf Bleibronze-Lagerausguß*

	Schichtdicke			Zusammensetzung		
	C	*D*	*E*	*C*	*D*	*E*
Laufschicht	40 μm	20 μm	20 μm	*Pb* Cu Sn 12	*Pb* Cu Sn 16	*Pb* Cu Sn 20
Diffusionsschicht (nach 3000 Std/170° C)	5—7 μm	6—8 μm	8—10 μm	$(Ni, Cu)_3Sn$	$(Ni, Cu)_3Sn_4$	$(Ni, Cu)_3Sn_4$
Sperrschicht (nach 3000 Std/170° C)	2—3 μm	1—2 μm	0	Ni	Ni	Ni
Sperrschicht (Neuzustand)	8 μm	8 μm	8 μm	Ni	Ni	Ni
Lagerausguß	0,7 mm	0,7 mm	0,7 mm	*Cu* Pb Sn	*Cu* Pb Sn	*Cu* Pb Sn

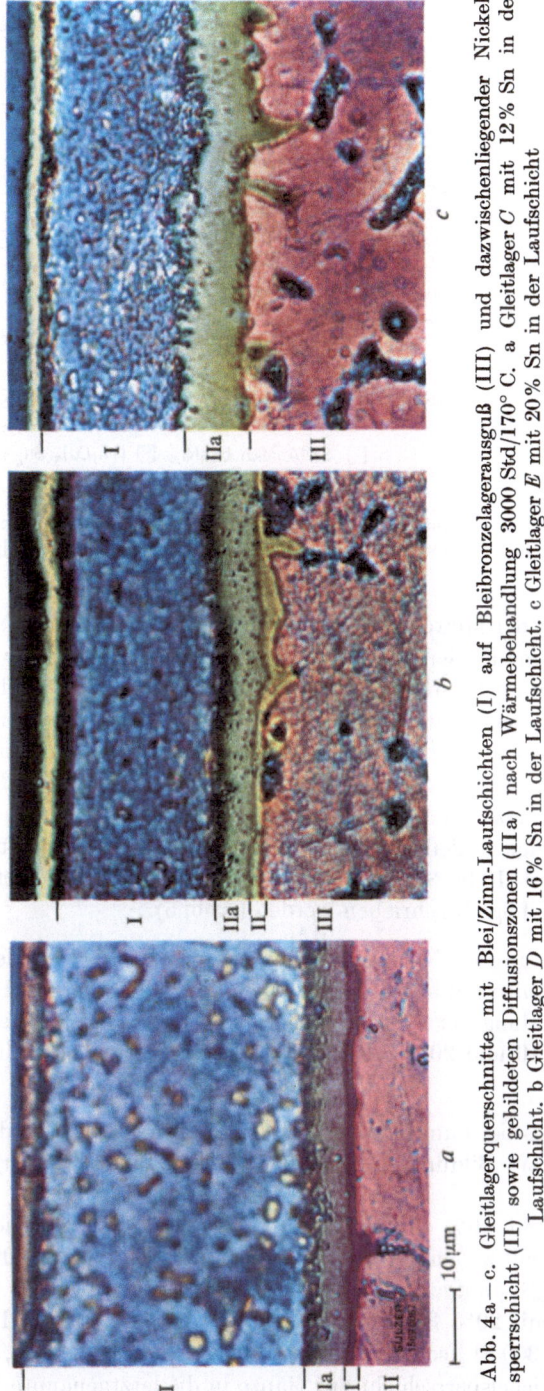

Abb. 4a—c. Gleitlagerquerschnitte mit Blei/Zinn-Laufschichten (I) auf Bleibronzelagerausguß (III) und dazwischenliegender Nickel-sperrschicht (II) sowie gebildeten Diffusionszonen (IIa) nach Wärmebehandlung 3000 Std/170° C. a Gleitlager C mit 12% Sn in der Laufschicht. b Gleitlager D mit 16% Sn in der Laufschicht. c Gleitlager E mit 20% Sn in der Laufschicht

Zur Optimierung des Zinngehaltes in der Laufschicht wurden an drei Gleitlagern C, D und E mit folgendem Aufbau im Ausgangszustand (Tabelle 3) Wärmebehandlungen bei 110° C und 170° C bis zu 3000 Std durchgeführt.

Der Einfluß des steigenden Zinngehaltes auf die Bildung der intermetallischen Ni/Cu/Sn-Phasen unter zunehmendem Abbau der Nickelsperrschicht kommt deutlich nach einer Wärmebehandlung bei 170° C/3000 Std zum Ausdruck (Abb. 4a—c).

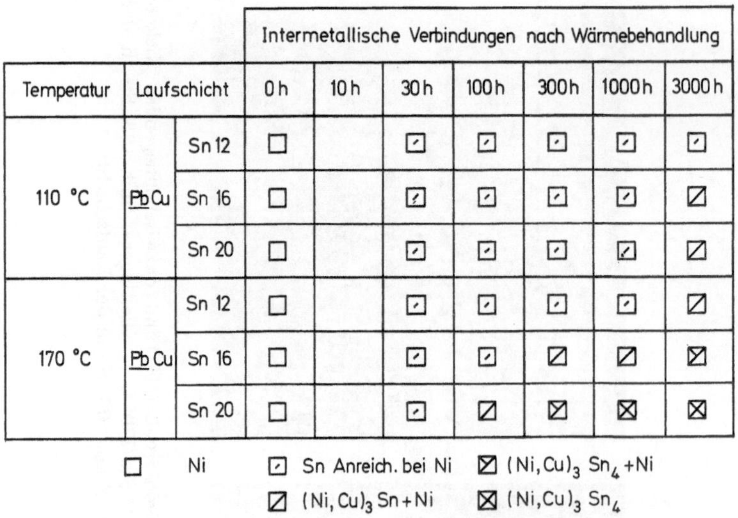

Temperatur	Laufschicht		Intermetallische Verbindungen nach Wärmebehandlung						
			0 h	10 h	30 h	100 h	300 h	1000 h	3000 h
110 °C	PbCu	Sn 12	□		☑	☑	☑	☑	☑
		Sn 16	□		☑	☑	☑	☑	◪
		Sn 20	□		☑	☑	☑	◪	◪
170 °C	PbCu	Sn 12	□		☑	☑	☑	☑	◪
		Sn 16	□		☑	☑	◪	◪	⊠
		Sn 20	□		☑	◪	⊠	⊠	⊠

□ Ni   ☑ Sn Anreich. bei Ni   ◪ $(Ni, Cu)_3 Sn_4 + Ni$

◪ $(Ni, Cu)_3 Sn + Ni$ ⊠ $(Ni, Cu)_3 Sn_4$

Abb. 5. Diffusionsvorgänge in Mehrstoff-Gleitlagern mit *Blei/Zinn-Laufschichten* auf *Nickelsperrschichten* unter Bildung intermetallischer Verbindungen in Funktion von Temperatur und Zeit bei der Wärmebehandlung

Untersuchungen mit der Mikrosonde bestätigten die unterschiedlich starke Zinnanreicherung neben Kupferanreicherung an der Nickelsperrschicht unter Bildung der zinnarmen intermetallischen Phase $(Ni, Cu)_3Sn$ im Gleitlager C und der zinnreicheren Phase $(Ni, Cu)_3Sn_4$ im Falle der Gleitlager D und E. Bei den Gleitlagern C und D war eine noch unveränderte Restnickel-Sperrschicht von 1 bis 3 μm feststellbar; hingegen war beim Gleitlager E mit dem höchsten Zinngehalt der Laufschicht im Neuzustand die Nickelsperrschicht vollständig in die an Zinn reiche intermetallische Phase umgesetzt worden.

Die Diffusionsvorgänge an den drei Gleitlagertypen C, D und E in Funktion von Temperatur (110 und 170° C) und Zeit (10 bis 3000 Std) können aufgrund der durchgeführten Untersuchungen mit der Mikrosonde wie folgt beschrieben werden (Abb. 5):

Bei 110 ° C können an den Nickelsperrschichten bis zu 1000 Std lediglich merkbare Zinnanreicherungen durch Diffusion des Zinns aus den Laufschichten mit 12, 16 und 20% Sn festgestellt werden. Die Bildung der intermetallischen Phase $(Ni, Cu)_3Sn$ ist bei Laufschichten mit höheren Zinngehalten (16 und 20% Sn) erst nach einer Wärmebehandlungsdauer von 3000 Std zu beobachten.

Bei 170° C kommt es bei Laufschichten mit 12% Sn unter 1000 Std zu Zinnanreicherungen und bei etwa 3000 Std zur Bildung der intermetallischen Phase $(Ni, Cu)Sn$ an der Nickelsperrschicht.

Bei Laufschichten mit 16% Sn erfolgt die Bildung dieser intermetallischen Phase schon bei etwa 300 Std, und mit zunehmender Wärmebehandlungsdauer (3000 Std) wird die an Zinn reichere Phase $(Ni, Cu)_3Sn_4$ aufgebaut; eine Restnickel-Sperrschicht ist jedoch noch vorhanden.

Bei Laufschichten mit 20% Sn kommt es bereits bei etwa 100 Std zur Bildung der Phase $(Ni, Cu)_3Sn$, welche bei 300 Std schon in die an Zinn reichere Phase $(Ni, Cu)_3Sn_4$ übergeht. Nach etwa 1000 Std ist die Nickelsperrschicht zur Gänze in die letztgenannte Phase umgesetzt.

4.2. Gleitlager mit Blei/Indium-Laufschicht auf Lagerausguß

In einem 1000stündigen Versuchsdauerlauf eines Lokomotiv-Dieselmotors wurde der Gleitlagertyp F mit einer Blei/Indium-Laufschicht direkt auf Bleibronze-Lagerausguß auf seine Einsatzfähigkeit geprüft.

Im Neuzustand betrug der Indiumgehalt der Laufschicht durchschnittlich 13 Gew.-%. Mikrosondenuntersuchungen an Mikroquerschnitten (Abb. 6) zeigten, daß sich Indium bei Abwesenheit einer Diffusionssperrschicht schon bei der Lagerfabrikation während der Diffusions-

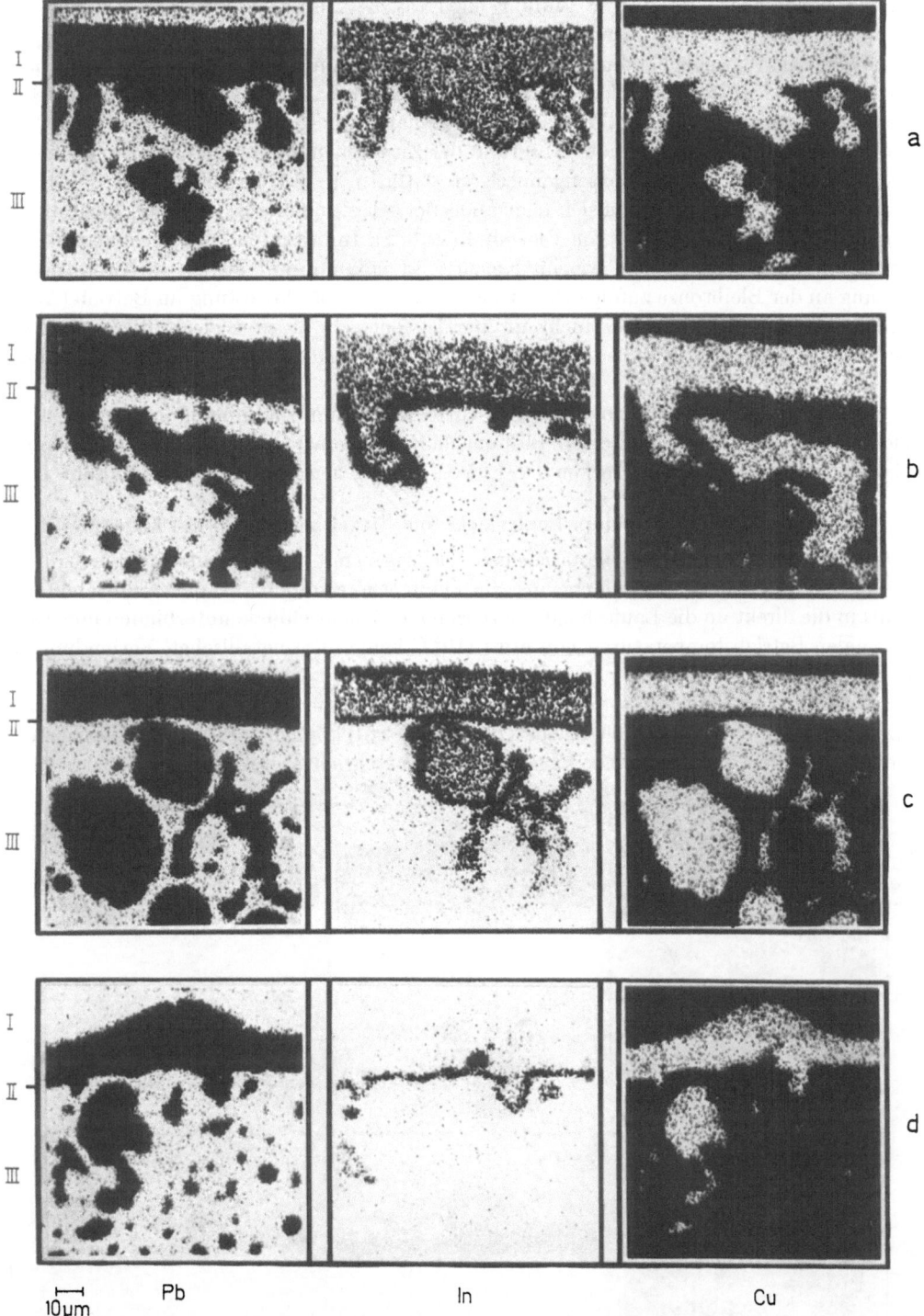

Abb. 6a—d. Querschnitte des Gleitlagers *F* mit Blei/Indium-Laufschichten (I) auf Bleibronzelagerausguß (III) und gebildeten Diffusionszonen (II); Mikrosonde-Rasterbilder mit Verteilung der Elemente Pb, In, Cu. a Neuzustand. b 500 Std/170° C. c 1000 Std im Betrieb (intakter Oberflächenbereich). d 1000 Std im Betrieb (Oberflächenbereich mit schwarzen Flecken)

wärmebehandlung (80° C/12 Std) gleichmäßig über die 20 μm dicke Laufschicht und auch die direkt angrenzenden Bleieinschlüsse der Bleibronze verteilt (Abb. 6a). Im Übergangsbereich der Laufschicht zum Lagerausguß durchgeführte Linienanalysen mit der Mikrosonde ergaben

Indiumanreicherungen, die auf die Bildung einer Cu/In-Phase schließen lassen, welche jedoch lichtoptisch noch nicht erkennbar war.

Eine Langzeit-Wärmebehandlung (500 Std/170° C) beschleunigt die Indiumdiffusion in Richtung Lagerausguß. Die lichtoptisch deutlich sichtbare, 1—2 μm breite Diffusionszone wies bei der Mikrosondenuntersuchung (Abb. 6b) einen Cu-Gehalt von durchschnittlich 70 Gew.-% und eine starke Indiumanreicherung auf. Aufgrund der Zusammensetzung und des Zweistoffsystems für Cu/In muß es sich um ein Phasengemisch Cu + Cu_9In_4 ($\alpha + \delta$) handeln.

An dem Gleitlager F befanden sich nach 1000 Betriebsstunden auf der Laufschichtoberfläche beider Kanten schwarze Flecken von Bleisulfid (Abb. 2). Im intakten Oberflächenbereich liegen praktisch dieselben Verhältnisse wie im Neuzustand vor, nur mit etwas stärkerer Indiumanreicherung an der Bleibronze zufolge 1000stündiger Temperatureinwirkung im Betrieb (Abb. 6c). Im Gegensatz dazu ist die Laufschicht im Bereich der schwarzen Bleisulfidflecke vollständig an Indium verarmt und weist lediglich noch die Cu/In-Verbindung in der Diffusionszone auf (Abb. 6d).

Die zerstörungsfreie röntgenographische Gitterkonstantenbestimmung ergab im intakten Bereich $a = 4{,}92$ Å, entsprechend einem Indiumgehalt von etwa 9 Gew.-% im Pb/In-Mischkristall, und im Bereich der schwarzen Flecken $a = 4{,}95$ Å, entsprechend Reinblei ohne Indium.

4.3. Gleitlager mit Blei/Indium-Laufschicht und Nickelsperrschicht auf Lagerausguß

Zum Verbessern der vorhin besprochenen Gleitlager mit Pb/In-Laufschichten baute man neuerdings eine Diffusionssperrschicht aus Nickel ein. Diese soll erstens das Abdiffundieren des Indiums in die direkt an die Laufschicht angrenzenden Bleieinschlüsse unterbinden und zweitens bei normalen Betriebstemperaturen von etwa 110° C keine intermetallischen Verbindungen mit Indium bilden.

Die Diffusionsvorgänge an den zwei Gleitlagertypen G und H (Tabelle 1) in Funktion von Temperatur (110, 170 und 220° C) und Zeit (10 bis 3000 Std) können aufgrund der durchgeführten Untersuchungen mit der Mikrosonde folgendermaßen beschrieben werden (Abb. 7, 8):

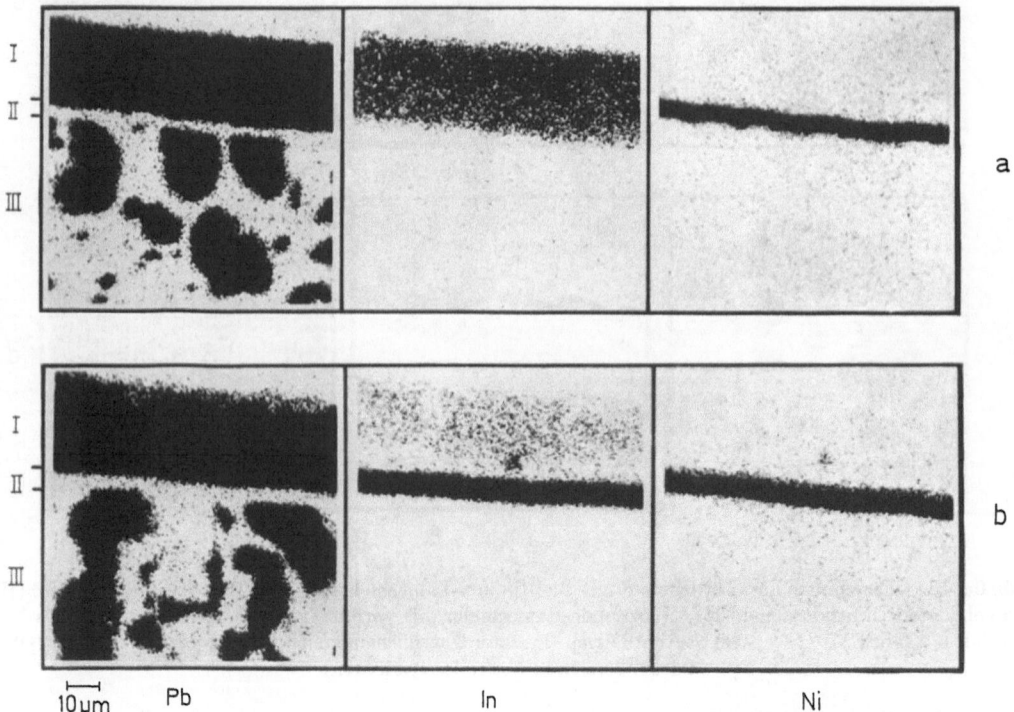

Abb. 7a u. b. Querschnitte des Gleitlagers H mit Blei/Indium-Laufschichten (I) auf Bleibronzelagerausguß (III) und dazwischenliegender Nickelsperrschicht (II) sowie gebildeter Diffusionszone; Mikrosonde-Rasterbilder mit Verteilung der Elemente Pb, In, Ni. a Neuzustand; b 300 Std/170° C

Temperatur	Laufschicht		Intermetallische Verbindungen nach Wärmebehandlung						
			0 h	10 h	30 h	100 h	300 h	1000 h	3000 h
110 °C	Pb	In 10	□		□	□	□	◪	◪
		In 20	□		□	□	□	◪	◪
170 °C	Pb	In 10	□		◪	⊡	⊡	◪	◪
		In 20	□		◪	◪	⊠	⊠	⊠
220 °C	Pb	In 10	□	⊡	⊡	⊡	◪	◪	◪
		In 20	□	⊡	◪	⊠	⊠	⊠	⊠

□ Ni ⊡ In Anreich. bei Ni ⊠ Ni In + Ni

◪ Ni₃ In + Ni ▨ Ni In

Abb. 8. Diffusionsvorgänge in Mehrstoff-Gleitlagern mit *Blei/Indium-Laufschichten* auf *Nickelsperrschichten* unter Bildung intermetallischer Verbindungen in Funktion von Temperatur und Zeit bei der Wärmebehandlung

Bei 110° C kommt es bis zu 300 Std sowohl im Falle der Laufschicht mit 10% In als auch mit 20% In weder zur Bildung intermetallischer Verbindungen noch zur Indiumanreicherung an der Nickelsperrschicht. Die letztgenannte Anreicherung von Indium ist erst nach einer Wärmebehandlungsdauer von 1000 und 3000 Std zu beobachten.

Bei 170° C können bei Laufschichten mit 10% In bis zu 300 Std schon geringfügige Indiumanreicherungen als Vorstufe zur Bildung der nach 1000 und 3000 Std aufgebauten intermetallischen Verbindung Ni₃In an der Nickelsperrschicht festgestellt werden.

Bei Laufschichten mit 20% In kommt es an der Nickelsperrschicht schon nach 100 Std zur Bildung der intermetallischen Verbindung Ni₃In, welche bei zunehmender Wärmebehandlungsdauer in die an Indium reichere Phase NiIn übergeht (Abb. 7a, b); neben dieser neugebildeten Zone ist allerdings noch bis zu 3000 Std eine Restnickel-Sperrschicht vorhanden.

Bei 220° C können bei Laufschichten mit 10% In bis zu 100 Std Wärmebehandlungsdauer nur zunehmende Indiumanreicherungen an der Nickelsperrschicht festgestellt werden. Ab 300 Std bildet sich hierauf die intermetallische Verbindung Ni₃In, wobei wiederum bis zu 3000 Std eine Restnickel-Sperrschicht erhalten bleibt.

Bei Laufschichten mit 20% In kommt es bereits nach 30 Std zur Bildung der Phase Ni₃In, welche nach 100 Std zufolge reichlichem Indiumnachschubes in die Phase NiIn übergeführt wird; eine Restnickel-Sperrschicht ist noch bis zu 3000 Std vorhanden.

5. Zusammenfassung und Schlußfolgerungen

Die Untersuchung über Diffusionsvorgänge an verschiedenen Kombinationen von Bleibasis-Laufschichten mit Nickelsperrschichten in Gleitlagern läßt folgende Schlußfolgerungen zu:

1. Bei Temperaturen von 110° C ist im Falle von PbSn-Laufschichten mit 16 und 20% Sn der Aufbau einer intermetallischen Phase (Ni, Cu)₃Sn erst bei 3000 Std Wärmebehandlungsdauer zu beobachten. Als Vorstufe dazu erfolgen merkbare Zinnanreicherungen an der Nickelsperrschicht unabhängig vom Sn-Gehalt der Laufschicht schon nach relativ kurzer Zeit (10—30 Std).

Entsprechende Indiumanreicherungen an der Nickelsperrschicht im Falle von PbIn-Laufschichten mit 10 und 20% In sind erst nach 1000 Std beobachtbar, so daß nach 3000 Std noch kein Aufbau einer intermetallischen Phase erfolgt ist.

2. Bei Temperaturen von 170° C, entsprechend erhöhten Betriebstemperaturen, ist der Einfluß des Sn- bzw. In-Gehaltes der Laufschicht auf die Bildungsgeschwindigkeit der intermetallischen Zwischenzonen schon stark ausgeprägt.

Bei Sn- bzw. In-Gehalten von 10% in der Laufschicht erfolgt die Bildung der intermetallischen Zonen $(Ni, Cu)_3Sn$ bzw. Ni_3In erst nach 1000—3000 Std, wobei Restsperrschichten aus Nickel noch erhalten bleiben.

Eine Erhöhung der Gehalte auf 20% bewirkt bereits nach 100 Std die Bildung der eben erwähnten intermetallischen Phasen, welche bei 300 Std schon in die an Sn- bzw. In-reicheren Phasen $(Ni, Cu)_3Sn_4$ bzw. NiIn übergehen. Bei den PbSn 20-Laufschichten ist die Nickelsperrschicht nach etwa 1000 Std zur Gänze in die intermetallische Phase umgesetzt, währenddem eine Restsperrschicht bei den PbIn 20-Laufschichten auch nach 3000 Std noch vorhanden ist.

3. Bei Temperaturen von 220° C, entsprechend örtlich extrem stark erhöhten Betriebstemperaturen, erfolgt bei PbIn 10-Laufschichten die Bildung der Phase Ni_3In erst nach 300 Std. Eine Restsperrschicht aus Nickel ist neben dieser Phase auch noch nach 3000 Std beobachtbar.

Im Falle der PbIn 20-Laufschichten bildet sich aus der nach 30 Std aufgebauten Phase Ni_3In die an In-reichere Phase NiIn, wobei die Restsperrschicht noch bis zu 3000 Std erhalten bleibt.

4. Eine Erhöhung des Sn- bzw. In-Gehaltes von 10 auf 20% in der Laufschicht bewirkt allgemein bei 110° C keinen wesentlichen Einfluß auf die Bildungsgeschwindigkeit beim Aufbau intermetallischer Zwischenzonen an der Nickelsperrschicht.

Bei höheren Temperaturen von 170 und 220° C führt ein Sn- bzw. In-Gehalt von 20% in der Laufschicht zum frühzeitigen Aufbau von Sn- bzw. In-reichen intermetallischen Phasen unter vollständigem Abbau der Nickelsperrschicht im Falle von PbSn-Laufschichten.

5. Die PbIn 10-Laufschichten bieten gegenüber den PbSn 10-Laufschichten den Vorteil, daß die Abdiffusion des Indiums in die Nickelsperrschicht bei normaler Betriebstemperatur von 110° C beträchtlich langsamer erfolgt als die von Zinn aus PbSn 10-Laufschichten. Für erhöhte Betriebstemperaturen bis zu 220° C wurden bereits Sperrschichten gefunden, welche eine stärkere diffusionshemmende Wirkung als Nickelschichten besitzen.

Literatur

1. SCHLÄPFER, O.: Eine neue Serie von Sulzer-Lokomotivmotoren mit V-Anordnung der Zylinder. Tech. Rundschau Sulzer Bd. **47**, Nr. 2, 65—79 (1965).
2. SCHLESER, H.: Atom-Absorptions-Spektrophotometrie. Z. Instrumentenk. **73**, Nr. 2, 1—9 (1965).
3. TYZAK, C., and G. V. RAYNOR: The lattice spacings of lead-rich substitutional solid solutions. Acta Cryst. No. 7, 505—510 (1954).
4. PEPPERHOFF, W.: Gefügeentwicklung durch Interferenz-Aufdampfschichten. Arch. Eisenhuettenw. **36**, Nr. 1, 269—273 (1961).
5. Glyco-Metallwerke, Daelen u. Loos GmbH: Das Hochleistungslager für Schiffsmotoren, Flugmotoren, Automobilmotoren, Kompressoren, S. 1—13. Wiesbaden-Schierstein 1958.
6. ANDERKO, K., F. W. RABENAU u. R. WEBER: Aluminium-Verbundlagerwerkstoffe, Aluminium **44**, Nr. 1, 41—46 (1968).

The Influence of Surface Treatment on the Diffusivity in the Surface Layers of Stainless Steels

T. ERICSSON and R. ROSÉEN

AB Atomenergi, Stockholm, Sweden

1. Introduction

It is well known that the rate of oxidation of stainless steels in steam at temperatures higher than about 400° C depends on the surface treatment of the material [1—4]. Ground specimens oxidize slower than electropolished or pickled ones although the effect at temperatures higher than about 700° C is of short duration. It has been suggested [1—4] that the lower oxidation rate of ground specimens is due to an increased diffusivity of the alloy components in the deformed surface layer giving a thicker and more protective chromium oxide.

However, little is known about the influence of surface treatment on the rate of diffusion. Russian workers [5] have studied the diffusion of an β-radioactive Ni isotope into specimens of a 20% Cr—80% NiTi-alloy (Kh20N80T3), pure Fe and pure Ni by the so-called absorption method. The specimens were either electropolished, vacuum annealed, ground or sand blasted. We have used the microprobe technique to study the interdiffusion between an electrolytic Ni layer and a 20% Cr — 35% Ni-steel subjected to various surface treatments.

2. Experimental Details

Discs, 4 mm thick, were cut from a round bar, 10 mm diameter, of a Sandvik steel 3 X R N 10, containing 19.5% Cr, 35% Ni, 44.5% Fe, 0.8% Mn and 0.2% Si. The specimens were mechanically polished and sealed into hydrogen filled quartz capsules. The capsules were annealed at 1150° C for 2 h and aircooled. One of the flat surfaces of each disc was given one of four treatments: as annealed (AN), electropolished in 95% acetic acid, 5% perchloric acid (EP), mechanically polished on a 1 μm diamond wheel (MP) or ground on 600 mesh paper (GR). The specimens were then lightly pickled for 5 minutes in 25% HCl + 1% HF and immediately transferred to an electrolytic bath containing water, 365 g/litre NiCl$_2$ · 6 H$_2$O and 75 g/litre HCl. Approximately 1.0 mg/cm² or a 1.2 μm thick layer of Ni was deposited on the specimens; this was measured by weighing. The specimens were again sealed into quartz capsules and diffusion annealed for various times at either 500° C or 800° C.

Before and after the diffusion anneals the specimens were analyzed for K_α of Fe and Cr in a CAMECA MS 46 microprobe. The analysis was carried out with the beam perpendicular to the Ni layer and with 20 kV accelerating voltage and appr. 20 nA beam current. In some cases 8, 10, 15 and 35 kV was also used to study the influence of the voltage. To minimize effects of small irregularities in the surface the beam was periodically scanned along a 10 μm long line with 50 c/s. The grain structure of the EP and AN specimens was well visible and the analysis was carried out in the interior of the grains except when the influence of the grain boundaries was studied. The intensity was measured in sixteen positions on each specimen. The ground specimens were analyzed with the scratches in several orientations. The effect of orientation was negligible. After correction for dead time and background, the ratios between the average intensities from each specimen and from pure standards of Cr, Fe or Ni were calculated.

3. Absorption and Fluorescence Calculations

To interpret the observed intensity ratios quantitatively a calculation of absorption and fluorescence effects has to be made. We have computed the absorption in basically the same way as Whittle and Wood [6]. For the X-ray intensity function for element A, $\varphi_A(\varrho z)$, we have used Philibert's [7] analytical expression

$$\varphi_A(\varrho z) = R_\infty (N/A)\,\psi_A\{\exp(-\sigma\varrho z) - [(1 - R_0/R_\infty)\exp[-\sigma(1 + 1/h)\varrho z]]\} \tag{1}$$

where R_∞ and R_0 are constants, N is Avogadro's number, A is the atomic weight, ψ_A is the ionization cross section for element A, σ is the Lenard coefficient, ϱz is the mass thickness and h is a function of the atomic number Z. Following Philibert [7] we have chosen $R_\infty = 4$, $R_0 = 1.48$, and $h = 3.5 A/Z^2$. According to Duncumb and Shields [8] we have put $\sigma = 2.39 \cdot 10^5/(E_0^{1.5} - E_c^{1.5})$ where E_0 is the applied voltage and E_c the critical excitation potential for element A. The observed intensity I of element A is given by the integral

$$I_A = \int_0^\infty C_A(\varrho z)\,\varphi_A(\varrho z)\exp(-\chi_A\varrho z)\,d(\varrho z) \tag{2}$$

where $C_A(\varrho z)$ is the weight per cent of element A as a function of the depth z under the surface and $\chi_A = (\mu/\varrho)_A \csc\theta = 3.24\,(\mu/\varrho)_A$ for the CAMECA instrument. Heinrich's [9] mass absorption coefficients (μ/ϱ) have been used. Notice that (μ/ϱ) and h are also functions of ϱz

$$(\mu/\varrho)_A(\varrho z) = \int_0^{\varrho z} \left(C_{\mathrm{Ni}}(\varrho y)\,(\mu/\varrho)_A^{\mathrm{Ni}} + C_{\mathrm{Fe}}(\varrho y)\,(\mu/\varrho)_A^{\mathrm{Fe}} + C_{\mathrm{Cr}}(\varrho y)\,(\mu/\varrho)_A^{\mathrm{Cr}}\right)d(\varrho y)$$

$h = 0.262 \cdot \bar{C}_{\mathrm{Ni}}(\varrho z) + 0.289 \cdot C_{\mathrm{Fe}}(\varrho z) + 0.315\,C_{\mathrm{Cr}}(\varrho z)$. The integral I_A has been solved numerically with the upper limit of integration being appr. $\varrho \cdot 6\,\mu\mathrm{m}$. One can distinguish several cases. For pure standards the integral has an analytical form, Philibert [7]. For coated specimens before the diffusion anneal the integral has been calculated in two parts, one in the Ni layer, the other in the alloy. For coated specimens after the diffusion anneal concentration profiles have to be chosen for the elements. We have assumed the density $\varrho = 8\,\mathrm{g/cm^3}$ and error function profiles with a diffusion constant D independent of composition, depth z under the surface and of element. This is a simplification but the measurements do not justify a more refined assumption. The computed intensity ratios for various values of $\sqrt{Dt}$ and the mass thickness are shown in Fig. 1 for Cr and Fe at 20 kV.

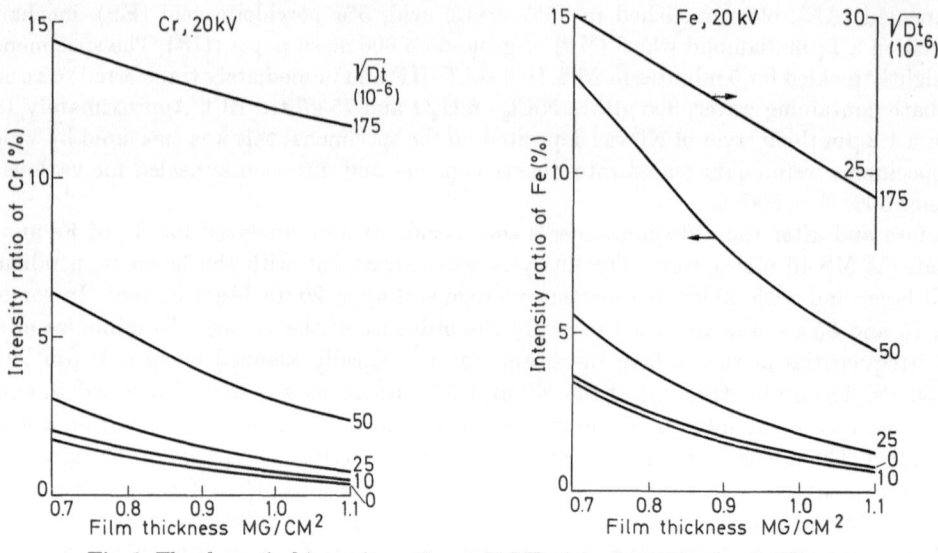

Fig. 1. The theoretical intensity ratios at 20 kV when absorption is considered

The NiK_α-radiation generated in the Ni-layer and reaching the alloy causes Fe and Cr fluorescence. If no diffusion has occured the NiK_α-intensity at the interface, ϱz_1, is given by

$$I_{\mathrm{Ni}} = \int\limits_0^{\varrho z_1} \varphi_{\mathrm{Ni}}(\varrho z) \int\limits_0^{\pi/2} \tfrac{1}{2} \cos v \exp\left[-\chi_{\mathrm{Ni}}^{\mathrm{Ni}} \varrho (z_1 - z)\, \mathrm{cosec}\, v\right] dv\, d(\varrho z)$$

where v is the angle between a certain beam travelling downward and the surface. The fluorescent Cr or Fe radiation is obtained according to BIRKS [10] by multiplying I_{Ni} with BIRKS's excitation efficiency E_{AB} and with the weight per cent of Cr or Fe in the alloy. The observed fluorescence from element A is

$$I_A^f = I_{\mathrm{Ni}} E_{A,\,\mathrm{Ni}} C_A^{\mathrm{alloy}} \exp(-\chi_A^{\mathrm{Ni}} \varrho z_1) \tag{3}$$

where $E_{\mathrm{Cr,\,Ni}} = 0.15$ and $E_{\mathrm{Fe,\,Ni}} = 0.22$. I_A^f has been divided by the intensity from the pure standard of element A, Eq. (2), to get the intensity ratio. This ratio contains the quantity $(\psi_{\mathrm{Ni}}/\psi_A) \cdot (A_A/A_{\mathrm{Ni}})$. For ψ an expression given by WEBSTER et al. [11] has been used

$$\psi_A = \mathrm{const}\, (1/E_c^2)\, U^{-m_2} \ln U$$

where $U = E_0/E_c$ and $m_2 = 0.837$. The final percental value for the fluorescence from Cr is 0.35 at 20 and 35 kV and from Fe 2.3 and 1.9 at 20 and 35 kV resp. As most of the electrons are stopped in the Ni layer this is the only fluorescence contribution considered.

4. Results and Discussion

The table gives the experimental conditions and the observed intensity ratios. The mass thickness of the Ni layer has been calculated from the intensity ratios before annealing by first substracting the fluorescence part and then using the curves for $\sqrt{Dt} = 0$ in Fig. 1. This so-called X-ray thickness is compared in the table with the mass thickness derived from the weight increase of the specimens. They agree within a few per cent in most cases. Analysis carried out at lower voltages has been of little value because in these cases the electrons do not or barely penetrate the Ni layer.

Knowing the mass thickness and the increase in observed intensity ratio the value of $\sqrt{Dt}$ can be derived from Fig. 1. $\sqrt{Dt}$ has usually been around $5 \cdot 10^{-5}$ cm which means that the concentration of Cr and Fe at the outer surface has been appr. 10% of the concentration in the alloy. We have therefore considered it reasonable to use the same fluorescence correction as for the case of no diffusion. The D-values are given in the table. The average D in cm²/s at 800° C is $7.4 \cdot 10^{-14}$ for EP specimens and $14 \cdot 10^{-12}$ for GR specimens and at 500° C $4.1 \cdot 10^{-16}$ for AN and EP specimens and $11 \cdot 10^{-14}$ for GR and MP specimens. Thus, grinding or polishing

Table. *A compilation of measured and calculated quantities based on Cr and Fe analysis*

Spec. No.	Surf. treatment	Voltage	Weighed thickness (mg/cm²)	X-ray thickness (mg/cm²)		Annealing		Intensity ratio %				Diff. constant (cm²/sec)	
						temp.	time	before ann.		after ann.			
				Cr	Fe	(°C)	(h)	Cr	Fe	Cr	Fe	Cr	Fe
1	EP	35	0.97	0.96	0.96	800	1	4.7	11.7	5.0	14.1	$2.8 \cdot 10^{-14}$	$7.8 \cdot 10^{-14}$
1	EP	20	0.97	0.97	1.03	800	1	1.0	3.1	1.4	4.1	$1.7 \cdot 10^{-14}$	$18 \cdot 10^{-14}$
1	EP	20	0.97	0.97	1.03	800	5	1.0	3.1	2.6	5.4	$8.0 \cdot 10^{-14}$	$6.4 \cdot 10^{-14}$
2	GR	35	1.00	0.96	0.97	800	1	4.7	11.3	13.1	31.8	$8.5 \cdot 10^{-12}$	$20 \cdot 10^{-12}$
2	GR	20	1.00	0.98	1.07	800	1	1.0	2.9	14.1	32.0	$10 \cdot 10^{-12}$	$16 \cdot 10^{-12}$
3	AN	20	0.96	0.92	0.92	500	344	1.2	3.9	1.5	4.6	$3.5 \cdot 10^{-16}$	$4.8 \cdot 10^{-16}$
4	EP	20	0.99	0.97	1.01	500	344	1.1	3.3	1.3	4.2	$3.5 \cdot 10^{-16}$	$4.5 \cdot 10^{-16}$
5	MP	20	0.88	0.93	0.97	500	8	1.2	3.6	4.0	8.9	$9.7 \cdot 10^{-14}$	$16 \cdot 10^{-14}$
6	GR	20	0.92	0.77	0.81	500	8	1.8	4.9	5.4	13.2	$7.3 \cdot 10^{-14}$	$9.0 \cdot 10^{-14}$

EP = electropolished, GR = ground or 600 mesh paper, AN = as annealed, MP = mechanically polished on 1 μm diamond wheel.

increases the D-value appr. 100 times compared to electropolishing or annealing at both 500 and 800° C. The increase of the intensity ratios observed over the grain boundaries in the EP and AN specimens has been about the same as for MP and GR specimens, indicating that the rate of grain boundary diffusion is comparable to the rate of diffusion in cold worked layers.

Our D-values agree reasonably well with the Russian ones [5] considering the simplifying assumptions in our analysis and the different techniques used. They found for the 20% Cr— 80% Ni-alloy at 800° C $D = 9.10^{-13}$ cm²/s for EP and $9.2 \cdot 10^{-12}$ for GR specimens and at 600° C $D = 4 \cdot 10^{-15}$ for EP and $12 \cdot 10^{-14}$ for GR specimens. Thus in their case grinding increased the diffusion rate 40 times at 600° C and 10 times at 800° C.

Acknowledgements. The authors thank AB Atomenergi for permission to publish this paper and Dr. G. Öst-berg for encouragement and support during the course of this work.

References

1. Warzee, M., J. Hennaut, M. Maurice, C. Son-nen, and J. Waty: J. Electrochem. Soc. **112**, 670 (1965).
2. Jansson, S., and B. Lehtinen: Metallurgie **7**, 1 (1967).
3. — W. Hübner, G. Östberg, and M. de Pour-baix: Brit. Corrosion J. **4**, 21, (1969).
4. Caplan, D.: Corrosion Sci. **6**, 509 (1966).
5. Bokhshtein, S. Z., M. A. Gubareva, and S. T. Kishkin: Diffusion processes, structure and properties of metals, New York: Consultants Bureau 1965.
6. Whittle, D. P., and G. C. Wood: Corrosion Sci. **6**, 397 (1966).
7. Philibert, J.: X-ray optics and X-ray micro-analysis. New York: Academic Press 1963.
8. Duncumb, P., and P. K. Shields: The electron microprobe. New York: J. Wiley & Sons, Inc. 1966.
9. Heinrich, K. F. J.: The electron microprobe. New York: J. Wiley & Sons, Inc. 1966.
10. Birks, L. S.: Electron probe microanalysis. New York: Interscience Publ. 1963.
11. See A. H. Compton and S. K. Allison: X-rays in theory and experiment. New York: Van Nostrand Inc. 1954.

Investigation of Hard Magnetic Materials by Small Angle Diffraction of Neutrons

G. Maier*

Institut für Festkörper- und Neutronenphysik der KFA Jülich, BRD

A microscopic description of magnetic materials must include direction and magnitude of local magnetic induction. Neutron diffraction crystallography has in many cases given this information on the scale of the unit cell. For unpolarized neutrons nuclear and magnetic scattering amplitudes do not interfere and the latter is proportional to the component of induction normal to the diffraction vector. Furthermore magnetic materials are characterized by their domain structure. The coercivity is known to be related to domain size, which is of the order of 10^2-10^4 Å in hard magnets and very much larger for soft ones.

A classification in domain size also is encountered in the theoretical description of the mechanism of neutron scattering.

Here it is of importance to distinguish whether or not an originally plane wave is substantially distorted on its passage through the specimen.

The magnetic index of refraction n in a medium of the induction $\vec{B}$ for neutrons of the magnetic moment μ_N and the energy $E = h^2/2m_N\lambda^2$ is:

$$n = 1 + \frac{\vec{\mu_N}\,\vec{B}}{2E} = 1 + \delta. \tag{1}$$

Thermal neutrons ($E = 25$ meV, $\lambda = 1,8$ Å) in iron ($B = 2.10^4$ gauss) give $\delta = 10^{-6}$. On a path D through a domain there will be a change of phase relative to vacuum of

$$\phi = 2\pi D\,\delta/\lambda = D\,m_N\,\vec{\mu_N}\,\vec{B}\,\lambda/h^2. \tag{2}$$

If two neighbouring domains are of opposite sign, a path of $D = 10^5$ Å will suffice to give a phase angle of 1 rad.

For $\phi > 1$, $D > 10^5$ Å the scattering of neutrons by a specimen must be regarded as the result of multiple refraction. The statistical nature of such a mechanism leads to a broadening of the primary beam proportional to the square root of the number of domains encountered by a single ray [1]. The angle of total reflection $\Theta_c = (4\delta)^{\frac{1}{2}}$ gives the maximum single deviation. Multiple refraction of neutrons by pure iron was already demonstrated 20 years ago by Hughes et al. [2]. On the other hand finely subdivided hard magnetic materials can be studied by small angle neutron diffraction, as the criterion $\phi < 1$ is fulfilled. It should be possible to augment the information given by electron microscopy by this means.

The scope of an investigation on these lines can be demonstrated with the example of TICONAL X. Anisotropy and coercivity of this system result from precipitation in an external magnetic field in [100] direction of the monophase single crystal. Electron micrographs by de Vos [3] show ferromagnetic rods in the preferred direction forming a quadratic pattern in its normal plane with a lattice spacing around 500 Å.

The corresponding diffraction pattern may be described in analogy to common notation. Then the Fourier transform of a rod is the form factor. In the magnetized state all precipitates

---

* Now at: Forschungslaboratorium am Goetheanum, CH-4143 Dornach, Switzerland.

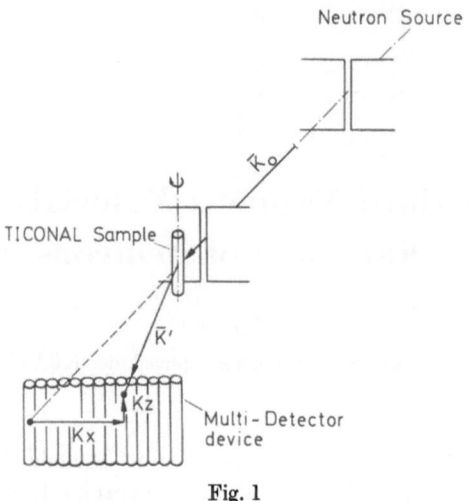

Fig. 1

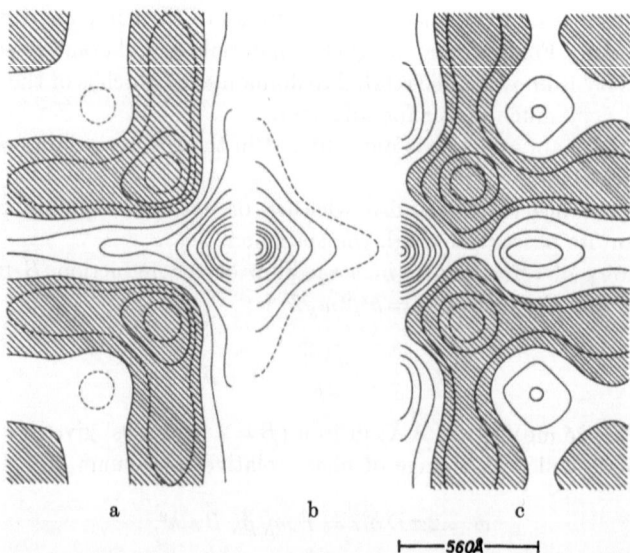

Fig. 2a—c. a magnetized; b demagnetized; c from micrograph. Contours of the correlation function for half
the plane. $G(0)$ is artificially normalized to 1. Full lines for step of 0.1. Negative values of $G(r)$ in the shaded
areas

will be magnetically alligned leading to the structure factor of a not very coherent quadratic
array. If the precipitates are uncorrelated in sign, the structure factor will be the Fourier transform of $\delta(r)$ and the precipitates will scatter independently.

The geometry of the scattering experiment ($K_0 = 2\pi/\lambda$, $\lambda = 4.2$ Å) is shown in Fig. 1. The
lengths for primary and secondary collimation are 2 m and 3.9 m respectively.

The rods lie in the axis of the 1.5 mm ∅ cylindrical specimen. As the form factor function
of a rod is very narrow in K_z-direction, integration in slit height is ensured. A section in the
direction K_x through the resulting two dimensional diffraction pattern is simultaneously recorded
by a set of fourteen, $\frac{1}{4}''$ inched, He_3 counters. By turning the specimen in steps of e.g. 3° this
function can be fully explored.

Fourier transformation leads to the section through the origin of the correlation function $G(r)$.
Comparing the contours for magnetized (a) and demagnetized state (b) as shown in Fig. 2 one
perceives that correlation between neighbouring precipitates has essentially gone lost on de-
magnetization as explained above. Fig. 2c is a correlation function generated by a computer

from an electron micrograph encoded as a quadratic pattern of ones and zeros. It is not surprising for (c) to give more contrast than (a) since it does not contain information on mutual correlation in direction of the rods. Zero level in the correlation functions is related to average scattering amplitude squared, as only inhomogenities lead to the scattering effect. Furthermore it is feasible to do scattering experiments while the specimen is in an external field, leading to differential information on the magnetization pattern at different points on hysteresis loops.

References

1. VINEYARD, G. H.: Phys. Rev. 85, 633 (1952).
2. HUGHES, B. H., M. T. BURGY, R. B. HELLER, and J. W. WALLACE: Phys. Rev. 75, 565 (1949).
3. VOS, K. J. DE: Diss. Eindhoven 1966. Z. Angew. Phys. 21, 381 (1966).
4. MAIER, G.: Z. angew. Phys. 27, 73 (1969).

Mesure d'homogénéité des mélanges de poudres par micro-analyse a sonde électronique

G. Chol, J. P. Auradon et F. Damay

Laboratoire Central de l'Eclairage, Paris, France

Introduction

L'homogénéité des mélanges de poudres destinées à être frittées influence beaucoup les qualités mécaniques, électriques ou magnétiques des matériaux formés. Ce fait est bien connu des fabricants de ferrites [1], de céramiques [2] ou de métaux frittés par exemple.

Quand on analyse un mélange de plusieurs phases $A_1, A_2, \ldots, A_i$, les échantillons comportant un nombre fini de particules donnent lieu à des fluctuations de compositions que l'on peut calculer dans le cas d'un mélange théoriquement homogène, connaissant le volume analysé ainsi que les proportions et les distributions en taille des particules de chaque phase.

L'analyse est effectuée à l'aide d'un micro-analyseur à sonde électronique en balayant l'échantillon et en enregistrant pour chaque élément les taux d'émission X qui nous permettent de calculer les fluctuations réelles de composition. Le rapport des écarts quadratiques moyens théoriques et mesurés $\chi = \sigma_0/\sigma$ constitue un critère quantitatif d'homogénéité pour chaque phase du mélange.

Nous présenterons ci-après les résultats relatifs à des mélanges de poudres d'oxydes constitutifs de ferrites doux de manganèse-zinc pour lesquels les tailles moyennes de particules sont faibles devant le volume analysé par la sonde électronique. Un autre cas sera également envisagé: celui d'un mélange de poudres métalliques dont les particules sont plus grosses que la section du faisceau électronique; nous montrerons qu'en ajoutant un balayage linéaire rapide à l'exploration de l'échantillon, on peut encore déterminer un critère quantitatif d'homogénéité.

Etude d'un mélange de particules fines

1. Analyse statistique

Nous considérons un mélange de particules fines par rapport aux dimensions du volume analysé par la sonde électronique.

Supposons que ce mélange soit constitué de deux phases 1 et 2, dont les proportions en volume sont v_1 et v_2: on a $v_1 + v_2 = 1$. En première approximation, nous supposerons que toutes les particules de chaque phase ont le même volume $= V_1$ et V_2 respectivement.

On calcule le volume moyen $\overline{V}$ des particules du mélange:

$$\frac{v_1}{V_1} + \frac{v_2}{V_2} = \frac{1}{\overline{V}} \tag{1}$$

ainsi que les proportions de chaque phase en nombre de particules:

$$n_1 = v_1 \frac{\overline{V}}{V_1}$$

et

$$n_2 = v_2 \frac{\overline{V}}{V_2}. \tag{2}$$

Dans le volume V analysé par la sonde électronique comportant un nombre limité N de particules, les proportions numériques n_1 et n_2 correspondant à un mélange supposé homogène présenteront des fluctuations statistiques dont la variance $\sigma^2(n)$ s'exprimera par:

$$\sigma^2(n_1) = \sigma^2(n_2) = \frac{n_1 \cdot n_2}{N}. \tag{3}$$

Les équations (1), (2) et (3) permettent de calculer les fluctuations statistiques de la composition en volume dans le volume analysé:

$$\frac{1}{v_1} = 1 + \frac{1 - n_1}{n_1} \frac{V_2}{V_1}.$$

Les variances des deux membres s'écrivent:

$$\sigma^2\left(\frac{1}{v_1}\right) = \left(\frac{V_2}{V_1}\right)^2 \frac{1}{n_1^4} \frac{n_1(1 - n_1)}{N}$$

d'où

$$\left[\frac{\sigma(v_1)}{v_1}\right]^2 = \frac{1}{V}\left[\left(\frac{1 - v_1}{v_1}\right)^2 v_1 V_1 + v_2 V_2\right]. \tag{4}$$

Il est facile de généraliser à un mélange d'un nombre quelconque de phases de proportion en volume v_i et dont les particules ont le volume $V_i =$

$$\left[\frac{\sigma(v_i)}{v_i}\right]^2 = \frac{1}{V}\left[\left(\frac{1 - v_i}{v_i}\right)^2 v_i V_i + \sum_{j \neq i} v_j V_j\right]. \tag{5}$$

Pour tenir compte de la répartition en taille des particules de chaque phase, caractérisée par σ_i, écart quadratique moyen, et $\overline{w_i}$, volume des particules de taille moyenne, nous utiliserons le calcul effectué par STANGE [3] qui a conduit à l'expression suivante:

$$\left[\frac{\sigma(v_i)}{v_i}\right]^2 = \frac{1}{V}\left[\left(\frac{1 - v_i}{v_i}\right)^2 v_i \overline{w_i}\left(1 + \frac{\sigma_i^2}{\overline{w_i}^2}\right) + \sum_{j \neq i} v_j \overline{w_j}\left(1 + \frac{\sigma_j^2}{\overline{w_j}^2}\right)\right]. \tag{6}$$

POOLE [4] a montré que les expressions telles que

$$\overline{w_i}\left(1 + \frac{\sigma_i^2}{\overline{w_i}^2}\right)$$

pouvaient se calculer connaissant la distribution en taille des particules et qu'elles représentaient le volume moyen dans la distribution en volume: $\overline{V_i}$. Dans le cas d'une phase constituée de particules sphériques obéissant à une loi de répartition log-normale, de diamètre moyen $\overline{d_i}$, le diamètre moyen de la répartition en volume $\overline{dv_i}$ est donné par:

$$\mathrm{Log}\,\frac{\overline{dv_i}}{\overline{d_i}} = \frac{3}{2}\,\mathrm{Log}^2\,\frac{\sigma_i}{\overline{d_i}}. \tag{7}$$

Le volume moyen $\overline{V_i}$ est alors:

$$\overline{V_i} = \frac{\pi}{6}\,\overline{dv_i}^3.$$

Il faut remarquer par ailleurs que le volume V intervenant dans l'équation (6) a été défini comme le volume occupé par les particules. Le volume V_0 de l'échantillon qui présente toujours une certaine porosité est plus grand que V. Soit ϱ_0 la masse spécifique du matériaux supposé massif et ϱ celle de la poudre pressée réellement obtenue, le volume de l'échantillon est:

$$V_0 = V\,\frac{\varrho_0}{\varrho}.$$

Nous sommes donc conduits à l'expression suivante pour les fluctuations de composition théoriques dans un volume V_0 d'un mélange homogène:

$$\left[\frac{\sigma(v_i)}{v_i}\right]^2 = \frac{\varrho_0}{\varrho\,V_0}\left[\left(\frac{1 - v_i}{v_i}\right)^2 v_i \overline{V_i} + \sum_{j \neq i} v_j \overline{V_j}\right]. \tag{8}$$

Lorsque l'on mesure ces fluctuations au microanalyseur, les caractéristiques de l'instrument interviennent: la constante de temps T du dispositif de comptage détermine le temps t, pendant lequel l'appareil intègre le taux de comptage instantané. La vitesse de défilement de l'échantillon $d\,x/dt$ conditionne le volume analysé pendant ce temps.

Si la sonde immobile analyse un volume vo que nous assimilerons à une demi-sphère pour calculer un ordre de grandeur de V_0:

Pour $T = 2$ s et $d\,x/dt = 20$ μm/mn ou $\frac{1}{3}$ μm/s, on obtient:

$$V_0 \sim 3,3\,v_0.$$

Enfin, le comptage lui-même donne lieu à des fluctuations s'ajoutant aux différences de composition:

a) D'une part des fluctuations qui peuvent provenir de défauts dans la surface de l'échantillon et qui se traduisent par une variation du courant absorbé. Ces fluctuations sont en général négligeables, sauf pour des échantillons particulièrement poreux. L'enregistrement du courant absorbé permet de les évaluer et d'éliminer, le cas échéant, les zones défectueuses.

b) D'autre part les fluctuations statistiques de comptage. Leur écart quadratique moyen s'exprime par:

$$\frac{\sigma(N)}{N} = 1/\sqrt{NT}$$

où N est le taux de comptage. Comme nous verrons par ailleurs que, dans le cas envisagé, le taux de comptage varie pratiquement linéairement avec la concentration, les fluctuations qu'enregistre l'appareil sont telles que:

$$\left[\frac{\sigma(N)}{N}\right]^2 = \left[\frac{\sigma(v_i)}{v_i}\right]^2 + \frac{1}{NT}. \tag{9}$$

2. Application au cas des oxydes métalliques

Lorsque nous avons étudié les mélanges de poudres d'oxydes métalliques destinés à la fabrication de ferrites doux, nous nous sommes trouvés dans le cas analysé ci-dessus. Le Tableau 1 donne les caractéristiques granulométriques, les concentrations et les différents paramètres conduisant au calcul des fluctuations théoriques de composition d'après les formules établies ci-dessus.

Les conditions expérimentales étaient les suivantes:

Constante de temps: 2 s
Vitesse de déplacement de l'échantillon: 20 μm/mn

Il est à remarquer que les particules d'oxyde de zinc, prismatiques, ne sont pas assimilables à des sphères, nous avons adopté pour $\overline{d_{zn}}$ la valeur ainsi calculée:

$$\overline{d_{zn}} = \sqrt[3]{l^2\,\overline{L}}$$

l et L étant la largeur et la longueur moyennes des particules.

Tableau 1

	Oxyde de Fer	Oxyde de Manganèse	Oxyde de Zinc
Proportion en poids	0,693	0,178	0,129
Densité	5,15	4,84	5,67
Proportion en volume v_i	0,693	0,190	0,117
Diamètre moyen numérique $\overline{d_i}$	0,057 μ	0,0515 μ	0,086 μ
$d_{\sigma i}/\overline{d_i}$	1,70	1,65	1,91
$\overline{d_{v i}}$	0,087 μ	0,075 μ	0,164 μ
$\overline{V_i}$	$0,357 \times 10^{-3} \mu^3$	$0,221 \times 10^{-3} \mu^3$	$2,31 \times 10^{-3} \mu^3$
$\sigma(v_i)/v_i$	0,0135	0,0254	0,0886

Tableau 2

	Oxyde de Fer	Oxyde de Manganèse	Oxyde de Zinc
Taux moyen de comptage	raie Fe K_α 3×10^4 cpm	raie Mn K_α 3×10^4 cpm	raie Zn K_α 6×10^3 cpm
$1/NT$	0,001	0,001	0,005
$[\sigma(v_i)/v_i]^2$ théorique	0,00018	0,00065	0,0079

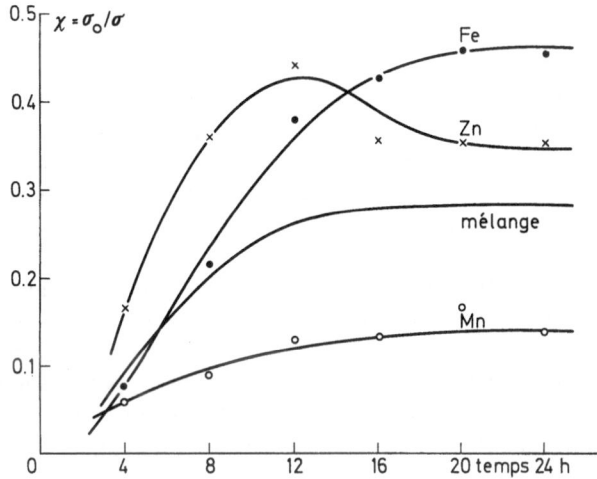

Fig. 1. Variation du rapport $\chi = \sigma_0/\sigma$ en fonction du temps de mélange dans le cas de poudres d'oxydes

Le Tableau 2 établit la comparaison, en fonction des taux moyens de comptage expérimentaux, entre les fluctuations statistiques de comptage et les valeurs limites des fluctuations de composition. On remarque que, pour un mélange parfaitement homogène, les fluctuations dues au comptage masquent les écarts de composition pour le fer; on pourra malgré tout détecter de faibles inhomogénéités, le rapport signal sur bruit restant au moins égal à 0,2. Les fluctuations de composition du manganèse pourront toujours être détectées avec un rapport signal sur bruit de 0,6 à la limite. Enfin, dans le cas du zinc, les fluctuations dues à la composition seront toujours prépondérantes, même dans le cas le plus homogène.

Pour calculer σ, écart type réel, d'après les enregistrements, on détermine la probabilité pour obtenir une concentration donnée C_i en mesurant la longueur de la portion de droite dont l'ordonnée correspond à la concentration C_i et qui est située sous la courbe enregistrée. On trace alors la courbe de probabilité pour que $C > C_i$ en fonction de C_i, en admettant que le taux de comptage varie linéairement avec la concentration, tout au moins au voisinage de la concentration moyenne. Cette approximation s'est trouvée justifiée par l'expérience. L'étalonnage est fourni par le planimétrage de la courbe enregistrée.

On a déterminé de cette façon l'influence du temps de mélange sur l'homogénéité en mesurant des échantillons prélevés dans un broyeur vibrant à circulation après des temps variables. La Fig. 1 donne les variations de $\chi = \sigma_0/\sigma$ pour le fer, le manganèse et le zinc en fonction du temps de mélange.

Pour avoir une idée de l'homogénéité de l'ensemble, on a également tracé la courbe relative à:

$$\sqrt[3]{\left(\frac{\sigma_0}{\sigma}\right)_{\text{Fe}} \left(\frac{\sigma_0}{\sigma}\right)_{\text{Mn}} \left(\frac{\sigma_0}{\sigma}\right)_{\text{Zn}}}$$

qui caractérise le mélange.

Il est visible qu'après 12 heures, l'amélioration devient très faible et que l'opération n'est plus rentable. La raison en est évidente: le zinc tend à se réagglomérer et l'on n'a pas intérêt à prolonger le mélange.

Etude des mélanges constitués de grosses particules

1. Méthode d'analyse

Nous allons considérer maintenant des mélanges de poudres constituées de particules dont le diamètre moyen est supérieur à 1 μm. Nous prendrons comme exemple les métaux destinés à la fabrication d'alliages magnétiques qui ont les caractéristiques suivantes:

$$\bar{d}_{Fe} = 3,7 \text{ à } 17 \text{ μm suivant les provenances}$$
$$d_{Ni} = 3,5 \text{ μm}$$
$$\bar{d}_{Mo} = 1,5 \text{ μm}$$
$$\bar{d}_{Cu} = 16 \text{ μm}$$

Le volume analysé par la sonde électronique immobile est alors inférieur au volume moyen d'une particule. Afin de conserver une analyse statistique convenable portant sur un nombre suffisamment grand de particules, il est nécessaire de modifier la quantité de matière vue instantanément par le microanalyseur. Ceci a été réalisé en utilisant un balayage linéaire rapide du spot électronique, d'amplitude L, qui s'ajoute au déplacement normal et plus lent de l'échantillon.

La profondeur de pénétration de la sonde est inférieure aux dimensions des particules et l'on procède, par conséquent, à une analyse en surface de l'échantillon. On calcule les fluctuations limites théoriques pour un mélange parfaitement homogène comme dans le cas envisagé plus haut en utilisant au lieu des volumes les surfaces interceptées par la sonde.

Soit s_i la proportion en surface de l'élément i, S_i la section d'une particule de i vue par la sonde, $\bar{S}_i$ la section moyenne dans la répartition en surface et S_0 la surface analysée. On sera conduit à la formule:

$$\left[\frac{\sigma(s_i)}{s_i}\right]^2 = \frac{1}{S_0}\left[\left(\frac{1-s_i}{s_i}\right)^2 s_i \bar{S}_i + \sum_{j \neq i} s_j \bar{S}_j\right] \tag{10}$$

et l'on calculera $\bar{S}_i$ comme dans le cas précédent:

$$\bar{S}_i = \frac{\pi}{4} \bar{d}_{S_i}^2$$

avec, dans le cas d'une répartition gaussienne:

$$\text{Log} \frac{\bar{d}_{s_i}}{d_i} = \text{Log}^2\left(\frac{\sigma_i}{d_i}\right).$$

Le diamètre qui intervient maintenant n'est plus le diamètre réel des particules mais celui qui est effectivement intercepté par la sonde. On peut calculer la correction à apporter au diamètre en évaluant le diamètre à la profondeur moyenne de pénétration du faisceau, soit 1 μm environ.

Cette correction n'est importante que pour les très grosses particules dont une partie se trouve cachée par les plus petites.

2. Application au cas des poudres de métaux

Les données expérimentales et les résultats des calculs relatifs à un type de poudres de métaux que nous avons étudiés sont donnés dans le Tableau 3.

Les enregistrements ont été effectués dans les conditions suivantes:

$$L = 400 \text{ μm}, \quad T = 5 \text{ s} \quad \text{et } \frac{dx}{dt} = 20 \text{ μm/mn}$$

ce qui correspond à:

$$S_0 \cong 1400 \text{ μm}^2.$$

L'importance des fluctuations statistiques de comptage est résumée dans le Tableau 4. On constate que dans tous les cas ces fluctuations sont très inférieures ou même tout à fait négligeables

Tableau 3

	Fer	Nickel	Molybdène	Cuivre
Proportion en moles	0,137	0,770	0,043	0,050
n_i proportion numérique	0,0725	0,467	0,459	$3,28 \times 10^{-4}$
Diamètre moyen numérique	3,7 μ	3,5 μ	1,5 μ	16,2 μ
$\overline{d_i}$ diamètre moyen intercepté	3,3 μ	3,2 μ	1,5 μ	10 μ
$d_{\sigma_i}/\overline{d_i}$	1,38	1,43	1,97	1,70
S_i correspondant à $\overline{d_i}$	8,55 μ^2	8,05 μ^2	1,76 μ^2	78,5 μ^2
s_i proportion en surface	0,119	0,719	0,157	0,0049
$\overline{S_i}$ (distribution en surface)	10,6 μ^2	10,4 μ^2	4,44 μ^2	138 μ^2
$\sigma(s_i)/s_i$	0,236	0,052	0,144	4,50

Tableau 4

	Fer	Nickel	Molybdène	Cuivre
Taux moyen de comptage	raie Fe K_α 5×10^3 cpm	raie Ni K_α 10^5 cpm	raie Mo K_α $1,5 \times 10^3$ cpm	raie Cu K_α 10^3 cpm
$1/NT$	0,0024	$0,12 \times 10^{-3}$	0,008	0,012
$[\sigma(s_i)/s_i]^2$ theorique	0,056	$2,7 \times 10^{-3}$	0,0206	20,3

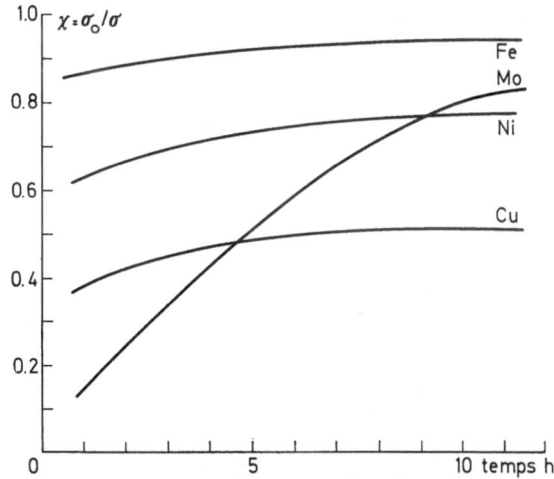

Fig. 2. Variation du rapport $\chi = \sigma_0/\sigma$ en fonction du temps de mélange dans le cas de métaux en poudre

devant les écarts de composition. Pratiquement, nous pourrons ne pas en tenir compte dans la lecture des enregistrements.

L'efficacité du mélange a été étudiée, comme dans le cas des poudres d'oxydes, en faisant des prélèvements successifs dans le mélangeur. Les courbes de la Fig. 2, représentant $\chi = \sigma_0/\sigma$ pour chaque métal en fonction du temps, permettent de fixer à 10 heures la durée optimale du mélange. On constate que les constituants les plus abondants: le fer et le nickel, sont mélangés plus rapidement: 5 heures suffisent, le molybdène, formé de particules fines est plus long à homogénéiser tandis que la répartition du cuivre, en grosses particules, ne s'améliore pas de façon sensible. On peut en conclure qu'il y a intérêt à utiliser des constituants dont les tailles de grains sont voisines.

Conclusion

La méthode que nous avons mise au point nous a permis de définir et de mesurer un critère quantitatif d'homogénéité. Les calculs théoriques justifient avec une excellente approximation toutes les hypothèses que nous avions émises. De plus, la mesure peut se faire de façon particulièrement facile, en évitant les problèmes d'échantillonnage rencontrés dans les méthodes décrites précédemment.

Résumé

On étudie en fonction des conditions de mélange les fluctuations de composition d'échantillons formés de poudres d'oxydes, de céramiques ou de métaux.

Les résultats expérimentaux, obtenus par enregistrement en déplaçant l'échantillon sous la sonde électronique, sont comparés aux valeurs théoriques calculées pour un mélange supposé parfaitement homogène. On définit de la sorte un critère quantitatif d'homogénéité pour chaque constituant analysé.

Deux exemples sont traités: dans le premier cas, les particules fines devant les dimensions de la sonde se prêtent à une détermination directe des fluctuations statistiques de composition. Dans le second cas, traitant de particules plus grosses, il nous faut tenir compte de la portion de chaque grain interceptée par le faisceau.

On peut ainsi comparer diverses méthodes de mélange et choisir un temps optimum pour cette opération.

Bibliographie

1. Strivens, M. A.: I.E.E., Conference on Magnetic Materials and their Applications, London 1967, No. 23, p. 168—172.
2. Poole, K. R.: Proc. Brit. Ceram. Soc. **3**, 43—48 (1965).
3. Stange, K.: Chem. Ing. Techn. **26**, 331—337 (1954).
4. Poole, K. R., R. F. Taylor, and G. P. Wall: Trans. Inst. Chem. Eng. (London) **42**, T 305—T 315 (1964).
5. Mugele, R. A., and H. D. Evans: Ind. Eng. Chem. **43**, 1317—1324 (1951).

Analyse des précipités extraits sur repliques, à l'aide d'un microanalyseur classique et d'un microanalyseur équipé d'un microscope électronique

M. ANCEY, G. HENRY, J. PHILIBERT et R. TIXIER

Irsid, St. Germain-en-Laye, France

Résumé

L'étude des alliages à hautes caractéristiques, de leurs propriétés mécaniques et de l'influence des traitements mécaniques et thermiques sur leur structure est liée à la détermination de la nature et de la composition des phases qui apparaissent sous forme de fins précipités dans le métal. On expose d'abord les techniques utilisées à l'Institut de Recherches de la Sidérurgie Française pour la préparation d'échantillons sous forme de répliques avec extraction. Puis on montre comment la microanalyse par sonde électronique, associée à la microscopie et à la diffraction électroniques, et éventuellement à la diffraction X, constitue un moyen de choix pour mener à bien l'identification complète des précipités extraits. On décrit la méthode utilisée pour l'analyse de ces échantillons minces et on cite quelques exemples d'applications.

I. Introduction

Les propriétés mécaniques à haute ou basse température, ainsi que certaines propriétés physiques des alliages à hautes caractéristiques sont souvent liées à la présence de phases qui apparaissent sous forme de précipités submicroscopiques dans le métal, et dont la nature et la composition peuvent varier suivant les traitements thermiques et mécaniques. L'identification de ces phases est un problème métallurgique important dont la résolution permet une meilleure compréhension des rôles respectifs des différents éléments d'alliages et des mécanismes de la précipitation. Cependant ces précipités sont généralement de taille nettement inférieure au micron, ce qui rend impossible leur observation et leur microanalyse dans les conditions habituelles, c'est-à-dire dans un échantillon massif. On est conduit à utiliser les préparations de microscopie électronique afin de les séparer de la matrice.

II. Techniques d'extraction

A. Intérêt des répliques avec extraction

La technique des lames minces est actuellement largement utilisée en microscopie électronique; elle apporte de précieux renseignements à la physique du métal. Cependant, pour déterminer la nature et la composition des phases précipitées, il est généralement préférable d'utiliser la technique des répliques avec extraction; celle-ci jointe à des méthodes d'attaque préférentielle, permet d'examiner au microscope électronique les précipités extraits, de les identifier par diffraction électronique ou, par diffraction des rayons X, à partir de plusieurs répliques avec extraction superposées, et enfin, de les analyser au moyen du microanalyseur par sonde électronique sans être gêné par la présence de la matrice [1—6].

B. Préparation des répliques

1. Attaque d'une surface métallique

Un réactif métallographique approprié permet la mise en évidence des précipités et, si la matrice est attaquée préférentiellement, leur extraction sur réplique. Cependant les réactifs sont rarement sélectifs et conduisent à une extraction globale des différents précipités présents, ce qui complique leur identification et leur analyse. Dans un grand nombre de cas, il est préférable d'effectuer une attaque électrochimique à l'aide d'un potentiostat, méthode beaucoup plus sélective [7].

2. Microfractographie

Une méthode d'extraction de nature complètement différente consiste à casser l'échantillon métallique à basse température ($-195°$ C). La rupture peut alors être fragile et fréquemment intergranulaire, ce qui permet l'extraction des micro-inclusions qui ont été à l'origine de la fracture. En particulier, lorsque celle-ci est intergranulaire, la méthode permet l'extraction sélective des précipités qui s'étaient formés dans les joints de grains [8].

3. Extraction des phases

a) *Support*. Les répliques sont généralement faites en carbone; cependant quand il est nécessaire de rechercher dans les précipités la présence de cet élément, nous employons le monoxyde de silicium[1].

Nous utilisons deux types de répliques [9].

b) *Répliques directes*. La pellicule de carbone ou de monoxyde de silicium déposée sur l'échantillon est détachée par dissolution anodique du métal sous-jacent, ou par dissolution chimique dans le brome-éthanol. Ce réactif ne dissout pas les carbures, la silice, l'alumine, les silicates, certains oxydes, sulfures ou nitrures. Il attaque le nitrure et le sulfure de fer, FeO, MnO et certains composés intermétalliques comme les nickelures.

c) *Répliques en deux temps*. Lorsqu'on veut éviter l'attaque des précipités au moment du décollement ou conserver intacte la matrice, on utilise une technique de réplique en deux temps: dépôt cathodique de nickel, décollement mécanique, puis dépôt de carbone ou de SiO sur l'empreinte obtenue. Le nickel est dissout dans le brome-éthanol; la dissolution est très rapide et les précipités restent intacts même s'ils sont solubles dans le brome [16].

Les répliques avec extraction obtenues ainsi sont déposées sur une grille support qui permet leur examen au microscope électronique et, sur le même échantillon, l'analyse des précipités au micro-analyseur par sonde électronique.

III. Microanalyse par sonde électronique des répliques avec extraction

Les précipités extraits sont généralement de taille inférieure au micron, mais ils se trouvent enrobés dans un film mince de carbone ou de monoxyde de silicium. On peut donc les analyser sans être gêné par la matrice. Il est possible de ce fait d'étudier des précipités bien plus petits que le diamètre de la sonde et d'analyser des volumes plusieurs dizaines de fois inférieurs aux quelques microns cubes excités dans un échantillon massif. La limite inférieure n'est imposée que par l'intensité minimale du signal X significatif détectable. La masse de matière analysée de façon non destructive dans ce cas est de l'ordre de 10^{-14} gramme et l'on est sensible à des quantités inférieures à 10^{-16} gramme d'un élément. Cependant de telles analyses présentent des difficultés plus grandes que celles d'un échantillon massif. Ces difficultés nous ont conduits à définir un mode opératoire particulier, de façon à manipuler sur un microanalyseur de conception classique, puis à réviser partiellement ce mode lorsque nous avons pu utiliser un microanalyseur équipé d'un microscope électronique.

---

1. Volatilisation sous vide, après préchauffage d'une dizaine de minutes entre 700 et 800° C.

A. Analyse avec un microanalyseur classique

1. Repérage

Le repérage direct de l'échantillon avec le microscope optique de la sonde étant pratiquement impossible, il est nécessaire de l'observer au préalable par microscopie électronique; nous utilisons des grilles supports dont les barreaux sont marqués, ce qui permet de retrouver par observation optique les zones choisies sur la préparation. D'autre part nous avons utilisé un écran fluorescent placé dans la colonne à environ vingt centimètres sous l'objet, et observable par un hublot. Cet écran permet un repérage assez grossier des précipités par microscopie électronique à ombre [10]; il a l'avantage de montrer directement si la sonde dérive sur l'objet au moment de la mesure.

2. Conditions d'analyse

L'échauffement de l'échantillon au point analysé peut conduire à la destruction du film support (fig. 1). On peut rarement dépasser une intensité de sonde de quelques dizaines de nanoampères. Si la préparation se charge trop, la sonde devient instable et on doit métalliser légèrement (100 Å, Cu) la face inférieure de la réplique.

L'émission X caractéristique est souvent faible, mais, si l'on évite d'analyser la réplique au voisinage d'un barreau de la grille support, le bruit de fond est généralement presque nul. Les conditions d'analyse ne seraient donc pas mauvaises si des difficultés pratiques ne les compliquaient pas.

La mise en place du point analysé de l'échantillon au point d'intersection des cercles de Rowland des spectromètres avec le plan du cercle de moindre confusion de la sonde est approchée puisque les conditions d'observation optique sont mauvaises. Il peut en résulter une

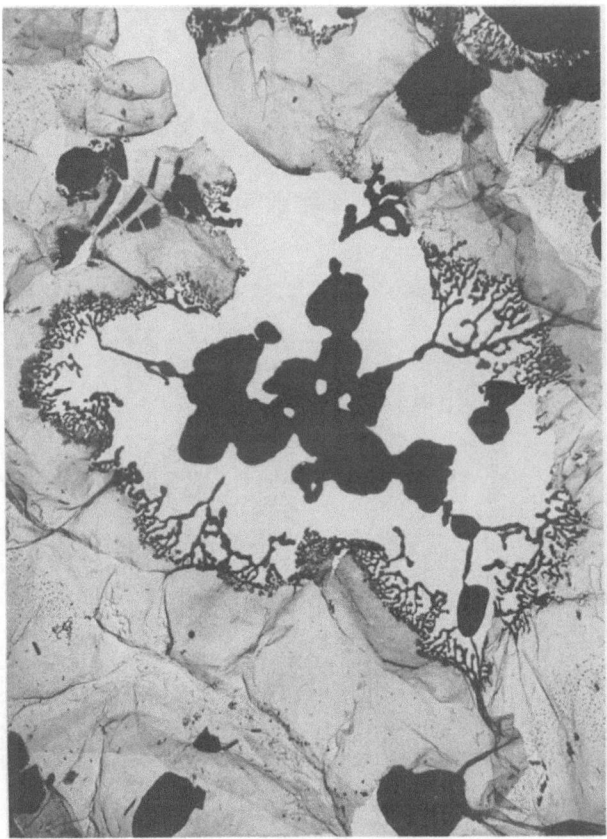

Fig. 1. Réplique partiellement détruite après analyse d'un précipité

défocalisation plus ou moins forte des spectromètres, ce qui fausse l'analyse. Les spectromètres à faible angle d'émergence et dont le plan contient l'axe du faisceau sonde sont les plus sensibles aux défauts de mise au point de l'échantillon. Les spectromètres à grand angle d'émergence, ou ceux dont le plan fait un angle assez grand avec l'axe du faisceau, sont beaucoup plus tolérants. Il faut donc souvent recaler les spectromètres sur la réplique, ce qui est rendu difficile par la faiblesse des intensités des rayonnements X cractéristiques. Il faut également recaler les amplificateurs des baies de comptage pour tenir compte du glissement de gain des compteurs dû à la grande variation d'intensité entre les témoins et l'échantillon.

3. Analyse qualitative

Il est parfois suffisant d'identifier les éléments présents et d'examiner leur répartition dans une zone de la réplique [2, 12]. On procède alors à des balayages à l'intérieur d'un carreau de grille support et on compare ceux-ci à une micrographie électronique par transmission de la

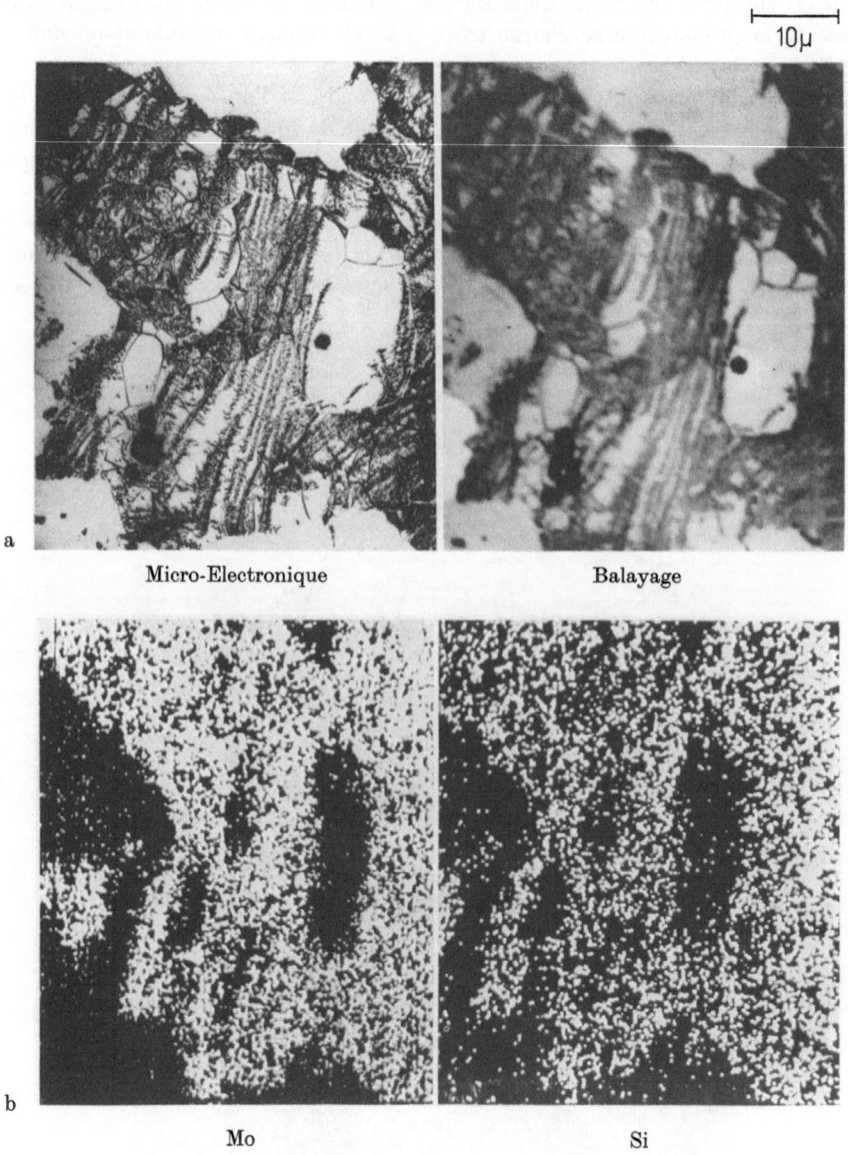

Fig. 2a—d. Analyse qualitative d'une réplique. Précipités extraits de la ferrite δ d'un acier inoxydable austénomartensitique

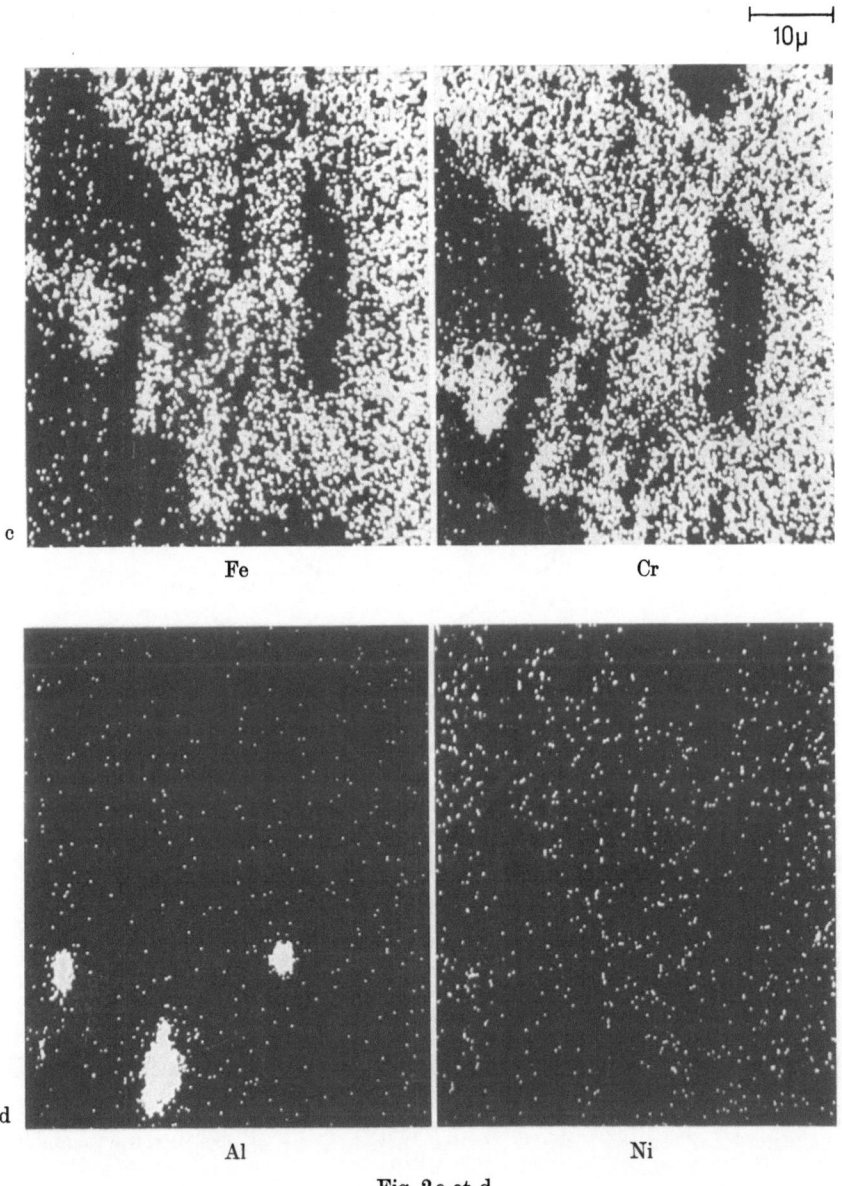

Fig. 2 c et d

même plage au même grandissement (environ 1000). Nous avons noté sur ces échantillons que les images électroniques obtenues, par balayage, à l'aide du courant échantillon, permettent dans certains cas de distinguer des détails bien plus petits que le diamètre de la sonde, en raison des conditions de contraste élevé de ces échantillons pour ce type d'images (fig. 2).

4. Analyse quantitative. Méthode

On ne peut pas envisager de disposer de témoins minces qui aient la même épaisseur massique que l'échantillon, ce qui supprimerait toute correction et permettrait des mesures absolues. Il nous a paru peu pratique d'utiliser des témoins sous forme de lames minces d'épaisseur connue en raison des difficultés de préparation de ces lames. Toutefois leur usage diminuerait l'importance des corrections et pourrait être intéressant dans certains cas. Nous utilisons plutôt les témoins massifs habituels.

La concentration apparente mesurée en un point, pour un élément, dépend de la masse volumique de l'échantillon en ce point. Elle varie considérablement puisque la densité en précipités de la réplique est loin d'être uniforme; de plus cette concentration n'a généralement pas de signification physique simple [11]. Par contre le rapport des concentrations apparentes en un point de deux éléments présents dans le précipité est significatif et permet d'obtenir le rapport réel des concentrations dans le précipité analysé. Les corrections à appliquer à la mesure font l'objet d'une autre communication à ce congrès.

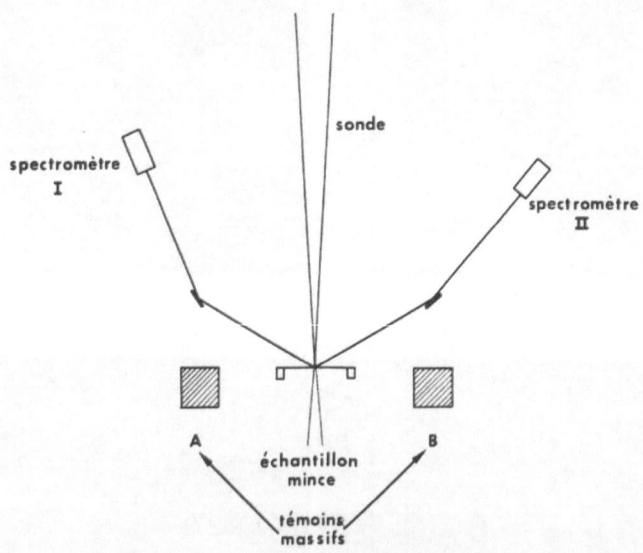

Fig. 3. Schéma de l'analyse d'une réplique

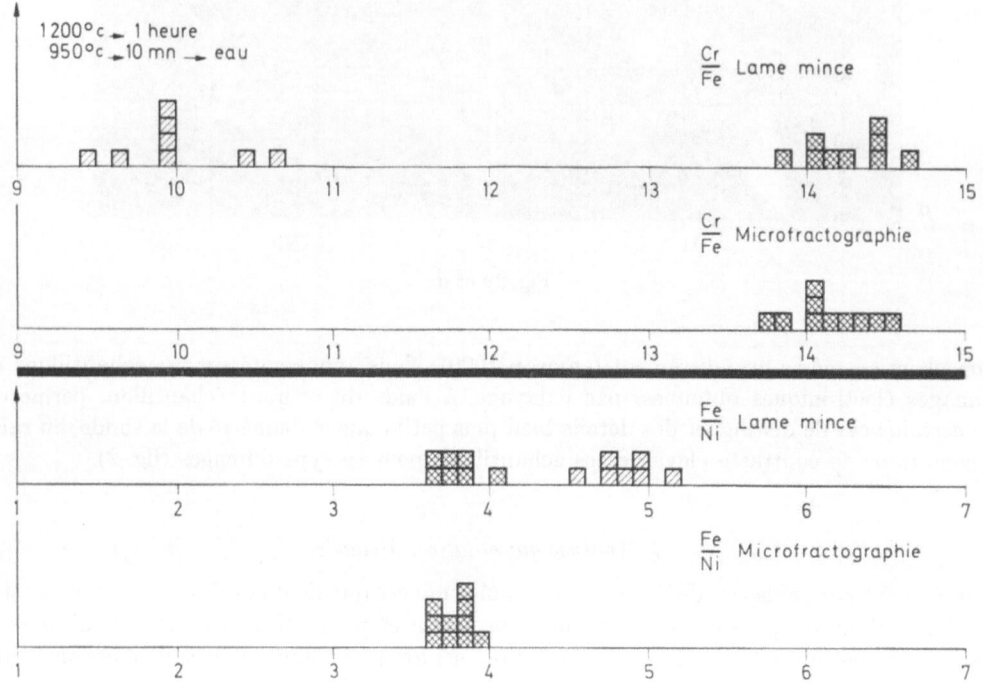

Fig. 4. Analyse quantitative des répliques. Histogramme des valeurs mesurées, carbures extraits d'un acier réfractaire au Cr—Ni

5. Mode opératoire

Il faut mesurer, en un même point de la cible, simultanément les concentrations apparentes de deux éléments. Pour celà (fig. 3) on cale les spectromètres I et II respectivement sur le rayonnement caractéristique des éléments A et B et, en un point donné de l'échantillon, on effectue les comptages simultanés sur les raies de A et B. Etant donné l'importance des variations statistiques, dues à la faiblesse des taux de comptage, on est conduit à répéter la mesure sur un assez grand nombre de points. On mesure également les bruits de fond par décalage des spectromètres; on amène ensuite les témoins de A, puis de B, à la place de l'échantillon, et on mesure les intensités caractéristiques de ces éléments. La tension d'accélération de la sonde est réglée à une valeur fixe, ce qui fait que souvent un seul des deux éléments est mesuré dans des conditions optimales d'excitation. Lors des mesures sur les témoins on peut réduire encore l'intensité, afin de minimiser les glissements de gain et les pertes par temps mort des compteurs.

Les rapports des concentrations apparentes mesurées sont dépouillés statistiquement à l'aide de l'histogramme des mesures. La fidélité de ces mesures a été vérifiée par le test suivant: on avait pu montrer par cette méthode de dépouillement l'existence dans une réplique de deux sortes de précipités bien distincts par les valeurs différentes des rapports de concentration chrome sur fer et fer sur nickel. Une autre extraction fut faite, mais cette fois-ci à partir d'une surface de rupture, de façon à éliminer les précipités intragranulaires. Effectivement les rapports de concentration mirent en évidence une seule sorte de précipités et ces rapports correspondaient avec une précision raisonnable à l'un de ceux déterminés sur les répliques antérieures (fig. 4).

B. Analyse avec un microanalyseur équipé d'un microscope électronique

1. Repérage

Le problème du repérage est évidemment résolu par le dispositif de microscopie électronique joint à la microsonde [13—15]. On observe une certaine plage de l'objet éclairé par la sonde légèrement défocalisée, l'objectif de la sonde servant alors de second condenseur du microscope électronique. On focalise ensuite la sonde au niveau de l'objet sur les précipités choisis; on peut d'ailleurs ajuster finement la position de la sonde en utilisant les plaques de déflexion électromagnétique. La microscopie électronique est caractérisée par une très grande profondeur de champ, en sorte que le déplacement en hauteur de l'échantillon peut défocaliser les spectromètres sans qu'on observe une diminution de netteté de l'image. On peut réduire considérablement cette profondeur de champ en utilisant le balayage rapide de la sonde («Wobbler»). Celui-ci permet une mise au point assez fine de l'image électronique par transmission à faible grandissement. On peut, d'autre part, en examinant l'image électronique par balayage, vérifier que la sonde est bien au point sur l'objet. Il n'en reste pas moins que le calage des spectromètres doit être vérifié avec soin, surtout lorsque ceux-ci ont un pouvoir séparateur élevé.

2. Méthode et mode de mesure

La méthode générale de mesure et le mode opératoire sont analogues à ceux décrits précédemment (microanalyseur classique); mais il est souvent commode de profiter, avant toute mesure quantitative, des facilités d'observation de l'échantillon soumis à un balayage de faible amplitude $(40 \times 40 \, \mu)$ simultanément sur l'écran du microscope électronique et sur les tubes cathodiques donnant les images X et électronique. Ceci permet de mettre en relation directe les formes et la disposition des précipités et les variations de composition. La fig. 5 montre un échantillon où l'on a ainsi trouvé que des inclusions très fines, extraites sur une surface de rupture, étaient constituées les unes de mélanges non-stoechiométriques et variés d'oxydes d'aluminium et de magnésium et les autres d'alumine. Une analyse quantitative suivie d'histogramme n'aurait pas donné de résultat net en raison des variations de composition d'un précipité à l'autre. Les conclusions de cette analyse ont été confirmées par diffraction de rayons X.

Micro-Inclusions
Réplique avec extraction

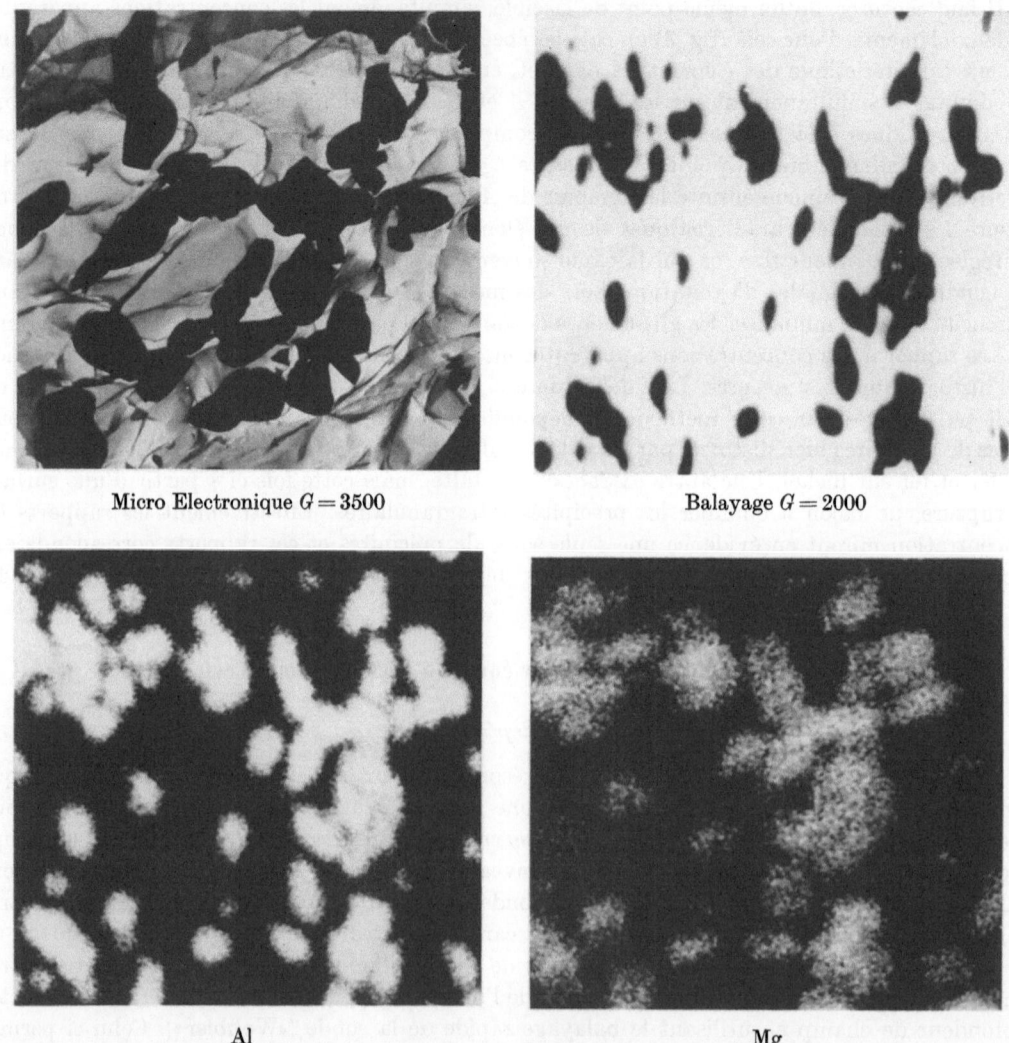

Micro Electronique $G = 3500$ Balayage $G = 2000$

Al   Mg

Fig. 5. Inclusions submicroscopiques extraites sur une surface de rupture

L'appareil dont nous disposons ne permet pas la diffraction électronique sur l'échantillon. Cependant le repérage de la réplique est suffisamment précis pour que la même plage puisse être examiné dans un microscope électronique classique, d'autant qu'il est possible de faire une micrographie électronique dans le microanalyseur. On peut d'ailleurs noter qu'un dispositif permettant la diffraction ne serait réellement utile que si l'on pouvait obtenir des tensions d'accélération bien supérieures à la limite actuelle de 40 kV.

3. Limitations pratiques

L'analyse des éléments assez lourds peut se faire à des tensions élevées, de l'ordre de 30 kV, pour lesquelles les répliques sont bien transparentes et facilement observables. Il n'en va pas de même quand la présence d'éléments légers conduit à réduire la tension d'accélération pour améliorer les conditions d'analyse quantitative et diminuer le bruit de fond. Nous avons pu montrer, sur des répliques avec extraction d'un acier inoxydable au bore, que deux rapports différents nickel sur fer correspondaient tantôt à un carbure, tantôt à un carbo-borure. Les répliques en monoxyde de silicium étaient presque opaques à 15 kV, tension qui conduit encore à un taux d'excitation tellement élevé qu'il rend peu vraisemblable une analyse quantitative des éléments très légers par cette méthode. Par contre, avec des répliques en carbone très fines, nous avons pu analyser des précipités d'un alliage aluminium-manganèse en les observant jusqu'à 10 kV.

IV. Applications

1. Composés intermétalliques

Nous avons recherché la composition des phases précipitées dans un acier inoxydable austéno-martensitique et identifié quelles que soient les conditions du traitement de vieillissement, le composé intermétallique NiAl. Cette analyse a été confirmée par diffraction électronique et par diffraction de rayons X. L'accord des trois mesures a montré la validité de la méthode de micro-analyse et du calcul de correction des rapports apparents. Nous procédons actuellement à une étude systématique de l'évolution de la composition d'un précipité intermétallique nickel-niobium en fonction des traitements thermiques. La microanalyse sur réplique est dans un tel cas pratiquement la seule méthode qui permette ce genre de mesures.

2. Composés semi-métalliques

L'identification des composés semi-métalliques constitue également un domaine prometteur. Nous avons montré que la substitution du bore au carbone entraînait un rapport chrome sur fer différent dans les carbo-borures et les borures. Nous envisageons des études systématiques de l'importance des substitutions du carbone, du bore et de l'azote, et en particulier l'étude des composés du type $M_{23}X_6$ ou M_7X_3, où X désigne un élément interstitiel. Malheureusement, il ne parait pas possible à l'heure actuelle de déterminer les taux de substitution des interstitiels (rapports C/N, C/B) en vue de les relier aux taux de substitution des éléments métalliques et aux paramètres réticulaires.

3. Micro-inclusions

Nous avons à plusieurs reprises appliqué cette technique à des cas d'analyse d'inclusions nombreuses mais dispersées et de taille inférieure au micron. Nous préparons alors une réplique avec extraction sur une surface de rupture obtenue à basse température. Les micro-inclusions favorisent la rupture fragile et se trouvent donc en quantité suffisante sur une réplique de la surface de rupture, ce qui permet leur extraction et leur analyse. Ainsi nous avons montré, dans un exemple cité plus haut (paragraphe B 2°) l'existence de mélanges d'oxydes d'aluminium et de magnésium dans des inclusions globulaires d'un diamètre moyen de 0,5 μ.

Remerciements. Les auteurs tiennent à remercier Monsieur A. QUENNEVAT et Mademoiselle D. BRYCKAERT pour de nombreuses mesures expérimentales et Madame S. GAUBE pour la préparation des échantillons.

Bibliographie

1. FISHER, R. M.: Electron probe microanalysis of submicron precipitates in steel. J. Appl. Phys. **28**, 1379—1380 (1957).
2. DUNCUMB, P.: Microanalysis with the X-ray scanning microscope. Proc. 4th Conference on Electron Microscopy 1958, p. 267—269. Berlin-Göttingen-Heidelberg: Springer 1960.
3. HENRY, G., J. PLATEAU et J. PHILIBERT: Utilisation du Brome en métallographie électronique. Compt. Rend. **246**, 2753—2756 (1958).
4. CASTRO, R., P. DEVORE et R. DEVIN: Le traitement thermique des aciers inoxydables au voisinage du solidus. Mém. Sci. Rev. Mét. **49**, No. 5, 343—356 (1962).
5. BAKER, T. N.: Nitride phase in silicon-killed carbon steel. J.I.S.I., 315—320 (1967).
6. RANZETTA, G. V. T., and V. D. SCOTT: Electron metallography of precipitates in irradiated and unirradiated stainless steel. Metals and Materials 146—150 (1967).
7. VOELTZEL, J., G. HENRY, J. MANENC et J. PLATEAU: Utilisation du potentiostat électronique pour l'attaque micrographique. Mém. Sci. Rev. Mét. **62**, No. 2, 129—134 (1965).

8. Weinryb, E., G. Henry, J. Philibert et J. Plateau: Analyse de précipités entraits sur répliques au moyen de la microsonde de Castaing. Proc. 5th Conference on Electron Microscopy K-7. New York: Acad. Press 1962.

9. Campos-Soares, R., J. Voeltzel et G. Henry: Déscription des techniques conduisant à l'examen, l'identification et l'analyse de phases précipitées très fines dans des matériaux métalliques. J. Microscopie 7, 17—38 (1968).

10. Boersch, H.: Das Elektronen-Schattenmikroskop I. Z. Techn. Physik 20, 346—351 (1939).

11. Philibert, J., et R. Tixier: Analyse quantitative des échantillons mines. Communication à ce congrès.

12. Campos-Soares, R., E. Weinryb et G. Henry: Identification et analyse des phases qui précipitent dans un acier inoxydable austéno-martensitique à durcissement secondaire. Compt. Rend. 264, 1165—1167 (1967).

13. Duncumb, P.: An electron-optical bench for microscopy, diffraction and X-ray microanalysis. Proc. 5th Congr. for Electron Microscopy, K.K-4. New York: Acad Press 1962.

14. Nixon, W. C., and R. Buchanan: An experimental electron-bench for electron microscopy and X-ray microanalysis. X-ray optics and microanalysis (1962), p. 441—444. New York: Acad. Press 1963.

15. Rouberol, J. M., M. Tong, C. Conty u. P. Deschamps: Studie und Anwendung eines kombinierten Gerätes „Microsonde-Elektronenmikroscop". Mikrochim. Acta, Suppl., II, 201—270 (1967).

16. Henry, G., et J. Plateau: Nouvelle méthode d'empreinte métallographique non-destructive. Rev. Nickel 28, 3, 3—8 (1963).

Diffusion chimique Zinc–Nickel.
Coefficients de diffusion intrinsèques

M. Andreani, P. Azou et P. Bastien

Institut de Physique des Métaux et de Métallurgie Ecole Centrale des Arts et Manufactures, France

Résumé

Une précédente étude a montré l'importance de l'interdiffusion zinc—nickel dans la fragilisation des aciers inoxydables austénitiques en présence de zinc. Nous envisageons cette interdiffusion dans un domaine de températures pour lequel le fer et le chrome gardent une concentration constante et nous mesurons l'énergie d'activation et le facteur de fréquence.

Parallèlement nous étudions le système binaire zinc—nickel, par des couples différentiels Ni pur—NiZn β. Nous montrons qu'on peut exprimer le coefficient de diffusion par la relation:

$$\tilde{D}_{\text{NiZn}} = D_1\, e^{-A c_{\text{Ni}}} e^{-\frac{Q}{RT}},$$

et nous établissons l'énergie d'activation et le facteur de fréquence pour les coefficients intrinsèques dans le plan de Kirkendall.

1. Introduction

L'étude de la diffusion chimique zinc—nickel présente un triple intérêt.

a) Un intérêt pratique immédiat, car, comme nous l'avons montré dans une précédente communication [1], la diffusion chimique zinc—nickel dans les aciers inoxydables austénitiques en contact avec le zinc contrôle la rupture fragile de ces aciers: au-dessus de 750°, dans ce système quaternaire Fe—Cr—Ni—Zn seule l'interdiffusion zinc—nickel intervient, provoquant la transformation $\gamma \rightarrow \alpha$,

b) un intérêt purement théorique, car la variation du coefficient de diffusion chimique avec la concentration présente une allure particulière, qu'il doit être possible de relier aux données thermodynamiques concernant la solution solide,

c) enfin cette étude nous a permis à l'aide des équations de Dehlinger et Darken de calculer les coefficients de diffusion intrinsèques pour la concentration dans le plan de Kirkendall, entre 600 et 1000°.

2. Diffusion du zinc dans les aciers inoxydables austénitiques

2.1. Diffusion à température supérieure à 750°

Il est nécessaire tout d'abord de rappeler et surtout, de compléter les résultats que nous avons obtenus précédemment concernant les aciers inoxydables austénitiques [1].

2.1.1. Etude qualitative

On observe que ces aciers sont très fragiles en présence de zinc au-dessus de 750°. Nous avons examiné à la microsonde électronique des échantillons d'acier immergés sans contrainte dans le zinc liquide, puis des éprouvettes de traction rompues en présence de zinc, entre 750 et 1100°. Seule l'interdiffusion zinc—nickel est importante, le fer et le chrome ne participent pas à la diffusion dans cet intervalle de températures et se comportent comme des solvants inertes dans la diffusion. L'étude des courbes concentration-pénétration a montré que l'inter-diffusion zinc—nickel provoque rapidement à température constante la transformation $\gamma \to \alpha$ de l'acier aux limites de pénétration du zinc. Ce fait est confirmé par une étude magnétique des échantillons.

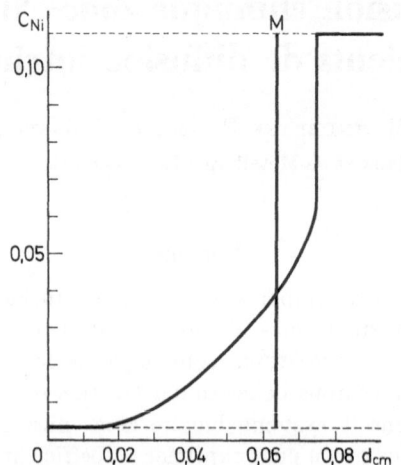

Fig. 1. Courbe de diffusion du nickel (1100° C)

L'allure d'une telle courbe de diffusion est donnée à la fig. 1. A droite de la discontinuité, du côté de la phase γ, le nickel garde, à la sensibilité de la sonde près, une concentration constante dans l'alliage jusqu'à la discontinuité due au changement de phase. Comme l'acier de départ est monophasé, il faut en conclure que la valeur du coefficient de diffusion chimique zinc—nickel apparaît négligeable dans la phase γ par rapport à sa valeur dans la phase α. Il en résulte que la transformation $\gamma \to \alpha$ se produit bien aux limites de pénétration du zinc. La mobilité des atomes lors de cette transformation, ainsi que l'importante variation du volume atomique, expliquent la propagation spectaculaire des fissures. On peut avoir une idée de l'importance des contraintes à l'interface $\alpha - \gamma$ par l'examen micrographique des échantillons immergés sans contrainte dans le zinc liquide: la phase α présente des fissures perpendiculaires à la surface qui pénètrent jusqu'au plan de changement de phase.

Pour assurer que cette transformation est bien responsable de la fragilité particulière des aciers inoxydables austénitiques en présence de zinc, nous avons fait les mêmes expériences sur un acier à 9,5% de nickel (austénitique à la température de l'expérience), pour lequel la trans-formation $\gamma \to \alpha$ ne se produit pas. Effectivement aucune fragilité importante n'est observée, en dehors de l'effet «normal» qui se produit pour la plupart des métaux au contact d'un métal liquide, et qui provient de la diminution de l'énergie superficielle.

2.1.2. Etude quantitative

Nous avons complété l'étude qualitative de cette diffusion par une étude quantitative: coefficient de diffusion chimique $\tilde{D}_{\text{NiZn}}$, énergie d'activation et facteur de fréquence.

Pour cela il a été calculé les corrections de fluorescence et d'absorption pour tous les alliages Fe—Cr—Ni—Zn. Nous avons utilisé les formules de Philibert [2]. Toutefois la valeur de la constante σ intervenant dans ces formules a été modifiée en tenant compte de la tension d'ex-citation E_K selon la formule proposée par Duncumb et Shield [3]

$$\sigma = \frac{2,39 \cdot 10^5}{(E_0)^{1,5} - (E_K)^{1,5}}$$

ainsi un meilleur accord est constaté avec l'analyse chimique.

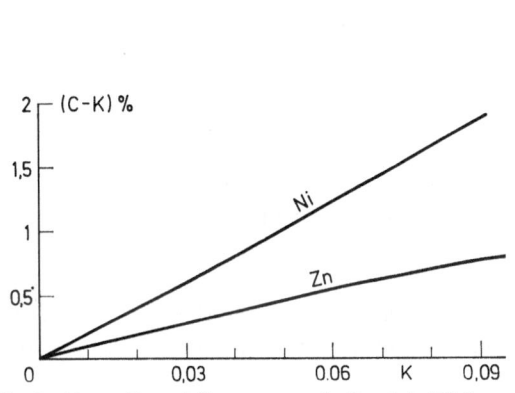

Fig. 2. Absorption et fluorescence du Zn et du Ni dans les alliages à 72% Fe et 17% Cr (20 kV)

Fig. 3. Variation du coefficient d'interdiffusion avec la concentration (1100° C)

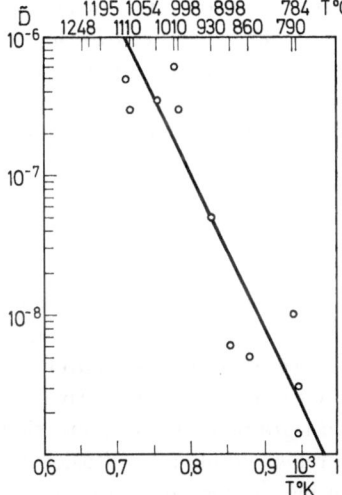

Fig. 4. $\tilde{D} = D_0 e^{-\frac{Q}{RT}}$, $C_{Ni} = 0,04$, $Q = 50$ kcal; $D_0 = 53$ cm³ s⁻¹

Les calculs, fort longs, ont été programmés sur ordinateur, et nous avons laissé les résultats complets sous forme d'un tableau qu'il n'est pas possible de présenter ici. La fig. 2 traduit graphiquement les corrections d'absorption et de fluorescence du zinc et du nickel dans les alliages à 72% de fer et 17% de chrome. C'est précisément notre cas puisque dans la zône de diffusion le fer et le chrome gardent cette concentration. On voit que ces corrections suivent une loi pratiquement linéaire, ce qui permet dans les comptages au spectromètre X de se rapporter à la concentration du nickel dans l'alliage de base, qui a été déterminée avec précision. Les corrections de numéro atomique et de fluorescence par le fond continu peuvent être valablement négligées dans notre cas.

Les durées de diffusion étaient limitées à 30—60 minutes, pour éviter une trop grande dissolution de l'échantillon dans le bain de zinc. Pour augmenter la précision sur la mesure des temps, l'échantillon est plongé dans le bain préalablement mis en température, puis trempé à l'eau à la fin de l'expérience. Les courbes concentration-pénétration ont été ainsi établies pour des températures échelonnées entre 750 et 1150°. Comme nous l'avions indiqué [1], il se forme dans les joints de grains une solution solide nickel—zinc, et les mesures ont été faites en évitant les joints.

Puis nous avons dépouillé les courbes par la méthode de MATANO [4]. Celle-ci est la seule technique rigoureuse dans notre cas, du fait de la discontinuité, et de la variation importante du coefficient de diffusion avec la concentration. Au sein d'un même échantillon les différentes courbes expérimentales donnant $\tilde{D}_{\mathrm{NiZn}}$ en fonction de la concentration présentent une allure semblable (fig. 3), mais on observe une dispersion sur $\tilde{D}_{\mathrm{NiZn}}$ de l'ordre de 50%. Nous pensons que cette dispersion provient soit d'un mouillage imparfait de certaines parties de l'échantillon au début de l'expérience, soit de l'existence éventuelle de «court-circuits de diffusion» par les joints de grains en présence d'une contrainte thermique.

Il faut donc considérer avec réserve les résultats quantitatifs concernant dans ce cas la diffusion chimique. D'ailleurs la fig. 4 montre une dispersion assez grande des points expérimentaux par rapport à la loi d'Arrhenius. Dans cette figure, chaque point représente une moyenne sur cinq courbes réalisées sur un même échantillon. La valeur de $\tilde{D}_{\mathrm{NiZn}}$ est prise pour la concentration moyenne de 4% de nickel. La droite, placée par la méthode des moindres carrés donne une énergie d'activation Q de 50 kcal/mole et un facteur de fréquence D_0 de l'ordre de 53 cm² s⁻¹.

2.2. Diffusion entre 420 et 750°

La diffusion du zinc dans cet intervalle de température présente un caractère complexe et se prête mal à une étude quantitative. Entre 420 et 570° la vitesse de dissolution l'emporte sur la vitesse de diffusion, et l'on ne constate aucun changement de composition de la matrice. Entre 570 et 750° la concentration des quatre éléments présents varie simultanément et le changement de phase n'est pas observé. Parallèlement, les éprouvettes de traction rompues en présence de zinc ne font pas apparaître de fragilité.

2.3. Diffusion à température inférieure à 420°

Comme précédemment, les quatre éléments présents participent à la diffusion, qui de plus est très lente. Il n'est pas possible de faire une étude complète. On ne peut absolument pas négliger les coefficients croisés dans les relations d'Onsager.

On peut cependant faire certaines remarques qualitatives. Comme dans la diffusion ternaire Fe—Ni—Co [5], sous l'influence du gradient de concentration des autres éléments (fig. 5), le nickel présente une redistribution remarquable (ou diffusion inverse). Il apparaît que la teneur en nickel tombe à une valeur voisine de zéro près du plan de soudure, dans une zône où la concentration en fer et en chrome reste importante. On peut donc penser que se produit le

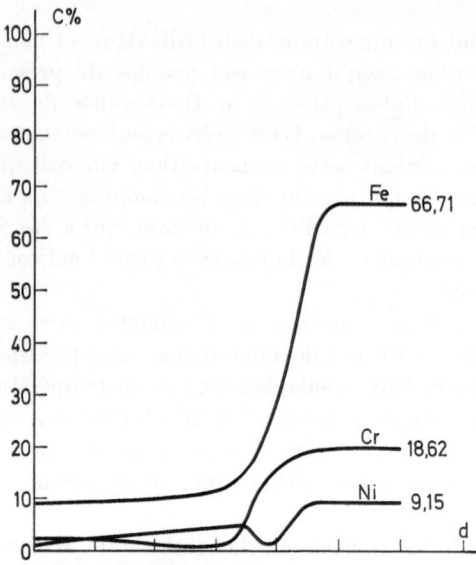

Fig. 5. Allure des courbes de diffusion à 350° C

même phénomène qu'au-dessus de 750°, et que la matrice γ est localement transformée en ferrite α. Cette hypothèse est parfaitement vérifiée, la micrographie du couple en présence de magnétite Fe_2O_3 (technique de BITTER) révèle une étroite bande de ferrite α le long du plan de soudure.

Ici encore apparaît donc une possibilité de fragilisation de l'acier par changement de phase en surface. Des cas de fissuration après frottement sur des pièces en zinc ont été observés dans l'industrie; ils peuvent s'expliquer de cette façon.

3. Diffusion chimique zinc-nickel

La problème théorique de la diffusion à quatre corps avec changement de phase est d'une complexité et d'une lourdeur insurmontables. Dans les trois domaines de température que nous avons envisagés, il en est cependant un où le problème se simplifie: au-dessus de 750°, où seule l'interdiffusion zinc—nickel intervient, et où nous avons calculé l'énergie d'activation et le facteur de fréquence de la diffusion. C'est pourquoi il nous a semblé intéressant de rapprocher les résultats ainsi obtenus en présence du solvant inerte fer—chrome de ceux qu'on obtient dans le système binaire simple nickel—zinc.

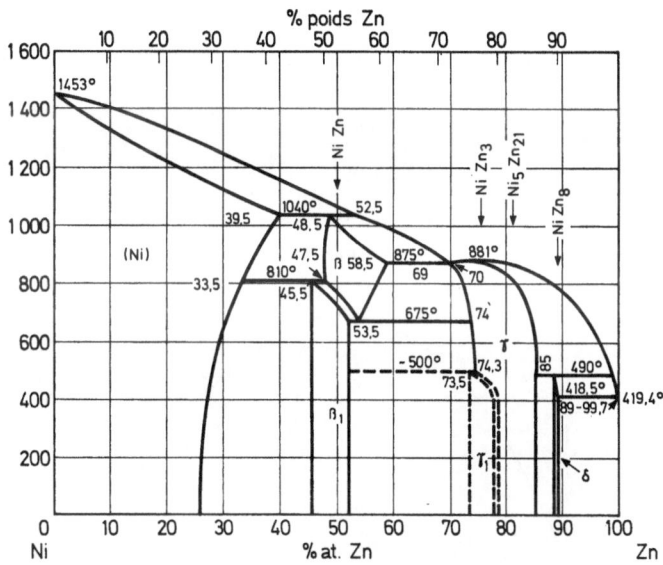

Fig. 6. Diagramme Nickel—Zinc

La diffusion chimique zinc—nickel a été étudiée par la méthode des couples semi-infinis différentiels. Le diagramme nickel—zinc est assez complexe (fig. 6), et comme la solution solide (Ni) est surtout intéressante dans notre cas, nous avons réalisé nos diffusions entre le nickel pur et la phase β (ou β_1 selon la température).

3.1. Techniques expérimentales

Le nickel utilisé a une pureté de 99,95%. Comme la température d'ébullition du nickel est bien inférieure à celle du point de fusion du nickel, l'alliage β a été réalisé par diffusion du nickel dans un bain de zinc à 99,999%, sous argon purifié, en élevant progressivement la température. Ainsi la pression partielle de la vapeur métallique était maintenue à une valeur négligeable devant la pression d'argon, jusqu'à l'obtention de la phase liquide homogène. Nous avons ensuite vérifié l'homogénéité de l'alliage à la microsonde, puis contrôlé les paramètres par diffractométrie X.

Le couple de diffusion est constitué par l'assemblage de deux plaquettes de 8×8 mm, d'épaisseur 2 mm. Les faces en contact sont polies au diamant, puis soudées par maintien dans

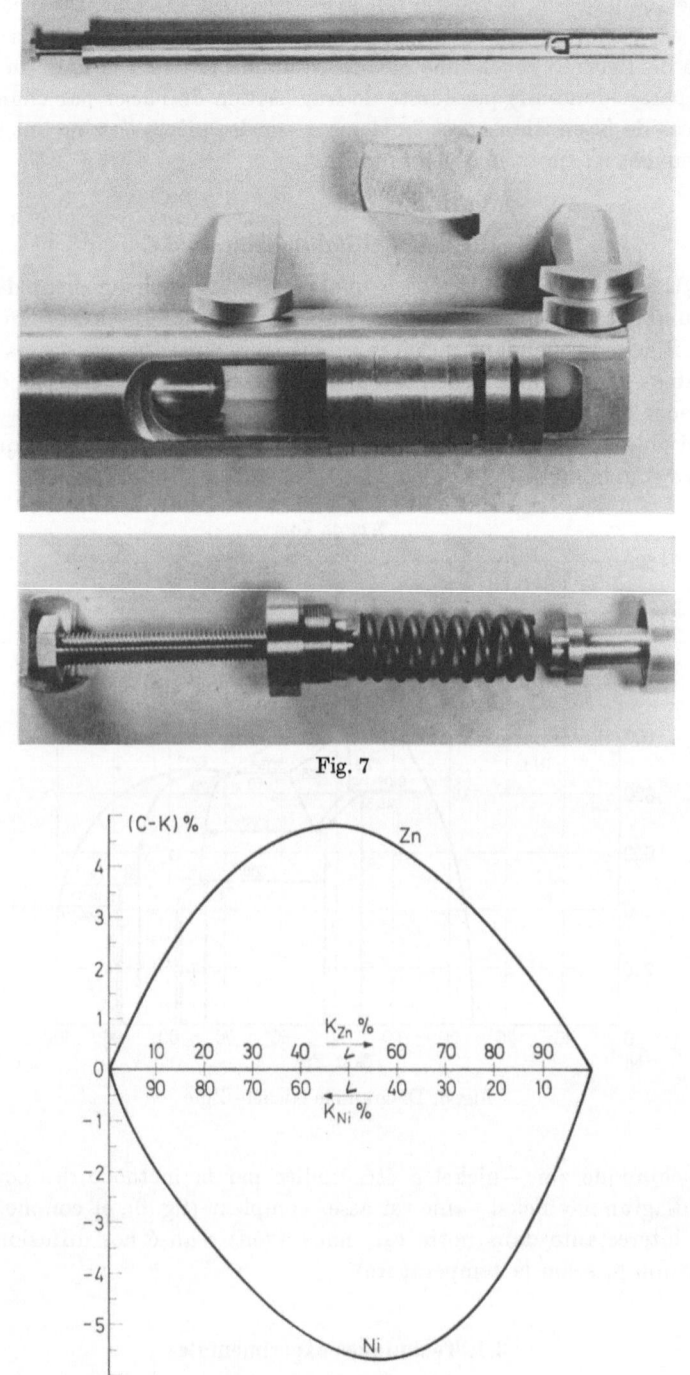

Fig. 7

Fig. 8. Corrections d'absorption et de fluorescence pour les couples Ni—Zn (20 kV)

une presse une heure à 600°. La presse à ressort utilisée (fig. 7) était réglée pour une pression de 3 kg/mm². Nous avons verifié à la sonde qué la diffusion due à ce traitement est insensible. Le recuit ultérieur est fait sous argon purifié. La température est stable à $\pm 2°$ près. La durée des recuits de diffusion varie entre 20 et 800 heures selon la température, et les «zônes de diffusion» ont des largeurs de 50 à 1000 μ. Après maintien en température, les couples sont sectionnés perpendiculairement à l'interface de soudure et les surfaces à analyser soigneusement polies:

Les enregistrements graphiques permettent de choisir une zône favorable, et dix courbes sont établies par comptage. La courbe concentration-pénétration finale est la courbe moyenne.

La fig. 8 montre les corrections de fluorescence et d'absorption calculées sur ordinateur selon les formules de PHILIBERT [2] modifiées par DUNCUMB et SHIELD [3]. Pour le nickel, la correction $(C-K)$ atteint 5,5% pour une concentration mesurée K de 45%. Cependant nous n'avons pas rapporté nos mesures au témoin pur, mais à l'alliage β dont la composition a été déterminée avec précision. On accroît ainsi la précision des mesures et la rapidité de l'opération, ce qui permet d'être moins tributaire des variations de caractéristiques de la sonde et des compteurs.

Nous avons constaté que le coefficient de diffusion chimique peut être multiplié par 100 lorsque la concentration en zinc passe de 0 à 40%. $\tilde{D}_{NiZn}$ a donc été calculé par la méthode de MATANO [4]. Rappelons que cette méthode reste valable lorsque la courbe concentration-pénétration présente des discontinuités; il suffit que la fonction représentative soit intégrable. Les surfaces ont été mesurées au planimètre, puis pour travailler avec une plus grande efficacité, nous avons mis au point un programme de calcul sur ordinateur. De cette façon à partir des mesures brutes fournies par la microsonde, l'ordinateur indique la position du plan de MATANO et la variation de $\tilde{D}_{NiZn}$ avec la concentration.

3.2. Résultats

On constate qu'en coordonnées semi-logarithmiques $(\log \tilde{D}_{NiZn} = f(C))$ nos points s'alignent compte tenu de la précision des mesures, c'est-à-dire à environ 3% près (fig. 9), et ceci pour toutes les températures. Toutefois entre 600 et 700° cette règle se vérifie moins bien, mais la précision est en même temps moins bonne. En effet, si on ne veut pas avoir des durées de diffusion exagérées (supérieures à 1000 heures) on doit se contenter d'une zône de diffusion plus étroite.

D'autre part, comme on le voit sur la fig. 10, la pente A de ces droites est sensiblement constante. De telle sorte que (fig. 11) les droites qui représentent les variations de $\log \tilde{D}_{NiZn}$ en fonction de $1/T$ pour des concentrations échelonnées sont sensiblement parallèles. L'énergie d'activation Q (fig. 12) varie peu avec la concentration. Il en résulte que dans un trièdre trirectangle si on porte les valeurs de $\log \tilde{D}_{NiZn}$ en fonction de C_{Ni} et de $1/T$, la surface moyenne représentative est un plan.

En résumé nos résultats sont avec une bonne approximation en corrélation avec la formule suivante:

$$\tilde{D}_{NiZn} = D_1 e^{-AC_{Ni}} e^{-\frac{Q}{RT}}$$

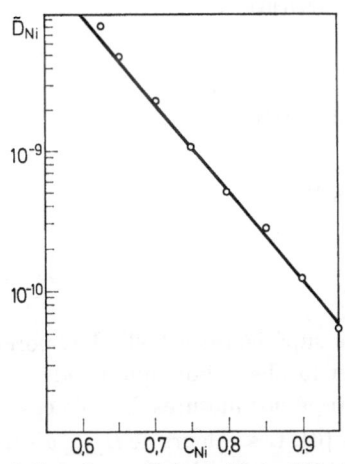

Fig. 9. Variation du coefficient d'interdiffusion avec la concentration (990° C)

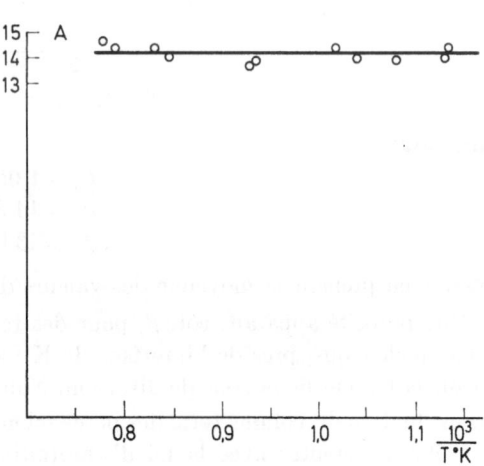

Fig. 10. Valeurs de A

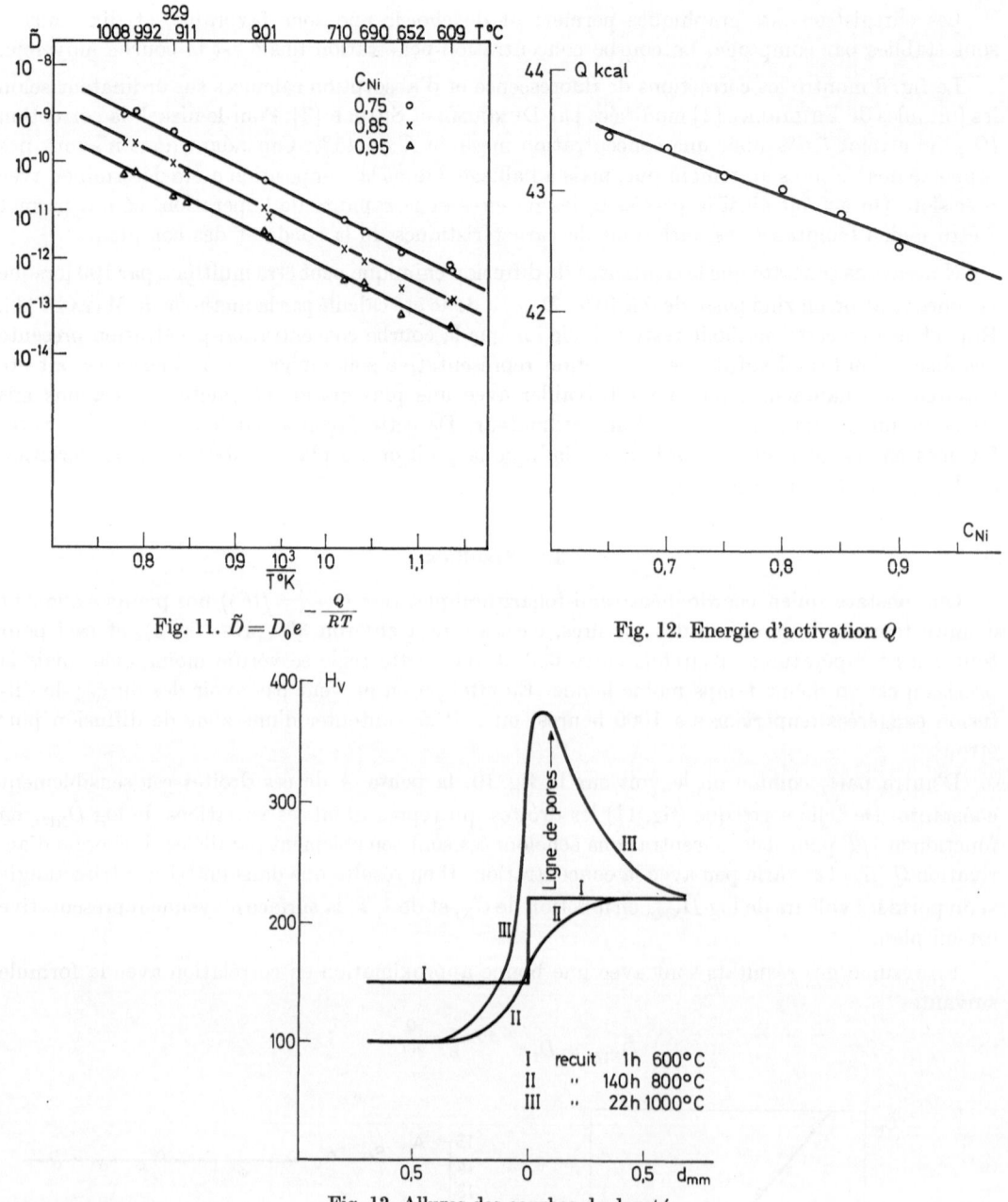

Fig. 11. $\tilde{D} = D_0 e^{-\frac{Q}{RT}}$　　　　　　　Fig. 12. Energie d'activation Q

Fig. 13. Allures des courbes de dureté

dans laquelle

$$D_1 = 1,05 \cdot 10^3 \text{ cm}^2 \text{ s}^{-1}$$
$$A = 14,24$$
$$Q = 43 \text{ kcal/mole}$$

obtenue en prenant la moyenne des valeurs de A et Q.

Une porosité apparaît, côté β, pour des températures supérieures à 950°. Les pores sont de forme quelconque, près de l'interface de Kirkendall, et localisés dans une bande assez petite devant la largeur de la zône de diffusion. Nous avons corrigé nos mesures dans ce cas en faisant une évaluation du volume total des pores selon Heumann [6]. Les valeurs de $\tilde{D}_{NiZn}$ ainsi obtenues sont plus cohérentes avec la loi d'Arrhenius. L'échantillon ne présente pas de gonflement. Cependant les courbes de microdureté ont été établies sur les trois couples de filiation suivants:

l'un simplement soudé une heure à 600°, le deuxième ayant subi un traitement de 140 heures à 800° et ne présentant pas de porosité, le troisième maintenu 24 heures à 1000° et présentant une ligne de trous. L'allure des courbes est donnée dans la fig. 13. On constate un accroissement considérable de la microdureté près de la ligne de trous, et on peut penser que les contraintes internes qui existent le long de cette ligne perturbent la diffusion. Nous envisageons d'étudier plus précisément ce phénomène.

3.3. Aspect thermodynamique

Il serait intéressant de relier ces variations particulières de log $\tilde{D}_{NiZn}$ en fonction de la concentration avec les données thermodynamiques sur la solution solide. Il semble en effet que la solution soit loin de l'idéalité, car elle s'écarte sensiblement de la loi de VEGARD. On peut le voir sur la fig. 14 où apparaissent les variations du volume atomique moyen avec la concentration. Notons au passage que toutes nos fractions atomiques mesurées devraient être corrigées en tenant compte de la variation du volume partiel atomique avec la concentration. Malheureusement toutes les mesures de paramètre dont nous disposons sont faites à température ambiante, et il n'est pas possible d'évaluer cette correction à la température de l'expérience.

Par ailleurs, on dispose actuellement de données sur les activités du zinc et du nickel en solution solide [7]. Mais il faudrait, pour utiliser les équations de DEHLINGER et DARKEN [8], mesurer les coefficients d'autodiffusion des isotopes dans une série d'alliages. Il est donc impossible actuellement dans notre cas de calculer un coefficient de diffusion chimique théorique.

En dehors du fait que le champ de forces dans lequel se déplacent les atomes est certainement très variable dans l'espace et dans le temps (puisque le gradient chimique varie au cours du temps), on peut se demander si le changement de phase n'est pas pour une part responsable de la variation de $\tilde{D}_{NiZn}$. Comme on le voit sur la fig. 15, l'allure des courbes concentration-pénétration est très particulière, la pente étant très faible au voisinage de la discontinuité. Effectivement, sur les droites log $\tilde{D}_{NiZn} = f(C)$, le coefficient de diffusion le plus élevé est obtenu pour la concentration limite à la discontinuité.

Nous avons donc cherché à savoir si le déplacement du plan de discontinuité pouvait provoquer un effet thermique, qui expliquerait le profil de variation de $\tilde{D}_{NiZn}$. En effet si l'on suppose que le plan de discontinuité constitue une source de chaleur, la température au sein de l'échantillon décroît linéairement de chaque côté en régime permanent, et donc $\tilde{D}_{NiZn}$ décroît selon une loi proche de la loi exponentielle observée.

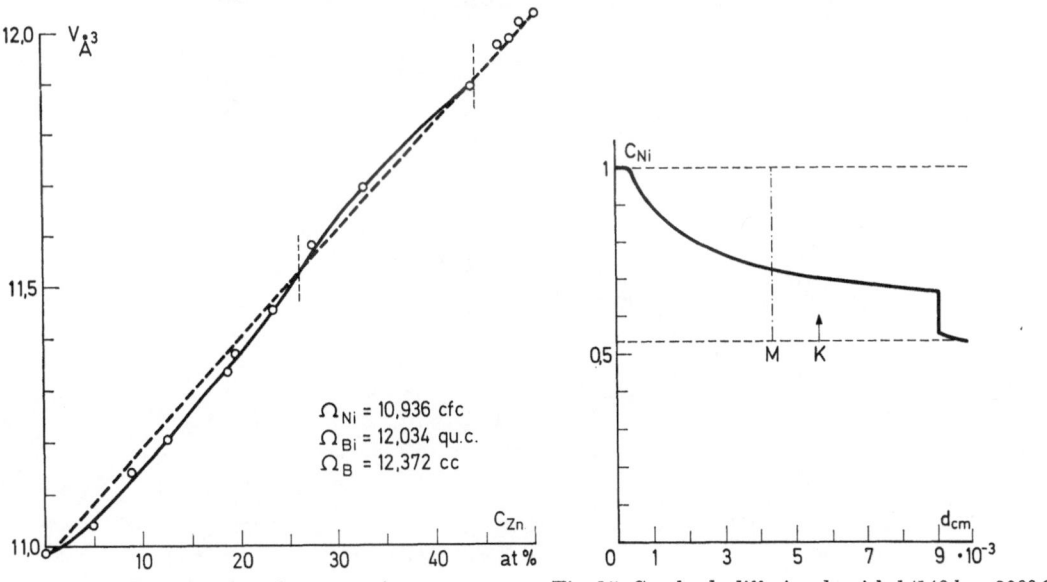

Fig. 14. Variation du volume atomique moyen   Fig. 15. Courbe de diffusion du nickel (140 h — 800° C)

528 M. ANDREANI, P. AZOU et P. BASTIEN:

Pour chercher à vérifier cette hypothèse, nous avons fait des expériences de diffusion à 800° dans un microcalorimètre. Nous avons constaté qu'il y a effectivement un dégagement de chaleur pendant la diffusion, mais ce dégagement est très faible: de l'ordre de quelques dizaines de calories pour 24 heures à 800°. Moyennant certaines hypothèses sur les échanges thermiques, on en déduit que l'élévation de température dans le plan de discontinuité n'est pas supérieure à une fraction de degré. C'est donc tout-à-fait insuffisant pour expliquer la variation rapide de $\tilde{D}_{\mathrm{NiZn}}$ avec C_{Ni}.

4. Coefficients de diffusion intrinséques

4.1. Techniques expérimentales

Sur les mêmes échantillons, dans le but d'évaluer les coefficients de diffusion intrinsèques à la concentration de l'interface de KIRKENDALL, nous avons mesuré les déplacements de cet interface. Il est repéré par des fils de platine lorsque c'est nécessaire, mais le plus souvent il apparaît à l'attaque par la forme des grains, comme on le voit sur la fig. 16.

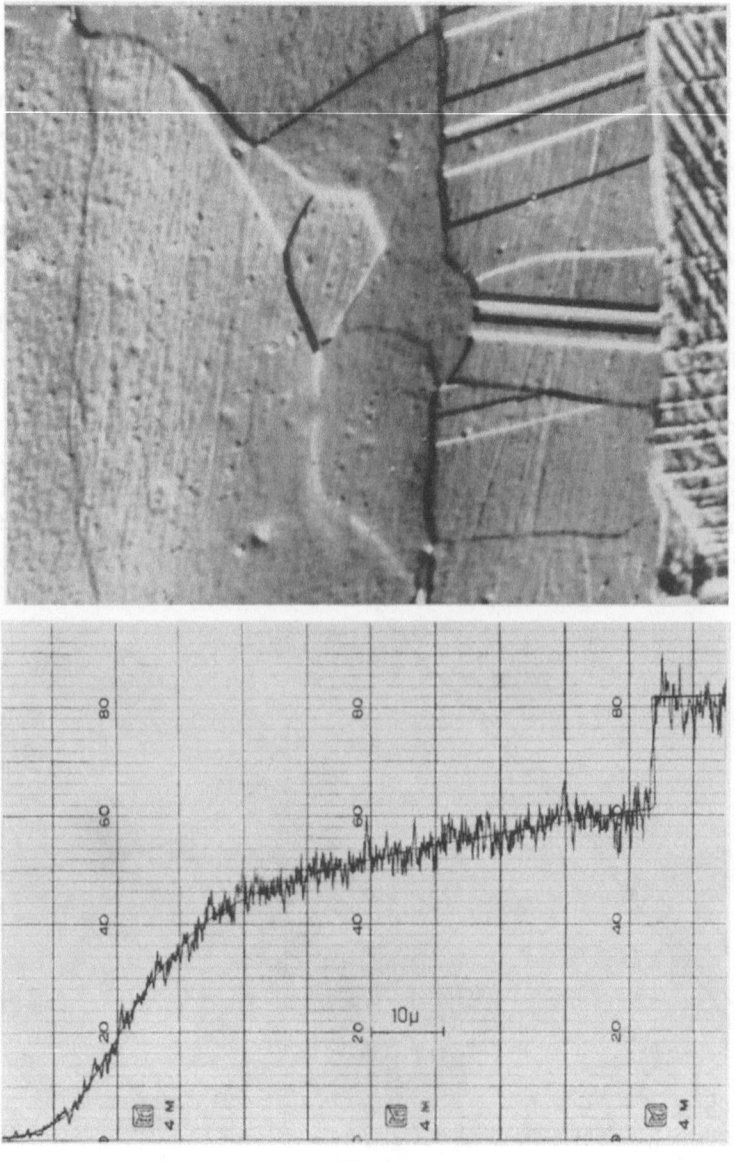

Fig. 16

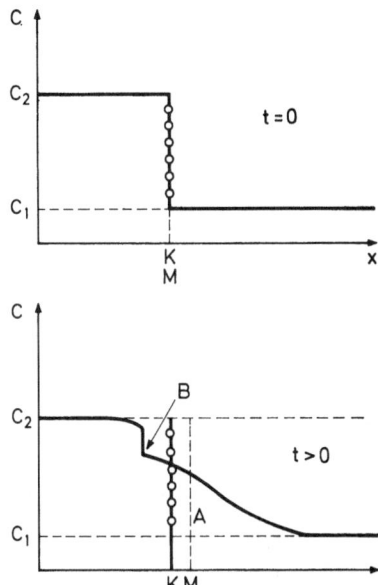

Fig. 17. Calcul des coefficients de diffusion intrinsèques dans le plan de KIRKENDALL

On sait (fig. 17) qu'en plaçant ce plan sur la courbe concentration-pénétration, on peut suivant la méthode de HEUMANN calculer le rapport des coefficients intrinsèques par le rapport des aires A et B:

$$\frac{D_{Zn}}{D_{Ni}} = \frac{A}{B}. \tag{1}$$

Cette relation suppose satisfaites deux hypothèses de départ:

a) l'absence de sursaturation de lacunes,

b) la conservation du nombre de sites cristallographiques.

Il faut une deuxième équation pour calculer D_{Zn} et D_{Ni}: on utilise généralement la relation donnant la vitesse de déplacement des fils repères. Dans notre cas, aux basses températures, lorsque la zône de diffusion est étroite, la précision sur cette vitesse est médiocre. Il nous est plus facile de mesurer la distance séparant le plan de discontinuité du plan de KIRKENDALL, et nous avons utilisé la relation de DARKEN:

$$\tilde{D}_{NiZn} = N_{Zn} D_{Ni} + N_{Ni} D_{Zn} \tag{2}$$

N_A désignant la fraction molaire de l'élément A dans le plan de KIRKENDALL.

4.2. Résultats

La fig. 18 montre la variation de la concentration dans le plan de KIRKENDALL. Selon la température, elle passe de 66 à 72%, écart de 6% non négligeable.

Nous avons cependant reporté les valeurs de log D_{Ni} et log D_{Zn} ainsi obtenues en fonction de l'inverse de la température, en négligeant sur ces 6% la variation de D_{Ni} et D_{Zn} avec la concentration. Il faudrait pour préciser ce point étudier l'effet KIRKENDALL sur toutes les concentrations, en réalisant des couples du type «mille-feuilles».

Comme on peut le constater à la fig. 19 l'alignement des points est satisfaisant, plus favorable que dans le cas de la diffusion chimique.

Les énergies d'activation et les facteurs de fréquence déduits de ces mesures ont pour valeur:

$$Q_{Zn} = 45,5 \text{ kcal/mole}, \quad D_{Zn}^0 = 0,176 \text{ cm}^2 \text{ s}^{-1},$$
$$Q_{Ni} = 43,5 \text{ kcal/mole}, \quad D_{Ni}^0 = 8,30 \cdot 10^{-2} \text{ cm}^2 \text{ s}^{-1}.$$

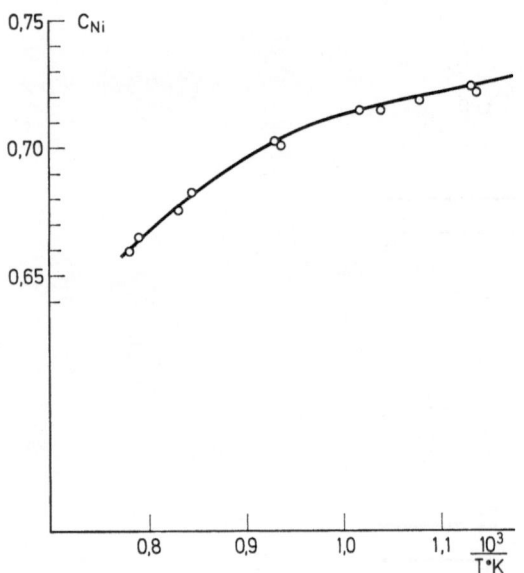

Fig. 18. Concentration au plan de Kirkendall

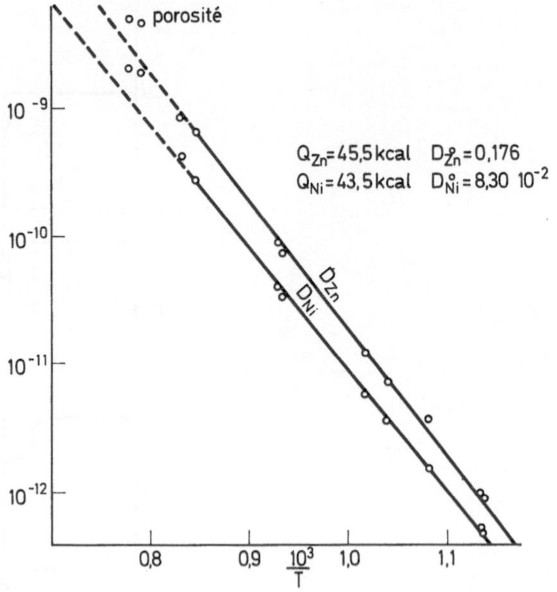

Fig. 19. Coefficients de diffusion intrinsèques $D_{Zn} D_{Ni}$

Sur la fig. 19, les points qui correspondent à l'apparition de la porosité s'écartent nettement de la droite bien que, comme précédemment indiqué, une correction ait été réalisée sur le déplacement des plans par évaluation du volume des pores [6].

La correction apparaît comme étant encore insuffisante, bien que sans correction l'écart à la loi d'Arrhenius soit beaucoup plus important. Les relations (1) et (2) sont donc très affectées par l'apparition de sursaturations de lacunes, mais sont appliquées avec succès lorsque les hypothèses de départ sont satisfaites.

5. Conclusion

Nous pouvons donc penser que l'interdiffusion zinc—nickel dans les aciers inoxydables, c'est-à-dire en présence de fer et de chrome, fait appel à un mécanisme de diffusion analogue à celui de la diffusion dans le couple binaire simple. Cependant la diffusion est beaucoup plus rapide et plus activée thermiquement: non seulement le solvant fer–chrome inerte dans la diffusion ne freine pas la migration des atomes, mais il crée un champ de forces favorable à cette diffusion.

Bibliographie

1. Andreani, M., P. Azou et P. Bastien: Compt. Rend. 263, 1041—1043 (1966).
2. Philibert, J.: Met. Cor. Ind. 465, 158—176; 466, 216—240; 469, 325—342 (1964).
3. Duncumb, P., and P. K. Shield: Tube Invest. Res. Lab., Techn. Rep. No. 181 (1964).
4. Matano, C.: Japan. J. Phys. 8, 109 (1933).
5. Sabatier, J. P., et A. Vignes: J. de Micr. 6, 108 (1967).
6. Heumann, T.: La diff. dans les métaux. Bibl. techn. Philips, 59 (1957).
7. Chart, T. G., J. K. Critchley, and R. Williams: J. Inst. Metals 96, 7, 224 (1968).
8. Darken, L. S.: Trans. AIME, 175, 184 (1948).

Electron Probe Microanalysis of Ti(C, N) and Zr(C, N) in Steel

T. SHIRAIWA and N. FUJINO

Central Research Laboratories, Sumitomo Metal Industries, Ltd., Amagasaki, Japan

Introduction

In steels containing titanium and zirconium, the angular precipitates (non-metallic inclusion) are observed by an optical microscope. They have been called nitride or carbide, and sometimes called carbonitride because it is suggested that they are uniform solid solutions of nitride and carbide.

In the present work, these precipitates in various types of steel are analysed by means of an electron probe microanalyser and an optical microscope.

Experiment

The compositions of steel examined in the present work are shown in Table 1. The electron probe microanalyser is the ARL-Shimadzu type. In X-ray spectroanalysis of nitrogen in titanium nitride, the $\mathrm{Ti}L_l$ line overlaps on $\mathrm{N}K_\alpha$ lines, as shown by the spectra in Fig. 1. In the present work, $\mathrm{N}K_\alpha$ is detected at the shorter wavelength than the peak of $\mathrm{N}K_\alpha$ as shown by an arrow in Fig. 1, where the intensity of $\mathrm{Ti}L_l$ line is negligible.

Table 1. *Composition of steels and recognized precipitates*

No.	Type of steel	Composition (wt. %)						Solid solution	Core structure	
		C	Cr	Ni	Mo	Ti	Zr	TiN : TiC	core	envelope
1	Low alloy steel	0.10	0.14	0.29	—	0.08	—	TiN		
2	Medium carbon steel	0.38	0.05	—	—	0.60	—	1:0.24		
3	High carbon steel	0.68	—	—	—	0.07	—	1:0.14		
4	Cr—Mo cast steel	1.99	0.87	—	0.27	0.12	—	(Ti, Mo)C	TiN	(Ti, Mo)C
5	18-8 stainless steel	0.06	17.65	12.23	0.06	0.42	—		TiN	TiC
6	20-30 heat resisting steel	0.08	20.80	32.50	0.10	0.33	—		TiN	TiC
7	Low carbon steel	0.17	0.41	—	—	—	0.07	Zr(C, N)	ZrN	ZrC

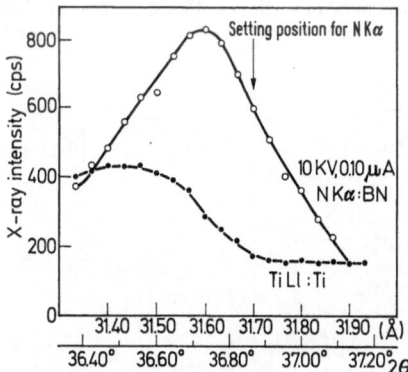

Fig. 1. Spectra of $\mathrm{N}K\alpha$ from boron nitride and $\mathrm{Ti}L_l$ from pure titanium

Results

1. Titanium Nitride, TiN. The titanium precipitates observed in the low alloy steel are found to be nearly pure titanium nitride which is shown in Fig. 2 a. The colour of these precipitates is orange yellow. The chart recordings across the precipitate are shown in Fig. 3.

2. Titanium Carbonitride Ti(C, N). Precipitates observed in the medium carbon steel and the high carbon steel are solid solution of titanium carbide and nitride which are usually called titanium carbonitride Ti(C,N). The line analyses of the precipitate in the medium carbon steel is given in Fig. 4. The colour of the precipitate is similar to the titanium nitride.

3. Titanium-Molybdenum Carbide (Ti, Mo)C. The titanium carbide which does not contain nitrogen is not observed in examined steels but the titanium carbide containing molybdenum is found in Cr−Mo cast steel. The colour is bluish white. Fig. 5 shows the line analyses of the precipitate, where nitrogen is scarcely detected. This also means that the separation between $\mathrm{Ti}\,L_l$ and $\mathrm{N}\,K_\alpha$ is successful.

4. Core Structure (Ti, Mo)C + TiN and TiC + TiN. In the Cr−Mo cast steel, 18-8 stainless steel and 20−30 heat resisting steel, the precipitate of core structure which consists of carbide

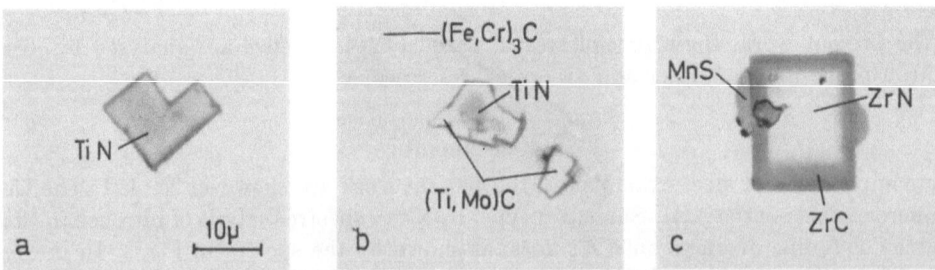

Fig. 2a−c. Microphotograph of precipitates. a) titanium nitride in low alloy steel; b) core structure type precipitate in Cr−Mo cast steel; c) core structure type precipitate in low carbon steel

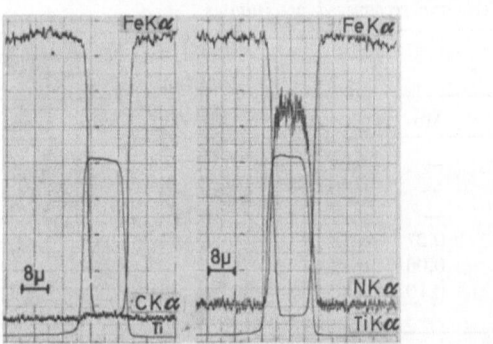

Fig. 3. Line analyses of titanium nitride in low alloy steel

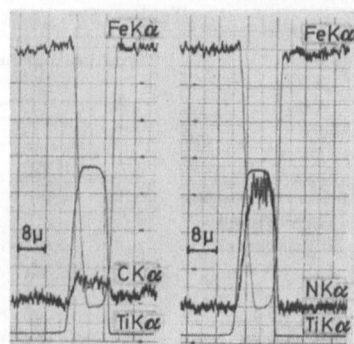

Fig. 4. Line analyses of titanium carbonitride in medium carbon steel

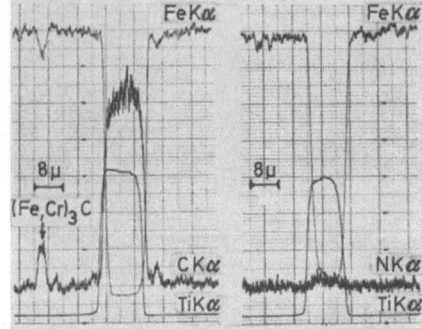

Fig. 5. Line analyses of titanium-molybdenum carbide in Cr−Mo cast steel

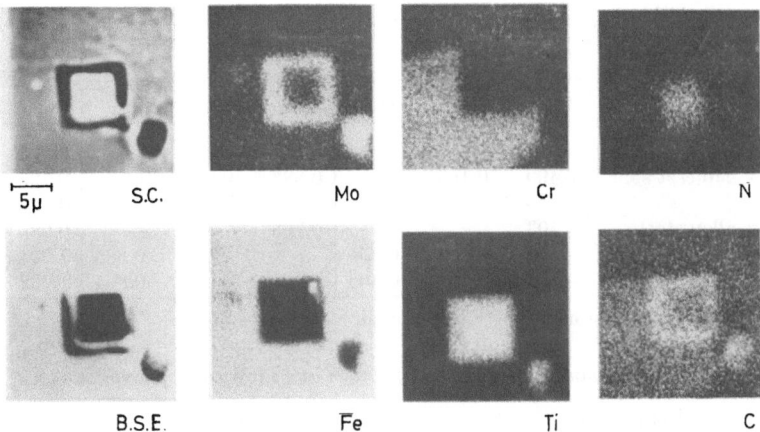

Fig. 6. E.B.S. images of core structure type precipitate in Cr—Mo cast steel

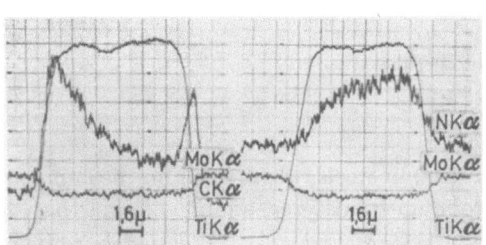

Fig. 7. Line analyses of the core structure type precipitate in 18-8 stainless steel

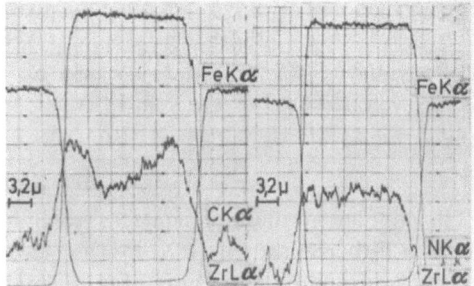

Fig. 8. Line analyses of core structure type precipitate in low carbon steel

and nitride is discovered where the nitride always concentrates in the centre of the precipitate and the carbide surrounds the nitride core [1].

In the Cr—Mo cast steel, the surrounding carbide contains molybdenum, but in the stainless steel and the heat resisting steel, the carbide is titanium carbide only. Fig. 2b is the microphotograph and Fig. 6 is the E.B.S. images of the core structure type precipitate in the Cr—Mo cast steel. The core and the envelope can be distinguished by their colour. Fig. 7 shows line analysis of the precipitate in the 18-8 stainless steel.

5. Zirconium Carbide, ZrC; Zirconium Nitride, ZrN and Core Structure, ZrC + ZrN. In the low carbon steel containing zirconium, zirconium carbide, zirconium nitride, and the core structure type precipitate, ZrC + ZrN, are observed. In the core structure precipitate, zirconium nitride exists in the core of the precipitate and zirconium carbide surrounds the nitride.

The colour of the zirconium carbide is orange and that of the nitride is yellowish white, and the core structure precipitate shows the colours corresponding to their constituents. Fig. 2c shows the microphotograph of the core structure type precipitate and Fig. 8 is the line analysis.

6. Quantitative Analysis. Results of the quantitative analysis of metallic elements in precipitates are shown in Table 2. The atomic number corrections given by POOLE and THOMAS [2] and the modified Philibert's absorption correction [3, 4] are applied, where the mass absorption coefficients are based on Birks' table [5]. Corrected results agree well with the stoichiometric value. Table 3 is the quantitative analysis results of carbon and nitrogen in solid solution type precipitates. They are obtained by above mentioned corrections where the standard samples of carbon and nitrogen are synthetic titanium carbide and titanium nitride in the low alloy steel. The mass absorption coefficients for CK_α and NK_α are referred to Henke's data [6]. The deficiency of the total values in Table 3 is due to the inaccuracy of theoretical correction for carbon and nitrogen.

Table 2. *Quantitative analysis of metallic elements in precipitates*

Precipitate	Matrix	I_A^{AB}/I_A^A [a]			Corrected			Total[b]
		TiK_α	MoI_α	ZrL_α	Ti	Mo	Zr	
TiC	synthetic	0.734	—	—	0.798	—	—	0.998
(Ti, Mo)C	Cr—Mo cast steel	0.490	0.166	—	0.583	0.247	—	$0.729 + 0.278 = 1.007$ [c]
								$0.729 + 0.262 = 0.991$ [d]
TiN	low alloy steel	0.703	—	—	0.765	—	—	0.988
ZrC	low carbon steel	—	—	0.854	—	—	0.885	1.001
ZrN	low carbon steel	—	—	0.856	—	—	0.894	1.032

[a] Acc. volt. 30 kV.   [b] As stoichiometric compound.   [c] TiC + MoC.   [d] TiC + Mo$_2$C.

Table 3. *Quantitative analysis of carbon and nitrogen in precipitates*

Matrix	Observed			Corrected[c]				Mol.
	I_{Ti}/I_{Ti}^{Ti} [a]	I_C/I_C^{TiC} [b]	I_N/I_N^{TiN} [b]	Ti	C	N	Total	TiC/TiN
Medium carbon steel	0.720	0.10	0.67	0.784	0.020	0.165	0.969	0.14
High carbon steel	0.724	0.15	0.58	0.788	0.030	0.146	0.964	0.24

TiC Synthetic; TiN Precipitate in sample No. 1.

[a] Acc. volt. 30 kV.   [b] Acc. volt. 15 kV.   [c] Weight fraction.

Conclusion

The angular titanium precipitates (carbide or nitride) in steels, which have not been precisely identified hitherto, are analysed by the electron probe microanalyzer, and it is found that their structures depend on the composition of matrix steel.

In the solid solution type precipitate, the carbide concentration increases with the carbon content in the matrix, and it suggests the equilibrium at the precipitation.

The appearance of the core structure precipitate means that the precipitation temperature is different for nitride and carbide, and nitride precipitates at an early stage in cooling and solidification of molten steel. But the reason why the core structure is observed in the Cr—Mo cast steel and the high chromium and nickel alloys cannot be interpreted, and it is necessary to investigate the effects of alloying elements on the precipitation.

In the Cr—Mo cast steel, molybdenum is observed in titanium carbide. This precipitate is suggested to be a solid solution of Mo$_2$C in TiC which has been reported by Elemenko et al. [7].

The colour of titanium carbide is bluish white and nitride is orange yellow, and they are distinguished by optical microscopic observation.

The similar modification of carbide and nitride is observed in zirconium precipitate in the low carbon steel.

The authors would like to acknowledge with sincere gratitude the encouragement given them by Dr. Motoo Sumitomo, Director of Central Research Laboratories, Sumitomo Metal Industries. The authors are also grateful to Dr. Gunji Shinoda, Emeritus Professor of Osaka University for his critical discussions.

References

1. Shiraiwa, T., and N. Fujino: Sumitomo Metals **18**, 509 (1966) [in Japanese].

2. Poole, D. M., and P. M. Thomas: J. Inst. Metals **90**, 228 (1961—1962).

3. Philibert, J.: Proc. Symposium on X-ray Optics and X-ray Microanalysis, Stanford, p. 379. New York: Academic Press 1963.

4. Adler, J., and J. Goldstein: Proc. Symposium on X-ray Optics and X-ray Optics and X-ray Microanalysis, Paris, p. 210. Paris: Hermann 1966.

5. Birks, L. S.: Electron probe microanalysis. New York: Interscience Publishers 1963.

6. Henke, B. L., R. L. Elgin, R. E. Lent, and R. B. Ledingham: AROSR No. 67-1254, U.S. Air Force, Office of Scientific Research, Washington, 1967.

7. Elekenko, V. N., T. Ya. Velikanova, and S. V. Shavanova: Tr. Soveshch. po Metallokhim., Metallaved. i Primeneniyu Titana i ego Splavov, 6th Moscow **1965**, 11—19.

Bestimmung der Grenzflächenenergie
zwischen Zementit-Partikeln und Ferrit-Matrix
mit Hilfe der Ostwald-Reifung

S. F. Dirnfeld

Technion-Israel Institute of Technology, Dept. of Materials, Haifa, Israel

Die Messung der Grenzflächenenergie zwischen den festen Phasen ist eine der wichtigsten Fragen, mit der sich die Untersuchung der Niederschläge befaßt.

Sind in einer gesättigten Lösung fein verteilte Teilchen einer zweiten Phase vorhanden, so besteht die Tendenz, daß infolge eines Überganges des gelösten Stoffes von Teilchen zu Teilchen die kleineren Teilchen sich auflösen und die größeren wachsen. Das Stadium des Wachstums großer Teilchen auf Kosten kleinerer (Ostwald-Reifung) wird durch die Grenzflächenenergie — σ bestimmt. Die treibende Kraft rührt dabei von der sich insgesamt ergebenden Verringerung der freien Energie der Zwischenflächen her. Folglich kann die Kinetik des Wachstums der Teilchen zur Bestimmung von σ herangezogen werden. Dieser Prozeß wurde von Carl Wagner [1], G. W. Greenwood [2] und R. A. Oriani [3] quantitativ wie auch theoretisch behandelt. In den Arbeiten von C. Wagner und G. W. Greenwald wurden Gleichungen für die zeitliche Änderung der Größenverteilungsfunktion der mittleren Teilchengröße bei Diffusions-Stoffübergang bestimmt. Für die größeren Teilchen von 10^{-7} cm wurde bestimmt [1], daß sich der mittlere Radius relativ zu der dritten Wurzel der Anlassungszeit vergrößert, wenn der Volumenbruchteil der dispersen Phase während des Prozesses konstant bleibt. In dieser Arbeit wurde die Wachstumskinetik der gleichachsigen Zementit-Partikeln in einer ferriten Matrix von eutektoidischem Stahl untersucht, welcher direkt von Austenit isothermisch bei verschiedenen Temperaturen (zwischen $600-700°$ C) angelassen worden war. Die Zusammensetzung des Stahls: C 0,89%; Al 0,28%; Si 0,18%; Mn 0,2%.

Abb. 1 zeigt, wie sich der laminäre Perlit (a) mit der Zeit in einem globularen Perlit zersetzt (b) und sich danach die Zementit-Teilchen unter Einfluß der verschiedenen Temperaturen vergrößerten. Oben sieht man eine Mikrophotographie nach einer Ätzung in 2% Nital (b) und unten eine ähnliche (c) nach einer Ätzung mit einem speziellen Reagenten von Beraha [4], der den Vorteil hat, daß die Grenzen zwischen den weißgefärbten Zementit-Partikeln und der schwarzen ferriten Matrix scharf gekennzeichnet sind. Von jedem Muster wurden 8 Mikrographe 3000mal vergrößert photographiert und das Ausmaß der Dispersion von den Karbidteilchen bei 5 verschiedenen Temperaturen und verschiedenen Anlaßzeiten nach den Cahn-Fulman [5]- und Saltykov [6]-Methoden analysiert. Die Resultate wurden mit einem Elliot-Computer-503 ausgearbeitet. Die folgenden Resultate ergaben sich als Funktionen der Anlassungszeit bei verschiedenen Temperaturen:

$N =$ die Gesamtzahl der Teilchen je Volumeneinheit,
$\bar{d} =$ der mittlere Durchmesser der Teilchen,
$S =$ die gesamte Oberfläche der Teilchen je Volumeneinheit,
$\delta =$ der mittlere freie Spielraum zwischen den Teilchen,
$V =$ der Volumenbruchteil der dispersen Teilchen.

Um festzustellen, nach welcher Funktion sich $\bar{d}$ während der Anlassungszeit t verändert, wurden Kurven für alle 5 Temperaturen gezeichnet (Abb. 2).

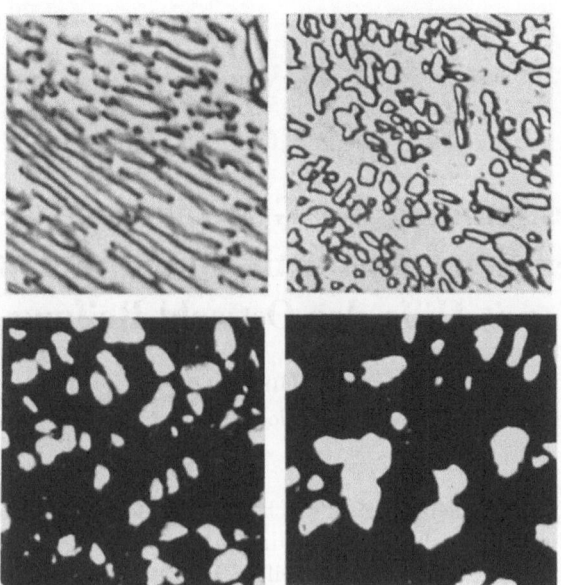

Abb. 1. Laminarer und globularer Perlit ($\times$ 3000)

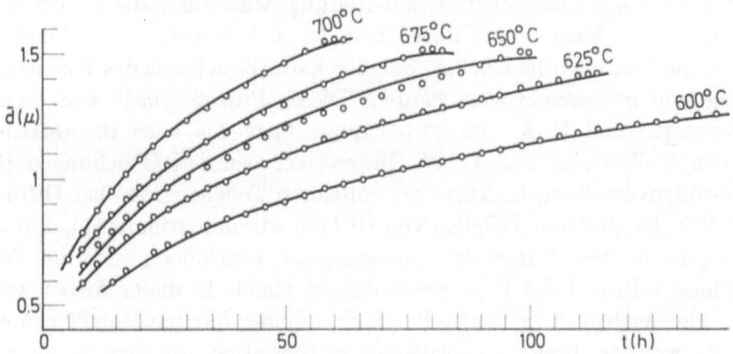

Abb. 2. Der mittlere Durchmesser $\bar{d}$ als Funktion der Anlaßzeit t

Es stellt sich heraus, daß $\bar{d}$ sich mit der Anlassungszeit t exponentiell vergrößert gemäß der Gleichung:

$$\bar{d} = b\, t^{a}, \tag{1}$$

in der a und b von der Temperatur abhängige Konstanten sind.

Um die Werte von a und b zu gewinnen, wird die Gl. (1) folgendermaßen geschrieben:

$$\log \bar{d} = \log b + a \log t. \tag{2}$$

Aus der Neigung der Linien in Abb. 3 lassen sich die Werte von a bestimmen, welche um 0,32 liegen.

Nun kann man die Gl. (1) in folgender Form schreiben:

$$\bar{d} = b\, t^{\frac{1}{3}}. \tag{3}$$

V bleibt während des Prozesses bei allen Temperaturen konstant. Informationen über N, S und δ sind in den folgenden Abb. 4—6 enthalten.

Die Forscher O. BANNYH und MODIN [7] stellten ebenfalls dieselbe Beziehung zwischen $\bar{d}$ und t fest, als sie mit einem Elektronen-Mikroskop das Wachsen der Zementit-Partikel in Sorbit für eutektoidischen Stahl untersuchten.

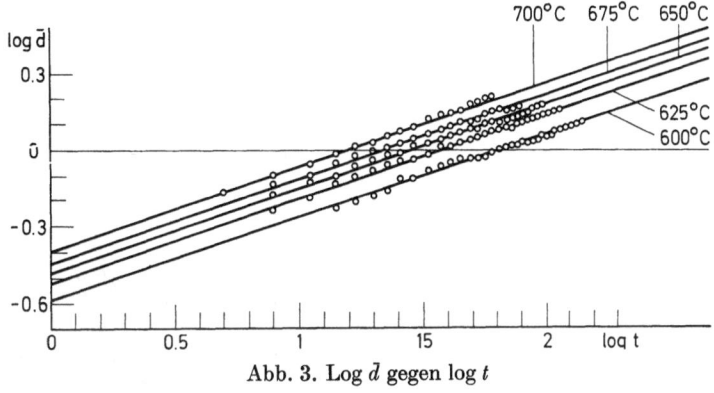

Abb. 3. Log d gegen log t

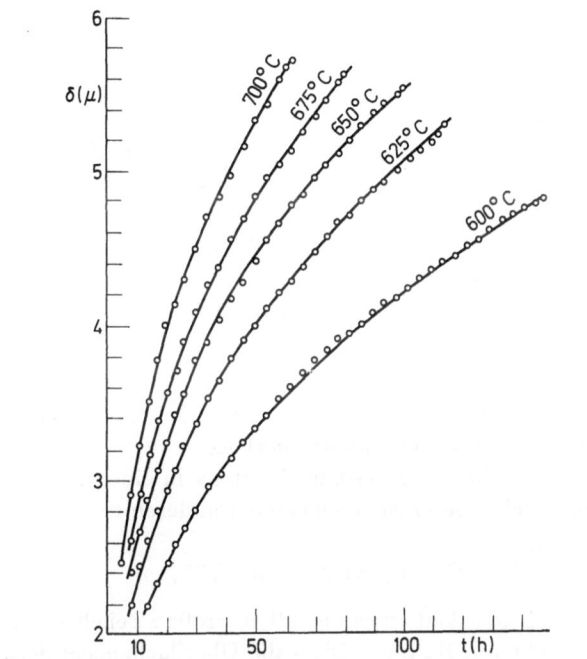

Abb. 4. Der mittlere freie Spielraum δ als Funktion der Anlaßzeit

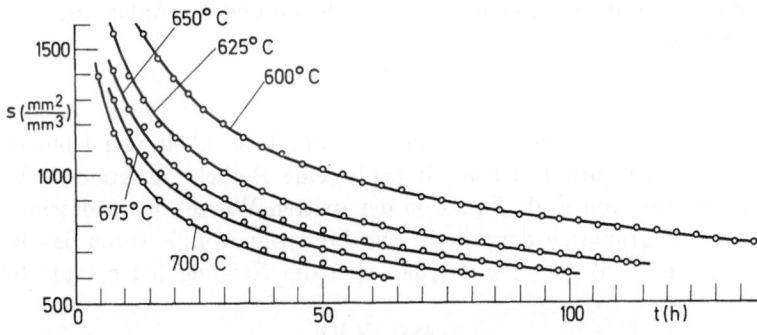

Abb. 5. Die gesamte Oberfläche S als Funktion der Anlaßzeit

Die Geschwindigkeit des Anwachsens der karbidischen Oberfläche wurde bestimmt, und daraus wurde mit Hilfe der bekannten Methode die Aktivierungsenergie Q für diese Reaktion bestimmt. Daraus folgte, daß $Q \approx 20\,000$ cal/mol war und z.B. von derselben Größe wie die

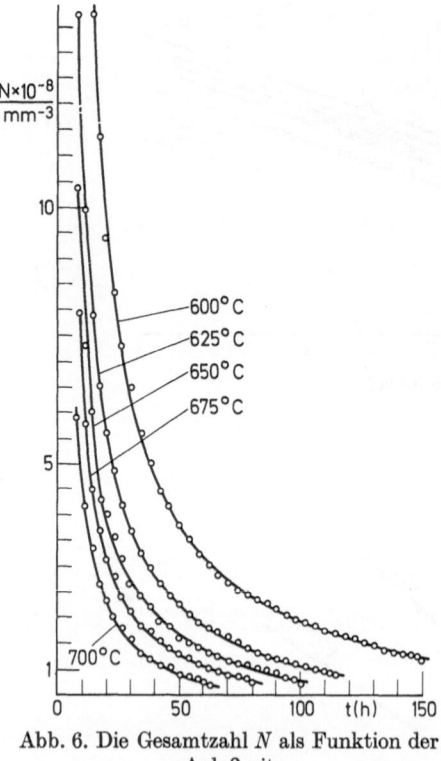

Abb. 6. Die Gesamtzahl N als Funktion der Anlaßzeit

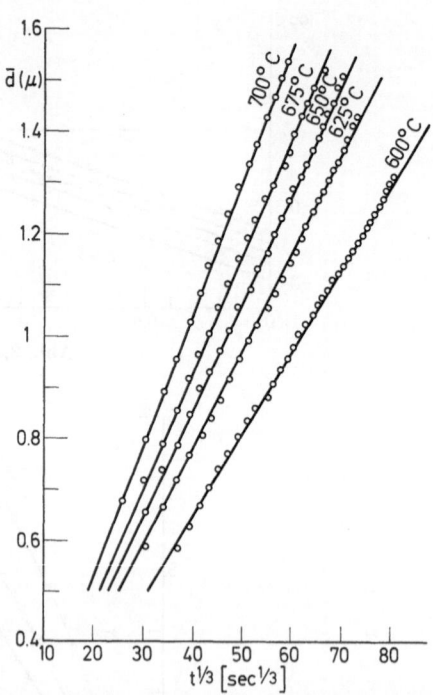

Abb. 7. Der mittlere Durchmesser $\bar{d}$ als Funktion der dritten Wurzel der Anlaßzeit t

Aktivierungsenergie der Kohlenstoffdiffusion durch Ferrit. Noch ein Beweis, daß dieser Prozeß durch diffusionsbestimmten Stoffübergang kontrolliert ist.

Wenn wir die bekannte Gleichung von Gibbs-Thomson in einer analogen Form für die Löslichkeit kleiner Festkörperteilchen in einer umgebenden anderen festen Phase schreiben [1], so bekommen wir die Gleichung:

$$C_r = C_0 \exp(2\sigma V_m / r \cdot RT),\tag{4}$$

wobei C_0 der Grenzwert der Löslichkeit eines unendlich großen Teilchens und C_r die Löslichkeit eines Teilchens mit dem effektiven Radius r ist, σ die Oberflächenenergie, V_m das Molvolumen, R die Gaskonstante und T die absolute Temperatur.

Mit einer Greenwoods [2] ähnlichen Entwicklungstheorie, angepaßt für Zementit-Teilchen in einer Ferrit-Matrix, wurde die Abhängigkeit zwischen $\bar{d}$ und der Anlassungszeit T theoretisch folgendermaßen bestimmt:

$$\bar{d} = \left(\frac{48\,D_c^\alpha\,C_\alpha^e\,M\,\sigma}{RT\,\varrho^2}\right)^{\frac{1}{3}} t^{\frac{1}{3}},\tag{5}$$

wobei D_c^α und C_c^α die Diffusion und der Grenzwert der Löslichkeit von Kohlenstoff in Ferrit sind, ϱ die Dichte der Zementit-Teilchen. Gl. (5) ist eine ähnliche funktionelle Verbindung wie Gl. (1). Die Abhängigkeit von $\bar{d}$ als Funktion der dritten Wurzel der Anlassungszeit bei allen 5 Temperaturen wurde graphisch dargestellt. Die Linien in Abb. 7 waren das Resultat.

Aus den Gln. (1) und (5) geht hervor, daß sich die Neigung jeder Linie folgendermaßen ausdrücken läßt:

$$b = \left(\frac{48\,D_c^\alpha\,C_c^\alpha\,M\,\sigma}{RT\,\varrho^2}\right)^{\frac{1}{3}}.\tag{6}$$

Die aus den Linien entstandenen b-Werte wurden in der oberen Tabelle zusammengefaßt. Aus Gl. (6) wird nun der σ-Wert zwischen Zementit-Teilchen und Ferrit ermittelt:

$$\sigma = \frac{b^3\,RT\,\varrho^2}{48\,D_c^\alpha\,C_c^\alpha\,M}.\tag{7}$$

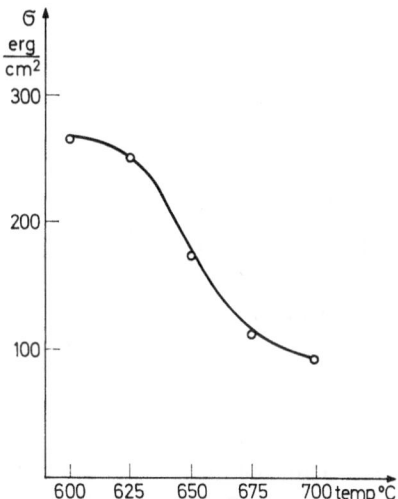

Abb. 8. Die Grenzflächenenergie σ als Funktion der Temperatur

Die in der unteren Tabelle enthaltenen Werte für die 5 Temperaturen wurden eingefügt und es ergaben sich die σ-Werte, welche in Abb. 8 zusammengefaßt wurden: $\sigma_{600} = 266{,}2$ erg/cm², $\sigma_{625} = 250$ erg/cm², $\sigma_{650} = 173{,}6$ erg/cm², $\sigma_{675} = 114{,}4$ erg/cm² und $\sigma_{700} = 92{,}1$ erg/cm².

Tabelle

Temp. (°C)	b (cm sec$^{-\frac{1}{2}}$) $\times 10^6$
700	2,56
675	2,27
650	2,13
625	1,94
600	1,65

Temp. (°C)	D_c^α (cm²·sec^{-1})	C_c^α (Gew.-%)	C_c^α (mol·cm^{-3})
700	$8{,}7 \times 10^{-7}$	$1{,}69 \times 10^{-2}$	$1{,}1 \times 10^{-4}$
675	$6{,}4 \times 10^{-7}$	$1{,}245 \times 10^{-2}$	$0{,}82 \times 10^{-4}$
650	$4{,}5 \times 10^{-7}$	$0{,}944 \times 10^{-2}$	$0{,}62 \times 10^{-4}$
625	$3{,}1 \times 10^{-7}$	$0{,}706 \times 10^{-2}$	$0{,}46 \times 10^{-4}$
625	$3{,}1 \times 10^{-7}$	$0{,}706 \times 10^{-2}$	$0{,}46 \times 10^{-4}$
600	$2{,}2 \times 10^{-7}$	$0{,}518 \times 10^{-2}$	$0{,}34 \times 10^{-4}$

Die Werte für D_c^α und C_c^α wurden mit Hilfe von SMITH's [8] Resultaten ermittelt. ORIANI [3] errechnete mit BANNYHs [7] Versuchsresultaten σ bei einer Temperatur von 700° C ($\sigma = 5750$ erg pro cm²). Er bemerkt dabei, daß dieser Wert 10mal größer sei (z.B. wegen der Schwierigkeiten, die sich bei der Messung der Durchmesser ergaben).

σ-Werte derselben Größenordnung ergaben sich auch bei den Untersuchungen von KRAMER und MEHL [8], deren Berechnungen nach anderen Methoden gemacht worden waren.

Literatur

1. WAGNER, C.: Z. Elektrochem. **65**, 581 (1961).
2. GREENWOOD, G. W.: Acta Met. **4**, 243 (1956).
3. ORIANI, R. A.: Acta Met. **12**, 1399 (1964).
4. BERAHA, E.: J.I.S. Inst. **203**, 454 (1965).
5. CAHN and FULLMAN: Trans. AIME **206**, 610 (1956).
6. SALTYKOV, S. A.: Stereometric metallography, 2nd ed. Mascow: Metallurgizdat 1958.
7. BANNYH, O., H. MODIN, and S. MODIN: Jernkonter **146**, 774 (1962).
8. SMITH, R. P.: Trans. Met. Soc. AIME **224**, 105 (1962).
9. KRAMER, J. J., G. M. POUND, and R. F. MEHL: Acta Met. **6**, 763 (1958).

Die Röntgenröhrenspannung als Einflußgröße bei der quantitativen RFA

H. Ebel und T. Katinger

Institut für angewandte Physik der Technischen Hochschule Wien, Österreich

Bei der quantitativen RFA mit äußerem Standard wird die Fluoreszenzausbeute der zu untersuchenden Probe mit der einer Eichprobe verglichen. Wird als Eichprobe das entsprechende Reinelement verwendet, so ergibt die Abhängigkeit des Quotienten aus den Fluoreszenzintensitäten — der im allgemeinen kleiner als eins ist — keinen linearen Verlauf über der Konzentration. Derartige Eichkurven werden neben der Zusammensetzung auch von der Spannung an der Spektroskopieröhre beeinflußt. SEEMANN, SCHMIDT und STAVENOV haben darauf bereits hingewiesen [1]. Im folgenden soll versucht werden, durch Anwendung der theoretischen Ansätze für die Fluoreszenzausbeute [2] und der näherungsweisen Beschreibung der Wellenlängenabhängigkeit der weißen Strahlung und der Schwächungskoeffizienten gemäß KRAMERS [3], die Spannungsabhängigkeit theoretisch zu erfassen und nach einer Diskussion der möglichen Fälle eine Anwendung der Theorie auf konkrete experimentelle Beispiele zu geben.

Der Intensitätsquotient errechnet sich zu

$$r_i = \alpha_1 \cdot \frac{\int\limits_{\lambda_0}^{\lambda_K} \frac{\lambda \cdot (\lambda - \lambda_0)}{\lambda^3 + \alpha_2} \cdot d\lambda}{\int\limits_{\lambda_0}^{\lambda_K} \frac{\lambda \cdot (\lambda - \lambda_0)}{\lambda^3 + \alpha_3} \cdot d\lambda}$$

$\alpha_{1,2,3}$ von der Elementkombination und der Versuchsgeometrie abhängige Proportionalitätsfaktoren.

Dieser Ausdruck enthält bereits die gesuchte Spannungsabhängigkeit, da die kurzwellige Grenze λ_0 des Kontinuums durch $h \cdot c/e \cdot V$ zu ersetzen ist. Es sind darin folgende Voraussetzungen enthalten:

1. Das Fehlen einer selektiven Anregung.

2. Das Fehlen einer Anregung durch charakteristische Linien des primären Spektrums.

3. Die Kramersschen Näherungen.

Die Lösung der Integrale erfolgt durch zweimalige partielle Integration. Mit Hilfe eines elektronischen Rechners kann dann die zahlenmäßige Abhängigkeit der relativen Intensität von der Röhrenspannung ermittelt werden.

Für ein binäres System sind fünf Fälle zu unterscheiden.

a) Das Zusatzelement ist in geringer Konzentration vertreten.

b) Die beiden Elemente sind im periodischen System benachbart, wobei jene Strahlung zur Analyse gelangt, für die die Schwächungskoeffizienten nicht durch eine Absorptionskante getrennt sind.

c) Wesentlich leichteres Zusatzelement.

d) Wie b), jedoch mit Absorptionskante.

e) Im periodischen System weit auseinander liegende Elemente.

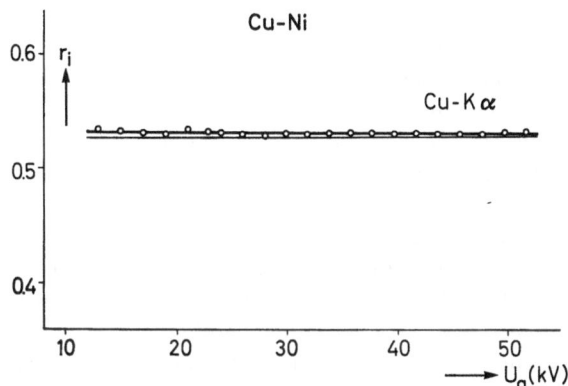

Abb. 1. Relative Fluoreszenzintensität in Abhängigkeit von der Spannung an der Spektroskopieröhre
(System Cu—Ni, Cu $K\alpha$-Strahlung)

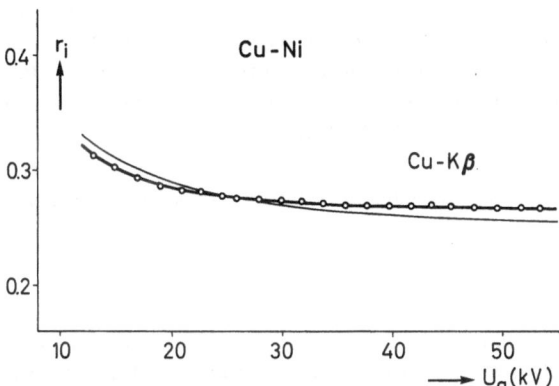

Abb. 2. Wie Abb. 1 (System Cu—Ni, Cu $K\beta$-Strahlung)

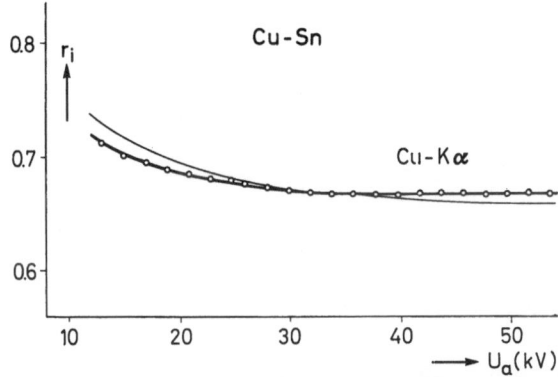

Abb. 3. Wie Abb. 1 (System Cu—Sn, Cu $K\alpha$-Strahlung)

Für a) bis c) ist nur eine geringe und meßtechnisch nicht feststellbare Spannungsabhängigkeit
zu erwarten. Die Fälle d) und e) führen zu einer Abhängigkeit, die experimentell einwandfrei
nachgewiesen werden kann. Da a) und c) nicht von besonderer Bedeutung sind, wurden die
Elementkombinationen entsprechend b), d) und e) theoretisch und experimentell behandelt.
b) wurde mit Hilfe einer Cu—Ni-Legierung und Cu K_{α}-Strahlung verifiziert, d) mit derselben
Legierung unter Verwendung der Cu K_{β}-Strahlung und e) mit einer Cu—Sn-Probe. Die Abb. 1—3
zeigen die gemessenen relativen Intensitäten und zum Vergleich die errechneten Kurven. Die
Abweichungen zwischen den Kurven sind nahe der Absorptionskante auf die Kramerssche

Näherung und bei höheren Spannungen auf die Vernachlässigung der zusätzlichen Anregung durch die WL-Linien zurückzuführen.

Aus den Ergebnissen ist zu ersehen, daß die relative Intensität über der Röhrenspannung bis zu 10% schwankt, wobei die Abhängigkeit bei höherer Spannung geringer wird. Eichkurven, die die relative Intensität in Abhängigkeit von der Zusammensetzung wiedergeben, sollten daher die Spannung als Parameter enthalten.

Der bei höheren Spannungen weitgehend konstante Intensitätsquotient weist darauf hin, daß die Eindringtiefe der weißen Strahlung wesentlich größer ist als die Fluoreszenztiefe. Berechnungen einer das weiße Spektrum ersetzenden Strahlung effektiver Wellenlänge haben aber ergeben, daß diese auch bei höheren Röhrenspannungen nicht als konstant anzusetzen ist. Die auf einer monochromatischen Ersatzstrahlung aufbauenden quantitativen Näherungsverfahren sind aus diesem Grunde eher problematisch. Schließlich hängt die effektive Wellenlänge in einem Mehrstoffsystem im Gegensatz zu der Annahme von Beattie und Brissey [4] noch zusätzlich von der zahlenmäßigen Zusammensetzung ab.

Bei der quantitativen RFA mit variablem Beobachtungswinkel [5] gelangt ebenfalls eine mittlere Wellenlänge zur Anwendung, doch ist das Analysenergebnis spannungsunabhängig, da der die Zusammensetzung charakterisierende Grenzwert der Relativintensität einem schleifenden Beobachtungswinkel der Fluoreszenzstrahlung entspricht, wodurch die Fluoreszenztiefe gegen Null geht und damit die Eindringtiefe und die tiefenabhängige Variation der spektralen Intensitätsverteilung der Primärstrahlung bedeutungslos werden.

Der Ludwig-Boltzmann-Gesellschaft danken wir für die Förderung der Untersuchungen.

Literatur

1. Seemann, H. J., G. Schmidt u. F. Stavenov: Z. Naturforsch. 16a, 25 (1961).
2. Müller, R. O.: Spektrochemische Analysen mit Röntgenfluoreszenz, S. 56 u. 65. München: R. Oldenbourg 1967.
3. Blochin, M. A.: Physik der Röntgenstrahlen, S. 82 u. 141. Berlin: VEB Verlag Technik 1957.
4. Beattie, H. J., and R. M. Brissey: Anal. Chemistry 26, 980 (1954).
5. Ebel, H.: Z. Met. 57, 454 (1966).

Messungen der Dichte dünner Aufdampfschichten

H. Ebel, A. Wagendristel und H. Judtmann

Institut für angewandte Physik der Technischen Hochschule Wien, Österreich

Die Dichte dünner Schichten liegt, bedingt durch Fehlordnungen und Poren, unter der des kompakten Materials. Die Abweichungen sind durch die Aufdampfgeschwindigkeit, die Güte des Vakuums, unter dem die Bedampfung erfolgte, sowie durch eine anschließende Wärmebehandlung zu beeinflussen. Das Ziel der gegenständlichen Untersuchungen war es, den relativen Unterschied $\Delta \varrho / \varrho$ der Dichten zu bestimmen.

Die gravimetrische Messung der Dichte versagt bei dünnen Aufdampfschichten. Der Weg über die Messung der Massenbelegung m/F und der geometrischen Dicke d führt hingegen zur Dichte ϱ_{sch} der Schicht

$$\varrho_{\mathrm{sch}} = \frac{1}{d} \cdot \frac{m}{F}.$$

Mit der Dichte ϱ des kompakten Materials errechnet sich der Dichteunterschied $\Delta \varrho$ zu

$$\Delta \varrho = \varrho_{\mathrm{sch}} - \varrho.$$

Zur Bestimmung der Massenbelegung können verschiedene Verfahren herangezogen werden. Es kann dies auf chemischem Wege oder mittels einer Vergleichsmessung während des Aufdampfvorganges an einem Schwingquarz sowie röntgenfluoreszenzanalytisch erfolgen. Die Röntgenfluoreszenzanalyse ist auf zwei verschiedene Arten einsetzbar. Entweder durch Ablösen der Schicht — Flüssigkeitsanalyse — oder aber durch Anwendung eines Absolutverfahrens [1], das die Massenbelegung der Schicht direkt zu messen gestattet. Die Dicke der Schicht ist durch optische Interferenzmessung zu erhalten. Sind Schicht und Trägermaterial unterschiedlich, so muß zur Erzielung eines gleich großen Phasensprungs bei der Reflexion die Kante zwischen Träger und Schicht mit einer einheitlichen Substanz überdampft werden. Die andere Variante, die geometrische Dicke der Schicht mit Hilfe des Kiessig-Verfahrens [2] zu bestimmen, ist vergleichsweise langwieriger und kann nur auf Schichtdicken von maximal 600 bis 1000 Å ausgedehnt werden.

Im folgenden werden Versuchsergebnisse gebracht, aus denen die Dichte der dünnen Schichten folgt. Die Massenbelegung der Schichten wurde nach dem röntgenfluoreszenzanalytischen Absolutverfahren [1] bestimmt und die geometrische Dicke durch optische Interferenz. Zu dem röntgenographischen Verfahren ist zu bemerken, daß dieses für die Schichtdicke nur dann den richtigen Wert ergibt, wenn die Dichte ϱ_{sch} der Schicht und nicht die des kompakten Materials (ϱ) in die Rechnung einbezogen wird. Durch Verwendung von ϱ weicht die röntgenographische Dicke um denselben Prozentsatz von der optischen oder besser geometrischen Dicke ab wie die Dichte ϱ_{sch} von ϱ.

In der Tabelle sind von den untersuchten Proben die optisch und röntgenographisch gefundenen Dicken angegeben. Außerdem ist in dieser Tabelle die Aufdampfgeschwindigkeit und schließlich der Dichteunterschied angegeben. Die Messungen wurden an Kupferschichten, die auf Glasträger gedampft worden waren, ausgeführt. Zur Analyse gelangte die Kupfer-K_α-Linie des Schichtmaterials. Als Dichtewert des kompakten Materials wurde $\varrho_{\mathrm{Cu}} = 8{,}932$ g/cm³ und für den Massenschwächungskoeffizienten der Kupfer-K_α-Strahlung in Kupfer $\mu/\varrho = 49{,}61$ cm²/g der Literatur entnommen. Der Strahlenuntergrund wurde durch Ausmessung des gesamten CuK_α-Profils ermittelt, so daß die Genauigkeit der Messungen möglichst günstig war.

Tabelle

Probe	D_0 (Å)	D_R (Å)	v (Å/min)	$\Delta\varrho/\varrho$ (%)
1	754	680	600	9,8 ± 6,5
2	508	577	670	—11,4 ± 15,6
3	1084	1019	720	6,0 ± 5,1
4	1149	1013	1150	11,8 ± 4,5
5	578	622	1150	—7,6 ± 15,2
6	1120	973	1150	13,1 ± 5,2
7	1117	907	1800	18,8 ± 5,1
8	1016	946	2030	6,9 ± 3,3
9	1326	1154	2650	13,0 ± 7,1
10	1692	1530	1130	5,9 ± 3,5
11	2417	2161	1610	10,6 ± 4,5
12	1671	1336	1670	20,1 ± 5,2
13	1989	1658	2270	16,7 ± 4,1
14	1960	1822	2610	7,0 ± 3,2
15	2658	2197	3540	17,3 ± 4,0
16	1868	1685	3730	9,8 ± 4,6
17	3386	2861	1930	15,5 ± 2,2
18	5425	4478	2170	17,4 ± 2,0
19	4547	3920	2270	13,8 ± 2,0
20	2926	2512	2340	14,2 ± 1,9
21	3389	2827	3390	16,6 ± 3,0
22	3566	3116	3570	12,6 ± 2,8
23	4058	3193	4060	21,3 ± 2,3
24	3086	2609	4100	15,5 4,0

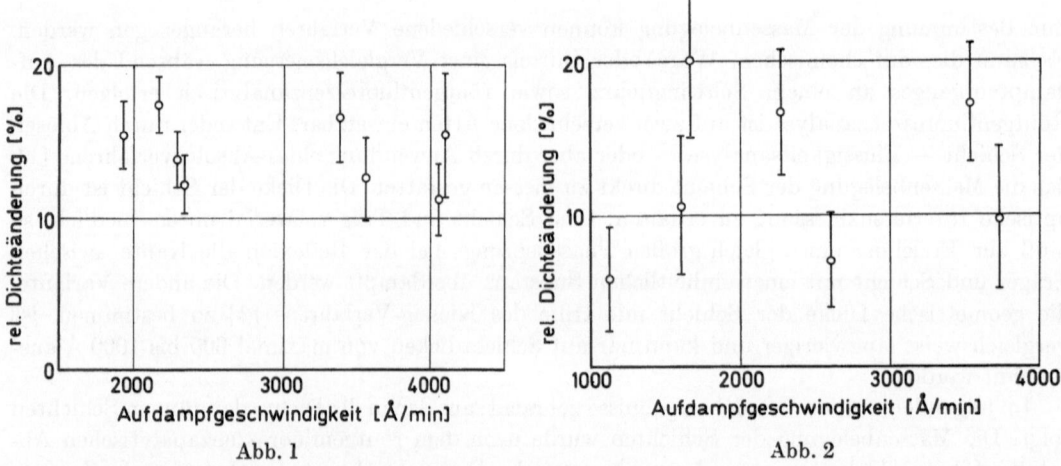

Abb. 1   Abb. 2

Abb. 1. Relative Abweichung der Dünnschichtdichte von der Dichte des kompakten Materials in Abhängigkeit von der Aufdampfgeschwindigkeit. Die Dicke der Schichten liegt zwischen 3000 und 6000 Å

Abb. 2. Wie Abb. 1. Die Schichtdicke variiert zwischen 1500 und 2000 Å

Abb. 1 zeigt den relativen Dichteunterschied in Abhängigkeit von der Aufdampfgeschwindig-keit. Die Dicke der Schichten liegt hier zwischen 3000 und 6000 Å. Der Dichteunterschied beträgt etwa 15%. Bei dünneren Schichten, entsprechend der Abb. 2, sind bereits größere Streuungen der Versuchsergebnisse zu erkennen. Die mittlere Abweichung der Dichte liegt etwas unter 15% und scheint nach niederen Aufdampfgeschwindigkeiten hin abzunehmen. Die Dicke dieser Schichten liegt zwischen 1500 und 2000 Å. In der Abb. 3 schließlich sind die Ergebnisse für dünne Schichten — 500 bis 1000 Å — eingetragen. Die versuchsbedingten Streuungen liegen wieder über jenen der Abb. 2, doch ist aus den Ergebnissen zu ersehen, daß mit abnehmender Auf-

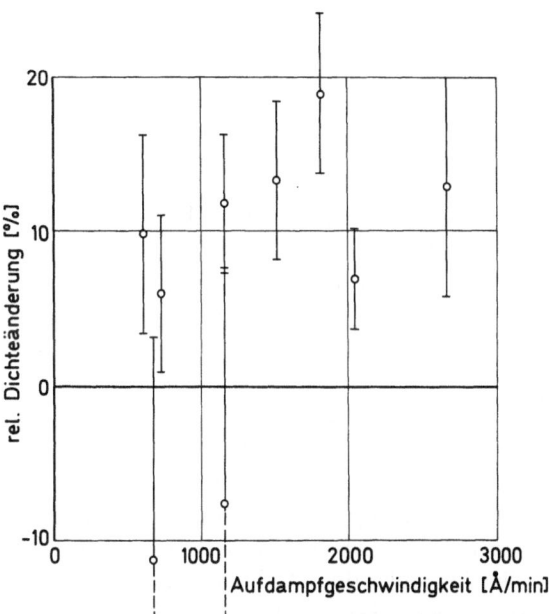

Abb. 3. Wie Abb. 1. Hier liegt die Schichtdicke zwischen 500 und 1000 Å

dampfgeschwindigkeit der Dichteunterschied der Schicht gegenüber dem kompakten Material zusehends geringer wird.

Sämtliche Ergebnisse wurden an Schichten gewonnen, die unter einem Vakuum von etwa $2-5 \cdot 10^{-6}$ hergestellt worden waren. Eine Variation des Vakuums bis zu etwa $5 \cdot 10^{-4}$ brachte keine meßbare Beeinflussung der Schichtdichte.

Aus den gezeigten Ergebnissen ist zu erkennen, daß die Dichte dünner Aufdampfschichten mit zunehmender Aufdampfgeschwindigkeit von der des kompakten Materials abweicht, wobei die Dichte der Schicht stets kleiner, bestenfalls gleich der des kompakten Materials ist. Bei dünnen Schichten ist die Dichteabweichung geringer, als bei dicken Schichten. Sie strebt mit zunehmender Schichtdicke einem Durchschnittswert von etwa 15% relativer Dichteabweichung zu, der dann unabhängig von der Schichtdicke und auch der Aufdampfgeschwindigkeit ist.

Allerdings besteht die Möglichkeit, daß eine Bedampfung unter einem besseren Vakuum als $2-5 \cdot 10^{-6}$ zu geringeren Abweichungen als 15% führt. Diese Experimente konnten mit den vorhandenen Geräten nicht realisiert werden.

Der Ludwig-Boltzmann-Gesellschaft danken wir für die Förderung der Arbeiten.

Literatur

1. EBEL, H.: Z. Met. **56**, 802 (1965).   2. KIESSIG, H.: Ann. d. Phys. **10**, 715, 769 (1931).

Biological Applications

Biological Work Using Microfluorescence Analysis

H. O. E. Röckert

Department of Histology, University of Göteborg, Sweden

There is a great need of simple micro-methods for elementary analyses on the cellular level. A number of physiological and pathological mechanisms concerning the electrolyte transport through the cell membrane are unknown so far. Not only the lack of important fundamental data makes the need for elementary analysis obvious but also a check of the treatment of patients in clinical medicine makes analyses of the electrolyte content in cells and tissue fluids necessary. The common analyses of blood samples do not always reflect what has happened in tissues and organs in various parts of the body [1]. Local tests can be made on the value of various therapeutic measures for diseases affecting the electrolyte transport.

This can best be done by using a good isolation technique for cells and by using a method for collecting tissue fluid normally and during pathological conditions like extracellular oedema in tissues.

Most of the methods used are performed at a macro-level and assume that the blood values are also true for other tissues. Attempts have been made to calculate the electrolyte composition of tissue fluids from indirect measurements [2], but direct measurements are lacking or unsatisfactory on the whole.

Ordinary flame-photometry has not the sensitivity to analyse elements in the amounts which occur in single cells. Neutron activation may work in theory for analysing potassium in big cells but has in practice turned out to be unsatisfactory. The use of autoradiography is not the best method for these problems among other things due to difficulties in quantitation and time consumption. In these indirect methods for estimating the intracellular electrolyte content, corrections have to be made for fat, collagen, blood and extracellular space before the probable intracellular parameter could be calculated.

The use of the microprobe technique will offer a sensitivity beyond the picogram region and would make it possible to study the distribution of elements within the cell. For total cell values an integration procedure would be necessary. For single cell analysis and analysis of e.g. microdrops of tissue fluid the preparation technique is difficult. The cost of the microprobe equipment is considerably higher than for any other instrument for elementary analysis within the micro-region.

X-ray fluorescence micro-analysis has decided potential for obtaining elementary information on the single cell level both for cells and tissue fluids. An alternative is given by ultramicro-flame photometry. The advantages and disadvantages of the two methods in biological work are compared and discussed in a separate paper during the present symposium.

X-ray fluorescence micro-analysis has the advantage of being a non destructive method. There is no continous spectrum excited by fluorescence because the primary X-rays cannot lose their energy in a continous fashion analogous to the deceleration of electrons in the target of the X-ray tube. If an analysing crystal is used about half the background intensity is caused by scattering of the fluorescent radiation from the analysing crystal. The other half of the background is caused by primary radiation scattered by the specimen. The half scattered from the crystal may be eliminated by pulse amplitude discrimination. The half representing primary radiation scattered from the specimen, and diffracted by the crystal according to wavelength, cannot be eliminated because it will have the same wavelength as the part of the spectrum on which it is superimposed [3].

If a broad beam is used for exciting radiation, difficulties in localization of the analysed area will occur. For analysing microareas a reasonable intensity of the primary beam will be required. A Cosslett-Nixon X-ray microscope will produce a small focal spot. By using collimating apertures in the primary beam it is possible to obtain a lateral resolution in the range of 10—50 μ.

The main physical problems to be overcome in the practice of the technique are the collimation of the primary beam by an aperture system which produces little or no scattered radiation, and the efficient collection and discrimination of the radiations excited in the specimen. For the latter, the alternative of a nondispersive proportional counter system, and a curved crystal spectrometer are available. The limiting factors which determine the sensitivity of the method are twofold: first, the intensity of scattered radiation of a wavelength close to that of the fluorescence line: and second, the intensity in the fluorescence spectrum. Thus, even though a crystal spectrometer may result in a much better rejection of scattered radiation, the very low efficiency will partly offset this advantage.

The nondispersive system has been chosen by us and is the one with which we have most experience. It has turned out to function well. The deflection coils mounted in the place of the pole piece of the magnetic lens can deflect the electron beam so as to bring the focal spot on the axis of the aperture system. The latter consists of two Siemens-type electron microscope apertures mounted immediately above the copper-foil target so as to illuminate an area of the specimen 100 or 200 μ in diameter. The specimen is mounted on a thin Mylar foil on top of the last aperture. The space between specimen and counter window is evacuated to about 10 mm Hg in order to reduce scattering and to eliminate the argon K fluorescence radiation, whose pulse-height distribution would overlap that of potassium.

One of the practical difficulties in the use of the point-focus X-ray tube is the instability of the output. Both filament shift and pitting of the target contribute to fluctuations in X-ray intensity. One possible solution to this difficulty is to monitor the direct beam with a second counter. An alternative method, adopted here, is to include with the specimen a constant amount of a second element not normally present, and to refer the intensity due to the unknown element to that of the internal standard. In the case of potassium excited by Cu K radiation, iron forms a very convenient standard, since its pulse-height distribution is well separated from that of potassium.

The fluorescence radiation from the standard iron-bearing foil and from the specimen were recorded simultaneously by a single proportional counter whose output pulses, after amplification, were passed to a multichannel pulse-height analyzer so as to produce a spectrum containing peaks due to both iron K and potassium K radiation. Since both the specimen and its support are very thin, there is no appreciable attenuation of the radiation from the standard.

The sensitivity and linearity of the technique can be determined by using standards prepared by evaporating dilute solutions of potassium hydroxide of known concentration. The weights of potassium in individual standards were determined by adding ^{131}I to the solution and comparing the activity of each evaporated standard with that of an aliquot of the bulk by means of a scintillation counter. The activities of the standards are very low and no increase in the background of the proportional counter of the X-ray apparatus was detectable. Using the set up described it is possible to determine potassium in amounts down to 10^{-11}—10^{-12} g under optimal conditions [4]. If the geometry of the aperture system could be more carefully determined and the beam of the X-ray microscope more easily focussed and measured, the sensitivity could probably be increased by a factor of 10. Generally for single cell analysis the potassium amount has been about the order of 10^{-10} g, which makes the present arrangement sufficient for its purpose. It has also the advantage of using commercial electron microscope apertures.

Some types of specimens are better suited for analyses than other. When studying sections the difficulties depend of course on what element is to be analysed. In paraffin embedded sections quite a number of elements will be dissolved during preparation and a number of chemical and morphological artifacts will occur. Freezed dried sections are of course more favorable, but localization in a freezed dried section is generally very difficult. Firstly it is difficult to be sure that the area to be analyzed is immediately on top of the aperture. Secondly it may be difficult to ensure intimate contact of the area to be analyzed with the supporting foil, or the top surface of the aperture. A separation of one or a few microns above the aperture will result in the analyses of a bigger area than is wanted, since the primary beam is divergent. Also the holes in electron microscope apertures are not strictly cylindrical. When using them in this method, the varying energy of the primary beam results in varying penetration through the aperture edges, thus producing a field of varying diameter.

Generally it is safer to isolate the specimen which is to be analysed and then irradiate the entire specimen.

In many cases it is necessary to dissect out cells from fresh material. Cells down to about $30-40$ μ in diameter can be dissected out by hand with fine needles under a binocular microscope. If the cells are smaller, a group of cells can be excised in the same way, and their number counted or the specimen's mass can be determined by X-ray absorption after the X-ray fluorescence analysis has been completed. When single cells with a size smaller than $25-30$ μ have to be isolated a micromanipulator generally has to be used. If a fluid dissection medium is employed the composition of this medium is most important. One has to be absolutely sure that no uncontrolled leakage occurs through the cell membrane in any direction, because this might influence the analytical result. Also one has to be sure that if the dissection medium contains ions of the same sort as are being analysed in the cell, that the excessive fluid outside the cell is carefully removed when the cell has been transported on to the supporting foil. This can be checked by using radioactive isotopes in the medium and the activity on a specimen measured by a scintillation counter after the excessive fluid has been sucked away from the mylar foil. Sometimes a cell can be dissected out from a tissue directly without any artificial medium around it.

It is also important to check the thickness of the specimen. When sections are analysed the microtome will produce sections of suitable thickness. If dissected specimens or cells are used a thickness control is necessary, not for each specimen but for the type of specimen. Most single cells after being dissected out become flat when placed on a mylar foil. Groups of cells on the other hand have a tendency to arrange themselves as a hemisphere on a supporting foil. If the specimen is too thick there is still a way around the problem. If the element to be analysed can be extracted the problem is not too difficult; e.g. intracellular potassium can be extracted in a drop of water. The drop can then be placed in the X-ray beam. In practice such an extraction is best performed under liquid potassium free paraffin and the microdrop sucked up with a micropipette sealed off with liquid paraffin in both ends to prevent evaporation. Accurate thickness determination can be made with the method of HALLÉN. To just use the focussing device and read the scale of the micrometer screw on an ordinary microscope is not accurate. HALLÉN's method is based on the principle of measuring the movement of the microscope tube with an accurate microcator when the top and bottom surface are focussed separately [5] with the aid of the shadow of a hair.

X-ray microfluorescence can also be used *in vivo*. With the arrangement already described it is possible to determine iron in haemoglobin in tissue as a measurement of the size of the vascular bed. For example in the frogs web or in the cheek pouch of the guinea pig where it is possible to obtain transmission of the tissue. Even if the thickness causes a systematic error it is constant, and the arrangement could be used for registering relative changes in the size of the vascular bed provided the exposure time is short enough not to produce any irreversible changes by X-irradiation [6].

X-ray fluorescence micro-analysis has so far proved to be a convenient method for quantitative work when working with electrolyte problems of three different types of tissues, vis: muscular tissue, nervous tissue and dental tissue.

During shock a proper knowledge of the pathological and physiological reactions of single cells is necessary for the evaluation of correct therapeutic measures. The skeletal muscle cell is of great importance to the electrolyte equilibrium, especially with regard to potassium, as muscular tissue constitutes about 40 to 50% the bulk of the cellular tissue of the body. Potassium determinations would serve as a vitality test for the cells as it mainly has an intracellular position, while sodium is mainly located extracellularly. The sodium pump will partly be damaged by insufficient oxygen supply to the cell during shock as the aerobic cellular metabolism is affected. With the present method it is possible to see if the effect of the treatment will be on the cell or the tissue fluid around the cell. Previous methods have the disadvantage that they can not distinguish without approximation between intra- and extracellular elements. Biopsies were performed on dogs, in which haemorrhagic shock was being induced so that muscle cells could be isolated before and after the bleeding. The dogs were bled to 45 mm Hg bloodpressure for 135 minutes. Parts of single cells were dissected out in an oxygenated dissection solution. The cells were placed on a mylar foil and allowed to dry under dustfree conditions. Pieces from the central part of each undamaged muscle fiber were analysed for potassium separately so making it possible to detect any maltreatment of the pieces of muscle cell. To obtain the amount of potassium per mass unit of each specimen the dry mass was determined by using X-ray absorption measurements. Individual variations showed a potassium loss between 10 to 50% (mean 26 per cent) [7, 8]. At the same time it was noticed that each cell showed a characteristic pattern of shifts of the potassium content in vitro in relation to time. Analyses from patients suffering from essential hypokalemia have shown the method to be useful for clinical diagnoses. The patients are not caused any more discomfort by a small muscle biopsy than from an ordinary blood test.

Separate analyses of potassium in surrounding tissue fluid sucked up in small glass capillaries showed a potassium increase resulting from leakage of intracellular potassium. In cases with a large reduction blood volume, the amount of potassium in the plasma did not reflect the changes in the tissue fluid due to the impaired blood supply, and the decreased fluid exchange between these two compartments [1].

In the studies of nervous tissue X-ray fluorescence micro-analyses have been used as a test of damages after X-irradiation. Insufficiency in the sodium pump is discovered earlier this way than are morphological changes. In a series of experiments on the rabbit brain given 3,000 r in single doses the maximum response occurs 4 to 6 days after irradiation. The damage is reversible and is reastablished in 2 to 3 weeks like other measured parameters such as total organic mass and succinoxidase activity [9—11].

In the application to dental tissues, X-ray fluorescence has been used for analyses of enamel and dental fluids. Only microdrops can be obtained from this material and a new field of research in dentistry is about to be born, as the content of these fluids may reflect physiological and pathological changes in the dental pulp. It is otherwise impossible to obtain samples from this tissue without causing non physiological damage on the way into the pulp chamber. For these fluids, however, the amount of potassium and calcium seem to be of the same order of magnitude, thus causing difficult interference problems when using a non dispersive system for analysing the fluorescent radiation [12]. Maybe a dispersive system would contribute to give useful information in the field of dental fluid research.

References

1. HAGBERG, S., H. HALJAMÄE, and H. RÖCKERT: Ann. Surg. 168, 259 (1968).
2. MANERY, J. F.: Physiol. Rev. 34, 334 (1954).
3. BIRKS, L. S.: X-ray spectrochemical analysis. New York and London: Interscience Publ. Inc. 1959.
4. LONG, J. V. P., and H. O. E. RÖCKERT: Proc. 3rd int. symp. on X-ray optics and X-ray microanalysis, p. 513, 1963. Ed. PATTEE, COSSLETT and ENGSTROM. New York: Academic Press.
5. HALLÉN, O.: Acta Anat., Suppl. 25 (1956).
6. RÖCKERT, H.: Experientia 21, 169 (1965).
7. HALJAMÄE, H.: Acta Chir. Scand. 133, 259 (1967).
8. HAGBERG, S., H. HALJAMÄE, and H. RÖCKERT: Acta Chir. Scand. 133, 265 (1967).
9. HAMBERGER, A., and H. RÖCKERT: J. Neurochem. 11, 757 (1964).
10. RÖCKERT, H. O. E., and B. H. O. ROSENGREN: Proc. IVth int. symp. on X-ray optics and X-ray microanalysis, p. 650, 1966. Ed. CASTAING, DESCHAMPS and PHILIBERT. Paris: Herman.
11. HALLÉN, O., A. HAMBERGER, B. ROSENGREN, and H. RÖCKERT: J. Neuropathol. Exptl. Neurol. 26, 327 (1967).
12. BERGMAN, G., L. Å. LINDÉN, and H. RÖCKERT: Advances in fluorine research and dental caries prevention 4, 163 (1966).

Biological Applications of Projection X-Ray Microscopy

R. L. de C. H. Saunders

Department of Anatomy, Sir Charles Tupper Medical Building, Dalhousie University, Halifax, Nova Scotia, Canada

Projection X-ray microscopy is a recently developed technic which provides the biologist with the means of obtaining enlarged images directly with X-rays, and of studying the internal structure of optically opaque objects in the dead and living state. Hitherto most papers have dealt with the physical and instrumental features of the X-ray projection microscope. Fewer papers have reported its use for biological studies in any depth, presumably because of the limited number of instruments yet available.

Specifically, projection X-ray microscopy offers a new and promising approach to the study of the microcirculation, under normal, pathological or experimental conditions. While X-ray microscopy by point projection cannot compete with the electron microscope, owing to the linked problem of spot size and X-ray intensity, it is nevertheless also providing a new approach to histological research.

The X-Ray Projection Microscope

This method of microscopy was suggested by SIEVERT (1936) [1], von ARDENNE (1939) [2] and others, and subsequently developed by COSSLETT and NIXON (1952) [3] at Cambridge. Briefly the X-ray projection microscope consists of an electron gun followed by a demagnifying system consisting of a condenser and objective lens. This system is used to focus electrons into a microbeam that strikes a metal foil target, so producing a point source of X-rays. This point source is used to project an enlarged image of a nearby object on to a phosphor viewing screen or distant photographic plate.

The resolution of the instrument is determined by the size of the point source, and at its best is about 1,000 Å. The magnification provided is determined by the ratio of the target-specimen and target-plate distances, so that for a constant plate position the closer the object is brought to the target the greater the magnification obtained. Because the target forms the end wall of the microscope, biological objects can be positioned directly over the point source. Most biological studies can be performed over the microscope under atmospheric conditions, but when soft X-rays are used a vacuum camera is fitted to the microscope target to eliminate their absorption by air. The 5—30 kV range of the low voltage microscope (XM 30) meets most biological needs, but the 60 kV of the high voltage model (XMPJ) [4] has proved useful for special problems requiring greater X-ray penetration.

Biological Applications

Biological applications of the X-ray projection microscope date historically just before the first conference on X-ray microscopy held at Cambridge in 1956 [5]. About that time a variety of specimens, chiefly of an entomological nature, were used as test objects by COSSLETT and NIXON [6], LE POOLE [7], ONG SING POEN [8], and others, to determine the resolution and performance of the X-ray projection microscope, as well as to devise ways of improving specimen contrast. Their work showed that the instrument could provide sharp, contrasty images, and also be used for stereomicroscopy because of its great depth of field.

These projection studies covered a variety of insects, such as the green fly or aphid (Aphis fabae), mosquito (Culex), fruit fly (Drosophila melanogaster), and cicada (Platypleura capitata). They demonstrated that external features such as antennae, body divisions, and wing structure, could be accurately recorded in remarkable detail. Internal structures, otherwise difficult to dissect, such as the limb musculature, ovaries and gut, could be shown intact. For example, the V-shaped tymbal muscles responsible for the distinctive note of the cicada were beautifully demonstrated stereographically by NIXON (1954) [9].

Microangiography

Microangiography, or the X-ray study of microscopic blood vessels filled with contrast medium, was one of the earliest biological applications of the projection X-ray microscope SAUNDERS and FRYE (1958) [10]; SAUNDERS (1959) [11]. Pioneer work by BELLMAN (1953) [12] had demonstrated the value of microangiography performed by contact mircoradiography, especially in obtaining wide field views of the microcirculation. The projection X-ray microscope however increased the potential of the technic, in that it not only provides magnified and sharper images, but all the blood vessels remain in focus owing to its great depth of field. The contact and projection methods of microangiography complement one another, and both have been practised with the X-ray microscope with marked success.

The skin of the rabbit ear has been used extensively for studies of the microcirculation with the X-ray microscope [13]. Contrast has been produced by the intravascular injection of a radiopaque of colloidal dimension, capable of flowing through the smallest blood vessels, such as 25% Micropaque made up in warm saline, or Thorotrast. Injection, performed close to the heart to avoid inducing vascular spasm or shock in the tissue field under study, has then been followed by serial exposures. In this way it has been possible to record vascular reaction to experimental treatment, such as trauma, heat and cold.

Such studies showed the small terminal arteries, capillaries, and draining venules of the microcirculation in the skin, with great detail and clarity. They also revealed patterns of microvascular organisation, such as the coarse distributor networks formed by interconnecting arteries and the finer nutritive networks of capillaries enclosed by them. Vascular shunts, or S-shaped arterio-venous anastomoses, that connect small arteries to adjacent cutaneous veins were also recorded, and rendered more obvious by chilling the ear. Such shunts have a caliber of 0.05 — 0.1 mm, and length of 0.5—1 mm. Even the flow of contrast medium through the living capillary bed could be recorded in this way. Such was the complexity of the small vessels forming the microcirculation, that it was found helpful to record the same vessels after animal sacrifice to interpret the rapid and sometimes bizarre flow patterns observed in the living animal.

Successful application of projection X-ray microscopy to microangiography led to study of the small blood vessels and microvascular patterns within the human skin [14], lung [15], tooth [16], intestine [17], and brain [18].

In the case of human skin, projection studies showed that all parts of the complex vascular system of fetal or adult skin could be readily demonstrated. Examination of the whole thickness of the skin of the human palm or foot, following injection with contrast medium, revealed that the pattern of the skin's friction ridges or finger and toe prints, were repeated by the underlying blood vessels (Fig. 1). The papillary capillaries thus repeated the basic patterns of arch, loop or whorl, so giving a "vascular print". Where a skin ridge "forked", it was noted that the underlying vessel pattern did likewise. In studies at higher power it was possible to identify the thick and thin limbs of the cutaneous capillary loops. Such studies have understandably interested dermatologists and surgeons concerned with the problems of skin grafting.

Projection X-ray microscopy has proved particularly useful in the field of cardio-pulmonary research, where it has been used to study the small blood vessels within the lung, and assess the revascularisation of the heart muscle following the implantation of arterial grafts therein.

Projection studies of the injected human lung removed at autopsy demonstrated that the pulmonary arterial tree could be traced diminuendo to capillary level within a single plate,

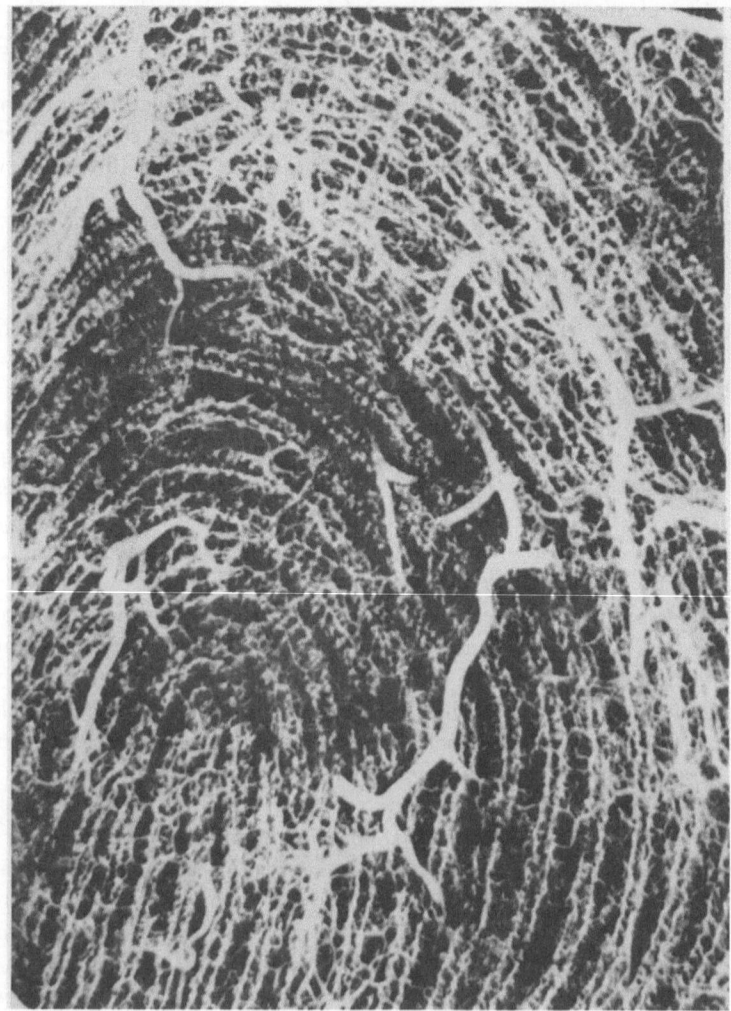

Fig. 1. Projection X-ray micrograph of whole thickness of human skin showing that the capillary pattern repeats the pattern of the overlying friction ridges. Skin from the second toe. Micropaque injection. Mag. × 24

a distinct advantage over the light microscope, whose limited depth of focus forces the use of thin microtome sections with consequent destruction of the vascular patterns under study. In general the terminal branches of the pulmonary artery were seen to follow and imitate the ramification of the respiratory bronchioles and small air or alveolar ducts. The lung capillaries were observed to arise abruptly from the sides of saccular branches or precapillary arterioles, and so form an almost continuous sheet of capillaries surrounding the air sacs. The honeycomb appearance of the polygonal air sacs and their interalveolar septa was a conspicuous feature.

Projection micrographs at high magnification showed that the mesh spaces of the pulmonary capillary net were often smaller than the diameter of its capillaries, and that it was possible to perform capillary counts on some of the interalveolar septa. X-ray microscopy clearly opens the way to detailed studies of the intrapulmonary vessels in health and disease, and of physical factors that may influence pressure and flow within the pulmonary circuit.

In cardiac research much attention has been given to finding ways of improving the blood supply in ischemic heart disease. The Vineberg operation, in which the internal mammary artery is implanted in a tunnel within the heart muscle or myocardium, has had some acceptance. But some doubt has remained regarding its effectiveness in perfusing the myocardial circulation.

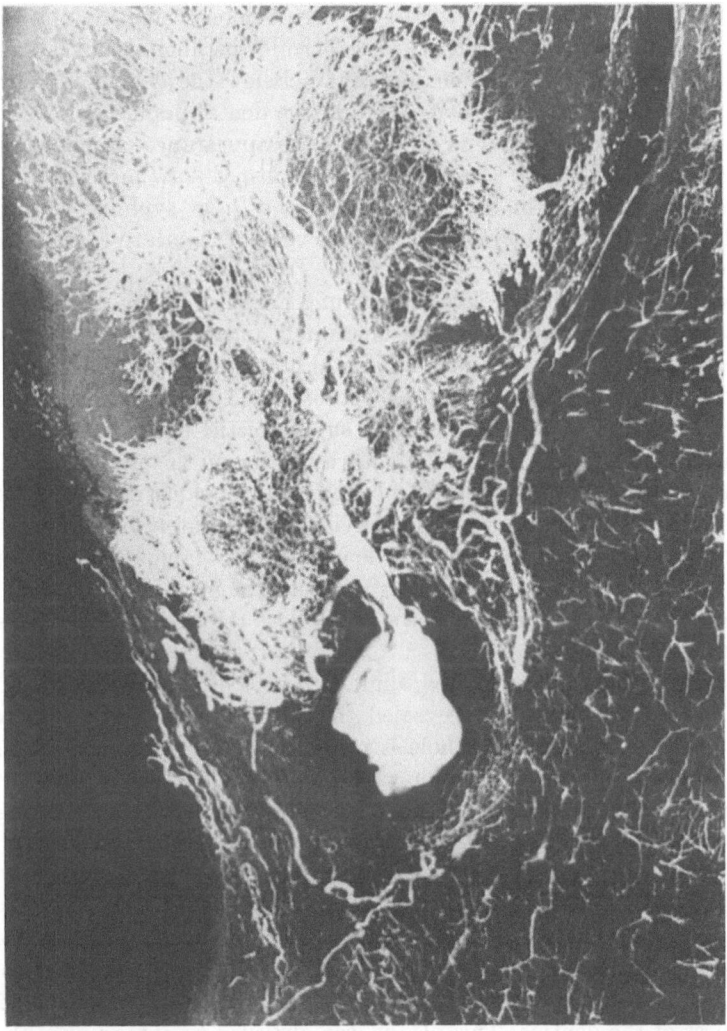

Fig. 2. Projection X-ray micrograph of a dog heart showing a cross section of an implanted artery with a side branch feeding the capillary bed. Micropaque injection. Mag. × 18

Clinically it has been difficult to obtain an objective assessment, for although standard radiography has recently provided some information on the coronary arteries, the resolution is limited.

An experimental technic was therefore developed to assess objectively the myocardial vascularisation that occurs from an implanted artery [19]. The Vineberg operation was performed on dogs and six or more weeks later the implanted artery was exposed in the living animal and injected with contrast medium. X-ray microscopy was then used to assess the vasculature of the heart. This clearly showed that the arterial implant gave off functioning side branches, and that consistent filling (perfusion) of the adjacent microcirculation follows suitable injection of an internal mammary implant (Fig. 2). X-ray microscopy is the only method available whereby the microvascular bed of the injected heart wall can be adequately visualised, revealing such features as the transmyocardial vessels and blood supply of the papillary muscles which control the heart valves.

In the field of dental research, projection X-ray microscopy has been used to study the blood supply of the jaws and teeth, in both animals and man [20, 21]. Study of these vessels has hitherto suffered from severe technical difficulties imposed by the hard and opaque surrounding tissues.

The blood supply of the developing tooth, has been demonstrated by X-ray microscopy both within the human fetal jaw and the isolated tooth germ. A peridental plexus was observed about the dental sac, closely applied to the enamel organ. Below the developing crown was seen an intradental plexus of blood vessels, situated within the dental papilla or precursor of the tooth pulp. This intradental plexus was found to precede the appearance of enamel and the mineralisation of the crown. These plexuses are believed to transport raw materials for the developing enamel and dentine, and explain how infectious agents (e.g. syphilis, German Measles) and certain antibiotics (e.g. tetracycline) can enter the developing tooth to cause either malformation or discoloration of the tooth.

Projection studies of the injected jaw of the young Rhesus monkey proved particularly useful for demonstrating the blood supply of the unerupted and erupted teeth. They showed very clearly how the inferior dental artery gave off dental branches to the deciduous dentition above, and also germinal branches that supplied the plexuses in and around the germs of the permanent successors below. Thus revealing that the blood supply of the two dentitions is shared, in that there is a common vascular path. This possibly has a bearing on the mechanism of tooth eruption.

Studies of the blood supply of erupted monkey teeth demonstrated that the apical artery on entering the root of an erupted tooth, subdivided immediately to form a vascular bundle. On ascending the root canal, this vascular bundle was seen to "fan out" within the pulp chamber of the tooth to supply the subdentinal capillary plexus. The interdental arteries could be traced between the teeth, and connections demonstrated between the periodontic vessels and capillaries of the gum.

X-ray microscopy of freshly extracted adult human teeth injected with contrast medium by the suction-injection technic [22], revealed the blood vessels within the tooth pulp with marked clarity. In this technic a drill hole is made in the tooth crown to permit the cutting of several capillary loops in the pulp, after which suction is applied to the tooth crown, while keeping the root of the tooth submerged in contrast medium. Such studies revealed the pattern of the dental pulp vessels in both single and multi-rooted teeth.

For example single rooted teeth such as the adult human bicuspid showed that the larger centrally located vessels of the pulp were surrounded by smaller paraxial vessels, which subdivided to form a rich peripheral or subdentinal capillary plexus that lay in close contact with the dentine of the tooth. High magnifications proved that the subdentinal plexus was an anastomotic network, with an irregular contour due to its capillary loops extending into small bays on the deep aspect of the dentine. This plexus was observed to extend over the whole coronal part of the tooth pulp and also down into the root canal. Its richness may be regarded as an index of pulpal metabolic activity. A similar vessel pattern was observed in multi-rooted teeth, such as the molars, except that vascular bridges united the pulpal vessels across the lower part of the pulp chamber. Further work is required regarding the "take up" of nutrients and minerals from this pulpal vascular system, and their outward diffusion or transdental flow in both the growing and mature tooth.

In the fields of neuroanatomy and neurosurgery, X-ray microscopy has provided a new approach to the study of the small blood vessels within the animal and human brain. The arterial and venous components of the human cerebral circulation have been described, and while the full significance of these vessel patterns in health and disease have not yet been determined, certain neurosurgical applications have been quickly recognised [23].

Lack of precise information of the flow and distribution of blood in these small vessels presently limits our understanding of cerebral vascular disorders in general. Patterns of blood flow have recently been examined in the surface vessels of the surgically exposed brain, following intracarotid injection of certain dyes and radioisotopes [24]. Radioisotopic flow curves have been measured directly by placing miniature gamma detectors or counters on the brain surface during operation. Consequently it became important to determine the nature of the vascular territory "seen" by such counters. X-ray microscopy was able to supply this information by showing the typical pattern of blood vessels within a brain convolution or gyrus corresponding in size to the counter field [25].

Projection X-ray microscopy has successfully demonstrated the blood supply of the spinal cord and spinal nerves. It has made possible the study of both the peripheral system of vessels surrounding the cord, and the brush-like central arteries that penetrate it to supply the grey matter. The radicular arteries could be traced along the spinal nerve roots, and the capillary bed within the spinal root ganglia was a striking feature [26].

General Biology and Historadiography

In general biology, projection X-ray microscopy has proved useful for studying diatoms, plankton and foraminifera [27—29], providing sharp pictures of the complex inner structures of these lower forms of life. Pictures of these are sometimes difficult to secure because of the limited focal depth of the optical microscope and density of the specimen. In addition, it has provided the biologist with a practical method of examining many foraminifera populations. BAEZ's (1957) [30] demonstration of a trematode by X-ray microscopy suggests that the technic might have a role in parasitology.

Botanical material lends itself to study by X-ray microscopy since thin plant structures often require little or no preparation, other than the removal of water to improve image contrast. For example, the microstructure and crystals of rose leaves, and their seasonal variations have been studied by ELY [31] using the projection method. Plant histology was one of the first applications of the X-ray microscope because it was found that the marked X-ray absorption of plant cells provided contrasty pictures of cell features, mitotic division (Allium cepa) and cellular patterns of the phloem and xylem (Helianthus) [32].

Sections of various woods, such as obeche, pine and mahogany have been successfully studied by projection, revealing the fibres, parenchymal and sap cells, and crystals of calcium salts [33]. Recently JONGEBLOED [34] has pointed out the value of projection X-ray microscopy for scientific and technological research on wood. His beautiful micrographs of tropical and other woods show that anatomical details can be studied in thin sections (30—50 microns) and that problems concerned with conservation, impregnation and the painting of wood can be investigated.

Historadiography, or the study of tissues and cells by X-ray methods dates from LAMARQUE and TURCHINI (1936) [35]. They performed such studies by contact microradiography using very soft X-rays generated at about 2 kV. Since then, ENGSTROM and GREULICH (1956, 1960) [36, 37]. LINDSTROM (1955) [38] and others of the Swedish school have played a great part in the development of methods of obtaining pictures of high resolution and of extracting quantitative information from histological material by the contact method. The contact method yields excellent results, but as is well known, is limited by registration at unit magnification, specimen thickness and emulsion grain size. The primary magnification, great depth of field, and higher resolution afforded by the point projection method are of distinct advantage in histological research. Fortunately the X-ray projection microscope can be used to obtain either contact or projection images.

Methods of preparing specimens for historadiography by projection X-ray microscopy have been described by LE POOLE (1957) [39], ONG SING POEN (1959) [40], and more recently by JONGEBLOED (1965) [41]. Briefly these include direct staining (e. g. with alcoholic iodine), selective staining with a heavy element (e.g. osmium tetroxide, thorotrast), shadow casting (e.g. with gold), replication, and the administration of contrast agents (e.g. barium sulfate, lead acetate). All of these methods are applicable to biological material, and in passing it might be noted that gold shadowed bull sperm on X-ray microscopy clearly show the 0.3 micron wide tail, so indicating the resolution obtainable [42].

The first study of bone by projection X-ray microscopy revealed that the diploic veins within the tables of the dog skull could be demonstrated in about 5 minutes, in contrast with the days required by conventional technics [43]. Histological sections of mineralised tissues such as developing teeth, ovarian dermoid teeth [44], calcified cartilage, and bone provide contrasty projection micrographs, even when decalcified. Such micrographs provide a striking demonstration of the epiphyseal and metaphyseal growth zones of young bone such as seen in

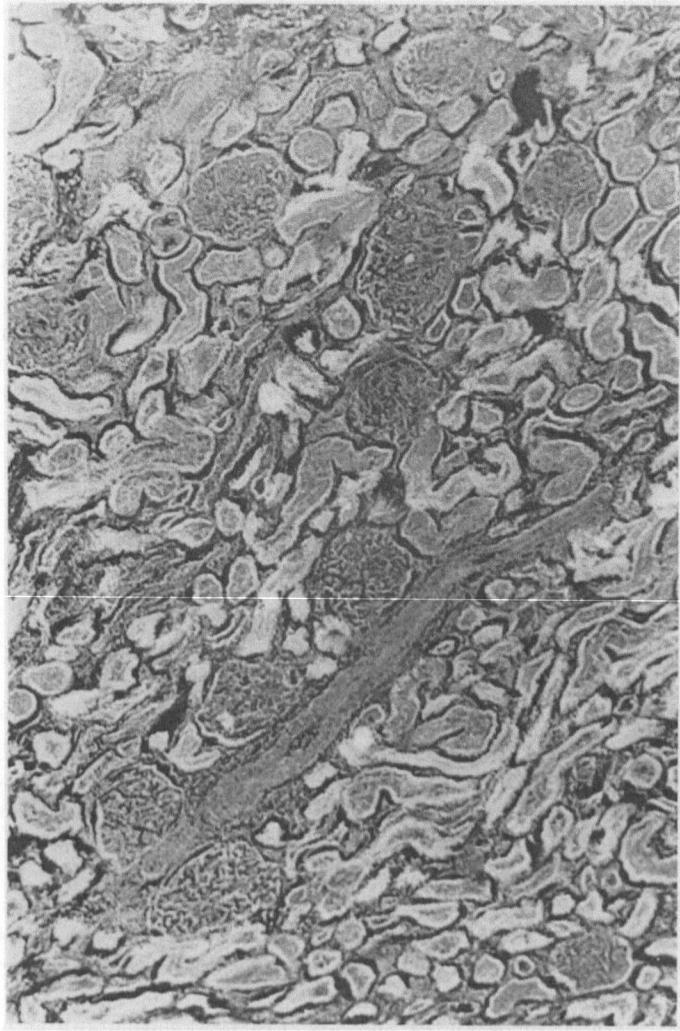

Fig. 3. Projection X-ray micrograph of unstained, freeze dried, rat kidney showing the renal corpuscles and cortical labyrinth. Note the marked contrast between the glomeruli and the more X-ray absorbent proximal convoluted tubules. Mag. × 242

the rabbit tibia. The concentric Haversian systems with their lamellae, entrapped osteocytes, and central canals, as well as the columns of proliferating and maturing cartilage cells with their calcifying intercellular matrix, can all be seen to advantage. The metaphyseal area distinctly shows the calcified trabeculae, osteoblasts lining the dark marrow spaces, and the areas of new bone deposition. High power views show details such as the lacunae that contain the osteocytes, and the collagen fibrils which separate the rows of maturing cartilage cells, as well as other interesting features.

Turning to soft tissues, Oderr (1964) [45] studied the internal architecture of the human lung prepared by inflation-fixation using only alcohol and air. Sections of lung (0.5—2 mm thick) examined with a specially designed microscope of the projection type, revealed a lace-like pattern of interlocking respiratory units. High power studies of these respiratory units showed that the air sacs or alveoli were arranged in spiral tiers, forming helices about the alveolar ducts. Gas exchange was considered in the light of the geometric figures so discovered. Cunningham (1960) [46] used the projection method to study pathology sections of various lung diseases such as asbestosis, and among other things showed bronchial cartilage undergoing calcification, and the bronchial cilia with the characteristic phospholipid line at their bases. X-ray microscopy has

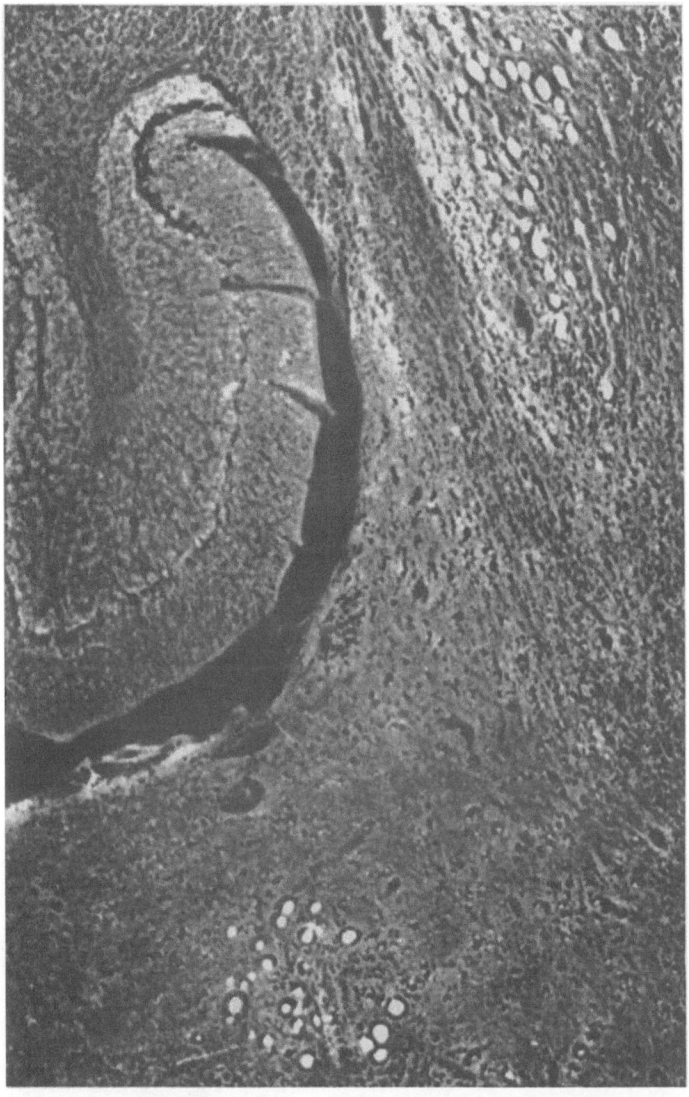

Fig. 4. Projection X-ray micrograph of sectioned rat brain, stained with lead for acid phosphatase enzyme
activity, showing two groups of neurons (brain stem nuclei). Mag. × 83

also been used to demonstrate the location of tin within a tin miner's lung [47], and also the
characteristic perivascular disposition of iron particles from a case of pulmonary siderosis in
an electrical welder [48].

Chemicals used for fixing, embedding and staining, have deleterious effects on soft tissues
which can be avoided in X-ray microscopy by using cryostat sections. Ways of applying projection X-ray microscopy to the histology of unstained, freeze dried, animal and human tissues
were explored by SAUNDERS and VAN DER ZWAN (1960) [49]. Many tissues such as the skin,
kidney, thyroid, salivary and sebaceous glands were found to give contrasty projection X-ray
micrographs with soft aluminium radiation even when unstained.

For example, sections of unstained human scalp showed the cellular organisation of the outer
and inner root sheaths of the hair follicles with almost diagrammatic clarity, such was their
X-ray absorption. From without inwards it was possible to identify the vitreous membrane, cells
of the outer sheath, Huxley's and Henle's layers. The process whereby these inner layers undergo
conversion to soft keratin could also be clearly seen.

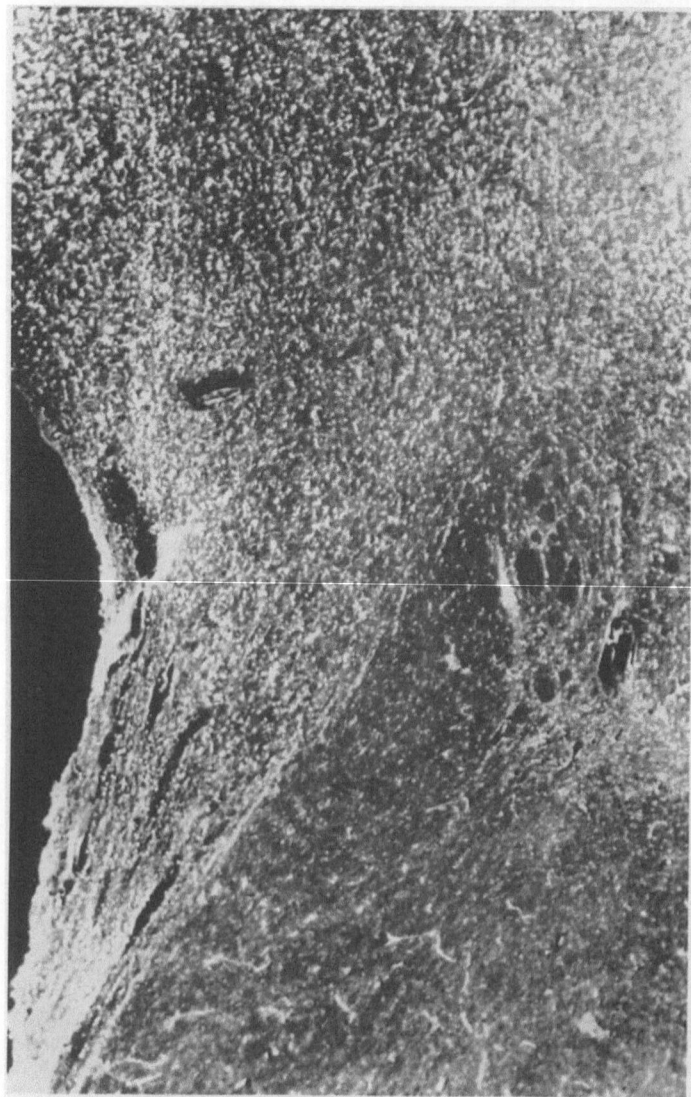

Fig. 5. Projection X-ray micrograph of sectioned human hypothalamus, stained with lead for acid phosphatase activity, showing the contrast between the unstained optic tract area with the markedly X-ray absorbent cells of the supraoptic area. Mag. × 36

The sebaceous glands related to the hair follicles, although unstained, also provided interesting X-ray micrographs. High power studies showed the gland lobules and their cells with striking clarity. The cell membranes were markedly X-ray absorbent and hence stood out in contrast with the accumulation of dark, radiolucent, lipid droplets within the sebaceous cells. Some of the sebaceous cells could be seen to contain discret droplets of lipid, whereas others showed the cell nucleus surrounded by confluent droplets, which accordingly gave the cell a black appearance. Other micrographs showed the duct of a sebaceous gland entering a hair follicle, while the centre of the gland acinus demonstrated cells packed with lipid and surrounded by the debris of broken down cells.

Projection X-ray studies of unstained rat kidney also provided much interesting information, showing details of the cellular structure of the renal corpuscles and the cortical labyrinth. The interlobular arteries, vascular pole and capillary tuft of the glomeruli could be readily identified, as well as Bowman's capsule and space and its continuity with the proximal convoluted tubule.

The glomeruli contrasted markedly with the surrounding and more X-ray absorbent proximal convoluted tubules (Fig. 3). The reason for this radiopacity differential is unknown, but is possibly related to the selective absorption that occurs in these tubules. Other X-ray micrographs showed that the distal and collecting tubules could be readily identified by virtue of their numerous and radiolucent nuclei. The rat kidney was used by Mosley, Scott and Wyckoff (1957) [50] as a test object for projection studies with the soft K radiation (10 Å) of magnesium.

In neurology, X-ray microscopy can be used to study sections of nervous tissue stained by metallic impregnation following the classical technics of Golgi and Cajal. For example, X-ray micrographs of the cerebellar cortex stained by the Cajal method clearly show the pear-shaped Purkinje cells with their dendrites branching within the molecular layer, and subjacent to them the small round cells of the granular layer. But while X-ray microscopy can quickly and accurately record the general and detailed features of tissue sections prepared in this way, such staining technics are both time consuming and capricious.

Recently it has been found that the modern precisely controlled histochemical technics, or modifications of these, can be used effectively with X-ray microscopy to produce radiopacity in the desired location. For example, use of the phosphatase reactions has made it possible to demonstrate various collections of nerve cells (or nuclei) and fibre patterns within the brain stem of the rat (Fig. 4), as well as within the human hypothalamus (Fig. 5). The cells of the human supraoptic nucleus can be so illustrated with a contrast rarely obtained by photomicroscopy, the X-ray micrograph even showing the minute corpora amylaceae sometimes seen in the surrounding hypothalmic brain tissue. The organisation of the blood vessels and nerve cells within the cortex of the human brain as well as their interrelationships have recently been studied in detail by this method [51].

In conclusion, it will be evident that projection X-ray microscopy has manifold biological applications. Because the image obtained is totally different in character to the light microscope and is caused by a pure absorption phenomenon, the information is more specific and informs us about the type of material present in the specimen [52]. Hence it is to be expected that projection X-ray microscopy will make a significant contribution to biological research, especially when coupled with modern X-ray microanalytical technics.

Acknowledgements. I gratefully acknowledge the contributions of two departmental associates — Dr. C. Edwin Kinley for Fig. 2, and Miss Mary A. Bell for Fig. 4 and 5. Thanks are also due Mrs. Pat Troyer and Mr. Victor R. Carvalho for help with X-ray microscopy and photography, and Miss Lesley Gentry for library research.

References

1. Sievert, R.: Acta Radiol. 17, 299 (1936).
2. Ardenne, M. v.: Naturwissenschaften 27, 485 (1939).
3. Cosslett, V. E., and W. C. Nixon: Proc. Roy. Soc. (London), Ser. B 140, 422 (1952).
4. Saunders, R. L. de C. H., and R. V. Ely: X-ray optics and microanalysis, p. 642. Paris: Hermann 1965.
5. Cosslett, V. E., A. Engstrom, and H. H. Pattee: X-ray microscopy and microradiography. Proceedings of a Symposium held at the Cavendish Laboratory, Cambridge, 1956. New York: Academic Press 1957.
6. —, and W. C. Nixon: Nature 170, 436 (1952).
7. Le Poole, J. B., and Ong Sing Poen: X-ray microscopy and microradiography, p. 91. New York: Academic Press 1957.
8. Ong Sing Poen: Microprojection with X-rays. The Hague: Martinus Nijhoff 1959.
9. Nixon, W. C.: Proceedings of Internat. Conference on Electron Microscopy, held at London, July 1954, p. 307. London: Roy. Mic. Soc. 1956.
10. Saunders, R. L. de C. H., and R. M. Frye: Microscopy Symposium, p. 67. Chicago: Walter C. McCrone Associates 1958.
11. — J. Anat. Soc. India 8, 1 (1959).
12. Bellman, Sven: Acta Radiol., Suppl. 102, 7 (1953).
13. Saunders, R. L. de C. H.: J. Roy. Microscop. Soc. 83, 55 (1964).
14. — Blood vessels and circulation, p. 38. London: Pergamon Press 1961.
15. —, and V. R. Carvalho: X-ray optics and X-ray microanalysis, p. 109. New York: Academic Press 1963.
16. —, and H. O. E. Röckert: Structural and chemical organisation of teeth, vol. 1, p. 199. New York: Academic Press 1967.
17. — Experentia 17, 1 (1961).
18. — X-ray microscopy and microanalysis, p. 244. Amsterdam: Elsevier 1960.
19. Kinley, C. E., and R. L. de C. H. Saunders: Can. J. Surg. 11, 27 (1968).

20. SAUNDERS, R. L. DE C. H.: Oral Surg. **22**, 503 (1966).
21. — J. Can. Dental Assoc. **33**, 245 (1967).
22. KRAMER, I. R. H.: Anat. Rec. **111**, 91 (1951).
23. SAUNDERS, R. L. DE C. H., W. H. FEINDEL, and V. R. CARVALHO: Med. and Biol. Illus. **15**, 108, 236 (1965).
24. FEINDEL, W., Y. L. YAMAMOTO, and C. P. HODGE: Can. med. Ass. J. **96**, 1 (1967).
25. FEINDEL, W. H., Y. L. YAMAMOTO, M. A. BELL, and R. L. DE C. H. SAUNDERS: Correlation of radiation detector field and vascular territory for studies on cortical blood flow. J. Nucl. Med. (in preparation).
26. SAUNDERS, R. L. DE C. H.: X-ray microscopy and microanalysis, p. 244. Amsterdam: Elsevier 1960.
27. DUPOUY, G., F. PERRIER, and P. VERDIER: Compt. Rend. **250**, 3083 (1960).
28. JONGEBLOED, W. L.: X-ray optics and microanalysis, p. 636. Paris: Hermann 1965.
29. HOOPER, K.: X-ray microscopy and microanalysis, p. 216. Amsterdam: Elsevier 1960.
30. BAEZ, A. V.: The X-ray microscope, p. 39. California: University of Redlands 1956.
31. ELY, R. V.: The encyclopedia of microscopy, p. 638. New York: Rheinhold 1961.
32. SAUNDERS, R. L. DE C. H.: X-ray microscopy and X-ray microanalysis, p. 305. Amsterdam: Elsevier 1960.
33. JACKSON, C. K.: X-ray microscopy and microradiography, p. 487. New York: Academic Press 1957.
34. JONGEBLOED, W. L.: Norelco Rep. **12**, 93 (1965).
35. LAMARQUE, P., and T. TURCHINI: C. R. Ass. Anat. **31**, 341 (1936).
36. ENGSTROM, A., and R. GREULICH: Exp. Cell Res. **10**, 251 (1956).
37. GREULICH, R.: X-ray microscopy and microanalysis, p. 273. Amsterdam: Elsevier 1960.
38. LINDSTROM, B.: Acta Radiol., Suppl. **125**, 9 (1955).
39. LE POOLE, J. B., and ONG SING POEN: X-ray microscopy and microradiography, p. 91. New York: Academic Press 1957.
40. ONG SING POEN: Microprojection with X-rays, p. 103. The Hague: Martinus Nijhoff 1959.
41. JONGEBLOED, W. L.: X-ray optics and microanalysis, p. 636. Paris: Herman 1965.
42. LE POOLE, J. B.: The X-ray microscope, p. 39. California: University of Redlands 1956.
43. HEWES, C. L., W. C. NIXON, A. V. BAEZ, and O. F. KAMPMEIER: Science **124**, 129 (1956).
44. RÖCKERT, H., and R. L. DE C. H. SAUNDERS: Experientia **14**, 59 (1958).
45. ODERR, C.: Am. Rev. Resp. Diseases **90**, 401 (1964).
46. CUNNINGHAM, G.: Tools of biological research, p. 155. Oxford: Blackwell 1960.
47. SAUNDERS, R. L. DE C. H.: Unpublished observations 1957.
48. ANGERVALL, L., G. HANSSON, and H. RÖCKERT: Acta Pathol. Microbiol. Scand. **49**, 373 (1960).
49. SAUNDERS, R. L. DE C. H., and L. VAN DER ZWAN: X-ray microscopy and X-ray microanalysis, p. 305. Amsterdam: Elsevier 1960.
50. MOSLEY, V. M., D. B. SCOTT, and R. W. G. WYCKOFF: Biochim. Biophys. Acta **23—24**, 235 (1957).
51. SAUNDERS, R. L. DE C. H., M. A. BELL, and V. R. CARVALHO: Proceedings of the Fifth Congr. X-Ray Optics and Microanalysis. Berlin-Heidelberg-New York: Springer 1969, p. 569.
52. ANDERTON, H.: Sci. Progr. (London) **55**, 337 (1967).

Comparison between X-Ray Fluorescence and Ultra-Micro Flame Photometric Analyses of the Electrolyte Content in Single Cells

H. HALJAMÄE and H. RÖCKERT

Department of Histology, University of Göteborg, Sweden

Ordinary flame photometric methods are usually used for the determination of tissue electrolytes in macrosamples. Corrections must be made for fat, collagen, blood and extracellular space before the probable intracellular concentrations of electrolytes can be calculated [1—4]. Analyses of single cells are not possible with ordinary flame photometry due to limited sensitivity (10^{-8} to 10^{-7} g). Neutron activation analysis, a more sensitive approach, has been used for analyses of small samples of tissue taken with biopsy-needles [5, 6]. In theory this method is sensitive enough for analyses of single cells; the main advantage being that all elements present will be activated and from the radiation spectra several elements in the sample can be quantitated. The phase distribution of radioactive isotopes using dilution methods has been applied to the problem of intracellular electrolyte content [7, 8]. These methods, however, are indirect approaches to intracellular electrolyte content and for phase distribution the factors used in the calculations of the intracellular concentrations under control conditions may not be valid under different pathological conditions. The "normal" distribution of electrolytes between the intravascular compartment, the capillary ultrafiltrate, the local tissue fluid, and the intracellular phase is still not completely known. As stated by THORN [9], there might exist a more fixed phase of fluid in the tissues with a composition different from the experimentally calculated tissue fluid. The existence of local macromolecular and fibrillar components may affect the distribution and exchange of electrolytes between the intra- and extracellular compartments. Especially during pathological conditions with profound changes in tissue perfusion (e.g. shock conditions), it may not be completely possible to reveal the extent of changes in local cellular electrolyte metabolism using the indirect approach of ordinary macro methods.

The authors have been studying changes in cellular electrolyte metabolism of single skeletal muscle cells caused by induced haemorrhagic chock. Micro-techniques raise many new difficulties due to problems of isolation of single cells, sample handling, and also technical problems with the equipment used for the micro-analyses. To overcome some of these problems, two different analytic procedure have been used: X-ray fluorescence microanalysis and ultra-micro flame photometry.

X-Ray Fluorescence Micro-Analysis

In Fig. 1, a diagram of the technical setup is given. For these studies a Cosslett-Nixon type of tube served as X-ray source and has the advantage of giving a high X-ray intensity over a small spot. Over the target two electron microscope apertures were placed, 100 and 200 μ in diameter. Above the aperture a thin mylar foil coated with iron, Fe_2O_3, was attached. The specimen on another thin mylar foil was then placed over the iron coated foil. The mylar foils contain H, N, O and C and do not disturb the measurements.

When the X-ray beam hits the specimen, most of the energy passes straight through as the transmitted beam. A minor portion excites secondary radiation, i.e., X-ray fluorescence. The

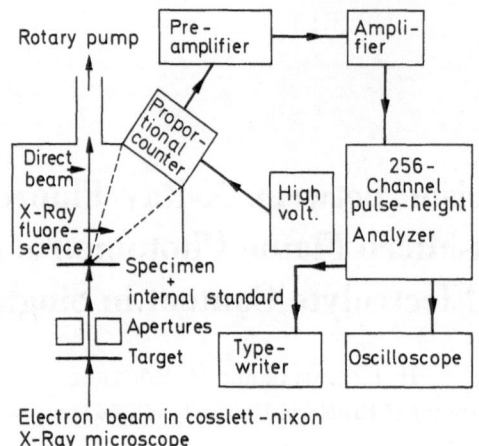

Fig. 1. Block diagram of the X-ray fluorescence set up

wave lengths of this radiation depend on the elements present in the specimen and the fluorescent intensities are directly proportional to the amounts of the constituent elements. A constant sector of the fluorescent radiation was registered by a proportional counter, which together with a multichannel pulse height analyser sorted out the different wave lengths. The high voltage used on the X-ray tube was 20 kV.

In order to decrease the absorption in air of the fluorescent radiation, the specimen was surrounded with an evacuated brass chamber which had a hole for the window of the proportional counter. A moderate vacuum, about 10 mm Hg, was used. Since the argon content of air produces a peak close to that of potassium in the energy spectrum, the evacuation chamber also reduced the argon interference considerably. The iron foil over the top aperture served as an internal standard in order to make it possible to correct for variations in radiation per unit time of the primary beam. Standard curves were obtained by analysing micro-drops of a mixture of a potassium hydroxide solution of known concentration and a $Na^{131}I$ solution of known activity. The volume of the micro-drops analysed, was obtained by measuring the radioactivity in a scintillation counter. (For further details [10].)

Ultra-Micro Flame Photometry

A detailed description of the dual-channel ultra-micro flame photometer used has been published previously [11]. A diagram of the equipment is given in Fig. 2.

Briefly the flame was supplied with city gas and air at constant pressure from cylinders. A known nanoliter volume of the sample was applied onto the tip of a 100 μ platinum-iridium wire, attached to a brass sample holder. The sample holder was inserted into a guidance slit leading to the burner house compartment. In the slit the sample was completely dried at a distance of 4 mm from the flame. The sample holder was then connected to an electromagnet, which when activated pushed the sample holder into "analysing position", that is with the tip of the thin Pt—Ir-wire in the lower part of the flame. Constant sectors of the light emitted, when the sample was burnt off, were focused by two collector lens systems onto two separate photo-cells, arranged on either side of the flame. Interference filters, one for each channel, were placed between the collector and the detector units. The filter for the potassium channel had a T_{max} at 762 mμ, that for the sodium channel had T_{max} at 584 mμ. Both filters had a half widht of 20 mμ. The photomultiplier used for potassium light detection had S-8 response and that of sodium light detection S-4 response. Each channel had separate amplifiers, integrators and indicators. From the indicators, the integrated values of the potassium and sodium content of the samples were obtained. A special control unit synchronized the onset of integration with the insertion of the sample into the flame. The electro-magnet together with a timer, which can

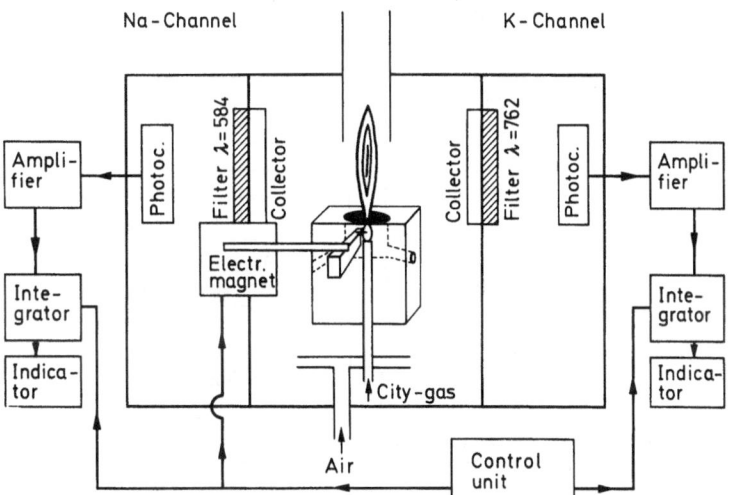

Fig. 2. Schematic diagram of the ultra-micro flame photometer

be preset for the needed time period for analysis (0.5—2 sec), and the integrators were controlled by this unit. The high voltage supply for the photomultipliers and the low voltage supply for the amplifiers and the control unit were common for both channels.

To increase the stability of the flame photometer, certain precautions were taken. The burner, collector, and detector units were built into a metal box, the inside surfaces of which were blackened to prevent light reflection. The only outlet of this box was a narrow chimney, and by having a high flow of filtered air at constant pressure enter the box, room air was prevented from entering the flame. The slight instability of the tip of the flame was prevented from reaching the detector systems by having it hidden in the chimney. A special diaphragm with a small central aperture sealed off from the detectors the lower part of the flame, into which the Pt—Ir-wire was inserted at analysis. The glow from the wire could not reach the detectors and variations in back-ground light emission at analysis were minimized.

Preparation of Single Cells and Analytic Procedures

Single skeletal muscle cells from dogs were isolated as previously described [12, 13]. The procedure is outlined in Fig. 3. Dissections were performed in a buffered (HCO_3^-), oxygenated (O_2 95%, CO_2 5%), Krebs-Ringer solution, pH 7.3—7.4. The dissection of single cells were continued for 40—50 minutes during which time 15—20 cells could be isolated. Fresh solution in the dissection dish was exchanged every 5 min to prevent changes in electrolyte composition and pH due to leakage from damage cells. With the B-procedure (see Fig. 3) several single cells could be isolated during the same time period. Cells were examined under phase contrast microscopy to ascertain if the cells were usable for further analysis [12, 13]. Since cellular damage was always present at the cut ends of the single fibers, only central parts of cells, which had a "normal" remaining cross-striation, no retraction nodes, and which were not too flattened out on the mylar foil, were used. To be able to exclude contamination of a sample during the different steps at the analyses, 3—4 pieces from each cell were analysed separately.

For X-ray fluorescence analysis the pieces of skeletal muscle cells were placed over the top aperture of the Cosslett-Nixon X-ray microscope and duplicate analyses of each piece followed by duplicate analyses of background fluorescence were performed. After the fluorescence analysis the dry mass of each cell was determined by X-ray absorption [14].

For ultra-micro flame photometry, after dry mass was determined, the pieces of muscle cell were then placed on a clean quartz-glass in a Petri dish and covered with liquid paraffin. Under the liquid paraffin cover, each piece was extracted for 2 hrs with 70.8 nanoliters of quartz

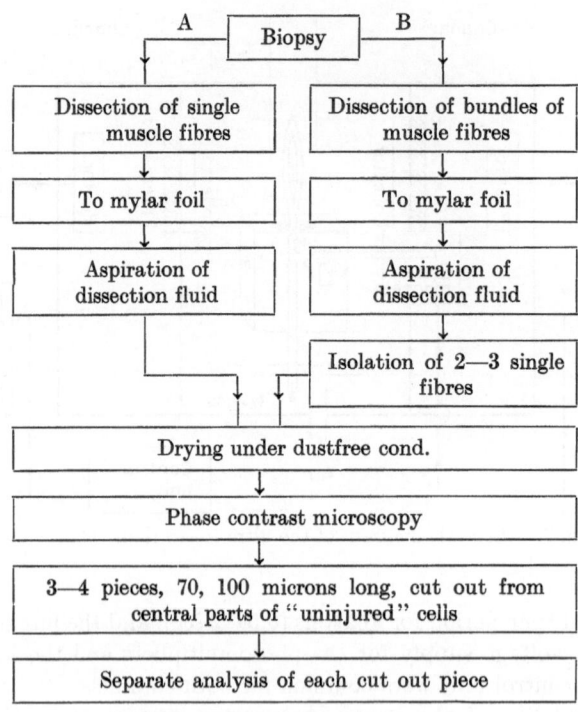

Fig. 3. Dissection procedures used

redistilled water. Duplicate samples of 21.8 nl of the extract were analysed with the ultra-micro flame photometer. The calibrated nl-pipettes used were made according to Keesey [15] and calibrated according to Prager et al. [16]. Only pipettes with a variation of less than 2 per cent were used.

From standard curves made up from solutions of known potassium concentrations, the potassium content in each piece of cell could be calculated and given as mEqK/100 g dry muscle cell.

Results with Single Cells

In Table 1 results for the potassium content of single skeletal muscle cells from two different dogs are given.

The cells from one experiment were analysed by X-ray fluorescence microanalysis and those from the other by ultra-micro flame photometry. The cells from dog A were dissected according to A-procedure (Fig. 3) and those from dog B according to the B-procedure, so that in the latter case two or more cells from the same time period in the dissection solution could be obtained.

The data showed that there is a great discrepancy between the values obtained in mEqK/100 g dry muscle cell with the two methods. The values obtained with X-ray fluorescence microanalyses are markedly higher than those obtained with ultra-micro flame photometry. This difference in the obtained average value for the potassium content of each cell is mainly due to different interference errors of the two methods.

With the X-ray fluorescence equipment it was not possible to separate completely the fluorescence radiation of chlorine, potassium and calcium. As shown in Fig. 4, it was possible to separate the main peaks of chlorine and potassium fluorescence radiation but there was some interference due to overlapping of the outer parts of these peaks. Part of the error may be attributed to the intracellular muscle cell chlorine content as well as the small amount of chlorine in the dissection fluid adhering to the cell membrane. Since the manipulations of the dissection and isolation procedures are well standardized, the fraction of dissection fluid adhering to the cells will be a relatively constant error. A substitution for the Krebs-Ringer solution devoid of potassium,

Table 1. *For each cell 2—4 cut out pieces were analysed separately and each of these values are given. The time given for each cell is elapsed time from the taking of the biopsy until the cell was removed to the mylar foil*

X-ray fluorescence microanalysis						U-M flame photometry			
dog A cell	time (min)	mEqK/100 g dry cell				dog B cell	time (min)	mEqK/100 g dry cell	
		obtained	average	corrected	average			obtained	average
1	11	131.97		69.82		1a	13	29.10	
		145.01		77.28				35.90	
		69.31		36.83				42.40	35.80
		83.63	107.48	44.25	54.54	1b	13	37.60	
2	12	84.65	84.65	44.76	44.76			31.10	
3	13	98.98		52.43				33.00	33.90
		91.05	95.02	48.34	50.39	2	17	42.20	
4	16	26.34		14.07				41.70	41.95
		47.31		25.06		3a	20	17.10	
		42.71	38.79	22.76	20.63			12.30	
5	17	47.06	47.06	25.06	25.06			11.90	13.77
6	24	83.12		43.99		3b	20	10.50	
		95.91		50.90				11.50	
		97.70		51.66				11.50	11.17
		77.49	88.56	41.18	46.93	4	23	32.50	
7	34	41.94		22.25				30.80	31.65
		91.82		48.59		5	25	31.30	
		63.43	65.73	33.50	34.78			23.00	
8	38	74.94		39.64				24.10	26.13
		50.38		26.60		6a	30	47.20	
		119.44		63.43				40.90	
		75.70	80.12	40.15	42.46			36.60	41.57
						6b	30	45.30	45.30
						7a	33	34.30	
								31.30	32.80
						7b	33	31.50	31.50

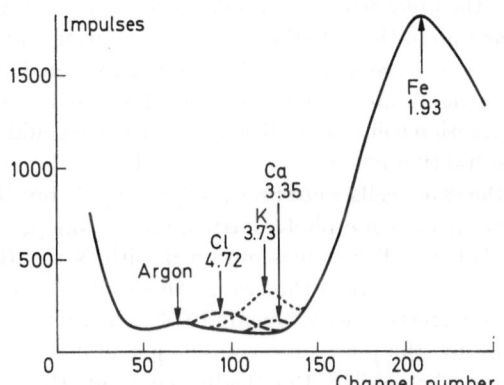

Fig. 4. Fluorescent energy levels of the elements discussed

chloride and calcium could not be used since that would deleteriously affect the in vitro electrolyte metabolism of the cells.

Small quantities of potassium and calcium will adhere to the cells but the concentration of these electrolytes is low in the dissection fluid compared to the intracellular potassium content so that this error will be small. The sensitivity of the X-ray fluorescence equipment for chlorine is markedly lower than for potassium [12], thus reducing the magnitude of chlorine interference problem slightly. When the values for K/Fe-peak minus Background/Fe-peak are calculated for the cell and the corresponding absolute amount of potassium is obtained from the standard curve, the potassium content per unit dry weight of the cell will be higher than the expected.

Table 2. *Some specimens analysed with the two methods*

Cell	Dry weight $\times 10^{-9}$ g	Potassium				U-M flame ph./ X-ray fluorescence
		X-ray flame		U-M flame ph.		
		$\times 10^{-11}$ g	mEq/100 g	$\times 10^{-11}$ g	mEq/100 g	
1	23.3	115	126.3	51.5	56.5	0.45
2	25.6	113	112.8	44.1	44.0	0.39
3	21.8	70	82.1	46.8	55.0	0.67
4	21.1	112	135.8	47.2	57.3	0.42
					Average	0.48

In addition the interference of chlorine with the potassium peak, the fluorescent radiation energies of potassium and calcium are so closely related that an acceptable separation of these peaks was not possible (see Fig. 4). The calcium content of single cells, however, is low compared to the potassium content, but it will contribute to a slightly higher obtained potassium value than expected.

To correct these errors the average volume of adhering dissection solution was determined by using isotopes in the dissection solution and calculating a correction factor. This factor was as high as 0.53 [12].

The lower values for the potassium content of skeletal muscle cells obtained with ultra-micro flame photometric analysis were closer to the expected. Chloride did not interfere with the determination. There were no cross-effects on the separate indicator readings when analysing standard solutions with different concentrations of potassium and sodium. By comparing results from analysis of reference solutions containing potassium chloride and sodium chloride and reference solutions to which different ions occuring in biological material were added, the possible interference effects of these different substances were determined. When calcium ions were added (1 mg/l) the average indicator readings for potassium were 96 per cent of those for the control reference solution. Phosphate (0.1 mM) decreased the indicator readings 2—4 per cent, magnesium (1 mg/l) 2—6 per cent and bicarbonate (8 mg/l) 3 per cent [11, 15]. Therefore when analysing single skeletal muscle cells, the obtained value would be 5 to 10 per cent lower than the absolute. The small amounts of potassium in the adhering dissection solution will however also be included, causing the obtained values to seem closer to the values expected. If, during the extraction of the single cells for micro flame photometer analysis all the potassium content were not free to equilibrates with the extraction solution, a slightly lower value could be obtained. No existence of such a bound potassium fraction could be demonstrated.

In some experiments, the same cells were first analysed by X-ray fluorescence, the dry mass determined, and then ultra-micro flame photometric analysis was performed. Results from this comparison are shown in Table 2. The values obtained with X-ray flourescence analysis were significantly higher and the ratio between the two methods (0.48) was similar to the correction factor calculated from the interference experiments (0.53; see above).

Theoretically, the values for the potassium content per unit dry weight of each piece from the same muscle cell ought to be similar. Practically, however, there are several difficulties to overcome when working with single cells affecting these results. Cells may be contaminated with dust from the air or from the thin needles with which they are handled. Loss of fragments from the cells may occur during manipulations. Slight variations in the reproducibility of the dry mass determined by the X-ray absorption will also affect the final results. The variations between the values for the potassium content of the different pieces from the same cells seen in Table 1 include all these possible errors, and so do not reflect directly the reproducibilities of the two methods. When considering the above facts both methods had a good reproducibility with biological material. Usually the potassium content of the different pieces from the same cell were of the same magnitude (Table 1). This was also observed for the average potassium content of cells isolated after the same time under in vitro conditions (dog B: cell 1a—1b, 3a—3b, 6a—6b, 7a—7b) or cells isolated within a time interval of only 1 minute (dog A: cell 2 and 3,

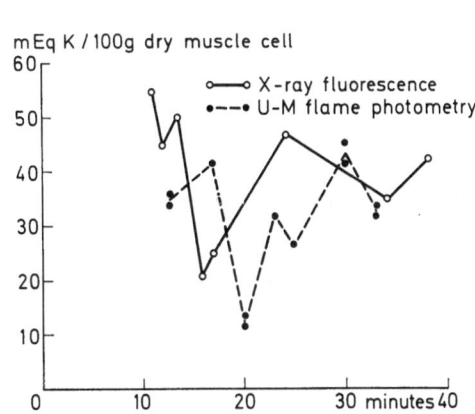

Fig. 5. Potassium content of single skeletal muscle cells in relation to time in vitro

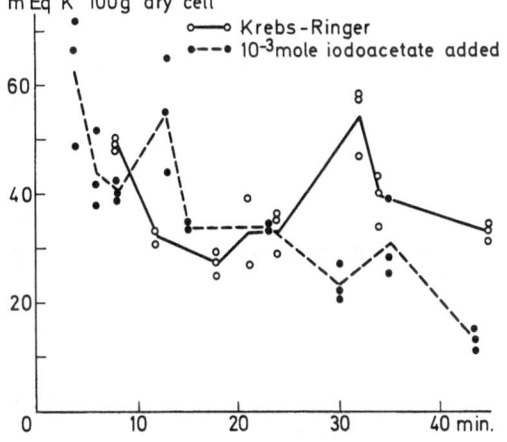

Fig. 6. The effect of iodoacetate on the reaccumulation of intracellular potassium

4 and 5). The overall variations for the values for the different pieces from the same cells were, however, higher by X-ray fluorescence analysis. The difference in sensitivity between the two methods probably contributes to this variation. The absolute potassium content of the cells analysed was usually in the 10^{-10} g range. Within this range the variations by X-ray fluorescence microanalyses for analyses of standard solutions were $\pm 2.95 \times 10^{-11}$ g [12], while the average variation for ultra-micro flame photometry was only $\pm 1.2\%$ [11].

When the average potassium value for each cell is plotted against the time under in vitro conditions, a curve for skeletal muscle electrolyte shifts taking place under the in vitro conditions is obtained. As shown in Fig. 5, the potassium content of the cells is high at the beginning, followed by a decrease. After 15 to 20 minutes the cells start to accumulate potassium again and after 25 to 30 minutes the potassium content was approximately as high as that of the originally isolated cells. The average values for cells isolated after the same time (dog B: 1a—1b, 3a—3b, 6a—6b, 7a—7b) were very similar with an average difference of 2.38 mEqK/100 g. This makes it possible to exclude that the obtained cellular electrolyte shifts are due to variations in cellular damage during isolation, since the same degree of cellular damage or contamination of each sample seems unlikely. Further proof of the active nature of these shifts could be obtained by adding different metabolic inhibitors to the dissection solution [13]. Iodoacetate 10^{-3} mole was added to the Krebs-Ringer solution to inhibit active glycolysis [17] and the contralateral muscle was dissected in ordinary Krebs-Ringer. The control dissection showed the same pattern of potassium shift as described (Fig. 6). With added iodoacetate, however, there was a continuous decrease of the potassium content with no accumulation of intracellular potassium at a later time period.

Conclusions

X-ray fluorescence is a non-destructive method and subsequent analyses of the tissue can be performed. Micro-flame photometry is a quicker and easier method for routine analyses of many samples of potassium and sodium. It is technically less complicated and the risk of technical breakdowns is small. If several elements together with sodium and potassium need to be analysed, the X-ray fluorescence method may be used, provided that the interference problems are carefully assessed. Other elements apart from potassium and sodium in or around a single cell are too low to be analysed accurately by X-ray fluorescence except for mineralized tissues or certain pathological samples. This difficulty could be overcome by the isolation of several cells, pooling them, and then extracting the element or elements under liquid paraffin as described above. It has been possible to carefully check the errors and judge the reliability of the information obtained by using both methods for analyses of the same tissue. The sensitivity of the

X-ray fluorescence method could be increased more by using optimal aperture systems in order to minimize the scattered radiation.

The possibility of obtaining microradiagrames at the same time as the microanalysis takes place is a promising approach under development in Cambridge.

References

1. MANERY, J. F.: Physiol. Rev. 34, 334—417 (1954).
2. CONWAY, E. J.: Physiol. Rev. 37, 84—132 (1957).
3. DARROW, D. C., and S. HELLERSTEIN: Physiol. Rev. 38, 114—137 (1958).
4. USSING, H. H., in: The alkali metal ions in biology. Handbuch der experimentellen Pharmakologie. Berlin-Göttingen-Heidelberg: Springer 1960.
5. REIFFEL, L., and C. A. STONE: J. Lab. clin. Med. 49, 286—291 (1957).
6. BERGSTRÖM, J.: Scand. J. Clin. Lab. Invest. 14, Suppl. 68 (1962).
7. LEVITT, M. F., and M. GANDINO: Amer. J. Physiol. 159, 67 (1949).
8. DEANE, N., and H. W. SMITH: J. clin. Invest. 31, 197—199 (1952).
9. THORN, N. A., in: The alkali metal ions in biology. Handbuch der experimentellen Pharmakologie. Berlin-Göttingen-Heidelberg: Springer 1960.
10. LONG, J. V. P., and H. O. E. RÖCKERT, in: X-ray optics and x-ray microanalysis. New York: Acad. Press 1963.
11. HALJAMÄE, H., and S. LARSSON: Chem. Instrum. (in press).
12. — Acta Chir. Scand. 133, 259—263 (1967).
13. — Unpublished reports.
14. ROSENGREN, B. H. O.: Acta Radiol., Suppl. 178 (1959).
15. KEESEY, J. C.: J. Neurochem. 15, 547—562 (1968).
16. PRAGER, D. J., R. L. BOWMAN, and G. G. VUREK: Science 147, 606 (1965).
17. WEBB, L. J., in: Enzyme and metabolic inhibitors, chapt. 1, vol. III. New York and London: Acad. Press 1966.

X-Ray Histochemistry Used for Simultaneous Demonstration of Neurones and Capillaries in the Human Brain

R. L. de C. H. Saunders, M. A. Bell, and V. R. Carvalho

Department of Anatomy, Sir Charles Tupper Medical Building, Dalhousie University, Halifax, Nova Scotia, Canada

Histochemistry has been found to extend the usefulness of X-ray microscopy, because many of the classical histochemical technics developed for optical microscopy deposit metal in the tissues. Modifications of these, and new technics, can be developed to produce radiopacity in the desired location. The added metal may increase a small amount naturally present, or introduce an entirely new element.

In this study, histochemical reactions were used to increase the contrast of nerve cells and capillaries in the human brain, to permit their study by contact and projection X-ray microscopy. The relations between nerve cells and blood vessels within the cortex of the brain are important because it is known that the cortical neurones cannot survive oxygen deprivation for 3 minutes or less. Owing to their great metabolic activity these neurones require a constant supply of oxygen and glucose.

Material and Preparation

Nineteen post mortem human brains were used, ranging from $7^1/_2$ foetal months to 90 years. Small blocks were removed from the motor cortex and fixed (15—24 hours) in a cold (4° C) formalin-saline or formalin-calcium fixative (10%). Sections of 35—100 microns in thickness were cut on a freezing microtome mounted on glass slides with a gelatin adhesive, air dried and stained. After staining, the sections were dehydrated, cleared and mounted with DPX mounting medium in the usual way for optical microscopy and photography. For X-ray study, the sections were then uncovered, filmed with DPX and removed from the glass slide with a razor blade [1]. X-ray microscopy was performed with a Philip's contact microradiography unit (CMR-5) and a Cosslett-Nixon projection x-ray microscope (XM 30). Both units were equipped with a vacuum camera. The contact unit, which has a tungsten target, was operated at 4 kV 2 mA, while the projection microscope was operated at 15 kV 50 μA using an 8 micron aluminium target. Micrographs were recorded on either Kodak high resolution (HR) or spectroscopic (0—649) plates, processed in either Kodak Dektol or Ilford Caustic Hydroquinone developer. After X-ray studies it was possible to return the sections to a glass slide and coverslip if desired for storage or further optical control studies.

Staining Methods and Results

Previous X-ray microscopy of the human brain following intravascular injection of a radiopaque such as Micropaque or Chromopaque, had shown the pattern of the small blood vessels that form the epicerebral and intracerebral circulations [2]. The larger cortical branches of the cerebral arteries and their fine pial branches give rise to numerous short cortical and long transcerebral arteries, which plunge from the brain surface into the substance of the underlying gyrus.

The short cortical arteries (Fig. 1) on penetrating the grey matter or cortex of the brain form a palisade of small vessels which supply its nerve cells or neurones. The longer transcerebral arteries course in graceful curves through the cortex, and traverse the subjacent white matter where they terminate in a periventricular plexus about the lateral ventricle [3]. Such is the density of the short cortical and transcerebral arteries that they give the appearance of a cascade.

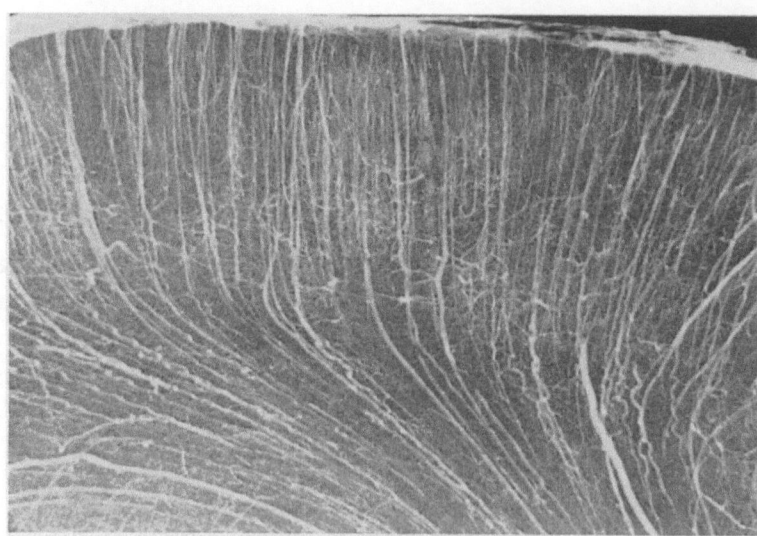

Fig. 1. Injected human cortex showing the short cortical arteries breaking up within the grey matter and transcerebral arteries entering the subjacent white matter. Micropaque injection. (Mag. × 20)

The slender, short cortical arteries terminate in a grapnel-like leash of pre-capillary arterioles
within the cortex. These arterioles spread outwards and upwards to supply the capillary bed
of the cortex. The cortical capillary bed is a continuous three-dimensional net, which in the
X-ray micrograph appears to enclose dark radiolucent spaces that contain the nerve cells.

Since microangiography showed the cortical capillary bed in this way, it seemed feasible
to add a picture of the stained neurones to the injected capillary network and thus demonstrate
both simultaneously. However, careful study and comparison of microtome sections, thick and
thin (2 mm to 10 microns) by optical and X-ray microscopy, revealed that the capillary bed
was seldom completely injected. This was not surprising since no injection method can guarantee
complete filling of a microcirculatory bed. It was therefore decided to selectively stain the cortical
capillary vessels specially for X-ray microscopy.

Small arteries and veins and the capillary vessels of the brain (Figs. 2 and 3) contain the
non-specific alkaline phosphatase enzyme in the endothelial lining of their walls (where it is
assumed to play a role in the transport of nutrients from the blood to the brain tissue); hence
to demonstrate these vessels a modification of the classic Gomori stain [4] for alkaline phosphatase
activity was used, resulting in the deposit of a black sulfide compound within the vessel walls.
Optical microscopy made it clear that stained brain material provided a more complete capillary
picture than injected material. The slight modification of the original method substituted a final
reaction product of lead sulfide for the more usual but less radiopaque cobalt sulfide so that
the sections could also be effectively studied by X-ray microscopy. The penetration and great
depth of field provided by the X-ray microscope permitted study of thicker sections; all the blood
vessels consequently remained in focus, so preserving the three dimensional character of the
cortical capillary bed.

X-ray micrographs (Figs. 4 and 5), when compared with microphotographs of the same
tissue areas, demonstrate that X-ray microscopy not only shows the same general vascular
features, but also accentuates the cortical capillaries so that they stand out in contrast with the
granular background of the brain tissue. Such micrographs clearly show how the short cortical
arteries terminate in the capillary bed. It was these X-ray micrographs that first drew our
attention to the additional horizontal pattern formed by the arterioles and venules as they
leave and join the larger vertical vessels that were emphasized previously by the injection
techniques.

Because the small arteries and veins exhibit a different degree of phosphatase activity in
the brain there is a difference in their radiopacity. The arteries stain more darkly and conse-

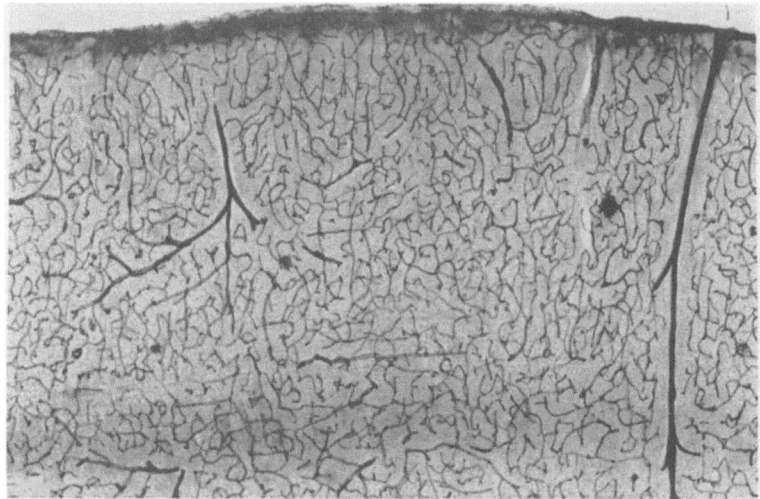

Fig. 2

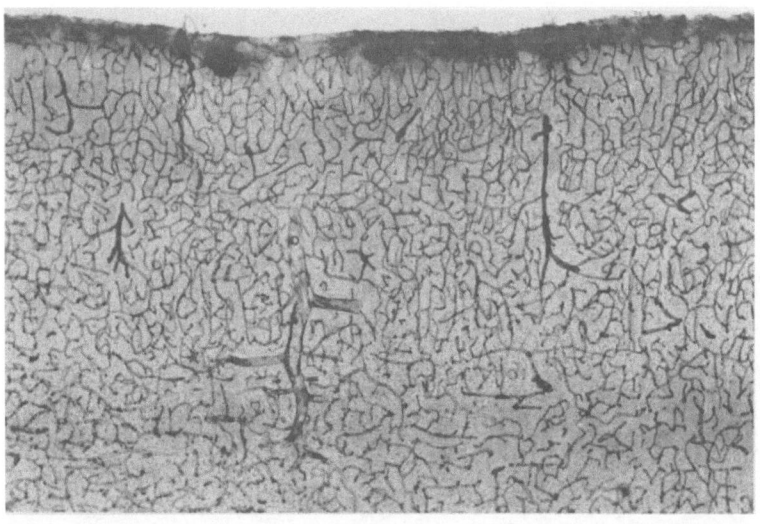

Fig. 3

Figs. 2 and 3. Microphotographs of human cortex demonstrating the intracortical arteries, veins and capillaries as stained by the alkaline phosphatase reaction. Note that the veins appear paler than arteries of corresponding size. (Mag. × 50)

quently are more X-ray absorbent. Hence they can be distinguished from the fainter, less X-ray absorbent, veins. The radiopacity difference between artery and vein is evident in both Figs. 4 and 5. Fig. 6 shows that the intensity of the phosphatase reaction likewise lessens in the venous terminals of the capillary bed, which accordingly are fainter on X-ray.

X-ray micrographs, such as Figs. 7 and 8, show the completeness with which the finest capillaries in the human cortex can be demonstrated by X-ray microscopy following phosphatase staining. They also show the nature of the cortical capillary bed and as a corollary that of the neuronal areas.

After obtaining satisfactory X-ray visualisation of the cortical capillaries, we turned to the problem of adding the nerve cell population to the picture. The majority of nerve cells contain the enzyme acid phosphatase, presumably associated with lysosomal systems for intracellular digestion, autolysis and pigment accumulation [5]. Therefore, a second technique, also developed

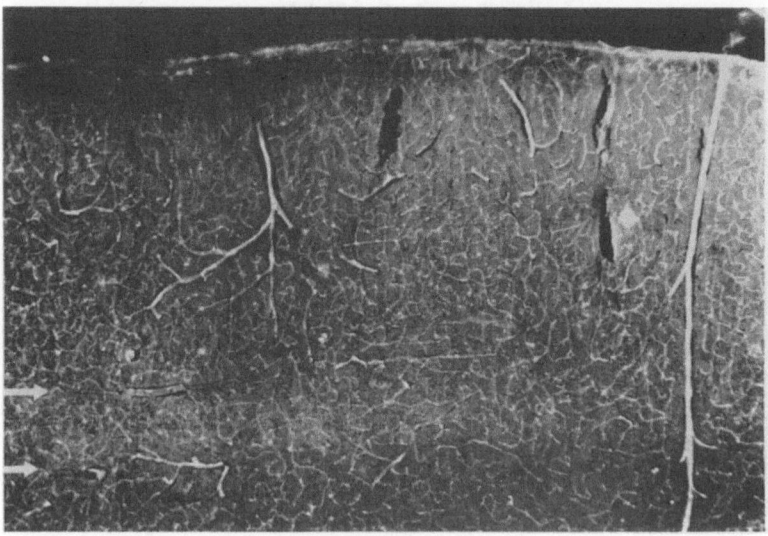

Fig. 4

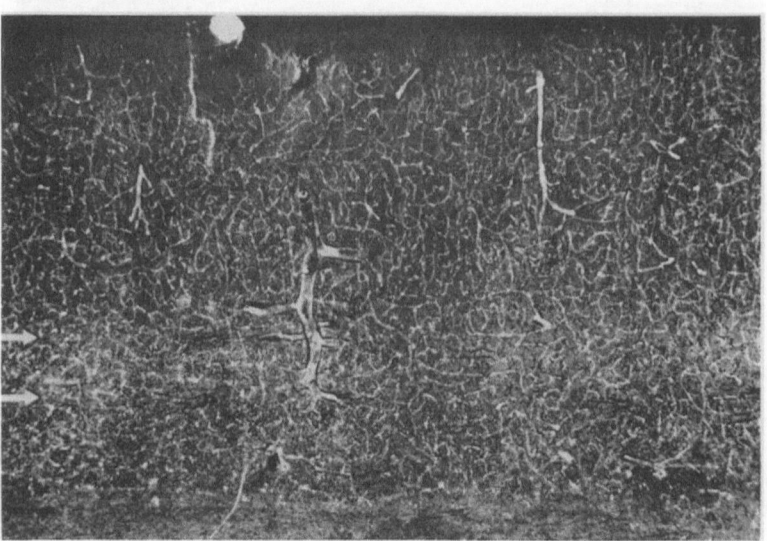

Fig. 5

Figs. 4 and 5. X-ray micrographs showing the same fields as in Figs. 2 and 3. Note the main vertical vessels, the definite horizontal orientation of the arterioles and venules deep in the cortex (in line with the arrows), and the conspicuous cortical capillaries. (Mag. × 50)

by Gomori [6] and also resulting in the deposition of lead sulfide at the site of activity of this enzyme, was used to produce an image of the nerve cells alone. It should be noted that in all our applications of these enzyme techniques we have overstained by using incubation times ranging from 3 to 18 hours in order to obtain sufficient metal deposit to produce a useful X-ray image; in this way we have sacrificed the delicate localisation of enzyme activity for which the techniques were originally developed.

An X-ray micrograph (Fig. 9) of a single transected gyrus (approx. 1 cm in width) stained for acid phosphatase illustrates the density of the nerve cells or neurones in the human cortex, estimated to be about 50,000 per sq.mm [7]. The human cortex is about 3 mm thick. It is immediately evident that the elementary organisation of the human motor cortex consists of

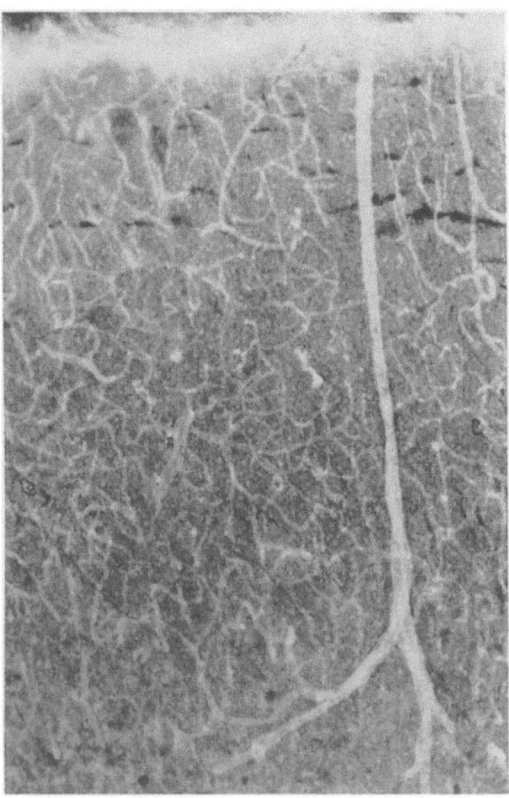

Fig. 6. Human cortex showing a branching short cortical artery. Compare the X-ray absorption of this artery with that of capillaries draining into a venous fragment at left centre. Alkaline phosphatase reaction. (Mag. × 130)

columns of nerve cells oriented vertically to the cortical surface. This vertical organisation of the nerve cells forms a beautiful radiation which we have named the aurora gyralis. The radiolucent spaces between the columns of nerve cells are now known to be largely occupied by vessels, but of course are also shared by nerve cell fibres and glial cells.

This X-ray evidence of vertical organisation contrasts with traditional teaching of a horizontal organisation of the neurones into layers or cortical laminae. It brings strong morphological support to the concept that the cortical neurones are basically and functionally organized in vertical patterns at right angles to the neurone layers. This concept arose from microelectrode studies performed on the cortex of the cat [8].

Projection studies of the motor cortex at higher power (Fig. 10) clearly demonstrated that in thickness the individual columns of neurones consisted of one or more pyramidal nerve cells. Many of the pyramidal cells seen here show apical dendrites that taper from the top of the cell body into fine processes.

After first recording the cortical blood vessels, and then the nerve cells alone, the next step taken was to evoke the acid and alkaline phosphatase reactions sequentially in that order in the same brain section, so that the vessels and neurones could be studied together. Details of the modifications required to combine the 2 technics will appear in a forthcoming thesis [1].

The necessarily prolonged staining procedure both darkened and increased the density of the brain sections. This consequently reduced their value for optical microscopy, making it difficult to obtain satisfactory black and white or colour microphotographs. Microphotographs of brain sections stained by both phosphatase methods therefore showed a lack of contrast between the blood vessels and nerve cell columns, although the latter were obvious.

On the other hand, the same area of the same section on examination by X-ray microscopy showed marked contrast between the patterns of blood vessels and nerve cells (Fig. 11). The

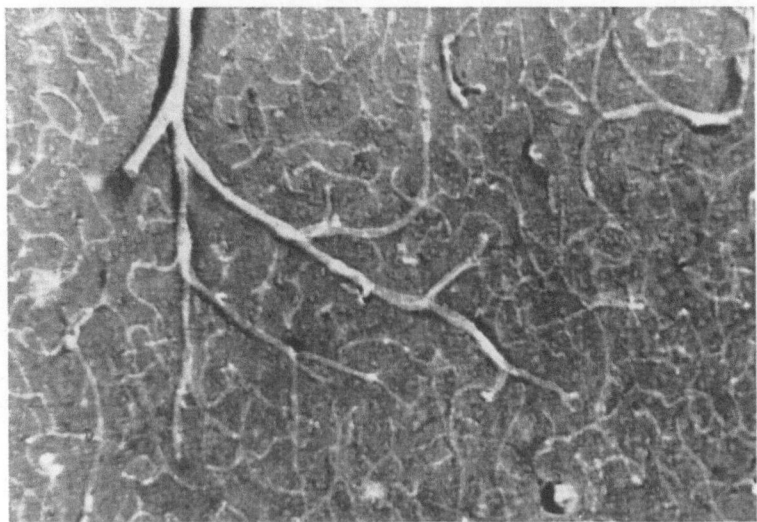

Fig. 7

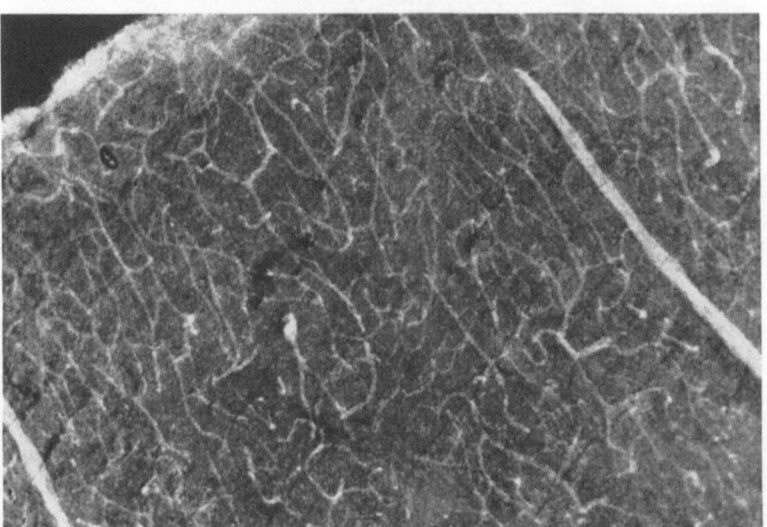

Fig. 8

Figs. 7 and 8. X-ray micrographs of human cortex showing the capillary bed and neuronal areas. Alkaline phosphatase reaction. (Mag. × 130)

nerve cell columns are surrounded by the cortical capillary net, which somewhat obscures their columnar pattern. The close relationship that exists between the nerve cells and capillaries should be noted.

X-ray projection micrographs at higher power (Figs. 12 and 13) demonstrate the clarity with which nerve cells and capillaries can be recorded by this technique. They illustrate the intimate nerve cell — capillary or neurono — vascular relationship. Since the brain capillaries measure about 5 microns in diameter, it is evident that the cortical nerve cells lie within a 30 micron radius of a capillary vessel, often so close as to appear in contact with it. Fig. 12 shows a large pyramidal nerve cell lying within this radius of a curved capillary vessel.

This combined technique also provided information about the nerve cells and their processes. X-ray micrographs, such as Fig. 13, show short processes, known as basal dendrites, arising

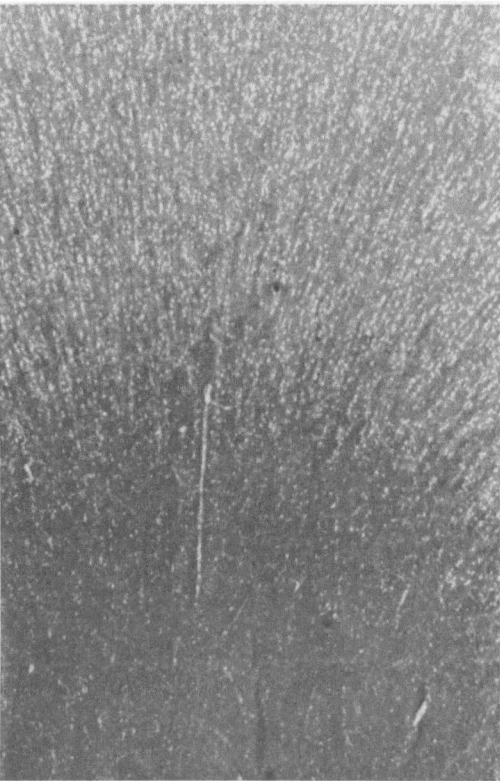

Fig. 9. X-ray micrograph of a transected human gyrus showing the radiating columns of neurones that form the aurora gyralis. Acid phosphatase reaction. (Mag. × 40)

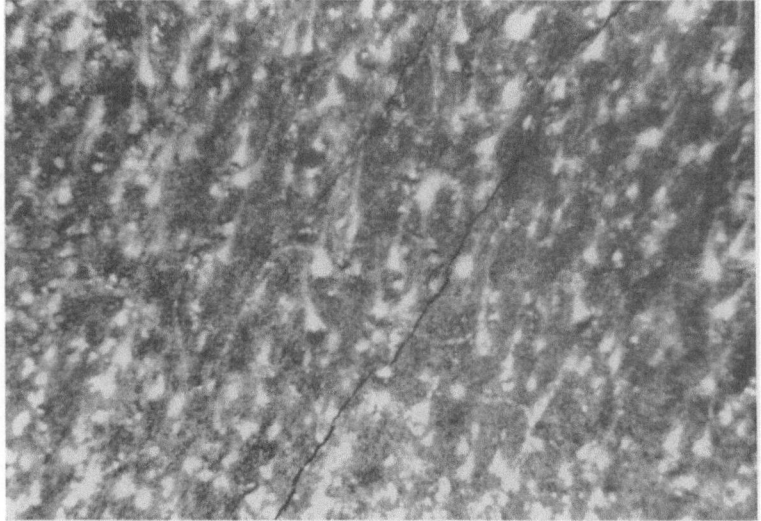

Fig. 10. High power view of the neuronal columns seen in Fig. 9. Acid phosphatase reaction. X-ray micrograph. (Mag. × 250)

from the large pyramidal cells. The tops of other pyramidal cells show the long tapering processes known as apical dendrites. A vertical organisation of nerve cell fibres is evident in some of the nerve cell columns. Surrounding the nerve cells can be seen neuroglial nuclei, which appear as white spots.

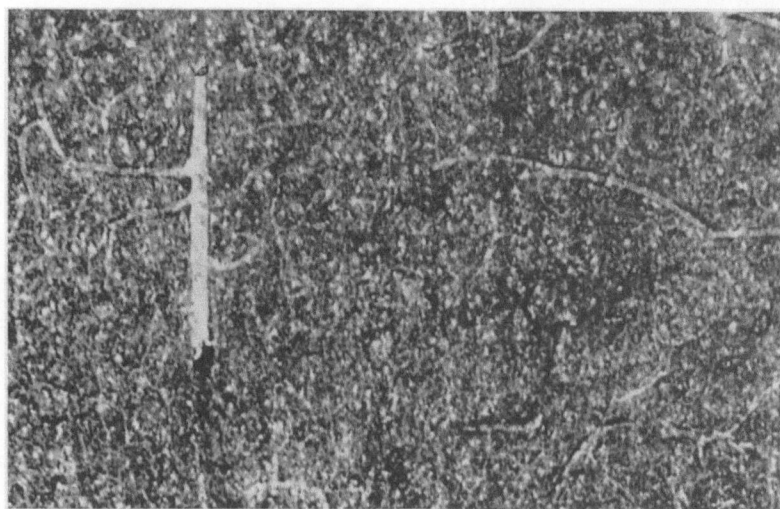

Fig. 11. X-ray micrograph of human cortex following combined staining for acid and alkaline phosphatase showing nerve cell columns, cortical artery and capillary bed. (Mag. × 96)

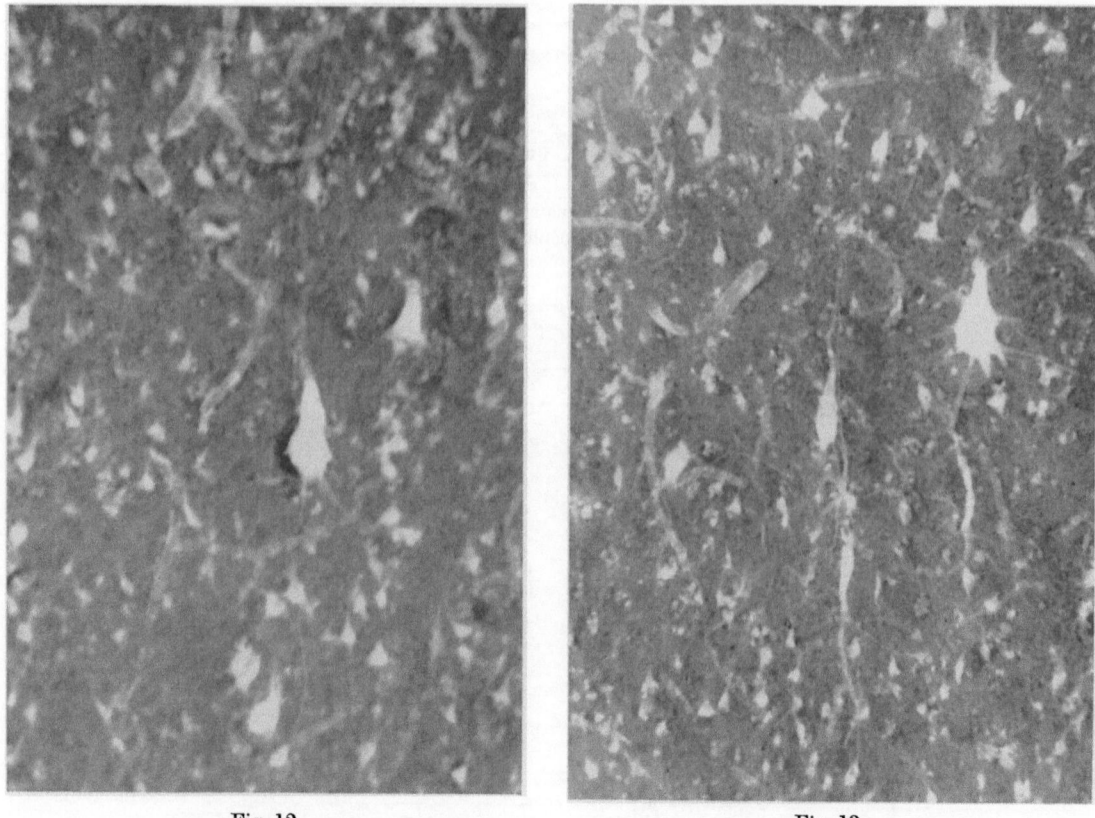

Fig. 12   Fig. 13

Figs. 12 and 13. X-ray micrographs, following combined staining for acid and alkaline phosphatase, showing neurono-vascular relationships. Note that many cell processes can be traced. (Mag. × 250)

Microphotography (Fig. 14) of a large, phosphatase stained pyramidal nerve cell shows its tapering apical branch or dendrite arising from the top of the cell. At the bottom of the cell emerges the long slender nerve cell fibre or axon process. The enlargement of this process after an initial thin segment may be related to the beginning of a protective myelin sheath.

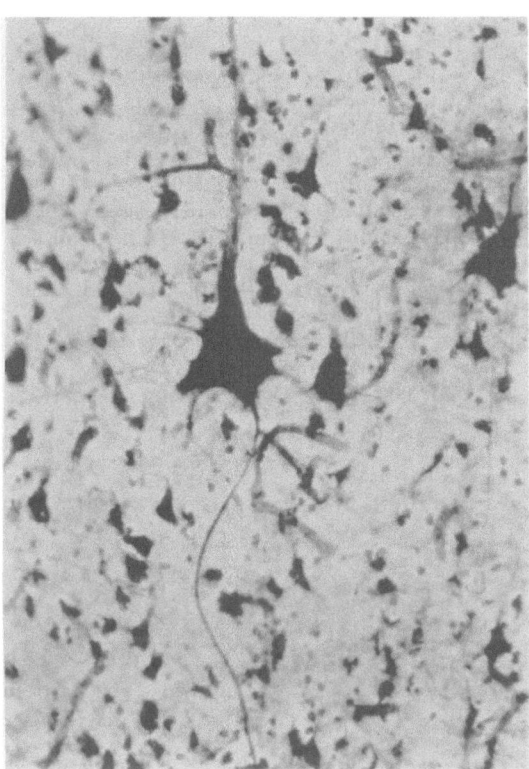 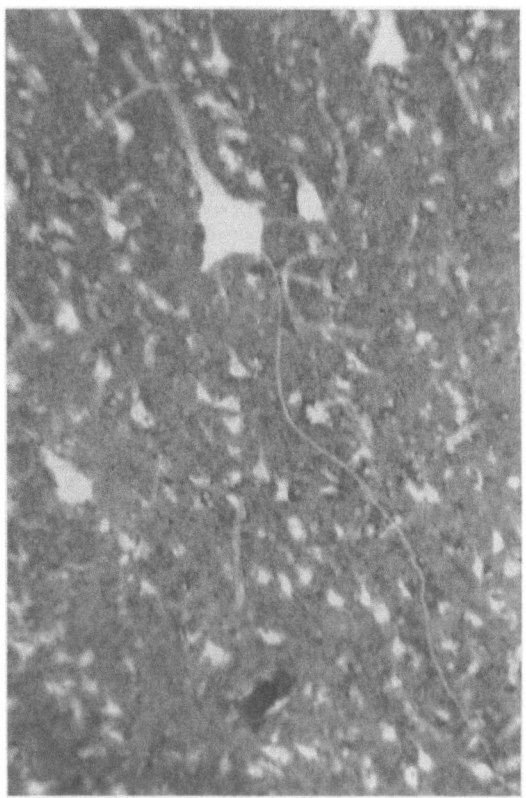

Fig. 14   Fig. 15

Fig. 14. Microphotograph of a large pyramidal nerve cell showing the apical dendrite crossing a forked capillary, and the long axonal process showing an increase in diameter that may be related to its myelin sheath. Acid and alkaline phosphatase reactions. (Mag. × 290)

Fig. 15. X-ray micrograph of the large pyramidal nerve cell shown in Fig. 14. Acid and alkaline phosphatase reactions. (Mag. × 250)

A high power X-ray projection micrograph (Fig. 15) of the same pyramidal cell shows the same features, and even on X-ray it is possible to identify the point where the axon process increases in diameter. Such processes may reach 1 meter in length and have 1,000 times the bulk of the cell body, and they presumably derive their nutrition from the capillary network of the white matter. The large cell (80 microns across) was situated in the paracentral lobule of the human brain, or the uppermost end of the motor cortex.

Conclusions

It will be apparent that X-ray microscopy combined with histochemical technics provides a new approach to the study of vascular and cellular organisation. In the human brain it has 1. demonstrated the pattern of the intra-cortical vessels and revealed both vertical and horizontal components; 2. strikingly demonstrated the vertical organisation of the cortical neurones, thereby bringing strong morphological support to modern physiological concepts; 3. demonstrated simultaneously both the capillaries and neurones of the cortex, thereby throwing light on the nature of the important neurono-vascular relationships.

Acknowledgements. We wish to thank Miss JEAN McK. MORRISON of Edinburgh, Scotland for library research, and Dr. N. KERENYI of the Pathological Institute, Dalhousie University, for providing autopsy material. Thanks are also due Miss LESLEY GENTRY and Mrs. AUDREY JACKSON for secretarial assistance, and Mrs. PAT TROYER for expert technical assistance.

References

1. BELL, MARY A.: Thesis leading to the degree of master of science, Dalhousie University, Halifax, Nova Scotia, 1969.
2. SAUNDERS, R. L. DE C. H., W. H. FEINDEL, and V. R. CARVALHO: X-ray microscopy of the blood vessels of the human brain. Medical and biological illustration. Parts 1 and 2, vol. XV, p. 108—122 and 234—246 (1965).
3. — Microangiography of the brain and spinal cord. In: X-ray microscopy and X-ray microanalysis, p. 244—256. Amsterdam: Elsevier 1960.
4. GOMORI, G.: Cit. by T. BARKA and P. J. ANDERSON, Histochemistry: Theory, practice and bibliography, p. 214—226. New York: Hoeber Medical Division, Harper and Row 1963.
5. ANDERSON, P. J., and S. K. SONG: Acid phosphatase in the nervous system. J. Neuropathol. Exptl. Neurol. 21, 263—283 (1962).
6. GOMORI, G.: Cit. by J. F. A. McMANUS and R. W. MOWRY, Staining methods: Histologic and histochemical, p. 162—164. New York: Hoeber International Reprint, Harper and Row 1960.
7. ECCLES, SIR JOHN: The neurophysiological basis of mind. In: Human machine mechanisms. The living universe, an encyclopedia of the biological sciences, p. 75—81. New York: Thomas Nelson 1965.
8. NOBACK, C. R.: The human nervous system, p. 232. New York: McGraw-Hill 1967.

Histochimie par spectrographie des RX

A. DUPREZ

Laboratoire d'Anatomie Pathologique Faculté de Médecine, Nancy, France

La détection par les méthodes histochimiques usuelles des éléments contenus dans les structures histologiques n'est possible que dans quelques cas particuliers. Le plus souvent, les atomes font partie intégrante des molécules biologiques et perdent ainsi leurs propriétés chimiques caractéristiques. Celles-ci ne peuvent être retrouvées que par une analyse élémentaire destructive presque toujours dénuée d'intérêt pour les études topographiques que sont les déterminations histochimiques. Dans cette perspective on conçoit que l'histochimie élémentaire soit encore rudimentaire.

Pour être efficaces et générales, ces caractérisations et les mesures devraient faire appel à des techniques non destructives dont la spécificité de détection ne soit pas liée à la réactivité chimique de l'élément à analyser. La spectrographie des RX répond à ces deux critères fondamentaux et devrait dans l'avenir être une méthode de choix en ce domaine. Dans cette communication nous voudrions simplement montrer comment nous utilisons trois techniques de spectrographie X en histopathologie humaine courante. Mais l'étude critique de ces méthodes doit d'abord situer le système de références de ces analyses.

Les references de la quantification en histochimie

Pour comparer les teneurs, la référence en poids de la zone dans laquelle la mesure d'un atome a été réalisée, semble à priori le système le plus simple. En réalité, les masses explorées sont si petites que toute détermination pondérale directe est irréalisable. Mais s'il en était autrement, il serait encore nécessaire de corriger ce poids par rapport à l'épaisseur qui conditionne l'auto-absorption.

L'utilisation du volume est en pratique beaucoup plus aisée puisque l'épaisseur de la coupe est connue et assez constante et que la surface est toujours planimétrable quelle que soit la petitesse de la zone explorée. Mais la densité des diverses structures histologiques et leur proportion varient d'un tissu à l'autre et pour un même tissu selon qu'il est normal ou pathologique. Le volume n'a donc qu'une signification mineure pour la quantification à l'échelle histologique.

En spectrographie X, le meilleur système de référence nous parait être la mesure dans la même zone que l'atome étudié, d'un autre élément représentatif de tout tissu. Parmi les 4 éléments dont la présence est ubiquitaire: H, C, N, O, les trois derniers sont détectables. L'azote ne peut toutefois servir de référence que si l'atome étudié a partie liée avec les molécules protéïques ou les acides nucléïques seuls détenteurs d'azote dans leur structure. Le carbone et l'oxygène se retrouvent, en revanche, dans les trois familles biochimiques des divers tissus. Notre choix s'est porté sur le carbone mais des mesures comparatives nombreuses seront nécessaires avant de le préférer définitivement. Quoiqu'il en soit, le principe de chiffrer les analyses élémentaires des tissus par rapport à l'un de ses atomes parait être la meilleure solution au problème de l'expression des résultats en histopathologie.

Methodes

Nous avons fait appel à deux types de sources d'émission X: la fluorescence X et l'émission primaire X (macro et microsondes). Les divers appareils que nous avons utilisés sont tous des dispositifs commercialisés: fluorescence X Philips P.W. 1540 — macro-sonde J.P. X₃ Jéol — micro-sonde Cameca.

A. La fluorescence X

Technique. Des coupes à la paraffine de 7,5 μ sont étalées, collées sur du mylar de 6 μ d'épaisseur, puis déparaffinées, réhydratées et enfin soigneusement séchées. L'analyse sous vide est alors entreprise par balayage ou enregistrement d'une raie particulière. Les préparations sont ensuite colorées (le mylar restant incolore) puis montées entre lame et lamelle.

Une autre modalité encore plus simple consiste en un avivement au microtome du bloc de paraffine jusqu'à ce que le tissu apparaisse. Sans autre préparation, le bloc est alors analysé. Un bloc témoin de paraffine sert de blanc. L'analyse d'autres blocs de tissu sain homologue permet d'établir des comparaisons. Enfin, la première coupe est montée et observée.

Avec l'un ou l'autre de ces procédés, nous avons détecté [1, 2] du fer dans le foie (hémochromatose) du cuivre dans le foie et le système nerveux (maladie de Wilson), de la silice dans le poumon (silicose pulmonaire), du calcium et du soufre dans de l'os et le cartilage (ostéochondrome), de l'iode dans la thyroïde (adénome), du phosphore et du potassium dans tous les tissus analysés.

Critique. La simplicité et la rapidité de détection sont des avantages auxquels s'ajoutent une bonne sensibilité et une excellente spécificité. Comme l'intensité des excitations restent toutefois modeste avec les dispositifs conventionnels, on ne peut analyser efficacement que d'assez grandes surfaces (5 à 10 mn de diamètre). Ceci exclut le plus souvent les analyses sur prélevements biopsiques. Mais l'inconvénient majeure réside dans l'impossibilité actuelle de détecter les éléments légers essentiels en biologie comme système de références. Des tubes sans fenêtre feront peut-être évoluer ce problème. Actuellement, les domaines d'application de la fluorescence X en histopathologie se limitent aux études systémiques d'éléments supérieurs au numéro 12 répartis sur d'assez grandes surfaces c'est-à-dire provenant de pièces d'exérèse ou d'autopsie. Quant à l'aspect quantitatif, il se cantonne dans des comparaisons avec un tissu sain homologue.

B. La macro-sonde électronique

Technique. Des coupes à la paraffine de 5 ou 7,5 μ d'épaisseur sont étalées puis collées selon la technique ordinaire sur un bloc d'aluminium pur (99,99%) parfaitement plan et rendu réfléchissant par un soigneux polissage métallurgique. Après séchage, les coupes sont déparaffinées, réhydratées puis à nouveau séchées. Sans métallisation, les coupes sont introduites dans l'appareil. Après l'analyse, les préparations sont colorées, montées sous une résine (eukitt) et observées ou photographiées à l'aide d'un microscope par réflexion.

Avec cette technique, nous avons détecté du fer et du cuivre dans le foie (hémochromatose et maladie de Wilson), de la silice (silicose pulmonaire) mais surtout du carbone dans divers tissus.

Critique. Cette technique simple et très spécifique permet l'étude précise de toutes les surfaces entre 0,5 et 16 mn de diamètre. Cette échelle de dimension a l'avantage d'être utilisable au niveau de tous les prélevements biopsiques ou non, donc applicable à la routine. De plus et surtout, la détection des éléments légers permettra l'élaboration d'un système de références satisfaisant. Le premier inconvénient réside dans l'échauffement qui détermine des brulures du tissu. On peut toutefois le minimiser dans de nombreux cas en réduisant l'intensité du courant sonde ou en diminuant le temps d'analyse. Des dispositifs de refroidissement sont en voie d'élaboration. Le second facteur défavorable est constitué par la nécessité d'observer par réflexion la coupe, ce qui est esthétiquement moins agréable pour le morphologiste. Mais l'utilisation d'un support en quartz avec évaporation d'or devrait permettre de revenir à l'observation par transmission. La macro-sonde parait être actuellement l'appareil le plus satisfaisant pour résoudre la majorité des problèmes d'analyse élémentaire en histopathologie humaine.

C. La micro-sonde électronique

Technique. Les préparations sont soit des frottis sanguins [3] réalisés selon la technique usuelle mais sur aluminium et séchés très rapidement, soit des coupes histologiques de 5 µ ou 7,5 µ montées suivant le procédé envisagé pour la macro-sonde. Les détections sont faites point à point avec comptage ou enregistrements graphiques ou balayage suivant un axe. Dans ce dernier cas, la préparation est déplacée à faible vitesse (10 à 20 µ/mn) alors que le papier se déroule à une vitesse moyenne (quelques cm/mn). Après la mesure, les coupes sont colorées et montées. Le trait de sonde, de teinte brunâtre est bien repérable sur les microphotographies prises par réflexion. Les enregistrements sont photographiés. Les deux séries de photos sont enfin agrandies dans le même rapport et mises en contact pour objectiver les corrélations morphoquantitatives [4].

Avec ces méthodes, nous avons étudié la répartition du phosphore nucléaire de quelques tissus, les variations du potassium dans les globules rouges humains au cours du «lavage des hématies», le fer contenu dans certaines mélanines humaines. Enfin, nous avons exploré la répartition du fer dans l'intestin grêle en utilisant la technique du balayage.

Critique. Cette technique d'une extrême finesse topographique et quantitative est à la macro-sonde ce que le microscope électronique est à la microscopie optique. La détermination point par point a un intérêt considérable à l'échelle cytologique et en particulier pour l'étude des pertubations ioniques mais nécessitent de nombreuses mesures pour être significative. Le balayage technique beaucoup plus simple car automatique, fournit des renseignements morpho-quantitatifs extrêmement intéressants pour l'étude de la répartition dans les tissus. Le trait de sonde est pour l'histologiste un facteur favorable puisqu'il permet un repérage aisé du lieu d'analyse sans pour autant gêner l'observation. La brulure ne semble pas, du moins dans les cas que nous avons étudiés, modifier les comptages puisque nos séries de mesures en un même point furent toujours stables quel que soit le temps d'exposition aux électrons. De toute façon, si un déplacement de matières se produit, son importance devient biologiquement négligeable lorsqu'on réalise des analyses simultanées de deux atomes dont un est le système de référence.

L'utilisation de ces diverses techniques et d'autres encore plus élégantes [5—8] dont nous n'avons pas l'expérience devraient faire de la spectrographie X une méthode de premier plan pour les analyses élémentaires en histologie normale et pathologique.

Bibliographie

1. DUPREZ, A., A. WITTMANN, G. RAUBER et P. FLORENTIN: Compt. Rend. Soc. Biol. 161, 871—874 (1967).
2. KISSEL, P., J. SCHMITT et A. DUPREZ: Rev. neurol. 118, 379—386 (1968).
3. DUPREZ, A., et A. VIGNES: Compt. Rend. Soc. Biol. 161, 1358—1360 (1967).
4. —, et G. RAUBER: Ann. Anat. Pathol. 13, 373—378 (1968).
5. GALLE, P.: Rev. Franc. Etudes Clin. Biol. 7, 1084—1086 (1962).
6. HALL, T. A., J. A. HALE, and V. R. SWITSUR: In electron microprobe, p. 805—833. New York: J. Wiley 1966.
7. LONG, J. V., and H. O. RÖCKERT: In III. internat. symposium on X ray optics and X ray microprobe, p. 513—521. Stanford: Academic Press 1963.
8. TOUSIMIS, A. J.: In III. internat. symposium X ray optics and X ray microprobe, p. 539—557. Stanford: Academic Press 1963.

The Application of Microprobe Analysis to Biology

T. A. HALL

Cavendish Laboratory, Cambridge, England

H. J. HÖHLING

Institut für medizinische Physik der Universität Münster, BRD

Microprobes have been used for biological research to an extent which may surprise this audience; it certainly surprised the speaker when he compiled a bibliography. The number of publications amounted to 82. Furthermore, most of the papers now report actual biological research instead of the evaluation of possibilities which prevailed in the early days.

In most of the published work, the instrumental techniques are quite familiar. Thus a wide field of biological work requires no special technology. Since this audience is presumably interested at present in the instrumental problems posed by biological applications, we shall give little attention to these relatively straightforward studies. But to ignore them completely would distort the description and leave the impression that the field is dominated by technical problems, so we shall start with a survey of some applications. They may be classed as follows:

1. The Analysis of Minute Foreign Bodies or Particles. Lung is the most common site, and a good example is a study of asbestos fibres in the lung reported at this conference [1].

2. The Localisation and Analysis of Abnormal Elemental Concentrations Built up by Ingestion. Such accumulations occur often in the kidney, and especially rewarding studies have been carried out with a combination transmission electron microscope-microanalyser [2].

3. The Localisation and Analysis of Abnormal Elemental Concentrations Caused by Metabolic Defects. Examples are the accumulation of copper in Wilson's disease [3] and of iron in the liver in haemachromatosis and haemosiderosis [4, 5]. A recent report [6] of abnormal concentrations of zinc and titanium in leucocytes in leukemia may belong in this category.

4. The Analysis of Calcium and Phosphorus and Also the Localisation and Analysis of other Elements in Bone and Teeth. This field is very important medically and is the subject of a great deal of probe work [7—17]. Fortunately the tissues are relatively tough and the concentrations of the interesting elements are generally fairly high. The technical problems are much like those in mineralogical work.

5. The Confirmation of the Performance of Histochemical Stains. The probe can sometimes establish directly the degree of correspondence between a visible stain and the tissue-component which it is intended to display. Examples are the correlation of the visible compound cobalt sulphide with the target of the stain, namely phosphate released by the enzyme alkaline phosphatase [18], the correlation of chromium in the stain chrome gallocyanin with the target in the tissue, nucleic acid [19]; and the correlation of the silver of the von-Kossa stain with the target element calcium.

Fig. 1 shows silver and calcium signals recorded during a line-scan through a mineralising band in von-Kossa stained turkey epiphyseal cartilage [20]. The silver was well localised to calcium, but deposited only at high concentrations of calcium (well mineralised zones). The lack of staining at lower concentrations, observable in Fig. 1, can be documented more fully by static-probe measurements.

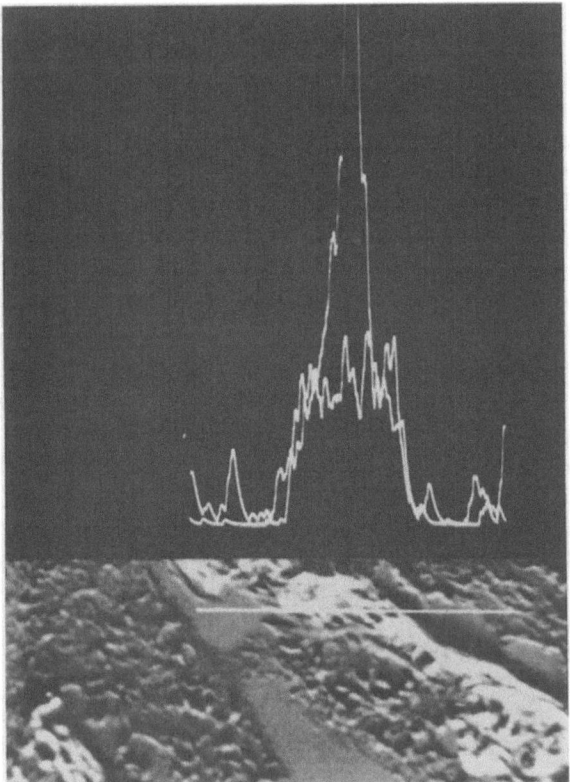

Fig. 1. Intensities of calcium (higher peak) and silver (flatter peak) along a line scan in von-Kossa stained mineralising cartilage. Image by backscattered electrons

While some studies in each of the above five categories have been carried out successfully without concern for instrumental problems associated especially with biological tissues, for an even wider range of interesting work instrumental problems are acute and decisive. We shall now consider these problems.

1. Damage to the Specimen Inflicted by the Probe. "Soft" (i.e., non-calcified) biological tissues not only conduct heat poorly, but in comparison with most probe specimens they are much less able to tolerate elevated temperatures. While conventional evaporated coatings of 100 or 200 Å conduct heat away to a useful extent, the maximum safe probe current is nevertheless often in the range 20—50 nanoamperes. Fortunately modern probes are so sensitive that higher currents are usually not required, but they may be necessary when one must deal with low elemental concentrations in thin specimens. While it may not help much to increase the thickness of the evaporated coating on a thick specimen, for thin specimens (100—600 μg/cm²) it has been shown that currents up to 1 microampere may be tolerated if the specimen is coated on both sides with aluminium up to a total thickness of 1,000 Å [21].

Among the embedding media commonly used to prepare tissue sections, paraffin has been found unsatisfactory under the probe (it usually warps and flows), while hard media like the epoxy resins are stable. (Of course, paraffin-embedded preparations can be deparaffinized before analysis.)

Another problem in thick specimens is the possible *displacement* of ions, presumably by electric fields produced by accumulated charge [22]. This danger also should be less in thin specimens, where the ions cannot migrate to a depth beyond the reach of the probe.

The obvious way to test for probe damage is to see if the X-ray counting rates are steady in time. According to this test, the problem of probe damage is not often worrisome, although quite thick conducting coatings are sometimes necessary.

2. Spatial Resolution. The density of specimens of soft tissue varies from about 0.1 g/cc for certain tissues after freeze-drying to somewhat more than 1 g/cc for embedded tissues. The electron stopping-power is consequently so low that, if spatial resolution of the order of one micron is to be preserved in thick specimens, the probe voltage E_0 must not exceed the ionization energy E_x of the analysed element by more than several kilovolts. But low overvoltage ratios E_0/E_x produce low X-ray intensities, and also widely disparate excited volumes when several elements must be studied. The best procedure for thick specimens, as stressed by ANDERSEN [10, 23], is to use low probe voltages and to observe only the efficiently excited soft radiations, switching from K to L radiations in the neighbourhood of atomic number 23. In this way, you can have both a good overvoltage ratio E_0/E_x and a small overvoltage difference $(E_0 - E_x)$.

The alternative means of preserving spatial resolution is to study thin specimens, such as tissues sections or cell smears a few microns or less in thickness. Such preparations are anyhow the natural objects of study in many areas of biology. For probe voltages above 30 kV, the lateral spread of the beam within thin specimens is usually unimportant. However, it is essential to mount the specimens on very thin supports to prevent excitation over a wide area by backscattered electrons.

In the analysis of *thin* specimens, there seems to be little advantage to using low kilovoltages and soft radiations [24]. In ANDERSEN's tabulations of minimal measurable amounts [10, 23], operation at low kilovoltages seems superior but this is because, in a thick specimen, more energetic electrons would spread over a larger volume containing more of the element. Thus his tabulation does not apply to thin specimens.

3. Histological Correlation. One needs an image of the specimen, either optical or formed by electrons, showing enough structure to enable one to position the probe as desired. Unfortunately, contrast in soft tissues is intrinsically very weak because there are only slight variations in atomic number, and cellular structures do not vary much in density. This is why one must resort to stains in the optical microscopy of dry tissue sections, and why electron images of soft tissues usually reveal merely boundaries where there are abrupt changes in density. The difficulty is fundamental and applies equally to scanning images formed by backscattered electrons, by secondary electrons or by current retained in the specimen. The identification of cytological entities like nuclei or mitochondria is usually difficult or impossible in unstained sections in the probe. (However, in some tissues nuclei may stand out because of what seems like a difference in surface *texture*, a difference which may be accentuated in preparative procedures like frozen-fracturing followed by freeze-etching.)

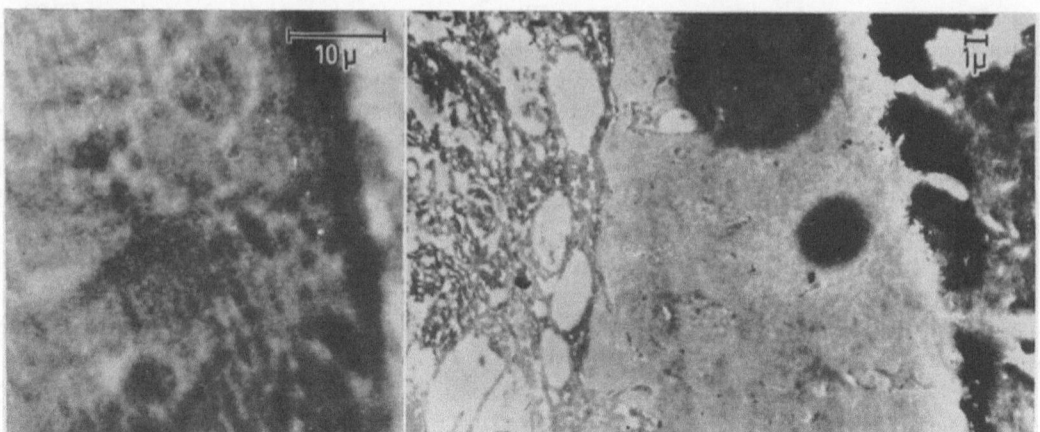

Fig. 2. Micrographs of unstained sections of embryonic rat incisor. Left: Scan image, backscattered electrons, 6-μm section of frozen tissue. Right: Transmission electron image, 1,000-Å section. The mineralising border is at the right of each image; it appears dark by transmission and bright by backscattering. The dark circles in the pre-dentine of the transmission image are contamination deposited during static-probe measurements

Of course, one can resort to stains if necessary. In doing this, one may have to renounce one of the important potentials of the microprobe technique: the possibility of chemical analysis *in situ* without the risk of displacing elements in the preparative procedure. We shall return to this point later. However there is no doubt that stains can be used satisfactorily for some studies at least without putting the results in question.

There remains one method of image formation possessing superior sensitivity to slight gradations within a specimen and hence providing superior contrast, namely transmission electron microscopy. Fig. 2 shows the difference between a scanning electron micrograph of an unstained 6-μm section of a rat incisor tooth in an embryonic stage, and a transmission electron micrograph of an unstained ultrathin section of the same tissue. While the mineralising dentinal border is obvious in both images, the band of adjacent cell-free matrix can be reliably distinguished from the inner odontoblast cells only in the transmission image, where the distinction is quite plain. The possibility of improved histological correlation through higher contrast is probably the main contribution offered to biology by combination transmission electron microscope-microanalysers.

4. Quantitation. It should be pointed out first that very accurate quantitation is not often needed in biological probe work. Even when precise analyses are available, it may be difficult to exploit a high degree of precision in drawing conclusions, because phases are rarely encountered in a pure state. For example, consider bone, a relatively well characterised and simple tissue. Here one might want to distinguish between the crystalline substances apatite (with a calcium/ phosphorus atomic ratio of 10/6) and octacalcium phosphate (ratio 8/6). But bone contains a considerable amount of calcium and phosphorus as amorphous calcium phosphate and in other forms with unknown ratios Ca/P, so that the local measurement of the ratio Ca/P, no matter how precise, cannot determine in itself whether the crystalline substance is apatite or octacalcium phosphate.

But with the question of precise analysis set aside, even relatively crude quantitation is complicated by another feature typical of biological tissues, an inhomogeneous distribution of dry mass per unit volume. The resulting problems are:

a) Dry mass per unit volume is in itself one of the most important things to know about the structure of a tissue. One wants to quantitate not only in terms of weight-fractions, which may be measureable by means of the conventional theory of quantitation, but also in terms of amounts per unit volume.

b) If a tissue is merely dried, even when it is essentially homogeneous in atomic number there may be such variation in local dry mass per unit volume that absorption corrections cannot be made reliably in thick specimens.

c) If the dry tissue is infiltrated with material of matched absorption coefficient [11], one can correct for absorption, but the price is that the measured weight-fractions refer to the entire mass, tissue plus embedding medium, not to the tissue alone. (Of course, *ratios* of elemental weight-fractions should be unaffected and may be successfully measured after the embedding.)

For thin[1] specimens of soft tissue a separate theory of quantitation has been developed, based on the use of bremsstrahlung intensity as a measure of total mass per unit area. An additional detector is needed at the probe column, as indicated in Fig. 3. In the simplest form of the theory, discussed earlier in this conference [25], the bremsstrahlung intensity is taken as a direct measure of relative mass per unit area in a thin specimen (which is mounted on a very thin support). Since, in a section of uniform thickness, mass per unit area is related to mass per unit volume by a constant factor, one has here the measure of amount per unit volume which was sought above. Furthermore, since the intensity of a characteristic radiation generated in a thin specimen is proportional to the amount per unit area of the corresponding element,

---

1. The term "thin specimen" is now meant to imply specifically that the probe electrons lose only a small fraction of their energy as they pass completely through the specimen, while "thick specimens" are understood to be electron-opaque. (Thus a given section of tissue may be regarded as thick with respect to low-voltage probes, and thin with respect to high-voltage probes.) We shall adhere to this meaning in the remainder of this paper.

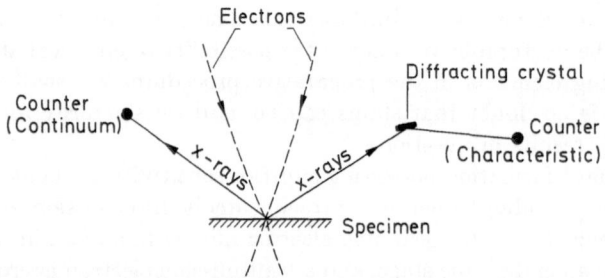

Fig. 3. Disposition of detectors for the simultaneous measurement of
characteristic and continuum radiation

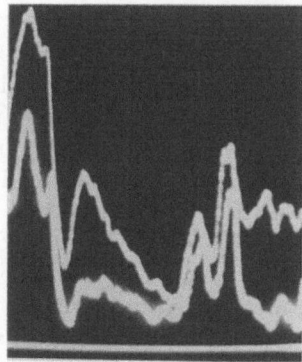

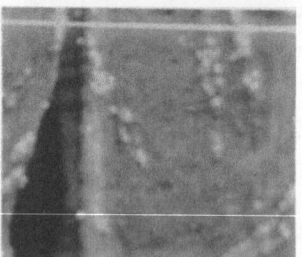

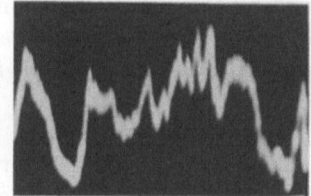

Fig. 4. Sulphur weight-fraction and mass per unit volume in layers
of the skin. Centre: Backscattered-electron image of a 5-μm transverse section of skin, showing a line of scan. Local variations along
the line of scan are shown by the three traces which represent, from
top to bottom, total dry mass per unit area, sulphur per unit area,
and sulphur per unit dry mass. The corneum is at the left of the
fissure; underlying layers of the skin are to the right

Fig. 4

the ratio of characteristic and bremsstrahlung intensities is a relative measure of the weight-
fraction of the element.

The use of this method to determine relative mass per unit area and relative weight-fractions
is illustrated in a recent study of the distribution of sulphur and dry mass per unit area in layers
of the skin [26]. The two upper traces in Fig. 4 are proportional to the amount of sulphur per
unit area and the amount of dry mass per unit area along the indicated line of scan in the transverse section of skin. The bottom trace, the ratio of the upper two, is proportional to the sulphur
weight-fraction. The technique shows clearly that as the inner boundary of the corneum is
approached from the interior, there is an abrupt increase in the amounts per unit area with no
increase in the sulphur weight-fraction.

The theory has been extended to provide absolute values of weight-fractions by introducing
the known physics of the production of continuum radiation [27]. The yield per probe electron
of a characteristic radiation generated in a thin specimen can be expressed (with notation defined
directly below) in the form

$$n_A \, dx = w_A \, S_A \, N_A \, dx. \tag{1}$$

The yield of continuum radiation can be expressed in the form

$$n_w \, dV \, dx = \frac{k}{E} \, \frac{dV}{V} \left(\sum_r N_r Z_r^2 B_r \right) dx. \tag{2}$$

In the comparison of the intensities from the specimen and from a standard, most of the constants
in the equations cancel, and the division of Eq. (1) by Eq. (2) gives the following equation for

the ratio of characteristic and continuum intensities:

$$\frac{\text{(characteristic count/white count)}_{\text{specimen}}}{\text{(characteristic count/white count)}_{\text{standard}}} = \frac{\left(N_A \big/ \left(\sum_r N_r Z_r^2 B_r\right)\right)_{\text{specimen}}}{\left(N_A \big/ \left(\sum_r N_r Z_r^2 B_r\right)\right)_{\text{standard}}} . \tag{3}$$

The definitions of the symbols in these equations are:

$n_A \, dx =$ the number of quanta of a characteristic radiation of element A generated in path length dx;

$w_A =$ fluorescence yield;

$S_A =$ ionization cross section for radiation A, cm²/atom;

$N_A =$ atoms/cc of element A;

$n_w \, dV \, dx =$ the number of continuum quanta in the quantum-energy range between V and $(V+dV)$ generated in path length dx;

$k =$ a constant;

$E =$ energy of the probe electrons;

$Z =$ atomic number;

$r =$ index running over all of the constituent elements of the specimen (or standard);

$B =$ a factor which is unity if one accepts KRAMER's expression [28] for the intensity of bremsstrahlung, or is the logarithmic term written by BETHE as $\ln 2E/J$ in his stopping formula if one accepts the expression for bremsstrahlung intensity used by MARSHALL and HALL [27]. The choice between these two forms of B has little effect on the numerical results in practice, especially with specimens of soft tissue.

The only requirements on the standard in Eq. (3) are that it must contain element $\underline{A}$; its composition must be known; and it must be thin in the sense defined above. Except for this specification, there is no need to know its thickness.

Straightforward algebraic manipulation of Eq. (3) [29] gives a practical equation for the measurement of weight-fractions in thin specimens:

$$C_A = Q_A G_A A_A \left[\sum_u (Q_u G_u A_u) + H \left(1 - \sum_u Q_u G_u Z_u^2 B_u\right) \right]^{-1}, \tag{4}$$

with

$$H^{-1} = \sum_m \frac{C_m'}{A_m} Z_m^2 B_m \quad \text{and} \quad C_m' = \frac{C_m}{\sum_m C_m}.$$

The symbols not already defined in Eq. (4) are

$C_A =$ the weight-fraction of element A;

$Q_A = \dfrac{\text{(characteristic count/white count)}_{\text{specimen}}}{\text{(characteristic count/white count)}_{\text{standard}}}$;

$G_A = N_A / (\sum N Z^2 B)$ in the standard for element A;

$A_A =$ the atomic weight of element A.

The subscript m is an index which runs over those constituent elements in the specimen which are part of a matrix whose composition is sufficiently well known. (Only the composition of the matrix is assumed, not its concentration in the specimen. The method does not require the postulation of a matrix in general. If no matrix is postulated, the last term in Eq. (4) becomes *nil* and the equation is still valid.) The subscript u includes all of the elements to be assayed and any other important elements which are not part of the postulated matrix.

Eq. (4) provides absolute values of weight-fractions in thin specimens. To use the equation, for each major constituent *either* the intensity of its characteristic radiation must be measured *or* it must be part of a matrix of known composition. The equation is especially suitable for the very common case where an element of interest is present in a matrix containing no major constituents besides protein. The measurement of a weight-fraction can then be carried out by observing only the one characteristic intensity plus the continuum, since the atomic composition of protein is adequately constant and known.

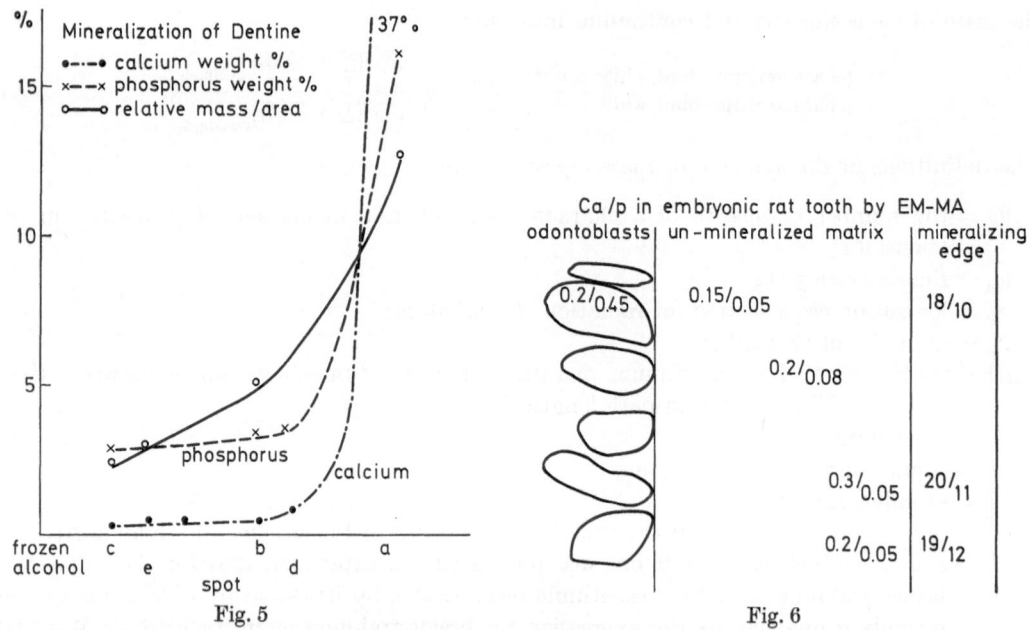

Fig. 5. Static probe measurements of calcium and phosphorus weight-fractions in 5-μm sections of embryonic rat incisor. Odontoblast cells at the left; mineralising border at the right; cell-free matrix in between

Fig. 6. Static probe measurements of calcium and phosphorus weight-fractions in an ultrathin section of embryonic rat incisor. The positions of the numbers indicate approximately where the measurements were made in the section

We wish to make two remarks about this method.

a) Because it is a ratio method and the characteristic and continuum radiations are produced in parallel as the electrons traverse the specimen, there is no need for corrections for electron backscatter or "penetration". Furthermore, absorption and fluorescence are usually negligible in the thin specimens to which it applies. Finally, as a ratio method it is independent of fluctuations in beam intensity.

b) It seems especially attractive to use the method for the very thin specimens studied in combination transmission electron microscope-microanalysers, but it is not yet clear how successfully one can reduce or correct for the high background of extraneous continuum radiation in these instruments.

The use of the continuum-method for absolute measurements may be exemplified by data from an extensive set of studies [30—34] on the progressive accumulation of calcium, phosphorus and dry mass per unit volume in very early stages or pre-stages of mineralisation in several tissues (chiefly aorta, bone and embryonic teeth). Fig. 5 gives the results of some static-probe measurements on sections of embryonic rat incisor like the one in Fig. 2a, while Fig. 6 gives corresponding results for an ultrathin section like the one in Fig. 2b [35]. It should be noted that "weight-fraction" and "total mass per unit area" must refer to the specimen as presented to the probe, including embedding material if any is present. The thicker sections were either cut from frozen, unembedded tissue or freed of embedding before measurement, but in the case of the ultrathin sections the embedding material had to remain, inevitably depressing the observed weight-fractions to some extent and leaving little significance to the measurement of a total mass per unit area. (The preparative procedure for ultrathin sections introduces other serious problems of quantitation as well, as we shall see shortly, but the disadvantages are offset by the superior histological correlation. In the thicker sections, one had to work back blindly from the mineralising border, relying for histological interpretation on prior knowledge of the width of the cell-free band.)

5. The Measurement of Low Concentrations. When we discussed above the types of probe study which require no special instrumental technique, most of the categories involved elemental concentrations far above the usual levels of interesting elements in normal tissues. Many important biological studies are ruled out simply because the concentrations are below the limit of detectability.

The limits for thick specimens have been worked out in great detail by ANDERSEN [10, 23], in terms of both minimal absolute amounts and minimal weight-fractions. The absolute minima are most often in the range $10^{-15}-10^{-16}$ g, and the weight-fractions are most often in the range $10^{-3}-10^{-4}$ (i.e., $0.1-0.01\%$).

For thin specimens (a few microns in thickness) we shall not quote from the theory which has been developed [36] (although the theory is simpler than for thick objects), but shall merely state from the experience of many analyses, our own and others, that in fact the limits are about the same. This may seem surprising since, in the case of thin specimens, only a small part of the energy of the probe electrons is available for X-ray production. But this disadvantage is compensated by the fact that higher probe voltages can be used without loss of resolution, and higher probe currents are then available and tolerated [24].

For ultrathin sections (1,000 Å and less, for transmission electron microscopy) the minimal absolute amounts remain the same, but the weight-fractions must be higher in order to have enough of the element under the probe. The range of minimal weight-fractions is commonly shifted unfavourably by a factor of the order of ten from the range quoted above.

We have already mentioned that maximum sensitivity can be realised in thin specimens only if a heavy conducting coating is applied. It is also necessary to mount such specimens on very thin supports (like the Formvar or collodion films used in electron microscopy, or similarly prepared Nylon films) in order to reduce background from the support.

Given the quoted limits, what are the implications for the study of normal tissues? We can see from Table 1 that there is enough worthwhile work to keep many hands busy, although it is also true that many important subjects are beyond microprobe analysis.

Table 1. *Normal average concentrations of certain elements in some tissues*

Tissue	Element	Weight per cent (g of element/ 100 g of dry tissue)
Muscle	K	1.0
	Ca	0.01
	Mg	0.06
	Zn	0.02
Liver	Zn	0.01
	Cu	0.002
Aorta	Ca	0.2
Prostate gland	Zn	0.05
Sperm cells (some species)	Zn	0.2
Pancreas	Zn	0.008
Soft tissues, average	S	0.6
	P	0.6
	Cl	0.3

6. Specimen Preparation and Artifacts. We have already mentioned that bone and teeth may be prepared by the techniques common to mineralogical specimens, that paraffin-embedded soft tissues must be deparaffinized because paraffin generally does not withstand a probe, and that thin specimens should be heavily coated and mounted on thin supports if they are to be studied with high probe currents. Fig. 7 is an illustration of a thin specimen mounted on a thin support, before coating in the evaporator. The point to be stressed now is that the procedures most frequently used for preparing tissue sections involve a very great danger of the displacement

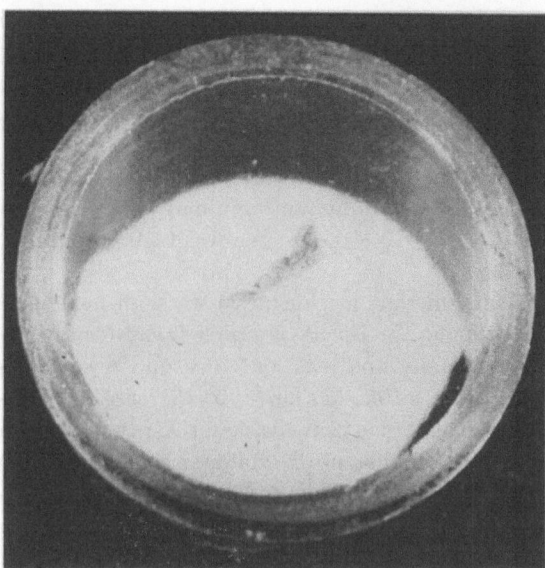

Fig. 7. A section of skin mounted on a thin support; thickness of section: 5 μm; thickness of Nylon support: 2,000 Å; diameter of aluminium tube: 1 cm

or loss of the elements to be studied, because the specimen is taken through a series of liquid baths in the course of fixation and embedding (and staining, if a stain is applied). The safest procedure, instead, seems to be to freeze a block of tissue quickly, to section it frozen, to put the frozen section on a cold support and to dry it by sublimation. The risk in other procedures has been demonstrated by ANDERSEN [10], who compared nucleated red blood cells prepared in different ways. Among other effects he observed that the concentrations of sodium and phosphorus were reduced by approximately 90% or more in the cytoplasm of the cells prepared by "wet" methods.

The problem is especially awkward with the ultrathin sections prepared for transmission electron microscopy because it is extremely difficult to cut ultrathin sections of frozen tissues. To prepare ultrathin sections, conventionally tissues are fixed in liquid baths, embedded and cut; and the sections drop from the knife to a water bath from which they are recovered. The consequences, for example in ultrathin sections of rat incisor fixed in alcohol and embedded in methacrylate, are shown in Table 2. During preparation phosphorus was almost entirely removed from the pre-dentinal band.

Table 2. *Calcium and phosphorus weight-fractions in the pre-dentine of rat incisor in frozen and in ultrathin sections*

	Ca (%)	P (%)
Frozen section, 4 μm	0.2—0.4	1—4
Ultrathin section, fixed in alcohol, embedded in methacrylate	0.2—0.4	approx. 0.06

Thus, while transmission electron microscopy seems to offer the best hope for good histological correlation, the associated conventional preparative procedures seem most threatening to the integrity of the specimens. However, the position is not hopeless. Recently we have studied ultrathin sections prepared by one of the established "dry" methods. The block of tissue is rapidly frozen, fixed by freeze-drying and then infiltrated with the embedding medium and sectioned; the sections are transferred directly from the knife to the thin supporting film. There is no contact with liquid except for the embedding medium, and according to our limited experience so far, this method seems to avoid the loss of elements.

A similar procedure has been used [37] for another technically formidable problem, the localisation of readily diffusable electrolytes (K, Cl and Na) in tissues of muscle and mucosa. Notably, the block of tissue was exposed to osmium tetroxide *vapour* after freeze-drying and before embedding. This vapour serves both as a fixative and as an electron-stain, and should disrupt the tissue much less than liquid fixatives or stains. Three-micron sections were studied. The procedure succeeded in the very difficult requirement of retaining a marked difference between intracellular and extracellular concentrations of potassium. It would be hopeless to study electrolyte concentrations in sections prepared in more usual ways, and there is no doubt that in general preparative procedures may be crucial to obtaining meaningful results.

In overall summary, we may say that much worthwhile biological work can be done with non-specialised probe techniques, and that much more can be done with techniques that have now been developed under the impetus of the special requirements of biology.

References

1. ROSENSTIEL, A. P. v., and H. B. ZEEDIJK: This conference.
2. GALLE, P.: Proc. 3rd Intern. Congr. of Nephrology, p. 306. Basel and New York: Karger 1967.
3. TOUSIMIS, A. J., and I. ADLER: J. Histochem. Cytochem. 11, 40 (1963).
4. GUEFT, B., Y. KIKKAWA, and J. MOSKAL: J. Appl. Phys. 35, 3077 (1964).
5. MARSHALL, D. J.: Ph. D. Thesis, Cavendish Laboratory, University of Cambridge (1967).
6. CARROLL, K. G., and J. L. TULLIS: Nature 217, 1172 (1968).
7. MELLORS, R. C.: Lab. Invest. 13, 183 (1964).
8. BAUD, C. A., S. KIMOTO and H. HASHIMOTO: Experientia 19, 524 (1963).
9. TOUSIMIS, A. J., in: X-ray optics and X-ray microanalysis (editors PATTEE, COSSLETT and ENGSTROM), p. 539. New York: Academic Press 1963.
10. ANDERSEN, C. A., in: Methods of biochemical analysis (editor DAVID GLICK), vol. 15, p. 147. New York: Interscience Publ. 1967.
11. FRANK, R. M., M. CAPITANT, and J. GONI: J. Dental Res. 45, 672 (1966).
12. ROSSER, H., A. BOYDE, and A. D. G. STEWART: Arch. Oral Biol. 12, 431 (1967).
13. SÖREMARK, R., and P. GRØN: Arch. Oral Biol. 11, 861 (1966).
14. FRAZIER, P. D.: Arch. Oral Biol. 12, 25 (1967).
15. SAFFIR, A. J., and R. E. OGILVIE: Trans. of the 2nd National Conference on Electron Microprobe Analysis, Boston (1967).
16. WEI, S. H. Y., and M. J. INGRAM: Trans. of the 3rd National Conference on Electron Microprobe Analysis, Chicago (1968).
17. ROSENSTIEL, A. P. v., J. VAHL, and J. KYSELOVA: This conference.
18. HALE, A. J.: J. Cell Biol. 15, 427 (1962).
19. SIMS, R. T., and D. J. MARSHALL: Nature 212, 1359 (1966).
20. GARDNER, D. L., and T. A. HALL: J. of Pathology (in press). Thanks are due to the Cambridge Scientific Instrument Company for the use of their Microsan V microprobe.
21. HALL, T. A., A. J. HALE, and V. R. SWITSUR, in: The electron microprobe (editors McKINLEY, HEINRICH and WITTRY), p. 805. New York: John Wiley & Sons 1966.
22. VASSAMILLET, L. F., and V. E. CALDWELL: Trans. of the 3rd National Conference on Electron Microprobe Analysis, Chicago (1968).
23. ANDERSEN, C. A.: Brit. J. Appl. Phys. 18, 1033 (1967).
24. HALL, T. A., in: Quantitative electron probe microanalysis (editor K. F. J. HEINRICH), p. 269. National Bureau of Standards Special Publ. 298, Washington D.C. (1968).
25. —, and P. WERBA: This volume.
26. SIMS, R. T., and T. A. HALL: J. Cell Sci. 3, 563 (1968).
27. MARSHALL, D. J., and T. A. HALL: Brit. J. Appl. Phys. 1, 1651 (1968).
28. KRAMERS, H. A.: Phil. Mag. 46, 836 (1923).
29. Equation (4) is a more general form of the equation derived in reference number 24, where the derivation from equation (3) is described in detail.
30. HÖHLING, H. J., T. A. HALL u. R. W. FEARNHEAD: Naturwissenschaften 54, 93 (1967).
31. — — et al.: Naturwissenschaften 54, 142 (1967).
32. — — u. A. BOYDE: Naturwissenschaften 54, 617 (1967).
33. — — et al.: In: Les Tissus Calcifiés (editors G. MILHAUD, M. OWEN et H. J. J. BLACKWOOD), p. 323. Paris: Société D'Edition D'Enseignement Supérieur 1968.
34. — — —, and A. P. v. ROSENSTIEL: Calcified Tissue Research 2 (August Suppl.), 5 (1968).
35. We must thank our colleague A. BOYDE for the preparation of the tissue sections used for Fig. 5, and we must thank the staff at the Tube Investments laboratory, in particular P. DUNCUMB, C. J. COOKE, and P. HUNNEYBALL, for the use of their electron microscope-microprobe, the original "EMMA", to obtain the data of Fig. 6.
36. HALL, T. A.: Manuscript in preparation.
37. HOGBEN, C. A. M., and M. J. INGRAM: Trans. of the 2nd National Conference on Electron Microprobe Analysis, Boston (1967).

Electron Microprobe Studies on the Mineralization Process of Tooth and Bone

S. Suga

Department of Pathology, Nippon Dental College, Tokyo, Japan

The electron microprobe analyses on the biological hard tissues have mainly been performed on their ground surfaces [1, 2, 6]. In this instance, the biological specimens give rise to difficulties in connection with the interpretation of quantitative data. The degree or depth of electron beam penetration into the specimens depends upon the density of tissues in the region. For this reason, in the regions of lower tissues density, the electron beam will penetrate further and excite production of characteristic X-ray at a greater volume. In order to resolve the problem mentioned above, the author in the present study has tried to use the thin undemineralized microtom sections of the developing enamel, dentin and bone to the analyses. And, in the case of developing enamel and dentin, the cryostat sections were used, in order to avoid the removal of elements from the specimens and the contamination of specimens, which occure during specimen preparation.

Materials and Methods. For the analyses of the developing enamel and dentin, the upper molar germ of the 6—16 days rats, and, for those of the developing bones, the small fragments of human alveolar bone were used respectively. A part of the rat molar germs were deep freezed with chilled N-hexan (−60° C) immediately after sacrificing, and, the undemineralized cryostat sections (3—4 μ) were subsequently made. Another part of tooth germs and bones were fixed with 10% formol, and, then, the undemineralized paraffine sections (3—4 μ) were made. Both the paraffine and cryostat sections were placed directly on the surface of formval films covered on the metal rings. During this procedure for the cryostat sections, the use of water was avoided. The paraffine was removed by the immersion into xylol after the sections were dried.

Before analysing by electron microprobe, the sections mounted on the formval film had been previously microradiographed by soft X-ray. The adjoining serial sections were also stained with the various histological and histochemical methods.

The specimen support metal rings were placed on the surface of specimen holder and fixed using an electric conductive paint.

The electron microprobe apparatus used is Shimadzu-ARL, type EMX-2 (Shimadzu Seisakusho, Kyoto, Japan). The apparatus was operated at rather high accelerate voltage (30 kV) with sample current, ranging from 0.05 to 0.08 μA. By this arrangement, the surface analyses which required over 100 minutes could be carried out without the destruction of sections.

The curved single crystals used are 4 inch LiF for the detection of Zn, 4 inch ADP for Ca, P and S, and 4 inch KAP for Mg.

The elements determined were Ca, P, Mg, S and Zn. On the line scan analyses at the present study, only the gradients of distributions of elements were investigated.

Findings:

1. The noticable difference were not found on the elemental distribution patterns revealed by both the line scan and surface analyses, between the cryostat and paraffine sections, at least, in the cases of Ca, P, Mg, S and Zn.

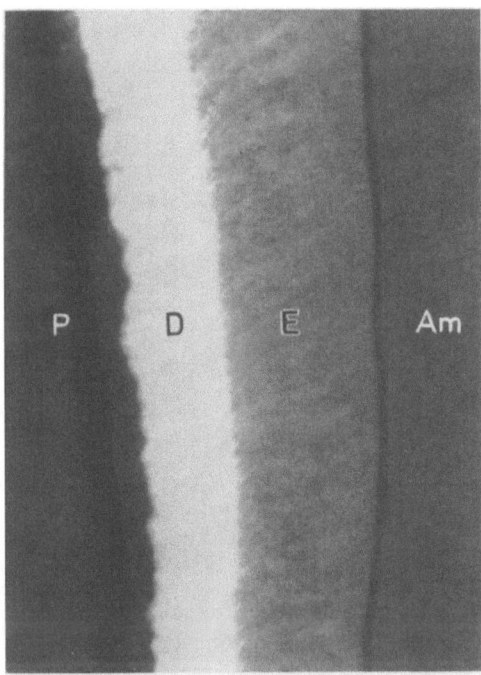

Fig. 1. Microradiogram of a longitudinal undemineralized cryostat section of upper first molar germ of 8 days rat. *D* dentin, *E* enamel, *P* dental pulp, *Am* ameloblasts

2. Developing enamel and dentin. Fig. 1 indicates the microradiogram of a longitudinal undemineralized cryostat section of upper first molar germ of 8 days rat. The enamel matrix here is mineralized slightly immediately after the matrix is laid down and, then, its mineralization degree increases very gradually towards the amelodentinal junction. On the other hand, the mineralization degree of dentin is higher than that of developing enamel, and appears to extend to the final rate soon after the mineralization began. It is recognized that the X-ray images by CaK_α and PK_α bring out rather distinctly a difference of mineralization degree between the two tissues and the gradient of mineralization increase in the developing enamel, as observed by microradiography.

According to the line scan analyses performed on the cryostat sections across the enamel and dentin layers in the direction perpendicular to the ameloidentinal junction, Ca and P (Fig. 2a) represent almost the same gradienat of distribution. In the enamel, Ca and P concentrations which showed lowest degree at the outermost narrow layer (the pre-enamel) increase their degree gradually as it approaches the ameloidentinal junction. Finally, the intensities of CaK_α and PK_α emissions at the enamel along the ameloidentinal junction measure more than 8 times higher than that at the pre-enamel (to about $5\,\mu$ inside of the surface). However, they cover about one-half of them in the dentin adjacent to the ameloidentinal junction. In the dentin, as looked from the direction of the predentin, the intensities of CaK_α and PK_α emissions begin to increase very rapidly and appear to reach the almost final degree in the width of about $4\,\mu$. The fluctuation of intensities of these emissions observed during the scanning across the dentin layer is probably caused by the closing of electron beam over the dentinal tubules.

MgK_α emission intensity in the enamel increases rapidly in the range of about $4\,\mu$ in width, after the mineralization begins. But, during the increase of Ca concentration, its gradient becomes slower than that of CaK_α emission. Thus, the intensity at the innermost layer of enamel is to be regarded as about 2 times as high as that at the pre-enamel. On the other hand, MgK_α emission intensity in the dentin shows abrupt increase proportional to an increase of CaK_α emission intensity and forms a peak at the outermost narrow layer adjacent to the dentin-predentin junction, after which it begins to decrease towards the amelodential junction (Fig. 2b). The

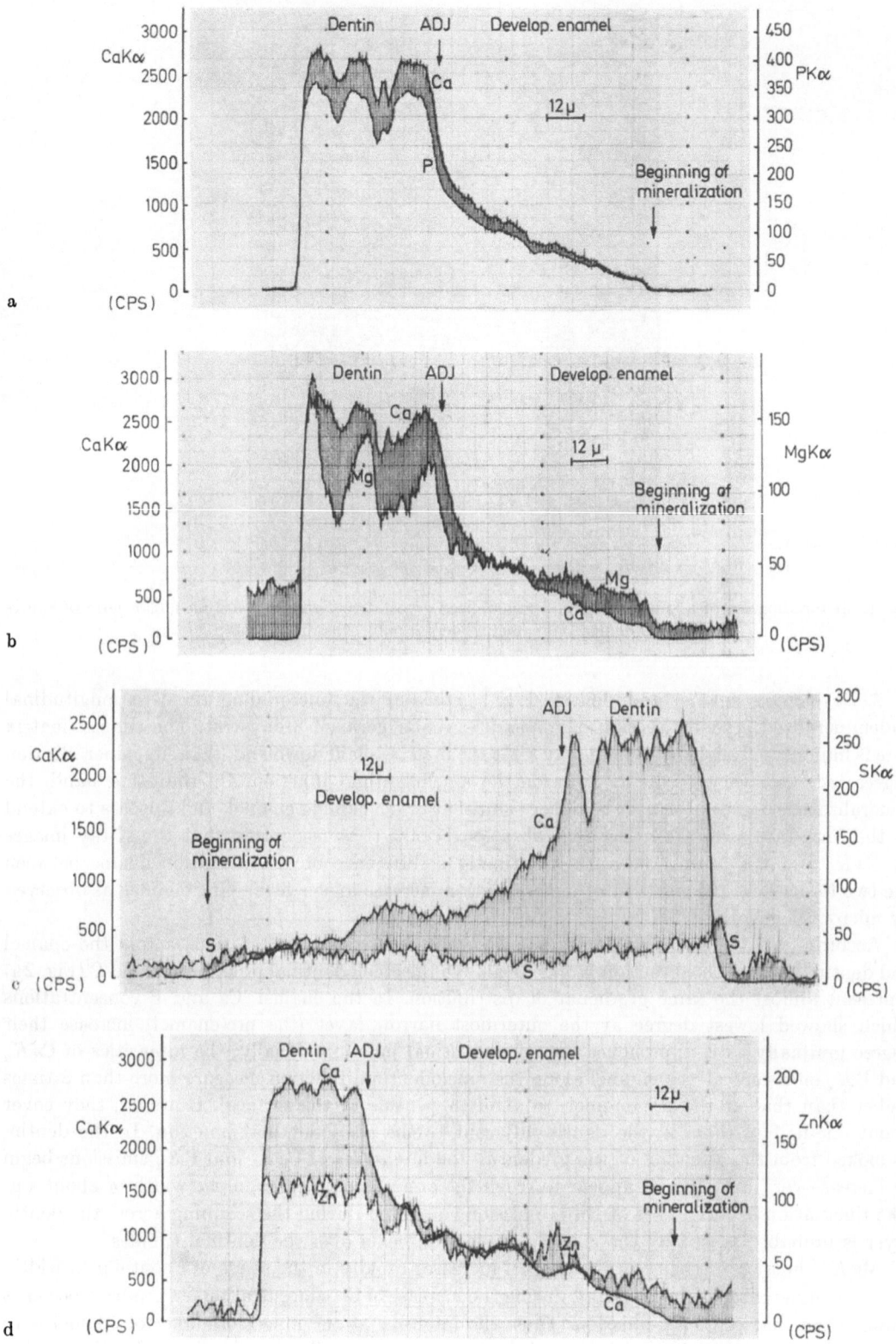

Fig. 2a—d. Line scan analyses of Ca, P, Mg, S and Zn, performed on the cryostat sections of rat developing enamel and dentin. The enamel is at the stage of matrix formation. In each case of analyses, two elements were simultaneously analysed. The difference of emission intensities between two elements does not reveal the actual difference of concentrations between them

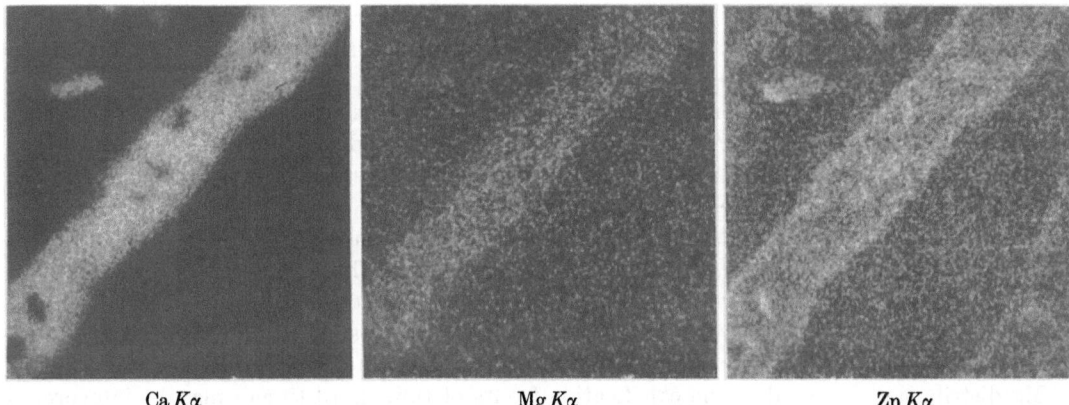

CaKα   MgKα   ZnKα

Fig. 3. The patterns of surface analyses by CaK_α, ZnK_α and MgK_α emissions, obtained from the paraffine
section of the formol fixed human alveolar bone

intensity of SK_α emission observed at the ameloblastic layer becomes 2 times higher soon after
the onset of enamel matrix formation, however, during the mineralization process, this intensity
shows tendency to decrease very slowly. In the predentin, SK_α intensity increases with steep
gradient from the juxtaodontoblastic layer towards the dentin-predentin junction. When the
mineralization begins, the gradual decrease of SK_α intensity becomes observable as it moves
towards the deeper layer (Fig. 2c). Zn concentration is slightly higher in the dentin than in the
developing enamel, and, the gradient of its concentration increase in the developing enamel is
very much slower than that of Ca concentration (Fig. 2d).

3. Developing bones. Fig. 3 shows the patterns of elemental distributions obtained by the
surface analyses on the paraffine section of human alveolar bone. The pattern of Ca distribution
resembles closely the microradiographic X-ray absorption image obtained from the same section,
especially in the gradient of increase of its intensity, from the periphery to the core of the bone
trabeculae. Mg concentration is higher in the trabeculae than in its surrounding soft tissues.
The difference of intensity of MgK_α emission between the periphery and the core of bone trabeculae is not so noticeable as is the case with CaK_α image. Zn shows a characteristic pattern of
distribution. Its concentration is highest at the peripheral layer including both the unmineralized
osteoid and the thin mineralized layers, and, then, it becomes slighter towards the core of
trabeculae. The presence of this element at the osteocytic lacunae should also deserve our
attention.

The results of line scan analyses performed across the trabeculae showed the patterns of
elemental distributions correspond to the patterns by the surface analyses. Especially, it was
interesting to note that ZnK_α emission intensity shows very steep increase and soon reaches a peak
at the peripheral part where Ca emission intensity shows still a very rapid increase. At the layer
where Ca intensity forms peak, Zn emission intensity has begun to decrease.

Discussion. The hard tissues are composed of the mineral salts, namely the hydroxyapatite,
the protein matrix and the ground substances. Furthermore, it contains the various trace
elements, such as Na, F, Mg, Zn, Al, Sr, Pb, Cu, Si, Ag, Fe, Sn and Mn [4]. It is generally
recognized that some of them plays very important rôle in the process of progressive mineralization and gives the characteristic physico-chemical properties to the regions of matured hard
tissues [4]. The pattern of mineralization increase differs among three hard tissues in the mode
and the gradient, as established already by microradiography and autoradiography. The mineralization of dentin appears to be accomplished soon after the mineralization begins. The bone
mineralizes with slower gradient than does the dentin. The enamel mineralization accomplished
through at least two stages which differ in the mode and gradient of mineralization increase.
At the stage of matrix formation, the organic matrix mineralizes very slightly after it was laid
down. Then, very slow increase of mineralization degree along the incremental pattern is

observed. However, once in the stage of maturation, the mineralization of enamel becomes to increase its degree very rapidly throughout the whole layer with much steeper gradient than that in the former stage [9, 10]. In the present investigation, the microprobe analyses were done only on the section of enamel at the stage of matrix formation.

The gradient of $Ca K_\alpha$ and $P K_\alpha$ emission intensities, revealed by the line scan analyses and the characteristic X-ray images, showed the almost same characteristic patterns of mineralization increase as observed in the three tissues by microradiography.

It has been supposed that Mg places itself mainly on the surface of apatite crystal and is not in the lattice position. Thus, the concentration of Mg in the hard tissues may correlate with the total surface of apatite crystal contained in the tissue region [4, 13]. It is estimated that the increasing rate of total surface of apatite crystals becomes slower as the crystal growth progresses and, finally, it will begin to decrease. Thus, the characteristic behaviours of gradient of Mg distribution observed at the calcification fronts of three hard tissues may be interpreted in regard to the different increasing rate of total surface of the crystals, accompanied by the crystal growth [8, 11].

It seems that Zn is combined both with the protein matrix and the apatite crystals. And, Zn competes with Ca on the apatite crystal surface [3]. It is a well known fact that Zn is a co-factor of some sort of metallo-enzyme, namely carbonic anhydrase. The histochemical reaction of Zn by dithizone method is intensively positive only at the site of active mineralization, of dentin and bone, and the uptake of ^{65}Zn is, also, demonstrated at the mineralization site of bone [5, 12]. While ^{45}Ca uptake is observed not only at the active mineralization site, but, also, throughout the already formed osteon whose mineralization is not completed yet, the uptake of ^{65}Zn is observed only at the site of active mineralization [12]. The findings on the distribution of Zn in the present study are probably compatible with the findings by the autoradiography. And, in the developing bone, it seems likely that a part of deposited Zn is released from the tissue during the progress of mineralization. The discrepancy between the distribution pattern obtained by electron microprobe and histochemical reaction pattern, of Zn, is probably due to the reducing of its histochemical reactivity by so-called "masking effect" and it suggests the technical limitations of the ordinary stain histochemistry.

The main sulfur containing chemical composition in the biological tissues seems to be derived from sulfur-containing amino acids, chondroitin sulfate and sulfatide. Chondroitin sulfate plays important role as a seeding matrix for apatite crystals. The almost even distribution throughout the unmineralized and mineralized layers is the characteristic feature of this element. According to the comparison between the characteristic X-ray image of S and the histochemical reaction patterns of several kinds of sulfur containing substances, such as SH groups, SS groups and chondroitin sulfate, no histochemical reaction which reveals the similar pattern with $S K_x$ emission image was found.

References

1. Boyde, A., V. R. Switsur, and R. W. Fearn-head: J. Ultrastruct. Res. 5, 201 (1961).
2. — —, and A. D. G. Stewart, from: Advances in fluorine research and dental caries prevention, edit. by J. L. Hardwick et al., p. 185. Oxford: Pergamon Press 1963.
3. Brudevold, F., L. T. Steadman, M. A. Spinelli, B. H. Admur, and P. Grøn: Arch. Oral Biol. 8, 135 (1963).
4. —, and R. Söremark: from "Structural and chemical organization of teeth, edit. by A. E. W. Miles, vol. 2, p. 247. New York: Academic Press 1967.
5. Fiore-Donno, G., and L.-J. Baum: Helv. Odont. Acta. 10, Suppl., 139 (1966).
6. Frank, R. M., M. Capitant, and J. Goni: J. Dental Res. 45, 672 (1966).

7. Haumont, S.: J. Histochem. Cytochem. 9, 141 (1961).
8. Nylen, M. U., E. D. Eanes, and K.-Å. Omnell: J. Cell Biol. 18, 109 (1963).
9. Suga, S., and G. Gustafson, from: Advances in fluorine research and dental caries prevention, edit. by J. L. Hardwick et al., p. 223. Oxford: Pergamon Press 1963.
10. —, and Y. Murayama: Odontology (Tokyo) 53, 154 (1965).
11. Takuma, S., from: Structural and chemical organization of teeth, edit. by A. E. W. Miles, vol. 1, p. 325. New York: Academic Press 1967.
12. Vincent, J.: Clin. Orthop. 19, 161 (1962).
13. Wood, N. V.: Science 105, 531 (1947).

Electron Probe Microanalysis of Filled Human Teeth

D. BAX

Analytisch Chemisch Laboratorium der Rijks Universiteit te Utrecht, Utrecht, Netherlands

L. W. J. VAN DER LINDEN

Tandheelkundig Instituut der Rijks Universiteit te Utrecht, Utrecht, Netherlands

By studying radiographs of amalgam filled human teeth sometimes small areas are found just beneath the filling, that have a higher than normal absorption of X-rays. From these findings it may be concluded that in these areas the average atomic number is higher than in areas that do not show this increased absorption.

In the literature there has been some controversy about the nature of the phenomenon: Some authors assumed that mercury might have diffused into the dentin [1], others assumed that hypercalcified dentin is responsible for the higher absorption [2]. Since neither theory has been proved in a satisfactory way it was our purpose to investigate the composition of the area in order to explain the observed phenomenon.

Radiographs of large numbers of extracted filled human teeth have been studied to select suitable specimens. Eight specimens were cut with a diamond saw in such a way that the areas under investigation were in the plane of the cut surface. The cut teeth were not sectioned, but mounted as a whole in brass rings with a diameter of 25 mm and a height of 12 mm, which are the standard specimen holders for the Norelco microprobe. The teeth were mounted in Araldite epoxy resin, and polished first with silicon carbide paper and finally with diamond pasta. Light optical pictures were made of the polished specimens, so they could be correlated with the radiographs, and the specimens were then coated with a thin layer of carbon.

The investigation was carried out partly with a Norelco Microprobe, and partly with a Norelco Microprobe fitted with a "Minicolumn", designed and manufactured by the Institute of Technical Physics in Delft (Holland). This column has been described at this conference [3].

In order to understand what process may have taken place, it is necessary to give the composition of some materials:

Material	Element	Approximate composition (%)
Amalgam	Hg	50
	Ag	35
	Sn	13
	Cu	0—3
	Zn	0—1
Dentin	Ca	27
	P	13
	C, H, N, O	60
Zinc phosphate cement ZnO/H_3PO_4	Zn	45

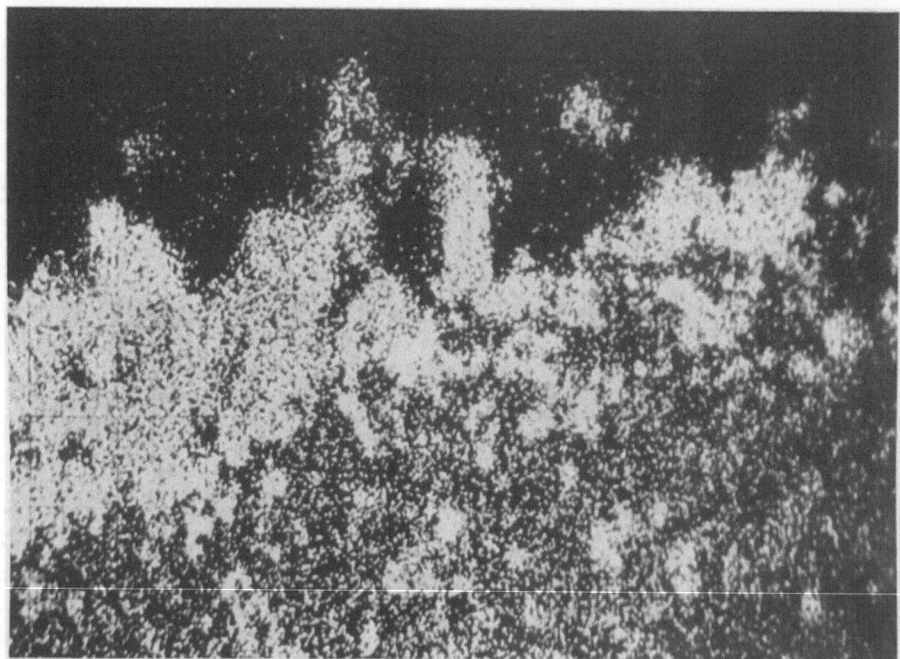

Fig. 1. X-ray scan picture ($Sn L_\alpha$), showing the distribution of tin along the dentino-amalgam junction

In filling teeth it is preferable to keep the composition of amalgam between certain limits, since a small deviation causes a rapid decreases in mechanical qualities [4]. Between the amalgam and the dentin sometimes a cement base is inserted in order to thermally insulate the pulp. This cement is made by mixing zinc oxyde and phosphoric acid.

Guided by literature we first searched for the presence of mercury or an increased calcium concentration. The results being negative, we ran a number of spectra. These spectra showed that the areas contained no detectable amounts of mercury silver or copper. The calcium and phosphorus concentrations were up to 10% lower than in normal dentin. The high absorption was found to be caused by the presence of either tin or zinc.

Zinc was found in those cases when the area investigated was in contact with a layer of zinc phosphate cement. The concentration was then approximately 1%. Tin was found in areas close to amalgam where no cement had been used. The concentration of tin was also about 1%. The difference in concentration of zinc in zinc phosphate cement (45%) and of tin in amalgam (13%) does not explain the findings of equal concentrations of these elements in the areas investigated. We therefore also investigated the chemical composition of the amalgam along the dentino-amalgam junction. The tin concentration along the border varied and was at times as high as 60—70%. This phenomenon is shown very clearly on the X-ray scan pictures (Fig. 1). A number of mechanical line scans were made while the zinc K_α or the tin L_α radiation was monitored on a recorder. In this way some idea of the concentration of zinc or tin as a function of the place could be obtained. At the transition of normal to abnormal dentin, a sharp rise in zinc or tin concentration was apparent (Fig. 2a—c).

To show that zinc or tin in these low concentrations could indead cause the observed absorption, we made some artificial dentin [5], and mixed this with tin and zinc compounds to a metal concentration of 1%. We compared the resultant X-ray absorption with that of pure artificial dentin. As the radiograph shows, the resultant difference in absorption agrees reasonably with the specimens (Fig. 3).

These findings permitted us to make the following hypothesis about the origin of the phenomenon. We supposed that the primary cause was the presence of residual caries, which was not excavated properly. Moisture possibly penetrated between the dentin and the amalgam, and

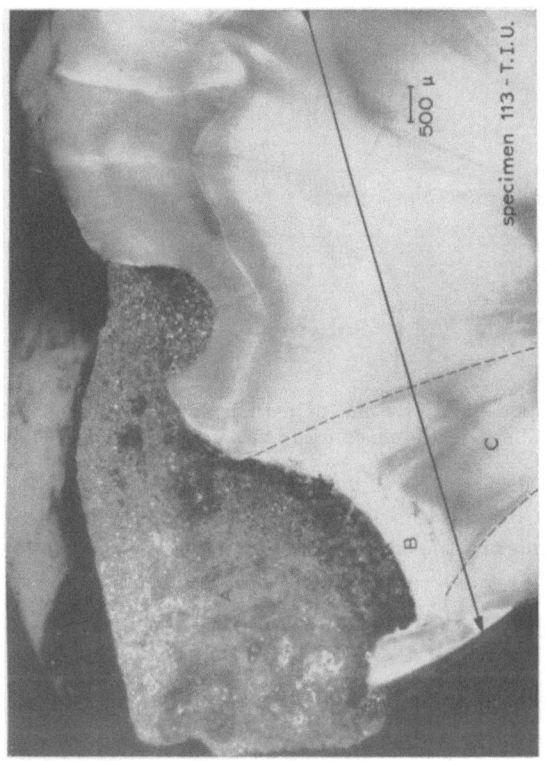

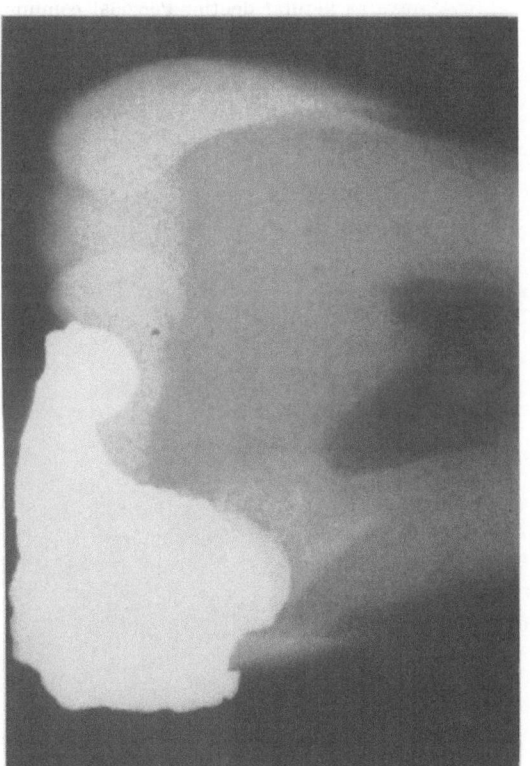

b

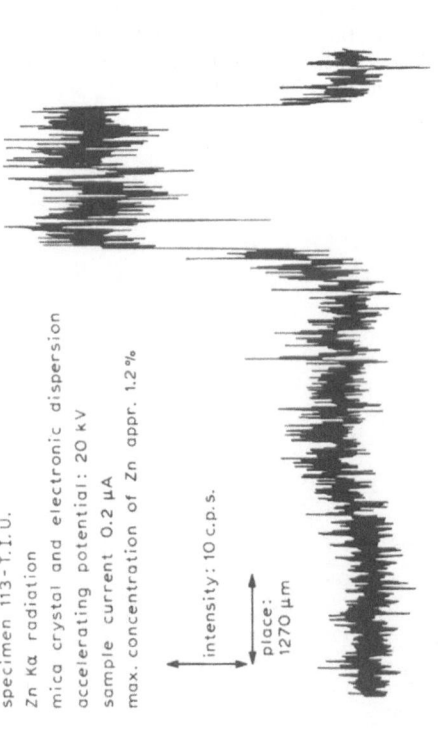

a

specimen 113 - T.I.U.
Zn Kα radiation
mica crystal and electronic dispersion
accelerating potential: 20 kV
sample current 0.2 μA
max. concentration of Zn appr. 1.2 %

intensity: 10 c.p.s.

place:
1270 μm

c

Fig. 2a—c. An amalgam filled tooth with a zinc phosphate cement layer between the amalgam and the dentin. a Radiograph, showing an area with higher absorption. b Light optical picture, showing the track of the mechanical line scan. A amalgam, B zinc phosphate cement, C area with higher absorption. c Recording of the $Zn\,K_\alpha$ radiation along the track mentioned in Fig. 2b

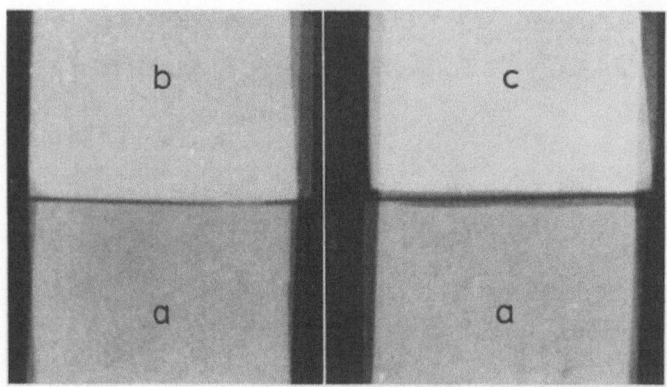

Fig. 3. Radiograph showing the difference in absorption between 4 mm of pure artificial dentin (a) and 2 mm of pure artificial dentin combined with 2 mm artificial dentin mixed with 1% ZnO (b) and 1% SnO (c)

dissolved the metal, in the case of tin probably by an electrochemical process. The metal ions then diffused into the dentin. Since tin and zinc are the most reactive metals present in the filling, it is not unreasonable to suppose that they would also go into solution first.

Quite recently a part of our findings has been confirmed by other investigators [6].

References

1. Massler, M., and T. K. Barber: Action of amalgam on dentin. J. Am. Dental Assoc. **47**, 415—422 (1953).
2. Huysen, G. van, W. F. Bale, and H. C. Hodge: Comparative study of X-ray absorption properties of normal and pathological dentin by densitometric and ionization methods. J. Dental Res. **14**, 168—169 (1934).
3. Fontijn, L. A.: The T.P.D. Electron probe X-ray analyser. This conference.
4. Skinner, E. W., and R. W. Phillips: The science of dental materials, 6th ed., chap. 23, p. 366—383. Philadelphia: W. B. Saunders Co. 1967.
5. Aken, J. van: A solid mixture of polymers and additives which has the same absorption of X-rays as natural dentin. Personal communication, Utrecht (1967).
6. Wei, S. H. Y., and M. J. Ingram: Electron microprobe analysis of the silver amalgam-tooth interface. Third National Conference on Electron Microprobe Analysis, Chicago, August 1968, paper No. 44.

Zerstörungsfreie Analyse von Zahnhartsubstanzen mit der Elektronenstrahlmikrosonde

H. Wörner

Klinik und Poliklinik für Zahn-, Mund- und Kieferkrankheiten der Universität Tübingen, BRD

H. Wizgall

Institut für angewandte Physik der Universität Gießen, BRD

Die Frage ob Qualität und Widerstandsfähigkeit der Zahnhartsubstanzen gegenüber äußeren Schädlichkeiten vom Mineralisationsgrad abhängig sind, ist schon seit langer Zeit Gegenstand wissenschaftlicher Diskussionen.

Die Mineralisationsstörungen der Zähne haben unterschiedliche Ursachen. Jede schwere Erkrankung und Stoffwechselstörung kann während der Zahnbildungsperiode solche Störungen hervorrufen. Die Mineralisationsstörungen des Schmelzes sind im Gegensatz zu Veränderungen bei der Knochenbildung später nicht mehr reparabel, da es nach dem Durchtritt des Zahnes keine schmelzbildenden Zellen mehr gibt.

Zähne sind feste Körper. Schmelz besteht zu 96% und Zahnbein zu 72% aus anorganischem Material. Die relative Härte des Schmelzes beträgt 250 Vickers-Einheiten und ist mit der des Stahles vergleichbar. Zahnbein ist weicher, seine Härte liegt aber immerhin noch bei 70 Einheiten. Hiermit sind günstige Voraussetzungen dafür gegeben, Zähne zu Proben für die Elektronenstrahlmikroanalyse zu verarbeiten.

Die qualitative Zusammensetzung der Zahnhartsubstanzen ist durch herkömmliche Analysenverfahren weitgehendst bekannt. Darüber hinaus erlaubt es die Röntgenmikroanalyse, die Zusammensetzung der Probe in einem nur wenige μm^3 großen Volumen zu ermitteln. Vorläufig beschränkten wir unsere Untersuchungen, die wir mit der JEOL JXA-3A-Mikrosonde durchführten, auf die Bestimmung der Elemente Calcium, Phosphor und Fluor, die für die Qualität der Zahnhartsubstanzen von ausschlaggebender Bedeutung sind.

Als Ausgangsmaterial verwendeten wir die Molaren von Albinoratten. Die Herstellung und Verarbeitung der Proben erfordert besondere Sorgfalt. Die Tiere werden durch Narkose getötet, der Unterkiefer herausgelöst. Mit der von Janssen angegebenen Serienschliffmaschine legen wir einen vertikalen Längsschliff durch den Unterkiefer, und zwar so, daß dieser Schliff durch den medialen und distalen Kontaktpunkt des 2. Molaren verläuft. Die entstandenen Hälften werden mit ihrer Schlifffläche auf einer Glasplatte fixiert und in selbsthärtendem Kunststoff eingebettet. Die Polymerisation des Kunststoffes erfolgt im Drucktopf. Mit dieser Maßnahme erreichen wir eine innige Verklebung zwischen Probe und Einbettmittel. Eine Luftpolsterbildung an den Berührungsflächen wird vermieden. Die Gefahr, daß die planparallele Oberfläche der Probe durch deren Ablösung aus der Einbettmasse verlorengeht, wird wesentlich verringert. Aus diesem Grunde sind wir vom Dentalamalgam als Einbettmittel wieder abgekommen.

Die Politur der Probenoberfläche wird mechanisch mit einer speziell hierfür angefertigten Maschine durchgeführt. Nach der Politur ist eine sorgfältige Reinigung mit einer weichen Bürste und Aqua dest. erforderlich, um die Reste der Schleif- und Poliermittel — wir verwendeten Carborundpulver und Chromoxyd — zu entfernen. Vor dem Aufdampfen der leitenden Schicht — wir entschieden uns für Reinaluminium — müssen die Präparate noch 24 Std an der Luft getrocknet werden.

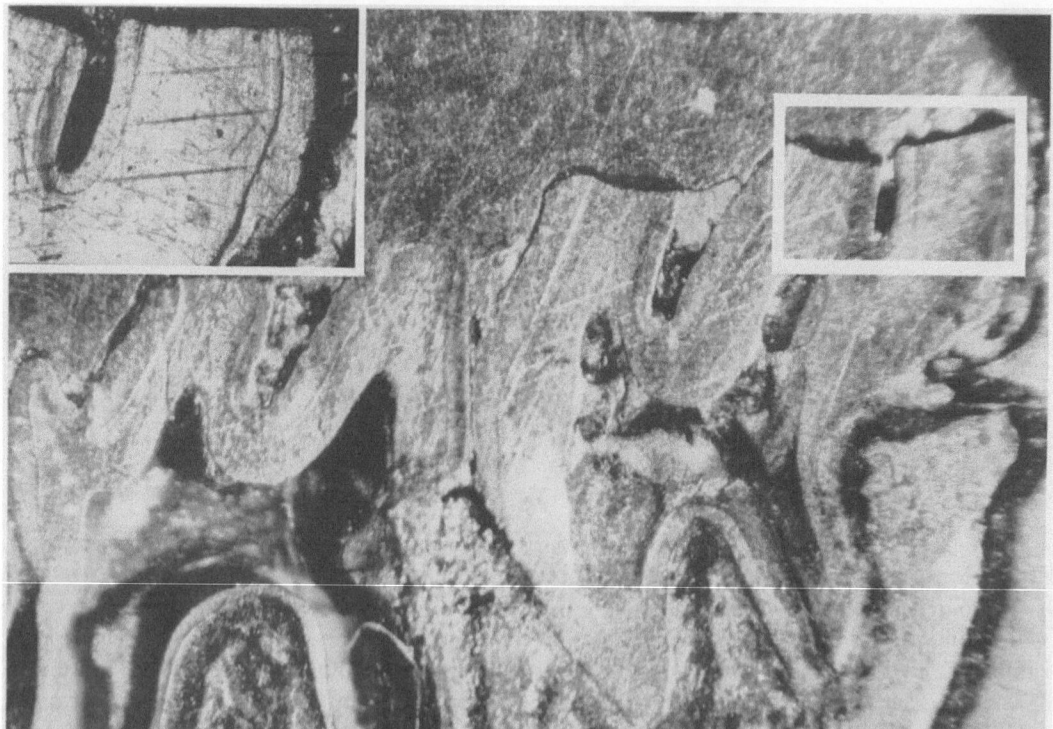

Abb. 1. Längsschliff durch den Unterkiefer der Ratte mit M_1 und M_2 nach Al-Beschichtung. Links oben Ausschnittsvergrößerung des Mesialhöckers von M_1. Im Abstand von 125 μ drei in sich parallele Skanninglinien (Originalfoto) nur im Dentin sichtbar, das gegenüber Elektronenbeschuß weniger widerstandsfähig ist. Vergrößerung ≈ 300fach, Ausschnitt ≈ 566fach

In Voruntersuchungen hatte sich gezeigt, daß eine Anregungsspannung von 20 kV und eine Stromstärke von 2×10^{-7} A für alle drei Elemente gut auswertbare Intensitätswerte ergibt. Wir haben diese Bedingungen daher bei allen Untersuchungen beibehalten. Die Kenntnis der genauen absoluten Zusammensetzung ist für vergleichende Untersuchungen zunächst nicht erforderlich. Angaben über die relative Änderung und Verteilung der Elemente in den einzelnen Zahn-abschnitten, die durch die linienförmige Abtastmethode zu gewinnen sind, geben uns wertvolle Hinweise, wie wir an einem Beispiel zeigen wollen.

Abb. 1 zeigt einen Längsschliff durch den Unterkiefer einer 33 Tage alten Ratte. Ausschnitts-vergrößerung des mesialen Höckers des 1. Molaren mit drei in sich parallelen Skanninglinien im Abstand von 125 μ. Sie beginnen am äußersten mesialen Schmelzrand und ziehen mit 15° Neigung zur Zahnachse nach distal. Die Skanninglinien kamen durch Verschieben der Probe mit einer Geschwindigkeit von 100 μ/min zustande. Entlang dieser Linien wurde die jeweilige Konzentra-tion von Calcium und Phosphor aufgezeichnet (Abb. 2). Der höhere prozentuale Anteil von Calcium und Phosphor im Schmelz ist eindeutig zu erkennen. Auch der deutliche Abfall beider Werte an der Schmelzdentingrenze ist deutlich zu sehen. Die Calciumkonzentration nimmt dann langsam zu, während die Phosphorwerte ständig kleiner werden. Die Phosphorwerte liegen außer-dem in der schmelznahen Zone im Bereich der Abtastlinie 1 und 2, d.h. in der früher gebildeten Dentin-Substanz höher als bei der Abtastlinie 3, die jüngerem Zahnbein entspricht.

Während die Aufzeichnung der Konzentrationsunterschiede entlang einer Skanninglinie wert-volle Informationen liefert, ist die Darstellung der Oberflächenverteilung der Elemente Calcium, Phosphor und Fluor weniger aufschlußreich. Auch hierfür ein Beispiel. Die Abb. 3 und 4 zeigen die Oberflächentopographie eines Zahnabschnittes mit den Hartsubstanzschichten Schmelz und Dentin. Abb. 3 wurde durch rückgestreute Elektronen gewonnen. Verschiedentliche Umzeich-nungen von Röntgenstrahlenintensitätsmessungen ergaben die dazugehörige Verteilung von Cal-cium, Phosphor und Fluor (Abb. 4). Eine charakteristische Anordnung dieser Elemente, wie sie

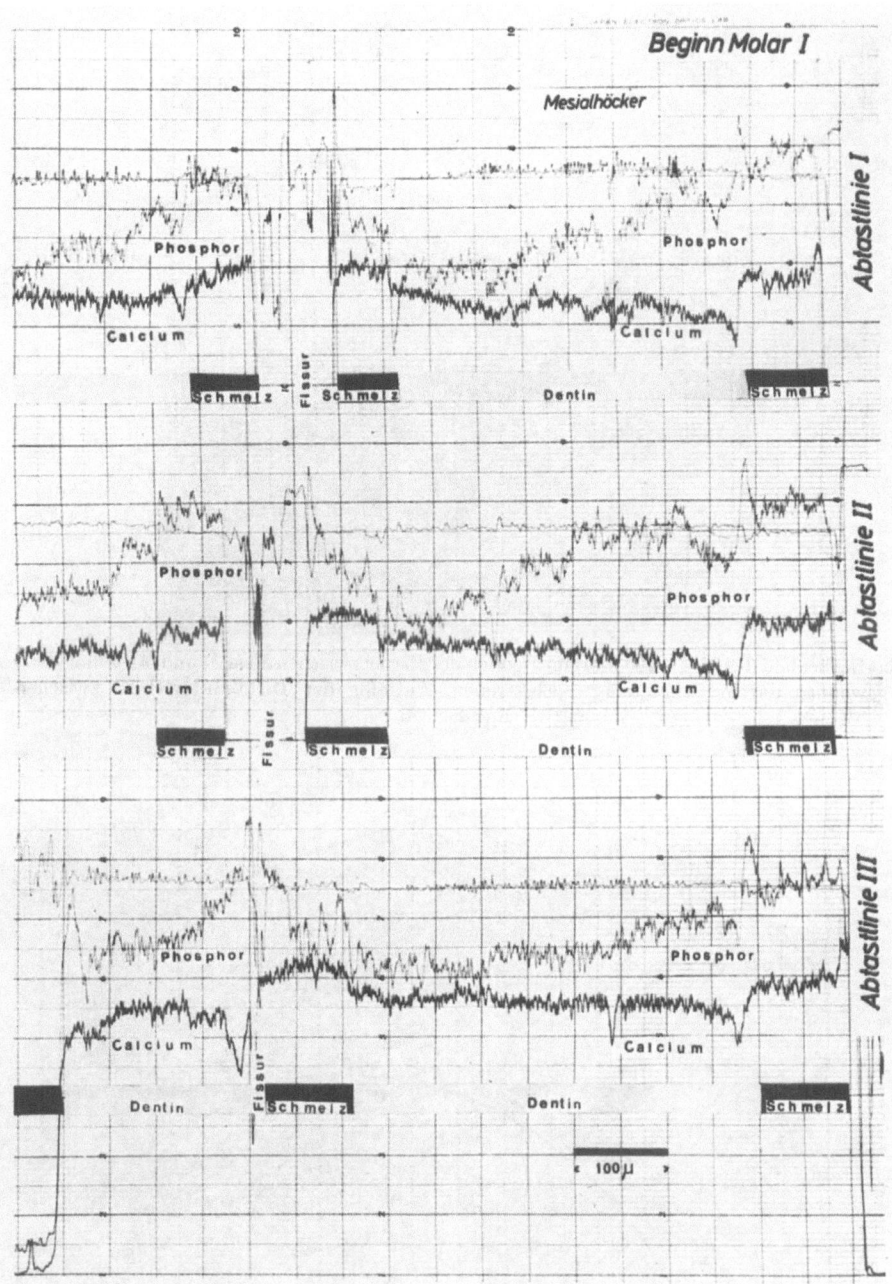

Abb. 2. Wiedergabe der Konzentrationsverhältnisse von Calcium und Phosphor entlang der im Ausschnitt Abb. 1 dargestellten Skanninglinien. Anregungsspannung 20 kV. Stromstärke 2×10^{-7} A

eigentlich auf Grund der chemischen Bindung zu erwarten wäre, läßt sich aber nicht erkennen. Dies erklärt sich wohl dadurch, daß die Zonen, die jeweils zur Strahlungsemmission angeregt werden, statistischen Auswahlregeln folgen. Außerdem reicht das Auflösungsvermögen der Sonde nicht aus, um in diese molekularen Bereiche vorzudringen.

Um tatsächliche, durch Umwelteinflüsse hervorgerufene Änderungen im Mineralisationsverhalten der Zahnhartsubstanzen zu erfassen, ist es erforderlich, in den voneinander abzugrenzenden Bereichen zahlreiche Punktanalysen durchzuführen und die Mittelwerte mit statistischen Methoden zu vergleichen. Ein derartiger Vergleich ist ebenfalls ohne genaue Kenntnis der

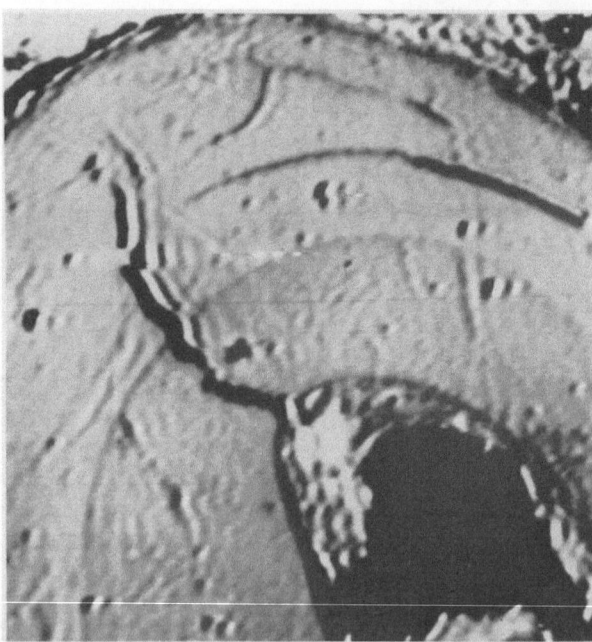

Abb. 3. Oberflächenbild (0,166 × 0,166 mm) im Bereich der Fissur zwischen Mesial- und Mittelhöcker des zweiten Molaren, gewonnen durch rückgesteuerte Elektronen. Auffällig der Dichteunterschied zwischen Schmelz und Dentin

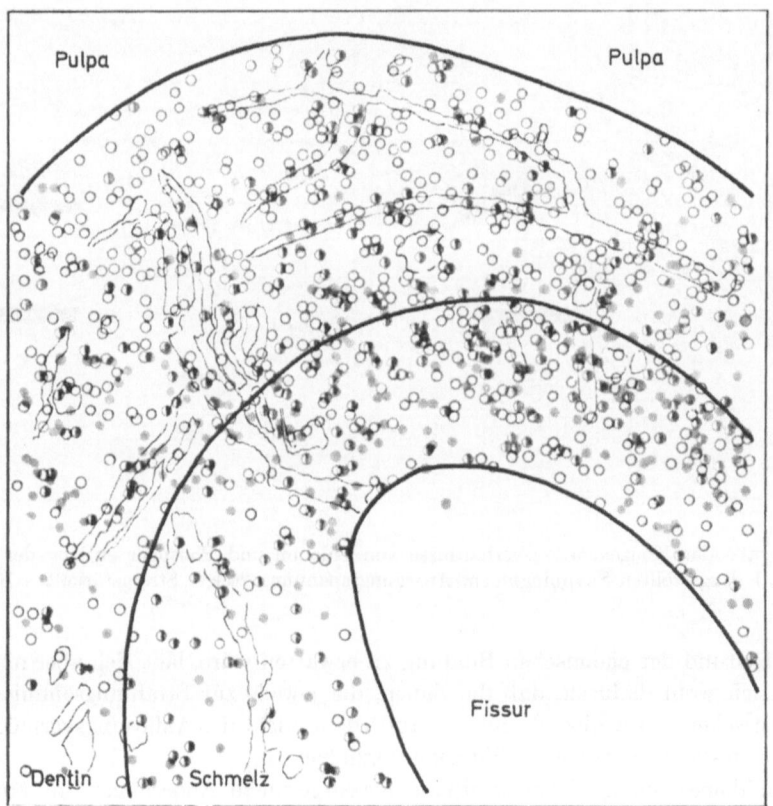

Abb. 4. Anordnung von Calcium, Phosphor und Fluor des in Abb. 3 dargestellten Bereiches. Darstellung entstand durch maßstabgetreue Umzeichnung der durch Röntgenemission charakterisierten Bezirke. Ein ordnendes Prinzip ist nicht zu erkennen. Anregungsspannung 10 kV. Stromstärke 0,2 × 10⁻⁷ A

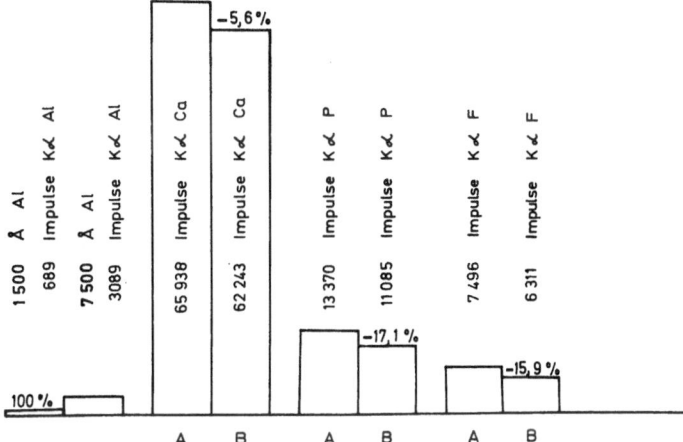

Abb. 5. Die von Ca, P und F ermittelten Strahlungsintensitäten werden durch Veränderung der aufgedampften Al-Schicht unterschiedlich beeinflußt

Absolutwerte für die Zusammensetzung der zu prüfenden Hartsubstanz möglich, aber aus den verschiedensten Gründen recht unpraktisch. Mit Hilfe eines Eichstandards gelang uns die Umrechnung in besser verständliche Prozentwerte. Als Eichstandard für die quantitative Analyse verwendeten wir einen in der Natur vorkommenden Fluor-Apatit-Kristall, der entsprechend seiner naßchemischen Untersuchung 37,2% Calcium, 17,4% Phosphor und 2,15% Fluor enthielt. Wir wählten Fluor-Apatit, da dieses Mineral mit den Kristalliten im Schmelz und Dentin größtmögliche Ähnlichkeit besitzen soll.

Der Eichstandard wurde derselben Vorbehandlung unterzogen wie das Untersuchungsmaterial. Da nicht alle Proben zugleich mit dem Eichstandard bedampft werden können, ergab sich die Aufgabe, die Schichtdicke des Aluminiumniederschlags auf den einzelnen Proben zu bestimmen und zu prüfen, inwieweit eine veränderte Schichtdicke die Intensität der K_α-Strahlung von Calcium, Phosphor und Fluor verändert. Die Dicke der aufgedampften Al-Schicht ergibt sich in etwa aus den Intensitätsverhältnissen der K_α-Strahlung für Al auf der Probe gegenüber einem Prüfkörper aus Rein-Al. Hierbei ist die für die Röntgenemissionsanregung mögliche Schicht des reinen Al-Standards mit der Eindringtiefe in dieses Material gleichzusetzen. Diese Eindringtiefe beträgt nach den Angaben von WITTRY bei einer Anregungsspannung von 20 kV und einer Stromstärke von 2×10^{-7} A etwa 8,2 µ. Hieraus ergibt sich der Wert für die Bedampfungsschicht unseres Eichstandards mit 1500° A.

Um zu prüfen, inwieweit die aufgedampfte Al-Schicht die Intensitätsverhältnisse beeinflußt, haben wir den Eichstandard nach Abschluß der Untersuchungen nochmals — und zwar sehr intensiv — bedampft und die Schichtdicke wiederum bestimmt. Sie belief sich jetzt auf ca. 7500° A. Stellen wir die Intensitätsverhältnisse der K_α-Strahlung von Ca, P und F nach der ersten und zweiten Bedampfung vergleichsweise gegenüber, so ergibt sich unter der Voraussetzung einer linearen Abhängigkeit zwischen Schwächung und zunehmender Schichtdicke: Eine Vermehrung der Al-Schicht von 1000° A vermindert die Calciumwerte um 0,9%, die Phosphorwerte um 2,8% und die Fluorwerte um 2,6% (Abb. 5). Praktisch bedeutet dies, daß wir den Fehler, der bei unseren Untersuchungen durch eine unterschiedliche Aufdampfschicht in den Grenzen von ± 150° A entstand, vernachlässigen können, zumal bei statistischer Betrachtungsweise diese Abweichungen wieder ausgeglichen werden.

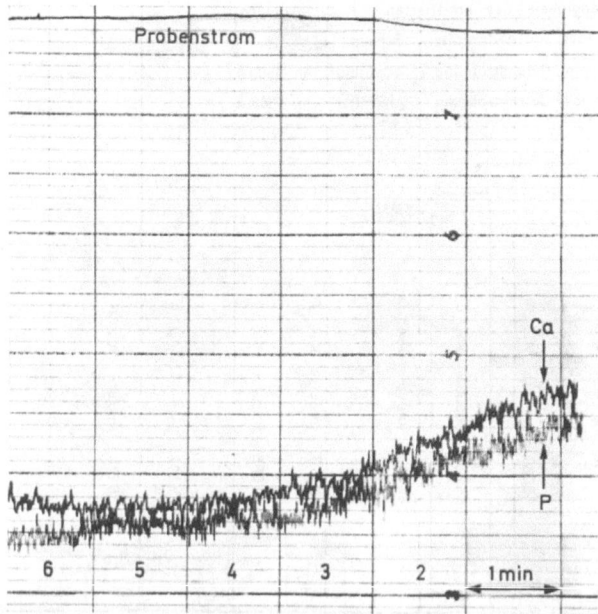

Abb. 6. Veränderung der Dentinsubstanz durch Elektroneneinfall an einem einzigen Punkt über einen längeren Zeitraum. Die aufgefangenen Intensitätswerte nehmen bis zu 25% ab

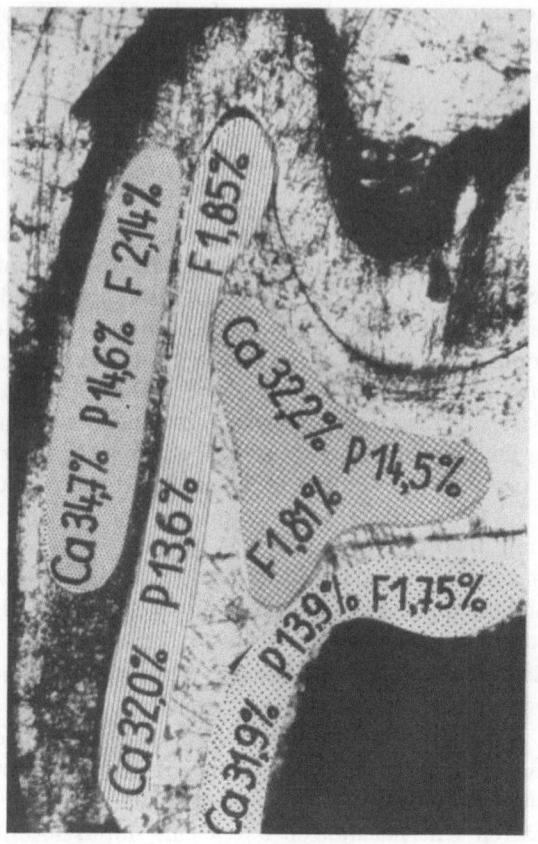

Abb. 7. Übersichtsdarstellung der Calcium-, Phosphor- und Fluoranteile im distalen Höckerbereich von M_2 einer 33 Tage alten Ratte. Zonen sind unterteilt in Schmelz, schmelznahes, mittleres und pulpanahes Dentin

Von größerem Einfluß auf unsere Untersuchungen ist die Tatsache, daß das Zahnmaterial beim Elektronenbeschuß einer definierten, ca. 1 μ^2 großen Stelle nur für ca. 30 sec unverändert bleibt (Abb. 6). Bei längerer Dauer der Einstrahlung ist eine deutliche Intensitätsabnahme der erzeugten Röntgenstrahlung zu verzeichnen. Von einer wirklich zerstörungsfreien Analyse können wir somit nur dann sprechen, wenn die Analysenwerte in sehr kurzer Zeit ermittelt werden. Diese Schwierigkeit haben wir dadurch umgangen, daß die Probe während der Analyse mit konstanter Geschwindigkeit von 10 μ/min bewegt wurde. Auf diese Weise wird der Bezirk, aus dem die Einzelmessungen stammen, zwar auf etwa 20 μ vergrößert, er ist aber für unsere vergleichenden Untersuchungen noch hinreichend klein.

Unter Berücksichtigung all dieser Gegebenheiten fanden wir bei einer 33 Tage alten Ratte im Schmelz: 34,7% Ca, 14,6% P, 2,4% F.

In der äußersten Dentinschicht: 32% Ca, 13,6% P, 1,85% F.

In der mittleren Dentinschicht: 32,2% Ca, 14,5% P, 1,81% F und in der pulpanahen Dentinschicht: 31,9% Ca, 13,9% P und 1,75% F (Abb. 7).

Literatur

Birks, L. S.: Electron probe microanalysis. New York: Interscience Publ. 1963.

Fromme, H. G., H. Riedel, J. Vahl: Elektronenmikroskopische und elementanalytische Untersuchungen zur Differenzierung der Odontoblasten und zur Dentinbildung. Deut. Zahnaerztl. Z. 5 (1968).

Höhling, H. J.: Die Bauelemente von Zahnschmelz und Dentin aus morphologischer, chemischer und struktureller Sicht. München: Carl Hanser 1966.

Stickler, R.: Einführung in die Grundlagen und Arbeitsmethoden der Elektronenstrahlmikroanalyse. München: Selbstverlag Kontron GmbH & Co. 1967.

Theisen, R.: Electron beam microanalysis. Berlin-Göttingen-Heidelberg: Springer 1965.

Tousimis, A. J.: Electron probe X-ray microanalysis of medical and biological specimens. ASTM Spec. Techn. Publ. 349, 193—206 (1963).

The Correlation of Bone Mineral
with Body Build and Bone Turnover

N. J. D. SMITH

King's College Hospital Dental School, London, England

M. H. HOBDELL

The Anatomy Department, King's College, London, England

The literature concerning the dynamics of normal and pathological bone tissue contains many references to the age changes in bone. Surprisingly, there is scant reference to any quantitative relationship between body build and bone structure.

It is the purpose of this paper to report a preliminary study which was carried out in an attempt to correlate the *in vivo* determination of bone mineral density with body stature. The same method, developed for the *in vivo* determination of bone mineral content, was then used on a small series post-mortem and compared with the bone activity, measured by microradiographic techniques, of autopsy specimens from the same cadavers.

The mineral density of the calcaneus was determined in a series of 54 young adult European volunteers in the 20—29 years age group (14 females and 40 males), using a 50 kV XX 90 X-ray microscope and two parallel scintillation counting circuits (SMITH, 1968). The calcaneus was selected for a number of reasons, among which was the fact that MAINLAND (1957) has stated that this bone does not undergo the progressive demineralization with age that both he and other workers have found to occur in most other bones. At present our series is not large enough to confirm this, and so for the moment it is assumed that this is so. The results, expressed as grams of calcium chloride equivalent per square centimetre, showed that the spread of the values so obtained was large, the ratio between the least mineralized and the most heavily mineralized being of the order of 1:8. The distribution of the determinations would appear to approximate to the expected normal curve (Fig. 1). This wide range of values is in keeping with the bone mineral determinations obtained by other methods at other sites, MAYO (1957) having obtained a ratio of least mineralization to greatest mineralization of the order of 1:10 in the ulna.

A series of standard anthropometric measurements was taken on each subject immediately after the bone mineral content had been determined in order to see if any correlation could be found between bone mineral content and body stature. The measurements which were taken were the height, weight, bi-acromial width and bi-iliac width. The anthropometric methods were those described by TANNER (1964) except that the height was read from a boxwood scale attached to the weighing machine, and that the women were wearing light clothing.

A plot of bone density against height $\times$ (bi-acromial width + bi-iliac width) would indicate that some degree of correlation does exist between bone mineralization and body size (Fig. 2). This plot was then used to normalize the experimentally derived bone mineral determinations to a man of "standard" stature. The "standard" stature chosen was one where the height $\times$ (bi-acromial width + bi-iliac width) gave a product of 11,250 cm². The histogram so obtained from these normalized values was more compact than that obtained from the uncorrected values, and the ratio of minimum mineralization to maximum mineralization dropped to the order of 1:5, further showing the probability of a relationship between bone density and body stature (Fig. 3).

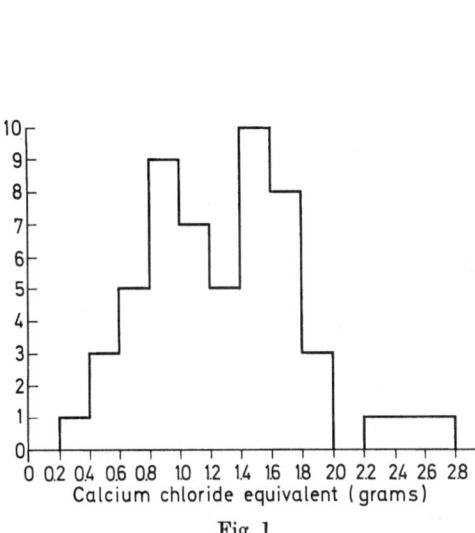

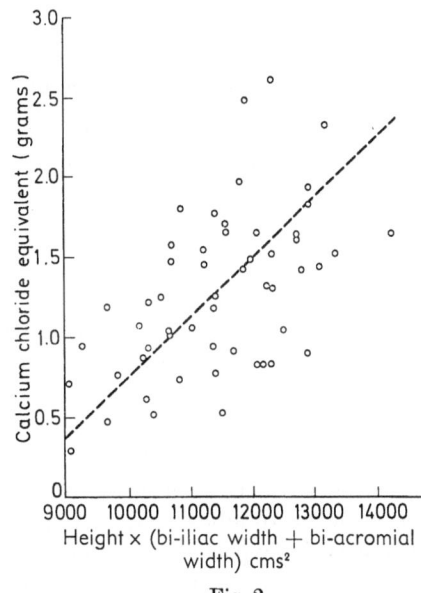

Fig. 1

Fig. 2

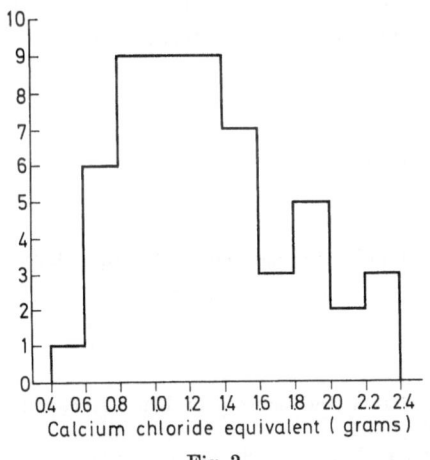

Fig. 3

Fig. 1. This histogram shows the distribution of the bone mineral determinations of 54 subjects in the 20—29 year age group

Fig. 2. The bone mineral density expressed as grams of calcium chloride equivalent per square centimetre is plotted against a measurement derived from the stature of the subject. (The dotted line is drawn by eye.)

Fig. 3. This histogram illustrates how the spread of the bone mineral determinations is reduced after a correction has been applied to allow for the stature of the subject

It is hoped that subsequent computer analysis of a larger sample, of which these determinations will form part, may show a rather better correlation between the degree of mineralization and body stature since it is clear that the relationship is a complicated one.

The second stage of this investigation was to use the method described for the *in vivo* bone mineral determinations to assess the density of the calcaneus of five subjects post-mortem. The anthropometric measurements were repeated on the cadavers and the mineral density determinations were then normalized as described for the *in vivo* study. Autopsy specimens of the anterior iliac crest and of the vertebral column were taken from the same cadavers, and plane parallel 100 micron thick undemineralized sections were cut after methyl methacrylate embedding. These were micro-radiographed using a copper target at 20 kV and 1 mA tube current, on Kodak Experimental V 6028 film.

The morpho-histological technique of JOWSEY et al. (1965) was modified by measuring both the forming and resorbing surfaces in a photographic enlargement (×30) of the microradiograph, and expressing this length of surface as a percentage of the total bone surface. This was done to provide a measure of the total surface activity of the specimen, rather than a measure of either formation or resorption.

Table

Patient No.	Mineral uncorrected	Mineral corrected	Activity (%)
A	1.7	1.7	26
B	1.1	1.1	28
C	1.0	0.8	25
D	0.7	0.6	11
E	0.6	1.0	31

At present a comparison between the bone mineral density and the activity of the bone surfaces is limited in value, because of the small size of the sample. The table shows the bone mineral figures (both before and after correction) ranked from the highest to the lowest and the corresponding bone activity for the same subject.

It is known that there are large errors in the bone activity measuring techniques (Jowsey et al., 1965, op. cit.), errors of up to 10% have been shown to exist in the repeated estimations from the same microradiograph, so that where there are small differences in corrected bone mineral it may be expected that the errors in bone activity measurements may well be too great to discern between different degrees of mineralization. However, where there is a large difference in bone mineral it should be possible to demonstrate a change in bone activity.

The limitation of this technique seems to be the microradiographic interpretation of bone activity rather than the bone mineral density determination. However, the techniques of Hobdell and Boyde (this volume) will enable a more accurate microradiographic assessment of the bone activity to be made.

This preliminary work is being expanded in two directions. Firstly, by study of both density and activity in the same bone specimen, which will be done post-mortem, and secondly, by repeating the present study in vivo using iliac crest biopsies.

Acknowledgements. We would like to acknowledge the help and encouragement of Mr. R. V. Ely and of the M.R. Research Trust during the preparation of this paper.

References

Hobdell, M. H., and A. Boyde: The correlation between microradiography and scanning electron microscopy of bone section surfaces. This Volume p. 611.

Jowsey, J., J. P. Kelly, B. L. Rigg, A. J. Bianco Jr., A. Scholz, and J. Gershon-Cohen: Quantitative microradiographic studies of normal and osteoporotic bone. J. Bone Joint Surg. 47 A, 4785 (1965).

Mainland, D.: A study of age differences in the X-ray density of the adult human calcaneus-variation and sources of bias. J. Gerontol. 12, 53 (1957).

Mayo, K. M.: Quantitative measurement of bone mineral content in normal adult bone. Brit. J. Radiol. 34, 693 (1961).

Smith, N. J. D.: The measurement of bone mineral in vivo using two parallel scintillation counting circuits. Proceedings of the Symposium Ossium of the European Assoc. of Radiol. (in the press) (1968).

Tanner, J. M.: The physique of the Olympic athlete. London: George Allen and Unwin 1964.

The Correlation between Microradiography and Scanning Electron Microscopy of Bone Section Surfaces

M. H. HOBDELL

Anatomy Departments, King's College London

A. BOYDE

University College London, England

Microradiographs were prepared of carefully washed 200 µ ground sections of human mandibular cortical bone: these sections were then extracted with hot 1.2 ethane-diamine in a soxhlet apparatus, washed with alcohol, dried, coated with carbon and gold in the vacuum evaporator, and photographed in Cambridge Scientific Instruments Stereoscan MKI scanning electron microscope operated at 10 kV.

The interpretation of the microradiographic image (following JOWSEY et al., 1965), was found to be in good correlation with the interpretation of the scanning electron microscopic image of bone surfaces given by BOYDE and HOBDELL (1969). Thus forming surfaces present as diffuse edges of low density regions, showing a gradual increase in density towards the surrounding lamellae in the microradiographs. In the scanning electron micrographs, forming surfaces present as a mineralizing front in which the collagen fibre bundles are not completely impregnated with mineral and therefore show as short or irregular segments. There is a distinct change in the pattern of forming free bone surfaces after the removal of the organic matrix with 1.2 ethane-diamine.

Resting surfaces show as the smooth borders of more perfectly mineralized regions in microradiographs and as smooth bundles of completely impregnated collagen fibres in the scanning electron micrographs: the pattern of the collagen fibres is unchanged by the 1.2 ethane-diamine extraction in this case.

Resorbing surfaces show easily recognised Howship's lacunae in both types of image. In the microradiographs these present as scalloped profiles of the edge of densely mineralized bone with normally a narrow low density seam. The border lines between adjacent lacunae show as prominent white lines in scanning electron micrographs.

A new type of forming surface has been encountered in the depths of previous resorption lacunae in which new bone formation has just commenced. One cannot recognise any pattern of collagen fibre bundles in this situation — rather, the mineral presents as microcalcospheritic nodules.

Poorly mineralized osteons appear brighter in the scanning electron microscope image of 1.2 ethane-diamine extracted sections: the same regions also show a pattern of broad, low, flat ridges, corresponding to the dimensions of the lamellae proper, and narrow grooves which would be attributed to the interlamallar planes. We feel it to be most probable that these two effects are interrelated and that interruptions in the continuity of the conducting coating at cracks located at the interlamellar planes cause the isolation and insulation of "islands" which charge up under the electron beam.

This communication has been submitted for publication in full in the „Zeitschrift für Zellforschung und Mikroskopische Anatomie" (which is published by Springer-Verlag).

The X-ray microscope used in these studies was provided through the personal generosity of Mr. R. V. ELY and The M. R. Research Trust and the Ely-Webster Trust. The *Stereoscan* scanning electron microscope was provided by the Science Research Council (U.K.). This work was also supported by the Medical Research Council.

References

BOYDE, A., and M. H. HOBDELL: Scanning electron microscopy of lamellar bone. Z. Zellforsch. **93**, 213—231 (1969).

JOWSEY, J., P. KELLY, B. L. RIGGS, A. J. BIANCO, D. A. SCHOLZ, and J. GERSHON-COHEN: Quantitative microradiographic studies of normal and osteoporotic bone. J. Bone Joint Surg. **47** A, 785—806 (1965).